Research Advancements in Pharmaceutical, Nutritional, and Industrial Enzymology

Shashi Lata Bharati
North Eastern Regional Institute of Science and Technology, India

Pankaj Kumar Chaurasia
LS College Muzaffarpur, India

A volume in the Advances in Medical Technologies and Clinical Practice (AMTCP) Book Series

IGI Global
DISSEMINATOR OF KNOWLEDGE

Published in the United States of America by
IGI Global
Medical Information Science Reference (an imprint of IGI Global)
701 E. Chocolate Avenue
Hershey PA, USA 17033
Tel: 717-533-8845
Fax: 717-533-8661
E-mail: cust@igi-global.com
Web site: http://www.igi-global.com

Library of Congress Cataloging-in-Publication Data

Names: Bharati, Shashi Lata, 1983- editor. | Chaurasia, Pankaj Kumar, 1987- editor.
Title: Research advancements in pharmaceutical, nutritional, and industrial enzymology / Shashi Lata Bharati and Pankaj Kumar Chaurasia, editors.
Description: Hershey PA : Medical Information Science Reference (an imprint of IGI Global), [2018] | Includes bibliographical references.
Identifiers: LCCN 2017038144| ISBN 9781522552376 (hardcover) | ISBN 9781522552383 (ebook)
Subjects: | MESH: Enzymes--pharmacology | Chemistry, Pharmaceutical--methods
Classification: LCC RM666.E55 | NLM QU 135 | DDC 615.3/5--dc23 LC record available at https://lccn.loc.gov/2017038144

This book is published in the IGI Global book series Advances in Medical Technologies and Clinical Practice (AMTCP) (ISSN: 2327-9354; eISSN: 2327-9370)

British Cataloguing in Publication Data
A Cataloguing in Publication record for this book is available from the British Library.

All work contributed to this book is new, previously-unpublished material. The views expressed in this book are those of the authors, but not necessarily of the publisher.

For electronic access to this publication, please contact: eresources@igi-global.com.

Advances in Medical Technologies and Clinical Practice (AMTCP) Book Series

Srikanta Patnaik
SOA University, India
Priti Das
S.C.B. Medical College, India

ISSN:2327-9354
EISSN:2327-9370

Mission

Medical technological innovation continues to provide avenues of research for faster and safer diagnosis and treatments for patients. Practitioners must stay up to date with these latest advancements to provide the best care for nursing and clinical practices.

The **Advances in Medical Technologies and Clinical Practice (AMTCP) Book Series** brings together the most recent research on the latest technology used in areas of nursing informatics, clinical technology, biomedicine, diagnostic technologies, and more. Researchers, students, and practitioners in this field will benefit from this fundamental coverage on the use of technology in clinical practices.

Coverage

- Clinical Data Mining
- Neural Engineering
- Patient-Centered Care
- Nutrition
- Biomechanics
- Medical informatics
- Telemedicine
- Medical Imaging
- Biomedical Applications
- Clinical High-Performance Computing

The Advances in Medical Technologies and Clinical Practice (AMTCP) Book Series (ISSN 2327-9354) is published by IGI Global, 701 E. Chocolate Avenue, Hershey, PA 17033-1240, USA, www.igi-global.com. This series is composed of titles available for purchase individually; each title is edited to be contextually exclusive from any other title within the series. For pricing and ordering information please visit http://www.igi-global.com/book-series/advances-medical-technologies-clinical-practice/73682. Postmaster: Send all address changes to above address.

Titles in this Series

For a list of additional titles in this series, please visit: www.igi-global.com/book-series

Multifunctional Nanocarriers for Contemporary Healthcare Applications
Md. Abul Barkat (K.R. Mangalam University, India) Harshita A.B. (K.R. Mangalam University, India) Sarwar Beg (Jubilant Generics, India) and Farhan J. Ahmad (Jamia Hamdard, India)
Medical Information Science Reference • copyright 2018 • 335pp • H/C (ISBN: 9781522547815) • US $265.00 (our price)

Biomedical Signal and Image Processing in Patient Care
Maheshkumar H. Kolekar (Indian Institute of Technology Patna, India) and Vinod Kumar (Indian Institute of Technology Roorkee, India)
Medical Information Science Reference • copyright 2018 • 312pp • H/C (ISBN: 9781522528296) • US $265.00 (our price)

Next-Generation Mobile and Pervasive Healthcare Solutions
Jose Machado (University of Minho, Portugal) António Abelha (University of Minho, Portugal) Manuel Filipe Santos (University of Minho, Portugal) and Filipe Portela (University of Minho, Portugal)
Medical Information Science Reference • copyright 2018 • 286pp • H/C (ISBN: 9781522528517) • US $245.00 (our price)

Innovative Research in Thermal Imaging for Biology and Medicine
Ricardo Vardasca (University of Porto, Portugal) and Joaquim Gabriel Mendes (University of Porto, Portugal)
Medical Information Science Reference • copyright 2017 • 340pp • H/C (ISBN: 9781522520726) • US $210.00 (our price)

Internet of Things and Advanced Application in Healthcare
Catarina I. Reis (Polytechnic Institute of Leiria, Portugal) and Marisa da Silva Maximiano (Polytechnic Institute of Leiria, Portugal)
Medical Information Science Reference • copyright 2017 • 349pp • H/C (ISBN: 9781522518204) • US $210.00 (our price)

Integrating Biologically-Inspired Nanotechnology into Medical Practice
B.K. Nayak (K.M. Centre for Post Graduate Studies, India) Anima Nanda (Sathyabama University, India) and M. Amin Bhat (Sathyabama University, India)
Medical Information Science Reference • copyright 2017 • 394pp • H/C (ISBN: 9781522506102) • US $190.00 (our price)

IGI Global
DISSEMINATOR OF KNOWLEDGE

701 East Chocolate Avenue, Hershey, PA 17033, USA
Tel: 717-533-8845 x100 • Fax: 717-533-8661
E-Mail: cust@igi-global.com • www.igi-global.com

Table of Contents

Foreword xvi

Preface xvii

Acknowledgement xx

Chapter 1
Enzyme Replacement Therapy: Therapeutic Enzymes Used for the Treatment of Hereditary Deficiency Diseases 1
Abderrezak Khelfi, National Center of Toxicology, Algeria

Chapter 2
Therapeutic Enzymes Used for the Treatment of Cardiovascular Diseases and Coagulation Disorders 23
Abderrezak Khelfi, National Center of Toxicology, Algeria

Chapter 3
Therapeutic Enzymes Used for the Treatment of Non-Deficiency Diseases 46
Abderrezak Khelfi, National Center of Toxicology, Algeria

Chapter 4
Colonic Bacterial Enzymes: Pharmaceutical Significance and Applications 71
Srushti M. Tambe, Bombay College of Pharmacy, India
Namita D. Desai, Bombay College of Pharmacy, India

Chapter 5
Human Immunodeficiency Virus Reverse Transcriptase (HIV-RT): Structural Implications for Drug Development 100
Anuradha Singh, University of Allahabad, India
Ramendra K. Singh, University of Allahabad, India

Chapter 6
The Use of Liposomes in Enzymes and Drug Design: Liposomes Drug Delivery System 128
Mahmoud Balbaa, Alexandria University, Egypt
Doaa Awad, Alexandria University, Egypt

Chapter 7
Enzymatic Research Having Pharmaceutical Significance 141
Ishan H. Raval, Central Salt and Marine Chemicals Research Institute, India
Arvind Kumar Singh Chandel, Central Salt and Marine Chemicals Research Institute, India

Chapter 8
Enzyme-Triggered Hydrogels for Pharmaceutical and Food Applications 159
Lakshmishri Upadrasta, Techno India, India
Vijay Kumar Garlapati, Jaypee University of Information Technology, India
Nafisa Lakdawala, Techno India, India
Rintu Banerjee, IIT Kharagpur, India

Chapter 9
Laccase Catalysis: A Green Approach in Bioactive Compound Synthesis 178
Helina Patel, Natubhai V. Patel College of Pure and Applied Sciences, India
Akshaya Gupte, Natubhai V. Patel College of Pure and Applied Sciences, India

Chapter 10
Laccase-Mediated Treatment of Pharmaceutical Wastes 213
Hamid Forootanfar, Kerman University of Medical Sciences, Iran
Shokouh Arjmand, Kerman University of Medical Sciences, Iran
Mina Behzadi, Kerman University of Medical Sciences, Iran
Mohammad Ali Faramarzi, Tehran University of Medical Sciences, Iran

Chapter 11
Laccase From White Rot Fungi Having Significant Role in Food, Pharma, and Other Industries 253
Ankita Kushwaha, University of Allahabad, India
Shweta Maurya, University of Allahabad, India
Ravi Kant. Pathak, University of Allahabad, India
Sonam Agarwal, University of Allahabad, India
Pankaj Kumar Chaurasia, MNNIT Allahabad, India & and LS College Muzaffarpur, India
M. P. Singh, University of Allahabad, India

Chapter 12
The Potential Application of Peroxidase Enzyme for the Treatment of Industry Wastes 278
Sonam Agarwal, University of Allahabad, India
Krishna Kumar Gupta, University of Allahabad, India
Vivek Kumar Chaturvedi, University of Allahabad, India
Ankita Kushwaha, University of Allahabad, India
Pankaj Kumar Chaurasia, MNNIT Allahabad, India & and LS College Muzaffarpur, India
M. P. Singh, University of Allahabad, India

Chapter 13
Biotechnology of Microbial Xylanase: Overview 294
Hooi Ling Ho, UCSI University, Malaysia

Chapter 14
Functional and Antioxidant Properties of Protein Hydrolysates From Ricinodendron Heudelotii(Bail.) Flours From Cameroon 326
Ngangoum Eric Serge, National Advanced School of Industrial Sciences, Cameroon
Mezajoug Kenfack Laurette Blandine, University of Ngaoundere, Cameroon
Sanjit Kumar, Vellore Institute of Technology, India
Mookambeswaran Vijayalakshmi, Vellore Institute of Technology, India
Tchiégang Clergé, University Institute of Technology, Cameroon

Chapter 15
Enzyme Use and Production in Industrial Biotechnology 341
Subir Kumar Nandy, Technical University of Denmark, Denmark

Chapter 16
Trends and Challenges in Enzymatic Bioengineering of Natural Products to Industrially Valuable Products: Status and Future of Industrial Enzymology 351
Sujata Sinha, Indian Institute of Technology, India

Chapter 17
Industrial Enzyme Technology: Potential Applications 375
Michael Bamitale Osho, McPherson University, Nigeria

Chapter 18
Role of Enzymes From Microbes in the Treatment of Recalcitrant From Industries 395
Veena Gayathri Krishnaswamy, Stella Maris College, India

Chapter 19
Moving into Nanotechnology Roles to Mimic and Boost Enzyme Activity 421
Maria Laura Soriano, University of Cordoba, Spain

Compilation of References 441

About the Contributors 541

Index 546

Detailed Table of Contents

Foreword xvi

Preface xvii

Acknowledgement xx

Chapter 1
Enzyme Replacement Therapy: Therapeutic Enzymes Used for the Treatment of Hereditary Deficiency Diseases 1
Abderrezak Khelfi, National Center of Toxicology, Algeria

Enzyme replacement therapy is a therapeutic approach in which the specific enzyme that is absent or inactive in affected individuals is replaced with a functional enzyme molecule derived from biological sources or produced by biotechnology. A large number and variety of enzyme defects have been identified in humans. Over 40 hereditary deficiency diseases were reported. The common feature is that enzyme deficiency leads to the accumulation of undegraded molecules and lysosomal storage, resulting in organ dysfunction. Crude enzyme preparations are often unsuitable for therapeutic uses because of their potential contamination and antigenicity. Advances in gene identification and cloning led to the subsequent production and demonstration of equal efficacy of recombinant human enzyme. The adverse events recorded range from boxed warnings for severe allergic reactions. This chapter summarizes therapeutic enzymes used in clinical practice, with particular reference to those obtained from biological sources and biotechnology processes.

Chapter 2
Therapeutic Enzymes Used for the Treatment of Cardiovascular Diseases and Coagulation Disorders 23
Abderrezak Khelfi, National Center of Toxicology, Algeria

The successful introduction of enzyme replacement therapy opened the way for the use of enzymes, first as crude preparations and later as highly purified enzymes for use in cardiovascular diseases, clotting disorders, etc. Elimination of blood clot is the key factor in thrombolytic therapy and fibrinolytic enzyme therapy can be practiced to remove the clot. Based on the mechanism of action, they are of two types of enzymes: plasminogen activators and plasmin-like enzymes. Plasma products are usually employed as a source of several enzymes used for the treatment of coagulation disorders. While these products have traditionally been purified from blood donations and obtained as foreign proteins obtained from heterogeneous sources, most are now produced by biotechnology. The therapeutic enzymes reviewed in this chapter are used for the treatment of cardiovascular diseases and hereditary diseases leading to coagulation disorders. Enzyme preparations obtained by direct fractionation from a naturally producing source and recombinant enzymes are considered in this chapter.

Chapter 3
Therapeutic Enzymes Used for the Treatment of Non-Deficiency Diseases 46
Abderrezak Khelfi, National Center of Toxicology, Algeria

Therapeutic enzymes have a broad variety of specific uses and clinical applications, particularly as antineoplasic agents, wound debridement therapeutics, and anti-inflammatory drugs, etc. These enzymes can elicit immune response, contributing allergic reactions. Newer drugs with improved stability and less antigenicity have been developed. Covalent modification of enzymes is used to circumvent this immunogenicity. Advancements in drug delivery have revolutionized enzyme therapy. Microencapsulation and artificial liposomal entrapment are some of the techniques used to increment the stability and half-life of enzyme drugs. Several enzymes are now used as prodrug that metabolizes inactive substances into active metabolites through bioactivation process. This approach comprises a suit of techniques that allow activation of drugs locally and at the site of action. This chapter gives an outline of clinical uses of therapeutic enzymes used in non-deficiency diseases. Developments of these enzymes are reviewed with a particular focus on bioengineering applied to the native proteins.

Chapter 4
Colonic Bacterial Enzymes: Pharmaceutical Significance and Applications 71
Srushti M. Tambe, Bombay College of Pharmacy, India
Namita D. Desai, Bombay College of Pharmacy, India

This chapter reviews various enzymes produced by the colonic microflora and their utilization in the development of pharmaceutical dosage forms to achieve colon-specific drug delivery. This chapter discusses the applications of colonic bacterial enzymes in order to surrogate colonic conditions in vivo so as to evaluate in vitro drug release from microbially triggered/enzymatically triggered colon-specific drug delivery systems. This chapter also discusses different methods to produce colonic bacterial enzymes as well as use of probiotics as a source to produce colonic bacterial enzymes.

Chapter 5
Human Immunodeficiency Virus Reverse Transcriptase (HIV-RT): Structural Implications for Drug Development 100
Anuradha Singh, University of Allahabad, India
Ramendra K. Singh, University of Allahabad, India

Reverse transcriptase (RT) is a multifunctional enzyme in the life cycle of human immunodeficiency virus and represents a primary target for drug discovery against HIV-1 infection. Two classes of RT inhibitors, the nucleoside and the non-nucleoside RT inhibitors, are prominently used in the highly active antiretroviral therapy in combination with other anti-HIV drugs. This chapter deals with the salient features of HIV-RT that make it an attractive target for rational drug design and chemotherapeutic intervention in the management of acquired immunodeficiency syndrome. Further, the role of RT in the viral life cycle, the ways the drugs act to inhibit the normal functions of RT, and the mechanisms that the virus adapts to evade the available drugs have been discussed. Computational strategies used in rational drug design accompanied by a better understanding of RT, its mechanism of inhibition and drug resistance, discussed in this chapter, shall provide a better platform to develop effective RT inhibitors.

Chapter 6
The Use of Liposomes in Enzymes and Drug Design: Liposomes Drug Delivery System 128
Mahmoud Balbaa, Alexandria University, Egypt
Doaa Awad, Alexandria University, Egypt

Liposomes are phospholipid vesicles that share many of membranes properties. Liposomes can be easily prepared in a range of sizes. They are able to improve the unfavorable properties of many free drugs such as increasing the amount of drugs delivered to various diseased sites in addition to decreasing the drug toxicities. Encapsulation of enzymes and food ingredients, as well as antioxidants in liposomes also received a lot of awareness. Moreover, an increase for drugs delivered to various diseased tissues was achieved by encapsulating drugs in the liposomes. The topics of encapsulation of enzymes and food ingredients as well as antioxidants in liposomes were highly investigated.

Chapter 7
Enzymatic Research Having Pharmaceutical Significance.. 141
Ishan H. Raval, Central Salt and Marine Chemicals Research Institute, India
Arvind Kumar Singh Chandel, Central Salt and Marine Chemicals Research Institute, India

The enzymes' biocatalysts act by lowering the activation energy without getting consumed in the reaction. The immense number of enzymes acts as a correctly matched orchestra to ensure that enormously complex life mechanisms and processes occur in a right direction. Sufficient quantity and accurate function of enzymes results in proper functional maintenance of body. The enzymes play a major role in the diagnosis, curing, biochemical investigation, and monitoring of many dreaded diseases of the century. The development of recombinant DNA technology had a significant impression on production levels of enzymes. Around 50% of the enzyme market is covered by recombinant enzymes. Because of development in molecular biology tools, several pharmaceutically enzymes have been identified and are being actively used in the pharmaceutical industry either for diagnostic or treatment. Information on this topic is very insufficient, and thus, the present chapter is an attempt to compile information on the sources, properties and applications of important therapeutic enzymes.

Chapter 8
Enzyme-Triggered Hydrogels for Pharmaceutical and Food Applications 159
Lakshmishri Upadrasta, Techno India, India
Vijay Kumar Garlapati, Jaypee University of Information Technology, India
Nafisa Lakdawala, Techno India, India
Rintu Banerjee, IIT Kharagpur, India

Enzyme-mediated polymeric hydrogels are drawing considerable attention in pharmaceutical and food sectors owing to their superior biocompatibility and process controllability under physiological conditions. Enzymes play a significant role in polymeric hydrogel formation through different mechanisms. Oxidases (e.g., horseradish peroxidase and tyrosinase) have demonstrated to drive the crosslinking of gel precursors by oxidizing the phenolic or acrylic moieties to free radicals. Transferases and hydrolases catalyze elongation of biopolymer chains which gradually self-assemble into hydrogels. Still more certain enzymes also participate in hydrogel formation by releasing gelation factors. Enhancement of the desired properties of certain hydrogels through the interior and exterior post-modifications has also been demonstrated by certain enzymes. Hence, in this chapter, the authors explore the different mechanisms of enzyme-mediated hydrogels preparations and its fabrication towards pharmaceutical and food sectors along with the discussion of recent trends and further prospects.

Chapter 9
Laccase Catalysis: A Green Approach in Bioactive Compound Synthesis 178
Helina Patel, Natubhai V. Patel College of Pure and Applied Sciences, India
Akshaya Gupte, Natubhai V. Patel College of Pure and Applied Sciences, India

The search for cost-effective and environmental benign practices for the production of bioactive compounds has gained considerable attention since last decade, due to increasing demand of eco-friendly processes. Many industries have started adopting routes for the development of green chemistry by employing enzymatic approaches to overcome the limitations of physico-chemical methods and environmental concerns. Laccase is one such enzyme which has gained considerable attention in recent years as a biocatalyst in organic synthesis. Laccases possess versatile biochemical properties and the reactions catalyzed by laccase require only molecular oxygen with concomitant release of water as a byproduct. They have been widely used for reactions such as dimerization, polymerization, coupling, and grafting reactions and for antibiotic modifications. This chapter summarizes the advances that have been made in developing technologies based on laccase mediated reactions in the field of medicine, agriculture, food, and pharmaceuticals.

Chapter 10
Laccase-Mediated Treatment of Pharmaceutical Wastes.. 213
Hamid Forootanfar, Kerman University of Medical Sciences, Iran
Shokouh Arjmand, Kerman University of Medical Sciences, Iran
Mina Behzadi, Kerman University of Medical Sciences, Iran
Mohammad Ali Faramarzi, Tehran University of Medical Sciences, Iran

Laccases are versatile multi-copper enzymes belonging to the superfamily of oxidase enzymes, which have been known since the nineteenth century. Recent discoveries have refined investigators' views of the potential of laccase as a magic tool for remarkable biotechnological purposes. A literature review of the capabilities of laccases, their assorted substrates, and their molecular mechanism of action now indicates the emergence of a new direction for laccase application as part of an arsenal in the fight against the contamination of water supplies by a number of frequently prescribed medications. This chapter provides a critical review of the literature and reveals the pivotal role of laccases in the elimination and detoxification of pharmaceutical contaminants in aquatic environments and wastewaters.

Chapter 11
Laccase From White Rot Fungi Having Significant Role in Food, Pharma, and Other Industries 253
Ankita Kushwaha, University of Allahabad, India
Shweta Maurya, University of Allahabad, India
Ravi Kant. Pathak, University of Allahabad, India
Sonam Agarwal, University of Allahabad, India
Pankaj Kumar Chaurasia, MNNIT Allahabad, India & and LS College Muzaffarpur, India
M. P. Singh, University of Allahabad, India

Laccases (E.C. 1.10.3.2 benzenediol: oxygen oxidoreductase) are an interesting group of N glycosylated multicopper blue oxidase enzymes and the widely studied enzyme having a broad range of substrate specificity of both phenolic and non-phenolic compounds. They are widely found in fungi, bacteria plant, insects, and in lichen. They catalyze the oxidation of various phenolic and non-phenolic compounds, with the concomitant reduction of molecular oxygen to water. They could increase productivity, efficiency, and quality of products without a costly investment. This chapter depicts the applications of laccase enzyme from white rot fungi, having various industrial (such as textile dye bleaching, paper and pulp bleaching, food includes the baking, it also utilized in fruit juice industry to improve the quality and stabilization of some perishable products having plant oils), pharmaceutical (as it has potential for the synthesis of several useful drugs such anticancerous, antioxidants, synthesis of hormone derivatives because of their high value of oxidation potential) significance.

Chapter 12
The Potential Application of Peroxidase Enzyme for the Treatment of Industry Wastes.................. 278
Sonam Agarwal, University of Allahabad, India
Krishna Kumar Gupta, University of Allahabad, India
Vivek Kumar Chaturvedi, University of Allahabad, India
Ankita Kushwaha, University of Allahabad, India
Pankaj Kumar Chaurasia, MNNIT Allahabad, India & and LS College Muzaffarpur, India
M. P. Singh, University of Allahabad, India

Environmental pollution is becoming one of the major threats around the world because of the release of toxic and hazardous substances from food, pharmaceutical, and other industries as well. These wastes are mainly dumped indiscriminately which ultimately reached water bodies, thereby affecting marine ecosystem. Therefore, effective effluent treatment is an important step which can help in conserving our water resources. White rot fungus (WRF) have been shown to degrade and mineralize a wide variety of wastes because of their nonspecific extracellular lignin mineralizing enzymes (LMEs). These enzymes are used for the decolorization of synthetic dyes. They help in the degradation of pesticides, polycyclic aromatic hydrocarbons (PAHs), and pharmaceuticals wastes like- anti-inflammatory, lipid regulatory, antiepileptic drugs, endocrine disrupting chemicals, etc. They also help in degrading the food waste and convert them into useful products which can be used as food, feed, fodder; some of these wastes are lignocellulosic waste, viticulture waste, olive mill waste, molasses waste, etc.

Chapter 13
Biotechnology of Microbial Xylanase: Overview 294
Hooi Ling Ho, UCSI University, Malaysia

Xylanases are inducible enzymes responsible for the complete hydrolysis of xylan into xylose. Both solid state fermentation (SsF) and submerged fermentation (SmF) are used in the production of xylanase. SsF has become a popular approach due to its economic value. In fact, higher biomass and lower protein breakdown are among the factors involved in determining the production of xylanases in SsF. Agricultural extracts which are abundantly available in the environment such as rice bran and wheat bran are commonly used as the potential carbon source in xylanases production. Xylanase is indeed one of the valuable enzymes which show immense potential in vast industrial applications. The demand for xylanase is increasing because of its prodigious utilization in pulp and paper, bakery, food and beverage, detergents, textile, and animal feed. Xylanase has therefore become one of the important commercial enzymes in recent years.

Chapter 14
Functional and Antioxidant Properties of Protein Hydrolysates From Ricinodendron Heudelotii(Bail.) Flours From Cameroon 326
Ngangoum Eric Serge, National Advanced School of Industrial Sciences, Cameroon
Mezajoug Kenfack Laurette Blandine, University of Ngaoundere, Cameroon
Sanjit Kumar, Vellore Institute of Technology, India
Mookambeswaran Vijayalakshmi, Vellore Institute of Technology, India
Tchiégang Clergé, University Institute of Technology, Cameroon

Ricinodendron heudelotii kernels were defatted and used as substrate to produce protein hydrolysatesusing papain (PHP), trypsin (PHT), proteases from Abrus precatorius (PHAp)and B. enneandra(PHBe). The degree of hydrolysis (DH), antioxidant (DPPH method), and functional properties of hydrolysates were performed. The DH value, whatever hydrolysis time, was highest with PHP. The water holding capacity decreased with the hydrolysis time from 21.50 ± 0.44% to 5.20 ± 0.07%. After 6h of hydrolysis, PHAp exhibited maximum solubility value (70.17 ± 2.15%) while PHBe had lower solubility value (18.43 ± 0.12%). The highest value of emulsifying activity index was found at pH 9 with 0.25% (w/v) hydrolysate concentration. Within the range of pH used (4-9), the best foam capacity and foam stability were exhibited by PHBe. PHP, PHAp, and PHT inhibited DPPH radical at 83.30 ± 0.46, 75.07 ± 0.15, and 56.78 ± 0.40%, respectively, at 6h of hydrolysis.

Chapter 15
Enzyme Use and Production in Industrial Biotechnology 341
Subir Kumar Nandy, Technical University of Denmark, Denmark

This chapter demonstrates the bioprocess strategies involved in the application and production of enzymes from an industrial view point. Moreover, bottlenecks in enzyme production and novel strategies to overcome the barriers are demonstrated here. Enzymes are produced from different sources of microorganisms and mostly all biological reactions happen due to the help of enzymes within a very short time. The different uses of enzymes are discussed in this chapter.

Chapter 16
Trends and Challenges in Enzymatic Bioengineering of Natural Products to Industrially Valuable Products: Status and Future of Industrial Enzymology 351
Sujata Sinha, Indian Institute of Technology, India

Bioengineering of natural products to useful products are trending rapidly. These products are cost effective and eco-friendly and fitting into consumer demands of natural and organic. Wastes from industries, agriculture, fishery, dairy, etc. are being investigated for transformation to useful biomolecules for other industries like cosmetics, food supplements/preservation, dairy, etc. Biocatalytic transformation looks promising in the present scenario, but needs intensive research looking for novel enzymes/process and their optimization. Immobilization and scale up is also required for taking this process up to industrial level. Process improvement and downstream processing research for product purification is going on. Some of the areas which look promising are metagenomic screening of novel biocatalysts, gene cloning for overexpression and purification, etc. Bioreactor designing for scale up and simultaneous production and purification of desirable products are also being emphasised. Exploring biological activities after enzymatic reaction is one of the main areas of research nowadays.

Chapter 17
Industrial Enzyme Technology: Potential Applications... 375
Michael Bamitale Osho, McPherson University, Nigeria

Biotechnology, being the application of biological organisms and their components in pharmaceutical and other industrial processes, has emerged as the basic transformation tool for starch hydrolysis enzyme. Several advantages over chemical catalysts under mild environmental conditions with efficiency and high specificity have been accrued to this fact. Such include ingredient substitution through continuous fermentation, increased products yield and plant capacity, processing aid substitution, more efficient processing, less undesirable products with improved products. This chapter reports on the molecular properties of thermostable enzymes such as alpha-amylases, alpha-glucosidases, glucoamylases pullulanases as relates to pharmaceutical industries; highlights various technology development, continuous solid-state fermentation, metabolic engineering, sol-gel immobilized enzyme arrays often use in enzyme industries. The new modern biotechnology leads to improvement in the effects of various physiological conditions which may allow various industrial processes to carry out lower energy consumption, harmless to the environment, high efficiency, and the product's properties enhancement.

Chapter 18
Role of Enzymes From Microbes in the Treatment of Recalcitrant From Industries......................... 395
Veena Gayathri Krishnaswamy, Stella Maris College, India

The limited availability of fresh water is a global crisis. The growing consumption of fresh water due to anthropogenic activities has taken its toll on available water resources. Unfortunately, water bodies are still used as sinks for waste water from domestic and industrial sources. Azo dyes account for the majority of all dye stuffs, produced because they are extensively used in the textile, paper, food, leather, cosmetics, and pharmaceutical industries. Bacterial degradation of azo dyes under certain environmental conditions has gained momentum as a method of treatment, as these are inexpensive, eco-friendly, and can be applied to wide range of such complex dyes. The enzymatic approach has attracted much interest with regard to degradation of azo dyes from wastewater. The oxido-reductive enzymes are responsible for generating highly reactive free radicals that undergo complex series of spontaneous cleavage reactions, due to the susceptibility of enzymes to inactivation in the presence of the other chemicals. The oxidoreductive enzymes, such as lignin peroxidase, laccases, tyrosinase, azoreductase, riboflavin reductive, polyphenol oxidase, and aminopyrine n-demethylase, have been mainly utilized in the bacterial degradation of azo dye. Along with the reductive enzymes, some investigators have demonstrated the involvement in some other enzymes, such as Lignin peroxides and other enzymes. This chapter reviews the importance of enzymes in dye degradation.

Chapter 19
Moving into Nanotechnology Roles to Mimic and Boost Enzyme Activity 421
Maria Laura Soriano, University of Cordoba, Spain

A new tendency toward the design of artificial enzymes based on nanostructures (nanodots, nanofibers, mesoporous materials) has emerged. On one hand, nanotechnology bestows self-catalytic nanoparticles with a specific activity to achieve efficient reactions with low number of by-products. On other hand, the nanoparticles may behave as nanometric scaffolds for hosting enzymes, promoting their catalytic activity and stability. In this case, enzyme immobilization requires the preservation of the catalytic activity by preventing enzyme unfolding and avoiding its aggregation. These approaches render many other advantages like hosting/storing enzymes in nanotechnological solid, liquid, and gel-like media. This chapter focuses on the most up-to-date approaches to manipulate or mimic enzyme activity based on nanotechnology, and offers examples of their applications in the most promising fields. It also gives new insight into the creation of reusable nanotechnological tools for enzyme storage.

Compilation of References 441

About the Contributors 541

Index 546

Foreword

In the present time, science is running very fast because of hasty developments of various technologies and research in order to make human life more comfortable, luxurious and healthy. Pharmaceuticals, food research and industries are the three main pillars upon which human life is mainly dependent. The faster growth of development and industries related to the human based on any types of materials and technologies will also generate several human health issue along with it. In order to improve the human health there are significant role of medicines/pharmaceutics in their life. Nutritional food, harmless medicines (medicines without side effects or with lesser side effects), eco-friendly industrial activities create the environment of a real healthy, green and flourishing atmosphere. From past few years, a lot number of researches are going on very fast on enzymology and enzyme based green technologies to develop the medicines with lesser or no side effects and healthy food products with enzymatic removal of serious pollutants, harmful food additives and food based by-products produced from related industries. Enzymes are the valuable large protein molecules that play a significant role in living system along with their role in food industries, pharmaceutical industries, synthetic chemistry, paper, pulp and dyes industries, bioremediation study, green technology developments etc.

The present book entitled *Research Advancements in Pharmaceutical, Nutritional and Industrial Enzymology* may be a milestone in the direction of enzymes based research and developments. This is unique collection of 19 chapters written by experts of enzymatic field belonging to different countries. Each chapter is focused on the valuable studies based on pharmaceutical, food and industrial enzymology. Many chapters of this book contain significant assessment of pharmaceutical role of enzymes while others illustrate the significant role of enzymes in food industries in the form of quality enhancements as well as bioremediation of pollutants generated from food, pharmaceutical and other industries. This book fully justifies its novel contents and materials.

The writing style, scientific significance, easy and comprehensible language and nice subject based detailed discussions make this book highly noteworthy for all scientific readers

Sudha Yadava
DDU Gorakhpur University, India

Preface

Enzyme is the most renowned term for the scientific as well as social society. It is involved in human life from the birth till end. Every system either natural or living based have large varieties of enzymes inside and their different types of role. They are macromolecules of proteins generally act as biocatalysts in speeding up the reactions occurring inside body without being used up. They play a variety of role in living organisms, plants, insects, bacteria etc. They have several valuable biological roles in metabolism, in control of activity, in diseases and diseases control, pharmaceutical industries, food industries, synthetic industries etc. They have variable applications in the field of biofuel industry (cellulases and ligninases), biological detergent (proteases, amylases, mannanases), brewing industry (amylase, glucanases, proteases, β-glucanases, pullulanases and acetolactate decarboxylase), food processing (laccases, amylases, proteases, trypsin, cellulases and pectinases), dairy industry (renin, lipases), culinary uses (papain), personal care (proteases), paper industry (xylanases, hemicellulases, lignin peroxidases, laccases), molecular biology (nucleases, DNA ligases, polymerases, etc.) (https://en.wikipedia.org/wiki/Enzyme). Few examples of enzymes of medicinal importance are Asparaginase (use in leukaemia), Glutaminase (use in leukaemia), Hyaluronoglucosaminidase (heart attack), Lysozyme (bacterial cell wall hydrolysis, antibiotic), thiosulphate sulfurtransferase (in cyanide poisoning), β-lactamases (penicillin allergy), Streptokinase (in blood clots), Trypsin (in protein hydrolysis, inflammation), Ribonuclease (in RNA hydrolysis, antiviral), uricase (in gout), urokinase (in blood clots) (http://www1.lsbu.ac.uk/water/enztech/medical.html).

Aforementioned applications of enzymes are only like a trailer of a long movie. There are huge known and unknown applications of enzymes. Keeping the different valuable fields and applications of enzymes, editors have best tried to produce such types of book that can provide novel information to research, academic and scientific audience. This book is dedicated to the pharmaceutical, nutritional and industrial enzymology. Different types of novel chapters are compiled in this book. Each chapter has been written by the experts of their own fields. Enzyme based contributions of authors of different countries like India, Spain, Denmark, Malaysia, Cameroon, Tehran, Nigeria and Egypt in this book on smoldering topic make this book novel.

This book entitled *Research Advancements in Pharmaceutical, Nutritional, and Industrial Enzymology* is made up of total 19 chapters which have been contributed by the experts of their fields. All the selected chapters contain promising information regarding enzymatic studies and their roles in different fields like pharmaceutical/medicines, nutrition and industries. All the chapters in this book may be highly handy in growth of enzyme based knowledge and it will attract a huge audience working in the field of science and technology, biochemistry, biotechnology, microbiology, enzymology, bio-synthetic chemistry and medicinal chemistry. Subjects, contents and languages of chapters of this book are highly simplified, justified, improved and easy to understand so that its advantageous position could be easily reached up to research scholars, scientists, academic graduate and post graduate students and industries.

This book starts with Chapter 1 written by the Abderrezak Khelfi which has detailed valuable manifestation and discussions on the therapeutic enzymes and their roles in the cure of hereditary deficiency diseases. In this chapter Khelfi discussed and summarized the therapeutic enzymes used in clinical practice, with particular reference to those obtained from biological sources and biotechnology processes.

Chapter 2 written by Abderrezak Khelfi have detailed novel discussions on enzyme replacement therapy and he has compiled various information and research works based on use of therapeutic enzymes in the treatment of cardiovascular diseases and coagulation disorders. The therapeutic enzymes reviewed in this chapter by Khelfi are used for the treatment of cardiovascular diseases and hereditary diseases leading to coagulation disorders. Enzyme preparations obtained by direct fractionation from a naturally producing source and recombinant enzymes are considered in this chapter.

Chapter 3 written by Abderrezak Khelfi is also based on therapeutic roles of enzymes. They describe very adequately the role of enzymes in treatment of non-deficiency diseases. This chapter gives an outline of clinical uses of therapeutic enzymes used in non-deficiency diseases. Developments of these enzymes are reviewed with a particular focus on bioengineering applied to the native proteins.

Chapter 4 written by Tambe and Desai on colonical bacterial enzymes and their pharmaceutical significance and application. This chapter reviews the importance of colonic bacterial enzymes, their production methods as well as their utilization to develop various formulations for achieving colon-specific drug delivery.

Chapter 5 has been written by Singh and Singh on "Human Immunodeficiency Virus Reverse Transcriptase (HIV-RT): Structural Implications for Drug Development". They showed and discussed the salient features of HIV-RT that make it an attractive target for rational drug design and chemotherapeutic intervention in the management of acquired immunodeficiency syndrome. Further, they have also conferred the role of RT in the viral life cycle, the ways the drugs act to inhibit the normal functions of RT and the mechanisms that the virus adapts to evade the available drugs.

Chapter 6 has been written by Awad and Balbaa on topic "The Use of Liposomes in Enzymes and Drug Design: Liposomes Drug Delivery System". They discussed the role of liposomes in enhancing the effectiveness of therapeutic agents including enzymes along with the clarification of the importance of liposomes in improving the limitations of drugs.

Chapter 7 written by Raval and Chandel presents the extensions of enzyme based research for the pharmaceutical application and studied that enzymes play a major role in the diagnosis, curing, biochemical investigation and monitoring of many dreaded diseases of the century. They compiled the information on the sources, properties and applications of important therapeutic enzymes.

Chapter 8 written by Upadrasta et al shows the significance of "Enzyme Triggered Hydrogels for Pharmaceutical and Food Applications". In this chapter they have put forth the different mechanisms of enzyme-mediated hydrogels preparations and its fabrication towards pharmaceutical and food sectors along with the discussion of recent trends and further prospects.

Chapter 9 written by Patel and Gupte deals a green approach in bioactive compound synthesis via laccase catalysis. This chapter summarizes the advances that have been made in developing technologies based on laccase mediated reactions in the field of medicine, agriculture, food and pharmaceuticals.

Chapter 10 has been written by Forootanfar et al on "Laccase-Mediated Treatment of Pharmaceutical Wastes". In this chapter they have discussed the role of laccase in the removal of pharmaceutical wastes.

Their chapter provides a critical review of the literature and reveals the pivotal role of laccases in the elimination and detoxification of pharmaceutical contaminants in aquatic environments and wastewaters.

Chapter 11 written by Kushwaha et al deals the discussion on the role of laccase obtained from white rot fungi in food, pharmaceutical and other types of industries. This chapter deals the laccase role in food industry in respect of wine stabilization, beer stabilization, fruit juice processing, baking, improvement in food sensory parameters and sugar beet pectin gelation. They also showed the role of laccase in compound detection in beverages. They have also discussed the various other role of laccase in textile industries and pharmaceutical industry.

Chapter 12 written by Sonam et al was focused on the potential applications of peroxidase enzymes in treatment of industry wastes. This book chapter is focused on the remediation and degradation potential of lignin peroxidase (LiP), Manganese peroxidase (MnP) and versatile peroxidase (VP), whether alone or in combination, on different types of wastes which are toxic to environment if left untreated.

Chapter 13 written by Ho involve the discussion on biotechnology of microbial xylanase enzyme which are inducible enzymes responsible for the hydrolysis of xylan into xylose and have significant role in pulp and paper, bakery, food and beverage, detergents, textile and animal feed as important commercial enzymes.

Chapter 14 was written by Serge et al that involve the study on the "Functional and Antioxidant Properties of Protein Hydrolysates From Ricinodendron Heudelotii (Bail.) Flours From Cameroon". They defatted the *Ricinodendron heudelotii* and used it as substrate to produce protein hydrolysates using papain (PHP), trypsin (PHT), proteases from *Abrus precatorius* (PHAp) and *B. enneandra* (PHBe). They performed the degree of hydrolysis (DH), antioxidant (DPPH method) and functional properties of hydrolysates.

Chapter 15 deals the compiled discussion on the enzyme use and production in industrial biotechnology which has been written by Nandy. This chapter demonstrates the bioprocess strategies involved in the application and production of enzymes from industrial view point.

Chapter 16 written by Sujata and Tripathi on "Trends and Challenges in Enzymatic Bioengineering of Natural Products to Industrially Valuable Products: Status and Future of Industrial Enzymology". They summarized the natural products being investigated for the industrially valuable product as a result of enzymatic hydrolysis and covered the trends and challenges in biochemical production of these molecules and their yield improvement.

Chapter 17 has been written by Osho who presents a chapter on "Industrial Enzyme Technology: Potential Applications". This chapter deals the properties and other of thermo-stable enzymes such as alpha-amylases, alpha-glucosidases, glucoamylases pullulanases as relates to pharmaceutical industries. In this chapter, authors have lengthily explored the valuable values of starch hydrolyzing enzymes as prospective tool for the many biotechnological opportunities.

Chapter 18 has been prepared by Krishnaswamy on topic "Role of Enzymes From Microbes in the Treatment of Recalcitrant From Industries" in which he has discussed the significance of microbial enzymes in treatment of industrial recalcitrant. This chapter mainly reviews on the importance of enzymes in dye degradation.

Chapter 19 written by Soriano Dotor is based on the topic "Moving Into Nanotechnology Roles to Mimic and Boost Enzyme Activity" in which they have focused to discuss on the most up-to-date approaches to manipulate or mimic enzyme activity based on nanotechnology, and offers examples of their applications in the most promising fields. This chapter also gives new insight into the making of re-usable nano-technological tools for enzyme storage.

Acknowledgement

I, Dr Shashi Lata Bharati, Editor of the book entitled *Research Advancements in Pharmaceutical, Nutritional and Industrial Enzymology* am cordially thankful to the Department of Chemistry, NERIST, Nirjuli, Arunachal Pradesh, India for providing necessary facility at its level. I am very much thankful to Prof H. S. Yadav (Department of Chemistry, NERIST, Nirjuli, Arunachal Pradesh, India) for his kind and continuous support at each level of my academic problems. I am also thankful to my research scholar Miss Chandana Sarma, younger sisters Pusp Lata, Sneh Lata and Hem Lata.

I am dedicating this book to my parents Shri Om Prakash and Smt Ramanti Devi who are the lodestar of my life.

I, Dr Pankaj Kumar Chaurasia, Editor of the book entitled *Research Advancements in Pharmaceutical, Nutritional and Industrial Enzymology* am cordially thankful to the Department of Biotechnology, Motilal Nehru NIT, Allahabad for providing required facilities during the work and also thankful to the Department of Chemistry, University of Allahabad as intial place for the work as Guest Faculty. I am too much grateful to CSIR-HRDG New Delhi for providing the fund support in the form of Research Associate [file No. 09/1032(0011)2K17]. I am very much thankful to Dr. Ashutosh Mani (Assistant Professor, Department of Biotechnology, Motilal Nehru NIT, Allahabad) for his continuous moral support to me. I am grateful to Prof. MP Singh (Centre of Biotechnology, University of Allahabad) and Prof. Sudha Yadava (Department of Chemistry, DDU Gorakhpur University, Gorakhpur) for their relevant and valuable suggestions. I am very much thankful to my younger brother Er. Neeraj Kumar Chaurasia and cousin brother Dr Ajay Kumar Chuarsia for their moral help during the every stage of my life.

I am dedicating this book to my lovely parents Shri Hari Narayan Chaurasia and Smt Sumitra Devi.

Shashi Lata Bharati
North Eastern Regional Institute of Science and Technology, India

Pankaj Kumar Chaurasia
LS College Muzaffarpur, India

Chapter 1
Enzyme Replacement Therapy:
Therapeutic Enzymes Used for the Treatment of Hereditary Deficiency Diseases

Abderrezak Khelfi
National Center of Toxicology, Algeria

ABSTRACT

Enzyme replacement therapy is a therapeutic approach in which the specific enzyme that is absent or inactive in affected individuals is replaced with a functional enzyme molecule derived from biological sources or produced by biotechnology. A large number and variety of enzyme defects have been identified in humans. Over 40 hereditary deficiency diseases were reported. The common feature is that enzyme deficiency leads to the accumulation of undegraded molecules and lysosomal storage, resulting in organ dysfunction. Crude enzyme preparations are often unsuitable for therapeutic uses because of their potential contamination and antigenicity. Advances in gene identification and cloning led to the subsequent production and demonstration of equal efficacy of recombinant human enzyme. The adverse events recorded range from boxed warnings for severe allergic reactions. This chapter summarizes therapeutic enzymes used in clinical practice, with particular reference to those obtained from biological sources and biotechnology processes.

1. INTRODUCTION

Enzymes are biomolecules that catalyze and accelerate chemical reactions. Almost all biological processes require enzymes in order to maintain the homeostasis of cells. They are extremely selective for their substrates, and their activity is highly regulated by substrate concentration, pH, and temperature.

Enzyme replacement therapy (ERT) is a therapeutic approach in which the specific enzyme that is absent or inactive in affected individuals is replaced with a functional enzyme molecule derived from biological sources or produced by DNA technology. The concept of ERT has been around for at least 50 years. For example, de Duve described enzymes as part of therapeutic protocols for genetic deficiencies in the 1960s (Vellard, 2003). A large number and variety of enzyme defects have been identified in

DOI: 10.4018/978-1-5225-5237-6.ch001

humans, many leading to diseases produced by altered amounts of metabolites resulting from enzyme deficiency. In the early 1960s, the first metabolic storage disease was identified. Since then over 40 such diseases have been reported. Metabolic storage disorders are life-threatening diseases caused by insufficient activity of enzymes required for the catabolism of biological materials that arise from the normal turnover of body constituents. The severity of symptoms depends on whether the deficiency is partial or complete. The so-called storage diseases provided both the best early examples of a clear association of enzyme deficiency and disease as well as candidates for a successful therapeutic approach to treatment. Lysosomal storage disorders (LSD) are strong candidates for the development of specific innovative therapies. The common feature is that enzyme deficiency leads to accumulation of undegraded macromolecules and lysosomal engorgement, resulting in organ dysfunction.

The importance of ERT in treating LSD was recognized in the early 1990s when an effective enzyme replacement was first developed for Type I Gaucher disease using highly purified placenta-derived β-Glucocerebrosidase. It markedly improved hematological indices and reduced hepatosplenomegaly in patients (Barton, Furbish, Murray, Garfield, & Brady, 1990). Advances in gene identification and cloning led to the subsequent production and demonstration of equal efficacy of recombinant human enzyme produced in Chinese hamster ovary (CHO) cells. The introduction of Cerezyme ® led to the withdrawal of the placental product (Grabowski, Barton, Pastores, Dambrosia, Banerjee, McKee, & Brady, 1995).

The successes of ERT in Gaucher disease established a viable medical and industrial model for the development of ERTs for other LSD. In fact, ERT has been successfully done in other enzyme deficiency disorders like Fabry disease, Pompe disease, mucopolysaccharidosis (MPS), severe combined immunodeficiency (SCID), etc (Table 1). However, even with an early introduction of treatment there are still several limitations of ERT such as delivering therapeutic levels of lysosomal enzymes across the blood-brain barrier and despite efforts, ERT mostly slows down disease progression and attenuates symptom expression.

Research and drug developments fostered under Orphan Drug Product Development Programs have greatly assisted the introduction of efficient and safe enzyme-based therapies for several rare disorders. The Orphan Drug Act was passed in 1983 in the USA to encourage biopharmaceutical companies to develop treatments for these rare disorders affecting only small number of people (<200,000).

Adverse events recorded during ERT range from boxed warnings for severe allergic reactions, including anaphylaxis, to rare effects on the heart, lungs, liver, and blood as well as a variety of mild to moderate commonly seen responses such as gastrointestinal symptoms, headache, and mild skin reactions.

Crude enzyme preparations are often unsuitable for therapeutic uses because of their potential contamination with endotoxins and antigenicity. When the enzyme used is a foreign protein, it can elicit an immune response that alters the clearance rate or induces severe allergic reactions in the host. In the field of modern innovative biotechnology, a relatively small number of recombinant enzymes have been developed as the main active substances of therapeutics.

Advances in the knowledge of therapeutic uses of enzymes have heightened interest in the manufacture and processing of these macromolecules. Moreover, gene identification and cloning led to the subsequent production and demonstration of equal efficacy of recombinant human enzyme produced in CHO cells and various optimization strategies have overcome the current limitations of ERT.

This chapter summarizes therapeutic enzymes used in clinical practice, with particular reference to those obtained from biological sources and biotechnology processes. Some of the pathophysiological features of hereditary deficiency diseases are described. Some of the key advances that have led to successful commercial production of enzymes for ERT will be addressed.

2. ENZYMES USED FOR THE TREATMENT OF HEREDITARY DEFICIENCY DISEASES

2.1. Enzymes Used for the Treatment of Lysosomal Storage Diseases

2.1.1. Alglucerase/Imiglucerase

Type 1 Gaucher disease is a hereditary metabolic disorder characterized by a functional deficiency in β-glucocerebrosidase activity (Weinreb, Barranger, Charrow, Grabowski, Mankin, & Mistry, 2005). This 60-kDa enzyme (β-D-glucosyl-N-acylsphingosine glucohydrolase, E.C. 3.2.1.45) is a lysosomal glycoprotein, which catalyzes the hydrolysis of the glycolipid glucocerebroside or glucosylceramide to glucose and ceramide. Deficiency of this enzyme results in the accumulation of lipid glucocerebroside primarily in peripheral tissue macrophages, which become engorged and known as Gaucher cells. This gives rise to the most prevalent LSD known as Gaucher disease, characterized by swollen organs (due to accumulation of affected macrophages therein), bone pain, anemia, and sometimes neuronal damage. The most common form of the disease is caused by a point mutation, resulting in an asparagine to serine substitution at amino acid residue 370 of the 497 amino acid mature protein. This mutation is inherited in an autosomal recessive way with an incidence of 1/600 to 1/2,500 in Ashkenazi Jews and has an estimated worldwide prevalence of around 1/75,000 births (Grabowski, 2005).

Gaucher disease can be treated by ERT using a β-Glucocerebrosidase. In 1991, Alglucerase (Ceredase ® mannose-terminated-β-glucocerebrosidase purified from human placenta) was the first ERT for Type 1 Gaucher disease. Concerns relating to cost, source availability and accidental transmission of disease leaded to the development of a recombinant glucocerebrosidase form. In fact, the human enzyme has been replaced by recombinant glucocerebrosidase named Imiglucerase (Cerezyme ®) produced in engineered CHO cell line. This was critical for overcoming the limitations of availability in view of 20,000 placentas were required to provide a one-year supply for a single patient. Imiglucerase was regarded as the 'gold-standard' against which all other products are compared even though the disorders being treated are diverse.

Imiglucerase differs from the native glycoprotein in two important respects. First, there is a single amino acid substitution in the primary sequence, which arose in the cloning process. Second, the glycosyl moiety of the protein is deliberately modified after purification. More precisely, Imiglucerase (MW 60.4 kDa) contains 497 amino acids with four N-linked glycosylation sites, and differs from the wild-type glucocerebrosidase amino acid sequence by a single amino acid substitution at position 495, where arginine being substituted by histidine. This amino acid change does not have any effect on enzyme biological activity. The oligosaccharide chains at the glycosylation sites are modified after purification to terminate in α-mannose. The enzyme-based processing step uses an Exoglycosidase to remove outer sugar residues that cap the oligosaccharide side chains of the β-glucocerebrosidase. This exposes mannose residues underneath, facilitating specific uptake by macrophages via macrophage cell-surface mannose receptors. In this way, the product is specifically targeted to the cell type most affected by the disease. Indeed, unmodified glucocerebrosidase, if administered, is quickly removed from the bloodstream by the liver.

As the native enzyme, Imiglucerase catalyzes the hydrolysis of glucocerebroside to glucose and ceramide preventing secondary hematologic, spleen, liver, and skeletal complications. More than 5,600 patients in 90 countries have been treated with Cerezyme ® (Das & Goyal, 2014). Imiglucerase (Cerezyme ®) was approved in 1994 by the U S Food and Drug Administration (FDA) and in 1997 by the

European Medicines Agency (EMA) for use as long-term ERT in patients with confirmed diagnoses of non-neuropathic (Type 1) or chronic neuropathic (Type 3) Gaucher disease. The therapy is ineffective in neuropathic Gaucher disease even if instilled directly into the cerebrospinal fluid. It has also relatively poor efficacy against pre-existing bone and lung disease.

Cerezyme ® is supplied as a pure, lyophilized product available in 400-U vials and is reconstituted in sterile water for clinical use. The majority of patients treated with Imiglucerase are prescribed 15 to 60 units/kg/dose every 2 weeks, administered intravenously over 2 to 3 hours that can be lowered as therapy (and clinical response to treatment) progresses. Its application leads to a marked improvement in the clinical manifestation of the Gaucher disease. In fact, injection of Imiglucerase into patients leads to elevated levels of the enzyme in serum and reduction in the accumulation of Glucocerebroside which reduces anemia, thrombocytopenia, spleen and liver size, and decreased cachexia (Pastores, Weinreb, Aerts, Andria, Cox, Giralt, & Tylki-Szymanska, 2004). General side effects, such as fatigue, headache, fever, abdominal pain, chills, dizziness, and backache, have been reported in only 1.5% patients (Das & Goyal, 2014). Other related adverse events were reported such as chest discomfort, pruritus, rash, urticaria, dyspnea, nausea, diarrhea, headache, tachycardia, and flushing. Overall, adverse events can be classified into three categories: general disorders and administration site reactions (nausea, vomiting, pyrexia, chills, and chest discomfort); skin and subcutaneous tissue disorders (pruritus, rash, and urticaria); respiratory, thoracic, and mediastinal disorders (dyspnea, cough, and throat irritation). They are predominantly self-limited and effectively managed by slow and gradual increasing infusion or pretreatment with antipyretics and antihistamines.

The most troublesome side effect is life threatening hypersensitivity (hypotension, dyspnea, pruritus, urticaria, angioedema, and anaphylaxis). The seroconversion and IgG antibody formation can be observed with their disappearance over time. Very rarely, IgE antibodies form can cause serious adverse events. Because humans can develop antibodies that may affect treatment efficacy, it is recommended that patients should be monitored for any allergic reaction during their first year of treatment. After this period, if no IgE antibodies have developed, patients can receive infusions at home instead of in a medical setting.

2.1.2. Velaglucerase-Alfa

Velaglucerase-alfa (VPRIV ®) is a human β-glucocerebrosidase produced in a human cell line used also for long-term treatment of Gaucher disease. It is a 63 kDa monomeric glycoprotein with 497 amino acids containing five potential N-linked glycosylation sites that targets macrophages via mannose receptors, and acts to degrade accumulated glucocerebroside within the macrophages. Unlike Imiglucerase, the amino acid sequence of Velaglucerase-alfa is identical to the human wild-type enzyme. Another distinguishing structural feature is that Velaglucerase-alfa has higher mannosyl N-linked glycans content than Imiglucerase for target cell recognition.

Velaglucerase-alfa is available in 200-or 400-U vials. The dose recommended is 60 units/kg every 2 weeks, administered over 1 hour.

The adverse effects related to Velaglucerase-alfa are hypersensitivity including anaphylaxis and allergic dermatitis, infusion reactions, headache, pyrexia, arthralgia, nasopharyngitis, dizziness, and bone pain.

2.1.3.Taliglucerase-Alfa

Taliglucerase-alfa (Elelyso ®) has been developed using genetically engineered carrot cells for the treatment of Gaucher disease (Aviezer, Brill-Almon, Shaaltiel, Hashmueli, Bartfeld, Mizrachi, & Galun, 2009). This enzyme has a prolonged half-life compared with Imiglucerase. Plant-cell expression system is more efficient and produces safe enzymes that are free from animal-derived components (Shaaltiel, Bartfeld, Hashmueli, Baum, Brill-Almon, Galili, & Aviezer, 2007; Zimran, Brill-Almon, Chertkoff, Petakov, Blanco-Favela, Munoz, & Aviezer, 2011). Compared with Imiglucerase, amino acid sequence of Taliglucerase-alfa (MW 60.8 kDa) contains two additional amino acids at the N-terminus and seven additional amino acids at the C-terminus. As Imiglucerase, Taliglucerase-alfa differs from the human glucocerebrosidase at residue 495, where there is a substitution between arginine and histidine. This enzyme is also up taken by target cells via mannose receptors. It catalyzes the hydrolysis of glucocerebroside to glucose and ceramide preventing secondary hematologic, spleen, liver, and skeletal complications.

Taliglucerase-alfa is supplied as a lyophilized powder and is reconstituted in sterile water for clinical use. It is available in 200-U vials. The recommended dose is 60 units/kg administered every 2 weeks as a 60 to 120-min intravenous infusion.

The list of adverse events induced by Taliglucerase-alfa is similar to the lists for the other two β-Glucocerebrosidases with anaphylaxis, allergic and infusion reactions, headache, pharyngitis, arthralgia, as well as back and extremity pain.

2.1.4. Agalsidase

Agalsidase is used in ERT for Fabry disease. Fabry disease is a X-linked, recessively inherited glycosphingolipid LSD which leads to the abnormal metabolism and progressive lysosomal accumulation of neutral glycosphingolipids (mainly globotriaosylceramide (GL-3)) in many tissues and cell types, such as kidney, heart, skin, and cells lining the blood vessels. The incidence of this disease is 1 in 40,000 to 1 in 60,000 worldwide, mostly in Caucasians (Spada, Pagliardini, Yasuda, Tukel, Thiagarajan, Sakuraba, & Desnick, 2006). It primarily affects the vasculature and results in renal failure, pain, and corneal clouding. The disease is caused by a genetically encoded defect of the lysosomal α-galactosidase A (E.C. 3.2.1.22).

Two recombinant α-Galactosidases (Agalsidase alfa or Replagal ® produced in a human cell line and Agalsidase beta or Fabrazyme ® obtained by recombinant DNA technology in a CHO cells) have been shown to mobilize and clear the sphingolipid deposits. The FDA approved Agalsidase beta in 2003 for long-term ERT in patients with Fabry disease. It has the same amino acid sequence as the native enzyme. As a recombinant equivalent of naturally occurring α-galactosidase, recombinant human α-galactosidase was intended to provide an exogenous source of α-galactosidase A and to limit GL-3 accumulation in tissue, thereby ultimately to alleviate the clinical consequences of GL-3 accumulation. This enzyme drug contains two subunits (MW~100 kDa) with 398 amino acids and hydrolyzes globotriaosylceramide (GL-3, ceramide trihexoside) and lipids with terminal α-D-galactose in the lysosome thus reducing GL-3 deposition in capillary endothelia.

Agalsidase beta has some side effects such as difficulty in breathing, choking of throat, hives, rashes, itching, and fever. However, the infusion reaction is the most common adverse event seen during administration of Agalsidase beta. Cardiac disorders, when they occur, are commonly tachycardia and gastrointestinal reactions (nausea, vomiting, and diarrhea).

Other adverse reactions have been reported such as immunogenicity and allergic reactions (anaphylaxis, urticaria, pruritus, and rash), stroke, ataxia, and pain.

2.1.5.Alglucosidase Alfa

In 2006 recombinant Alglucosidase alfa received approval by the FDA and the EMA for the treatment of Pompe disease. Pompe disease is a glycogen storage disease Type II and an autosomal recessive genetic disorder arising from the deficiency of the lysosomal enzyme acid α-Glucosidase. The estimated incidence of Pompe disease range from 1/40,000 to 1/300,000 (Phupong, Shuangshoti, Sutthiruangwong, Maneesri, Nuayboonma, & Shotelersuk, 2005). α-Glucosidase or acid Maltase (E.C. 3.2.1.3) catalyses the hydrolysis of glycogen and other natural α-glucans by hydrolyzing α-1,4- and α-1,6-glucosidic bonds. Therefore, the deficiency of this enzyme results in the intracellular accumulation of glycogen in various tissues, with largest impacts on respiratory, cardiac, and skeletal muscles.

Pompe disease can manifest in two different forms, which depend mainly on the age of onset, and include infantile, juvenile, and adult onset.

The infantile form is characterized by an almost total lack of α-Glucosidase activity and has a very poor prognosis. It is characterized by prominent hypotonia, muscle weakness, motor delay, feeding problems, hypertrophic cardiomyopathy, respiratory and cardiac insufficiency that lead to death in the first year of life in the majority of affected patients. Juvenile, adult or late onset forms of the disease retain residual enzyme activity, therefore clinical symptoms are less severe and they generally exhibit only moderate cardiac involvement. However, the disease progresses with age, leading to deterioration of the respiratory and skeletal muscles (often mistaken for limb girdle muscular dystrophy). This can result in early mortality, generally from respiratory failure. The combined incidence of all forms of Pompe disease is estimated to be 1/40,000 (Disease, Kishnani, Steiner, Bali, Berger, Byrne, & Watson, 2006).

The recombinant α-Glucosidase contains 952 amino acids with an apparent MW of 110 kDa. There are seven glycosylation sites, and some of the polysaccharide chains are phosphorylated. In addition to glycosylation and phosphorylation, the maturation of α-Glucosidase involves proteolytic processing at both the N- and C-terminal ends.

The final recombinant acid α-Glucosidase is a glycoprotein with a MW of 109 kDa and 883 amino acid residues. CHO-cell-produced enzyme was found to be effective in reducing glycogen granules in affected tissues.

There are two approved commercial preparations of Alglucosidase alfa by the FDA, Myozyme ® and Lumizyme ®.

Lumizyme ® is approved by the FDA as ERT for late-onset Pompe disease without evidence of cardiac hypertrophy. Myozyme ® is approved for infantile-onset Pompe disease.

ERT with Alglucosidase alfa has been shown to significantly improve survival, quality of life, and motor outcomes in most patients with infantile Pompe disease. It is clear that ERT is effective in infantile form, especially on reducing the cardiac hypertrophy and prolonging the survival. Additionally, the potential central system involvement in Pompe patients deserves attention because the recombinant enzyme currently used in treating Pompe disease does not cross the blood-brain barrier hence cannot counteract the CNS damage effectively.

The dosage of enzyme required to effect an improvement in clinical condition in Pompe disease is usually 20 mg/kg/dose.

Both preparations, Myozyme ® and Lumizyme ®, carry warnings for severe allergic reactions including life-threatening anaphylaxis and cardiorespiratory failure. Serum IgE antibodies to Alglucosidase can be detected.

Lumizyme ® is responsible for a wide range of reactions including pyrexia, flushing, hyperhidrosis, headache, dizziness, hypertension, rash, and urticaria while for Myozyme ®, infusion reactions (urticaria, fever, and decreased oxygen saturation), infections such as pneumonia and respiratory syncytial virus, respiratory distress/failure, and gastrointestinal problems predominate.

2.1.6. Laronidase

Laronidase (α-L-iduronidase (IDUA)) is intended for the treatment of MPS Type I. MPS I is an autosomal recessive LSD that results in the deficiency of IDUA (E.C. 3.2.1.76), a glycosyl hydrolase necessary for the cleavage of terminal α-iduronic acid residues from glycosaminoglycans (GAG), heparan sulfates, and dermatan sulfates that remain stored in lysosomes.

The estimated incidence of MPS I in British Columbia and Northern Ireland is approximately 1/76,000 and 1/300,000 newborns (Kang & Stevens, 2009).

Depending on the severity of the mutation, the consequent accumulation of GAG in many tissues, particularly connective tissue results in a spectrum of multisystemic clinical symptoms, ranging from mild clinical phenotype (Scheie syndrome), to intermediate phenotype (Hurler-Scheie syndrome), and the most severe clinical presentation, termed Hurler syndrome with an incidence of 1/115,000 (Kresse, 2004). Patients can experience frequent respiratory, ear infections, cardiac disease as well as enlarged liver and spleen, joint stiffness, excessive head growth, hernias, and malformation of the spine. This disease can also result is developmental delay in more affected cases.

The development of Laronidase began with the cloning of the human IDUA cDNA in 1991 in CHO cells. The cDNA codes for a protein of 653 amino acids of 76 kDa containing six potential N-linked oligosaccharide sites, all of which are utilized. The recombinant enzyme undergoes signal peptide cleavage and post-translational addition of the N-linked oligosaccharide chains as well as mannose 6-phosphate addition on glycosylation sites 3 and 6 to produce the mature monomeric form. This results in a recombinant variant of 628 amino acids with MW of 83 kDa.

The FDA and EMA approved the recombinant form of IDUA, Aldurazyme ®, in 2003 for use in the treatment of MPS I.

The most common adverse events observed with Laronidase treatment were upper respiratory tract infection, rash, hyperreflexia, paresthesia, pyrexia, and injection site reaction. Injection site reaction was offset by slowing the infusion rate, temporarily stopping the infusion, and/or administering additional antipyretics and/or antihistamines.

Hypersensitivity responses were observed and antibodies to Laronidase were detected in some patients during infusions. In general, these responses declined or resolved with continued weekly therapy.

2.1.7. Iduronate-2-Sulfatase

Iduronate-2-sulfatase or Idursulfase (IDS) (E.C. 3.1.6.13) is a member of the sulfatase family that is responsible for the cleavage of sulfate group from the 2-position of L-iduronic acid present on GAG such as dermatan sulfate and heparan sulfate. MPS Type II (Hunter syndrome) is a rare X-linked recessive disease caused by deficiency of this lysosomal enzyme. This disease is characterized by progressive accumulation of GAG in nearly all cell types, tissues, and organs. The incidence of Hunter's syndrome ranges from 1/132,000 in the United Kingdom (Kang & Stevens, 2009). As with several other forms of LSD, Hunter's syndrome may present in a spectrum of phenotypic severity, largely dependent on the mutations involved. Clinical manifestations include severe airway obstruction, skeletal deformities, cardiomyopathy and, in most patients, neurological decline with mental retardation. Multiorgan and tissue involvement can potentially lead to fatality. Death usually occurs in the second decade of life in severe cases.

IDS ERT is of significant benefit in reducing symptoms of MPS II. The only enzymatic treatment modality available for Hunter's syndrome is a recombinant form of IDS (Elaprase ®), which has been approved for use as an ERT (FDA: 2006; EMA: 2007). This form is a human IDS that contains 525 amino acids with MW of 76 kDa and 8 Asp-linked glycosylation sites. Cysteine amino acids are modified to formylglycine.

The most common adverse events with this treatment are infusion-based reactions, headache, nasopharyngitis, abdominal pain, arthralgia, pruritus, rash, flushing, as well as urticaria, pyrexia, dyspepsia, diarrhea, vomiting, cough, anxiety, and chest pain.

IDS carries an FDA black box warning of the risk of hypersensitivity after/during infusion of the drug. In a recent Korean study, anti-IDS IgE antibodies were detected in 34 patients taking the drug for MPS II (Kim, Park, Kim, Lee, Maeng, Cho, & Jin,, 2013).

2.1.8.Galsulfase

Galsulfase is a recombinant human N-acetylgalactosamine 4-sulfatase used for the treatment of MPS VI. MPS VI or Maroteaux-Lamy syndrome is a rare autosomal recessive (incidence 1/360,000), clinically heterogeneous LSD that develops due to a deficiency in the enzyme N-acetylgalactosamine 4-sulfatase (E.C. 3.1.6.12) called also Arylsulfatase B (ARSB) (Kang & Stevens, 2009). This enzyme is normally required for the breakdown of certain complex carbohydrates known as GAG, ant its deficiency results in the accumulation of partially degraded GAG in lysosomes of many cell types, tissues, and organs throughout the body and urinary excretion of dermatan sulfate, chondroitin sulfate, and other GAG. Indeed, Arylsulfatase B is responsible for hydrolysis of the sulfate moiety of the GAG chondroitin 4-sulfate and the GAG dermatan sulfate.

Patients with MPS VI usually appear normal at birth, but may present a wide range of symptoms that may progress slowly or rapidly. The following signs characterize this disease: dolicocephaly, wide forehead, spinal abnormalities, bony stenosis, meningeal thickening, and short stature. As the disease progresses they may develop facial infiltration, hepatosplenomegaly, joint contractures, degenerative joint disease, enlargement of heart valves, cardiomyopathy, sinusitis, otitis, pulmonary impairment, ocular abnormalities (corneal clouding, glaucoma and papilledema with optical atrophy), neurological involvement (hydrocephalus, medullar compression; mental retardation is not a common occurrence), obstructive sleep apnea syndrome and umbilical and inguinal hernias.

Galsulfase is a monomeric glycoprotein with a MW of 66 kDa. Amino acid sequence of the recombinant enzyme is identical to the native human enzyme. It contains 495 amino acids with six glycosylation sites and four mannose-6-phosphate. Cys53 residues are modified to formylglycine.

As the native enzyme, Galsulfase catalyzes the cleavage of the sulfate ester from terminal N-acetylgalactosamine 4-sulfate residues of GAG chondroitin 4-sulfate and dermatan sulfate. Increased catabolism of GAG in turn reduces systemic dermatan sulfate accumulation, thereby reducing the primary symptoms of MPS VI. This drug was approved by the FDA and EMA in 2005 and 2006 respectively for the treatment of patients with MPS VI. It has been granted orphan drug status. Naglazyme is a formulation of Galsulfase, which is a purified human enzyme that is produced by recombinant DNA technology in a CHO cell line.

Clinical trials have demonstrated that Galsufase provides clinically important benefits for MPS VI patients, specifically improved endurance and walking ability, reduction in hepatosplenomegaly, urinary GAG levels, and improvements in pulmonary function (Harmatz, Giugliani, Schwartz, Guffon, Teles, Miranda, & Swiedler, 2006). However, cardiac valve disease, joint disease, skeletal disease, cervical spinal cord compression, as well as ocular and nervous disorders remained largely unaffected in the course of ERT (Giugliani, Lampe, Guffon, Ketteridge, Leao-Teles, Wraith, & Harmatz, 2014; Harmatz, Garcia, Guffon, Randolph, Shediac, Braunlin, & Decker, 2014; Horovitz, Magalhaes, Acosta, Ribeiro, Giuliani, Palhares, & Llerena, J. C., Jr , 2013). Starting treatment at an earlier age may be beneficial for growth, and some aspects of the syndrome may be modified by longer periods of treatment.

Galsulfase is generally considered a relatively safe drug although some patients developed antibodies against the enzymes, leading to allergic reaction. Typical signs of an allergic reaction include shock, laryngeal headache, joint pain, and eye redness, edema, apnea, pyrexia, urticaria, pruritus, respiratory distress, and angioedema. Other adverse effects were reported such as Infusion-related reactions, rash, chills, headache, nausea, vomiting, visual abnormalities, anxiety, dyspepsia, and cough, as well as limb, chest, ear and abdominal pain.

2.1.9. Arylsulfatase A

Metachromatic leukodystrophy is a rare autosomal recessive metabolic disorder, arising from the deficiency of Arylsulfatase A (ARSA). The overall incidence of metachromatic leukodystrophy is generally low and ranges from 1/150,000 to 1/2,500 according to ethnic origin, Eskimos and Navajo Indians being the most affected populations (Kang & Stevens, 2009). Metachromatic leukodystrophy is characterized by the intralysosomal accumulation of the cerebroside 3-sulfate substrate that results in the progressive demyelination of the central and peripheral nervous system, leading to severe neurological impairment.

Although a recombinant form of ARSA is currently under clinical trials (Metazym) as an ERT, no treatment modality is currently available for this fatal metabolic condition.

2.1.10. Acid Sphingomyelinase

Niemann-Pick disease is a lipid storage disorder that results from the partial or complete absence of the lysosomal enzyme acid sphingomyélinase (E.C. 3.1.4.12), resulting in lysosmal accumulation of sphingomyelin in numerous tissues (spleen, liver, lungs, bone marrow, and brain), but particularly in macrophage and neurons.

Clinically patients suffer from progressive hepatosplenomegaly with reduced appetite, abdominal distension, pain, and thrombocytopenia. Accumulation of sphingomyelin in the CNS results in ataxia, dysarthria, and dysphagia. Accumulation of sphingomyelin results also in many succumbing to pulmonary complications and pneumonia in late childhood to middle age.

On 5 December 2016, orphan designation (EU/3/01/056) was granted by the European Commission for a recombinant product of the human acid Sphingomyelinase (also known as Olipudase alfa) for the treatment of Niemann-Pick disease.

2.1.11. Elosulfase Alfa

Elosulfase alfa (Vimizim ®) is a CHO cell-derived recombinant N-acetylgalactosamine-6-sulfatase (rhGALNS) monomer with 496 amino acids and MW of 55.4 kDa. It contains two glycosylation sites and the Cys53 situated in the active site is modified to formylglycine. It is indicated for the treatment of MPS IV Type A also known as Morquio A syndrome, an autosomal recessive disorder. This enzyme increases the catabolism of GAG that remain incompletely catabolized causing skeletal dysplasia and other abnormalities. Elosulfase alfa was granted marketing approval by the FDA and the EMA in 2014. The reported adverse reactions related to the use of Elosulfase alfa are anaphylaxis, urticaria, dyspnea, flushing, pyrexia, vomiting, nausea, headache, chills, and abdominal pain.

2.1.12. Sebelipase Alfa

Sebelipase alfa (Kanuma ®) is a recombinant human lysosomal acid lipase with a MW of 55 kDa produced in egg white of genetically engineered chickens. It has the same amino acid sequence as the native enzyme and six N-linked glycosylation sites. Sebelipase alfa is used for the treatment of patients with lysosomal acid lipase deficiency. This LSD is an autosomal recessive disease characterized by lysosomal accumulation of cholesteryl esters and triglycerides in organs including liver, intestine, and blood vessel walls. This results in increased liver fat and liver disease, lipid accumulation in intestinal walls leading to malabsorption, growth failure, and dyslipidemia with elevated LDL-C and triglycerides together with HDL-C.

Sebelipase alfa binds to cell surface receptors via its glycan structures and is internalized into lysosomes where it hydrolyses cholesteryl esters and triglycerides to free cholesterol, glycerol, and free fatty acids.

The reported adverse reactions related to the use of Sebelipase alfa are hypersensitivity to eggs or egg products. Other adverse reactions can appear such as fever, headache, nausea, vomiting, diarrhea, constipation, anemia, cough, nasopharyngitis, and fever.

2.2. Dornase Alfa

Bovine DNase I was first purified from pancreatic bovine extracts in the late 1940s. In 1958, a bovine pancreatic Deoxyribonuclease I (DNase I) preparation was used as a mucolytic agent for human use. However, severe adverse reactions like bronchospasms and asthmatic reactions after a repeated administration (high immunogenicity and high content of Trypsin and Chymotrypsin) led to the withdrawal of this preparation from the market, and therefore, clinical interest in DNase I diminished. DNase I received much attention once again when Shak, et al. cloned the human version of it in 1990 (Shak, Capon, Hellmiss, Marsters, & Baker, 1990). In 1993, this led to the marketing of recombinant human DNase I (Dornase alfa) under the trade name Pulmozyme ®.

Dornase alfa is a recombinant human DNase I (rhDNase) (E.C. 3.1.21.1), an extracellular glycoprotein produced in human pancreas, salivary glands stomach, and small intestine, which selectively cleaves DNA. This enzyme is present in many biological fluids such as urine, serum, and amniotic and cerebral fluid. rhDNase is used for the treatment of cystic fibrosis, which is a lethal autosomal recessive disorder caused by a defective cystic fibrosis transmembrane conductance regulator and a chloride channel protein, leading to improper salt balance and thick tenacious secretions of exocrine glands. The frequency of occurrence varies among populations, with persons of northern European extraction being most at risk. Within such populations approximately 1 in 2,500 newborns are affected (Walsh & Walsh, 2015). In patients affected by cystic fibrosis, expression of the aberrant gene results in compromised function of a number of tissue types, including the pancreas and sweat glands. However, pulmonary defects are the major reason for mortality in cystic fibrosis patients. The major clinical symptom of cystic fibrosis is undoubtedly the production of extremely viscous mucus in the respiratory tract leading to airway obstruction, lung inflammation, and bronchiectasis. Sputum becomes thick due to dead neutrophils.

The pathophysiology of this disease is as follow; defective chloride channel protein leads to a decrease in chloride transport from the lung epithelial cells into the surrounding respiratory mucus, and in an associated increased movement of sodium ions in the reverse direction. This leads to an increased level of water absorption from the mucus into the epithelial cells, increasing subsequently viscosity and elasticity of the mucus. This in turn decreases clearance of the mucus, allowing any inhaled pathogens to remain in the lung. Infections trigger an immune response in which large numbers of neutrophils are attracted to the site of infection. Ingestion and destruction of bacterial populations by such white blood cells results in the liberation of large amounts of DNA. This released DNA interacts with a variety of additional extracellular substances present in the infected lung, thus further increasing mucus viscosity. Viscous extracellular DNA accumulates in sputum of patients contributing to reduced pulmonary function and frequent pulmonary infection. Mucus containing significant amounts of extracellular DNA from degenerating leukocytes is also a problem in other pathological conditions such as bronchiectasis patients, patients with respiratory syncytial virus bronchiolitis, and patients with atelectasis.

Dornase alfa is produced by genetically engineered CHO cell lines and contains 260 amino acids identical to natural enzyme with an approximate MW of 37 kDa. It contains at its N terminus a 22-amino acid-long sequence encoding a secretion signal, four cysteine residues (amino acids 101, 104, 173, and 209) and two disulfide bridges. The first disulfide bridge (C101–C104) is not essential for enzyme activity, while the second one (C173–C209) is essential for the structure and activity of the enzyme. The recombinant enzyme contains two potential N-linked glycosylation sites, N18 and N106 and depending on the expressing cell type, one or both sites become glycosylated. The half-life of rhDNase in the lungs of cystic fibrosis patients is estimated to be between 2 and 5 hours.

Dornase alfa breaks down nucleic acid materials present in sputum/mucus of cystic fibrosis patients and reduces sputum viscosity in the airways and adhesiveness of lung secretions, promoting improved clearance of secretions. It converts extracellular DNA to 5'-phosphonucleotide end products by hydrolysis of P-O 3' bonds, without affecting intracellular DNA. The enzyme does not appear to affect sputum in the absence of an inflammatory response to infection nor does it affect the sputum of healthy individuals.

Clinical trials and observation studies on the efficacy of Dornase alfa inhalatory therapy have shown improvement in lung function and a decrease of respiratory exacerbations in patients with cystic fibrosis and moderate lung disease as well as a decrease in cystic fibrosis-related morbidity, including the risk of respiratory tract infectious exacerbations (Suri, Metcalfe, Lees, Grieve, Flather, Normand, & Wallis, 2001). Pulmozyme ® received one of the fastest approvals by the FDA (1993) under the orphan drug status.

This drug is well tolerated but some side effects were reported. The most common adverse effects reported after inhalation of rhDNase I are voice alterations (hoarseness), pharyngitis, sore throat, chest pain, rash, pharyngitis, laryngitis, and conjunctivitis. Antibodies against rhDNase I have been found in the serum of 3% of cystic fibrosis patients (Niek, Stefaan, & Demeester, 2006).

Urticaria, abdominal pain, nausea, vomiting, dyspnea, dyspepsia, pyrexia, and headache have also been reported following Dornase alfa therapy.

2.3. Pegademase

Pegademase, is adenosine deaminase enzyme (ADA) derived from bovine intestine, which breaks down adenosine and 2′-deoxyadenosine into inosine and 2′-deoxyinosine respectively.

This enzyme is used for treating SCID associated with the deficiency of ADA (E.C. 3.5.4.4). It is a primary immune deficiency caused by several genetic mutations affecting the immune system, with at least 12 genes implicated. SCID is characterized by the deficiency in lymphocyte functions (T, B, and NK lymphocytes). It is referred to as 'combined' because both cell-mediated and humoral immunity are affected.

The deficiency of ADA leads to the accumulation of adenosine, 2'-deoxyadenosine, and their metabolites, which are toxic to lymphocytes thereby leading to diminished immune function and maturation. A deficiency of T lymphocytes, particularly functional helper T cells and B lymphocytes, results in a markedly decreased production of antibodies and impairment of both arms of the adaptive immune response. This makes SCID patients susceptible to infections and extremely vulnerable due to lack of defense mechanisms.

The most common genetic defect in SCID is an X-linked mutation or the so-called gamma chain defect leading to decreased amounts of IgG. This form represents about 50% of cases, while defects involving the gene for ADA account for approximately 15% (Baldo, 2015).

Injection of unmodified ADA is not effective because of its short circulating life (less than 30 min) and the potential for immunogenic reactions to a bovine-sourced enzyme. The attachment of PEG to ADA increases its circulating life (extended plasma half-life of 48 to 72 hours) and masks epitopes on the surface of the ADA molecule to avoid immunogenic reactions. Hershfield et al. (1987) introduced the first PEG-ADA, which was prepared by conjugating mono-methoxypoly(ethylene glycol) (MW 5 kDa) to bovine intestinal enzyme ADA (Hershfield et al., 1987). As a side note, the experimental study of Bax et al. (2000) has shown that the efficient entrapment of native ADA in carrier erythrocytes also improves substantially the half-life of the enzyme (Bax, Bain, Fairbanks, Webster, & Chalmers, 2000).

Bovine ADA with polyethylene glycol is commercially available as Adagen ®. Adagen ® injection was the first successful application of ERT for an inherited disease and approved in 1990 by the FDA under the Orphan Drug Act for the treatment of patients with ADA-deficient SCID worldwide.

Therapy with Pegademase has been shown to diminish the frequency of opportunistic infections and relieve symptoms of diarrhea and dermatitis.

The possible side effects of this drug are allergic reactions like difficulty in breathing, choking of throat, swelling of the lips, tongue, face, hives, and signs of infection such as sore throat, fever, or congestion. Small range of adverse events, including injection site reactions, headache, urticaria, and pain, were noted in some clinical trials. Hemolytic anemia, autoimmune hemolytic anemia, and thrombocytopenia have also been reported.

2.4. α1-Antitrypsin

α1-Antitrypsin is a proteolytic enzyme that hydrolyzes proteins. It is a 52-kDa serum glycoprotein with 394 amino acid residues and three glycosylation sites. α1-Antitrypsin is synthesized in the liver and constitutes over 90% of the α1-globulin band observed on electrophoresis of serum. It is normally present in serum at a concentration of 2 g/L and represents the major serine protease inhibitor present in mammalian serum. It serves as a potent inhibitor of the protease Elastase and as such prevents damage to lung tissue by neutrophil Elastase. In lungs, the alveoli are chronically exposed to low level of neutrophil Elastase released from activated and degenerating neutrophils. This proteolytic activity can destroy the elastin in the alveolar walls if unopposed by the inhibitory action of α1-Antitrypsin.

Genetic deficiencies of α1-Antitrypsin results usually in life-threatening emphysema. These deficiencies are particularly prevalent in persons of northern European descent, with the disorder affecting in the region of 1 in 2,500 such individuals (Walsh & Walsh, 2015). This results from uncontrolled damage to lung tissue by neutrophil Elastase. The deficiency can be reversed by weekly intravenous administration of α1-Antitrypsin. This enzyme is also used for the treatment of cystic fibrosis.

α1-Antitrypsin preparations used medically are normally obtained from pooled plasma fraction. The possibility of accidental transmission of disease via infected source material has encouraged the development of alternative recombinant sources. α1-Antitrypsin has been successfully produced in various recombinant systems, including in the milk of both transgenic mice and sheep, although no recombinant product has thus far been approved for general medical use.

2.5. Phenylalanine Hydroxylase/Phenylalanine Ammonia Lyase

Phenylalanine hydroxylase (PAH) (E.C. 1.14.16.1) is an enzyme that catalyzes the hydroxylation of phenylalanine to generate tyrosine. Mutations in the PAH gene cause an autosomal recessive disorder called phenylketonuria with the decrease or loss of PAH activity. Deficiency in PAH results in the accumulation of phenylalanine, which is eventually converted into neurotoxic phenylpyruvate, leading to intellectual disability, seizures, and other serious medical problems. Disease control is made possible by ERT. ERT with PAH or enzyme substitution with Phenylalanine ammonia-lyase (PAL) is practiced.

Because of the complex structure of native PAH, truncated forms of PAH are often used in order to stabilize the structure and increase the activity. Additionally, PEG conjugation is also employed to lower the immune response of PAH and improve the pharmacokinetics, which shows promise as an alternative treatment to phenylketonuria.

PAL (E.C. 4.3.1.24) is an enzyme that catalyzes the biotransformation of phenylalanine to trans-cinnamic acid and ammonia and therefore, has the potential to manage the accumulation of phenylalanine. Since it exists widely in plants, yeasts, and fungi, PAL is much easier to obtain than PAH and no complicated cofactors are required for PAL to function.

An oral treatment, Phenylase, is being developed based on the use of recombinant yeast PAL. However, proper protections are required because PAL is highly immunogenic and vulnerable to proteolytic degradation. Another common approach is achieved by conjugating PAL with PEG. Studies have shown that conjugation of PAL with a 20 kDa linear PEG could abolish its immunogenicity without compromising its activity. PEG-PAL is less immunogenic and PEG-PAL formulations with more stability are in development for oral intake (Longo, Harding, Burton, Grange, Vockley, Wasserstein, & Sile, 2014).

On 28 January 2010, orphan designation (EU/3/09/708) was granted by the European Commission to pegylated recombinant PAL for the treatment of hyperphenylalaninemia.

2.6. Porphobilinogen Deaminase

Recombinant human porphobilinogen deaminase (rhPBGD, E.C. 4.3.1.8), an enzyme involved in the heme-synthesizing pathway. It catalyzes the head to tail condensation of four porphobilinogen molecules into the linear hydroxymethylbilane while releasing four ammonia molecules.

The enzyme obtained orphan designation in 2002 by the EMA for the treatment of acute intermittent porphyria (AIP), a clinical manifestation of an autosomal dominant genetic disorder caused by mutation in the PBDG gene and resulting in the reduction of erythrocyte activity with the accumulation of porphobilinogen in the cytoplasm. This may lead to an increase in blood and urinary concentrations of heme precursors such as porphobilinogen and a deficit in heme synthesis. Onset of AIP typically occurs during puberty or later. Individuals may experience acute episodes of neuropathic symptoms. Common symptoms include severe abdominal pain, peripheral neuropathy, and psychiatric symptoms. Crises may be precipitated by a broad range of medications (including barbiturates and sulfa drugs). Administration of PBDG is expected to decrease plasma concentrations of heme precursors.

2.7. Asfotase Alfa

Asfotase alfa (Strensiq ®) is a recombinant glycoprotein that contains the catalytic domain of the tissue-nonspecific alkaline phosphatase. It is a metallo-enzyme glycoprotein of two identical polypeptide chains each containing 726 amino acids (MW ~161 kDa) linked by two disulfide bonds and produced in CHO cells. Each chain is made up of the enzyme catalytic domain, the human IgG1 Fc domain, and an aspartate decapeptide used to target bone. The enzyme reduces elevated substrate levels including inorganic pyrophosphate, which inhibit bone mineralization.

Asfotase alfa is indicated for the treatment of hypophosphatasia (HPP), a rare inherited bone disorder caused by mutations in the tissue-nonspecific alkaline phosphatase, an enzyme that plays a role in bone mineralization, which leads to malformed bones, bone fractures, and loss of teeth. In the absence of alkaline phosphatase, the enzyme's substrates including inorganic pyrophosphate accumulate, blocking hydroxyapatite crystal growth and inhibiting bone mineralization. This results a multi-systemic syndrome with rickets, bone deformation, osteomalacia, and muscle weakness.

Asfotase alfa is the first compound to be approved in the USA and the EU since 2015 for long-term ERT in patients with pediatric-onset HPP to treat the bone manifestations of the disease.

This treatment was associated with skeletal, respiratory and functional improvement in perinatal, infantile and childhood-onset HPP. Patients with life-threatening perinatal and infantile HPP treated with Asfotase alfa had substantially improved bone mineralization, growth, fine and gross motor function, agility and decreased pain. Asfotase alfa rapidly improve rickets severity as reflected in improvements in bone mineralization, respiratory function, gross motor function, cognitive development, muscle strength and ability to perform activities of daily living, and catch-up height-gain. The most common adverse effects of the treatment included injection site reactions (pain, itching, erythema, etc), headache, limb pain, and hematoma.

Warnings for the use of this enzyme cover hypersensitivity reactions including anaphylaxis, localized lipodystrophy including lipoatrophy and lipohypertrophy at injection sites, and ectopic calcifications of the eye (including the cornea and conjunctiva) and the kidneys. Less commonly occurring adverse reactions are hypocalcemia, renal stones, chronic hepatitis, and decreased vitamin B6 levels.

2.8. Enzymes as Digestive Aids

Various enzymatic preparations may be used as digestive aids. Most of these enzymes are depolymerases catalysing the enzymatic breakdown of a number of dietary components including polysaccharides, proteins, and lipids. Some enzyme preparations consist of a single enzyme that catalyses the degradation of a specific dietary substance. Others contain multiple enzymatic activities that exhibit broad digestive ability.

These supplemental enzymes can be used to complement normal digestion or to provide an additional digestive capacity. Enzyme supplementation is also used to manage some digestive disorders as they are mostly due to hereditary lack of digestive enzymes. The majority of enzyme supplementation derived from animal (pancreatic extract), microbial (e.g., Aspergillus oryzae), or plant (e.g., barley) sources.

2.8.1. Glutenases

Celiac disease is intolerance to dietary protein gluten in genetically susceptible individuals. It is an autoimmune disorder resulting in an inflammatory response to gluten and leading to nutrient malabsorption. Gluten is a protein found in different nutrients such as wheat, barley and rye. Treatment of this digestive disorder with oral enzyme preparation containing some gluten-specific proteases (Glutenases) is promising. Glutamine-specific Endopeptidase from barley (EP-B2), prolyl oligopeptidase from *Flavobacterium meningosepticum* (FM-POP), prolyl endopeptidase from *Sphingomonas capsulata* (SC-PEP), prolyl endoprotease from *Aspergillus niger* (AN-PEP), etc. are some examples of Glutenases with high potential to detoxify gluten prior to reaching the small intestine. Different combinations of these Endopeptidases are currently under clinical trials.

2.8.2.α-Amylase

α-Amylase (E.C. 3.2.1.1) plays an important digestive role in higher animals. It catalyses the hydrolysis of the α1-4 glycosidic linkages of polysaccharides such as starch and glycogen to yield dextrins, oligosaccharides, maltose, and glucose, which are easier to use by cells. The enzyme may be isolated from saliva and pancreatic tissue. Various amylase preparations have been administered orally to aid the digestion of dietary carbohydrate. This is also the case for extracts from *Aspergillus oryzae* containing cellulases, proteases, as well as α- and β-Amylases.

2.8.3. Cellulase

Cellulase (E.C. 3.2.1.4) is not produced by humans and is administered as a digestive supplement to alleviate flatulence and to improve overall digestion, especially in high-fiber diets. Cellulase hydrolyzes the β(1-4) glycosidic bonds of cellulose, an indigestible plant polysaccharaide, releasing therefore glucose.

2.8.4. Lactase

β-D-galactoside-galactohydrolase (E.C. 3.2.1.23), commonly known as β-galactosidase or "lactase," catalyses the hydrolysis of the disaccharide lactose, the principal sugar of milk, forming glucose and galactose.

Lactose is digested in the gastrointestinal tract by lactase-phlorizin hydrolase, a membrane-bound enzyme of the small intestinal epithelial cells. Intestinal lactase insufficiency results in a metabolic disorder and one of the most common gastrointestinal disorders called lactose intolerance (maldigestion with clinical symptoms). Intestinal β-galactosidase insufficiency can be a result of down regulation in the expression of the β-galactosidase gene, or by injury to the intestinal mucosa. β-Galactosidase-deficient populations have difficulty in consuming milk and other lactose-containing products, as ingestion of lactose can result in abdominal pain, diarrhea, and flatulence. The incidence of this disorder has been reported to be large. β-galactosidase enzymes obtained from various sources can be used for this disorder. The sources include yeasts such as *Kluyveromyces fragilis* and *Kluyveromyces lactis*, fungi such as *Aspergillus niger* and *Aspergillus oryzae*, and bacteria such as *Bacillus coagulans*.

Preparations of lactase may be employed as a digestive aid to alleviate symptoms associated with lactose intolerance. Some of the most popular products are Lactaid ®, Dairycare ®, Lacteeze ®, Lifeplan ®, Lactrase ®, and SureLac ®. The majority of these digestive aids contain *Aspergillus*-derived β-Galactosidases. The β-galactosidase enzyme derived from *Aspergillus oryzae* is an extracellular protein. It is a glycosylated homodimer with a MW of 105 kDa.

Clinical studies assessing the efficacy of lactase-based treatments for infantile colic have reported favorable results in alleviating the condition (Kanabar, Randhawa, & Clayton, 2001).

Another approach to overcome lactose intolerance is supplementation of dairy along with probiotics. In probiotics some of the micro-organisms producing β-galactosidase are present such as *Lactobacillus, Bifidobacterium, Streptococcus,* and *Lactococcus*.

2.8.5. Pancreatin

Pancreatic enzymes are essential for digestion of macromolecules such as protein, fats, and carbohydrates in the small intestine. Pancreatin is a proteinaceous preparation extracted from the pancreas. It contains amylase, protease, lipase and nuclease activities, and may be administered orally to patients suffering from conditions caused by deficient secretion of pancreatic enzymes. These conditions include chronic pancreatitis, pancreatic carcinomas, partial or total pancreatic ectomy, gastrogenic maldigestion, and cystic fibrosis (characterized partly by the blockage of pancreatic ducts by thick mucus). These conditions require not only replacement of proteolytic, but also of amylolytic and lipolytic activities. Therefore, Pancreatin, a hog or bovine pancreatic powder containing Trypsin (E.C. 3.4.21.4), Chymotrypsin (E.C. 3.4.21.1), α-Amylase (E.C. 3.2.1.1), and Lipase (E.C. 3.1.1.3) activities, is frequently used. Protease digests proteins in food to oligopeptides; amylase digests starch and other carbohydrates into oligo and disaccharides; while lipase hydrolyses fat into fatty acids and monoglycerides.

Pancreatin and Pancrelipase are similar except that Pancrelipase has higher lipase activity than Pancreatin.

Pancreatin preparations are administered to restore nutrient digestion and proper intestinal function in exocrine pancreatic insufficiency. They certainly improve the quality of the patient's life by improving fat absorption and reduction of steatorrhea, weight loss, and other symptoms associated with malabsorption.

Pancreatin and many other enzymatic digestive aids, display a pH versus activity profile, which renders them maximally active in the upper portion of the small intestine. Such enzymes, when administered orally, must pass through the stomach in order to reach their site of action. Inactivation of a large proportion of these enzymes can occur in the stomach, due to the low pH values encountered therein, and due to their proteolytic degradation by stomach pepsin. To protect the Pancreatin enzymes against the acid pH of the stomach, they are administered preferentially as enterocoated formulations, which release the enzymes at the alkaline pH of the duodenum. More recently, it has been shown that pegylating some enzymes can potentially protect them from acid inactivation. Overdose of Pancreatin and Pancrelipase can cause diarrhea and stomach upset.

Some of the main Pancreatin/Pancrelipase preparations are marketed by the brand names Creon, Pancrease, Cotazyme, Zenpep, Ultresa, Pertyze, Nutrizym, Pancrex, and Viokase.

2.8.6. Other Enzymes

- **Pepsin (E.C. 3.4.23.1):** Is used for support of gastric function to increase the digestive capacity of gastric juice. In combination with bismuth complexes, it is also used for treatment of gastric ulcers. Pepsin is secreted naturally in the stomach of most animals, where it catalyses the proteolytic degradation of dietary protein to shorter polypeptides. Pepsin preparations employed as digestive aids are obtained by extraction from the mucous membranes of the stomach of various slaughterhouse animals.
- **Invertase:** Is utilized as a digestive aid in alleviating the symptoms of sucrose-isomaltase deficiency. Invertase alleviates symptoms by hydrolyzing sucrose to glucose and fructose, which are absorbed to portal blood. The enzyme is extracted from *Aspergillus* and *Saccharomyces* species.
- **Papain:** Is a highly active plant protease obtained from the juice of the fruit of the tropical plant *Carica papaya*. It has a broad range of proteolytic activity.
- **Bromelain:** Is a complex natural mixture of enzymes derived from pineapple plant (*Ananas comosus*). Bromelain is used as a digestive aid to promote healthy digestion by assisting in the hydrolysis of dietary protein. It is typically derived from either the fruit (E.C. 3.4.22.33) or stem (E.C. 3.4.22.32) of the plant, with most commercial sources being derived from the stem. Bromelain includes a grouping of sulfhydryl proteolytic enzymes. In addition to the proteolytic portion, Bromelain contains peroxidase, and acid phosphatase. Bromelain can also be taken for treating cardiovascular and skin disorders as well as for treating osteoarthritis.

3. CONCLUSION

Advancements in biotechnology over past years have allowed pharmaceutical companies to produce safer enzymes with enhanced specificity. Along with these advances, changes in orphan drug regulations and new initiatives by the FDA and EMA have been effective in facilitating efforts to develop enzyme drugs. This synergy has had a beneficial effect on the development of treatments for rare diseases like LSD.

ERT provided to some degree of benefit in all patients with LSD. While some of the enzymes discussed have been used or are currently in clinical testing, others remain under development in the laboratory or have been discontinued. The extent of improvement varies greatly according to the disease itself and the stage of evolution. The full potential of ERT has not yet been realized because the majority of patients treated either in clinical trials or on therapy have had advanced disease. Because the initiation of ERT early in life before the development of secondary organ damage may be crucial, attention should be given to early diagnosis and even screening for LSD in early age. Drug delivery to the brain remains a major obstacle for treatment of LSD with CNS disorders and represents one of the most important limitations of ERT. The efforts should be aimed at improving targeting to brain and avoiding adverse health effects. In addition, the recognition and management of infusion reactions to ERT are critical to ensure patient safety during and after ERT administration. Cost remains an issue for many countries, so that ERT remains impossible to obtain for many affected patients.

One of the most common uses of enzymes is modern preparations containing digestive enzymes that should offer a good efficacy and better absorption of nutrients. Though therapeutic enzyme development is not a new approach, only few enzymes available for clinical application have received FDA and EMA approval.

Successful complementation of such enzyme defects will enhance significantly patients' quality of life, not only with regard to the reduction of symptoms but also with regard to less daily restrictions for healthy lifestyle. Furthermore, progression in the latest technology will enable manufacturers to make safer and better therapeutic enzymes for these patients.

REFERENCES

Aviezer, D., Brill-Almon, E., Shaaltiel, Y., Hashmueli, S., Bartfeld, D., Mizrachi, S., & Galun, E. (2009). A plant-derived recombinant human glucocerebrosidase enzyme--a preclinical and phase I investigation. *PLoS One*, *4*(3), e4792. doi:10.1371/journal.pone.0004792 PMID:19277123

Baldo, B. A. (2015). Enzymes approved for human therapy: Indications, mechanisms and adverse effects. *BioDrugs*, *29*(1), 31–55. doi:10.100740259-015-0116-7 PMID:25648140

Barton, N. W., Furbish, F. S., Murray, G. J., Garfield, M., & Brady, R. O. (1990). Therapeutic response to intravenous infusions of glucocerebrosidase in a patient with Gaucher disease. *Proceedings of the National Academy of Sciences of the United States of America*, *87*(5), 1913–1916. doi:10.1073/pnas.87.5.1913 PMID:2308952

Bax, B. E., Bain, M. D., Fairbanks, L. D., Webster, A. D., & Chalmers, R. A. (2000). In vitro and in vivo studies with human carrier erythrocytes loaded with polyethylene glycol-conjugated and native adenosine deaminase. *British Journal of Haematology*, *109*(3), 549–554. doi:10.1046/j.1365-2141.2000.02059.x PMID:10886203

Das, D., & Goyal, A. (2014). Pharmaceutical Enzymes. In S. K. Brar, G. S. Dhillon, & C. R. Soccol (Eds.), *Biotransformation of Waste Biomass into High Value Biochemicals* (pp. 367–387). New York, NY: Springer New York. doi:10.1007/978-1-4614-8005-1_15

Disease, A. W. G. o. M. o. P., Kishnani, P. S., Steiner, R. D., Bali, D., Berger, K., Byrne, B. J., & Watson, M. S. (2006). Pompe disease diagnosis and management guideline. *Genetics in Medicine, 8*(5), 267-288.

Giugliani, R., Lampe, C., Guffon, N., Ketteridge, D., Leao-Teles, E., Wraith, J. E., & Harmatz, P. (2014). Natural history and galsulfase treatment in mucopolysaccharidosis VI (MPS VI, Maroteaux-Lamy syndrome)-10-year follow-up of patients who previously participated in an MPS VI Survey Study. *American Journal of Medical Genetics. Part A*, *164a*(8), 1953–1964. doi:10.1002/ajmg.a.36584 PMID:24764221

Grabowski, G. A. (2005). Recent clinical progress in Gaucher disease. *Current Opinion in Pediatrics*, *17*(4), 519–524. doi:10.1097/01.mop.0000172702.33128.19 PMID:16012266

Grabowski, G. A., Barton, N. W., Pastores, G., Dambrosia, J. M., Banerjee, T. K., McKee, M. A., & Brady, R. O. (1995). Enzyme therapy in type 1 Gaucher disease: Comparative efficacy of mannose-terminated glucocerebrosidase from natural and recombinant sources. *Annals of Internal Medicine*, *122*(1), 33–39. doi:10.7326/0003-4819-122-1-199501010-00005 PMID:7985893

Harmatz, P., Garcia, P., Guffon, N., Randolph, L. M., Shediac, R., Braunlin, E., & Decker, C. (2014). Galsulfase (Naglazyme(R)) therapy in infants with mucopolysaccharidosis VI. *Journal of Inherited Metabolic Disease*, *37*(2), 277–287. doi:10.100710545-013-9654-7 PMID:24108527

Harmatz, P., Giugliani, R., Schwartz, I., Guffon, N., Teles, E. L., Miranda, M. C., & Swiedler, S. J. (2006). Enzyme replacement therapy for mucopolysaccharidosis VI: A phase 3, randomized, double-blind, placebo-controlled, multinational study of recombinant human N-acetylgalactosamine 4-sulfatase (recombinant human arylsulfatase B or rhASB) and follow-on, open-label extension study. *The Journal of Pediatrics*, *148*(4), 533–539. doi:10.1016/j.jpeds.2005.12.014 PMID:16647419

Hershfield, M. S., Buckley, R. H., Greenberg, M. L., Melton, A. L., Schiff, R., Hatem, C., ... Abuchowski, A. (1987). Treatment of adenosine deaminase deficiency with polyethylene glycol-modified adenosine deaminase. *The New England Journal of Medicine*, *316*(10), 589–596. doi:10.1056/NEJM198703053161005 PMID:3807953

Horovitz, D. D., Magalhaes, T. S., Acosta, A., Ribeiro, E. M., Giuliani, L. R., Palhares, D. B., & Llerena, J. C. Jr. (2013). Enzyme replacement therapy with galsulfase in 34 children younger than five years of age with MPS VI. *Molecular Genetics and Metabolism*, *109*(1), 62–69. doi:10.1016/j.ymgme.2013.02.014 PMID:23535281

Kanabar, D., Randhawa, M., & Clayton, P. (2001). Improvement of symptoms in infant colic following reduction of lactose load with lactase. *Journal of Human Nutrition and Dietetics*, *14*(5), 359–363. doi:10.1046/j.1365-277X.2001.00304.x PMID:11906576

Kang, T. S., & Stevens, R. C. (2009). Structural aspects of therapeutic enzymes to treat metabolic disorders. *Human Mutation*, *30*(12), 1591–1610. doi:10.1002/humu.21111 PMID:19790257

Kim, J., Park, M. R., Kim, D. S., Lee, J. O., Maeng, S. H., Cho, S. Y., & Jin, D. K. (2013). IgE-mediated anaphylaxis and allergic reactions to idursulfase in patients with Hunter syndrome. *Allergy*, *68*(6), 796–802. doi:10.1111/all.12155 PMID:23621439

Kresse, G. B. (2004). Nonindustrial Enzyme Usage: Therapeutic Enzymes. In Enzymes in Industry (pp. 295-317). Wiley-VCH Verlag GmbH & Co. KGaA.

Longo, N., Harding, C. O., Burton, B. K., Grange, D. K., Vockley, J., Wasserstein, M., & Sile, S. (2014). Single-dose, subcutaneous recombinant phenylalanine ammonia lyase conjugated with polyethylene glycol in adult patients with phenylketonuria: An open-label, multicentre, phase 1 dose-escalation trial. *Lancet*, *384*(9937), 37–44. doi:10.1016/S0140-6736(13)61841-3 PMID:24743000

Niek, N., Stefaan, C., & Demeester, J. (2006). Deoxyribonuclease I. In B. M & G. W (Eds.), Directory of thepraeutic enzymes. Taylor & Francis Group.

Pastores, G. M., Weinreb, N. J., Aerts, H., Andria, G., Cox, T. M., Giralt, M., & Tylki-Szymanska, A. (2004). Therapeutic goals in the treatment of Gaucher disease. *Seminars in Hematology*, *41*(4Suppl 5), 4–14. doi:10.1053/j.seminhematol.2004.07.009 PMID:15468045

Phupong, V., Shuangshoti, S., Sutthiruangwong, P., Maneesri, S., Nuayboonma, P., & Shotelersuk, V. (2005). Prenatal diagnosis of Pompe disease by electron microscopy. *Archives of Gynecology and Obstetrics*, *271*(3), 260–262. doi:10.100700404-004-0620-3 PMID:15146345

Shaaltiel, Y., Bartfeld, D., Hashmueli, S., Baum, G., Brill-Almon, E., Galili, G., & Aviezer, D. (2007). Production of glucocerebrosidase with terminal mannose glycans for enzyme replacement therapy of Gaucher's disease using a plant cell system. *Plant Biotechnology Journal*, *5*(5), 579–590. doi:10.1111/j.1467-7652.2007.00263.x PMID:17524049

Shak, S., Capon, D. J., Hellmiss, R., Marsters, S. A., & Baker, C. L. (1990). Recombinant human DNase I reduces the viscosity of cystic fibrosis sputum. *Proceedings of the National Academy of Sciences of the United States of America*, *87*(23), 9188–9192. doi:10.1073/pnas.87.23.9188 PMID:2251263

Spada, M., Pagliardini, S., Yasuda, M., Tukel, T., Thiagarajan, G., Sakuraba, H., & Desnick, R. J. (2006). High Incidence of Later-Onset Fabry Disease Revealed by Newborn Screening. *American Journal of Human Genetics*, *79*(1), 31–40. doi:10.1086/504601 PMID:16773563

Suri, R., Metcalfe, C., Lees, B., Grieve, R., Flather, M., Normand, C., & Wallis, C. (2001). Comparison of hypertonic saline and alternate-day or daily recombinant human deoxyribonuclease in children with cystic fibrosis: A randomised trial. *Lancet*, *358*(9290), 1316–1321. doi:10.1016/S0140-6736(01)06412-1 PMID:11684212

Vellard, M. (2003). The enzyme as drug: Application of enzymes as pharmaceuticals. *Current Opinion in Biotechnology*, *14*(4), 444–450. doi:10.1016/S0958-1669(03)00092-2 PMID:12943856

Walsh, G., & Walsh, G. (2015). Therapeutic Proteins: Blood Products, Vaccines and Enzymes. *Proteins*, 177–203.

Weinreb, N. J., Barranger, J. A., Charrow, J., Grabowski, G. A., Mankin, H. J., & Mistry, P. (2005). Guidance on the use of miglustat for treating patients with type 1 Gaucher disease. *American Journal of Hematology*, *80*(3), 223–229. doi:10.1002/ajh.20504 PMID:16247743

Zimran, A., Brill-Almon, E., Chertkoff, R., Petakov, M., Blanco-Favela, F., Munoz, E. T., & Aviezer, D. (2011). Pivotal trial with plant cell-expressed recombinant glucocerebrosidase, taliglucerase alfa, a novel enzyme replacement therapy for Gaucher disease. *Blood*, *118*(22), 5767–5773. doi:10.1182/blood-2011-07-366955 PMID:21900191

APPENDIX

Table 1. Therapeutic enzymes used for the treatment of hereditary deficiency diseases

Enzymes	Enzyme Commission Number	Sources	Brand Names	Indications	FDA Approval	EMA Approval
Glucosylceramidase (β-Glucocerebrosidase)/Alglucerase	3.2.1.45	Human placenta	Ceredase	Gaucher disease Type I	1991	-
Recombinant Glucosylceramidase (β-Glucocerebrosidase)/Imiglucerase	3.2.1.45	CHO cells	Cerezyme	Gaucher disease Type I	1994	1997
Recombinant Glucocerebrosidase/ Velaglucerase alfa	3.2.1.45	Human fibroblast cell line	VPRIV	Gaucher disease Type I	2010	2010
Recombinant Glucocerebrosidase/ Taliglucerase alfa	3.2.1.45	Carrot root cell	Elelyso	Gaucher disease Type I	2012	-
PEGβ-Glucocerebrosidase	3.2.1.45	Human placenta	Lysodase	Gaucher disease	-	-
α-Galactosidase, recombinant/ Agalsidase	3.2.1.22	Human cell line, CHO cells	Replagal, Fabrazyme	Fabry disease	Fabrazyme 2003	Replagal 2001 Fabrazyme 2001
Recombinant acid α-1,4-Glucosidase (acid Maltase)/Alglucosidase alfa	3.2.1.3	CHO cells	Myozyme, Lumizyme	Pompe disease	Myozyme 2006 Lumizyme 2010	Myozyme 2006
Recombinant α-L-Iduronidase/ Laronidase	3.2.1.76	CHO cells	Aldurazyme	Mucopolysaccharidosis I (Hurler, Hurler-Scheie, Scheie forms)	2003	2003
Recombinant Iduronate-2-sulfatase/ Idursulfase	3.1.6.13	Human cell line	Elaprase	Mucopolysaccharidosis II (Hunter's Syndrome)	2006	2007
N- Recombinant Acetylgalactosamine-4-sulfatase (Arylsulfatase B)/Galsulfase	3.1.6.12	CHO cells	Aryplase, Naglazyme	Mucopolysaccharidosis VI (Maroteaux-Lamy syndrome)	Naglazyme 2005	Naglazyme 2006
Recombinant Arylsulfatase A	3.1.6.1	-	Metazym	Metachromatic leukodystrophy	-	-
Recombinant Sphingomyelin phosphodiesterase 1 (acid Sphingomyelinase)/Olipudase alfa	3.1.4.12	CHO cells	-	Niemann-Pick disease	-	-
Elosulfase alfa	3.1.6.4	CHO cells	Vimizim	Mucopolysaccharidosis IVA (Morquio A Syndrome)	2014	2014
Sebelipase alfa	-		Kanuma	Lysosomal acid lipase deficiency	2015	2015
Recombinant Desoxyribonuclease, recombinant/ Dornase alfa	3.1.21.1	CHO cells	Pulmozyme	Cystic fibrosis Chronic bronchitis	Pulmozyme 1993	-
PEGadenosine deaminase/ Pegademase	3.5.4.4	Bovine intestine	Adagen	SCID with adenosine deaminase deficiency	1990	-
α1-Antitrypsin	-	Human plasma	-	Genetic deficiencies of α1-Antitrypsin resulting in emphysema Cystic fibrosis	-	-
Phenylalanine hydroxylase	1.14.16.1	-	-	Phenylketonuria	-	-
Phenylalanine ammonia-lyase	4.3.1.24	-	Phenylase	Phenylketonuria	-	-
Recombinant Uroporphyrinogen-1-synthase (Porphobilinogen deaminase)	4.3.1.8	*E. coli*	-	Acute intermittent porphyria	-	-

continued on following page

Table 1. Continued

Enzymes	Enzyme Commission Number	Sources	Brand Names	Indications	FDA Approval	EMA Approval
Asfotase alfa	3.1.3.12	CHO cells	Strensiq	Perinatal/infantile- and juvenile-onset hypophosphatasia	2015	2015
Glutenases	-	Barley, *Flavobacterium meningosepticum, Sphingomonas capsulata, Aspergillus niger*	-	Celiac disease	-	-
α-Amylase	3.2.1.1	Porcine pancreas	-	Digestion aid	-	-
β-Amylase	3.2.1.2	*Aspergillus oryzae*	-	Digestion aid	-	-
Pancreatin (mixture of pancreatic enzymes, e.g. Trypsin, Chymotrypsin, Lipase, α-Amylase)	-	Porcine pancreas	Cotazym, Kreon, Mezym F, Nutrizym, Pancrex, Pangrol, Pankreatan, Pankreon, Panzytrat	Digestion aid	Cotazym 1996	-
Protease extracts containing Proteases, Cellulases, RNases, α- and β-Amylases	-	*Bacillus subtilis, Aspergillus oryzae*	Travase, Nortase, Combiase	Debridement of wounds Digestion aid	1969 (discontinued)	-
Lipase, Amylase, and Protease	-		TheraCLEC-Total	Pancreatic insufficiency	-	-
Cellulase	3.2.1.4	*Trichoderma viride*	-	Digestion aid	-	-
β-Galactosidase (Lactase)	3.2.1.23	*Kluyveromyces fragilis, Aspergillus oryzae, Aspergillus niger, etc*	Lactaid, Lactrase, SureLac, Dairycare, Lacteeze, Lifeplan	Lactose intolerance	-	-
Pepsin	3.4.23.1	Porcine stomach	Enzymorn, Sentinel	Support of gastic function Gastric ulcers	-	-
Invertase	3.2.1.26	*Aspergillus* and *Saccharomyces* species	-	Sucrose-isomaltase deficiency	-	-
Papain	3.4.22.2	Carica papaya	Panafil	Digestion aid Debridement of wounds Reduction of edema after dentalsurgery	-	-
Bromelain	3.4.22.32	*Ananas comosus*	Traumanase, Phlogenzym, NexoBrid	Debridement of wounds Digestion aid Inflammation	-	NexoBrid 2012
Lipase	3.1.1.3	*Rhizopus arrhizus*	-	Digestion aid	-	-
Chymotrypsin	3.4.21.1	Bovine pancreas	-	Wound healing Digestion aid	-	-
Trypsin	3.4.21.4	Bovine pancreas	Leukase, Granulex	Debridement of wounds Digestion aid	-	-
Sacrosidase	-	*Saccharomyces cerevisiae*	Sucraid	Congenital sucrase-isomaltase deficiency	1998	-
T4 Endonuclease V	3.1.21.	Bacteriophage T4-infected *E. coli*	Dimericine T	Xeroderma pigmentosum	-	-

Chapter 2
Therapeutic Enzymes Used for the Treatment of Cardiovascular Diseases and Coagulation Disorders

Abderrezak Khelfi
National Center of Toxicology, Algeria

ABSTRACT

The successful introduction of enzyme replacement therapy opened the way for the use of enzymes, first as crude preparations and later as highly purified enzymes for use in cardiovascular diseases, clotting disorders, etc. Elimination of blood clot is the key factor in thrombolytic therapy and fibrinolytic enzyme therapy can be practiced to remove the clot. Based on the mechanism of action, they are of two types of enzymes: plasminogen activators and plasmin-like enzymes. Plasma products are usually employed as a source of several enzymes used for the treatment of coagulation disorders. While these products have traditionally been purified from blood donations and obtained as foreign proteins obtained from heterogeneous sources, most are now produced by biotechnology. The therapeutic enzymes reviewed in this chapter are used for the treatment of cardiovascular diseases and hereditary diseases leading to coagulation disorders. Enzyme preparations obtained by direct fractionation from a naturally producing source and recombinant enzymes are considered in this chapter.

1. INTRODUCTION

The application of biotechnology to pharmaceutical research, development, and manufacturing is a growing field. Over the last decade, the need of the biopharmaceutical industry for novel, more potent and stable enzyme therapeutics, has increased exponentially.

Enzyme therapies contribute a prominent share in clinical practice these days. Unlike common medicinal products, which can temporarily solve the particular health problems, pharmaceutical enzymes address the underlying cause of health problem and the patient can achieve permanent relief. Enzymes

DOI: 10.4018/978-1-5225-5237-6.ch002

can be thought of as protein molecules with a specific mission-to initiate and regulate countless biologic reactions in living organisms. Indeed, the main advantage of an enzyme drug is its specificity. Enzymes specifically bind to target molecules, which make enzymes stand out from any other class of drugs.

The successful introduction of enzyme replacement therapy (ERT) opened the way for the use of enzymes, first as crude topical and oral preparations with proteolytic and hydrolytic activity, and later as highly purified enzymes for use in cancer chemotherapy, metabolic deficiencies, cardiovascular diseases, clotting disorders, etc.

Development of enzyme therapeutics against cardiovascular diseases witnessed a tremendous explosion in the past four to five decades and resulted in the development of the first approved genetically engineered drug against cardiovascular diseases (Activase ®) in 1989. Later, many recombinant cardiovascular drugs have been developed and approved for clinical application (Table 1).

Elimination of blood clot or thrombus is the key factor in thrombolytic therapy and fibrinolytic enzyme therapy can be practiced to remove or lyse the clot. Enzymes are used as thrombolytic agents for the treatment of myocardial infarction, thromboembolic strokes, or deep vein thrombosis. Enzymes like Streptokinase, Urokinase, Alteplase, Reteplase, etc are used in the treatment of thrombosis to restore perfusion of the affected tissue. Thrombolytic therapy using fibrinolytic enzyme has an advantage over classical treatments, as the enzymes could act upon the existing clot. Based on the mechanism of action they are of two types; plasminogen activators (e.g. tissue plasminogen activator (tPA) and Urokinase) and plasmin-like enzymes (e.g., Nattokinase and Lumbrokinase) (Table 1). The thrombolytic enzymes are mainly administered by injection or infusion, but some enzymes are used for systemic therapy after oral administration.

Plasma is usually employed as a source of several therapeutic enzymes particularly those used for the treatment of coagulation disorders. The use of whole blood for such purposes would be wasteful if the specific purified component required is already available. Furthermore, fractionation procedures used to produce specific purified blood products considerably reduce the risks of accidental transmission of disease from contaminated blood donations. Most such proteins have been commercially available for many years. While these products have traditionally been purified from blood donations and obtained as foreign proteins obtained from heterogeneous sources, most are now produced by recombinant DNA technology. Despite the advantages of recombinant coagulation factors only a few enzymes were approved by the Food and Drug Administration (FDA). Some adverse events associated with these enzyme drugs is the major setback for their development.

ERT are often complicated by immune responses to the therapeutic enzymes that may cause adverse clinical effects by neutralizing product activity, altering biodistribution and leading to rapid removal and inactivation of enzyme drugs. There may be also problems associated with severe hypersensitivity reactions and immunological reactions. It is for this reason that these enzyme products are administered to patients in very small dose to avoid possible side effects.

The therapeutic enzymes reviewed in this chapter are used for the treatment of cardiovascular diseases and hereditary diseases leading to coagulation disorders. They either have been employed as therapeutic agents in the past or are at a developmental or clinical trial stage as new therapeutics. Enzymes from venomous animals are also reviewed. Enzyme preparations obtained by direct extraction from a naturally producing source and recombinant enzymes are considered in this chapter. This chapter addresses also the issues facing development of ERT for the treatment of some hereditary disorders with inhibitory antibodies.

2. ENZYMES FOR TREATMENT OF CARDIOVASCULAR DISEASES

2.1. Plasminogen Activators

2.1.1. Streptokinase

Streptokinase is a secreted microbial plasminogen activator. The fibrinolytic activity of Streptokinase was first demonstrated in 1933 when it was found that filtrates of broth cultures of certain strains of Streptococcus bacteria (β-hemolytic streptococci; *Streptococcus haemolyticus*) could dissolve a fibrin clot (Sikri & Bardia, 2007).

Streptokinase has a MW of about 47 kDa with a single chain of 414 amino acids without intramolecular disulfide bonds. It contains three structural domains designated α (residues 1 to 150), β (residues 151 to 287), and γ (residues 288 to 414), which all have different associated functional properties. Each domain binds plasminogen, although none can activate plasminogen independently. Streptokinase is a potent plasminogen activator but, unlike other plasminogen activators, Streptokinase is not a protease.

Streptokinase protein creates an active complex that catalyses the conversion of inactive plasminogen to fibrinolytic plasmin. In fact, the fibrinolytic activity of Streptokinase, mediated by generation of plasmin from plasminogen, is initiated by binding to the *β*-chain, ie, Val 561-Asp 790, of circulatory plasminogen forming Streptokinase-plasminogen activator complex. This results in a conformational change in the plasminogen molecule, which in turn, acts as a specific protease in activating a second plasminogen molecule to plasmin by cleaving the Arg 560/Val 561 bond. To achieve optimal thrombolysis, the ratio of Streptokinase to plasminogen should be approximately 1:10. Since Streptokinase generates free circulating plasmin, its use is associated with a systemic lytic state.

Streptokinase in some cases has been shown to be more effective than tPA. It has also a better half-life compared to tPA. The initial plasma half-life ($t_{1/2}$ α) of Streptokinase is about 18 min, and the β half-life of about 83 min.

Streptokinase is obtained by the purification of bacterial proteins elaborated by group C β-hemolytic *streptococci.* Many Streptokinase-based thrombolytic drugs are available in the market such as Streptase ®, Kabikinase ®, Eskinase ®, Thrombosolv ®, Zykinase ®, Varidase ®, Kinalysin ®, and Prokinase ®. Heberkinasa ® is a recombinant Streptokinase obtained through the isolation and cloning of the Streptokinase gene of a strain of *Streptococcus equisimilis* group C and its expression in *E. coli.*

Streptokinase was initially used in combating fibrinous pleural exudates, hemothorax, and tuberculous meningitis. In 1958, Streptokinase was used in patients with acute myocardial infarction. It was also used in the treatment of pulmonary embolism, deep vein thrombosis, arterial thrombosis or embolism, and occlusion of arteriovenous cannulae.

The most important side effect of Streptokinase-based therapy is related to the strong antigenic potential of Streptokinase. Since the enzyme is obtained from bacterial sources, it can induce immune responses from allergy (fever and rash) to life-threatening anaphylactic shock (flushing, itching, rash, dyspnea, bronchospasm, hypotension, angioedema, urticaria, and anaphylaxis) (0.1% in one study) (Jayaram, Ahluwalia, & Cooney, 2000).

The thrombolytic efficacy of Streptokinase treatment may be compromised by the presence of antibodies to the enzyme in the patient's blood. These neutralizing antibodies may arise because of a prior streptococcal infection, or prior Streptokinase treatment.

Since Streptokinase has no clot-selective property, its therapeutic levels cause extensive activation of circulating plasminogen to plasmin leading to systemic lytic state that significantly increases the potential for hemorrhage, a serious hazard of thrombolytic therapy. Bleeding complications, including hemorrhagic strokes, have occurred, although infrequently, after administration of this enzyme. Retroperitoneal, lingual, and uvula hematomas, and hematoma of the rectus muscle after Streptokinase for acute myocardial infarction have been described.

Hypotension commonly results if Streptokinase is given too quickly but this is usually overcome by slowing the infusion rate and giving fluids. There are reports of acute renal failure complicating a serum sickness like reaction following prolonged infusion of Streptokinase but also a few cases of serum sickness with renal impairment following a short infusion of the enzyme.

Leukocytoclastic vasculitis, another type III hypersensitivity response, has been described after intravenous Streptokinase.

Other possible adverse reactions are headache, fever, chills, back pain, myalgia, and asthenia.

Despite these substantial limitations with Streptokinase therapy, easy availability and lower cost make it a drug of choice for thrombolytic therapy in many underdeveloped countries.

2.1.2. Staphylokinase

Staphylokinase is a tPA protein secreted by some *Staphylococcus aureus* strains. Its ability to dissolve fibrin clots has been recognized since the 1940s (S, 2006). The processed mature protein has 136 amino acids, a MW of 16.5 kDa without disulfide bridges. In contrast to Streptokinase, Staphylokinase exhibits fibrin-specific activation of plasminogen. This provides Staphylokinase with an advantage for use in the treatment of thrombosis. Staphylokinase is devoid of catalytic activity, but like Streptokinase, it forms a 1:1 stoichiometric complex with plasmin, leading to the activation of other plasminogen molecules.

Natural strains of *Staphylococcus aureus* produce very low levels of Staphylokinase. As a result, many efforts have been made to clone and overexpress the gene. The gene has been cloned into *E. coli*, *Bacillus sp.*, and various other recombinant systems. Early recombinant forms of Staphylokinase are produced with reduced immunogenicity by site-directed mutagenesis. They show higher fibrin specificity and are more active toward platelet-rich arterial blood clots with fewer bleeding complications, but hemorrhage may still occur in Staphylokinase therapy. PEG-addition to Staphylokinase increases its circulating half-life, the extent of which appears to be proportional to the molecular weight of the PEG-conjugate. This extension leads to more rapid lysis of the thrombus and better patient outcome.

Recombinant forms of Staphylokinase have been evaluated for the treatment of myocardial infarction and peripheral arterial occlusion.

2.1.3. Anistreplase

The general drawbacks of Streptokinase therapy are its relative short plasma half-life (18 min) and its poor clot selectivity. These drawbacks have been addressed by making conjugates of the enzyme. Anistreplase is a human tPA that has more clot selectivity than Streptokinase alone and does not require prolonged infusion to achieve its thrombolytic effects 23. Anistreplase was found to be highly effective after a single intravenous dose of 30 units over a 5-min period compared to a 60-min infusion of 1.5 million units of Streptokinase. It has a considerably longer half-life than Streptokinase, ie, 90 min compared to 20 min.

Anistreplase is a complex of bacterial Streptokinase and purified human plasminogen that has been acylated to protect the enzyme's active site. When administered acyl group hydrolyses, freeing the activated Streptokinase-proctivator complex. This should delay the formation of plasmin but has no influence on the lysine-binding sites involved in the binding of the complex to fibrin. Moreover, unlike Streptokinase which is inactive until circulating plasminogen binds, this complex is a readily available plasminogen activator and activates plasminogen at the site of deposition of the thrombus. It cleaves the Arg-Val bond in plasminogen to form plasmin. Plasmin in turn degrades the fibrin matrix of the thrombus, thereby exerting its thrombolytic action (Figure 1).

Anistreplase is used to eliminate blood clots or arterial blockages that cause myocardial infarction. It is marketed by the name of Eminase ®. The side effects of Anistreplase appear to be similar to those of Streptokinase, including immune reactions and a systemic lytic state conductive to hemorrhage.

2.1.4. Urokinase

Fibrinolytic potential of human urine was first described in 1947. The active molecule is Urokinase (Urokinase-type plasminogen activator, u-PA, E.C. 3.4.21.73), which is secreted by the kidney and eliminated in the urine. The synthesis of Urokinase has been attributed to many renal cell types, two of which are tubular epithelial cells and glomerular visceral epithelial cells. Urokinase is also present at several other physiological locations, such as blood stream and extracellular matrix.

Urokinase is a Trypsin-like serine protease, which specifically cleaves the Arg-Val bond in inactive plasminogen to form active plasmin. Activation of plasmin triggers a proteolysis cascade that, depending

Figure 1. Coagulation cascade

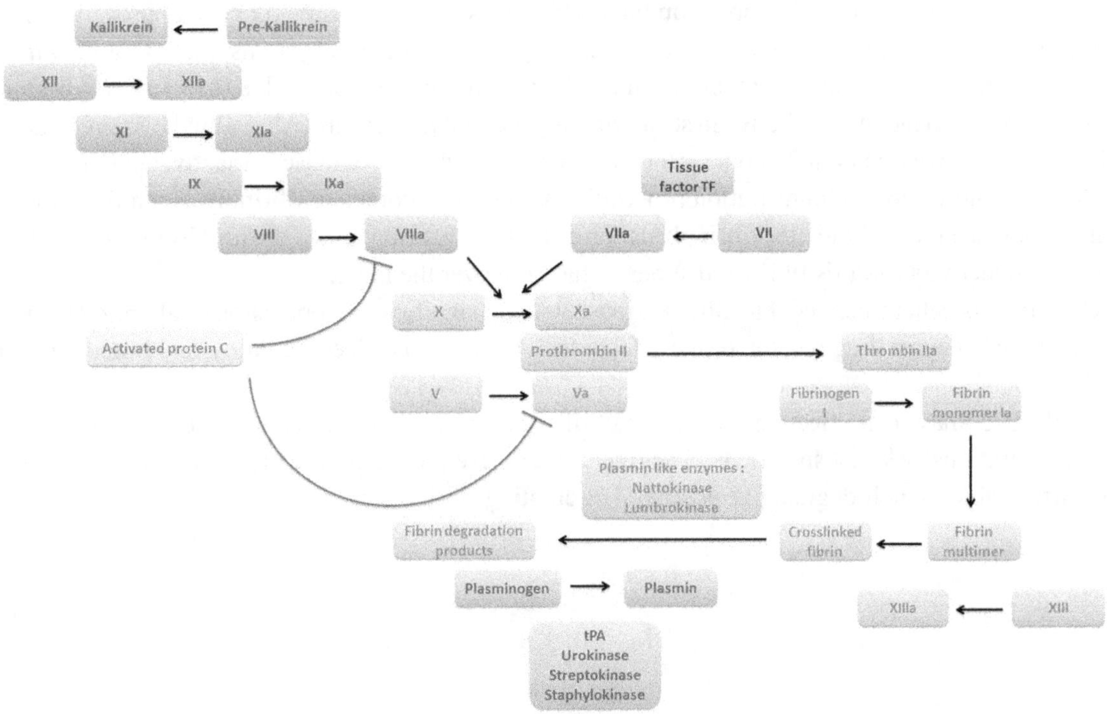

on the physiological environment, participates in thrombolysis and extracellular matrix degradation. This enzyme is used for treating cardiovascular diseases. It converts plasminogen to plasmin inside thrombus and thus, it is protected against circulating antiplasmins. In addition, the absence of Urokinase inhibitors in plasma makes it readily available at the site of action.

Urokinase has a half-life of about 10-15 min. It is a glycoprotein synthesized as a single-chain form with 411 amino acids and 12 disulfide bonds (scu-PA, prourokinase, MW 55 kDa), which is then processed at the Lys-158-Ile-159 bond into a two-chain derivative (tcu-PA or HMW Urokinase) by some proteases such as plasmin, kallikrein, and cathepsin B. Two-chain HMW Urokinase displays the catalytic activity and activates plasminogen efficiently. In plasma, in the absence of fibrin, scu-PA does not activate plasminogen; in the presence of the fibrin clot, scu-PA (but not tcu-PA) induces fibrin-specific clot lysis due to enhanced activity versus fibrin-bound plasminogen.

Further proteolytic cleavage of Urokinase result in a form with a MW of 33 kDa (LMW Urokinase). The low-molecular mass-Urokinase is derived by proteolytic cleavage between Glutamine 143 and Leucine 144 by matrix metalloproteinase pump-1. This fully active form possesses a short interdomain peptide, which is linked to the protease domain by the disulfide bridge between C148 and C279.

For several years, LMW-Urokinase obtained from human urine or from human neonatal kidney cell culture has been used for thrombolytic therapy. Since Urokinase is a naturally occurring enzyme present in human urine, plasma, and tissues, it does not cause any immunological complications. However, some manufacturers have run into problems with raw material supply (human urine) and uncertainties about possible viral contamination (human cell culture), leading to temporary unavailability of Urokinase in some countries.

Urokinase preparations are used for the treatment of acute myocardial infarction and pulmonary embolisms. There are available by the brand name of Abbokinase ®, Kinlytic ®, Actosolv ®, and Ukidan ®.

Recombinant versions of single-chain Urokinase (Saruplase and Prourokinase) have been in clinical studies for a number of years, but no recombinant Urokinase products have yet been approved for marketing. It has been produced by recombinant technology in a number of systems, including *E. coli* and yeast cells. Prolyse is a recombinant prourokinase (rpro- Urokinase) produced in the murine hybridoma cell line and was found to be effective in stroke and myocardial infarctions (Ali, Salim Hossain, & Islam, 2014). It is effective for about 24 hours after administration while Urokinase is inactivated within a few hours. Upon intravenous administration pro-Urokinase attained complete fibrinolysis within 1.5 hours while Urokinase took 3 hours for fibrinolysis. Prourokinase is also preferred over Urokinase due to its improved selectivity towards fibrin and superior half-life over the latter.

Urokinase is better tolerated than Streptokinase with no significant hypotension or allergic reactions. However, this enzyme drug has some side effects such as epistaxis, bleeding gums, and bloody or tarry stools.

While Urokinase can effectively induce clot dissolution in the majority of patients if given early, some preparations lack clot specificity. Treatment with these preparations can result in a systemic lytic state attributable to their degradative action on circulating fibrinogen.

2.1.5. Tissue Plasminogen Activators

2.1.5.1. Overview of Tissue Plasminogen Activator

tPA (E.C.3.4.21.68) is a naturally occurring serine protease found on endothelial cells and catalyzes the conversion of plasminogen to biologically active plasmin. It is the physiologic thrombolytic agent responsible for most of the body's natural efforts to prevent excessive thrombus propagation. It was first identified in 1966, but characterized until 1979 when a sufficient quantity of tPA was purified from human uterine tissue (Uxa, Baczyk, Kingdom, Viero, Casper, & Keating, 2010). A more extensive characterization of tPA was performed following its purification from a human melanoma cell line that produces large quantities of the molecule. Subsequent to these studies, human tPA was cloned and purified by recombinant DNA technology.

tPA is the main enzyme for breaking down the fibrin clots and allows blood flow to the affected areas, thus preventing tissue damage. As a protease, tPA plays a crucial role in regulating blood fibrinolysis, maintaining the homeostasis of extracellular matrix. tPA is found not only in the blood, but also in the central nervous system and plays an important role in modulating the post-translational activation of growth factors.

The MW of tPA is about 70 kDa, with 527 amino acids in its single-chain form. Two forms of tPA may be purified. Type 1 tPA is a single-chain polypeptide, whereas type II consists of two polypeptide chains connected by a disulfide linkage. The single-chain molecule is converted to its active two-chain form by hydrolysis of the Arg275-Ile276 peptide bond in the presence of plasmin linked by one inter-chain disulfide bond.

These two forms are both proteolytically active and comprise three N-glycosylation sites. The carbohydrate component is associated with rapid clearance from the bloodstream. For the correct folding of tPA, correct pairing of the 17-disulfide bonds in the molecule is required. tPA is made up of five structural domains, a looped "finger" domain near the N-terminal, a growth factor domain, the kringle 1 and kringle 2 domains as well as a serine protease domain. The finger and kringle 2 domains bind fibrin clots while the protease domain with its catalytic site at the C-terminus catalyzes the conversion of plasminogen to plasmin.

tPA has little plasminogen activator activity in the absence of fibrin and the presence of fibrin increases the catalytic activity of tPA. Thus, when introduced into the systemic circulation, tPA will initiate local fibrinolysis at the site of blood clots with only limited systemic proteolysis. Fibrin contains binding sites for both plasminogen and tPA. At the clot surface, tPA generates small quantities of plasmin; these then initiate the thrombolytic cycle by generating additional plasmin at the clot surface, resulting in dissolution. tPA activates plasminogen by cleaving a single Arg-Val bond. The initial half-life of exogenous tPA in humans is about 5 to 8 minutes.

Investigations were undertaken to provide a suitable high yielding, reliable source of the protein. This led to transfection of Chinese hamster ovary (CHO) cells with the tPA gene, extraction of the recombinant product from the culture medium and attempts to produce the protein from *E. coli*.

Recombinant tPA are available and approved for therapeutic usage. They are known as Alteplase, Reteplase, and Tenecteplase. They differ in their source, antigenicity, half-life and hemorrhagic potential. Recombinant tPA obtained from CHO cells was approved as a thrombolytic agent for the treatment of acute myocardial infarction and for acute massive pulmonary embolism. They are also used in clinical medicine to treat stroke or cerebrovascular accident (Del Zoppo, Saver, Jauch, & Adams, 2009).

In the 1990s, many approaches were reported to design tPA mutants with improved properties, e.g., higher fibrin specificity, greater zymogenicity (activity ratio of two-chain vs single-chain form), slower clearance from the circulation, and resistance to plasma proteinase inhibitors, but only some of these mutants have been clinically developed. The most frequently seen and potentially serious adverse event associated with these thrombolytic agents is bleeding, especially at intracranial, gastrointestinal, retroperitoneal, and pericardial sites.

2.1.5.2. Alteplase

Alteplase (Activase ®) was the first recombinant tPA product to be approved for therapeutic use. It is synthesized using the cDNA for natural human tPA obtained from a human melanoma cell line. The manufacturing process involves secretion of Alteplase into the culture medium by an established CHO cell lines into which the cDNA for Alteplase had been genetically inserted. The serine protease produced by this process is identical to the native tPA with 527 amino acids, MW of 59 kDa and different glycosylated part.

Alteplase binds to fibrin-rich clots via the fibronectin finger-like domain and the kringle 2 domain. The protease domain then cleaves the Arg-Val bond in plasminogen to form plasmin. Plasmin in turn degrades the fibrin matrix of the thrombus, thereby exerting its thrombolytic action. Alteplase also produces limited conversion of plasminogen in the absence of fibrin.

Alteplase is used for intravenous treatment of ST-elevation myocardial infarction, acute ischemic stroke, and acute massive pulmonary embolism.

Alteplase is associated with some deleterious effects such as hemorrhagic transformation during ischemic stroke, although it is efficient in treating the condition. Other side effects are hypotension, dizziness, sepsis, venous thrombosis, and anaphylaxis.

Because of its short plasma half-life (about 2 minutes), Alteplase must be administered via infusion, usually over periods of up to 90 min. The so-called second-generation recombinant tPA or the third-generation thrombolytic agent Reteplase and Tenecteplase are variants of tPA engineered for a longer half-life and resistance to inhibition.

2.1.5.3. Reteplase

Reteplase (sold under the trade names Ecokinase, Retavase, and Rapilysin) is considered a third-generation thrombolytic agent, genetically engineered to work more rapidly and to have a lower bleeding risk than Alteplase.

Reteplase is a recombinant single-chain non-glycosylated variant of human tPA, which has been modified to contain 355 of the 527 amino acids of the native human tPA (amino acids 1-3 and 176-527) and consisting only of the activity-related kringle 2 and the catalytic serine protease domains of human tPA with a MW of 39.5 kDa. The aim in developing Reteplase was to produce a faster thrombolytic agent with a longer effective half-life without increasing the risk of thrombosis.

In fact, Reteplase molecule no longer contains the structural elements responsible for the rapid clearance of tPA from the bloodstream (finger domain and glycocomponent). Indeed, it shows extended half-life in blood circulation (13-19 min as compared to 3-4 min for wild-type tPA), facilitating its convenient clinical administration via a single intravenous injection. Reteplase is produced by recombinant DNA technology in *E. coli*. The protein is expressed as inactive inclusion bodies from *E. coli*, converted into active form by an *in vitro* folding process and purified by chromatographic separation. Reteplase does not bind fibrin as tightly as native tPA because it lacks fibrin-binding domain (binding to fibrin is five times lower as compared to tPA), allowing the drug to diffuse more freely through the clot rather than binding only to the surface the way tPA does.

The plasminogenolytic activities of Reteplase in the absence of a stimulator do not differ, but in the presence of stimulating fibrinogen fragments, Reteplase activity is about four times lower as compared to tPA.

Reteplase was approved for treatment of acute myocardial infarction in 1998 and 1996 by the FDA and European Medicines Agency (EMA) respectively.

It works best to help people with strokes caused by clots (ischemic strokes) when it is given right away after the stroke symptoms begin.

Several large clinical studies have demonstrated that a clinical benefit of Reteplase over Alteplase could not be shown, and the two agents can be considered equivalent (Kresse, 2000). The reported adverse reactions are bleeding, allergic reactions, fever, infusion site reaction, nausea, vomiting, and cardiac disorders. Intracranial hemorrhage can be caused by Reteplase treatment.

2.1.5.4. Tenecteplase

Tenecteplase (TNK-tPA; Metalyse ® and TNKase ®) is another recombinant tPA produced in CHO cells with multiple point mutations. It is a glycoprotein that contains 527 amino acids developed by introducing mutations at three sites to the cDNA of the natural human tPA. It contains a substitution of threonine 103 with asparagine and a substitution of asparagine 117 with glutamine, both within the kringle 1 domain, as well as a tetra-alanine substitution at amino acids 296-299 (Lys296Ala, His297Ala, Arg298Ala and Arg299Ala) in the protease domain. The glycosylation site in kringle 1 was deleted by substitution of Asn-117 by Gln, whereas a new glycosylation site was introduced, at a different locus, by the replacement of Thr-103 by Asn. Overall, these mutations result in an extended half-life (17 min), resistance to plasminogen activator inhibitor (PAI)-1 (80-fold), fibrin specificity (14-fold), and thrombolytic potency against platelet-rich thrombi. The longer half-life makes it suitable for bolus administration. Tenecteplase is the most fibrin-specific of the tPA and is slightly more fibrin specific than tPA.

Like natural human tPA, the recombinant forms bind fibrin in clots via the fibronectin finger and kringle 2 domains before protease domain-mediated cleavage of a plasminogen Arg-Val bond to form plasmin, which in turn, exerts a thrombolytic action by degrading fibrin matrix of the thrombus, thereby exerting its thrombolytic action. This helps to remove blood clots and arterial blockages that cause myocardial infarction. As mentioned before, Tenecteplase has a long half-life; therefore, a single bolus injection is sufficient for the treatment.

This enzyme is the latest tPA approved for use in clinical practice. It was approved by the FDA as a fibrinolytic agent in 2000 for the treatment of acute myocardial infarction and lysis of intracoronary emboli.

Finally, fewer bleeding complications and dizziness were observed with Tenecteplase treatment compared to Alteplase. Other adverse reactions are allergic reactions, nausea, vomiting, fever, hypotension, and cardiac disorders.

2.2. Ocriplasmin

The use of plasmin to degrade fibrin polymers has been a priority in human therapy for more than 60 years. In the recombinant protein age, it was found that intracellular plasminogen activators made the production of plasminogen, the precursor of plasmin, difficult. This led to the production of short forms of plasmin that retained fibrinolytic activity. Ocriplasmin (also known as des-kringle 1-5 plasmin or microplasmin), a proteolytic enzyme and truncated recombinant form of human plasmin produced in yeast *Pichia pastoris* expression system, lacks all five kringle domains. This structure retains the fibrinolytic activity and does not alter the production of plasminogen. This truncated enzyme of 27.24 kDa MW is composed of 249 amino acids in the form of two polypeptide chains linked by two disulfide bonds joining residues 6 and 124 and 16 and 24. Four intrachain disulfide bonds stabilize the larger polypeptide chain of 230 amino acids.

Ocriplasmin has activity against the clinically relevant plasmin receptors, proteoglycans including fibronectin and laminin, components of the vitreoretinal interface. Vitreomacular adhesion may lead to traction and macular holes, but clinical trials have shown that intravitreal injection of Ocriplasmin and its proteolytic action against fibronectin and laminin can induce separation of the vitreous and macular surfaces, thereby resolving vitreomacular traction and closing macular holes (Mitchell & Miller, 2012; Stalmans, Benz, Gandorfer, Kampik, Girach, Pakola, & Haller, 2012; Syed & Dhillon, 2013). It is commercially available as Jetrea ®.

The reported adverse reactions related to ocriplasmin therapy are vitreous floaters, conjunctival hemorrhage, injection site pain, photopsia, dyschromatopsia, blurred vision, macular hole, reduced visual acuity, conjunctival hemorrhage, visual impairment, and retinal edema and detachment.

2.3. Plasmin-Like Enzymes

2.3.1. Nattokinase

Nattokinase or subtilisin NAT is a proteolytic enzyme derived from a Japanese food known as Natto, a preparation of soybeans that has undergone fermentation with a bacterium known as *Bacillus subtilis natto*. Later similar fibriolytic enzymes were obtained from other fermented food such as Chungkookjang soy sauce, doen-jang, douche, etc. Nattokinase can be secreted by other organisms like *B. amyloliquefaciens, B. licheniformis, B. amylosacchariticus, Flavibacterium, Pseudomonas sp.*, etc (Kurosawa, Nirengi, Homma, Esaki, Ohta, Clark, & Hamaoka, 2015).

Nattokinase is an extracellular enzyme with fibrinolytic potential. It is a cysteine-free serine protease with a MW of 28 KDa comprising 275 amino acids.

Nattokinase is a direct-acting fibrinolytic enzyme, which not only dissolves fibrin but also activates tPA to form plasmin, thereby increasing fibrinolysis mechanism. Further, it shows higher specificity towards fibrin. Nattokinase causes mild enhancement of fibrinolysis in plasma, as evidenced by its effect on fibrinolytic parameters. The fibrinolytic activity of Nattokinase is four-fold that of plasmin, the main fibrinolytic enzyme found in the body.

The stability of Nattokinase in the gastrointestinal tract makes it a more suitable candidate for oral administration as a potent cardiovascular disease neutraceutical. The main applications of Nattokinase include stroke, angina, deep-vein thrombosis, atherosclerosis, venous stasis, peripheral vascular disease, and claudication.

2.3.2. Lumbrokinase

Lumbrokinase is a preparation containing multiple enzyme fractions with fibrinolytic activities present in various species of earthworms and isolated first from *Lumbricus rubellus*. Later fibrinolytic enzymes were isolated from various species of earthworms such as *Lumbricus bimastus*, *Eisenia fetida*, *E. andrei*, etc. Six fibrinolytic enzymes were isolated from *L. rubellus* while seven enzymes were isolated from *E. fetida.*

Lumbrokinase is a group of proteolytic enzymes with molecular weight from 25 to 32 kDa, which includes plasminogen activator and plasmin. In 1991, Mihara et al. successfully extracted and characterized this group of fibrinolytic enzymes (Mihara, Sumi, Yoneta, Mizumoto, Ikeda, Seiki, & Maruyama, 1991). They can dissolve blood fibrin clots and reduce the platelets activation and aggregation as well as thinning of blood by reducing viscosity. They are more fibrin specific compared to other thrombolytics such as tPA and Urokinase.

Lumbrokinases can dissolve the fibrin itself or convert plasminogen to plasmin by inducing endogenous tPA activity to dissolve fibrin clots.

Lumbrokinases are regarded as potential therapeutics against thrombosis. They are widely used clinically as a thrombolytic agent in China to treat cerebral infarction, coronary heart disease, pulmonary heart disease, deep vein thrombosis, and angina pectoris. In Japan, Korea, Canada and USA, Lumbrokinases have been used as oral supplement to support and maintain healthy cardiovascular function.

Lumbrokinases are commercially available as Boluoke ®, which is an oral formulation enzyme supporting a healthier blood state, and it shows great promise in supporting a healthy coagulation/fibrinolysis balance in the body. It has been approved by the China Food and Drug Administration. However, it has not received FDA approval as a thrombolytic drug to date. The main advantage of Lumbrokinases is that they are a safer therapeutic enzyme with no bleeding disorders.

2.4. Proteases from Snake Venoms

Snake venoms contain enzymes such as Ancrod, isolated from the Malayan pit viper *Agkistrodon rhodostoma*, and Batroxobin, obtained from *Bothrops atrox moojeni* or *Bothrops jararaca* that act specifically on the A-α chain of fibrinogen without promoting the cross-linking of the resulting fibrin. Fibrin I, resulting from this action, stimulates the release of plasminogen activators resulting in its degradation by plasmin into non-coagulating degradation products. This leads to a decrease of fibrinogen plasma levels and to reduced coagulability of the blood with easily dispersible clots.

In addition, Ancrod stimulates the production of prostacyclin by endothelial cells, which inhibits platelet aggregation and formation of thrombi.

Ancrod and Batroxobin are used in clinical practice for the prevention of thromboses, supporting of drug-induced fibrinolysis, and for the production of chirurgic tissue glues. They are also use as anticoagulant agents in patients with heparin-induced thrombocytopenia.

The U.S. Stroke Treatment with Ancrod Trial (STAT) and the European Stroke Treatment with Ancrod Trial (ESTAT) proved the efficacy and safety of Ancrod in the treatment of acute stroke.

Loss of activity after prolonged therapy, resulting from inhibitory antibodies, is observed with this protease.

Indeed, repeated administration of natural Ancrod to patients suffering from thrombotic diseases results in a stimulation of the immune system, which counteracts the pharmacological activity of the enzyme and limits its clinical applicability.

Ancrod is a thrombin-like serine protease with 234 amino acids with a MW of 35 kDa. It is a glycoprotein containing five potential N-glycosylation sites. Immunogenicity of Ancrod may be due to the presence of its non-mammalian-type glycans. Recombinant production of the enzyme in *E. coli* and mouse fibroblast cells appears to have overcome the problem (Geyer, Jacobi, Linder, Stirm, Bialojan, Strube, & Geyer, 1996).

The recombinant protein is considered more effective for therapy than the natural protein owing to its reduced immunogenicity and reduced hemorrhagic complications on its administration.

Viprinex is a commercial preparation of Ancrod, which completed phase-II clinical trials in the USA and started phase-III trials in 2005.

Alfimeprase is a truncated form of fibrolase, a metalloproteinase isolated from the venom of the Southern copperhead snake *Agkistrodon contortrix contortrix*. Both Fibrolase and Alfimeprase have a direct proteolytic activity against the fibrinogen A-α chain. In has been shown that thrombolysis with Alfimeprase is up to six times more rapid than with plasminogen activators.

On 28 October 2005, orphan designation (EU/3/05/322) was granted by the European Commission for Alfimeprase for the treatment of acute peripheral arterial occlusion.

3. TREATMENT OF BLEEDING DISORDERS

3.1. Von Willebrand Factor

Factor VIII complex consists of two separate gene products. The smaller polypeptide exhibits coagulant activity and is often designated VIII:C. The FVIII gene codes for this polypeptide. The larger polypeptide, designated von Willebrand factor (VIII:vWB), is predominantly associated with platelet adhesion. On synthesis, individual von Willebrand factor polypeptides polymerize forming large multimeric structures. The product of the FVIII gene (VIII:C polypeptide) then associates with the multimeric VIII:vWB protein, forming the overall complex structure VIII:C-VIII:vWB, which may be co-purified from plasma.

Failure to synthesize VIII:C results in classical hemophilia A, while failure to synthesize VIII:vWB results in von Willebrand disease. Hemophilia is the most common congenital disorder of coagulation and affects approximately 1 in 10 000 males around the world (Kresse, 2000).

Patients suffering from von Willebrand disease actually synthesize normal factor VIII:C, but this polypeptide is rapidly degraded, as its stabilization requires association with the VIII:vWB polypeptide.

Plasma-derived coagulation factor VIII:vWB is obtained using high-resolution chromatographic techniques, including immunoaffinity chromatography. The final product is normally sterilized by filtration, filled into final containers and freeze-dried. The brand name of such preparations is Wilate ®, which has been approved by the FDA in 2009 to treat children and adults with von Willebrand disease and control of bleeding episodes and perioperative management of bleeding. Plasma-derived coagulation factor VIII:vWB preparations can induce hypersensitivity reactions, urticaria, dizziness, dyspnea, nausea, vomiting, and cough.

Patients with von Willebrand disease may potentially develop inhibitory antibodies to von Willebrand factor or factor VIII in von Willebrand factor/coagulation factor VIII complex preparations like Wilate ®. This may result in failure to obtain the desired plasma levels of von Willebrand factor. Being prepared from human plasma, von Willebrand Factor/Coagulation factor VIII complex must always be considered as a possible carrier of infectious agents.

Plasma-derived factor VIII:vWB preparations were replaced by recombinant factor VIII (rFVIII) using advancement in recombinant DNA technology. A recombinant preparation of von Willebrand factor (Vonicog alfa; Vonvendi ®) expressed in CHO cells was recently approved by the FDA. However, it is subject to a warning for thromboembolic reactions including disseminated intravascular coagulation, venous thrombosis, pulmonary embolism, myocardial infarction, and stroke. Other adverse events concern generalized pruritus and hypersensitivity reactions including anaphylaxis.

3.2. Factor VIII:C

Factor VIII:**C** is a single-chain protein of 2332 amino acids and 300 kDa of MW with several domains; A1-A2-B-A3-C1-C2.

A purified, dried, and pasteurized concentrate of coagulation factor VIII containing von Willebrand factor (Humate-P ®) is used for the treatment of hemophilia A as well as treatment and prevention of bleeding in von Willebrand disease. As with other human plasma-derived preparations, the possibility of microbial diseases needs to be kept in mind when using this preparation. Other adverse reactions can occur like anaphylaxis including urticaria, chest tightness, rash, pruritus, edema, and shock as well as thromboembolic events, hemorrhage, nausea, and pain.

Recombinant, human, full-length factor VIII is known as Octocog alfa and the brand names are Kogenate ®, Kovaltry ®, and Advate ®. It is indicated only for the treatment of hemophilia A. It is a recombinant glycoprotein of multiple peptides (80 kDa) prepared in BHK cells and associated in inactive form with von Willebrand factor. Pruritus, bleeding, cerebral hemorrhage, hemoarthrosis, headache, rash, pyrexia, pain, and injection site reactions were reported as adverse events related to the use of Octocog alfa. Moreover, some patients can develop inhibitory antibodies against factor VIII, which creates problems in the treatment of patients with antibodies as inhibitors.

Systemic hypersensitivity reactions to unmodified (full-length) factor VIII recombinant preparations were identified during its post-approval use. Reactions reported include facial swelling, hypotension, tachycardia, flushing, chest tightness, cyanosis, vomiting, and urticaria as well as nausea, rash, restlessness, and tingling.

It was found that deletion of a large section of the B domain, which has no known function, causes a large reduction in protein size and enhanced expression. This enabled the construction of a functional truncated enzyme in the form of the so-called B-domain deleted recombinant factor VIII protein and which retains six potential N-linked glycosylation sites at Asp residues 41, 239, 582, 1685, 1810, and 2118. This preparation, produced first produced in CHO cells, is also called BDDrFVIII or Moroctocog alfa and the marketed name is ReFacto ® approved by the FDA in 2000. The deleted B-domain of ReFacto ® was replaced by a short 14 amino acid linker and the entire enzyme contains 1438 amino acids with MW of 170 kDa (90 + 80 kDa form).

The use of ReFacto ® can induce the production of factor VIII inhibitors and hypersensitivity, which require stopping the infusion or slowing the rate. Other adverse events were reported such as vomiting, nausea, headache, arthralgia, vascular access complications, asthenia, bleeding, hematoma, and pyrexia.

Other recombinant BDD-modified factor VIII preparations are used as antihemophilic factors to treat hemophilia A. These are Turoctocog alfa, Simoctocog alfa, Susoctocog alfa, Lonoctocog alfa, and factor VIII Fc fusion protein, also known as Efmoroctocog alfa.

3.2.1. Turoctocog Alfa

Turoctocog alfa (NovoEight ®) is produced in CHO cells and licensed by the FDA in 2013. It is a BDDrFVIII with a truncated B-domain of only 21 amino acids linking the A1-A2 domains to the A3-C1-C2 domains. Turoctocog alfa contains four N-linked glycosylations and six tyrosine sulfation sites, two of which, Tyr346 and Tyr1664, are necessary for optimal interaction with thrombin. The half-life of Turoctocog alfa is 11-12 hours (c.f. Advate ® 11-13 hours). Adverse events probably related to Turoctocog alfa administration are injection site reactions, pyrexia, and hepatotoxicity.

3.2.2. Simoctocog Alfa

Simoctocog alfa (Nuwiq ®, MW 170 kDa) is a BDDrFVIII produced in human embryonic kidney (HEK) cells. It was approved by the FDA and the EMA in 2015 and 2014 respectively. Simoctocog alfa (MW 170 kDa) comprises the factor VIII domains A1-A2 plus A3-C1-C2 with a 16 amino acid linker between the A2 and A3 domains replacing the deleted B-domain. The octapeptide of arginine in the linker sequence is made to ensure similar proteolytic processing as the full-length factor VIII molecule. Tyrosine 1680 of Simoctocog alfa is sulfated (important for binding to von Willebrand factor) as occurs in plasma-derived factor VIII. The mean half-life of Simoctocog alfa is 15 hours in adults.

Important precautions when using Simoctocog alfa include the possibility of hypersensitivity reactions and the development of neutralizing anti-FVIII antibodies. Adverse events include paresthesia, headache, injection site reaction, dizziness, back pain, and dry mouth.

3.2.3. Susoctocog Alfa

Susoctocog alfa (Obizur ®) is a BDDrFVIII analog of porcine factor VIII (pFVIII) expressed in hamster kidney cells, in which the B-domain is replaced with a 24 amino acid linker and the global number of amino acids is 1448. Like Moroctocog alfa, Turoctocog alfa, and Simoctocog alfa, Susoctocog alfa has a MW of around 170 kDa with heavy (H) and light (L) chains of MWs 90 kDa and 80 kDa, respectively.

The reported adverse reactions are hypersensitivity reactions and the development of antibody inhibitors to pFVIII.

3.2.4. Factor VIII Fc Fusion Protein

The inconvenience of repeated intravenous injections, especially in children, has led to the desire for products that reduce the number of infusions. Efforts have been directed to produce factor VIII preparations that have an increased half-life.

In 2014, the FDA approved the first antihemophilic fusion proteins, a recombinant factor VIII Fc fusion protein (rFVIIIFc) (Efmoroctocog alfa/Efraloctocog alfa) commercially available as Eloctate ®, indicated for the treatment of adults and children with hemophilia A.

The fusion protein was developed by combining a single molecule of BDDrFVIII and Fc domain of human IgG1 expressed in HEK cells. It contains 1890 amino acids with a MW of 330 kDa.

Compared to rFVIII, the fusion protein (rFVIIIFc) has increased efficacy and prolonged half-life (Peters, Toby, Lu, Liu, Kulman, Low, & Pierce, 2013). The half-life of rFVIIIFc was found to be increased 1.5-2-fold to ~19.7 hours. This product was designed to allow the fixation to the neonatal Fc receptor FcRn. FcRn prolongs immunoglobulin half-life by delaying degradation in lysosomes and circulating the protein back into the plasma.

rFVIIIFc can induce a small production of antibodies against factor VIII, but not Fc that decline over the treatment. The most common adverse reactions are nasopharyngitis, arthralgia, headache, rash, and upper respiratory infections. FDA warnings for the approved factor VIII fusion protein cover the possibilities of hypersensitivity reactions.

3.3. Recombinant Factor IX

Coagulation factor IX is a zymogen of a vitamin K-dependent serine protease that plays a critical role in the intrinsic blood-clotting pathway. Deficiencies of factor IX result in a severe bleeding disorder called hemophilia B. Treatments for this life-threatening disorder became available in the latter part of the 20th century consisting mainly of prothrombin complex concentrates (PCCs) and subsequently with high purity plasma-derived factor IX (pdFIX) products obtained from pooled blood donations. The treatment can be given on demand during an emergency or prophylactically to prevent bleeding. Owing to the incidence of viral transmission associated with these products, serious safety concerns have been raised. Consequently, efforts have been made to develop safer treatments for hemophilia B. Departments of Health in some developed country announced that all plasma-derived blood products were to be phased out and replaced with recombinant factor IX (rFIX). However, many countries worldwide still rely on pdFIX concentrates and PCCs.

The recombinant version of factor IX (BeneFIX ®) was approved in the USA in 1997 and received EU approval in 1999.

BeneFIX ® product is synthesized in CHO cells, which provide the capacity for glycosylation and other post-translational modifications adaptable to large-scale culture, and can be grown in the absence of animal- and human-derived raw materials, thereby minimizing the risk of pathogen transmission.

The rFIX has an amino acid sequence identical to the Thr148 allelic form of factor IX derived from plasma and structural and functional properties similar to natural human factor IX. The molecular mass of rFIX was found to be around 55 kDa, which is higher than the predicted value of 47 kDa, reflecting the additional mass added by post-translational modifications. The major disadvantage of rFIX infusion is the development of antibodies in some patients.

3.3.1. Factor IX Fc Fusion Protein

Factor IX has a relatively short half-life of 14-34 hours. This prompted the development of the recombinant factor IX Fc fusion protein of 867 amino acids with single ligand specificity and a half-life extended to around 3.5 days with improved safety and efficacy.

The factor IX was fused with the Fc domain of human IgG1 to form a fusion protein rFIXFc of MW of 98 kDa. Production of the recombinant form in HEK cells allowed all the necessary post-translational modifications to be incorporated.

It is commercially available as Alprolix ® and was approved by the FDA and the EMA in 2014 and 2016 respectively.

Common adverse reactions related to the use of rFIXFc are headache and oral paresthesia.

3.3.2. Factor IX Albumin Fusion Protein

Factor IX albumin fusion protein is generated by combining recombinant human albumin with rFIX, producing a single protein designated rIX-FP containing the Thr148 allelic form of human plasma-derived factor IX. It s a glycoprotein of 1018 amino acids produced in CHO cell lines. Factor IX albumin fusion protein is marketed with the brand name of Idelvion ®, which obtained FDA and EMA approval in 2016.

The fusion protein is claimed to be effective for an extended period with dosing schedules of up to 14 days. Pharmacokinetic studies revealed a half-life of 87-93 hours.

Formation of neutralizing antibody inhibitors to factor IX and an association between the occurrence of inhibitors and allergic reactions may occur. Other possible adverse events are headache, dizziness, rash, eczema, dysgeusia, oral paresthesia, breath odor, fatigue, infusion site pain, palpitations, obstructive uropathy, hypotension, and potential thromboembolic complications.

3.4. Enzymes Used in Bypass Therapy

Prophylactic administration of factor VIII and factor IX reduces hemorrhage in hemophilic patients but the development of inhibitory antibodies often made the treatment less effective, or even ineffective. Now, the choice of treatment for hemophilia A and B is the so-called bypass approach avoiding the requirement for factor VIII or IX single preparations.

For both hemophilia A and B, the preparations of choice for patients with inhibitors are the PCCs, activated prothrombin concentrate, and recombinant activated factor VII (rFVIIa).

Finally, the bypass therapy is, by its nature, short acting and, if given too often, may itself increase bleeding or lead to excess clotting.

3.4.1. Prothrombin Complex Concentrates

Prothrombin complex concentrate (Kcentra®) is obtained from human plasma and contains non-activated four factors II (prothrombin), VII, IX, and X plus anti-thrombotic proteins C and S. Factor IX is the lead factor for potency. This concentrate is used for urgent reversal of acquired coagulation factor deficiency induced by vitamin K antagonist therapy (e.g., warfarin) in adults with acute major bleeding. Adverse reactions related to the use of this concentrate are thromboembolic events (stroke, pulmonary embolism, and deep vein thrombosis), nausea, vomiting, headache, arthralgia, and hypotension.

Another prothrombin complex concentrate containing non-activated factors II, IX, X, and low levels of factor VII derived from pooled human plasma is Profilnine ®. This concentrate is indicated for prevention and control o bleeding in patients with factor IX deficiency due to hemophilia B with inhibitory antibodies. Adverse related reactions are thrombosis, urticaria, fever, chills, nausea, vomiting, headache, somnolence, lethargy, flushing, and tingling.

3.4.2. Activated Prothrombin Concentrates

Activated prothrombin concentrates (Feiba NF ®) are used for prevention of bleeding, perioperative management and routine prophylaxis in hemophilic A and B patients with antibody inhibitors. They are prepared from human plasma and contain activated factor VII and mainly non-activated factors II and X. Moreover, these preparations contain small amounts of factor VIII and factor IX, which may provoke a weak reaction by antibodies directed against factor VIII in patients with hemophilia A and antibodies directed against factor IX in hemophilia B. Other possible adverse events are hypersensitivity reactions including severe allergic manifestations such as anaphylaxis, risk of thrombotic events, transmission of infectious agents, chills, chest pain, dizziness, dysgeusia, dyspnea, hypoesthesia, nausea, pyrexia, anemia, and somnolence.

3.4.3. Recombinant Activated Factor VII

Despite the advantages of infusion therapy with clotting factors, the major challenge encountered is the development of antibodies against clotting factors like factor VIII and factor IX. Indeed, infusion therapy by replacing factor VIII/IX is unsuccessful in patients with antibody inhibitors, increasing the bleeding risk and mortality. rFVIIa (Eptacog alfa) is a bypassing agent to manage bleeding disorders in patients with inhibitors. Indeed, it is used for the treatment of bleeding in hemophilic A or B patients. It is also used to treat patients with factor VII deficiency, which is a rare disorder occurring in 1 in 500,000 (Kresse, 2000).

rFVIIa is a vitamin K-dependent serine protease of 406 amino acids (MW 50 kDa) that can bypass factor VIII and factor IX in clotting.

In January 2004, the European Commission approved rFVIIa for the control of bleeding in patients with factor VII deficiency and Glanzmann's thrombasthenia refractory to platelet transfusions. In 1996, rFVIIa was approved by the EMA for on-demand treatment of bleeding complications in patients with inhibitors.

In the USA, rFVIIa has also been approved for the treatment of bleeding episodes in hemophilia A or B patients with inhibitors to factors VIII or IX in 1999. NovoSeven ® is the commercially available forms of rFVIIa.

rFVIIa has been cloned and expressed in mammalian cells and analyses have shown that the amino sequence of rFVIIa is identical to plasma-derived activated factor VII.

rFVIIa contains four domains, which are (from N- to C terminus, i.e., from the membrane surface) a γ-carboxyglutamic acid (Gla)-rich domain, two epidermal growth factor (EGF)-like domains, and a serine protease domain. It contains light and heavy chains linked by a disulfide bridge. Post-translational modifications are N-linked glycosylation of asparagines 145 and asparagines 322, and O-linked glycosylation of serine 52 and serine 60.

Patients treated with rFVIIa can develop serum antibodies to the protein. Other reported adverse reactions are thromboembolic events, anaphylaxis, pyrexia, hemorrhage, injection site reactions, arthralgia, headache, nausea, vomiting, hypertension, hypotension, edema, rash, and urticaria.

3.5. Fibrinogen Preparations

3.5.1. Fibrin Sealant

Fibrin Sealant (Tisseel ®) contains two components Sealant from pooled human plasma; sealer protein (fibrinogen plus aprotinin) and thrombin. It is used as an adjunct to hemostasis in patients undergoing surgery when conventional control of bleeding is ineffective or impractical. It is also used as an adjunct to surgery to prevent leakage from colonic anastomoses.

Reported adverse events are hypersensitivity including anaphylaxis, cardiac and vascular disorders, nausea, wheezing, bronchospasm, dyspnea, angioedema, flushing, erythema, urticaria, and pruritus.

3.5.2. Fibrin Sealant Patch

Fibrin Sealant Patch (Evarrest ®) is a human plasma-derived fibrinogen and thrombin embedded in a patch of oxidized regenerated cellulose and polygalactin fibers. It is used as an adjunct to hemostasis for soft tissue bleeding during open retroperitoneal, intra-abdominal, pelvic and non-cardiac thoracic surgery when control of bleeding by standard surgery is ineffective or impractical.

Some adverse reactions were reported such as abdominal distension, post procedural and intra-abdominal hemorrhage, and pulmonary embolism.

3.6. Factor XIII

3.6.1. Concentrate of Factor XIII

The concentrate of the human factor XIII is marketed by the name of Corifact ®. This concentrate is obtained from pooled human plasma consisting of two A-subunits and two B-subunits.

This preparation is indicated for routine prophylactic treatment and perioperative management of surgical bleeding in patients with congenital factor XIII deficiency. Several adverse effects were reported after its commercialization like hypersensitivity, joint inflammation, arthralgia, headache, rash, pruritus, erythema, and hematoma.

3.6.2. Factor XIII A-Subunit

Factor XIII A-Subunit (Tretten ®) is a recombinant human factor XIII-A_2 homodimer composed of two factor XIII A-subunits. The factor XIII A-subunit is a 731 amino acid chain with an acetylated N-terminal serine.

It is used for routine prophylaxis for bleeding in patients with congenital factor XIII A-subunit deficiency.

The reported adverse reactions to factor XIII A-subunit are headache, extremity pain, injection site reactions, and immunogenicity.

3.7. Kallikrein

Kallikrein (kallidinogenase, E.C. 3.4.21.34) was discovered in 1926 as a substance in human urine exhibiting hypotensive properties. Serum kallikrein plays an important physiological role in maintaining low blood pressure due to its ability to cleave L-kininogen, a serum α-globulin protein, and release vasoactive kinin peptides (bradykinin or lysyl-bradykinin).

Kallikrein is a serine protease synthesized as an inactive precursor, prekallikrein, which must undergo proteolytic processing to become activated. This proteolytic activation is facilitated by factor XIIa (Figure 1).

3.8. Protein C

The vitamin K-dependent glycoprotein protein C, also known as autoprothrombin IIA and coagulation factor XIV, is a zymogen for the activated protein C (APC) (E.C. 3.4.21.69).

APC plays a central role in the regulation of vascular function. It is an anticoagulant serine protease that inactivates the highly procoagulant factor Va and factor VIIIa, leading to a decrease in thrombin formation (Figure 1).

A deficiency of protein C may manifest as a significantly increased risk of venous thrombosis, purpura fulminans, and severe disseminated intravascular coagulation.

Human protein C concentrate (Ceprotin ®) was approved by the FDA (2007) and the EMA (2001) for the treatment of congenital protein C deficiency and thus the prevention and treatment of venous thrombosis and purpura fulminans.

The reported adverse reactions related to its use are rash, itching, lightheadedness, injection site reactions, hemothorax, and hypotension.

4. CONCLUSION

Enzymes are important therapeutic agents because they accelerate specific chemical reaction to produce a useful effect or product. Identifying pharmacological activity based upon an understanding of how enzymes work at the molecular level has enabled the pharmaceutical industry to discover many new groups of successful enzymes for the treatment of cardiovascular diseases. The increasing use of therapeutic enzyme against these diseases witnessed a tremendous success of such agents in the past few decades. The development of these efficacious and safe enzyme-based therapies has occurred hand in hand with some remarkable advances in the preparation of recombinant enzymes and the understanding of underlying mechanisms operative in different cardiovascular diseases. Nevertheless, often the cost of production and the high price are major setbacks for their development.

The therapeutic enzymes reviewed in this chapter either have been employed as therapeutic agents in the past or are at a developmental or clinical trial stage as new therapeutics. Despite their advantages, they are waiting for approval by the FDA and EMA. We believe that they all represent interesting examples of the challenges and opportunities for use of enzymes for therapeutic applications.

For the majority of patients the treatment will be safe but there is increasing concern about immune reactions leading to treatment failures in a significant minority of patients. Covalent modification of enzymes by molecules such as polyethylene glycol is used to circumvent the immunogenicity and extensive research is ongoing for evolving newer therapeutic enzymes and for improvement of the safety and potency of available drugs to all patients.

Building on the past successes and failures of engineered enzymes in medical applications, the future holds much promise for therapeutic enzymes to tackle these diseases. Further research is needed in this field so that more number of enzymes can be used in modern therapeutics for better patient care.

REFERENCES

Ali, M. R., Salim Hossain, M., & Islam, M. A. (2014). *Aspect of thrombolytic therapy: a review*. Academic Press.

Del Zoppo, G. J., Saver, J. L., Jauch, E. C., & Adams, H. P. Jr. (2009). Expansion of the time window for treatment of acute ischemic stroke with intravenous tissue plasminogen activator: A science advisory from the American Heart Association/American Stroke Association. *Stroke*, *40*(8), 2945–2948. doi:10.1161/STROKEAHA.109.192535 PMID:19478221

Geyer, H., Jacobi, I., Linder, D., Stirm, S., Bialojan, S., Strube, K. H., & Geyer, R. (1996). Glycosylation of recombinant ancrod from Agkistrodon rhodostoma after expression in mouse epithelial cells. *European Journal of Biochemistry*, *237*(1), 113–127. doi:10.1111/j.1432-1033.1996.0113n.x PMID:8620863

Jayaram, H. N., Ahluwalia, G. S., & Cooney, D. A. (2000). *Enzyme Applications, Therapeutic Kirk-Othmer Encyclopedia of Chemical Technology*. John Wiley & Sons, Inc.

Kresse, G.-B. (2000). Enzymes, 6. Enzymes in Therapy. In Ullmann's Encyclopedia of Industrial Chemistry. Wiley-VCH Verlag GmbH & Co. KGaA.

Kurosawa, Y., Nirengi, S., Homma, T., Esaki, K., Ohta, M., Clark, J. F., & Hamaoka, T. (2015). A single-dose of oral nattokinase potentiates thrombolysis and anti-coagulation profiles. *Scientific Reports*, *5*(1), 11601. doi:10.1038rep11601 PMID:26109079

Mihara, H., Sumi, H., Yoneta, T., Mizumoto, H., Ikeda, R., Seiki, M., & Maruyama, M. (1991). A novel fibrinolytic enzyme extracted from the earthworm, Lumbricus rubellus. *Japanese Journal of Physiology*, *41*(3), 461–472. doi:10.2170/jjphysiol.41.461 PMID:1960890

Mitchell, M. D., & Miller, D. M. (2012). Looking at ocriplasmin as a new option in eye disease. *Drugs of Today (Barcelona, Spain)*, *48*(8), 519–524. PMID:22916339

Peters, R. T., Toby, G., Lu, Q., Liu, T., Kulman, J. D., Low, S. C., & Pierce, G. F. (2013). Biochemical and functional characterization of a recombinant monomeric factor VIII-Fc fusion protein. *Journal of Thrombosis and Haemostasis*, *11*(1), 132–141. doi:10.1111/jth.12076 PMID:23205847

S, O. C. (2006). Additional Therapeutic Enzymes. In B. M & G. W (Eds.), *Directory of therapeutic enzymes*. Taylor and Francis.

Sikri, N., & Bardia, A. (2007). A History of Streptokinase Use in Acute Myocardial Infarction. *Texas Heart Institute Journal*, *34*(3), 318–327. PMID:17948083

Stalmans, P., Benz, M. S., Gandorfer, A., Kampik, A., Girach, A., Pakola, S., & Haller, J. A. (2012). Enzymatic vitreolysis with ocriplasmin for vitreomacular traction and macular holes. *The New England Journal of Medicine*, *367*(7), 606–615. doi:10.1056/NEJMoa1110823 PMID:22894573

Syed, Y. Y., & Dhillon, S. (2013). Ocriplasmin: A review of its use in patients with symptomatic vitreomacular adhesion. *Drugs*, *73*(14), 1617–1625. doi:10.100740265-013-0124-1 PMID:24062204

Uxa, R., Baczyk, D., Kingdom, J. C. P., Viero, S., Casper, R., & Keating, S. (2010). Genetic Polymorphisms in the Fibrinolytic System of Placentas with Massive Perivillous Fibrin Deposition. *Placenta*, *31*(6), 499–505. doi:10.1016/j.placenta.2010.03.013 PMID:20447686

APPENDIX

Table 1. Therapeutic enzymes used for the treatment of cardiovascular diseases and coagulation disorders

Enzymes	Enzyme Commission Number	Sources	Brand Names	Indications	FDA Approval	EMA approval
Streptokinase	-	Cβ hemolytic Streptococci (*Streptococcus equisimilis*)	Kabikinase, Streptase, Kinalysin, Varidase, Eskinase, Thrombosolv, Zykinase, Prokinase	Acute myocardial infarction, coronary artery thrombosis, gastric ulcers (together with ribonuclease), pulmonary embolism, deep vein thrombosis, arterial embolism	-	-
Recombinant Streptokinase	-	*E. Coli*	Heberkinasa	Acute myocardial infarction	-	-
APSAC (P-Anisoylated Derivative of Plasminogen Streptokinase Complex)/ Anistrepase	-	Human plasma (plasminogen)/ Hemolytic streptococci (Streptokinase)	Eminase	Acute myocardial infarction	-	-
Recombinant Staphylokinase	-	*E. Coli*	-	Acute myocardial infarction and peripheral arterial occlusion	-	-
Urokinase	3.4.21.73	Human urine or human kidney cell culture	Abbokinase, Actosolv, Alphakinase, Rheothromb, Ukidan, Kinlytic	Acute myocardial infarction, pulmonary embolism	Kinlytic (1978)	-
Recombinant Urokinase	3.4.21.73	*E. coli*	Saruplase	Acute myocardial infarction	-	-
Recombinant Tissue Plasminogen Activator/Alteplase	3.4.21.68	CHO cells	Activase, Actilyse, Cathflo activase	Acute myocardial infarction, pulmonary embolism, acute ischemic stroke	Activase 1987 Cathflo activase 1987	-
Recombinant Tissue Plasminogen Activator/Reteplase	3.4.21.68	*E. coli*	Rapilysin, Retavase, Ecokinase, Cleactor	Acute myocardial infarction	Retavase (1998)	Rapilysin 1996 Ecokinase 1996
Recombinant Tissue Plasminogen Activator/TNK-tPA/ Tenecteplase	3.4.21.68	CHO cells	TNKase, Metalyse, Tenecteplase Boehringer	Acute myocardial infarction	TNKASE (2000)	Metalyse 2001 Tenecteplase Boehringer 2001
Ocriplasmin	3.4.21.7	*Pichia pastoris*	Jetrea	Symptomatic vitreomacular adhesion	2012	2013
Nattokinase	-	*Bacillus subtilis, B. amyloliquefaciens, B. licheniformis, B. amylosacchariticus, Flavibacterium, Pseudomonas sp.*	-	Stroke, angina, deep-vein thrombosis, atherosclerosis, venous stasis, peripheral vascular disease, claudication	-	-
Lumbrokinase	-	*Lumbricus rubellus Lumbricus bimastus, Eisenia fetida, E. andrei,*	Boluoke	Cerebral infarction, coronary heart disease, pulmonary heart disease, deep vein thrombosis, angina pectoris	-	-
von Willebrand Factor/Coagulation Factor VIII Complex (Human)	-	Human plasma	Wilate	von Willebrand disease	2009	-
Recombinant von Willebrand Factor/ Vonicog Alfa		CHO cells	Vonvendi	von Willebrand disease	2015	-
Coagulation Factor VIII	-	Porcine plasma	Hyate C	Hemophilia A	-	-
Coagulation Factor VIII	-	Human plasma	Autoplex, Factorate, Profilate HS, Monoclate-P, Hemate P, Hemofil M, Monarc-M, Beriate P	Hemophilia A	-	-

continued on following page

Table 1. Continued

Enzymes	Enzyme Commission Number	Sources	Brand Names	Indications	FDA Approval	EMA approval
Coagulation Factor VIII (Containing von Willebrand Factor)	-	Human plasma	Alphanate, Humate-P, Koate DVI, Immunate STM, Voncento	Hemophilia A von Willebrand disease	Alphanate 2007 Humate-P 1999	Voncento 2013
Recombinant Coagulation Factor VIII	-	CHO cells	Recombinate, Bioclate, Advate,	Hemophilia A	Recombinate 2010 Advate 2003	Advate 2004
Recombinant Coagulation Factor VIII	-	BHK cells	Kogenate, Helixate, Kovaltry	Hemophilia A	Kogenate 2000 Kovaltry 2016	Kogenate2000 Helixate 2000 Kovaltry 2016
Recombinant Coagulation Factor VIII (B-Domain Deleted)	-	Embryonic kidney (HEK) cells	Nuwiq, Vihuma	Hemophilia A	Nuwiq 2015	Nuwiq 2014 Vihuma 2017
Recombinant Coagulation Factor VIII (B-Domain Deleted)- Fc Fusion Protein	-	Embryonic kidney (HEK) cells	Eloctate	Hemophilia A	Eloctate 2014	-
Recombinant PEGylated Full-Length FVIII	-	CHO cells	Adynovate	Hemophilia A	2015	-
Recombinant Coagulation Factor VIII (B-Domain Deleted)	-	CHO cells	ReFacto, Xyntha, Novoeight, Afstyla	Hemophilia A	ReFacto 2000 Xyntha 2008 Novoeight 2013 Afstyla 2016	ReFacto1999 Novoeight 2013 Afstyla 2017
Recombinant Coagulation Factor VIII (B-Domain Deleted)/ Susoctocog Alfa	-	BHK cells	Obizur	Hemophilia A	2014	2015
Coagulation Factor IX	3.4.21.22	Human plasma	Mononine, AlphaNine, Berinin HS, Immunine, Prothromplex-T, Nonafact	Hemophilia B	-	Nonafact 2001
Recombinant Coagulation Factor IX	3.4.21.22	CHO cells	BeneFIX, Ixinity, Rixubis	Hemophilia B	Benefix 1997 Ixinity 2015 Rixubis 2013	Benefix 1997 Ixinity (withdraw 2013) Rixubis 2014
Recombinant Coagulation Factor IX-Fc Fusion Protein	3.4.21.22	Embryonic kidney (HEK) cells	Alprolix	Hemophilia B	2014	2016
Recombinant Coagulation Factor IX GlycoPEGylated	3.4.21.22	CHO cells	Rebinyn	Hemophilia B	2017	-
Recombinant Coagulation Factor IX-Albumin Fusion Protein	3.4.21.22	CHO cells	Idelvion	Hemophilia B	2016	2016
Prothrombin Complex Concentrate (Human): Prothrombin, FVII, FIX, FX, Proteins C and S	-	Human plasma	Kcentra	Urgent reversal of acquired coagulation factor deficiency induced by vitamin K antagonist (VKA, e.g., warfarin) therapy in adult patients with acute major bleeding. Hemophilic A and B with inhibitory antibodies against factor FVIII or FIX	2013	-
Factor IX Complex: Prothrombin, FIX, FX, and Low Levels of FVII	-	Human plasma	Profilnine	Hemophilic A and B with inhibitory antibodies against factor FVIII or FIX	2010	-
Activated Prothrombin Concentrate: Prothrombin, FVIIa, FX, and Small Amounts of FVIII and FIX	-	Human plasma	Feiba	Hemophilic A and B with inhibitory antibodies against factor FVIII or FIX	2010	-
Recombinant Coagulation Factor VIIa/ Eptacog Alfa	3.4.21.21	BHK cells	NovoSeven	Hemophilic A and B with inhibitory antibodies against factor FVIII or FIX	1999	1996

continued on following page

Table 1. Continued

Enzymes	Enzyme Commission Number	Sources	Brand Names	Indications	FDA Approval	EMA approval
Coagulation Factor VII	3.4.21.21	Human plasma	Factor VII S-TIM 4	Factor VII deficiency	-	-
Human Fibrinogen / Human Thrombin	-	Human plasma	Raplixa, Evarrest, Evicel, TachoSil, Tisseel, Artiss, Fibryna, RiaSTAP	Mild to moderate bleeding in adults undergoing surgery when control of bleeding by standard surgical techniques is ineffective or impractical.	Raplixa 2015 Tisseel 1998 Artiss 2008 Evicel 2003 Evarrest 2012 TachoSil 2010 Fibryna 2017 RiaSTAP 2009	Raplixa 2015 Evarrest 2010 Evicel 2008 TachoSil 2004
Factor XIII Concentrate	-	Human plasma	Corifact	Congenital Factor XIII deficiency	2011	-
Coagulation Factor XIII	-	Human plasma	Fibrogammin-P	Bleeding, factor XIII deficiency, fibrin glues	-	-
Recombinant Coagulation Factor XIII A-Subunit	-	*Saccharomyces cerevisiae*	Tretten, NovoThirteen	Factor XIII deficiency	Tretten 2013	NovoThirteen 2012
Kallikrein	3.4.21.34	Porcine pancreas	Padutin, Bioactin, Prokrein, Onokrein P	Peripheral vascular and coronary disease, fertility disturbances	-	-
Protein C Concentrate (Human)	3.4.21.69	Human plasma	Ceprotin	Congenital Protein C deficiency Prevention and treatment of venous thrombosis and purpura fulminans	2007	2001
Recombinant Activated Protein C/ Drotrecogin Alfa	3.4.21.69	CHO cells	Xigris, Zovant	Severe sepsis	-	Xigris (2002)
Coagulation Factor X (Human)	3.4.21.6	Human plasma	Coagadex	Hereditary factor X deficiency	2015	2016
Recombinant Human Thrombin	3.4.21.5	CHO cells	Recothrom	Control of minor bleeding during surgery	2008	Withdraw 2009
Thrombin, Topical (Human)	3.4.21.5	Human plasma	Evithrom	Bleeding from capillaries and small venules accessible, control of bleeding by standard surgical techniques is ineffective	2007	-
Thrombin, Topical	3.4.21.5	Bovine plasma	Thrombin-JMI	Bleeding from capillaries and small venules accessible	2011	-
Ancrod	-	*Agkistrodon rhodostoma* venom	Arwin, Venacil	Improvement of blood rheology, lymphoblastic leukemia, lymphosarcoma	-	-
Batroxobin	-	*Bothrops atrox* venom	Defibrase, Botropase, Reptilase, Vivostat	Improvement of blood rheology, chirurgic tissue glues and hemostyptic	-	-
Recombinant Truncated Form of Fibrolase	3.4.24.72	Eukaryotic cells	Alfimeprase	Peripheral arterial obstructive disease	-	-
Brinase	-	*Aspergillus oryzae*	-	Fibrinolytic agent	-	-

Chapter 3
Therapeutic Enzymes Used for the Treatment of Non-Deficiency Diseases

Abderrezak Khelfi
National Center of Toxicology, Algeria

ABSTRACT

Therapeutic enzymes have a broad variety of specific uses and clinical applications, particularly as antineoplasic agents, wound debridement therapeutics, and anti-inflammatory drugs, etc. These enzymes can elicit immune response, contributing allergic reactions. Newer drugs with improved stability and less antigenicity have been developed. Covalent modification of enzymes is used to circumvent this immunogenicity. Advancements in drug delivery have revolutionized enzyme therapy. Microencapsulation and artificial liposomal entrapment are some of the techniques used to increment the stability and half-life of enzyme drugs. Several enzymes are now used as prodrug that metabolizes inactive substances into active metabolites through bioactivation process. This approach comprises a suit of techniques that allow activation of drugs locally and at the site of action. This chapter gives an outline of clinical uses of therapeutic enzymes used in non-deficiency diseases. Developments of these enzymes are reviewed with a particular focus on bioengineering applied to the native proteins.

1.INTRODUCTION

Enzyme therapies are becoming more prevalent in clinical medicine, with many producers targeting their advantages in disease treatment. In the last 100 years, enzymes have been increasingly used to treat various diseases. With the progression of enzymology, therapeutic applications of newly discovered enzymes were explored. They as therapeutics hold some advantages over non-enzymatic drugs with their amazing specificity towards targets as well as multiple substrate conversion. Enzymes are not only used for the treatment of topical disorders, but also for the therapy of systemic diseases. Today, therapeutic enzymes have a broad variety of specific uses and find clinical applications, particularly as antineoplasic agents, wound debridement therapeutics, and anti-inflammatory drugs, etc (Table 1). Additionally, there is a growing group of miscellaneous enzymes of diverse function obtained from micro-organisms. Some

DOI: 10.4018/978-1-5225-5237-6.ch003

enzyme preparations are isolated from animal and plant organisms. These enzyme preparations introduced upon intravenous infusion are generally considered as foreign substances by the body. This could elicit immune response, contributing severe allergic reactions and life-threatening conditions. Newer drugs with improved stability and less antigenicity have been developed. Advances in the knowledge of therapeutic uses of enzymes have heightened interest in the manufacture and processing of these macromolecules. Indeed, covalent modification of enzymes by molecules such as polyethylene glycol is used to circumvent the immunogenicity.

Advancements in drug development and delivery over the past few decades have revolutionized enzyme therapy. Microencapsulation and artificial liposomal entrapment are some of the techniques used to increment the stability and half-life of enzyme drugs. Some commercial enzymes are currently available in oral form. A number of chemical modifications have been employed and include binding to inert surfaces and encapsulation to increase the resistance of these intrinsically labile macromolecules to decomposition.

The development of efficacious and safe enzyme-based therapies has occurred hand in hand with some remarkable advances in the preparation of the often specifically designed recombinant enzymes. Approved recombinant enzymes are now used as therapy for chronic gout (Pegloticase), tumor lysis syndrome (TLS) (Rasburicase), leukemia (Pegasparaginase), detoxification of drugs (Glucarpidase), etc.

Other enzymes like Collagenase, Trypsin, Chymotrypsin, and Chondroitinase ABC are frequently constituents of products marketed as wound-debriding agents and anti-inflammatory agents. Enzymes like Hyaluronidase have been widely employed as a “spreading factor” for anesthesia and have a history of safe and effective usage.

Several enzymes are now used as prodrug, a drug that is administered in an inactive or significantly less active form, but once administered it is metabolized in vivo into an active metabolite through bioactivation process. This approach is called enzyme prodrug therapies (EPT), which comprise a suit of techniques that allow synthesizing the drugs locally and at the site of action.

This chapter gives an outline of clinical uses of therapeutic enzymes used in non-deficiency diseases. Developments of these therapeutic enzymes are reviewed with a particular focus on bioengineering and modifications applied to the native proteins. Additional modifications for functional improvements of proteins including post-translational modifications such as Pegylation are discussed. Together with information on the involved mechanisms, safety findings recorded so far on the various adverse events and problems of immunogenicity of these enzymes are presented. For better understanding, these enzymes were organized into different subcategories according to their clinical use.

2.ENZYMES USED FOR CANCER THERAPY

Cancer therapy involves inhibition of cancer cell proliferation without damaging the normal cells. Amino acid deprivation is the main methodology used in cancer enzyme based therapy and it consists in the induction of starvation of amino acids in tumor cells, which are auxotrophic to particular amino acids. This method often reduces tumor proliferation. Indeed, some tumor cells may require unusual or specific nutrients derived from the bloodstream, or may require a nutrient in higher concentration than normal cells. In this area, microbial derived L-Asparaginase has been approved for the treatment of some forms of leukemia, and glutaminase is also under study for treatment of neoplastic diseases. Furthermore, enzyme therapy for cancer treatment can be made by the use of EPT.

2.1.L-Asparaginase

One of the main examples for the successful enzyme therapy of cancer came from the observation that L-Asparaginase (E.C. 3.5.1.1) from guinea pig serum had striking antilymphoma activity. L-Asparaginase is a tetramer protein with a MW of 120 kDa. L-Asparaginase preparations used clinically may be purified from a wide variety of micro-organisms including yeast, fungi, and bacteria. Bacterial producers include *Erwinia chrysanthemi, E. coli, Serratia marcescens, Proteus vulgaris, Vibrio succinogenes*, etc. *E. coli* produces two Asparaginase isoenzymes of which only one is clinically effective. Plant sources include *Pisum sativum, Withania somnifera,* etc. L-Aspraginase has also been isolated from pancreas, spleen and kidneys of many animals.

This enzyme is a widely used antineoplastic agent and mostly used in the treatment of certain forms of childhood leukaemia. Specific for the non-essential amino acid L-Asparagine, some L-Asparaginase preparations are currently approved (Elspar ® produced in *E. coli* and Erwinase ® produced in *Erwinia chrysanthemi*).

L-Asparaginase decomposes L-asparagine to L-aspartic acid and ammonia. Depletion of the supply of L-asparagine leads to cell cycle arrest in the G1 phase, inhibition of protein synthesis and apoptosis of susceptible leukemic cell populations. While most normal tissues are able to synthesize this non-essential amino acid in amounts sufficient for protein synthesis, some types of lymphoid malignancies derive the required amino acid from plasma. In fact, use of the enzyme in leukemia is because susceptible leukemia cells cannot synthesize L-asparagine due to lack of the enzyme Asparagine synthase, which makes these cells dependant on the exogenous source of this non-essential amino acid for survival. Thus, L-Asparaginase activity makes tumor cells starved for the amino acid and eventually prevents cell proliferation without affecting normal cells.

Asparaginase is widely used in the treatment of childhood acute lymphoblastic leukemia (ALL). Intravenous administration of this enzyme results in the rapid depletion of serum asparagine levels, which normally range between 0.5 and 1.5 mg/dL. Protein synthesis in malignant cells incapable of synthesizing asparagine is thus severely compromised. In contrast, untransformed cells begin to synthesize their own L-asparagine. Plasma levels of L-Asparaginase following intravenous administration are correlated to the administered dose.

Pegylated Asparaginase has shown better results for the treatment of children with newly diagnosed standard-risk ALL than the native L-Asparaginase. PEGaspargase/PEGasparginase is a monomethoxypolyethylene glycol succinimidyl conjugate of *E. coli* L-Asparaginase. It shows increased half-life, increased stability, decreased immunogenicity and appears to be associated with improved outcomes when administered for ALL. In fact, the covalent attachment of poly(ethylene glycol) to L-Asparaginase yields a polymer-enzyme conjugate whose plasma half-life ranges from one to two weeks, with minimal immunogenicity. PEGaspargase is then safer for patients with allergy to native enzyme. In contrast, unconjugated bacterial L-Asparaginase preparations have plasma half-lives of approximately 26 hours and require repetitive dosing with the need for frequent intramuscular injection. They are strongly immunogenic with very high rate of allergic reaction, and their use is complicated by the rapid emergence of L-Asparaginase-resistant tumor cells. Given its microbial sources, it is not surprising that L-Asparaginase elicits an immune response when administered to humans. This imposes obvious limitations on the enzyme's long-term clinical efficacy.

In addition to its use in the treatment of ALL, PEGaspargase is also found to be useful in chronic lymphocytic leukemia, Hodgkin's lymphoma, multiple myeloma, and plasma cell leukemia. 18 PEGaspargase exhibits activity even in patients with non-Hodgkin's lymphoma, a disease ordinarily refractory to the unmodified enzyme.

Oncaspar ® is the commercially available forms of PEGaspargase, which received FDA approval in 1994 for treatment of ALL.

Vomiting, fever, pancreatitis, glucose intolerance, hyperglycemia, hepatotoxicity, nephrotoxicity, posterior reversible encephalopathy syndrome, CNS dysfunction, and thrombosis are listed as the most serious adverse events amongst multiple toxic effects induced by L-Asparaginase therapy.

Coagulopathy involving intracranial hemorrhage with thrombosis of the extremities, immune hemolytic anemia, and abnormal collagen-stimulated platelet aggregation were also mentioned as adverse reactions. Side effects include also allergic reaction, occasionally fatal, especially in children where allergic sensitivity can sometimes be avoided by switching from the *E. coli* to the *Erwinia* enzyme.

2.2. Other Amino Acid Deprivation Enzymes

In addition to L-asparagine, tumor cells require abundant supplies of L-glutamine, L-arginine, L-cysteine, L-tryptophan, and other amino acids for growth and survival. L-Glutaminase-Asparaginase preparation from *Acinetobacter,* which hydrolyzes L-glutamine and L-asparagine to L-glutamic acid and L-aspartic acid, respectively, shows activity against cultured tumor cells. In addition, the chemical modification of this enzymatic preparation with succinic anhydride prolongs its plasma half-life in humans in much the same way as the poly(ethylene glycol) modification.

Some cancers are also auxotrophic for arginine, which is a non-essential amino acid synthesized by Argininosuccinate synthetase and Argininosuccinate lyase from citrulline. Arginine deprivation is a novel approach to target tumors that lack Argininosuccinate synthetase expression. Tumors that typically do not express Argininosuccinate synthetase and require circulating arginine include melanoma, lymphoblastic leukemia, pancreatic carcinomas, hepatocellular carcinoma, some mesotheliomas and some renal cell cancers.

Arginase (E.C. 3.5.3.1) is an enzyme that catalyzes the hydrolysis of arginine to ornithine. This enzyme has inhibitory effect on the proliferation of arginine-requiring cancers like T-cell in ALL (T-ALL). The conjugated form of Arginase with soluble dextran has longer blood circulating half-life and significantly lowers blood arginine concentration compared to the native Arginase.

Recent studies indicated that the anti-T-ALL effects could be induced by the administration of PEGarginase and suggested a safe and effective therapeutic for T-ALL treatments (Hernandez, Morrow, Lopez-Barcons, Zabaleta, Sierra, Velasco, & Rodriguez, 2010; Morrow, Hernandez, Raber, Valle, Wilk, Majumdar, & Rodriguez, 2013).

Another arginine deprivation enzymatic method for cancer therapy is based on the use of Arginine deiminase (ADI) (E.C. 3.5.3.6).

ADI is a microbial enzyme from mycoplasma that converts arginine to citrulline and ammonia. Citrulline can be recycled back to arginine in normal cells which express Argininosuccinate synthetase, whereas tumor cells cannot. Thus, enzymatic action of ADI causes arginine deprivation in tumor cells that lack Argininosuccinate synthetase, thereby inhibiting tumor progression. The native form of ADI is strongly antigenic with a half-life of 5 hours. Conjugation to 20 kDa MW polyethylene glycol (ADI-PEG20) decreases antigenicity as well as increases serum half-life, allowing weekly administration that reduces plasma arginine to undetectable levels. The pegylated form of ADI (ADI-PEG20) has shown *in vitro* and *in vivo* activity against melanoma, pancreatic carcinoma, and hepatocellular carcinoma.

2.3. Rasburicase

A non-recombinant Urate oxidase product isolated from *Aspergillus flavus* has been used in some countries for many years for prophylaxis or treatment of hyperuricemia in patients with hematological malignomas. Several factors, including the complex manufacturing process and hypersensitivity reactions, have limited the availability of *Aspergillus flavus* Urate oxidase.

Rasburicase is the recombinant form of Urate oxidase enzyme (E.C. 1.7.3.3) used to treat lymphoid leukemia, non-Hodgkin's lymphoma, and acute myelogenous leukemia.

Rasburicase is derived from a cDNA code from a modified *Aspergillus flavus* strain and expressed in a modified yeast strain of *Saccharomyces cerevisiae*. It is almost identical to the natural Aspergillus enzyme, differing only by a modified reactive cysteine and by having a higher specific activity. It is a tetrameric protein with identical subunits of 301 amino acids and a global of MW of 34 kDa. Rasburicase converts existing uric acid to allantoin, which is 5-10 times more soluble in urine than uric acid.

Rasburicase has proved effective in managing TLS, which usually occurs 48-72 h after initiation of cancer therapy when large numbers of tumor cells undergo apoptosis in a short time, releasing their intracellular contents into the circulation. This is an oncologic emergency most commonly occurring spontaneously or after chemotherapy in patients with large tumor burdens such as acute leukemia, aggressive lymphoma, or bulky solid tumors. This reaction causes an ionic imbalance involving hyperkalemia, hyperphosphatemia, hyperuricemia, metabolic acidosis, and possibly acute renal failure and death. TLS life-threatening renal failure results from the crystallization of uric acid and calcium phosphate in the renal tubules leading to intraluminal tubular obstruction. The injection of Rasburicase reduces levels of uric acid and mitigates the toxic effects of chemotherapy induced tumor lysis. This enzyme drug is marketed by the name of Elitek ® or Fasturtec ® and approved by the FDA and the EMA in 2002 and 2001 respectively.

In pediatric patients with ALL, Rasburicase has proved to be the treatment of choice and superior to allopurinol. In a study of the efficacy of the preparation for the prevention and treatment of hyperuricemia during induction of chemotherapy for aggressive non-Hodgkin lymphoma in adults, results showed the enzyme to be a highly effective, fast-acting, and reliable therapy (Coiffier, Mounier, Bologna, Fermé, Tilly, Sonet, & Herbrecht, 2003).

The most common adverse reactions (incidence $\geq$ 20%) occurring in Rasburicase treated patients are vomiting, nausea, pyrexia, peripheral edema, nausea, vomiting, diarrhea, abdominal pain, constipation, diarrhea, rash, anxiety, and headache (Baldo, 2016).

Anaphylaxis to Rasburicase and hemolysis and methemoglobinemia are the subjects of an FDA black box warning for the enzyme. The latter two events are relevant to patients with glucose-6-phosphate dehydrogenase deficiency whose erythrocytes are subject to the oxidative stress of hydrogen peroxide produced during the Rasburicase-catalyzed conversion of uric acid to allantoin.

2.4. Chondroitinase

Chondroitinase is an enzyme able to act on chondroitin sulfate proteoglycans (CSPGs). This enzyme is useful in conditions where there is an increase in the level of CSPGs, namely cancer, spinal cord injury, and vitreous attachment.

CSPGs have number of biological applications. Apart from providing mechanical strength, they regulate cell migration including cell adhesion and proliferation. Several studies suggest that CSPGs play a prominent role in tumor development. Indeed, CSPGs are an important players in cell migration (metastasis) (Kasinathan, Volety, & Josyula, 2016). Therefore, removal of CSPGs by chondroitinase AC and, to a lesser extent, by chondroitinase B results in the loss of metastatic behavior of melanoma cells and their ability to invade the basement membrane, and inhibits tumor growth, neovascularization and metastasis.

2.5. Enzyme Prodrug Therapy

EPT uses antibody-conjugated enzymes, converting prodrug into cytotoxic drug at tumor cells and thereby killing tumor cells (Table 2).

Increased attention has been focused on developing novel strategies that can more effectively target tumor cells and combine the specificities of enzymes with those of monoclonal antibodies.

Antibody-directed enzyme prodrug therapy (ADEPT) was proposed as a means of restricting the action of cytotoxic drugs to tumor sites, thereby increasing their efficacy and reducing their normal tissue toxicity.

This concept was first envisioned by Philpott (Philpott, Bower, Parker, Shearer, & Parker, 1974) and termed ADEPT by Bagshawe in 1987 (Bagshawe, 1987). It was suggested that this could be achieved in a multistep procedure. In the first step, the antibody-enzyme directed at a tumor-associated antigen is delivered systemically and allowed to localize to the target tumor antigen in vivo. The second step, a nontoxic prodrug is administered and activated by the enzyme to the toxic drug at the tumor site. Next, the highly cytotoxic drug can diffuse to nearby cells, creating a local bystander effect where tumor is destroyed (Figure 1).

This approach is being utilized to discover and develop a class of cancer therapeutics based on tumor-targeted enzymes that activate prodrugs (Table 3).

Given the obvious problems with stability and immunogenicity when administering protein drugs, the concept of gene therapy approaches has been used, where the prodrug-converting enzyme is not directed to, but expressed by tumor cells. The idea to use targeted gene therapy to treat cancer was intensively followed since the 1990s. This system is based on the delivery of a gene for a prodrug-converting enzyme to tumor cells.

It is one of the most important and successful prodrug delivery approaches and has shown great promise in cancer therapy. Gene-directed enzyme prodrug therapy (GDEPT) utilizes transgenes that encode enzymes that convert prodrugs into active therapeutic metabolites (Table 4). There are several other names in the literature for this approach, including virus-directed enzyme prodrug therapy (VD-EPT), suicide gene therapy, and gene prodrug activation therapy, among others.

GDEPT usually comprises a three-component system: an inactive drug (prodrug), a gene coding for an enzyme that converts inactive prodrug to an active drug, and a carrier. Figure 1 illustrates the basic mechanisms of the GDEPT system for the treatment of cancer.

3.ENZYMES FOR THE TREATMENT OF INFECTIOUS DISEASES

Lysozyme (E.C. 3.2.1.17) was discovered in 1922 as a bacteriolytic substance present in many cells. Its antibacterial effect is based on hydrolysis of the bacterial cell walls by cleaving the β-(1,4) glycosidic bonds between N-acetylmuraminic acid and N-acetylglucosamine residues. Lysozyme also has chitinolytic activity, which contributes to the cell-wall lysis of numerous bacteria. This enzyme has also been shown to possess RNase activity, which selectively degrade viral RNA.

Lysozyme, obtained from hen egg white, is still used as an antibiotic for the treatment of infectious and inflammatory diseases, for the treatment of viral diseases, for eye therapy, and within the oral cavity and throat in anti-angina preparations.

Other antimicrobial enzymes are Chitinases (E.C. 3.2.1.14). As an element of the cell wall of various pathogenic organisms, including fungi, protozoa and helminths, chitin is a good target for antimicrobial Chitinases.

4. DEBRIDING ENZYMES

Debridement is the process of removing necrotic tissues from a wound to allow prompt healing. Enzymatic debriding-agent formulations have significant value in the treatment of necrotic wounds as an alternative to surgical debridement. The removal of purulent exudate, necrotic tissue, or hard and resistant eschar is a prerequisite for the effective cleaning and sterilization of burns, ulcers, traumatic and surgical wounds, and for the initiation of the healing process. The enzyme preparations used as debriding usually contain Collagenase, Papain, Sutilains, and Fibrinolysin.

4.1. Collagenase

Collagenase is an enzyme that catalyses the proteolytic hydrolysis of collagen. Collagen represents the major and rigid component of wound tissue. It provides the framework to hold necrotic cells to the soft tissue bed. Therefore, when collagen is degraded, necrotic tissue detaches and granulation can occur, providing the surface needed for proper epithelialization and wound healing. Although it may be isolated from culture extracts of various animal cells, Collagenase is obtained from *Clostridium histolyticum*. It becomes the first approved enzyme by the FDA in 2010 and by the EMA in 2011 for nonsurgical treatment for Dupuytren's contracture.

In December 2013, the FDA approved Collagenase through a restricted treatment program for Peyronie's disease with a palpable plaque and curvature deformity of at least 30 degrees.

Dupuytren's contracture and Peyronie's disease are often associated with trauma and deposition of collagen plaques. Injection of Collagenase into a Dupuytren's cord, which is mainly interstitial collagen, results in enzymatic cleavage and disruption of the cord.

Collagenase as Xiaflex ® (USA) and Xiapex ® (Europe) is a mixture of class I (AUX-I (MW 114 kDa)) and class II (AUX-II (MW 113 kDa)) *C. histolyticum* Collagenases in a required ratio. Each class hydrolyses collagen at different sites but they act in a complementary manner. Class I Collagenases (α, β, γ, and η), products of the colG gene, hydrolyze collagen near the amino and carboxy termini, generating large proteolytic fragments. Class II Collagenases (δ, ε, and ζ), products of the colH gene, cleave interior sites of the molecule, generating smaller fragments. Together, this different, but complementary, substrate specificity leads to effective degradation of the entire collagen molecule. All seven Collagenases, α, β, γ, δ, ε, ζ, and η are zinc proteinases functionally related to matrix metalloproteinases, which, among other activities, degrade the extracellular matrix.

Many of the adverse events recorded for Collagenase are associated with the use of the enzyme to treat Dupuytren's contracture and Peyronie's disease.

Adverse reactions include peripheral edema, contusion, injection site reactions, tendon rupture, ligament damage, pain in extremities, allergic reactions, and pruritus.

Other adverse reactions were reported such as penile hematoma, swelling, ecchymoses, contusion, injection site hemorrhage, erectile dysfunction, penile fracture, penile hematoma, and genital pruritus.

4.2. Sutilains

Sutilains are a mixture of non-specific serine proteases derived from *Bacillus subtilis* that can remove fibrin layers from wounds. They are employed in the debriding of wounds, skin ulcers, burns, and a variety of necrotic tissue types.

4.3. Papain-Based Debridement

Papaya juice has been used for wound debridement since ancient times. Papain is a 23-kDa single-polypeptide cysteine protease enzyme isolated from Papya tree *Carica papaya.* It breaks the fibrinous material in necrotic tissue and thus promotes the wound healing process. It is a non-selective enzyme that catalyses the hydrolysis of peptide bonds involving basic amino acids such as lysine, arginine, and histidine.

Papain requires activators and urea, which are used in combination with the enzyme. Urea helps to expose the enzyme to activators. Another advantage of urea is that it renders necrotic tissue and proteins more susceptible to the catalytic action of Papain, leading to more effective debriding of wound. Therefore, Papain/urea-based formulation was found to be more effective in wound debridement. Papain produces inflammatory response and thus induces pain. In order to relieve pain chlorophyllin copper complex along with Papain/urea is used in commercial formulations. It has been found to be useful in the treatment of pressure ulcers. Panafil ® is a commercially available Papain/urea formulation.

Papain has been also shown to produce marked reductions of obstetrical inflammation and swelling, and of edema resulting from dental surgery.

Chymopapain, a second proteolytic enzyme produced by the papaya tree, has also found medical application. Chemonucleolysis is the term used for the treatment of sciatica by injecting Chymopapain into damaged intervertebral discs. Sciatica, a medical condition characterized by back and leg pain, is caused by the degradation of an intervertebral disc. The disc tends to protrude laterally, compressing spinal root neurons. The injection of Chymopapain appears to speed up the underlying process of disc degradation, thereby more quickly reaching an end stage where it is stable and asymptomatic.

4.4.Trypsin/Chymotrypsin

The proteolytic enzymes Trypsin (E.C. 3.4.21.4) and Chymotrypsin can also be employed as debriding agents. Commercial preparations of these enzymes are typically derived from bovine pancreas. In fact, topical application of Trypsin is often implemented for wound healing. Trypsin cleaves serum albumin, which in turn initiates fibrocyte differentiation and thereby facilitates wound healing. Xenaderm ® and Granulex ® are two products containing Trypsin available for topical application.

4.5. Fibrinolysin/DNase

Fibrinolysin (E.C. 3.4.21.7) is an enzyme derived from plasma of bovine origin. It is used locally only and exclusively together with a Desoxyribonuclease (extracted from bovine pancreas). Fibrinolysin, a protease enzyme consisting of two polypeptide chains connected by two disulfide bonds, aids in debridement by dissolution of fibrin occurring in undesirable exudates like superficial wounds and burns. It also inhibits clotting and dilates blood vessels, while Desoxyribonuclease helps to break down nucleic acid materials and nucleoproteins present in exudates and necrotic tissue. Fibrin-degradation products stimulate macrophages to release growth factors into the wound bed.

The combination of the two enzymes has a synergistic effect on necrotic but not on living tissue. It provides enhanced wound cleaning and accelerates the healing process. It was commercially available as Elase ® and approved by the FDA in 1964. This product was discontinued because it contains chloramphenicol, which led to aplastic anemia and death. Infrequently, local reactions such as increased pain or a stitching/burning sensation can be noticed.

4.6.Chondroitinases

Chodroitinase cleaves chondroitin sulfate into unsaturated disaccharides. It could be used for the treatment of spinal injuries where they have been demonstrated to promote regeneration of injured spinal cord. Following a primary injury to the spinal cord, the glial and non-CNS cells invade the injured site, which results in the formation of a scar that prevents axons from regeneration. Initially, the scar is composed of cell debris released because of primary insult. As the macrophages start infiltrating, CSPG's are expressed by the injured site. Many studies attempted to remove the scar to facilitate axonal regeneration.

In 1999, Lemon et al. demonstrated that application of Chondroitinase in the treatment of spinal cord injury. The research showed that the level of CSPGs considerably increases after spinal cord injury, and this could be reduced through delivery of Chondroitinase ABC (E.C. 4.2.2.4) (Lemons, Howland, & Anderson, 1999). Subsequently, it was shown that the axons could be regenerated if they are treated with Chondroitinase. Hyaluronidase has a similar hydrolytic activity on chondroitin sulfate and may help in the regeneration of damaged nerve tissue.

4.7. Other Enzyme Preparations

- **Debrase Gel Dressing:** Comprising a mixture of enzymes extracted from pineapple, received clearance in 2002 from the US FDA for a Phase II clinical trial for the treatment of partial-thickness and full-thickness burns. This product also received orphan drug status in Europe.
- **Vibrilase (Recombinant Vibriolysin):** Is a proteolytic enzyme from the marine micro-organism *Vibrio proteolyticus.* This enzyme has been shown to be efficient against denatured proteins like those found in burned skin. The safety and tolerability of this enzyme topically applied for debridement of burns is currently tested.
- **Streptodornase:** Is an enzyme (DNAase) produced by hemolytic streptococci that is used medically, often in combination with streptokinase, to dissolve purulent or fibrinous secretions from infections. It is also used to treat ulcers in order to liquefy blood clots and dead tissue so that these can be easily removed.

5.ANTI-INFLAMMATORY AGENTS

5.1. Bromelain

Bromelain has been demonstrated to have an anti-inflammatory effect. It has been shown to remove T-cell CD44 molecules from lymphocytes, thereby affecting T-cell activation and leading to a reduction of inflammation (Munzig, Eckert, Harrach, Graf, & Maurer, 1994). Treatment with Bromelain resulted in a clear reduction in swelling, pain at rest and during movement, and tenderness. Bromelain has been used as an anti-inflammatory and analgesic agent in treating clinical symptoms of arthritis.

5.2. Trypsin/Chymotrypsin

Trypsin and Chymotrypsin are thought to induce anti-inflammatory effects. They have been used as anti-inflammatory agents in the treatment of bronchial asthma and burns.

The molecular mechanisms by which Trypsin and Chymotrypsin achieve an anti-inflammatory effect remain to be properly elucidated. However, it is likely that the mode of action of these proteases is based on their ability to degrade protein inflammatory mediators involved in promoting the synthesis of such mediators.

5.3. Serrapeptase

Serrapeptase/Serratiopeptidase/Serralysin is an enzyme preparation used as an anti-inflammatory agent. The enzyme is derived from the bacteria *Serratia* E-15 isolated from the intestines of the silkworm *Bombyx mori*. Serrapeptase was later identified in various species of *Serratia*, *Pseudomonas aeruginosa*, and *Aspergillus oryzae*. It is a zinc metalloprotease containing 470 amino acids with a MW of approximately 50 KDa. It helps eliminate inflammatory swelling, accelerates liquefaction of pus, and enhances the action of antibiotics. It is administered orally as a microencapsulated formulation to protect from gastric acid and has high bioavailability.

Reduction of inflammation is achieved through thinning and decreasing the amount of fluids accumulated in tissues and by helping to drain out the fluids. It also breaks down dead tissue around an injury and thus aids in the healing process.

The anti-inflammatory property of Serrapeptase makes the enzyme suitable for treatment of breast engorgement, reduction in swelling due to sports injuries, treatment of chronic sinusitis, thinning of bronchopulmonary secretions, and reduction of post-surgical swelling.

In patients with chronic airway disease (in which mucus production and removal are problematic), treatment with Serrapeptase result in changes in sputum. Weight, viscosity, elasticity, and neutrophil content are all decreased. Using Serrapeptase for treating chronic lung conditions in which sputum production is a problem (for example, cystic fibrosis) leads to improved lifestyle parameters. Another property of Serrapeptase is its analgesic effect by inhibiting pain inducing bradykinin like amines in inflamed tissues and its potential in clot lysis.

5.4. Superoxide Dismutase

Superoxide dismutase (SOD) (E.C. 1.15.1.1), the most important detoxifying enzyme present in cells, transforms the highly toxic superoxide anion (O_2^-) to moderately toxic hydrogen peroxide. This antioxidant enzyme is essential for cell protection from damage by oxygen free radicals. SOD is produced by most organisms. Under normal circumstances, formation of superoxide anions is kept under tight control by SOD. However, under conditions such as acute and chronic inflammation, superoxide anions are formed at a faster rate than the capacity of the endogenous SOD enzyme to remove them. This accumulation of superoxide anions results in superoxide mediated damage.

Superoxide anion is crucial starting points in the formation of other reactive oxygen species, which may cause oxidative damage to cells and tissue destruction. This is especially evident in several types of cells, of which appears to be involved in pathophysiological conditions such as myocardial infarction, hyperoxic lung damage, atherosclerosis, ischemia-reperfusion damage by initiation of lipid peroxidation, and other oxidative modifications. SOD is used therapeutically as a powerful antioxidant preventing from oxygen toxicity and as an anti-inflammatory agent. It has established its efficacy in the management of inflammatory diseases such as rheumatoid arthritis, osteoarthritis, and lung inflammation. SOD has also been shown to exert anti-inflammatory effects on uveitis and to protect human bone marrow progenitor cells from the effects of radiation. It proved to be therapeutically effective in the treatment of radiation cystitis and interstitial cystitis. It is also effective for elimination of side effects associated with chemo- and radiation therapy. Liposomal SOD has been used to treat radiofibrosis caused by excessive radiation and leading to fibrosis at a local target site. However, SOD is susceptible to the known limits of native protein administration such as unsatisfactory pharmacokinetics and immunogenicity, although the incidence of allergic and anaphylactic reactions seems to be low.

Therapeutic SOD has traditionally been derived from bovine liver by enzymatic digestion of an aqueous liver extract. Bovine Cu/Zn-SOD was marketed as an anti-inflammatory drug under the trade name Orgotein ®. However, due to adverse immunological side effects, it was withdrawn from the market, and its marketing license was suspended in 1994. The enzyme was marketed for use in animals in the USA as the active ingredient in Palosein ®, which is used for the alleviation of musculoskeletal inflammation in dogs and horses and as an anti-inflammatory agent with efficacy in the treatment of traumatic arthritis of horses.

Recombinant human SOD (Sudismase) preparations obtained from recombinant *E. coli* or yeast are presently in clinical development for treatment of amylotrophic lateral sclerosis, bronchopulmonary dysplasia in premature neonates, asthma, and respiratory distress syndrome.

A recombinant Cu/Zn-SOD (Oxsodrol ®) was registered with the FDA in 1991 as an orphan drug for the treatment of bronchopulmonary dysplasia in premature infants. SOD injection to patients with osteoarthritis of the knee is one of the most impressive examples of beneficial SOD therapy. However, this therapy suffers from drawbacks such as immunogenicity and targeting associated with its use as a therapeutic agent. To overcome or reduce these problems, several approaches have since been attempted. To improve the blood residence time and simultaneously reduce immunogenicity, SOD has been entrapped in liposomes. Another approach is used, which consists in masking of the protein surface by biocompatible natural or synthetic polymers such as PEG. This conjugate (PEGorgotein) indeed was found to have greatly increased clearance time that depends on the MW of the PEG used for modification and less immunogenicity. The therapeutic efficacy of PEGorgotein has been studied in several indications, and the product is presently studied for treatment of stroke and closed head injury.

Promising results were observed by intratracheal and local administration of liposome SOD/Catalase for the treatment of lung injuries (Barnard, Baker, & Matalon, 1993; Padmanabhan, Gudapaty, Liener, Schwartz, & Hoidal, 1985; Walther, David-Cu, & Lopez, 1995) as well as for wound healing purposes (Jubeh, Antler, Haupt, Barenholz, & Rubinstein, 2005; Vorauer-Uhl, Fürnschlief, Wagner, Ferko, & Katinger, 2001), respectively. PEG-based amphiphilic block copolymers were also employed to encapsulate SOD and Catalase, the resulting SOD/Catalase colloids exhibited enhanced stability against enzymatic degradation and prolonged blood circulation time in vivo (Hu & Tirelli, 2012; Onaca, Hughes, Balasubramanian, Grzelakowski, Meier, & Palivan, 2010; Simone, Dziubla, Arguiri, Vardon, Shuvaev, Christofidou-Solomidou, & Muzykantov Simone, 2009).

6. ADJUVANT ENZYMES: HYALURONIDASE

Hyaluronidase (E.C.3.2.1.35) is an enzyme that depolymerizes hyaluronan, an essential element involved in the protective barrier of tissues. Found in the extracellular matrix of most types of connective tissue, hyaluronan is a linear polysaccharide of a repeating disaccharides unit of D-glucuronic acid-β(1-3)-*N*-acetyl D-glucosamine-β(1-4). It is a high molecular mass glycosaminoglycan (GAG) that may be as large as 6-8 MDa.

Hyaluronan is widely distributed in the extracellular matrix of higher organisms as well as in the vitreous humor, cartilage, and other tissues. It can also form aggregates with proteoglycans, which contain GAG such as chondroitin and chondroitin sulfate.

Connective tissue, which acts as a barrier to the flow of fluid through the interstitial matrix, can become permeable after treatment with Hyaluronidase. This enzyme randomly hydrolyzes hyaluronan by splitting the glucosaminidic bond between C1 of the glucosamine moiety and C4 of glucuronic acid to yield even-numbered oligosaccharides with glucuronic acid residues at their non-reducing end. This action temporarily decreases the viscosity of the cellular cement and increases diffusion of injected fluids and localized transudates and exudates, facilitating the absorption of injected fluids and resorption of exudates. This increased permeability lasts for up to about 48 hours after which collagen remains unchanged with no signs of inflammation. This action allowed the use of this enzyme in conjunction with other drugs to speed the dispersion and delivery of medicines given by subcutaneous route by making tissue more permeable. It is also used as an adjunct in subcutaneous urography to improve resorption of radiopaque agent media when the intravenous route is precluded. For many years, Hyaluronidase given subcutaneously has been used for rehydration therapy.

Testicular Hyaluronidase was the principal form used as a therapeutic agent. The enzyme has been extracted from both bovine and ovine testicular tissue and semen. This enzyme is used to promote the absorption and dispersion in the subcutaneous or intramuscular injection of relatively large volumes of fluids. The enzyme preparation has been branded as Hydase ®, Amphadase ®, Hylase ® (derived from bovine testicular tissue), and Vitrase ® (derived from sheep testicular tissue).

Preparations of bovine testicular Hyaluronidase have been applied therapeutically in the fields of orthopedics, ophthalmology, and internal medicine for many years. A common field of application is its addition to local anesthetic agents, especially in ophthalmic anesthesia filed, to improve the rapidity of onset, dispersion, depth, and duration of local anesthesia.

It was first included in retrobulbar blocks. Nowadays, it is used routinely for that purpose but also in peribulbar and sub-Tenon's blocks. The ophthalmic injectable formulations have also been developed for use in the treatment of vitrous hemorrhage in diseases such as diabetic retinopathy.

Hyaluronidase extracts from animal testes have a number of undesirable features including low enzyme concentrations (typically <1% enzyme per unit weight of protein), the presence of contaminants (e.g., immunoglobulins, proteases, anticoagulants, and vasopermeability factors), and immunogenicity.

To overcome the presence of non-human, impure, animal derived extracts, a recombinant form of PH20, rhuPH20 (Hylenex ®), was developed in Chinese hamster ovary (CHO) cells.

It appears to be more effective and well tolerated without any inflammatory or immunogenic responses following repeated subcutaneous injections. Co-injection of the highly purified rhuPH20 permitted the administration of a volume up to five times greater than that normally given for drugs injected subcutaneously.

The FDA approved Hylenex ® in 2005 for use in removing the cumulus matrix surrounding oocytes in preparation for assisted reproductive technology procedures as well as for use as a spreading agent to facilitate the dispersion and absorption of other drugs. It is also under study for the management of heart attack.

There are six endo-β-*n*-acetylhexosaminidase hyaluronidase-like sequences in the human genome but only three of the gene products, Hyal1, Hyal2, and PH20, are known to have Hyaluronidase enzyme activity. However, only PH20 (sperm adhesion molecule 1 (SPAM1); zona pellucida binding) degrades GAG under physiological conditions and represent the predominant Hyaluronidase in mammalian testes. The human enzyme is a 509 amino acid glycoprotein attached to the outer cell surface membrane via a glycosylphosphatidylinositol linkage that facilitates sperm penetration and fertilization.

The recombinant enzyme, which lacks the glycosylphosphatidylinositol linkage, is a soluble single chain polypeptide with N-linked glycans with 447 amino acids and MW of 61 kDa. Like the animal-derived Hyaluronidase preparations, it hydrolyzes the glucosaminidic bonds between the β 1,4 linked N-acetyl-D-glucosamine and D-glucuronic acid sugars of hyaluronic acids.

In October 2014, the FDA granted Orphan Drug designation for pegylated recombinant human Hyaluronidase (PEGrhuPH20) for the treatment of pancreatic cancer. The EMA granted the same designation in December 2014.

The main recorded adverse events with recombinant Hyaluronidase are injection site reactions, edema in association with hypodermoclysis, and allergic reactions including urticaria, angioedema, and rarely anaphylaxis. Not surprisingly, unpurified preparations of bovine and ovine Hyaluronidase are much more likely to elicit immediate IgE-mediated allergic reactions, especially after repeat administrations. When unpurified preparations are used for ophthalmologic purpose, symptoms include swelling, angioedema, tenderness, inflammation, hypersensitivity, reduced visual acuity, periorbital edema with erythema, increased intraocular pressure, exophthalmos, eye pain, orbital and eyelid edema, conjunctival chemosis, proptosis, and restricted eye movement.

7. ENZYMES USED FOR THE TREATMENT OF GOUT: PEGLOTICASE

Uricase (Urate oxidase, E.C. 1.7.3.3, UOx) is an enzyme that catalyzes the oxidation of uric acid to 5-hydroxyisourate, which is a more soluble product that is readily excreted in urine.

The first Urate oxidase produced is called Rasburicase. As mentioned before, it is used for the treatment of TLS in pediatric cancer patients. It was also found to lower serum urate levels in patients with gout. The use of Rasburicase for this purpose was found to be limited and impractical by its immunogenicity and relatively short half-life. Pegylated form of Uricase was developed to overcome this drawback. Gout is characterized by hyperuricemia, which is an abnormally high level of serum uric acid that may crystallize at locations with low body temperature. Gout is a medical condition usually characterized by recurrent attacks of acute inflammatory arthritis, and eventually results in other serious syndromes due to bone erosion. Some patients affected by gout are intolerants to urate-lowering strategies and the recombinant Uricase may be used as an alternative therapeutic approach to alleviate hyperuricemia. In fact, treatment with administration of Uricase can lower the uric acid concentration significantly; however, direct use of Uricase results low therapeutic efficacy due to its low stability and high immunogenicity. Great efforts have been devoted to overcoming these limitations. One major type of approaches is to conjugate Uricase with polymers.

Pegylated to reduce its potential for immunogenicity and organ accumulation as well to increase its circulatory half-life, Pegloticase, a recombinant porcine/baboon variant Uricase (Urate oxidase) produced in *E. coli*, is used for therapy of gout previously refractory to conventional therapy with uricostatic (e.g., allopurinol) and uricosuric drugs (e.g., probenecid, colchicine, and sulfinpyrazone).

Pegloticase, which is absent in humans, catalyzes the oxidation of uric acid to 5-hydroxyisourate and hydrogen peroxide. The former compound is unstable and converts to racemic allantoin, which is much more soluble than uric acid and is readily excreted. In 2010, FDA approved Pegloticase (Krystexxa ®) as a clinical therapeutic for the management of chronic gout. In the meantime, other encapsulated Uricase has also been explored, such as by loading Uricase into human erythrocytes (Ihler, Lantzy, Purpura, & Glew, 1975; Ihler, Glew, & Schnure, 1973; Magnani, Mancini, Bianchi, & Fazi, 1992) or liposomes (Tan, Wang, Yang, Chen, Xiong, Zhang, & Zhang, 2010; Tan, Wang, Yang, Zhang, Liu, Chen, & Zhang, 2010).

Krystexxa ® is a Pegylated recombinant porcine-like Uricase with four identical non-covalently linked chains each 300 amino acids and MW of 136.8 kDa. Nine of the 30 lysines of this tetrameric peptide are Pegylated, giving a final MW of 545 kDa.

The FDA has issued a warning for anaphylaxis and infusion reactions during and after administration of Pegloticase. Other adverse reactions are gout flares, chest pain, contusion or ecchymosis, nausea, vomiting, nasopharyngitis, cardiovascular events, and constipation.

8. ENZYMES FOR HYPERGLYCEMIA

Glucose is an important source of energy, yielding mostly in form of ATP to support other cell processes. The metabolic disorders of glucose concentration in the body usually cause hyperglycemia and diabetes. Hexokinase (E.C. 2.7.1.1) and Glucose oxidase (E.C. 1.1.3.4) are the enzymes capable of catalyzing the conversion of glucose, which endows them the potential in reducing blood glucose level, particularly to patients with insulin resistant. Incorporation of Glucose oxidase in erythrocyte interior space was first demonstrated by Taverna and Langdon, indicating that glucose is capable to pass through erythrocyte membrane (Taverna & Langdon, 1973). Further attempts revealed that administration of Hexokinase/ Glucose oxidase encapsulated erythrocytes in diabetic mice was able to regulate blood glucose at physiological levels for several weeks (Rossi, Bianchi, & Magnani, 1992).

9. DETOXIFYING ENZYMES

9.1. Glucarpidase

Methotrexate (MTX) and other antifolates administered for cancer therapy are normally eliminated through urine. High-dose MTX is taken with precautions and improper elimination can lead to toxicities and renal failure. Patients with renal impairment given the drug may experience high plasma concentrations. Glucarpidase (Voraxaze ®), a recombinant Pseudomonas Carboxypeptidase G2 expressed in *E. coli*, is used clinically to hydrolyze MTX and other antifolates. It readily cleaves MTX into glutamate and 2,4-diamino-N10-methyl-pteroic acid largely excreted by the liver.

Glucarpidase is indicated in such patients, ensuring that MTX is eliminated enzymatically, mainly by hepatic mechanisms, and not by the kidneys. Toxic blood levels of the drug can be rapidly decreased by intravenous administration of Glucarpidase.

Leucovorin, a reduced folate and potential substrate for Glucarpidase, should not be given with the enzyme that degrades it. In addition, the drug competes with MTX for the enzyme.

The most common adverse events reported are flushing, paresthesia, headache, shaking, hypotension, nausea, vomiting, tingling fingers, burning of face and extremities, feeling of warmth as well as allergic reactions including anaphylaxis

9.2. Rhodanese

Rhodanese (E.C. 2.8.1.1) is one of the sulfurtransferases related to the biotransformation of cyanide to the less toxic thiocyanate in the presence of a sulfur donor. Liposomal encapsulated Rhodanese systems were developed as more practical cyanide antidotes (Petrikovics, Budai, Baskin, Rockwood, Childress, Budai, & Szilasi, 2009; Petrikovics, Jayanna, Childress, Budai, Martin, Kuzmitcheva, & Rockwood, 2011). Particularly to Rhodanese-based cyanide antidotes, liposome is still the preferred carriers, because most sulfur donors are small molecules and liposome is capable to co-encapsulate enzymes as well as small molecules.

9.3. Other Enzymes

- **Organophosphorus Compounds Detoxification:** Organophosphorus (OP) compounds are Acetylcholinesterase (AChE) inhibitors and exert their toxicity by causing an excessive accumulation of the neurotransmitter, acetylcholine, and subsequent disruption of cholinergic nervous transmission. In case of poisonous exposure, they may cause acute intoxication or even lethality.

Phosphotriesterases (E.C. 3.8.1.1) and diisopropyl-fluorophosphatase (DFPase, E.C. 3.8.1.2) are a family of enzymes that are highly efficient in hydrolysis of some OP compounds. Several different phosphotriesterases were stabilized by liposomal encapsulation, resulting in extended circulating half-life and enhanced protection against the lethal effect of paraoxon. However, the activity of encapsulated phosphotriesterases appears to be low due to the weak diffusion of OP compounds caused by the liposome.

- **Alcohol Detoxification:** Alcohol causes numerous human disorders and increases the risk of many harmful health conditions.

Alcohol detoxification can be done by Alcohol dehydrogenase (ADH, E.C. 1.1.1.1) and Acetaldehyde dehydrogenase (ALDH, E.C. 1.2.1.10). Since both ADH and ALDH are NAD-dependent enzymes, encapsulation systems like erythrocyte and liposome were usually employed in order to co-delivery NAD+ together with the enzymes to allow them function normally. Encapsulation with human erythrocytes and liposome enhance the stability of the enzymes without altering their catalytic activities, although ADH/ALDH-loaded erythrocytes display a short blood circulating half-life and rapid accumulation in liver.

An alternative enzyme potentially applicable to alcohol detoxification is Alcohol oxidase (AOx or AlOx, E.C. 1.1.3.13) without the need of co-factor (e.g., NAD+). However, during the oxidation of alcohol, AOx also produces a highly toxic element, hydrogen peroxide. This problem may be avoided by co-encapsulation of AOx and catalase in one nanocapsule to construct a dual enzyme nanocomplex, although AOx-catalase nanocomplex is still not an essentially safe therapeutic for alcohol detoxification due to the generation of acetaldehyde.

10.CONCLUSION

Therapeutic utility of enzymes as drugs is increasing day by day. Their use offers tremendous opportunities for the development of highly effective and specific therapies. The functional versatility of enzymes makes these biological macromolecules a powerful and almost limitless resource for clinical applications particularly in cancer treatment.

ADEPT, GDEPT and VDEPT all offer potential routes to the selective delivery of cytotoxic drugs to tumors and thus hold significant promise as new and improved tumor therapies. ADEPT is the more advanced approach and early clinical trials have shown encouraging results. It contains attractive molecules for different cancer treatments. The antibody component confers the agent with exquisite specificity for targeting both primary tumors and metastases. Many of the important questions that were raised when ADEPT was first proposed have been answered, but some remain. What has already been demonstrated makes it a strong candidate for ongoing clinical applications.

Gene therapy is a promising approach, still at the early stages of development. Some major problems remain to be solved before this new strategy becomes routinely adopted in the clinic. Current goals include identifying or creating vectors that afford tissue and tumor specificity, enhanced efficiency of transduction and conditional expression in human cells. Nevertheless, the clinical applications of such approach are encouraging and illustrate both feasibility and future promise for GDEPT as a cancer treatment.

There is a considerable potential for the use of enzymes in other pathological conditions. For example, inflammatory diseases and healing processes characterized by the accumulation of cytotoxic products and debris may be reduced by supplementation of proteolytic enzymes. Additionally, enzyme therapy can serve as a magnifying component of combinational clinical strategies designed for dispersion and delivery of medicines given by subcutaneous route by making tissue more permeable.

When used properly, enzymes can also be helpful for patients with suffering from different metabolic dysfunctions. Biotherapy for poisoning cases holds enormous promise as fast removal or neutralization of toxic substances is essential to restore the normal biochemical processes. Enzymes with high catalytic specificity and efficiency are therefore the most versatile detoxification agents. Various enzyme detoxification agents have been developed or under developments, some of which have been used clinically. With the rapid advances in the engineering technologies, better understanding of toxicity mechanisms and the development of more advanced carriers, one can expect that increasing number of enzyme detoxification agents will be developed in the near future. The immunogenicity issue is most acute with the microbial enzymes that, in terms of catalytic activity, have been found so useful. Progress has already been made in identifying the dominant epitopes on one enzyme that opens the way to reducing its immunogenicity.

REFERENCES

Bagshawe, K. D. (1987). Antibody directed enzymes revive anti-cancer prodrugs concept. *British Journal of Cancer*, *56*(5), 531–532. doi:10.1038/bjc.1987.237 PMID:3426915

Baldo, B. A. (2016). Enzymes Approved for Therapy. In B. A. Baldo (Ed.), *Safety of Biologics Therapy: Monoclonal Antibodies, Cytokines, Fusion Proteins, Hormones, Enzymes, Coagulation Proteins, Vaccines, Botulinum Toxins* (pp. 419–477). Cham: Springer International Publishing. doi:10.1007/978-3-319-30472-4_9

Barnard, M. L., Baker, R. R., & Matalon, S. (1993). Mitigation of oxidant injury to lung microvasculature by intratracheal instillation of antioxidant enzymes. *The American Journal of Physiology*, *265*(4 Pt 1), L340–L345. PMID:8238368

Coiffier, B., Mounier, N., Bologna, S., Fermé, C., Tilly, H., Sonet, A., & Herbrecht, R. (2003). Efficacy and Safety of Rasburicase (recombinant urate oxidase) for the Prevention and Treatment of Hyperuricemia During Induction Chemotherapy of Aggressive Non-Hodgkin's Lymphoma: Results of the GRAAL1 (Groupe d'Etude des Lymphomes de l'Adulte Trial on Rasburicase Activity in Adult Lymphoma) Study. *Journal of Clinical Oncology*, *21*(23), 4402–4406. doi:10.1200/JCO.2003.04.115 PMID:14581437

Hernandez, C. P., Morrow, K., Lopez-Barcons, L. A., Zabaleta, J., Sierra, R., Velasco, C., & Rodriguez, P. C. (2010). Pegylated arginase I: A potential therapeutic approach in T-ALL. *Blood*, *115*(25), 5214–5221. doi:10.1182/blood-2009-12-258822 PMID:20407034

Hu, P., & Tirelli, N. (2012). Scavenging ROS: Superoxide dismutase/catalase mimetics by the use of an oxidation-sensitive nanocarrier/enzyme conjugate. *Bioconjugate Chemistry*, *23*(3), 438–449. doi:10.1021/bc200449k PMID:22292618

Ihler, G., Glew, R. H., & Schnure, F. W. (1973). Enzyme Loading of Erythrocytes. *Proceedings of the National Academy of Sciences of the United States of America*, *70*(9), 2663–2666. doi:10.1073/pnas.70.9.2663 PMID:4354859

Ihler, G., Lantzy, A., Purpura, J., & Glew, R. H. (1975). Enzymatic degradation of uric acid by uricase-loaded human erythrocytes. *The Journal of Clinical Investigation*, *56*(3), 595–602. doi:10.1172/JCI108129 PMID:1159075

Jubeh, T. T., Antler, S., Haupt, S., Barenholz, Y., & Rubinstein, A. (2005). Local prevention of oxidative stress in the intestinal epithelium of the rat by adhesive liposomes of superoxide dismutase and tempamine. *Molecular Pharmaceutics*, *2*(1), 2–11. doi:10.1021/mp0499095 PMID:15804172

Kasinathan, N., Volety, S. M., & Josyula, V. R. (2016). Chondroitinase: A promising therapeutic enzyme. *Critical Reviews in Microbiology*, *42*(3), 474–484. PMID:25319196

Lemons, M. L., Howland, D. R., & Anderson, D. K. (1999). Chondroitin sulfate proteoglycan immunoreactivity increases following spinal cord injury and transplantation. *Experimental Neurology*, *160*(1), 51–65. doi:10.1006/exnr.1999.7184 PMID:10630190

Magnani, M., Mancini, U., Bianchi, M., & Fazi, A. (1992). Comparison of Uricase-Bound and Uricase-Loaded Erythrocytes as Bioreactors for Uric Acid Degradation. In M. Magnani & J. R. DeLoach (Eds.), *The Use of Resealed Erythrocytes as Carriers and Bioreactors* (pp. 189–194). Boston, MA: Springer US. doi:10.1007/978-1-4615-3030-5_23

Morrow, K., Hernandez, C. P., Raber, P., Valle, L. D., Wilk, A. M., Majumdar, S., & Rodriguez, P. C. (2013). Anti-leukemic mechanisms of pegylated arginase I in acute lymphoblastic T-cell leukemia. *Leukemia*, *27*(3), 569–577. doi:10.1038/leu.2012.247 PMID:22926702

Munzig, E., Eckert, K., Harrach, T., Graf, H., & Maurer, H. R. (1994). Bromelain protease F9 reduces the CD44 mediated adhesion of human peripheral blood lymphocytes to human umbilical vein endothelial cells. *FEBS Letters*, *351*(2), 215–218. doi:10.1016/0014-5793(94)00860-4 PMID:7521849

Onaca, O., Hughes, D. W., Balasubramanian, V., Grzelakowski, M., Meier, W., & Palivan, C. G. (2010). SOD antioxidant nanoreactors: Influence of block copolymer composition on the nanoreactor efficiency. *Macromolecular Bioscience*, *10*(5), 531–538. PMID:20112239

Padmanabhan, R. V., Gudapaty, R., Liener, I. E., Schwartz, B. A., & Hoidal, J. R. (1985). Protection against pulmonary oxygen toxicity in rats by the intratracheal administration of liposome-encapsulated superoxide dismutase or catalase. *The American Review of Respiratory Disease*, *132*(1), 164–167. PMID:4014861

Petrikovics, I., Budai, M., Baskin, S. I., Rockwood, G. A., Childress, J., Budai, L., & Szilasi, M. (2009). Characterization of liposomal vesicles encapsulating rhodanese for cyanide antagonism. *Drug Delivery*, *16*(6), 312–319. doi:10.1080/10717540903003711 PMID:19606945

Petrikovics, I., Jayanna, P., Childress, J., Budai, M., Martin, S., Kuzmitcheva, G., & Rockwood, G. (2011). Optimization of Liposomal Lipid Composition for a New, Reactive Sulfur Donor, and In Vivo Efficacy Studies on Mice to Antagonize Cyanide Intoxication. *Journal of Drug Delivery*, *2011*, 7. doi:10.1155/2011/928626 PMID:22187653

Philpott, G. W., Bower, R. J., Parker, K. L., Shearer, W. T., & Parker, C. W. (1974). Affinity Cytotoxicity of Tumor Cells with Antibody-Glucose Oxidase Conjugates, Peroxidase, and Arsphenamine. *Cancer Research*, *34*(9), 2159–2164. PMID:4858425

Rossi, L., Bianchi, M., & Magnani, M. (1992). Increased glucose metabolism by enzyme-loaded erythrocytes in vitro and in vivo normalization of hyperglycemia in diabetic mice. *Biotechnology and Applied Biochemistry*, *15*(2), 207–216. PMID:1586460

Simone, E. A., Dziubla, T. D., Arguiri, E., Vardon, V., Shuvaev, V. V., Christofidou-Solomidou, M., & Muzykantov, V. R. (2009). Loading PEG-Catalase into Filamentous and Spherical Polymer Nanocarriers. *Pharmaceutical Research*, *26*(1), 250–260. doi:10.100711095-008-9744-7 PMID:18956141

Tan, Q. Y., Wang, N., Yang, H., Chen, L., Xiong, H. R., Zhang, L. K., & Zhang, J. Q. (2010). Preparation and characterization of lipid vesicles containing uricase. *Drug Delivery*, *17*(1), 28–37. doi:10.3109/10717540903508953 PMID:22747073

Tan, Q. Y., Wang, N., Yang, H., Zhang, L. K., Liu, S., Chen, L., & Zhang, J. Q. (2010). Characterization, stabilization and activity of uricase loaded in lipid vesicles. *International Journal of Pharmaceutics*, *384*(1), 165–172. doi:10.1016/j.ijpharm.2009.09.036 PMID:19782743

Taverna, R. D., & Langdon, R. G. (1973). Glucose transport in white erythrocyte ghosts and membrane-derived vesicles. *Biochimica et Biophysica Acta (BBA) -. Biomembranes*, *298*(2), 422–428. doi:10.1016/0005-2736(73)90369-6 PMID:4719139

Vorauer-Uhl, K., Fürnschlief, E., Wagner, A., Ferko, B., & Katinger, H. (2001). Topically applied liposome encapsulated superoxide dismutase reduces postburn wound size and edema formation. *European Journal of Pharmaceutical Sciences*, *14*(1), 63–67. doi:10.1016/S0928-0987(01)00149-X PMID:11457651

Walther, F. J., David-Cu, R., & Lopez, S. L. (1995). Antioxidant-surfactant liposomes mitigate hyperoxic lung injury in premature rabbits. *The American Journal of Physiology*, *269*(5 Pt 1), L613–L617. PMID:7491979

APPENDIX

Table 1. Therapeutic enzymes used for the treatment of non-deficiency diseases

Enzymes	Enzyme Commission Number	Sources	Brand Names	Indications	FDA Approval	EMA Approval
L-asparaginase	3.5.1.1	*E. coli/Erwinia carotovora*	Crasnitin, Elspar, Kidrolase, Leunase, Erwinase	Acute lymphocytic leukemia	Elspar 1978 Erwinase 2011	Spectrila, 2016
PEGasparaginase/ Pegaspargase	3.5.1.1	*E. coli*	Oncaspar	Acute lymphocytic leukemia	1994	2016
L-arginase	3.5.3.1	Mycoplasma	-	Antineoplastic	-	-
Recombinant PEGarginine deiminase/ ADI-PEG 20	3.5.3.6	*E. coli*	Melanocid, Hepacid, Melanoid	Invasive malignant melanoma, hepatocellular carcinoma	-	-
L-glutaminase	3.5.1.2	-	-	Antineoplastic	-	-
L-tyrosinase	1.14.18.1	-	-	Antineoplastic	-	-
L-serine dehydratase	4.3.1.17	-	-	Antineoplastic	-	-
L-threonine deaminase	4.3.1.19	-	-	Antineoplastic	-	-
L-tryptophanase	4.1.99.1	-	-	Antineoplastic	-	-
Urate oxidase	1.7.3.3	*Aspergillus flavus*	Uricozyme	gout; hyperuricemia in TLS patients	-	-
Recombinant Urate oxidase/ Rasburicase	1.7.3.3	*Saccharomyces cerevisiae*	Elitek, Fasturtec	Hyperuricemia in TLS patients	Elitek, 2002	Fasturtec, 2001
Lysozyme	3.2.1.17	Hen's egg white		Infectious diseases (antibiotic)	-	-
Collagenase	3.4.24.3	*Clostridium histolyticum*	Iruxol, Biozyme C, Santyl, Xiaflex, Xiapex	Debridement of wounds, dermal ulcers, Dupuytren's disease, Peyronie's disease	Santyl, 1965; Xiaflex, 2010	Xiapex, 2011
Papain	3.4.22.2	*Carica papaya*	Panafil	Digestion aid, debridement of wounds, reduction of edema after dentalsurgery	-	-
Chymopapain	3.4.22.6	*Papaya latex*	Discase, Chymodiactin	Chemonucleolysis of prolapsed invertebral disk	1982 (discontinued)	-
Trypsin	3.4.21.4	Bovine pancreas	Leukase, Granulex	Debridement of wounds, digestion aid	-	-
Chymotrypsin	3.4.21.1	Bovine pancreas	-	Wound healing, digestion aid	-	-
Plasmin (Fibrinolysin)	3.4.21.7	Bovine or human plasma	Fibrolan, Lysofibrin, Actase, Elase	Debridement of wounds	-	-
Chondroitinase ABC	4.2.2.20	*Proteus vulgaris*	-	Lumbar disk herniation, treatment of patients undergoing vitrectomy	-	-
Desoxyribonuclease (Streptodornase)	-	*Streptococcus haemolyticus*	Varidase	Gastric ulcers (together with streptokinase)	-	-
Bromelain	3.4.22.32	*Ananas comosus*	Traumanase, Phlogenzym, NexoBrid	Debridement of wounds, digestion aid, inflammation	-	NexoBrid, 2012
Serrapeptase (serralysin)	3.4.24.40	*Serratia* E15, *Pseudomonas aeruginosa, Aspergillus oryzae*	Danzen	Inflammation	-	-
Superoxide Dismutase	1.15.1.1	Bovine liver or erythrocytes	Peroxinorm, Ontosein, Oxinorm, Orgotein	Inflammatory arthrosis, polyarthritis, asthma, oxygen toxicity in premature infants	-	-
Recominant Superoxide Dismutase/Sudismase	1.15.1.1	*E. coli* or yeast	Oxsodrol	Prevention of bronchopulmonary dysplasia in premature neonates	-	-

continued on following page

Table 1. Continued

Enzymes	Enzyme Commission Number	Sources	Brand Names	Indications	FDA Approval	EMA Approval
Recombinant PEGsuperoxide dismutase/ Pegorgotein	1.15.1.1	-	Dismutec	Closed head injury	-	-
Hyaluronidase	3.2.1.35	Bovine and ovine testes	Hylase, Vitrase, Amphadase, Hydase	Dispersion and absorption of injected drugs; sc fluid administration for hydration; sc urography for resorption of radiopaque agents	Amphadase, 2004 Vitrase, 2004; Hydase, 2005	-
Recombinant Hyaluronidase	3.2.1.35	CHO cells	Hylenex	Removing the cumulus matrix surrounding oocytes in preparation for assisted reproductive technology procedures, spreading agent to facilitate the dispersion and absorption of other drugs	Hylenex, 2005	-
Recombinant PEGurate oxidase/ Pegloticase	1.7.3.3	*E. coli*	Puricase, Krystexxa	Hyperuricemia in patients with severe gout	Krystexxa, 2010	Krystexxa, 2013
Recombinant *Pseudomonas sp.* carboxypeptidase G2 / Glucarpidase	3.4.17.11	*E. coli*	Voraxaze	Clearance of methotrexate intoxication due to kidney failure	2012	-
ButyrylcholinestErase	-	-	-	Clearance of cocaine during a drug overdose	-	-
Recombinant Platelet-Activating Factor Acetylhydrolaset	3.1.1.47	*E. coli*	Pafase	Severe sepsis, respiratory distress syndrome, pancreatitis	-	-
Ribonuclease	3.1.26.4	*Rana pipiens*	Onconase	Malignant mesothelioma, refractory breast cancer, renal cell cancer, antiviral	-	-
Neuraminidase	3.2.1.18	-	-	Antineoplastic	-	-
L-α-Arabinofuranosidase	3.2.1.55	-	-	Antineoplastic	-	-
Thrombin	3.4.21.5	Human plasma	Thrombinar, Velyn	Superficial bleeding; fibrin glues	-	-
Oryzin (Aspergillus Alkaline Protease)	3.4.21.63	*Aspergillus sp.*		Inflammation	-	-
Sfericase	-	*Bacillus sphaericus*		Chronic bronchitis, acute pneumonia, chronic sinusitis	-	-
Ananain, Comosain	-	Pineapple stem	Vianain	Enzymatic debridement of severe burns	-	-
Papain, Trypsin and Chymotrypsin	-	Bovine pancreas	Wobe-Mugos	Multiple myeloma	-	-
Protease Extracts Containing Proteases, Cellulases, RNases, α- and β-Amylases	-	*Bacillus subtilis, Aspergillus oryzae*	Travase, Nortase, Combiase	Debridement of wounds, digestion aid	1969 (discontinued)	-

Table 2. Enzymes used in enzyme prodrug therapy

Enzymes	Prodrug	Cytotoxic drug	Potential Targets
Carboxypeptidase G2	4-[bis-(2-chloroethyl)-amino] benzoyl-L-glutamic acid	4-[bis-(2-chloroethyl)-amino] benzoic acid	Colon cancer
Penicillin V Amidase	Doxorubicin-N-p-hydroxyphenoxyacetamide	Doxorubicin	Lung carcinoma
	Melphalan-N-p-hydroxyphenoxyacetamide	Melphalan	
β-Lactamase	5-fluorouracil-cephalosporin	5-fluorouracil	Colon cancer
Cytosine Deaminase	5-fluorocytocine	5-fluorouracil	Colon cancer
β-glucuronidase	Doxorubicin –glucuronide	Doxorubicin	Lung Carcinoma
	Epirubicin-glucuronide	Epirubicin	
	Phenol mustard glucuronide	Phenol mustard	
Carboxypeptidase A1	Methotrexate-alpha-phenylalanine	Methotrexate	Lung carcinoma
Alkaline Phosphatase	Etoposide phosphate	Etoposide	Colon cancer
	Mitomycin phosphate	Mitomycin	
Azoreductase	Azobenzene mustards	Phenylenediamine-mustards	**Liver Cancer**
DT Diaphorase	5-(Aziridine-l-yl)-2,4-dinitrobenzamide (CB 1954)	5-(Aziridine-l-yl)-4-hydroxylamino-2-nitrobenzarnide	Lung, breast, and colon cancer
	3-Hydroxymethyl-5- aziridinyl-l-methyl-2-(IH indole-4,7-dione)prop-β-en-a-ol (E09)	Reduced-E09 products	
Plasmin	Peptidyl-p-phenylenediamine-mustard	Phenylenediamine-mustard	B16 Melanoma
	Peptidyl-*p*-doxorubicin	Doxorubicin	

Table 3. Enzymes used in antibody-directed enzyme prodrug therapy (ADEPT)

Enzymes	Prodrug	Cytotoxic Drug	Antibody
Alkaline Phosphate	Etoposide phosphate	Etoposide	Monoclonal antibodies, L6 (anticarcinoma) and 1F5 (anti-B lymphoma)
	Mitomycin C phosphate	Mitomycin	
	Doxorubicin phosphate	Doxorubicin	
	Phenolmustard phosphate	Phenolmustard	
Carboxypeptidase G2	Benzoic mustard glutamates	Benzoic acid mustards	CEA scFv Fab'(2)-fragment of anti-carcinoembryonic antigen antibody anti-human c-erbB2 monoclonal antibody
	CMDA (4-([2-chloroethyl][2-(mesyloxy) ethyl]amino) benzoyl-l-glutamic acid)	Benzoic acid mustards	
	ZD2767P (bis-iodo-phenol mustard)	Benzoic acid mustards	
Carboxypeptidase A1	Methotrexate peptide	Methotrexate	scFv to γ seminoprotein
	Methotrexate alanine	Methotrexate	
Cytosine Deaminase	5-fluorocytosine	5-fluorouracil	scFv to gpA33 antigen in colon cancers
β-Lactamase	Cephalosporin mustards	Nitrogen mustards	L6 scFv to breast, colon, lung and ovarian carcinoma dsFv to p185HER2 scFv to melanotransferrin p97 scFv to melanoma Camelid VHH to CEA scFv to TAG72 carbohydrate epitope
	Doxorubicin-cephalosporin	Doxorubicin	
	C-Mel or GC-Mel	melphalan	
β-Glucuronidase	Phenol mustard glucuronide	Phenol mustard	CEA Fab CD20 scFv scFv to EpCAM Fab, chimeric tumor necrosis therapy antigen binding fragment Monoclonal antibody against pancarcinoma antigen epithelilal transmembrane glycoprotein L49-sFv
	Daunorubicin glucuronide	Daunorubicin	
	Glucuronide camptothecin	Camptothecin	
	Epirubicin-glucuronide	Epirubicin	
Nitroreductase	CB1954 (5-(Aziridine-I-yl)-2,4-dinitrobenzamide)	5-(azirdin-1-yl)-4-hydroxy-amino-2-nitrobenzamide	Murine monoclonal ***antibody A5B7*** targets the carcinoembryonic antigen (CEA)
Penicillin Amidase	Palytoxin-4-hydroxyphenylacetamide	Palytoxin	Monoclonal antibodies, L6 (anticarcinoma)
	Doxorubicin-phenoxyacetamide	Doxorubicin	
	Melphalan-phenoxyacetamide	Melphalan	
Carboxyesterases	CPT-11, irinotecan	SN38	scFv antibody C28 directed against the epithelial cell adhesion molecule EpCAM mAb IgG2a anti-p97 expressed on melanoma cells
Glycosidases	Glycosides	Daucormycin	-
α-Galactosidase	N-[4-(a-galactopyranosyl)-benyloxycarbonyl]-daunorubicin	daunorubicin	-
β-Galactosidase	17-AG-C2-Gal, Galactose-amine derivate of geldanamycin	17-AG-C2, toxic geldanamycin derivate	Anti-TAG 72 antibody fragment
β-Glucosidase	Amygdalin	Cyanide	Monoclonal antibody against polymorphic epithelial mucin
Ribonuclease A (Mutant)	Ribotide prodrug	Aniline mustard	-

Table 4. Enzymes used in gene-directed enzyme prodrug therapy (GDEPT)

Enzymes	Prodrug	Cytotoxic Drug
Thymidine Kinase	Ganciclovir	Ganciclovir Triphosphate
	6-Methoxypurine	Adenine
	arabinonucleoside	Arabinonucleoside triphosphate
Cytosine Deaminase	5-fluorocytosine	5-fluorouracil
Cytochrome P450	Cyclophosphamide	4-OH Cyclophosphamide
	Ifosfamide	4-OH ifosfamide
Nitroreductase	CB1954 (5-(aziridin-1-yl)-2,4-dinitrobenzamide)	5-(aziridin-1-yl)-4- Hydroxylamino-2-Nitrobenzamide
Carboxypeptidase G2	CMDA (4-([2-chloroethyl][2-(mesyloxy) ethyl]amino) benzoyl-l-glutamic acid)	Benzoic Acid Mustards
Glucose Oxidase	Glucose	Hydrogen Peroxide
Carboxypeptidase A1	Methotrexate-alanine	Methotrexate
Xanthine Oxidase	Hypoxanthine	Hydrogen Peroxide + Superoxide
α-Galactosidase	N-h4-(α-D-Galactopyranosyl) Benzyloxycarbonyl daunorubicin	Daunorubicin
β-Glucosidase	Amygdalin	Cyanide
Azoreductase	Azobenzene mustards	Phenylenediamine Mustards
β-Glucuronidase	Epirubicin-glucuronide	Epirubucin
β-Lactamase	Vinca-cephalosporin	4-Desacetylvinblastine- 3-carboxyhydrazide
Alkaline Phosphatase	Phenolmustard phosphate	Phenolmustard
	Doxorubicin phosphate	Doxorubicin
	Mitomycin phosphate	Mitomycin Alcohol
	Etoposide phosphate	Etoposide
Penicillin Amidase	Palytoxin-4-hydroxyphenyl- acetamide	Palytoxin
	Doxorubicin-phenoxyacetamide	Doxorubicin
	Melphalan-phenoxyacetamide	Melphalan
Thymidine Phosphorylase	5' -Deoxy-5-fluorouridine (5'-DFUR)	5-FU
Horseradish Peroxidase	indole-3-acetic acid	3-Methylene-2-Oxindoles
Purine Nucleoside Phosphorylase	fludarabine phosphate	Fludarabine
Tyrosinase	Hydroyphenylpropanol or N-acetyl-4-S-cysteaminylphenol	Quinones

Figure 1. Enzyme prodrug therapy

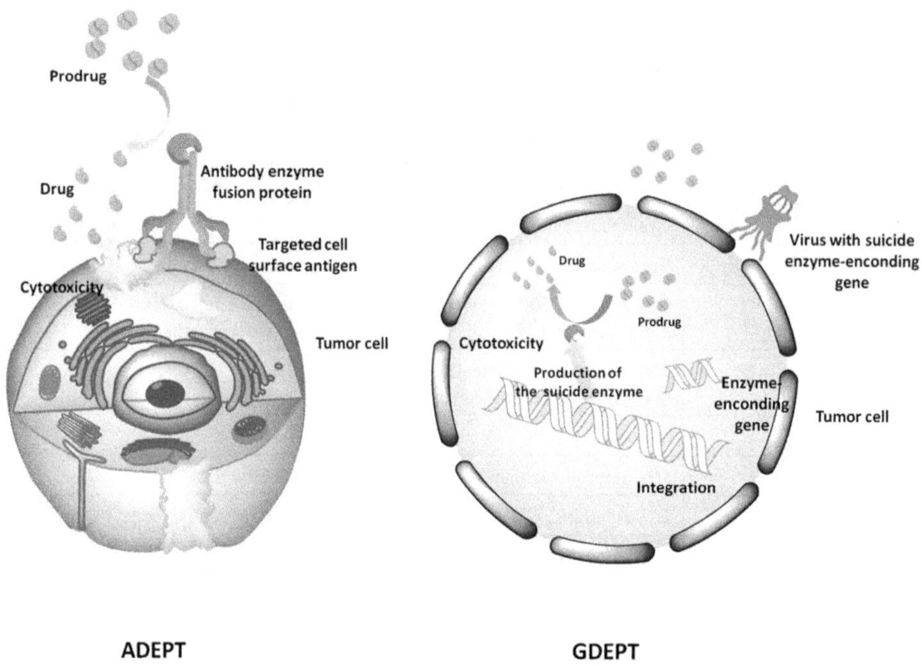

Chapter 4
Colonic Bacterial Enzymes:
Pharmaceutical Significance and Applications

Srushti M. Tambe
Bombay College of Pharmacy, India

Namita D. Desai
Bombay College of Pharmacy, India

ABSTRACT

This chapter reviews various enzymes produced by the colonic microflora and their utilization in the development of pharmaceutical dosage forms to achieve colon-specific drug delivery. This chapter discusses the applications of colonic bacterial enzymes in order to surrogate colonic conditions in vivo *so as to evaluate* in vitro *drug release from microbially triggered/enzymatically triggered colon-specific drug delivery systems. This chapter also discusses different methods to produce colonic bacterial enzymes as well as use of probiotics as a source to produce colonic bacterial enzymes.*

INTRODUCTION

The human colon comprises of a wide range of microflora representing approximately 10^{11}-10^{12} CFUs/ml with more than 400 bacterial species such as Bifidobacteria, Bacteroides, Clostridia etc. These bacterial species residing in the colon are capable of producing more than 500 different types of enzymes. The fermentation of various substrates that are left undigested in the small intestine after reaching the colon acts as a source of energy for the colonic bacteria in order to maintain their cellular functions. These substrates include disaccharides and trisaccharides such as lactulose, cellobiose, raffinose, stachyoseas well as partially fermented residues of polysaccharides such as pectins, galactomannans, xylans, etc. The colonic microflora produces wide range of bacterial enzymes such as α-D-galactosidase, β-D-glucosidase, β-xylosidase, β-arabinosidase, β-D-glucuronidase etc., to carry out fermentation of substrates (polysaccharides or dietary fibers). These bacterial enzymes are also capable of degrading the various polysaccharide-based coatings/matrices as well as break the bonds between the inert carrier and an ac-

DOI: 10.4018/978-1-5225-5237-6.ch004

tive drug moiety resulting in the release of active drug in the colon without being affected in the upper gastrointestinal tract (GIT). Taking this into consideration, several formulations have been developed to achieve colon targeting by using polysaccharides and carriers that are readily degraded by the colonic microflora but unaffected by the microflora present in the upper part of GIT. These systems, after reaching the colon are fermented (polysaccharide-based systems) or broken down (prodrug-based systems) by the colonic bacterial enzymes, thereby causing complete drug release in the colon. These approaches have been utilized in the treatment of various colonic conditions such as Crohn's disease, ulcerative colitis, diarrhea, constipation, amoebiasis and colorectal cancer. Additionally, these systems are utilized to avoid first pass metabolism of the drugs that are degraded in the upper gastric environment. Colonic bacterial enzymes have also found its applications in the *in vitro* release evaluation of colonic formulations serving as a discriminatory and biorelevant method to evaluate drug release from enzymatically/ microbially triggered colonic formulations. Colonic bacterial enzymes that are specific to the substrate e.g. pectinases for pectin-based colonic formulations or a mixture of several colonic bacterial enzymes representing the entire colonic microflora are employed during the dissolution studies under anaerobic conditions in order to simulate colonic conditions and assess drug release from the colonic formulations. This chapter reviews the importance of colonic bacterial enzymes, their production methods as well as their utilization to develop various formulations for achieving colon-specific drug delivery.

BACKGROUND

The human gastrointestinal tract (GIT) represents dynamic and ecologically diverse environment containing variety of microorganisms that contributes to its physiology and functions including its importance in the metabolism of ingested material. The upper part of the GIT including the upper region of the gastrointestinal tract, the stomach, duodenum, jejunum and upper ileum possess very small number of bacteria, mostly Gram-positive facultative organisms (Simon et al., 1982). However, the bacterial concentration increases significantly in the colon. The count of the colonic bacteria is 10^{11}-10^{12} colony forming units/mL (CFU/mL) (Moore, & Holdeman, 1975) and the anaerobic bacteria outnumber the aerobic bacteria by the factor of 1000 (Simon, Gorbach, & Bustos-Fernandez, 1983). This is because in the lower gut, the conditions are highly reducing and the oxygen tension is so low that even very-oxygen-sensitive (VOS) anaerobes can flourish and under these conditions facultative organisms compete poorly for nutrients (Hillman, 2004). The most predominant species of anaerobic bacteria in the colon are: Bacteroides, Bifidobacterium, Eubacterium, Peptococcus, Peptostreptococcus, Ruminococccus, Propionibacterium, Veillonella and Clostridium whereas; Escherichia coli and lactobacillus are important facultative bacteria in the colon (Simon, Gorbach, & Bustos-Fernandez, 1983; Drasar, & Hill, 1974). These bacteria produce wide range of hydrolytic and reductive metabolizing enzymes (Rowland, 1988). Various roles of the enzymes of bacterial origin in the colon are listed in Figure 1. The most widely exploited reductive colonic enzymes in drug delivery are azoreductases, nitroreductases, N-oxide reductases, sulfoxidereductases, deaminases and the most extensively investigated hydrolytic colonic enzymes are glucosidases, β-xylosidases, α-L-arabinosidases and glucuronidases (Scheline, 1973). These intestinal enzymes are used to trigger drug release from devices in various parts of gastrointestinal tract to degrade the polymeric coatings/matrices as well as to break bonds between an inert carrier and an active molecule (release of the drug using prodrug approach) (Tomlin, & Read, 1988; Prasanth, Jayaprakash, & Mathew, 2012). Such drug delivery systems are called as 'enzymatically triggered or

Figure 1. Role of colonic bacterial enzymes

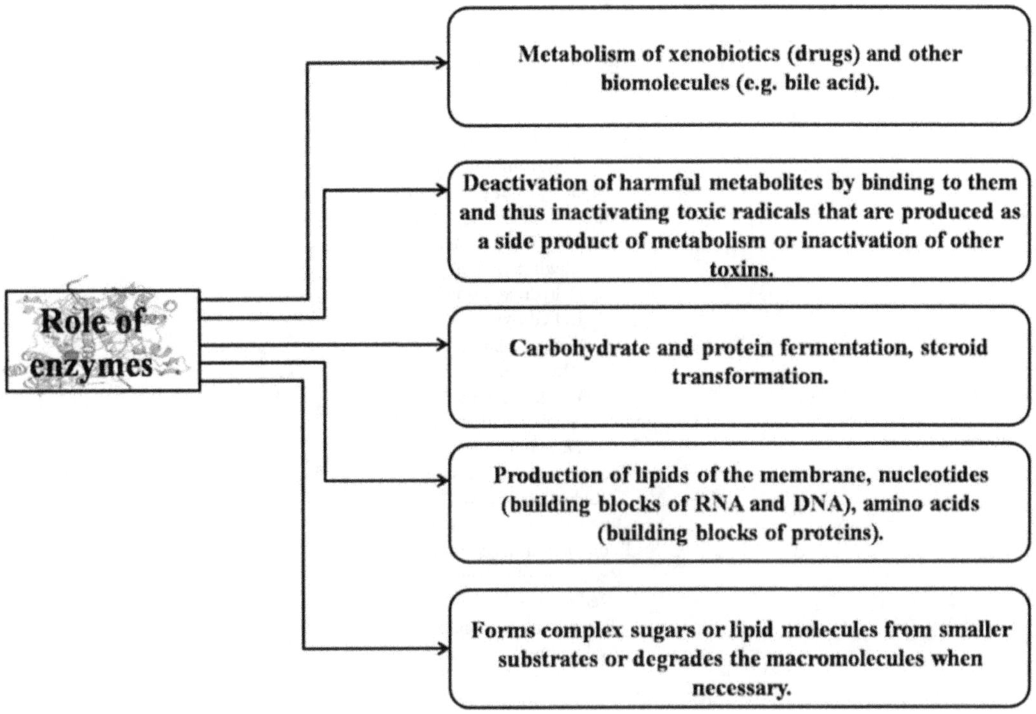

microbially triggered' colonic delivery systems for systemic or local drug action in the colon. In general, it can be said that reduction and hydrolysis are the major processes occurring in the lumen of the colon (Davaran, Hanaee, & Khosravi, 1999). The electron carriers (redox mediators) such as benzyl viologen and flavinmononucleotide, act as electron shuttles between the intracellular enzyme and the substrate (Jain, & Jain, 2008; Grim, & Kopecek, 1991). It was suggested that colon-specific drug delivery is a valid approach because of the low molecular weight electron mediators such as nicotinamide adenine dinucleotide phosphate (NADPH) that are present and are able to diffuse throughout a swollen polymeric matrix and not only because of particular organisms possessing specific azo-reductase exist in the colon (Lloyd, Martin, & Soozandehfar, 1994). Metabolic reactions carried out by colonic bacterial enzymes are shown in Table 1.

COLONIC BACTERIAL ENZYMES

Azoreductases

Azoreductases are the enzymes that are able to reduce the diazo bonds to azo bonds. The bacterial species in the colon that exhibit azoreductase activity are Escherichia coli, Streptococcus faecalis, Lactobacillus spp., Bacteroides fragilis, Eubacterium biforme, Bif. adolscentis, Cl.bifermentans, Cl.perfringens and Clostridium spp (Drasar, & Hill, 1974). Azo-compounds are the most common synthetic colored compounds used in foods, pharmaceuticals and cosmetics. Azoreductases produced by wide range of

Table 1. Metabolic reactions carried by colonic bacterial enzymes (Scheline, 1973; Van den Mooter & Kinget, 1995)

Hydrolysis of Glycosides
Glucuronides
Other Glycosides
Hydrolysis of sulphate esters
Hydrolysis of amides
Hydrolysis of esters
Hydrolysis of sulphamates
Hydrolysis of nitrates
Dehydroxylation
C Dehydroxylation
N Dehydroxylation
Decarboxylation
Dealkylation
O -Alkyl groups
Dehalogenation
Desamination
Heterocyclic ringfission
Reduction of C = C
Reduction of azo bonds
Reduction of nitro groups
Reduction of aldehydes
Reduction of ketones
Reduction of alcohols
Reduction of N-oxides
Acetylation
Esterification

colonic bacterial species are responsible for the reduction of the diazo linkages to the parent aromatic amines in the azo food colors, pharmaceuticals and cosmetics. The reducing activity of the bacterial azoreductases can be easily detected and quantified from the loss of pigment as the parent aromatic amines are colorless. Radomski et al., (1962) were able to demonstrate that the reductive fission of the azo groups in the colon was probably due to the action of wide range of bacterial species present in the colon. Azo reduction activity was reported by Fore et al., (1967) from the contents isolated from the distal portions of the small intestine and from the caecum of the rats, although no azo reducing activity was seen in stomach and duodenal contents. This study was able to conclude that the azo-reducing activity was of microbial origin in the lower part of the gastrointestinal tract. Drasar et al., (1974) observed that the reducing activity carried out by bacterial azoreductases required strict anaerobic conditions and an optimum pH around 7 for Enterococcus faecalis and pH value around 5 for Clostridium perfringens. It was also observed that no reducing activity took place by azoreductases when bacterial strains were incubated with diazo substrates in simple buffers even though strict anaerobic conditions were maintained. The azoreductase activity of Pseudomonas vulgaris was an NADPH-specific flavoprotein and was demonstrated by Roxon et al., (1967) whereas the azoreductase from Streptococcus faecalis was a flavoprotein that required NADH which was demonstrated by R. Scheline et al., (1970). The importance of flavins in the azo reduction process by bacterial azoreductases was showed by Gingell et al., (1971). The study showed that flavins act as two-electron shuttles between the flavoprotein and acceptor azo-bond. Table 2 shows the azoreductase activity of some colonic bacterial species. Various factors such as bile salts concentration, surface active and wetting agents, chelating agents affect the azoreductase activity in humans. A fivefold increase in azoreduction of tartarazine in the presence of bile salts was

Table 2. Azoreductase activity of colonic bacterial species.

BACTERIAL SPECIES	NUMBER OF STRAINS	AZOREDUCTASE ACTIVITY	
		MEAN	RANGE
Facultative anaerobes			
Escherichia coli	10	0.4	0.1-0.9
Micro-oesophilic			
Streptococcus faecalis	9	0.9	0.4-2.2
Lactobacilli spp.	10	13.0	0-50.7
Obligate anaerobes			
Bacteroides fragilis	9	0.2	0-0.8
Eubacterium biforme	5	0.3	0.2-0.5
Bifidobacterium adolscentis	4	0.8	0-3.1
Clostridium bifermentans	5	14.6	0-64.6
Clostridium perfringes	5	26.7	2.2-115.4

demonstrated by Allan et al., (1974). Azoreductase activity was also shown to be affected by the composition of the diet (Mallett, Wise, & Rowland, 1983; Wise, Mallett, & Rowland, 1982). The effects of diet on azoreductase activity are shown in Table 3.

Nitroreductases

Nitroreductase enzymes are mostly produced by the anaerobic gut bacteria such as Bacteroides thetaiotaomicron, Clostridium perfringens, Bifidobacterium infantis, Peptococcus anaerobius, Peptostreptococcus productus, etc. (Howard, Beland, & Cerniglia, 1983). Human nitroreductases are responsible to reduce several nitro compounds such as environmental contaminants generated during the combustion of fossil fuels like nitrobenzenes, nitrotoluenes, nitropyrenes etc. Also many nitro drugs like chloroamphenicol, nitroimidazoles, etc. are also reduced by bacterial reductases. (Rickert, 1987) isolated and identified nitroreductase producing bacteria. The bacterial isolates were identified as Clostridium spp., Clostridium leptum, Clostridium paraputrificium, Clostridium clostridiiforme as well as a Eubacterium spp. Eubacterium spp. was found to produce higher amount of nitroreductases and Clostridium clostridiiforme was found to produce the least. These species were found to be involved in the reduction of 4-nitrobenzoic acid. It was found that the conversion of 4-nitro benzoic acid to 4-aminobenzoic acid under anaerobic conditions occurred in all of the bacterial cultures. Moreover, nitroreductase was constitutive and extracellular and required FAD (Flavin Adenine Dinucleotide) for activity. Similarly, nitroreductase activity

Table 3. Effects of diet on azoreductase activity.

COMPOSITION OF DIET	EFFECT ON AZOREDUCTASE ACTIVITY
Pectin – 5% supplements	20% decrease
Cellulose – 15% supplements	20-fold increase
Carboxy Methyl Cellulose – 5% supplements	2.3-fold increase
From Grain-Based to Meat-Based Diet	2-fold increase

markedly increases with the addition of flavin (Rafil, Franklin, Heflich, & Cerniglia, 1991). It was also found that the nitroreductase activity from the bacterial strains were inhibited by both menadione and o-iodosobenzoic acid which binds to the sulfhydryl groups. This indicates that sulfhydryl groups are essential for the catabolic function of the enzyme and suggests that there may be cysteine residues in the active site. Bacterial reductases have several similar characteristics as that of nitrile reductases. The bacterial nitroreductases have low substrate specificity, acting on wide range of substituted nitrophenols. (Kiese, 1974) demonstrated that nitroreduction from E.coli required NAD which is a reducing agent and dicarboxylic acid like fumarate. Bacterial nitroreductases reduced nitro compounds, amino products with hydroxylamino and nitroso as intermediate compounds. These intermediate compounds are important in the methaemoglobinaemia produced by nitrobenzene. Nitroreductase produced by the colonic bacterial species are able to reduce nitropyrenes to produce potent mutagens (Howard, Beland, & Cerniglia, 1983). Nitropyrenes are widespread in urban air particles and exhaust fumes of diesel (Wang, Lee, King, & Warner, 1980). 1-nitropyrene was readily reduced to 1-aminopyrene by the action of bacterial reductases found in the colon on incubation with the rat colonic contents. Nitrotoluenes are also one of the environmental contaminants. It was found that 2-nitrotoluene is genotoxic and 2,6-dinitrotoluene is hepatotoxic in conventional rats (Doolittle, Sherrill, & Butterworth, 1983) but not in germ-free rats (Mirsalis, Hamm, Sherrill, & Butterworth, 1982).

Amino Decarboxylases

Berthelot, (1911) and Hanke et al., (1924) both demonstrated that the product formation after amino acid decarboxylation by bacterial decarboxylases are primary amines by using pure bacterial strains *in vitro* as well as *in vivo* studies in humans and animals. The product of amino acid decarboxylases are aliphatic amines, diamines, (product from basic amino acids) or omega-acid carboxylic acid (product from acidic amino acids). Tryosine on decarboxylation by bacterial decarboxylases gives octopamine which has similar pharmacological activity as that of noradrenaline. They further undergo portal transport to the liver followed by absorption in the colon and finally excreted in urine (Hanke, & Koessler, 1924). Bacterial decarboxylases are inducible enzymes that require a cofactor called pyridoxal phosphate. Histidine decarboxylase is an exception which does not require a cofactor for its activity. Gale, 2006 investigated decarboxylases in details and showed that decarboxylases are substrate specific, acting only on a single amino acid. In contrary, Haughton et al., (1961) demonstrated that decarboxylases are less substrate specific and act on wide range of amino acid substrates. Hayes et al., (1974) showed that decarboxylases are active over wide range of pH (4.0 - 7.5) and exhibit maximum activity at acidic pH close to 5.0.

Glycosidases

The most important glycosidases in the colon are β-glucosidase, α-glucosidase, β-glucuronidase, β-galactosidase and α-galactosidase. β-glucuronidase is most actively produced by *coliforms* and α-glucosidase, β-galactosidase and α-galactosidase are produced by Lactobacillus. β-glucosidase which degrades glycosides to release aglycones is highly produced by Enterococci (Hawksworth, Drasar, & Hili, 1971). Dabek et al., (2008) studied 40 different bacterial strains that represent the dominant bacterial species from human colon. It was found that most Bifidobacterium spp. and Bacteroides thetaiotaomicron carried β-glucosidase activity. β-glucosidases can exert beneficial as well as harmful effects which may be due to the aglycones formed by them with the help of a range of different plant glycosides that can lead to either mutagenic/toxic or health-promoting effects. Some plant glycosidases are glucuronized by serving as a substrate for β-glucuronidases when they reach the colon after being deconjugated by host β-glucosidases in the upper gut. The resulting aglycones of plant polyphenols are further degraded and undergo biotranformation by the colonic bacterial enzymes. Variation in the diet has shown to affect the activity of both β-glycosidases and β-glucuronidases. Consumption of soy has shown to increase the β-glycosidase activity in healthy volunteers. Faecal β-glucuronidases showed marked increase in rodents after the consumption of high protein/high fat diet whereas the activity was decreased after the consumption of carbohydrate-rich diet. β-glucuronidases liberate toxins and mutagens that have been glucuronated in the liver and excreted into the gut with the bile. This can in turn lead to high local concentration of carcinogenic compounds within the colon resulting in increased risk of carcinogenesis. Furthermore, enterohepatic circulation of xenobiotics compounds can occur due to the re-glucuronidation in the liver as well as reuptake of the deconjugated compounds from the gut which may increase their retention time in the body.α-galactosidases are enzymes that hydrolyze α-galactosidic bonds present in raffinose and stachyose which are α-galactosides of sucrose comprising three and four monomeric units, respectively. These oligosaccharides are non-digestible in upper GIT and small intestine due to the absence of α-galactosidases in human intestinal muscosa. However, they are readily metabolized in the lower intestine in the presence of bacterial α-galactosidases (Donkor, Henriksson, Vasiljevic, & Shah, 2007). Bacterial species that have been reported to have α-galactosidase activity are Lactobacillus spp., Bifidobacterium spp., Clostridium spp., etc. Deficiency of lysosomal α-galactosidase A activity can lead to Fabry's disease which is an X-linked inborn error of glycosphingolipid catabolism disease, progressive accumulation of globotriaosylceramide and related glycosphingolipids in vascular endothelial lysosomes of the kidneys, heart, skin, and brain leads to the main disease manifestations (Eng, Guffon, Wilcox, Germain, Lee, Waldek, . . . Desnick, 2001).

Applications of Colonic Bacterial Enzymes in Colon-Specific Drug Delivery Systems (CoDDS)

Targeting drugs and delivery systems to the colonic region of the gastrointestinal tract has received considerable interest in the recent years. Scientific endeavor in this area has been driven by the need to improve the treatment of local disorders of the colon such as inflammatory bowel disease (ulcerative colitis and Crohn's disease), irritable bowel syndrome and carcinoma. Also, colon is receiving significant attention as a portal for the entry of drugs into the systemic circulation. The most commonly used strategies to formulate colon-specific drug delivery systems include timed release systems, prodrugs, pH sensitive polymer coating/s and colonic micro- flora activated delivery systems (Drugs 2005; 1991-2007).

Polysaccharide Based Colon-Specific Drug Delivery Systems

In the human colon, wide spectrum of bacteria carry out the breakdown process of polysaccharides converting them to monosaccharides or disaccharides which are further fermented to produce a range of fatty acid end products. Several monosaccharides are part of the diet or can be released from dietary glycosides or polysaccharides. Some of the gut bacteria that utilize these monosaccharides are Peptococcus spp., Peptostreptococcus spp., Bacillus fragilis, Bacillus ovatus, Eubacterium rectale, Eubacterium lentum, Propionibacterium acnes, Lactobacillus plantarus, Bifidobacterium adolscentis. Polysaccharides are extensively investigated now-a-days for colon-specific drug delivery due to their unique proper-

Figure 2. General structure of Galactomannan, where, M = Mannose and G = Galactose

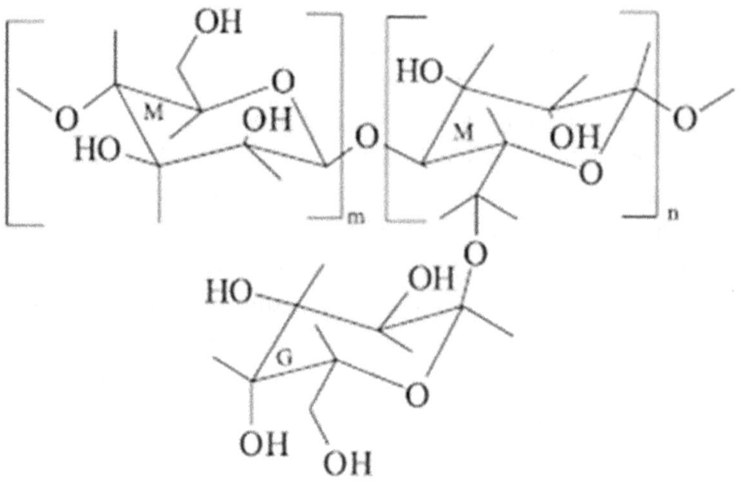

ties of being non-toxic, biodegradable, stability in acidic conditions and degradation by the intestinal anaerobic bacteria. Some examples of polysaccharides used as matrices/ coatings in drug delivery are mentioned below:

- **Guar Gum (Galactomannans):** Guar gum is a non-ionic, natural polysaccharide and it comprises of a linear chain of β-D- mannopyranosyl units linked (1→ 4) with single member α-D-galactopyranosyl units occurring at the side chains (Figure 2) (McCleary, & Matheson, 1986; Moreira, 2008). Although, humans and animals are incapable of digesting the galactomannan residues, they are completely fermented by bacterial enzymes to large amount of short chain fatty acids in the colon. Bacteroids ovatus and Ruminococcus albus have shown to participate in the degradation process of Guar gum. Various guar gum-based pharmaceuticals are evaluated by using enzymes such as α-galactosidase, galactomannanase, β-mannanase to study the degradation properties of the polysaccharides during *in vitro* drug release studies. Galactomannans exhibits greater potential as a carrier for colon specific drug delivery because they are readily fermented by various colonic bacterial enzymes without being affected in the upper gastrointestinal tract. Galactomannans such as guar gum and locust bean gum have been extensively studies as coating polymer/matrices to achieve colon targeting.
- **Degradation of Galactomannan Using Enzymes:** The galactomannandegrading enzymes are shown in Table 4. Additional enzymes are required to remove side-chain substituents that might be attached at various points on the mannan structure, creating more sites for subsequent enzymatic hydrolysis (Moreira, 2008). Enzymatic attack on galactomannans by different bacterial enzymes is shown in Figure 3. β-Mannanase also called as β-endomannanase, is an endo-type enzyme, which is responsible for the cleavage of β-1,4-linked internal linkages of the galacto (gluco) mannan backbone randomly to produce new chain ends (mannooligosaccharides) (McCleary, 1991).

Depending on the extent and pattern of substitution of the mannan backbone, β-mannanase affects the degradation process of galactomannan and galactoglucomannan.β-Mannanase has the lowest substitu-

Table 4. Galactomannan degrading enzymes and their EC numbers (Enzyme Commission number)

ENZYME	EC NUMBER
β-mannosidase (1,4-β-D-mannopyranoside hydrolase	EC 3.2.1.25
β-mannanase (1,4-β-D-mannanmannohydrolase	EC 3.2.178
β-glucosidase (1,4-β-D-glucosideglucohydrolase	EC 3.2.1.21
ADDITIONAL ENZYMES	
α-galactosidase (1,6-α-D-galactosidegalactohydrolase	EC 3.2.1.22
acetyl mannan esterase	EC 3.1.1.6

Figure 3. Enzymatic attack on galactoglucomannan structure (Puls, 1993)

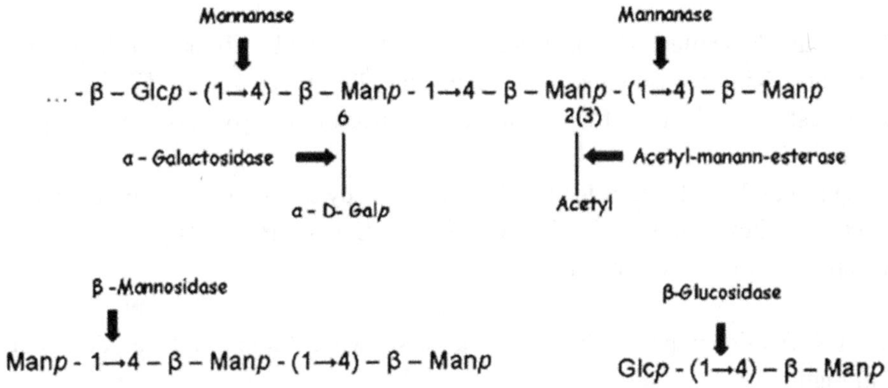

tion on the backbone and therefore. is the most active mannan-degrading enzyme (Civas, Eberhard, Le Dizet, & Petek, 1984). β-Mannosidase, an exo-type enzyme, releases mannose from the non-reducing end of mannans and mannooligosaccharides by cleaving β-1,4-linked mannosides (Coughlan, Tuohy Filho, EXF, Claeyssens, Vrsanska, & Hughes, 1993; Wong, & Saddler, 1993). Both β-mannanase and β-mannosidase have transglycosylation activity and can therefore be used for the synthesis of specific oligosaccharides (Moreira, 2008). Complete degradation of the galactomannan/glucomannan backbone to mannose by β-mannanase and β-mannosidase also depends on the action of β-glucosidase and α-galactosidase. β-glucosidase, an exotype enzyme, hydrolyzes 1,4-β-D-glucopyranose at the nonreducing end of the oligosaccharides released from glucomannan and galactoglucomannan by β-mannanase. α-Galactosidase, a debranching enzyme, catalyzes the hydrolysis of α-1,6-linked D-galactopyranosyl side chains of galactomannan and galactoglucomannan. Acetyl mannan esterase is a debranching enzyme which cleaves and releases acetyl groups from galactoglucomannan structure (Moreira, 2008). Mannan-degrading enzymes catalyze double displacement lysozyme-like reactions with retention of anomeric configurations. The mechanism involves stabilization of an oxocarbonium ion by electrostatic interaction with the carboxylate of an aspartate or glutamate residues at the active site. Moreover, this stabilization may occur by the formation of a covalent intermediate by nucleophilic attack of the aspartate or glutamate residues on C-1 of the reducing sugar (Coughlan, Tuohy Filho, EXF, Claeyssens, Vrsanska, & Hughes, 1993; M. Coughlan, 1991). The reaction is completed by the addition of hydroxyl group from water to the carbonium ion and a proton to the nucleophile. On the other hand, some few glycosyl hydrolases operate via single-displacement reactions with inversion of anomeric configuration. In this case, the reactions have the participation of a general acid (glutamate residue) and a general base (aspartate or glutamate residue) with nucleophilic attack by a molecule of water (Moreira, 2008).

- **Pectins (Polygalacturons):** Pectin is a class of polysaccharides found in the cell walls of higher plants and consists of D-galacturonic acid (GalA) units joined in chains by a means of α-(1→4) glycosidic linkage (homopolymer of (1→4) α-D-galactopyranosyluronic acid units with varying degrees of carboxyl groups methyl esterified along with some neutral sugars (Figure 4). Rhamnose is a minor component of the pectin backbone while neutral sugars such as arabinose, galactose and xylose occur in the side chains (Williams & Phillips 2009).

- **Pectin-Degrading Enzymes:** The pectinolytic enzymes may be divided in three broader groups as follows (Akiba, Kimura, Yamamoto, & Kumagai, 1995; Devi, & Rao, 1996):
- **Protopectinases**: They degrade the insoluble protopectin and give rise to highly polymerized soluble pectin.
- **Esterases**: They catalyze the de-esterification of pectin by the removal of methoxy esters.
- **Depolymerases**: They catalyze the hydrolytic cleavage of α-(1→4)-glycosidic bonds in the D-galacturonic acid moieties of pectin.

An extensive classification of pectinolytic enzymes is shown in Table 5 and their degradation mechanism is shown in Figure 5.

Protopectinases are enzymes that catalyze the solubilization of protopectin and the reaction is shown below:

$$Protopectin\ \left(insoluble\right)\ +\ H_2O \xrightarrow{PPase} Pectin\ \left(soluble\right)$$

Polygalacturonases (PGases) are the pectinolytic enzymes that catalyze the hydrolytic cleavage of the polygalacturonic acid chain with the introduction of water across the oxygen bridge.

Lyases (or transeliminases) perform non-hydrolytic breakdown of pectates or pectinates, characterized by a trans-eliminative split of the pectic polymer (Sakai, Sakamoto, Hallaert, & Vandamme, 1993). The lyases break the glycosidic linkages at C-4 and simultaneously eliminate H from C-5, producing a D 4:5 unsaturated product (Albersheim, Neukom, & Deuel, 1960; Codner, 1971). Lyases can be classified into following types on the basis of the pattern of action and the substrate acted upon by them (Table 5.).

Figure 4. Primary structure of pectins

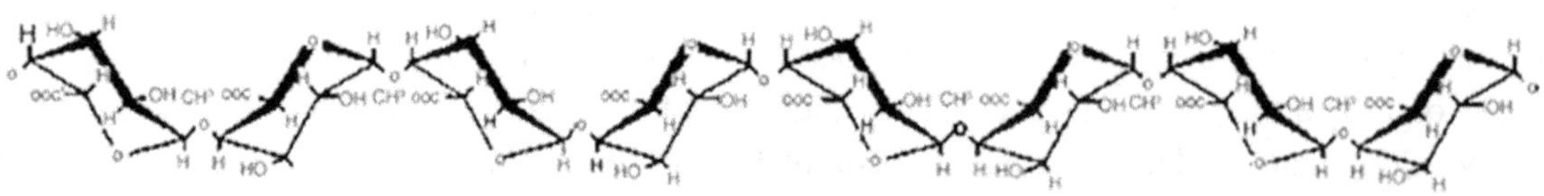

Figure 5. Mode of action of pectinases: (a) R = H for PG and CH_3 for PMG; (b) PE; and (c) R = H for PGL and CH_3 for PL. The arrow indicates the place where the pectinase reacts with the pectic substances. PMG, polymethylgalacturonases; PG, polygalacturonases (EC 3.2.1.15); PE, pectinesterase (EC 3.1.1.11); PL pectin lyase (EC-4.2.2.10) (Gummadi, S. N., & Panda, 2003).

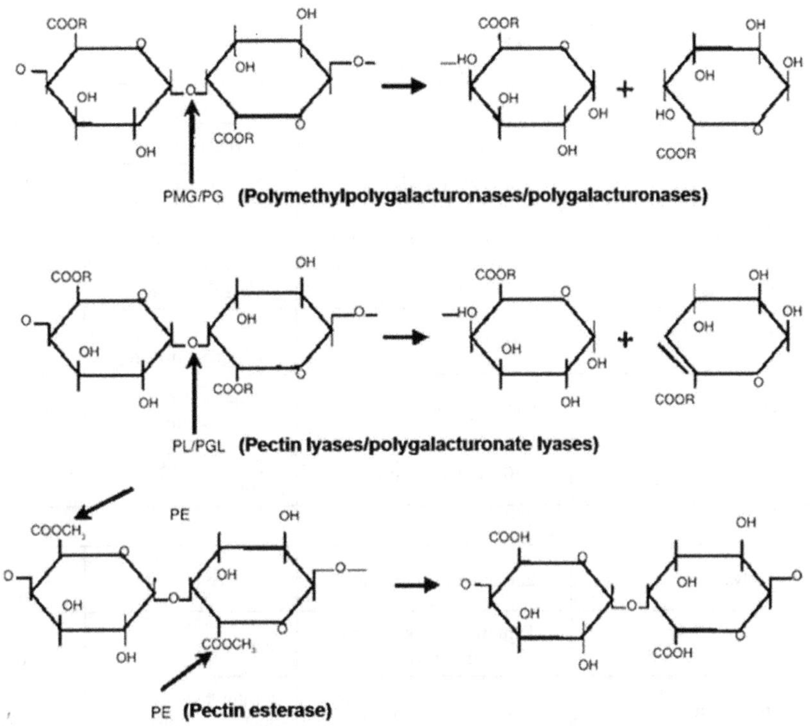

Pectinesterase (PE, Pectin pectlylhydrolase, EC 3.1.1.11) also called pectinmethylesterase, pectase, pectin methoxylase, pectin demethoxylase and pectolipase, is a carboxylic acid esterase and belongs to the hydrolase group of enzymes (Whitaker, 1984). It catalyzes the de-esterification of methyl ester linkages of galacturonan backbone of pectic substances to release acidic pectins and methanol (Cosgrove, 1997). The resulting pectin is then acted upon by polygalacturonases and lyases (Prade, Zhan, Ayoubi, & Mort, 1999; Sakai, Sakamoto, Hallaert, & Vandamme, 1993). The mode of action of PE varies according to its origin (Micheli, 2001). The reaction catalyzed by PE can be represented as follows:

$$Pectin + nH_2O \xrightarrow{PE} Pectate + nC_2H_5OH$$

PEs are highly specific enzymes and the activity is highest on 65–75% methylated pectin, since the enzyme is thought to act on methoxyl group adjacent to free carboxyl groups (Whitaker, 1984). PE action has a very little effect on viscosity of pectin containing solutions unless divalent ions are present, which increase viscosity due to cross-linking.

Depolymerases act on pectic substances by two different mechanisms, hydrolysis, in which they catalyze the hydrolytic cleavage with the introduction of water across the oxygen bridge and trans-elimination lysis, in which they break the glycosidic bond by a trans-elimination reaction without any participation of water molecule (Albersheim, Neukom, & Deuel, 1960; Codner, 1971).

Table 5. Classification of pectinolytic enzymes (Jayani, R. S., Saxena, S., & Gupta, 2005)

ENZYMES	MECHANISM AND PATTERN OF ACTION	SUBSTRATE	PRODUCT FORMED
Esterases Pectin Methyl Esterase	Random Hydrolysis	Pectin	Pectic Acid and Methanol
Depolymerizing Enzymes **1) Hydrolases**			
• Protopectinases	Random hydrolysis	Protopectin	Pectin
• Endopolygalacturonase	Random hydrolysis	Pectic acid	Oligogalacturonates
• Exopolygalacturonase	Terminal hydrolysis	Pectic acid	Monogalacturonates
• Exopolygalacturonan-digalacturono Hydrolase	Penultimate bonds hydrolysis	Pectic acid	Digalacturonates
• Oligogalacturonate Hydrolase	Terminal hydrolysis	Trigalacturonate	Monogalacturonates
• Δ 4:5 Unsaturated oligogalacturonate Hydrolases	Terminal hydrolysis	Δ 4:5(galacturonate)$_n$	Unsaturated monogalacturonates
• Endopolymethyl-galacturonases	Random hydrolysis	Highly esterified pectin	Oligomethylgalacturonates
• Endopolymethyl-galacturonases	Terminal hydrolysis	Highly esterified pectin	Oligogalacturonates
2) Lyases			
• Endopolygalacturonaselyase	Random trans-elimination	Pectic acid	Unsaturated oligogalacturonates
• Exopolygalacturonaselyase	Penultimate bond trans-elimination	Pectic acid	Unsaturated digalacturonates
• Oligo-D-galactosiduronatelyase	Terminal trans-elimination	Unsaturated digalacturonates	Unsaturated monogalacturonates
• Endopolymethyl-D-galactosiduronateLyase	Random trans-elimination	Unsaturated poly-(methyl-D-digalacturonates)	Unsaturated methyloligogalacturonates
• Exopolymethyl-D-galactosiduronatelyase	Terminal trans-elimination	Unsaturated poly-(methyl-D-digalacturonates)	Unsaturated methylmonogalacturonates

Prodrug Based Colon-Specific Drug Delivery Systems

Prodrug is defined as a pharmacologically inactive derivative of a parent drug molecule that requires spontaneous or enzymatic transformation *in vivo*, to release the active form of the drug. Colon-specific drug delivery via prodrug approach is highly based on the functional groups present on the drug molecule with chemical linkages. Prodrug approach for CoDDs is based on the release of active drug moiety from the carrier by the bacterial enzymatic action in the colon thereby protecting the drug release and absorption in the upper GIT.

- **Azo Linkages:** The metabolism of azo compounds by the colonic enzymes of bacterial origin is one of the most extensively investigated metabolic processes. In colon, the azo-bonds are cleaved by the azoreductases present in the colon and thus can be utilized for CoDDs (Saffran, Kumar,

Savariar, Burnham, Williams, & Neckers, 1986; Saffran, Bedra, Kumar, & Neckers, 1988). The use of azo compounds for colon targeting has been in the form of hydrogels as a coat over drug cores or as prodrugs. These azo bonds are stable in the upper gastrointestinal tract and are cleaved in the colon by azoreductases produced by the colonic microflora. Sulfasalazine, a prodrug which is used in the treatment of colonic inflammatory conditions has an azo bond between the active drug, 5-amino salicylic acid and the carrier, Sulfapyridine (Figure 6). After reaching the colon, the azo bond between them is cleaved by the bacterial azoreductases thereby releasing the active drug and carrier for local delivery to the colon (Khan, Piris, & Truelove, 1977). Other examples of azo-linked prodrugs include 5-amino salicylic acid (5-ASA) has been conjugated with various azo-conjugates like p-aminohippurate (Ipsalazide) (Chan, Pope, Gilbert, Sacra, Baron, & Lennard-Jones, 1983), 4-amino benzyol β-alanine (Balsalazide) (Chan, Pope, Gilbert, Sacra, Baron, & Lennard-Jones, 1983) with a reducible azo-linkages (Figure 7).Various azo-polymers are also being studied as a coating material over the drug cores. Examples include azo poly-

Figure 6. Mode of drug release from azo linked prodrugs

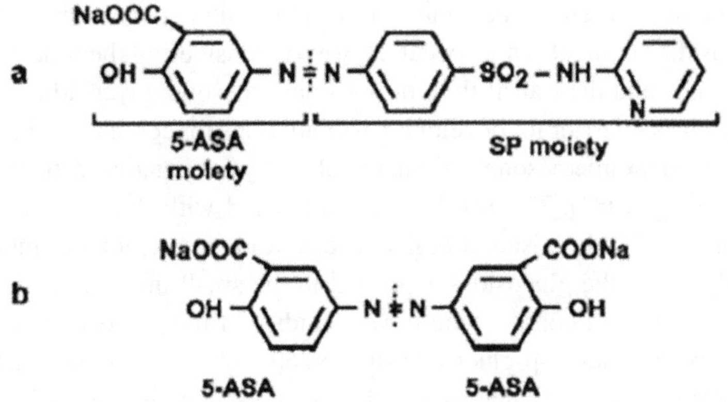

Figure 7. Enzymatic cleavage of Ipsalazide and Balsalazide

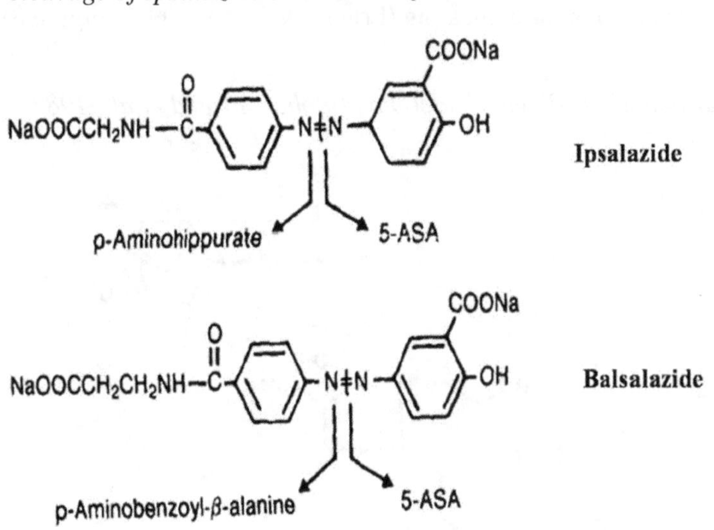

mers such as copolymers of styrene with 2-hydroxyethyl methacrylate coated over capsules of Vasopressin/Insulin (Saffran, Bedra, Kumar, & Neckers, 1988; Saffran, M., Field, J., Pena, J., Jones, R., & Okuda, 1991; Saffran, Kumar, Savariar, Burnham, Williams, & Neckers, 1986). 2-Hydroxyethyl methacrylate, methyl methacrylate and methacrylic acid and containing *N,N*-bis[methacryloyloxyethyl]oxy (carbonylamino)]azobenzene coated over Theophylline capsules is another example of azo polymeric coating over drug cores (Van den Mooter, & Kinget, 1995).

- **Glycosidic Linkages:** Sugar moieties such as glucose, galactose, cellobiose have also been conjugated to drug moieties to form glycosides that are very susceptible to intestinal glycosidases, galactosidases or cellobiosidases. Glycosides are hydrophilic and bulky in nature which makes them incapable to penetrate the biological membranes especially in the upper regions of the gastrointestinal tract with bacterial enzymatic activity is minimal (Pownall, Hickson, & Smith, 1983). Therefore, they are fermented by the bacterial glycosidases existing in the human colon. The major glycosidases produced by the wide range of colonic bacteria include β-D-galactosidase, α-L-arabinosidase, β-D-glucosidase and β-D-xylopyronisidase (Rubinstein, 1995). Glycosidic activity of the gastrointestinal tract is derived from the anaerobic intestinal activity or the sloughed or exfoliated cells of small intestine (Conchie, & Macdonald, 1959; Rubinstein, 1995). Presence of glycosidase in the small intestine may interfere with the colon-specific drug delivery as the hydrolysis of the glycosidic conjugates may occur. However, considering the time required for hydrolyzing glycosidic bonds and the transit time of the small intestine, glycosidic conjugate still carries potential to be used as a carrier for developing formulations for colon targeting. The colon-specific delivery of steroids, Dexamethasone and Prednisolone by conjugating with glucose was attempted by (Friend, & Chang, 1984). The results were compared with the same steroids free from any conjugation, both forms administered orally. The steroids free from any modifications were extensively hydrolyzed by the glucosidases present in the small intestine with only 1% of the oral dose reaching the colon. In contrast, the *in vivo* studies of the glucosidic conjugate of the dexamethasone, dexamethasone-21-β-glucoside showed 60% of the oral dose reached the colon in 4-5 hours whereas Predinosone conjugate showed poor site specificity with 15% of the administered oral drug reaching the colon in 4-5 hours. The animal studies of dexamethasone-21-β-glucoside showed that the prodrug was rapidly hydrolyzed by the hydrolytic enzymes (β-glucosidases) present in the caecum and colon of guinea pig (Friend, & Tozer, 1992) (Figure 8).

Figure 8. Dexamethasone-21- β -D-glucoside(David Robert Friend et al., 1984) (Arrow shows site of action of glucosidase)

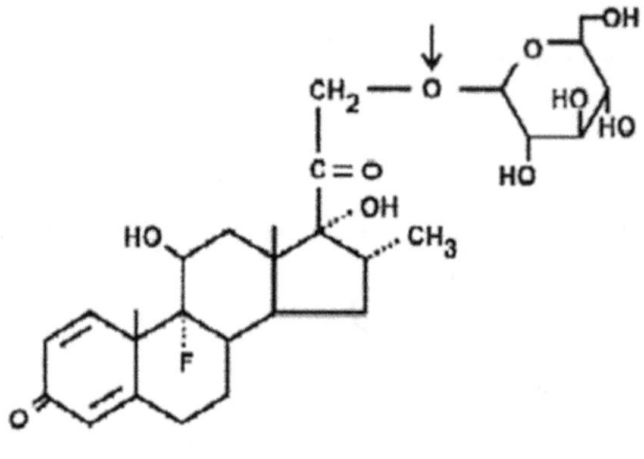

- **Glucuronide Linkages:** For many drugs, glucuronidation is the major mechanism for the inactivation and clearance of the drug from the body. Bacterial glucuronidases existing in the colon are capable of deglucuronidation of several drugs in the colon. Due of this advantage, many drugs are linked with glucuronide to achieve colon targeting. A glucuronide prodrug of dexamethasone namely dexamethasone-β-D-glucuronide by (Haeberlin, Rubas, Nolen III, & Friend, 1993) is shown in Figure 9. The study showed 30-fold increase in the luminal -β-D-glucuronidase activity between distal small intestine and caecum in normal rats. The glucuronidase activity gradient between distal small intestine and caecum is much lower, suggesting that the glucuronide prodrugs are capable of delivering drugs to the colon without being affected by the enzymes present in upper gastrointestinal tract. It was also observed that hydrolytic activity in the luminal contents of the colon in conventional rats was greater as compared to the colitic rats whereas, germ-free rats showed least luminal β-glucuronidase activity.
- **Dextran Linkages:** Dextran are polysaccharides of bacterial origin where monosaccharides are linked with each other by glycosidic linkages that can be hydrolyzed by molds, bacteria and mammalian cells (Sinha, & Kumria, 2001). The enzymes responsible for the hydrolysis of these linkages are called as dextranases. In the colon, dextran is hydrolyzed by the dextranases to give shorter prodrug oligomers, which are further split by the colonic esterases to release the active moiety from the carrier in the colon ((Sinha, & Kumria, 2001). Dextranase activity is absent in the upper gastrointestinal tract whereas high dextranase activity is found in the colon shown by gram negative anaerobic bacteria especially bacteroides. Bacteroides are present in abundance in the colon and constitute major part of the total colonic microflora (the concentration of bacteroides is as high as 10^{11} per gram contents in colon (Hehre, & Sery, 1952). These findings suggest that dextran can be used as a carrier to deliver drugs to the colon without being hydrolyzed in upper gastrointestinal tract. Dexamethasone linked with dextran using a spacer is shown in Figure 10. It was found that dexamethasone was released in the caecum and colon by the action of bacterial

Figure 9.Dexamethasone--β-D-glucuronide by Haeberlin et al., (1993)

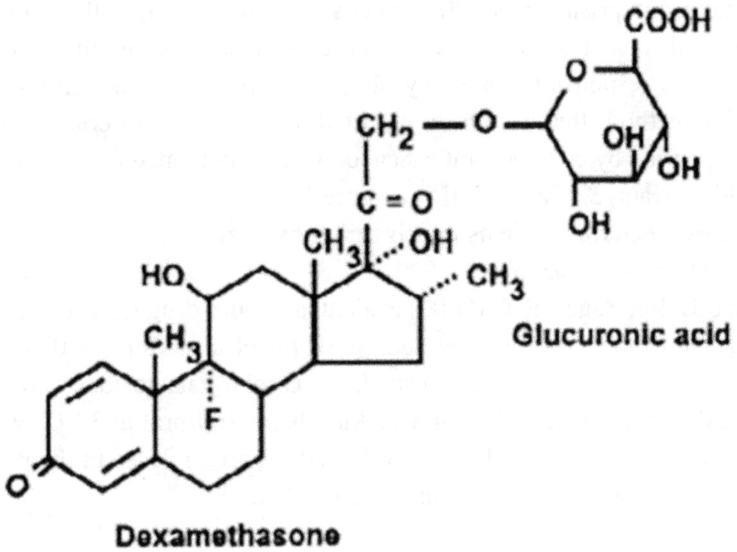

Figure 10. Dexamethasone-Dextran conjugates with succiniate spacer (McLeod, A. D., Friend, D. R., & Tozer, 1994). Arrow 1 shows site of action of dextranase.

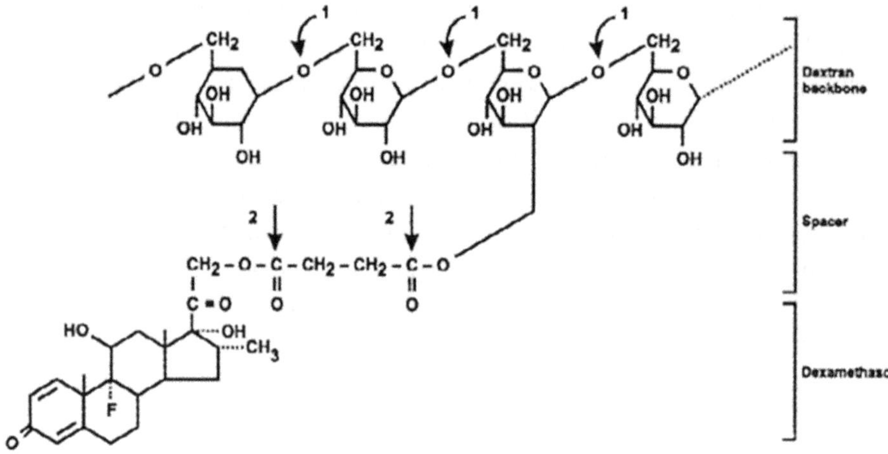

dextranases thereby treating the induced colitis in rats with little adrenal suppression (McLeod, Friend, & Tozer, 1994).

- ***In Vitro* Release Evaluation of Colon-Specific Drug Delivery Systems:** Dissolution testing is the most commonly employed *in vitro* evaluation technique for predicting the *in vivo* drug release kinetics from orally administered dosage forms including the modified release delivery systems. Novel dosage forms present unique problems in the development of *in vitro* release technology because of physicochemical properties of formulations and the unique physiological environment in which they should release their contents. Due to the distal location of colon, the colon targeted oral delivery systems traverse the longest path, coming across the most varied milieus. Designing a suitable biorelevant medium, therefore, has proven to be a complex task. Among the commonly used strategies, the microflora activated delivery systems have been found to be quite promising for delivering drugs to the colon. Human colon represents a dynamic and ecologically diverse environment, comprising over 400 distinct species of bacteria with a population of 10^{11}–10^{12} CFUs/mL of colonic contents. These bacteria produce wide spectrum of reductive and hydrolytic enzymes, which are responsible for many biorelevant processes like carbohydrate and protein metabolism. The methods that have been reported for evaluation of colon targeted delivery systems include triggering by enzymes, rat caecal contents and human faecal slurries (Singh, Yadav, Prudhviraj, Gulati, Kaur, & Vaidya, 2015) (Figure 11.).
- *In vitro* drug release determination using triggering by enzymes:
    - Mannan – Degrading Enzymes
        - Gliko-Kabir, Yagen et al. (2000) evaluated *in vitro* drug release of hydrocortisone from guar gum based hydrogel by adding 10 ml of a mixture of 0.2U/ml-of galactomannanase from Aspergillus niger) and 0.033U/ml α-galactosidase (from Escherichia coli) in sealed beakers mounted on a shaking bath (100rpm) at 37°C. Various galactomannan based colonic formulations evaluated for drug release in the presence of mannan-degrading enzymes is shown in Table 6.

Figure 11. Advantages and drawbacks of existing conventional methods for the evaluation of in vitro drug release of enzymatically triggered systems

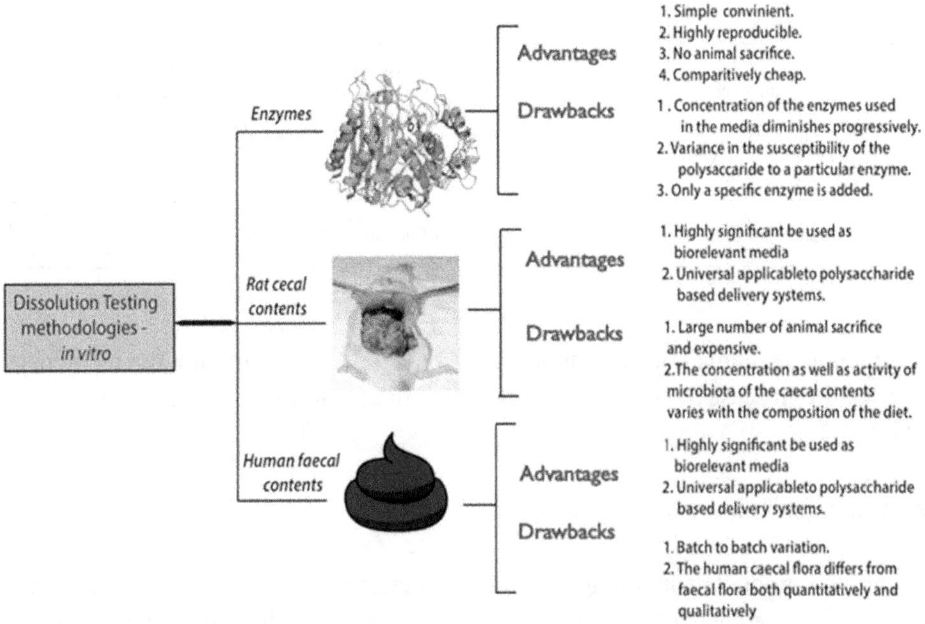

Table 6. Various galactomannan based colonic formulations evaluated for drug release in the presence of mannan-degrading enzymes.

FORMULATION	ENZYME	REFERENCES
Guar gum-alginate based Aceclofenacmicroparticles	α-galactosidase (33U/L) galactomannanase(200U/L)	Lakshmi et al., (2014)
Guar gum based hydrogel of Indomethacin and Budesonide	α-galactosidase(0.033U/ml) galactomannanase (0.175U/ml)	Gliko-Kabir et al., (1998)
Guar gum based hydrogel of Hydrocortisone	α-galactosidase (0.033U/ml) galactomannanase (0.2U/ml)	Gliko-Kabir et al., (2000)

- Pectinolytic Enzymes

Semdé et al., (2000) evaluated drug release from Theophylline coated pellets in the presence of 3 ml of Pectinex® SP-L Ultra (mixture of enzymes from *Aspergillus aculeatus*) in USP type-II paddle type apparatus. Some examples are mentioned in Table 7.

- *In vitro* drug release determination using rat caecal contents and human faecal slurries' triggering by enzymes

The dissolution medium that uses colon contents of rodents has been employed most frequently by the researchers. However, even this media suffers from a number of limitations like lack of reproducibility, cumbersomeness of procedure and most importantly, high cost and sacrifice of laboratory animals. The

Table 7. Various pectin based colonic formulations evaluated for drug release in the presence of pectin-degrading enzymes.

FORMULATION	ENZYME	REFERENCES
Tablets containing sodium fluroscein	The amount of pectinolytic enzymes used were 30, 925PG and 92, 775PG. (PG = milliequivalentsof reducing groups liberated from pectin per minute per unit of enzyme).	Ashford et al., (1994)
5-Amino Salicylic Acid tablets coated with pectin-hydroxypropylmethyl cellulose	3 ml Pectinex	Turkoglu et al., (2002)
Film coated tablets containingParacetamol	3 ml of Pectinex® SP-L Ultra	Wakerly et al., (1996)

dissolution method using human faecal slurries is the next best option, though the element of reproducibility is much less in this case as compared to the medium containing rat caecal content (Singh, Yadav, Prudhviraj, Gulati, Kaur, & Vaidya, 2015).

- *In vitro* drug release determination using probiotics

Probiotics are the products containing live micro-organisms that confer wide health advantages in human beings. There have been a few reports where the probiotic culture has been shown to mimic the colonic milieu in terms of the presence of polysaccharide metabolizing bacteria leading to perpetual generation of the relevant enzymes. This method utilizes probiotics as a source to generate gut bacteria and enzymes produced by them. In this method, probiotics capsules which are easily available commercially were inoculated in an anaerobic culture medium (Fluid thioglycollate medium) and incubated at 35°C for 48 hours under anaerobic conditions by using a CO_2 incubator. The probiotic capsules comprised of some of the dominant bacterial species found in the colon such as Lactobacillus acidophilus, Lactobacillus rhamnosus, Bifidobacterium longum, Bifidobacterium bifidum, Saccharomyces boulardi and Fructo Oligosaccharide (prebiotic). Dissolution testing of guar gum coated Sulfasalazine (SFZ) spheroids was carried out first in pH 1.2 (200 ml) for 2 hours and then the pH was changed to 6.8 by adding sodium hydroxide and volume was made up to 900 ml and the dissolution for carried out for further 3 hours. At the end of the 5 hours, the pH was adjusted to pH 7.4 using sodium hydroxide and the media was degassed using CO_2 followed by the addition of probiotic culture (approximately 10^{11}-10^{12} CFUs/ml representing the bacterial count in the human colon) to the dissolution medium and the testing was carried out till 18 hours. The probiotic culture method showed similar dissolution profiles as that of rat caecal content method and also significant increase in the drug release from the guar gum coated spheroids was seen in probiotic-assisted dissolution after the addition of the probiotic culture. The similarity factor method these two methods was found to be 50.66 showing that the dissolution profiles were statistically similar (Singh, Yadav, Prudhviraj, Gulati, Kaur, & Vaidya, 2015) (Figure 12.). This method is based on the fact that the probiotic culture containing the colonic bacterial species will leads to the perpetual generation of relevant enzymes that will carry out the degradation of the polysaccharide.

Probiotics-assisted dissolution method for the evaluation of microbially triggered drug delivery systems offers several advantages over conventional methods which are listed below (Singh, Yadav, Prudhviraj, Gulati, Kaur, & Vaidya, 2015):

Figure 12. Mean percentage cumulative release of drug from guar gum coated spheroids in probiotic culture medium, 4.5% w/v rat caecal contents, 5% w/v human faecal slurries and medium in absence of microbiota respectively.

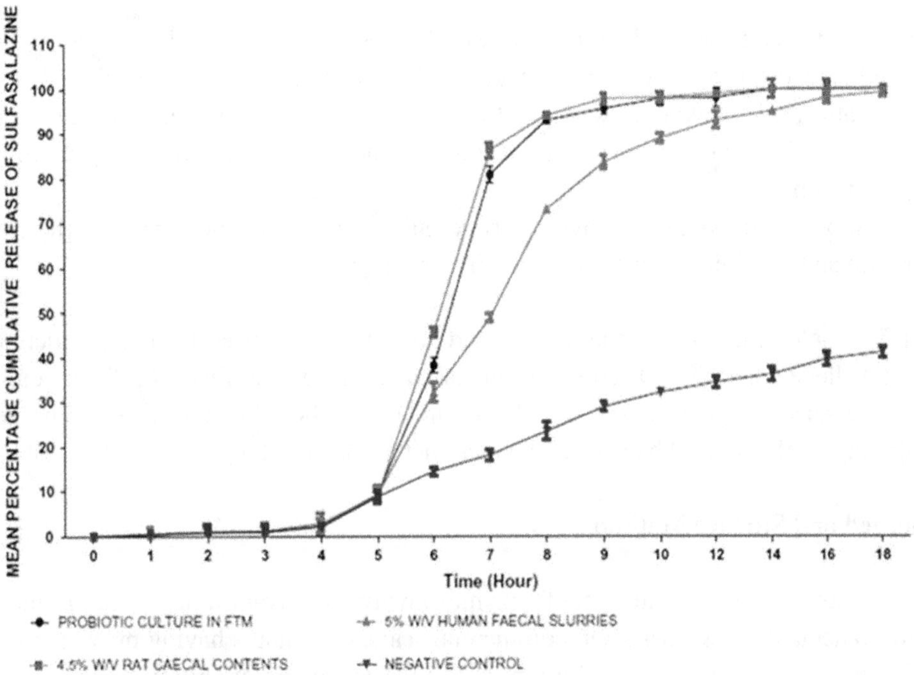

- No involvement of animal sacrifice, thus an animal sparing method.
- It overcomes the issues associated with enzyme addition method. This method involves incorporating live colonic bacterial species that will continuously produce colonic enzymes which are necessary to bring out the degradation process and the enzymatic activity will be replenished.
- They showed higher reproducible results as compared to other conventional dissolution testing methods.
- Closely resembles the colonic conditions *in vivo*.
- This *in vitro* dissolution method can be applicable to variety of polysaccharides, as polymers for colonic delivery.
- It is an inexpensive, cheap and convenient method and the sources are easily available.

PRODUCTION OF COLONIC BACTERIAL ENZYMES

Enzymes can be produced by various methods such as bran process, surface method, submerged method etc and some of them are discussed below:

- **Bran Process**

This process is the modification of the Takamine's original mold bran process where Takamine used wheat bran as a solid substrate for the production of fungal enzymes. In the modified mold bran process, other ingredients such as nutrient slats, acid or buffer to maintain/regulate pH, soy bean meal or beet cosettes to stimulate enzymes production can also be incorporated (Underkofler, Barton, & Rennert, 1958). The steps involved in one of the modified mold bran process are given below:

- Bran is first steamed for sterilization followed by cooling and is further inoculated with mold spores and spread out on trays for incubation (Underkofler, Severson, Goering, & Christensen, 1947; Forbath, 1957). Takamine's original mold bran process employed slowing rotating drums instead of trays for incubation. The use of trays for incubation gives more rapid growth and enzyme production.
- The incubation of the inoculated trays is carried out in chambers which are carefully controlled by temperature and humidity with the help of circulated air.

The modified mold bran process can also be and have been employed for the production of bacterial enzymes (Wallerstein, 1939). Different strains of bacteria e.g. Bacillus subtillis are cultivated in a special culture vessel as a pellicle on the surface of thin layer on liquid medium. Different media can be utilized depending on the desired bacterial enzymes such as bacterial amylases or bacterial proteases.

- **Submerged and Surface Method**

Submerged method for the production of enzymes involves cultivation of the molds and bacteria in shake flask or in aerated tubes/flasks. On commercial scale, deep tanks having provision for introducing sterile air and generating vigorous agitation are used. The amount of agitation, amount of sterile air and the degree of air dispersion are dependent variables. Optimum results are obtained when the air is dispersed in the form of very fine bubbles through the liquid culture medium (Underkofler, Barton, & Rennert, 1958). Submerged method has been employed for the production of both fungal and bacterial enzymes such as pectinase from Bacillus pumilus (Sharma, & Satyanarayana, 2006), protease from Aspergillus oryzae (Singhania, Sukumaran, Patel, Larroche, & Pandey, 2010), protease and amylase from Streptomyces rimosus (Yang, & Wang, 1999), pectinesterase and polygalacturonase from Aspergillus niger (Maldonado, & De Saad, 1998) etc. Surface method makes the use of trays for the cultivation and requires low power as well as low pressure air blower as compared to the submerged method. The differences between submerged method and surface method is discussed in Table 8. (Underkofler, Barton, & Rennert, 1958) and some commercial available enzymes and their sources are shown in Table 9.

- Production of Colonic Bacterial Enzymes by Using Probiotics

Steps involved in the production of colonic enzymes using probiotics are given below: -

Step 1: It involves the preparation of a suitable medium for the growth of anaerobic bacteria. Culture medium that can be used to cultivate anaerobes efficiently are Thioglycollate medium, Cooked meat broth, Anaerobic blood agar, Egg-yolk agar (EYA), Peptone-yeast extract glucose broth (PYG), Bacteroides bile esculin agar (BBE), Laked Kanamycin-vancomycin blood agar (LKV), Anaerobic phenylethyl alcohol agar (PEA) and Cycloserine cefoxitin fructose agar (CCFA). The selected media is sterilized by autoclaving at 121 degrees C for 15 minutes under 15 lbs pressure (Singh, Yadav, Prudhviraj, Gulati, Kaur, & Vaidya, 2015).

Table 8. The differences between submerged method and surface method.

SUBMERGED METHOD	SURFACE METHOD
Small flasks or tubes are used as fermenters.	Large amount of space is occupied due to the trays utilized for cultivation.
Requires minimum work	Comparatively, this process requires more hand labor.
High pressure air is required	Air blowers with lower pressure can be used.
Requires careful control	Minimum control over parameters is necessary
Less contamination	Frequent contamination
Recovery of the desired enzyme involves filtration or centrifugation and may also require evaporation and/or precipitation.	Recovery involves extraction with aqueous solution, filtration or centrifugation and may also require evaporation and/or precipitation.

Table 9. Some commercial enzymes and their sources (Underkofler, L., Barton, R., & Rennert, 1958)

SOURCE	ENZYME	MICROOGRANISM
Fungal	Amylases Glucosidases	Aspergillus oryzae, Aspergillus niger
	Pectinases Glucose oxidases	Aspergillus niger Penicillium notatum
Bacterial	Amylases	*Bacillus subtilis*
Yeast	Lactases	*Sacchromyces fragilis*

Step 2: The sterilized media is then inoculated with probiotic capsules which are the source of colonic bacteria. Commercially available probiotic mixture usually contains strains of Lactobacillus acidophilus, Lactobacillus rhamnosus, Bifidobacterium bifidus, Bifidobacterium longum, Saccharomyces boulardi, Lactobacillus salivarius, Lactobacillus paracasei, and Bifidobacterium lactis.

Step 3: After inoculation, the medium is incubated at 37°C for 48-72 hours under anaerobic conditions. Anaerobic conditions can be maintained by using various techniques such as CO_2 incubators, Gaspak® systems, anaerobic jars, anaerobic chambers, candle jar method, etc.

Step 4: After the incubation period, the dense colonies of anaerobic bacteria formed can be further processed. These bacteria will lead to perpetual generation of enzymes such as α-glucosidase, β-galactosidase, pectinases etc which can be later identified, separated and purified for different applications.

Schematic representation of the steps involved in the production of colonic bacterial enzymes by using probiotics is shown in Figure 13.

CURRENT CHALLENGES

Disease conditions, use of antibiotics, age, diet and several other factors influence the enzymatic activity in the colon. In colon cancer, there is an increase in the β-glucuronidase activity which may affect the drug release from formulation where β-glucuronidases play a major role in the degradation process. Use

Figure 13. Schematic representation of production of colonic enzymes by probiotic culture method.

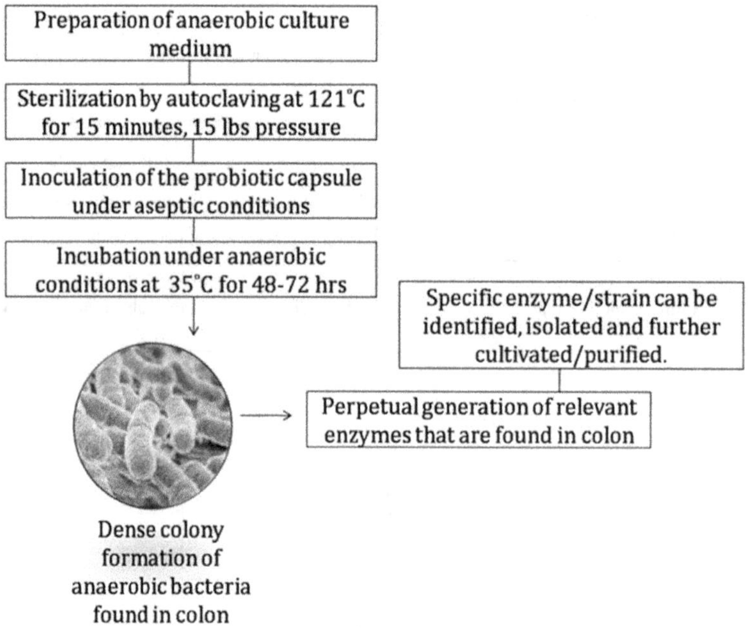

of antibiotics may inhibit the growth of some bacterial species which may affect the overall enzymatic activity in the colon thereby affecting the drug release from formulations reaching the colon. The enzymatic activity decreases with age in the colon resulting in slower degradation and drug release of colonic formulations and in turn delayed therapeutic action. Also, change in diet may alter the enzymatic activity in the colon causing retarding or accelerating therapeutic action of the drugs reaching the colon. Due to the several influencing factors, it also becomes difficult to surrogate colonic conditions to evaluate drug release from colonic formulations during formulation development. These circumstances highly affect the prediction of drug release during formulation development and therefore involve several risk factors.

SOLUTIONS AND RECOMMENDATIONS

Various probiotic formulations comprising of a mixture of bacterial species such as Lactobacillus acidophilus, Lactobacillus rhamnosus, Bifidobacterium longum, Bifidobacterium bifidum, Bifidobacterium breve, Bifidobacterium infantis, Lactobacillus plantarum, Lactobacillus paracasei, Lactobacillus bulgaricus, Streptococcus thermophilus etc. are developed in order to restore the disturbed enzymatic activity in the colon due to several conditions such as disease state, age, drugs such as antibiotics, age, diet etc. Some probiotics are also co-administered with colonic formulations developed for colon-specific drug delivery to overcome the problems associated with imbalanced colonic microflora and enzymatic activity (Prudhviraj, Vaidya, Singh, Yadav, Kaur, Gulati, & Gowthamarajan, 2015). Also, several models are being developed to simulate colonic conditions to evaluate *in vitro* drug release from colonic formulations such as dynamic colon model (Stamatopoulos, Batchelor, & Simmons, 2016) which provides realistic environment in terms of architecture of smooth muscle, motility patterns, and physical pressure occurring in the proximal colon as well as viscosity and probiotics-assisted dissolution method developed by (Singh, Yadav, Prudhviraj, Gulati, Kaur, & Vaidya, 2015) that makes the use of probiotics as a source of colonic bacterial enzymes to surrogate colonic conditions.

CONCLUSION

Oral CoDDS system has been the most popular and widely explored drug delivery system. Microbially triggered drug delivery systems that are activated by the wide-spread colonic bacterial species represent promising approach for the treatment of several local and systemic pathologies. Probiotics-assisted determination of *in vitro* drug release exhibits great potential for the development of inexpensive, highly reproducible, animal sparing dissolution medium for microbially triggered drug delivery systems. Production of colonic enzymes by using probiotics can also serve as a feasible method to produce colonic enzymes on large scale.

REFERENCES

Akiba, S., Kimura, Y., Yamamoto, K., & Kumagai, H. (1995). Purification and characterization of a protease-resistant cellulase from Aspergillus niger. *Journal of Fermentation and Bioengineering*, *79*(2), 125–130. doi:10.1016/0922-338X(95)94078-6

Albersheim, P., Neukom, H., & Deuel, H. (1960). Splitting of pectin chain molecules in neutral solutions. *Archives of Biochemistry and Biophysics*, *90*(1), 46–51. doi:10.1016/0003-9861(60)90609-3 PMID:13682085

Alkorta, I., Garbisu, C., Llama, M. J., & Serra, J. L. (1998). Industrial applications of pectic enzymes: A review. *Process Biochemistry*, *33*(1), 21–28. doi:10.1016/S0032-9592(97)00046-0

Allan, R., & Roxon, J. (1974). Metabolism by intestinal bacteria: The effect of bile salts on tartrazine azo reduction. *Xenobiotica*, *4*(10), 637–643. doi:10.1080/00498257409169766 PMID:4611051

Ashford, M., Fell, J., Attwood, D., Sharma, H., & Woodhead, P. (1994). Studies on pectin formulations for colonic drug delivery. *Journal of Controlled Release*, *30*(3), 225–232. doi:10.1016/0168-3659(94)90028-0

Berthelot, A. (1911). Recherches sur la flore intestinale. Isolement des microbes qui attaquent specialement les produits ultimes de la digestion proteiques. *Compt. rend. Acad. d. sc*, *153*, 306.

Brinton, C., Wichmann, H., Willaman, J., Wilson, C., & Dore, W. (1927). Definitions written by the Committee on Nomenclature of Pectin of the Agriculture-Food Division. *Journal of the American Chemical Society*, *49*, 38–40.

Chan, R. P., Pope, D. J., Gilbert, A. P., Sacra, P. J., Baron, J., & Lennard-Jones, J. E. (1983). Studies of two novel sulfasalazine analogs, ipsalazide and balsalazide. *Digestive Diseases and Sciences*, *28*(7), 609–615. doi:10.1007/BF01299921 PMID:6345112

Civas, A., Eberhard, R., Le Dizet, P., & Petek, F. (1984). Glycosidases induced in Aspergillus tamarii. Secreted α-d-galactosidase and β-d-mannanase. *The Biochemical Journal, 219*(3), 857–863. doi:10.1042/bj2190857 PMID:6331399

Codner, R. (1971). Pectinolytic and cellulolytic enzymes in the microbial modification of plant tissues. *Journal of Applied Microbiology, 34*(1), 147–160. PMID:5564376

Conchie, J., & Macdonald, D. (1959). Glycosidases in the mammalian alimentary tract. *Nature, 184*(4694), 1233–1233. doi:10.1038/1841233a0 PMID:13811463

Cosgrove, D. J. (1997). Assembly and enlargement of the primary cell wall in plants. *Annual Review of Cell and Developmental Biology, 13*(1), 171–201. doi:10.1146/annurev.cellbio.13.1.171 PMID:9442872

Coughlan, M. (1991). Towards an understanding of the mechanism of action of main chain-hydrolyzing xylanases. *Progress in Biotechnology*.

Coughlan, M. P., Tuohy Filho, M., Exf, P. J., Claeyssens, M., Vrsanska, M., & Hughes, M. (1993). Enzymological aspects of microbial hemicellulases with emphasis on fungal systems. Hemicellulose and Hemicellulases, 53-84.

Dabek, M., McCrae, S. I., Stevens, V. J., Duncan, S. H., & Louis, P. (2008). Distribution of β-glucosidase and β-glucuronidase activity and of β-glucuronidase gene gus in human colonic bacteria. *FEMS Microbiology Ecology, 66*(3), 487–495. doi:10.1111/j.1574-6941.2008.00520.x PMID:18537837

Davaran, S., Hanaee, J., & Khosravi, A. (1999). Release of 5-amino salicylic acid from acrylic type polymeric prodrugs designed for colon-specific drug delivery. *Journal of Controlled Release, 58*(3), 279–287. doi:10.1016/S0168-3659(98)00167-9 PMID:10099153

De Vries, J., Rombouts, F., Voragen, A., & Pilnik, W. (1982). Enzymic degradation of apple pectins. *Carbohydrate Polymers, 2*(1), 25–33. doi:10.1016/0144-8617(82)90043-1

Devi, N. A., & Rao, A. A. (1996). Fractionation, purification, and preliminary characterization of polygalacturonases produced by Aspergillus carbonarius. *Enzyme and Microbial Technology, 18*(1), 59–65. doi:10.1016/0141-0229(96)00055-5

Donkor, O. N., Henriksson, A., Vasiljevic, T., & Shah, N. (2007). α-Galactosidase and proteolytic activities of selected probiotic and dairy cultures in fermented soymilk. *Food Chemistry, 104*(1), 10–20. doi:10.1016/j.foodchem.2006.10.065

Doolittle, D. J., Sherrill, J. M., & Butterworth, B. E. (1983). Influence of intestinal bacteria, sex of the animal, and position of the nitro group on the hepatic genotoxicity of nitrotoluene isomers in vivo. *Cancer Research, 43*(6), 2836–2842. PMID:6850595

Drasar, B. S., & Hill, M. J. (1974). *Human intestinal flora: Academic Press (London) Ltd., 24/28 Oval Road*. London: NWI.

Eng, C. M., Guffon, N., Wilcox, W. R., Germain, D. P., Lee, P., Waldek, S., ... Desnick, R. J. (2001). Safety and efficacy of recombinant human α-galactosidase A replacement therapy in Fabry's disease. *The New England Journal of Medicine, 345*(1), 9–16. doi:10.1056/NEJM200107053450102 PMID:11439963

Forbath, T. (1957). Flexible processing keys enzymes' future. *Chemical Engineering (Albany, N.Y.)*, *64*, 226–229.

Fore, H., Walker, R., & Golberg, L. (1967). Studies on Brown FK. II. Degradative changes undergone in vitro and in vivo. *Food and Cosmetics Toxicology*, *5*, 459–473. doi:10.1016/S0015-6264(67)83147-X PMID:4295788

Friend, D. R., & Chang, G. W. (1984). A colon-specific drug-delivery system based on drug glycosides and the glycosidases of colonic bacteria. *Journal of Medicinal Chemistry*, *27*(3), 261–266. doi:10.1021/jm00369a005 PMID:6699871

Friend, D. R., & Tozer, T. N. (1992). Drug glycosides in oral colon-specific drug delivery. *Journal of Controlled Release*, *19*(1-3), 109–120. doi:10.1016/0168-3659(92)90069-4

Gale, E. F. (2006). The bacterial amino acid decarboxylases. *Advances in Enzymology and Related Areas of Molecular Biology*, *6*, 1–32.

Gingell, R., & Walker, R. (1971). Mechanisms of azo reduction by Streptococcus faecalis II. The role of soluble flavins. *Xenobiotica*, *1*(3), 231–239. doi:10.3109/00498257109033172 PMID:4341449

Gliko-Kabir, I., Yagen, B., Baluom, M., & Rubinstein, A. (2000). Phosphated crosslinked guar for colon-specific drug delivery: II. In vitro and in vivo evaluation in the rat. *Journal of Controlled Release*, *63*(1), 129–134. doi:10.1016/S0168-3659(99)00180-7 PMID:10640586

Gliko-Kabir, I., Yagen, B., Penhasi, A., & Rubinstein, A. (1998). Low swelling, crosslinked guar and its potential use as colon-specific drug carrier. *Pharmaceutical Research*, *15*(7), 1019–1025. doi:10.1023/A:1011921925745 PMID:9688054

Grim, Y., & Kopecek, J. (1991). Bioadhesive water-soluble polymeric drug carriers for site-specific oral drug delivery. Synthesis, characterization, and release of 5-aminosalicylic acid by Streptococcus faecium in vitro. *New Polymeric Materials, 3*(1), 49-59.

Gummadi, S. N., & Panda, T. (2003). Purification and biochemical properties of microbial pectinases—a review. *Process Biochemistry*, *38*(7), 987–996. doi:10.1016/S0032-9592(02)00203-0

Haeberlin, B., Rubas, W., Nolen, H. W. III, & Friend, D. R. (1993). In vitro evaluation of dexamethasone-β-d-glucuronide for colon-specific drug delivery. *Pharmaceutical Research*, *10*(11), 1553–1562. doi:10.1023/A:1018956232628 PMID:8290467

Hanke, M. T., & Koessler, K. K. (1924). Studies on proteinogenous amines XVII. On the faculty of normal intestinal bacteria to form toxic amines. *The Journal of Biological Chemistry*, *59*(3), 835–853.

Hawksworth, G., Drasar, B., & Hili, M. (1971). Intestinal bacteria and the hydrolysis of glycosidic bonds. *Journal of Medical Microbiology*, *4*(4), 451–459. doi:10.1099/00222615-4-4-451 PMID:5002686

Hayes, M., & Hyatt, A. (1974). The decarboxylation of amino acids by bacteria derived from human dental plaque. *Archives of Oral Biology*, *19*(5), 361–369. doi:10.1016/0003-9969(74)90177-0 PMID:4528022

Hehre, E. J., & Sery, T. W. (1952). Dextran-splitting anaerobic bacteria from the human intestine. *Journal of Bacteriology*, *63*(3), 424. PMID:14927577

Hillman, B. (2004). *Role of gut bacteria in human toxicology and pharmacology*. CRC Press.

Howard, P. C., Beland, F. A., & Cerniglia, C. E. (1983a). Reduction of the carcinogen 1-nitropyrene to 1-aminopyrene by rat intestinal bacteria. *Carcinogenesis, 4*(8), 985–990. doi:10.1093/carcin/4.8.985 PMID:6347427

Howard, P. C., Heflich, R. H., Evans, F. E., & Beland, F. A. (1983b). Formation of DNA adducts in vitro and in Salmonella typhimurium upon metabolic reduction of the environmental mutagen 1-nitropyrene. *Cancer Research, 43*(5), 2052–2058. PMID:6339047

Jain, S. K., & Jain, A. (2008). Target-specific drug release to the colon. *Expert Opinion on Drug Delivery, 5*(5), 483–498. doi:10.1517/17425247.5.5.483 PMID:18491977

Jayani, R. S., Saxena, S., & Gupta, R. (2005). Microbial pectinolytic enzymes: A review. *Process Biochemistry, 40*(9), 2931–2944. doi:10.1016/j.procbio.2005.03.026

Khan, A. A., Piris, J., & Truelove, S. (1977). An experiment to determine the active therapeutic moiety of sulphasalazine. *Lancet, 310*(8044), 892–895. doi:10.1016/S0140-6736(77)90831-5 PMID:72239

Kiese, M. (1974). *Methemoglobinemia: a Comprehensive Treatise: Causes, Consequences, and Correction of Increased Contents of Ferrihemoglobin in Blood*. CRC Press.

Lakshmi, P., & Pradeepthi, B. (2014). Formulation and in-vitro evaluation of colon targeted aceclofenac-guargum microspheres by ionic gelation process. *Pharmanest, 5*(3), 2163–2168.

Lloyd, A. W., Martin, G. P., & Soozandehfar, S. H. (1994). Azopolymers: A means of colon specific drug delivery? *International Journal of Pharmaceutics, 106*(3), 255–260. doi:10.1016/0378-5173(94)90011-6

Maldonado, M., & De Saad, A. S. (1998). Production of pectinesterase and polygalacturonase by Aspergillus niger in submerged and solid state systems. *Journal of Industrial Microbiology & Biotechnology, 20*(1), 34–38. doi:10.1038j.jim.2900470 PMID:9523455

Mallett, A. K., Wise, A., & Rowland, I. R. (1983). Effect of dietary cellulose on the metabolic activity of the rat caecal microflora. *Archives of Toxicology, 52*(4), 311–317. doi:10.1007/BF00316499 PMID:6307219

McCleary, B., & Matheson, N. (1986). Polysaccharides having a β-D-xylan backbone. *Advances in Carbohydrate Chemistry and Biochemistry, 44*, 158–164.

McCleary, B. V. (1991). *Comparison of endolytic hydrolases that depolymerize 1, 4-β-d-mannan, 1, 5-α-l-arabinan, and 1, 4-β-d-galactan*. ACS Publications. doi:10.1021/bk-1991-0460.ch034

McLeod, A. D., Friend, D. R., & Tozer, T. N. (1994). Glucocorticoid–dextran conjugates as potential prodrugs for colon-specific delivery: Hydrolysis in rat gastrointestinal tract contents. *Journal of Pharmaceutical Sciences, 83*(9), 1284–1288. doi:10.1002/jps.2600830919 PMID:7530303

Micheli, F. (2001). Pectin methylesterases: Cell wall enzymes with important roles in plant physiology. *Trends in Plant Science, 6*(9), 414–419. doi:10.1016/S1360-1385(01)02045-3 PMID:11544130

Mirsalis, J. C., Hamm, T. E., Sherrill, J. M., & Butterworth, B. E. (1982). Role of gut flora in the genotoxicity of dinitrotoluene. *Nature, 295*(5847), 322–323. doi:10.1038/295322a0 PMID:7057895

Moore, W., & Holdeman, L. (1975). Discussion of current bacteriological investigations of the relationships between instestinal flora, diet, and colon cancer. *Cancer Research, 35*(11 Part 2), 3418–3420.

Moreira, L., & Filho, E. X. F. (2008). An overview of mannan structure and mannan-degrading enzyme systems. *Applied Microbiology and Biotechnology, 79*(2), 165–178. doi:10.100700253-008-1423-4 PMID:18385995

Pokusaeva, K., Fitzgerald, G. F., & van Sinderen, D. (2011). Carbohydrate metabolism in Bifidobacteria. *Genes & Nutrition, 6*(3), 285–306. doi:10.100712263-010-0206-6 PMID:21484167

Pownall, H. J., Hickson, D. L., & Smith, L. C. (1983). Transport of biological lipophiles: Effect of lipophile structure. *Journal of the American Chemical Society, 105*(8), 2440–2445. doi:10.1021/ja00346a055

Prade, R. A., Zhan, D., Ayoubi, P., & Mort, A. J. (1999). Pectins, pectinases and plant-microbe interactions. *Biotechnology & Genetic Engineering Reviews, 16*(1), 361–392. doi:10.1080/02648725.1999.10647984 PMID:10819085

Prasanth, V., Jayaprakash, R., & Mathew, S. T. (2012). *Colon Specific Drug Delivery Systems: A Review on Various Pharmaceutical Approaches*. Academic Press.

Prudhviraj, G., Vaidya, Y., Singh, S. K., Yadav, A. K., Kaur, P., Gulati, M., &Gowthamarajan, K. (2015). Effect of co-administration of probiotics with polysaccharide based colon targeted delivery systems to optimize site specific drug release. *European Journal of Pharmaceutics and Biopharmaceutics, 97*, 164-172.

Puls, J. (1993). Chemistry of hemicelluloses: relationship between hemicellulose structure and enzymes required for hydrolysis. *Hemicellulose and Hemicellulases.*

Radomski, J., & Mellinger, T. (1962). The absorption, fate and excretion in rats of the water-soluble azo dyes, FD&C Red No. 2, FD&C Red No. 4, and FD&C Yellow No. 6. *The Journal of Pharmacology and Experimental Therapeutics, 136*(2), 259–266. PMID:14489780

Rafil, F., Franklin, W., Heflich, R., & Cerniglia, C. (1991). Reduction of nitroaromatic compounds by anaerobic bacteria isolated from the human gastrointestinal tract. *Applied and Environmental Microbiology, 57*(4), 962–968. PMID:2059053

Rowland, I. (1988). Factors affecting metabolic activity of the intestinal microflora. *Drug Metabolism Reviews, 19*(3-4), 243–261. doi:10.3109/03602538808994135 PMID:3068030

Rubinstein, A. (1995). Approaches and opportunities in colon-specific drug delivery. *Critical Reviews™ in Therapeutic Drug Carrier Systems, 12*(2-3).

Saffran, M., Bedra, C., Kumar, G. S., & Neckers, D. C. (1988). Vasopressin: A model for the study of effects of additives on the oral and rectal administration of peptide drugs. *Journal of Pharmaceutical Sciences, 77*(1), 33–38. doi:10.1002/jps.2600770107 PMID:3346821

Saffran, M., Field, J., Pena, J., Jones, R., & Okuda, Y. (1991). Oral insulin in diabetic dogs. *The Journal of Endocrinology, 131*(2), 267–278. doi:10.1677/joe.0.1310267 PMID:1744572

Saffran, M., Kumar, S., Savariar, C., Burnham, J. C., Williams, F., & Neckers, D. C. (1986). A new approach to the oral administration of insulin and other peptide drugs. *Science, 233*(4768), 1081–1085. doi:10.1126cience.3526553 PMID:3526553

Sakai, T., Sakamoto, T., Hallaert, J., & Vandamme, E. J. (1993). ⌈ Pectin, Pectinase, and Protopectinase: Production,⌈ Properties, and Applications. *Advances in Applied Microbiology, 39*, 213–294. doi:10.1016/S0065-2164(08)70597-5 PMID:8213306

Sakamoto, T., Hours, R. A., & Sakai, T. (1994). Purification, characterization, and production of two pectic transeliminases with protopectinase activity from Bacillus subtilis. *Bioscience, Biotechnology, and Biochemistry, 58*(2), 353–358. doi:10.1271/bbb.58.353 PMID:7764545

Scheline, R., Nygaard, R., & Longberg, B. (1970). Enzymatic reduction of the azo dye, acid yellow, by extracts of Streptococcus faecalis isolated from rat intestine. *Food and Cosmetics Toxicology, 8*(1), 55–58. doi:10.1016/S0015-6264(70)80223-1 PMID:4392274

Scheline, R. R. (1973). Metabolism of foreign compounds by gastrointestinal microorganisms. *Pharmacological Reviews, 25*(4), 451–523. PMID:4587548

Semdé, R., Amighi, K., Devleeschouwer, M. J., & Moës, A. J. (2000). Effect of pectinolytic enzymes on the theophylline release from pellets coated with water insoluble polymers containing pectin HM or calcium pectinate. *International Journal of Pharmaceutics, 197*(1), 169–179. doi:10.1016/S0378-5173(99)00465-2 PMID:10704804

Sharma, D., & Satyanarayana, T. (2006). A marked enhancement in the production of a highly alkaline and thermostable pectinase by Bacillus pumilus dcsr1 in submerged fermentation by using statistical methods. *Bioresource Technology, 97*(5), 727–733. doi:10.1016/j.biortech.2005.04.012 PMID:15936940

Simon, G. L., & Gorbach, S. L. (1982). Intestinal microflora. *The Medical Clinics of North America, 66*(3), 557–574. doi:10.1016/S0025-7125(16)31407-9 PMID:7043123

Simon, G. L., Gorbach, S. L., & Bustos-Fernandez, L. (1983). *Bacteriology of the colon. In Colon Structure and Function* (p. 103). Plenum Medical New York.

Singh, S. A., & Appu Rao, A. (2002). A simple fractionation protocol for, and a comprehensive study of the molecular properties of, two major endopolygalacturonases from Aspergillus niger. *Biotechnology and Applied Biochemistry, 35*(2), 115–123. doi:10.1042/BA20010077 PMID:11916453

Singh, S. K., Yadav, A. K., Prudhviraj, G., Gulati, M., Kaur, P., & Vaidya, Y. (2015). A novel dissolution method for evaluation of polysaccharide based colon specific delivery systems: A suitable alternative to animal sacrifice. *European Journal of Pharmaceutical Sciences, 73*, 72–80. doi:10.1016/j.ejps.2015.03.012 PMID:25829049

Singhania, R. R., Sukumaran, R. K., Patel, A. K., Larroche, C., & Pandey, A. (2010). Advancement and comparative profiles in the production technologies using solid-state and submerged fermentation for microbial cellulases. *Enzyme and Microbial Technology, 46*(7), 541–549. doi:10.1016/j.enzmictec.2010.03.010

Sinha, V., & Kumria, R. (2001). Colonic drug delivery: Prodrug approach. *Pharmaceutical Research, 18*(5), 557–564. doi:10.1023/A:1011033121528 PMID:11465408

Stamatopoulos, K., Batchelor, H. K., & Simmons, M. J. (2016). Dissolution profile of theophylline modified release tablets, using a biorelevant Dynamic Colon Model (DCM). *European Journal of Pharmaceutics and Biopharmaceutics*, *108*, 9–17. doi:10.1016/j.ejpb.2016.08.004 PMID:27531624

Tomlin, J., & Read, N. (1988). The relation between bacterial degradation of viscous polysaccharides and stool output in human beings. *British Journal of Nutrition*, *60*(3), 467–475. doi:10.1079/BJN19880119 PMID:3219318

Turkoglu, M., & Ugurlu, T. (2002). In vitro evaluation of pectin–HPMC compression coated 5-aminosalicylic acid tablets for colonic delivery. *European Journal of Pharmaceutics and Biopharmaceutics*, *53*(1), 65–73. doi:10.1016/S0939-6411(01)00225-9 PMID:11777754

Underkofler, L., Barton, R., & Rennert, S. (1958). Production of microbial enzymes and their applications. *Applied Microbiology*, *6*(3), 212. PMID:13534309

Underkofler, L., Severson, G., Goering, K., & Christensen, L. (1947). Commercial production and use of mold bran. *Cereal Chemistry*, *24*(1), 1–22.

Van den Mooter, G., & Kinget, R. (1995a). Oral colon-specific drug delivery: A review. *Drug Delivery*, *2*(2), 81–93. doi:10.3109/10717549509031355

Van den Mooter, G., Samyn, C., & Kinget, R. (1995b). In vivo evaluation of a colon-specific drug delivery system: An absorption study of theophylline from capsules coated with azo polymers in rats. *Pharmaceutical Research*, *12*(2), 244–247. doi:10.1023/A:1016283027139 PMID:7784340

Wakerly, Z., Fell, J., Attwood, D., & Parkins, D. (1996). Pectin/ethylcellulose film coating formulations for colonic drug delivery. *Pharmaceutical Research*, *13*(8), 1210–1212. doi:10.1023/A:1016016404404 PMID:8865314

Wallerstein, L. (1939). Enzyme Preparations from Microorganisms COMMERCIAL PRODUCTION AND INUSTRIAL APPLICATION. *Industrial & Engineering Chemistry*, *31*(10), 1218–1224. doi:10.1021/ie50358a012

Wang, C. Y., Lee, M.-S., King, C. M., & Warner, P. O. (1980). Evidence for nitroaromatics as direct-acting mutagens of airborne particulates. *Chemosphere*, *9*(2), 83–87. doi:10.1016/0045-6535(80)90093-4

Whitaker, J. R. (1984). Pectic substances, pectic enzymes and haze formation in fruit juices. *Enzyme and Microbial Technology*, *6*(8), 341–349. doi:10.1016/0141-0229(84)90046-2

Wise, A., Mallett, A. K., & Rowland, I. R. (1982). *Dietary protein and cecal microbial metabolism in the rat*. Academic Press.

Wong, K. K., & Saddler, J. N. (1993). Applications of hemicellulases in the food, feed, and pulp and paper industries. Hemicellulose and Hemicellulases, 127-143.

Yang, S.-S., & Wang, J.-Y. (1999). Protease and amylase production of Streptomyces rimosus in submerged and solid state cultivations. *Botanical Bulletin of Academia Sinica*, 40.

Chapter 5
Human Immunodeficiency Virus Reverse Transcriptase (HIV-RT): Structural Implications for Drug Development

Anuradha Singh
University of Allahabad, India

Ramendra K. Singh
University of Allahabad, India

ABSTRACT

Reverse transcriptase (RT) is a multifunctional enzyme in the life cycle of human immunodeficiency virus and represents a primary target for drug discovery against HIV-1 infection. Two classes of RT inhibitors, the nucleoside and the non-nucleoside RT inhibitors, are prominently used in the highly active antiretroviral therapy in combination with other anti-HIV drugs. This chapter deals with the salient features of HIV-RT that make it an attractive target for rational drug design and chemotherapeutic intervention in the management of acquired immunodeficiency syndrome. Further, the role of RT in the viral life cycle, the ways the drugs act to inhibit the normal functions of RT, and the mechanisms that the virus adapts to evade the available drugs have been discussed. Computational strategies used in rational drug design accompanied by a better understanding of RT, its mechanism of inhibition and drug resistance, discussed in this chapter, shall provide a better platform to develop effective RT inhibitors.

INTRODUCTION

Human immunodeficiency virus (HIV), a lentivirus belonging to the *Retroviridae* family, has been identified as an etiological agent of Acquired Immune Deficiency Syndrome (AIDS). In 1983, Luc Montagnier's group of Pasteur Institute, France, investigated Lymphadenopathy-associated virus (LAV). In 1984, Robert Gallo's group from National Institute of Health (NIH), USA, investigated a retrovirus, HTLV-III, first reported in 1981 in Los Angeles, New York and San Francisco, USA. In 1985, Jay Levy's group

DOI: 10.4018/978-1-5225-5237-6.ch005

from California University, San Francisco, USA, identified this virus as AIDS-Related Virus (ARV). In 1986, these three retroviruses (LAV, HTLV-III, ARV) were re-named as Human Immunodeficiency Virus (HIV) by an International Committee. In the same year, antigenic variants of HIV were designated as HIV-1 and HIV-2. For the discovery of HIV, Francoise Barre-Sinoussi and Luc Antoine Montagnier were awarded Nobel prize for physiology in 2008 (Gallo & Montagnier, 2003, Barre-Sinoussi, Chermann, Rey, Nugeyre, & Chamaret, 1983, Kumari & Singh, 2012).

Global HIV/AIDS Statistics

AIDS has been defined as the development of an immunocompromised state (Gottlieb et al., 1981) which continues to be a dreaded killer till date. It has caused a great burden to global wealth and health as 76.1 million [65.2 million–88.0 million] people have become infected with HIV since the start of the epidemic and 35.0 million [28.9 million–41.5 million] people have died so far because of AIDS-related illnesses. A global statistics have been shown in Table 1 (UNAIDS, 2016).

HIV/AIDS Status in India

According to a United Nations (UN) report, India has the third highest number of people living with HIV in the world with 2.1 million Indians, accounting for about four out of 10 people infected with the deadly virus in the Asia—Pacific region (Figure 1). As per the recently released data from National AIDS Control Organisation (NACO), HIV prevalence in Indian adult (15–49 years) is estimated at 0.26% (0.22% – 0.32%) in 2015 (0.30% among males and at 0.22% among females).

Among the States/ Union territories (UTs), in 2015, Manipur has shown the highest estimated adult HIV prevalence (1.15%), followed by Mizoram (0.80%), Nagaland (0.78%), Andhra Pradesh and Telangana (0.66%), Karnataka (0.45%), Gujarat (0.42%) and Goa (0.40%). Besides these States, Maharashtra, Chandigarh, Tripura and Tamil Nadu have shown estimated adult HIV prevalence greater than the national prevalence (0.26%), while Odisha, Bihar, Sikkim, Delhi, Rajasthan and West Bengal have shown an estimated adult HIV prevalence in the range of 0.21– 0.25%. All other States/UTs have levels of adult HIV prevalence below 0.20%. The adult HIV prevalence at national level has continued its steady decline from an estimated peak of 0.38% in 2001-03 through 0.34% in 2007 and 0.28% in 2012 to 0.26% in 2015 (NACO, 2015-2016).

Table 1. Global fact sheet 2016 (UNAIDS)

1. People living with HIV
In 2016, there were 36.7 million [30.8 million–42.9 million] people living with HIV.
34.5 million [28.8 million–40.2 million] adults
17.8 million [15.4 million–20.3 million] women (15+ years)
2.1 million [1.7 million–2.6 million] children (<15 years)
2. New HIV infections
Worldwide, 1.8 million [1.6 million–2.1 million] people became newly infected with HIV in 2016.
Since 2010, new HIV infections among adults declined by an estimated 11%, from 1.9 million [1.6 million–2.1million] to 1.7 million [1.4 million–1.9 million] in 2016.
New HIV infections among children declined by 47% since 2010, from 300 000 [230 000–370 000] in 2010 to 160 000 [100 000–220 000] in 2016.
3. AIDS-related deaths
AIDS-related deaths have fallen by 48% since the peak in 2005.
In 2016, 1 million [830 000–1.2 million] people died from AIDS-related illnesses worldwide, compared to 1.9 million [1.7 million–2.2 million] in 2005 and 1.5 million [1.3 million–1.7 million] in 2010.

Figure 1. Countries with most HIV cases

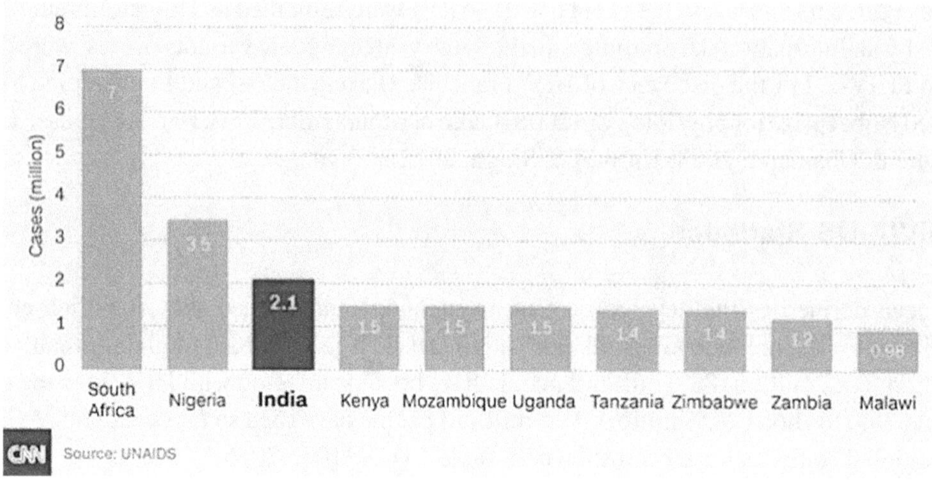

Need for Drug Development Against HIV

Fortunately, since the beginning of the epidemic, significant research has resulted in the development of more than 40 antiretroviral drugs that can help control the infection, when used in various combinations as shown in Table 2 and Table 3 (FDA, 2016; De Clercq & Li, 2016).

Despite major relief from current therapy, there are continued and concerted efforts in developing new agents for the treatment of HIV/AIDS for the following reasons:

- As of UNAIDS (2016), 19.5 million [17.2 million–20.3 million] people living with HIV were accessing antiretroviral therapy, up from 17.1 million [15.1 million–17.8 million] in 2015 and 7.7 million [6.8 million–8.0 million] in 2010. In 2016, around 53% [39–65%] of all people living with HIV had access to treatment. Some 54% [40–65%] of adults aged 15 years and older living with HIV had access to treatment, but just 43% [30–54%] of children aged 0–14 years had access. In 2016, around 76% [60–88%] of pregnant women living with HIV had access to antiretroviral medicines to prevent transmission of HIV to their babies. However, a good chunk of HIV infected people is yet to get access to ART.
- An effective vaccine or comparable preventative technique is unlikely to become available for years, and the epidemic will, therefore, sustain in endemic areas and increase in new regions.
- Low tolerance, less adherence, and more expensive treatments warrant further fruitful scientific interventions.
- Increasing resistance to existing drugs poses a serious challenge in ensuring effective treatment.

Thus, there is an urgent need for new anti-HIV drug candidates with increased potency, novel targets, improved pharmacokinetic properties, and reduced side effects.

Table 2. FDA approved anti-HIV drugs

	Name	Trade Name	Company	Launched
A	**Nucleoside or Nucleotide Reverse Transcriptase Inhibitors**			
1	Zidovudine(AZT, ZDV)	Retrovir	GlaxoSmithKline	March 19, 1987
2	Didanosine (ddI) enteric-coated didanosine, ddI EC)	Videx Videx EC (enteric-coated)	Bristol–Myers Squibb Co.	October 9, 1991 October 31, 2000
3	Zalcitabine (ddC)	HIVID	Roche	June 19, 1992
4	Stavudine(d4T)	Zerit	Bristol–Myers Squibb Co.	June 24, 1994
5	Lamivudine (3TC)	Epivir	GlaxoSmithKline,	November 17, 1995
6	Abacavir(ABC)	Ziagen	GlaxoSmithKline	December 17, 1998
7	Tenofovirdisoproxil fumarate(TDF)	Viread	Gilead	October 26, 2001
8	Emtricitabine (FTC)	Emtriva	Gilead	July 2, 2003
B	**Non-Nucleoside Reverse Transcriptase Inhibitors**			
9	Nevirapine extended-release nevirapine (NVP)	Viramune Viramune XR (extended release)	BoehringerIngelheim	June 21, 1996 March 25, 2011
10	Efavirenz (EFV)	Sustiva	Stocrin Bristol–Myers Squibb, Merck	September 17, 1998
11	Delavirdine(**DLV**)	Rescriptor	Agouron Pfizer	May 16, 2001
12	(ETR, TMC125)	Intelence	Tibotec Pharmaceuticals	January 18, 2008
13	(RPV, TMC278)	Edurant	Tibotec Pharmaceuticals	May 20, 2011
C	**Protease Inhibitors**			
14	Saquinavir(SQV)	Invirase	Hoffmann-La Roche	December 6, 1995
15	Indinavir(IDV)	Crixivan	Merck	March 13, 1996
16	Nelfinavir(NFV)	Viracept	Agouron, Pfizer	March 14, 1997
17	Amprenavir	Agenerase, Prozei	Vertex	1999
18	Lopinavir + ritonavir	Kaletra,	Aluvia Abbott	2000
19	Atazanavir(ATV)	Reyataz, Zrivada	Bristol–Myers Squibb Co.	June 20, 2003
20	Fosamprenavir(FPV)	Lexiva, Telzir	Vertex, GlaxoSmithKline	October 20, 2003
21	Tipranavir(TPV)	Aptivus	BoehringerIngelheim	June 22, 2005
22	Darunavir(DRV)	Prezista	Tibotec	June 23, 2006
D	**Fusion/ Entry Inhibitors**			
23	Enfuvirtide (T-20)	Fuzeon	Trimeris,Roche	March 13, 2003
24	Maraviroc (MVC)	Celsentri, Selzentry	Pfizer	August 6, 2007
F	**Integrase Inhibitors**			
25	Dolutegravir (DTG)	Tivicay	VIIV HLTHCARE	August 13, 2013
26	Elvitegravir (EVG)	Vitekta	Gilead Sciences, Inc.	September 24, 2014
27	Raltegravir (RAL)	Isentress Isentress HD	Merck	October 12, 2007 May 26, 2017
G	**Pharmacokinetic Enhancers**: Pharmacokinetic enhancers are used in HIV treatment to increase the effectiveness of an HIV medicine included in an HIV regimen.			
28	Ritonavir (RTV) *Although ritonavir is a PI, it is generally used as a pharmacokinetic enhancer	Norvir	Abbott, GlaxoSmithKline	March 1, 1996
29	Cobicistat (COBI)	Tybost GS-9350	Gilead Sciences, Inc.	September 24, 2014

Table 3. FDA approved Fix Dose Combination (FDC) HIV Medicines

S.No.	FDC	Components	Company	Approval Date
Two Drug Combinations				
1	Combivir	lamivudine and zidovudine (3TC / ZDV)	ViiV Healthcare Company	September 27, 1997
2	Kivexa/Epzicm	abacavir and lamivudine (ABC / 3TC)	ViiV Healthcare Company	August 2, 2004
3	Truvada	Emtricitabine and Tenofovirdisoproxilfumarate(FTC / TDF)	Gilead Sciences, Inc.	August 2, 2004
4	Evotaz	atazanavir and cobicistat (ATV / COBI)	Bristol–Myers Squibb Co.	January 29, 2015
5	Prezcobix	darunavir and cobicistat (DRV / COBI)	Janssen Pharmaceuticals, Inc.	January 29, 2015
6	Dutrebis	lamivudine and raltegravir (3TC/ RAL)	Merck & Co., Inc.	February 6, 2015
7	Descovy	emtricitabine and tenofoviralafenamide (FTC / TAF)	Gilead Sciences, Inc.	April 4, 2016
Three Drug Combinations				
8	Trizivir	abacavir, lamivudine, and zidovudine (ABC / 3TC / ZDV)	GlaxoWellcome	November 14, 2000
9	Atripla	efavirenz, emtricitabine, and tenofovir disoproxilfumarate(EFV / FTC / TDF)	Bristol-Myers Squibb Co. / Gilead Sciences,Inc.	July 12, 2006
10	Complera	emtricitabine, rilpivirine, and tenofovir disoproxilfumarate (FTC / RPV / TDF)	Gilead Sciences, Inc.	August 10, 2011
11	Stribild	elvitegravir, cobicistat, emtricitabine, and tenofovirdisoproxilfumarate (QUAD, EVG / COBI / FTC / TDF)	Gilead Sciences, Inc.	August 27, 2012
12	Triumeq	(ABC / DTG / 3TC)	ViiV Healthcare	August 22, 2014
13	Genvoya	elvitegravir, cobicistat, emtricitabine, and tenofoviralafenamidefumarate (EVG / COBI / FTC / TAF)	Gilead Sciences, Inc.	November 5, 2015
14	Odefsey	emtricitabine, rilpivirine, and tenofoviralafenamide(FTC / RPV / TAF)	Gilead Sciences, Inc.	March 1, 2016

BACKGROUND

Human Immunodeficiency Virus (HIV)

HIV can be classified into two types: HIV-1 and HIV-2, which are further divided into extensive groups, subtypes, and recombinant forms. HIV-2 is more than 55% genetically different from HIV-1. Due to this genetic difference, HIV-1 and HIV-2 antigens are distinct enough to the extent that if a test is developed to detect HIV-1, it will not reliably detect HIV-2. HIV-1 is identified as more virulent and prevalent than HIV-2 (De Clercq & Li, 2016).

Regarding the origin of HIV, it can be traced to West Central Africa in the late 19th or the early 20th century, when the butchering and consumption of primate bushmeat were widely practised (Sharp & Hahn, 2011). Due to multiple zoonotic transfers, HIV is known to be transmitted from chimpanzees (HIV-1 groups M and N), gorillas (HIV-1 groups P and O), and sooty mangabeys (HIV-2) to humans (Faria, Rambaut, Suchard, Baele, Bedford, Ward… & Posada, 2014; Tebitand Arts, 2011; D'arcet, Ayouba, Esteban, Learn, Boue, Liegeois… & Peeters, 2015; Moureze, Simon, & Plantier, 2013). As a blood-borne virus, HIV spreads mainly through HIV-contaminated blood or body fluids. Person can infect with HIV by sexual contact, needle sharing, blood transfusions, or maternal transmissions (Kumari & Singh, 2012). Monocytes having CD4 receptors are a prime target for HIV-1, which disturbs the cell-mediated immunity resulting in opportunistic infections leading to the death of the individual. CD4 receptors are protein molecules found predominantly on T-lymphocytes responsible for inducer or helper functions in the immature response (Veugelers & Schechter, 1997).

The Stages of HIV Infection

There are three stages of HIV infection (Veugelers & Schechter, 1997) as shown in Figure 2, which can be enumerated as:

1. **Acute HIV Infection:** Acute HIV infection is the earliest stage of HIV infection, and it generally develops within 2 to 4 weeks after a person is infected with HIV. The infected person develops flu-like symptoms-fever, headache, and rash. Rapid multiplication of the virus covers the whole body and destroys the CD4 cells of the immune system. This stage increases the risk of HIV transmission, as the viral load in the blood is quite high.
2. **Chronic HIV Infection:** The second stage of HIV infection is chronic HIV infection (also called asymptomatic HIV infection or clinical latency). The viral multiplication is very low at this stage. People with chronic HIV infection may not have any HIV-related symptoms, but they can still spread HIV to others. During chronic infection, the incubation period of HIV can be 8 to 11 years.

Figure 2. The Stages of HIV Infection

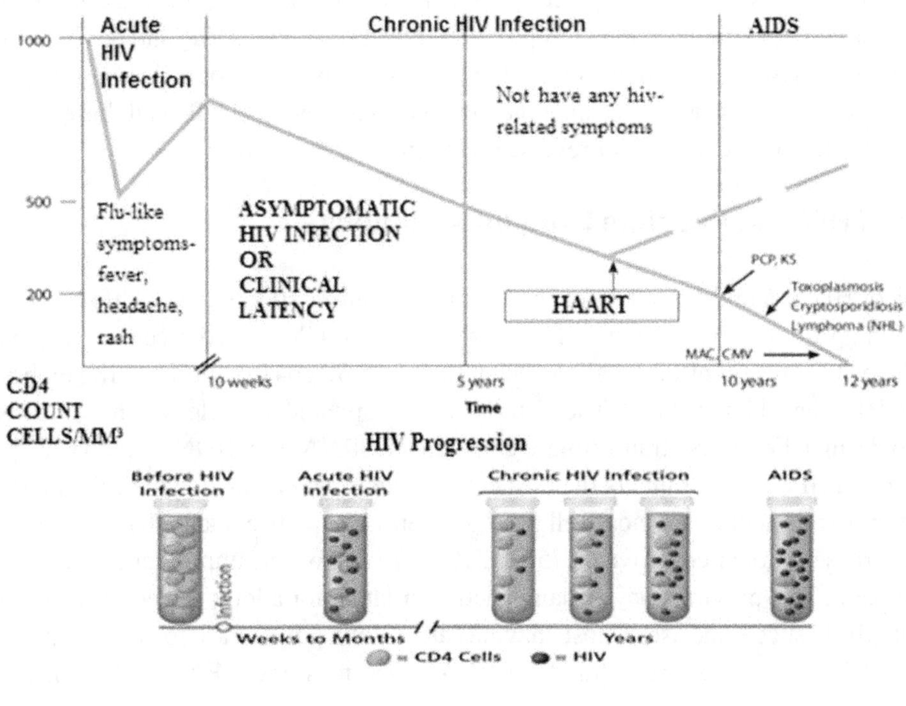

3. **AIDS:** AIDS is the final and the most severe stage of HIV infection. Because HIV has severely damaged the immune system, the body cannot fight off opportunistic infections. Opportunistic infections make the infected person more vulnerable to bacterial attack or cancer-related infections. People with HIV are diagnosed with AIDS, if they have a CD4 count less than 200 cells/mm^3 or if they have certain opportunistic infections.

Structure of HIV-1

The outer layer of the HIV consists of a lipid bi-layer, extracted from the host cell during budding of the newly formed virus, and carrying a number of proteins of both virus and host-cell origin. The host-cell major histocompatibility complex (MHC) proteins and actin remain embedded within the viral envelope. The envelope consists of the viral transmembrane protein gp41, which form a noncovalent complex with the viral outer membrane glycoprotein gp120. Protein gp120 may separate from the envelope and can be detected in the serum or within the lymphatic tissue of HIV-infected patients. The envelope protein is the most variable component of HIV. It is structurally divided into highly variable (V) and more constant (C) regions. The variability of V regions seems to be related to envelop functionality and may affect co-receptor use. Just beneath this outer envelope, a membrane-associated protein p18 is present. This p18 is essential for the virion integrity as it provides a matrix for the viral structure and surrounds a dense and cylindrical nucleoid having the capsid protein p24. Two identical RNA strands having p66/p55 component, the reverse transcriptase; p12 component, the integrase and p15 component and the protease are present in this nucleoid. Each strand contains the entire genetic blueprint coding for the structure and life cycle of HIV. Its genome is made up of 9800 nucleotides and encodes at least seventeen different proteins. It has genes encoding the Gag, Pol and Env polyproteins common to all retroviruses and carries open reading frames (ORFs) for several regulatory proteins. The transactivator (Tat) and regulator of virion expression (Rev) play a critical role in the HIV life cycle and are essential for replication. Viral protein R (Vpr) viral protein U (Vpu) viral infectivity factor (Vif) and negative factor (Nef), in contrast, are often described as 'non-essential' or 'accessory proteins' of HIV (Figure 3). At the ends of the genome are identical sequences called the long terminal repeats (LTR) and these contain regions that play a critical role in the process of reverse transcription (Chen, 2016).

Life Cycle of HIV and Possible Drug Interventions

The life cycle of HIV is a multi-step process as shown in Figure 4. HIV begins its life cycle by attaching its glycoprotein gp120 to CD4+T cells followed by CCR5 and CXCR4 co-receptors, depending on its tropism. This leads to fusion of the viral envelope, mediated by gp41, with the cell membrane and the release of the HIV capsid into the cell. The virus is now decapitated and releases the viral RNA into the host cell cytoplasm. HIV reverse transcriptase converts its ssRNA into dsDNA, known as the provirus. The newly formed HIV provirus enters the host cell's nucleus, where an enzyme called HIV integrase, integrates the provirus within the host cell's DNA. Here, host cell cofactor lens epithelium-derived growth factor/transcriptional co-activator 75 (LEDGF/p75), plays an important role and tethers integrase to chromatin. This provirus may remain inactive or latent for a long period of time. The provirus integrated into the host genome uses a host enzyme called RNA polymerase to create copies of the HIV genomic material, as well as shorter strands of RNA called messenger RNA (mRNA), which directs the synthesis of HIV proteins. An HIV enzyme called protease cuts the long chains of polypeptides into smaller individual functional proteins. These functional proteins combine with RNA genetic material, and thus assembled virus particles pushes out from the host cell. During budding, the new virus steals part of the cell's outer envelope studded with protein/sugar/lipid combinations and constitutes the HIV glycoproteins. The new copies of HIV can now move on to infect other cells.

Since the whole life cycle of this virus passes inside the host cell, it becomes really challenging to find an appropriate treatment. HIV infection can be inhibited or controlled through microbicidal compounds, which can be developed as down modulators of CD4 protein receptors, especially present over T-lymphocytes, at entry level, or as inhibitors of HIV reverse transcriptase (HIV RT)/integrase (IN) at replication level. The viable targets in the life cycle of HIV-1 can be exploited for the development of anti-HIV chemotherapy as shown in Table 4 (Kumari & Singh, 2012).

The retroviral RNA genome encodes for three enzymes essential for virus replication: reverse transcriptase (RT), integrase (IN) and protease (PR) that serve as important targets for therapeutic intervention. Targeting CD4 receptors is a little troublesome since the virus may escape identification at an early stage of infection. Similarly, developing microbicides as integrase inhibitors when proviral DNA has already emerged in the infected biological system may not prove much beneficial. The central point of replication of HIV is the process of reverse transcription in which the genomic RNA is converted into ds DNA by reverse transcriptase enzyme. Therefore, in the process of drug development against HIV, the most effective approach evolved is the synthesis of inhibitors of HIV Reverse Transcriptase (HIV RT). This enzyme, a heterodimer of p66/p51 units with a polymerase and RNase H activity, is essentially required for HIV proliferation and has no effect on normal cell replication. In fact, more than half of currently approved anti-HIV drugs target the RT enzyme (Table 2 and Table 3). With a better understanding of RT, its mechanisms of inhibition and drug resistance, it should be possible to develop more effective reverse transcriptase inhibitors (Geronikaki, Eleftheriou, & Poroikov, 2016).

Table 4. Possible HIV targets and potential interventions

Stages of HIV Life Cycle	Potential Intervention
Binding to the host cell	Antibodies to the virus or cell receptor
Entry to the host cell	Drugs blocking fusion
Reverse transcription	Reverse transcriptase inhibitors
Integration of DNA into the host genome	Integrase inhibitors
Expression of viral genes	Inhibitors of the tat protein
Production and assembly of viral components	Myristoylation, glycosylation and protease inhibitors
Budding of the virus	Interferons

Figure 3. Structure of HIV-1 genome

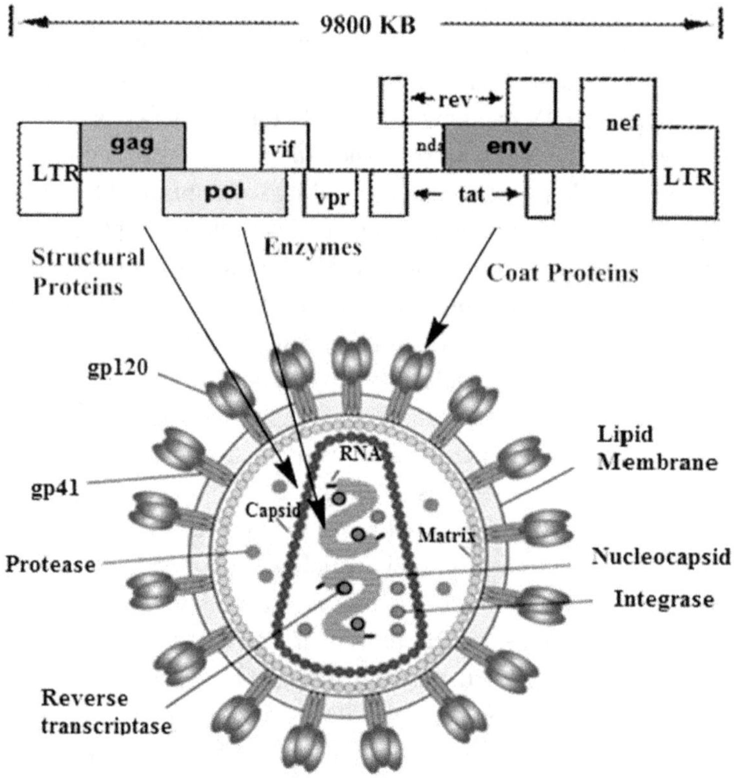

Figure 4. Life cycle of HIV

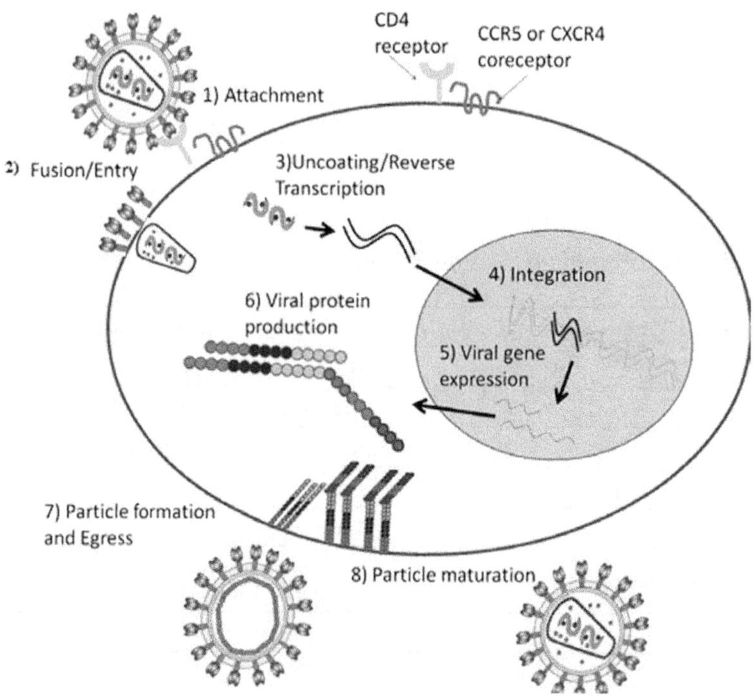

REVERSE TRANSCRIPTASE ENZYME AS A DRUG TARGET

Enzymes as a Good Drug Target

All human pathogens including viruses posses their functional enzymes and genetic material, which are suitable targets for drug intervention. More than half the drugs in clinical use today target the disease-associated or "druggable" portion of the genome dominated by enzymes (Hopkins & Groom, 2002).

The attractiveness of enzymes as target stems from their essential catalytic roles in many physiological processes that may be altered during the process of disease advancement. The structural determinants of enzyme catalysis lend themselves well to inhibition by small molecular weight drug-like molecules. As a result, there is growing interest in the study of enzymes with the aim of identifying inhibitory molecules that may serve as the starting point for drug discovery and development efforts. Conformational variation in binding sites that add catalysis offers various opportunities for drug interactions with the target molecule in a manner leading to the abolition of biological function. The biological functions can be modulated either by inhibiting their function with small molecules whose competitive binding affinity would be greater than their natural ligands that bind to the active sites (within the enzymes) or activating biomolecules (for normal functions) that are functionally deregulated in some diseases, such as cancer (Copeland, 2013).

For an enzyme inhibitor to become a practical drug, several criteria must be met (Goodman, 1996), like:

1. The inhibited biochemical pathway must be therapeutic to the patient.
2. The enzyme inhibitor must be specific.
3. The compound must have the pharmacokinetic characteristics of a practical drug.
4. The compound must have an acceptable toxicological profile.
5. The compound must survive a long and expensive clinical development process and ultimately be approved by regulatory agencies.
6. The compound must be economically viable in the market and compete successfully with other therapeutic alternatives.

Structure of HIV-1 Reverse Transcriptase

In 1975, Howard Temin, David Baltimore, and Renato Dulbecco shared the Nobel prize for Physiology for the discovery of the viral enzyme reverse transcriptase (RT), mainly associated with retroviruses. Their discovery challenged the prevailing central dogma of molecular biology, which states that DNA is transcribed into RNA, which is then translated into proteins (Baltimore, 1970; Temin & Mizutani, 2010).

HIV RT is a heterodimer of p66 (560 amino acids) and p51 (440 amino acids) is derived, from a Gag-Pol polyprotein synthesized by viral protease (PR). The larger subunit, p66, contains the active sites for both of the enzymatic activities of RT (polymerase and RNase H); and the smaller subunit plays a structural role. The p66 subunit folds into two domains, an N-terminal polymerase domain (440 residues) and a C terminal RNase H domain (120 residues). The polymerase domain is divided into four subdomains: the fingers (residues1-85, 118-155), the palm (residues 86-117, 156-237), the thumb (residues 238-318), and the connection (residues 319-426), analogous to the human right hand. The p51 subunit has only the polymerase domain with the same fingers, thumb, palm and connection subdomains. The polymerase domain of p66 folds into an open extended structure containing a large active site cleft while that of p51 is closed and compact (Figure 5) (Jonckheere & De Clercq, 2000).

The polymerase active site contains three conserved aspartic acid residues (Asp110, Asp185, Asp186) located in the p66 palm and lies at the base of the DNA-binding cleft. The DNA binding cleft is formed by the fingers, palm and thumb subdomains of the p66. The connection subdomains of both p66 and p51 form the floor of the DNA-binding cleft. The Asp185 forms an H-bond with the 3´-OH group of primer. Nearly all protein-DNA interactions involve atoms of the sugar-phosphate backbone of the nucleic acid. However, the phenoxyl side chain of Tyr 183, which is part of the conserved YMDD (Tyr-Met-Asp-Asp) motif, has H-bonding interactions with the nucleotide bases in the minor groove of the template-primer.

Within the palm domain, there are also two conserved Mg^{2+} ions, which assist polymerization of the incoming dNTP. During polymerization, the Mg^{2+} ions bind tightly, stabilize the pentavalent transition state on the alpha phosphate, provide a cooperative stabilization of the incoming dNTP, and facilitate the polymerization reaction (Hu & Hughes, 2012).

The Reverse Transcription Reaction

HIV-RT replicates the RNA using three main activities ((Jonckheere & De Clercq, 2000; Hu & Hughes, 2012):

1. RNA directed DNA polymerization (5’-->3’)- synthesizes DNA using RNA template
2. DNA directed DNA polymerization (5’-->3’)- synthesizes DNA using DNA template
3. RNase H activity-degrades RNA in the RNA:DNA hybrid

Figure 5. HIV-1 RT structure and its DNA polymerase and RNase H catalytic sites: Ribbon representation of the p66 subunit in blue and the p51 subunit in green. The template (red), the primer (orange) and the incoming dNTP (yellow) are represented by spheres. Polymerase binding site showing the location of catalytic residues Asp110, Asp185 and Asp186 (red spheres) and the side-chains of Lys65, Arg72, Tyr115 and Gln151 (blue sticks). The incoming nucleotide is represented with yellow sticks. (Right) RNase H active site showing the location of Asp443, Glu478, Asp498 and Asp549 (red spheres), and the coordinating metal ions (dot yellow surfaces).

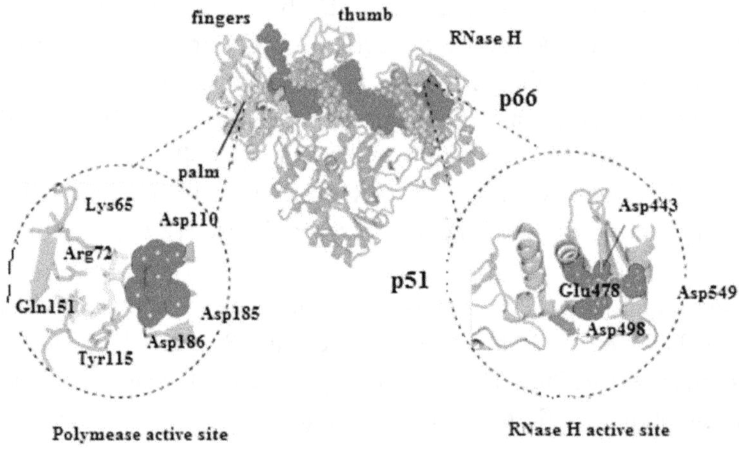

In addition to DNA elongation and RNA degradation activities of the enzyme, complete reverse transcription of genome requires two initiation events: one using tRNA3Lys as a primer for minus-strand DNA synthesis and other using RNase H resistant polypurine tract RNA as a primer for positive-strand DNA synthesis.

Reverse transcription as shown in Figure 6 has been described as:

Step 1: In HIV-1, a host tRNA3Lys (minus-strand, primer) hybridizes to the primer-binding site (PBS) located near the 5′ end of the viral genome (plus-strand, template) resulting in the synthesis of a

Figure 6. Reverse transcription process in HIV-1

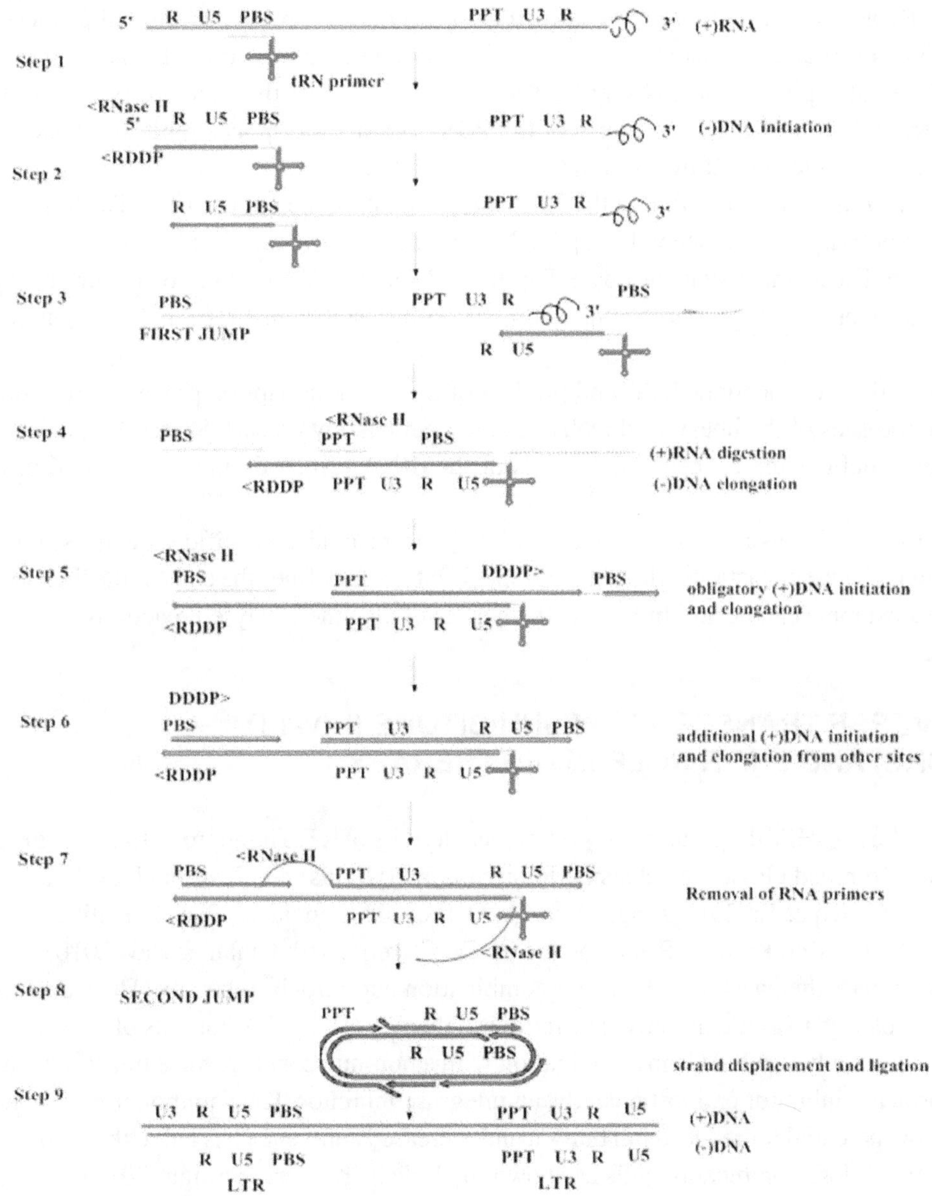

short stretch of single-stranded (-) DNA. PBS is complementary to 18 nucleotides at the 3' end of tRNAlys3.

Step 2: DNA synthesis is facilitated by RNase H digestion of the RNA portion of the RNA-DNA hybrid product, thus exposing the single-strand DNA product.

Step 3: This exposure facilitates hybridization with the R region at the 3' end of the same, or the second RNA genome, a strand-transfer reaction known as the "first jump".

Step 4: When minus-strand elongation passes a polypurine rich region called the polypurine tract (PPT) region, a unique plus-strand RNA primer is formed by RNase H cleavage at its borders. Plus-strand synthesis then continues back to the U5 region using the minus-strand DNA as a template.

Step 5: Meanwhile, minus-strand synthesis continues through the genome using the plus-strand RNA as a template, and removing the RNA template in its wake *via* RNase H activity.

Step 6: The RNase H digestion products formed are presumed to provide additional primers for plus-strand synthesis at a number of internal locations along the minus-strand DNA.

Step 7: PPT-initiated plus-strand DNA synthesis stops after copying the annealed portion of the tRNA to generate the plus-strand DNA form of the PBS, forming the "plus-strand strong stop" product. The tRNA gets cleared off by the RNase H activity of RT.

Step 8: This may facilitate annealing to the PBS complement on the minus strand DNA, providing the complementarity for the "second jump." DNA synthesis then continues.

Step 9: Strand displacement synthesis by RT to the PBS and PPT ends, and/or repair and ligation of a circular intermediate produces a linear duplex with long terminal repeats (LTRs) at both ends.

As has already been mentioned, the end product of reverse transcription process is the substrate for IN. As such, the ends of the linear viral DNA need to be relatively precise. RNase H cleavage removes the PPT primer defines the U3 end and removal of the tRNA primer defines the U5 end of the linear viral DNA.

Although RNase H has no mechanism that allows it to recognize specific sequences, it carries out these particular cleavage reactions with absolute specificity; and defines the ends of the linear viral DNA genome. This example can be highlighted as the beauty of enzyme catalysed reactions.

HIV-1 REVERSE TRANSCRIPTASE INHIBITORS (HIV-1 RT INHIBITORS) AND THEIR ROLE IN HIV THERAPY

Nearly half of the anti-HIV drugs target the polymerase activity of RT. The approved anti-RT drugs belong to one of the two broad classes: Nucleoside RT inhibitors (NRTIs) and Non-nucleoside RT inhibitors (NNRTIs) (De Clercq & Li, 2016; Singh, Miazga, Dąbrowska, Lipniacki, Piasek, Kulikowski, Shugar, 2014; Singh, Yadav, Rai, Kumari, Pannecouque, & De Clercq, 2010; Cihlar & Ray, 2010).

RT inhibitors are the backbone of current combination antiretroviral therapy. The standard care for HIV patients, referred to as highly active antiretroviral therapy (HAART), consists of three or more HIV drugs, most commonly two NRTIs in combination with a non-nucleoside reverse transcriptase inhibitor (NNRTI), protease inhibitor or, most recently, an integrase inhibitor. The common use of combinations of RTIs and the potential for reduced pill burden and increased adherence has led to the clinical development of the fixed-dose combination pills as shown in Table 3 (Kumari & Singh, 2013).

Nucleoside Reverse Transcriptase Inhibitors (NRTIs)

1. Structure of the Approved NRTIs

Zidovudine (AZT) has been identified as the first nucleoside inhibitor with *in vitro* anti-HIV activity. Since then, several NRTIs have been approved and in total, there are seven NRTIs approved for clinical use (Figure 7), viz. didanosine (ddI), zalcitabine (ddC), stavudine (d4T), lamivudine (3TC), abacavir (ABC), tenofovir disoproxil fumarate [TDF; prodrug for the oral delivery of the nucleotide analog tenofovir (TFV)] and, most recently in 2003, emtricitabine (FTC) (De Clercq & Li, 2016).

All NRTIs lack *3′* -OH group in the sugar moiety and are obligate chain-terminators of DNA elongation. Besides simply removing the *3′* -OH, as is the case for the *2′* , *3′* -dideoxy nucleosides ddI and ddC, AZT contains a replacement of the 3′-OH with a *3′* -azido functionality. Both d4T and ABC have unsaturation introduced into their ribose moieties resulting in *2′* , *3′* -dideoxy-*2′* , *3′* -didehydro

Figure 7. Structure of approved NRTIs for clinical use

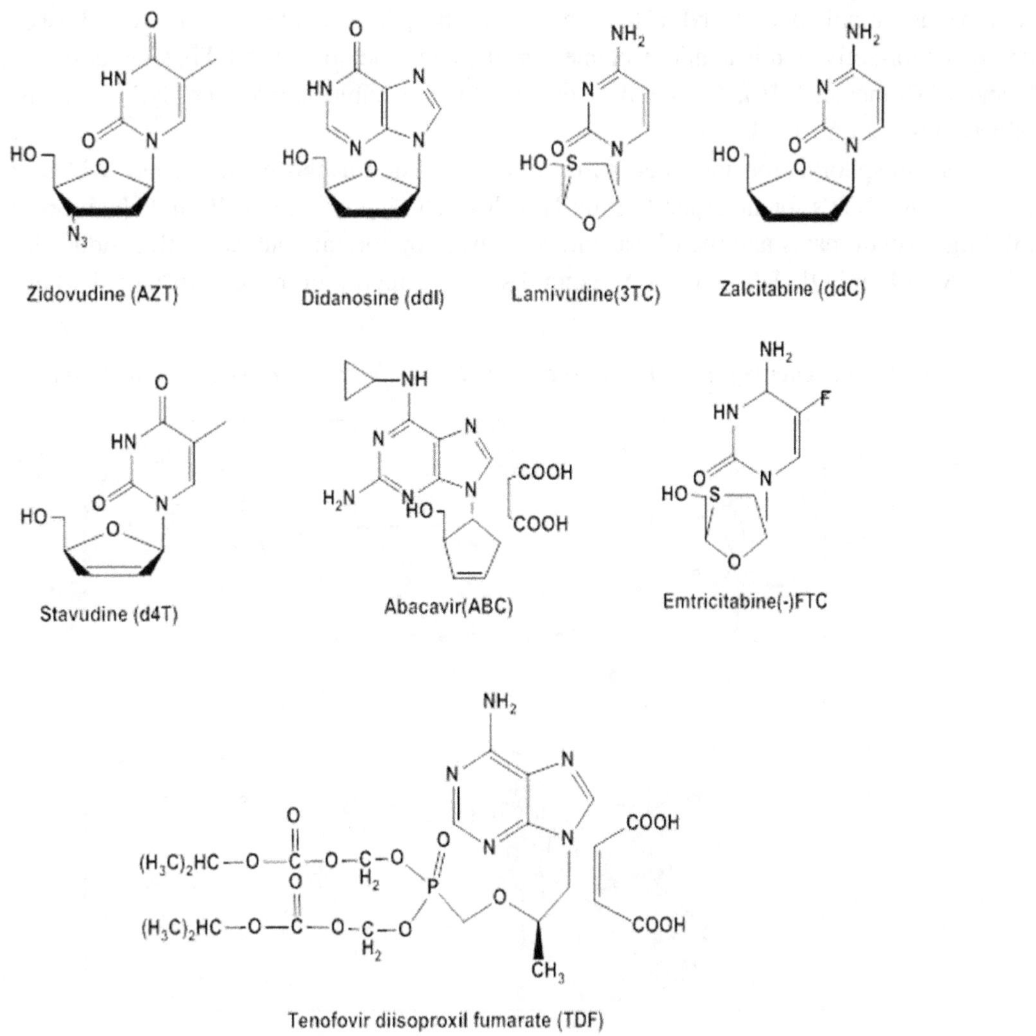

ribose ring analogs. Abacavir also replaces *O4′* with a carbon resulting in a carbocyclic ring and 3TC and FTC replace *C3′* with sulfur. The most marked ribose modifications are present in 3TC, FTC and TFV. In addition to the oxathiolane ring, 3TC and FTC have the unnatural L-enantiomeric ribose form. TFV, the only nucleotide analogue among approved NRTIs, has an acyclic linker attached to a modified phosphate moiety where a C–P phosphonate linkage replaces the normal O5′–P phosphate linkage. In contrast to the diverse chemical modifications of the ribose ring, very few base modifications are present in currently approved NRTIs. FTC contains a fluorine at the 5-position of its cytosine ring and ABC has a 6-modified diamino purine ring that serves as a prodrug to a guanine base; all the other approved NRTIs contain unmodified purine or pyrimidine bases (Bassetto, Massarotti, Coluccia, & Brancale, 2016).

2. Inhibition Mechanism of NRTIs

For NRTIs to be effective against HIV, they must enter and undergo phosphorylation to their active form, the NRTI triphosphate. The efficiency of this conversion to the active metabolite and the stability of NRTIs (and their triphosphates) in the presence of catabolic enzymes are important considerations in antiviral therapy because these factors help determine the concentration of the inhibitor in the bloodstream that is required for the NRTI to be effective. Thus, the absolute dependence on host cell enzymatic processes for activation is a unique element in the pharmacology of NRTIs (Bassetto, Massarotti, Coluccia, & Brancale, 2016). A general scheme for the reversible activation of NRTIs and its mechanism of action is presented in Figure 8.

After phosphorylation into their respective nucleoside triphosphate forms, NRTIs compete with endogenous dNTPs for incorporation by HIV RT. The incoming dNTP binds between the palm and the finger subdomains and the ribose and base make important contacts with residues including L74, Y115, M184 and Q151 (Figure 5). Structural studies support an induced fit model where proper base

Figure 8. General scheme for the reversible activation of NRTIs and its mechanism of action

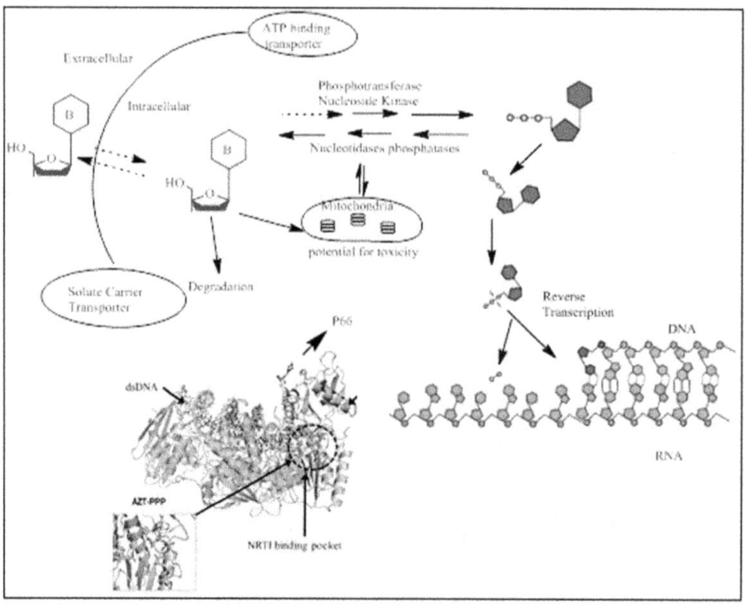

pairing by the incoming dNTP results in formation of a closed polymerase, primer/template, and dNTP complex with the incoming dNTP appropriately aligned to be attacked by the 3′ -OH at the terminus of the elongating primer strand. Kinetic studies showed a corresponding rate-limiting step that might reflect this conformation change. The dependence of the rate-limiting step on correct dNTP binding and base pairing forms the basis for the fidelity of polymerization. Despite the structural diversity present in NRTIs used clinically, their active triphosphates are able to mimic effectively the structural contacts of natural dNTPs in the HIV RT active site, allowing for efficient incorporation (Bassetto, Massarotti, Coluccia, & Brancale, 2016).

3. Limitations of Approved NRTIs
    a. Drug–Drug Interactions

NRTIs are metabolized by complex nucleotide metabolic pathways shared with the endogenous NTPs they compete with for activity, resulting in the potential for intra-class pharmacokinetic and pharmacodynamic drug interactions, for example, AZT and d4T as thymidine analogs, share common steps in their activation pathways including first, and second phosphorylation steps catalyzed by respective kinase. The strong affinity of AZT and AZT-MP for these enzymes likely explains the reduced levels of d4T-TP observed when the two NRTIs combined *in vitro* and their less than additive anti-HIV activity observed in a combination study in the clinic (Ray, 2005).

b. Resistance

The lack of a proofreading function of RT offers resistant mutant strains of HIV. Drug resistance remains a central challenge in HIV therapy because resistant viruses are easily transmitted and hence the prevalence of resistant viruses is increasing in untreated HIV-1 patients. For the virus to replicate and transmit, HIV-1 RT must be able to complete viral DNA synthesis, and NRTI-resistant RTs must retain the ability to incorporate normal dNTPs with reasonable efficiency. This means that NRTI resistance involves enhanced discrimination between normal nucleosides and NRTIs. Two basic types of NRTI-resistance mechanisms exist for HIV-1 RT (Garbelli, Riva, Crespan, & Maga, 2017):

- **Exclusion Mechanism:** It involves enhanced discrimination at the time the NRTI-TP is incorporated. The M184V/I mutations provide a clear example of the exclusion mechanism, where M184V/I selectively reduces the incorporation of 3TC and FTC by steric hindrance.
- **Excision Mechanism:** It involves the selective removal of the NRTI from the end of the viral DNA after it has been incorporated by RT. A well-studied example involves resistance caused by a set of mutations including M41L, D67N, K70R, L210W, T215F/Y, K219E/Q (these mutations will be collectively referred to as AZT resistant; they are also referred to as thymidine-analog mutations, TAMs, or excision-enhancing mutations, EEMs).

c. Adverse Effects

NRTIs therapy interferes with glycolysis, leading to lactic acidosis. During normal glycolysis, glucose is converted to pyruvate (in the cytosol), which is then transferred to the mitochondria and gets converted into *acetyl coenzyme A*, which, in turn, enters the tricarboxylic acid cycle to form NADH (the reduced form of nicotinamide adenine dinucleotide). The mitochondria use the NADH to produce adenosine triphosphate through oxidative phosphorylation. NRTIs inhibit DNA polymerase γ, which diminishes mitochondrial function, especially oxidative phosphorylation. This allows pyruvate and NADH to accumulate, enhancing the conversion of pyruvate to lactate. Impaired oxidation may also lead to a decrease in fatty acid oxidation. Free fatty acids then accumulate and metabolize to triglycerides. These excess triglycerides may accumulate in the liver, causing the characteristic hepatic serosis (Ho & Hitchcock, 1989; Edagwa, McMillan, Sillman, &Gendelman, 2017; Havlir, Tierney, Friedland, G.H., Pollard, R.B., Smeaton, L., Sommadossi, J.P., Fox, L., Kessler, Fife, & Richman, 2000).

Non-Nucleoside Reverse Transcriptase Inhibitors (NNRTIs)

1. Structure of the Approved NNRTIs

NNRTIs, chemically distinct from NRTIs, comprise of a small group of hydrophobic compounds (<600 Da) historically, originated from two classes of compounds discovered independently from each other: *1*-[(*2*-hydroxy-ethoxy) methyl]-*6*-phenylthiothymine (HEPT) analogues (Baba et al., 1989; Baba et al., 1994) and tetrahydro-imidazo [4, 5, 1- j k] [1, 4]-benzodiazepine-2(1H)-one and -thione (TIBO) analogues (Pauwels et al., 1994). FDA approved NNRTIs are compounds bearing a variety of heterocyclic rings, such as dipyrido [*1,4*] diazepine-6-one (nevirapine), benzoxazin-2-one (efavirenz), piperazine, and indolyl (delavirdine) moieties, pyrimidine (etravirine, rilpivirine) (Figure 9). NNRTIs contain at least one aromatic ring and based on their structures, NNRTIs reported to date can be divided into more than 50 classes (Singh & Singh, 2011; Kumari, Modi, Gupta, & Singh, 2011; Singh, Bhat, Verma, Kumawat,

Figure 9. Structure of FDA approved NNRTIs

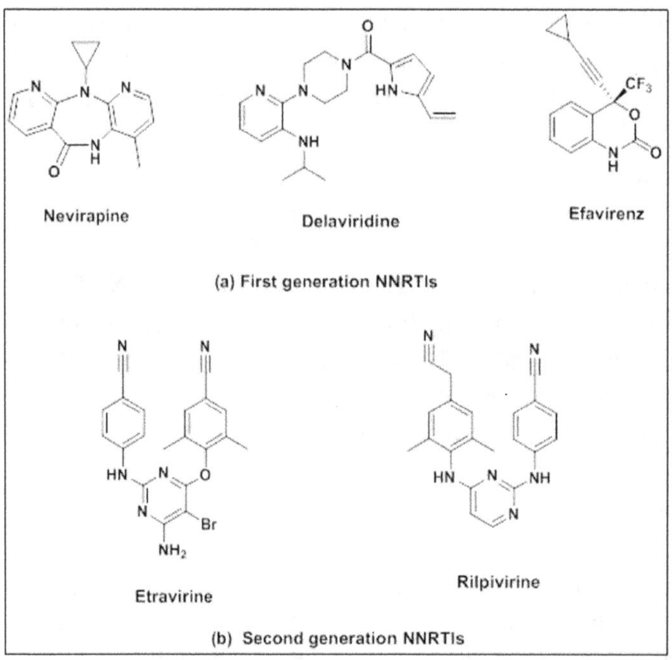

Kaur, Gupta, & Singh, 2013; Singh, Yadav, Yadav, Dhamanage, Kulkarni, & Singh, 2015; Singh, Yadav, Srivastava, Singh, Kaur, Gupta, & Singh, 2016; Kumari & Singh, 2016; Yadav, Singh, & Singh, 2017).

2. Inhibition Mechanism of NNRTIs

Despite the structural diversity of NNRTIs, they serve as non-competitive inhibitors that bind at allosteric site called non-nucleoside inhibitor binding pocket (NNIBP) located in palm domain of the p66 subunit of the RT, approximately 10Å away from the catalytic site of the enzyme. Binding of NNRTIs to HIV-1 RT does not prevent the binding of nucleic acid or nucleotide triphosphate substrates to the enzyme but blocks the chemical step of nucleotide incorporation Figure 10.

NNIBP is hydrophobic and flexible in nature and made up of aromatic (Y181, Y188, F227, W229, Y232), hydrophobic (P59, L100, V106, V179, L234, P236), and hydrophilic (K101, K103, S105, D132, E224) amino acids of the p66 subunit, and two amino acids of the p51 subunit (I135and E138). An important factor in the binding of the first generation NNRTIs, such as nevirapine, is the butterfly-like shape. Despite their configurational diversity, they assume very similar butterfly-like shape. Two aromatic rings of NNRTIs conform within the enzyme to resemble the wings of a butterfly. The butterfly structure has a hydrophilic center as a 'body' and two hydrophobic moieties representing the wings. Wing I is usually a heteroaromatic ring and wing II is a phenyl or allyl substituent. Wing I has a functional group at one side of the ring which is capable of forming hydrogen bonds with the main chain of the amino acids Lys-101 and Lys-103. Wing II interacts through π-π interactions within a hydrophobic pocket, formed in most part by the side chains of aromatic amino acids. On the main structure of the compound conforming to the butterfly body, a hydrophobic part fills a small pocket, formed by the side chains of Lys-103, Val-106 and Val-179. However, many other NNRTIs bind to RT in different modes. Second generation NNRTIs, such as etravirine, have a horseshoe-like shape with two lateral hydrophobic wings and a pyrimidine ring which is the central polar part (Figure 11).

3. HIV-1 RT and NNRTIs Interaction

Figure 10. (A) The ribbon representation of RT active domain of p66 monomer (B) HIV-1 RT complexed with dsDNA and nevirapine

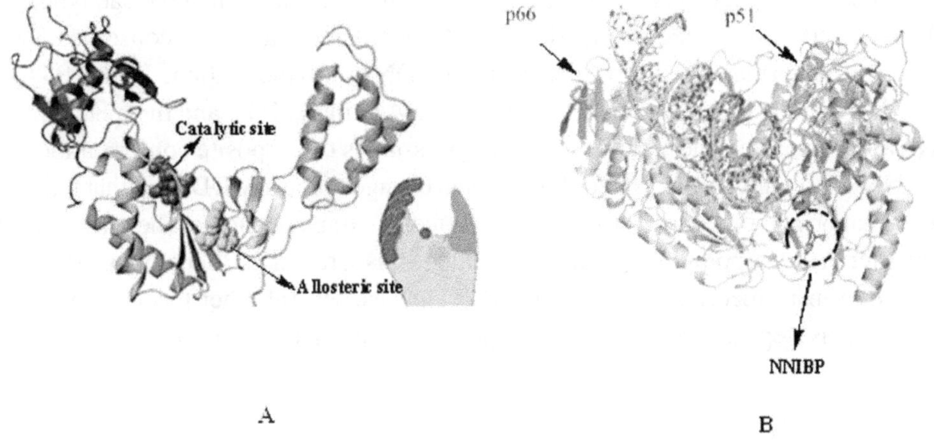

Figure 11. (A) Interaction of nevirapine in NNIBP (butterfly-like conformation) (B) Interaction of etravirine in NNIBP (horseshoe-like conformation)

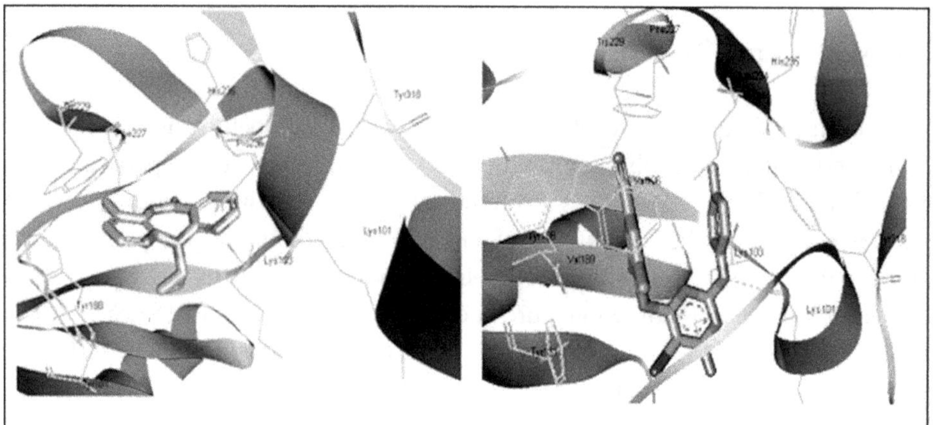

A B

Numerous co-crystal structures of HIV-1 RT unliganded, complexed with primer/template and dNTPs, or with bound NNRTIs have been resolved, providing a plethora of evidence for structural changes associated with NNRTI binding (deBéthune, 2010). The features of this interaction are:

- The NNIBP does not exist in native HIV-1 RT.
- The NNIBP forms by large structural rearrangements, particularly an extended conformation of the primer grip and rearrangements of the aromatic-ring containing residues Y181 and Y188, which locks the p66 thumb and fingers in their hyper-extended conformations.
- The NNIBP forms primarily from the p66 subunit; subunit p51, which contains amino acids identical to the polymerase domain of p66, does not have a second NNIBP.
- There are no metal co-factors bound at the active site with persuasive coordination geometry when NNRTI bind to RT.

The structural data suggest that NNRTI-binding distorts the alignment of the primer terminus with the polymerase active site, which could affect the chemical step of viral DNA synthesis. There is also structural evidence to suggest that NNRTI-binding affects the conformation of the catalytic carboxylates that bind the metal cofactors. Specifically, the β9–β10 loop that contains the conserved YMDD motif assumes different conformations during the course of DNA polymerization. In the RT/DNA/dNTP complex, the YMDD loop flexes and moves "down" by 2 Å to bind dNTP and the metal ions. However, in the RT/NNRTI complexes, the YMDD loop always assumes the opposite conformation ("up"). This suggests that NNRTI binding restricts a conformational change of the YMDD loop that facilitates metal-binding conformation and that this restriction could affect the translocation of the nucleic acid that normally occurs after the incorporation of each nucleotide (Sluis-Cremer & Tachedjian, 2008). In addition to this proposed primary mechanism of action, NNRTIs have several other mechanisms of action and interfere with various steps in the reverse transcriptase reaction. The inhibition of reverse transcription by the NNRTIs may be due to the effects on the RT RNase H activity and/or template/primer binding. Some NNRTIs interfere with HIV-1 Gag-Pol polyprotein processing and inhibit the late stage of HIV-1 replication (Zhou, Lin, & Madura, 2006).

4. Limitations of Approved NNRTIs
   a. NNRTI Resistance

Although NNRTIs are an essential ingredient of drug regimens used in HAART, they are not used as mono-therapeutic agents as they develop quick drug resistance. K103N and Y181C are the most frequently observed resistance mutations in patients treated with the approved NNRTIs. Other NNRTI resistance mutations observed in patients include L100I, K101E, V106A, V179D, Y188L, G190A and P236L. According to extensive crystallographic, molecular modelling, and biochemical studies, there are three broad classes of NNRTI-resistance mechanisms (Regina et al., 2010):

- **Loss/Change of Key Hydrophobic Interactions:** Specific residues Y181, Y188, and F227 located in the hydrophobic core of the NNIBP, have extensive interactions with NNRTIs. Mutations in some of the key residues (Y181C, Y188L, F227L) cause significant resistance through the loss of the aromatic ring interactions with NNRTIs. This causes high levels of resistance to the first generation NNRTIs, which is relatively rigid. Second generation NNRTIs have intrinsic flexibility (termed as "wiggling" and "jiggling") allowing compensatory interactions with RTs that have mutations that cause resistance to the first-generation NNRTIs.
- **Steric Hindrance:** Amino acid residues L100 and G190 are in the central region of the NNIBP. Mutations in either of these residues cause high levels of resistance to many NNRTIs. The L100I mutation confers resistance by changing the shape of the pocket (the amino acid is β-branched instead of γ-branched), whereas G190A introduces a bulge.
- **Pocket Entrance Mutations:** The K103N and K101E mutations are two NNRTI-resistance mutations that frequently cause resistance to first-generation NNRTIs. Amino acid residues K101 and K103 are located at the rim of the entrance to the NNIBP with their side chains pointing out. These mutations apparently cause resistance by interfering with the entry of NNRTIs into the pocket. Second generation NNRTIs overcome this problem, for example, rilpivirine is able to inhibit K103N mutant because it interacts with the side chain of the mutated N103 residue.

Table 5. Side Effects of NNRTIs

NNRTIs	Side Effects
Nevirapine	Elevations in liver function tests, hepatitis, liver failure
Delavirdine	Fatigue, Elevations in liver function tests, hepatitis, Nausea, diarrhea
Efavirenz	Abnormal dreams, drowsiness, dizziness, confusion, Mood changes, Elevations in liver function tests, Hyperlipidemia
Etravirine	Elevations in liver function tests
Rilpivirine	Insomnia, Depression, Elevations in liver function tests, Elevations in serum creatinine

ii. Side Effects of NNRTIs

All NNRTIs express significant drug-drug interactions. The use of NNRTIs is not without its side effects, and these can vary depending on the NNRTIs used (Table 5) (Reust, 2011).

Computational Strategies for Rational Design of HIV-1 RT Inhibitors

Presently, computational methods are an important part of the rational drug design process, known as computer-aided drug design (CADD). This approach relies on the knowledge of the three-dimensional structure of the biomolecular targets, such as enzymes and nucleic acids with an ultimate aim to reduce human power, cost, time and laboratory expenses in the drug discovery (Gashaw, Ellinghaus, Sommer, & Asadullah, 2012; Silverman & Holladay, 2014). The development of several approved anti-HIV drugs derive their origin in CADD process such as protease inhibitors; saquinavir, ritonavir and indinavir, integrase inhibitor; raltegravir, RT inhibitor; rilpivirine and fusion inhibitor; enfuvirtide (Wagner, Lee, Durrant, Malmstrom, Feher, & Amaro, 2016). CADD proceeds through the combination of molecular docking, molecular dynamics (MD) and quantitative structure–activity relationship studies (QSARs) Figure 12.

1. **Molecular Docking:** Molecular docking provides intricate details of interactions between a enzyme and a ligand. It is used for sampling the different poses of a ligand within its active site, based on the available structural data, in order to determine the best reciprocal orientation between the substituents of the ligand and the amino-acid side chains and water molecules present in the binding site. Earlier methods based on the lock-and-key theory of ligand-protein binding, treat both the structures as rigid bodies. However, the recent docking programs, like such as AutoDock, DOCK,

Figure 12. Computational strategies for the Rational Drug Design

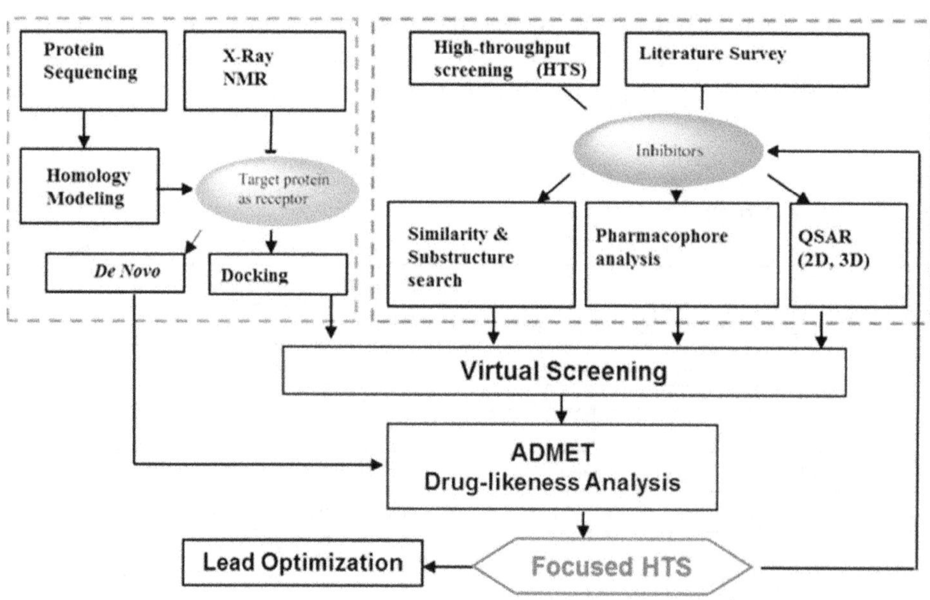

FlexX, Glide and GOLD consider ligands as flexibile units and protein as rigid bodies (Patani & LaVoie 1996; Congreve & Andrews 2017).

2. **Molecular Dynamics (MD):** MD is a powerful technique enabling the simulation of the atomic-scale movements of macromolecules over a milli- or nanosecond timescale. In this way, it is possible to determine the binding trajectories of a ligand and its subsequent adaptations to the interacting site (Congreve & Andrews 2017; Hansch, Sammes, & Taylor, 1989).
3. **Quantitative Structure–Activity Relationship Studies (QSARs):** QSAR, an approach to design lead molecules as the basis of their structural and physicochemical properties, uses Comparative Molecular Field Analysis (CoMFA) and Comparative Molecular Similarity Indices Analysis (CoMSIA) methods to derive active 3D QSAR models. This approach has also been utilised in developing HIV inhibitors up to certain extent (Gane & Dean, 2000).

Thus, the leads obtained from computer-aided screening exercises represent the starting points for further optimization of target affinity, target selectivity, biological effect, and pharmacological properties. Target affinity is commonly quantified using *in vitro* assays of enzyme activity. So, a lot of the information can be driven from the *in vitro* evaluation of enzyme–inhibitor interactions. Therefore, with the help of the key information obtained from rational drug design, one can answer the following questions:

- What opportunities for inhibitor interactions with enzyme targets arise from consideration of the catalytic reaction mechanism?
- How are inhibitors properly evaluated for potency, selectivity and mode of action?
- What are the potential advantages and liabilities of specific inhibition modalities with respect to efficacy *in vivo*?

Applications of Computational Strategies in Rationalization of NNRTIs

Rational design of NNRTIs based on *in silico* techniques utilises 'brute force' screening approach- a bunch of computational methods, because of diverse chemical scaffolds of NNRTIs. The availability of over 100 crystal structures of HIV-1 RT, either wild type (wt) or drug-resistant, in complex with approved and investigational NNRTIs, has allowed the identification of interactions between drugs and enzymes. This, in turn, has led to the possibility of rationally planning structural modifications in the NNRTIs scaffolds, which may escape the drug resistance. Nowadays, computer-aided drug design (CADD) is widely applied to the rationalization and speeding up the drug development process, starting from the promising hits obtained by high-throughput enzymatic assays or virtual screening. Examples of the application of these techniques have culminated the development of etravirine (ETR) and rilpivirine (RPV) as potential NNRTIs (Garbelli, Riva, Crespan, & Maga, 2017).

Importance of Conformational Flexibility of NNIBP in Drug Design

A good inhibitor is expected to occupy much of the available volume in the binding pocket of a drug target and to have favourable interactions within the pocket. For a rapidly mutating target, where the mutations affect the binding of an inhibitor, an effective drug candidate should have the ability to bind to the multiple related pocket conformations found in resistant mutants. This type of problem requires that an inhibitor be optimized against multiple related binding pockets in contrast to the classical approach where the optimization is carried out against a single binding pocket. An ideal inhibitor must accommodate with the physical and chemical changes of binding pocket due to mutations and its elastic limits. It should also adjust with different types of interactions with conserved and mutated amino acid residues in the pocket.

The conformational flexibility of an inhibitor may compensate for the effects of resistance mutations, if the inhibitor can be easily accommodated in the various binding pockets found in resistant targets. While the inhibitor should be flexible so that it can bind in the modified pockets of a mutant target, important inhibitor–target interactions should not be adversely affected by inhibitor flexibility. This consideration is important to avoid an entropic penalty upon the inhibitor binding. Hydrophobic pockets appear to be more suitable targets for designing such inhibitors. Inhibitor–target polar interactions are critical and, therefore, need to be treated appropriately. The second binding mode of an inhibitor should preserve the polar interactions present in its first binding mode, or compensate for the loss of those interactions by developing new interactions with the pocket. The flexibility of highly potent DAPY analogs and easy interconversions among related conformations (Figure 13) may help explain their improved resilience to resistance mutations when compared with other NNRTI drugs and drug candidates. The concept of exploiting conformational degrees of freedom to offset the effects of resistance mutations may have broader implications for drug design. Further, The NNIBP is elastic and its conformation depends on the size, shape, specific chemical composition and binding mode of an NNRTI. The limit of the pocket flexibility is not fully understood. Therefore, accurate prediction of the structure of HIV-1 RT/NNRTI

Figure 13. Flexible inhibitor is more effective than a rigid inhibitor in overcoming the effects of a resistance mutation (courtesy: Regina, Coluccia, & Silvestri, 2010)

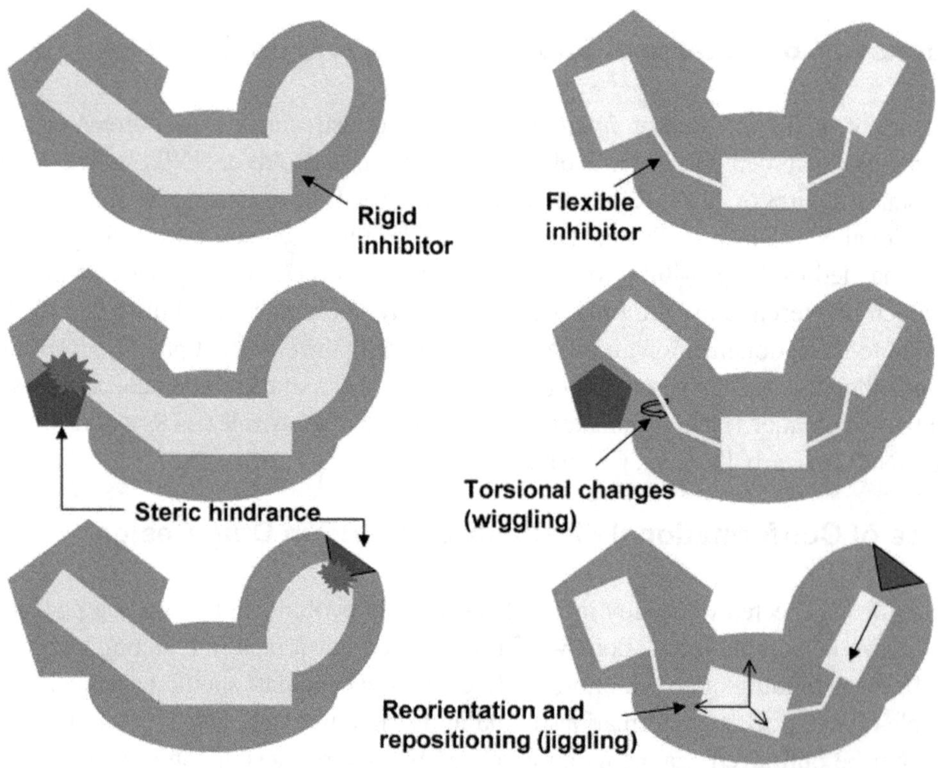

complexes by molecular modelling is a very challenging problem. Reliable molecular modelling of HIV-1 RT/NNRTI complexes requires experimentally determined structure of related NNRTIs in complexes with HIV-1 RT. The binding of an NNRTI also has long-range effects on the relative arrangement of the RT segments. These long-range effects also vary according to the nature of the bound NNRTI (Regina, Coluccia, & Silvestri, 2010).

CONCLUSION

Reverse transcription process is the hallmark of retroviruses and is highly significant in drug development against HIV/AIDS. The crystal structure of the HIV-1 RT, particularly those that reveal the structure of complexes with bound nucleic acids and incoming dNTPs, have allowed one to understand how the enzyme works at a molecular level. Structure of RT with bound anti-RT drugs provided better understanding of drug action and drug resistance. Complementary biochemical and genetic experiments have helped complete this picture. Although HIV-1 RT is an extremely validated target and widely studied, the discovery of novel allosteric sites and alternative mechanisms to this enzyme provide insights to develop new therapeutic classes of inhibitors. The high suitability of HIV-RT has resulted in the development of NRTIs and NNRTIs as the most effective inhibitors used in HAART against HIV/AIDS. However, further investigation of HIV-RT structure and its role can lead to the development of effective inhibitors targeted against HIV reservoirs in the body and mutant strains as well.

ACKNOWLEDGMENT

Financial assistance in the form of WOS-A fellowship (SR/WOS-A/CS-1053/2014) to Anuradha Singh by the DST, New Delhi is sincerely acknowledged.

REFERENCES

Baba, M., Shigeta, S., Yuasa, S., Takashima, H., Sekiya, K., Ubasawa, M., ... De Clercq, E. (1994). Preclinical evaluation of MKC-442, a highly potent and specific inhibitor of human immunodeficiency virus type 1 in vitro. *Antimicrobial Agents and Chemotherapy*, *38*(4), 688–692. doi:10.1128/AAC.38.4.688 PMID:7518216

Baba, M., Tanaka, H., De Clercq, E., Pauwels, R., Balzarini, J., Schols, D., ... Miyasaka, T. (1989). Highly specific inhibition of human immunodeficiency virus type 1 by a novel 6-substituted acyclouridine derivative. *Biochemical and Biophysical Research Communications*, *165*(3), 1375–1381. doi:10.1016/0006-291X(89)92756-3 PMID:2575380

Baltimore, D. (1970). Rna-dependent dna polymerase¡ n virions of rous sarcoma virus. *Nature*, *226*(121), 1. PMID:16057031

Barr-Sinoussi, F., Chermann, J. C., Rey, F., Nugeyre, M. T., & Chamaret, S. (1983). Isolation of a T-lymphotropic retrovirus from a patient at risk of acquired immune deficiency syndrome (AIDS). *Science*, *220*, 868. doi:10.1126cience.6189183 PMID:6189183

Bassetto, M., Massarotti, A., Coluccia, A., & Brancale, A. (2016). Structural biology in antiviral drug discovery. *Current Opinion in Pharmacology*, *30*, 116–130. doi:10.1016/j.coph.2016.08.014 PMID:27611878

Chen, B. (2016). HIV capsid assembly, mechanism, and structure. *Biochemistry*, *55*(18), 2539–2552. doi:10.1021/acs.biochem.6b00159 PMID:27074418

Cihlar, T., & Ray, A. S. (2010). Nucleoside and nucleotide HIV reverse transcriptase inhibitors: 25 years after zidovudine. *Antiviral Research*, *85*(1), 39–58. doi:10.1016/j.antiviral.2009.09.014 PMID:19887088

Congreve, M., & Andrews, S. (2017). Perspective in Medicinal Chemistry: Structure-Based Drug Design. *Current Topics in Medicinal Chemistry*, *17*(2), 93–94. doi:10.2174/15680266170216112220425 1 PMID:28093978

Copeland, R. A. (2013). *Evaluation of enzyme inhibitors in drug discovery: a guide for medicinal chemists and pharmacologists*. John Wiley & Sons.

Boesch, F. H., Madinda, C., Robbins, N. F., Gray, M. M., Cournil, M., Ooms, A., ... Peeters, M. (2015). Origin of the HIV-1 group O epidemic in western lowland gorillas. *Proceedings of the National Academy of Sciences of the United States of America*, *112*(11), E1343–E1352. doi:10.1073/pnas.1502022112 PMID:25733890

De Clercq, E., & Li, G. (2016). Approved antiviral drugs over the past 50 years. *Clinical Microbiology Reviews*, *29*(3), 695–747. doi:10.1128/CMR.00102-15 PMID:27281742

De Béthune, M. P. (2010). Non-nucleoside reverse transcriptase inhibitors (NNRTIs), their discovery, development, and use in the treatment of HIV-1 infection: A review of the last 20 years (1989–2009). *Antiviral Research*, *85*(1), 75–90. doi:10.1016/j.antiviral.2009.09.008 PMID:19781578

Edagwa, B., McMillan, J., Sillman, B., & Gendelman, H. E. (2017). Long-acting slow effective release antiretroviral therapy. *Expert Opinion on Drug Delivery*, 1–11. doi:10.1080/17425247.2017.1288212 PMID:28128004

Faria, N. R., Rambaut, A., Suchard, M. A., Baele, G., Bedford, T., Ward, M. J., ... Posada, D. (2014). The early spread and epidemic ignition of HIV-1 in human populations. *Science*, *346*(6205), 56–61. doi:10.1126cience.1256739 PMID:25278604

Gallo, R. C., & Montagnier, L. (2003). The discovery of HIV as the cause of AIDS. *The New England Journal of Medicine*, *349*(24), 2283–2285. doi:10.1056/NEJMp038194 PMID:14668451

Gane, P. J., & Dean, P. M. (2000). Recent advances in structure-based rational drug design. *Current Opinion in Structural Biology*, *10*(4), 401–404. doi:10.1016/S0959-440X(00)00105-6 PMID:10981625

Garbelli, A., Riva, V., Crespan, E., & Maga, G. (2017). How to win the HIV-1 drug resistance hurdle race: Running faster or jumping higher? *The Biochemical Journal*, *474*(10), 1559–1577. doi:10.1042/BCJ20160772 PMID:28446620

Gashaw, I., Ellinghaus, P., Sommer, A., & Asadullah, K. (2012). What makes a good drug target? *Drug Discovery Today*, *17*, S24–S30. doi:10.1016/j.drudis.2011.12.008 PMID:22155646

Geronikaki, A., Eleftheriou, P., & Poroikov, V. (2016). *Anti-HIV Agents: Current Status and Recent Trends*. Academic Press.

Goodman, L. S. (1996). *Goodman and Gilman's the pharmacological basis of therapeutics* (Vol. 1549). New York: McGraw-Hill.

Gottlieb, M. S., Schroff, R., Schanker, H. M., Weisman, J. D., Fan, P. T., Wolf, R. A., & Saxon, A. (1981). Pneumocystis carinii pneumonia and mucosal candidiasis in previously healthy homosexual men: Evidence of a new acquired cellular immunodeficiency. *The New England Journal of Medicine*, *305*(24), 1425–1431. doi:10.1056/NEJM198112103052401 PMID:6272109

Hansch, C., Sammes, P. G., & Taylor, J. B. (1989). *Comprehensive medicinal chemistry: the rational design, mechanistic study & therapeutic applications of chemical compounds* (Vol. 5). Pergamon Pr.

Havlir, D. V., Tierney, C., Friedland, G. H., Pollard, R. B., Smeaton, L., Sommadossi, J. P., ... Richman, D. D. (2000). In vivo antagonism with zidovudine plus stavudine combination therapy. *The Journal of Infectious Diseases*, *182*(1), 321–325. doi:10.1086/315683 PMID:10882616

Ho, H. T., & Hitchcock, M. J. (1989). Cellular pharmacology of 2',3'-dideoxy-2',3'-didehydrothymidine, a nucleoside analog active against human immunodeficiency virus. *Antimicrobial Agents and Chemotherapy*, *33*(6), 844–849. doi:10.1128/AAC.33.6.844 PMID:2764535

Hopkins, A. L., & Groom, C. R. (2002). Opinion: The druggable genome. *Nature Reviews. Drug Discovery*, *1*(9), 727–730. doi:10.1038/nrd892 PMID:12209152

Hu, W. S., & Hughes, S. H. (2012). HIV-1 reverse transcription. *Cold Spring Harbor Perspectives in Medicine*, *2*(10), a006882. doi:10.1101/cshperspect.a006882 PMID:23028129

Jonckheere, H., Anné, J., & De Clercq, E. (2000). The HIV-1 Reverse Transcription (RT) Process as Target for RT Inhibitors. *Medicinal Research Reviews*, *20*(2), 129–154. doi: PMID:10723025

Kumari, G., & Singh, K, R. (. (2013). Anti-HIV drug development: Structural features and limitations of present day drugs and future challenges in the successful HIV/AIDS treatment. *Current Pharmaceutical Design*, *19*(10), 1767–1783. doi:10.2174/138161281131999990295 PMID:23092282

Kumari, G., & Singh, R. K. (2012). Highly active antiretroviral therapy for treatment of HIV/AIDS patients: Current status and future prospects and the Indian scenario. *HIV & AIDS Review*, *11*(1), 5–14. doi:10.1016/j.hivar.2012.02.003

Kumari, G., & Singh, R. K. (2016). Molecular Modeling, Synthesis, and Anti-HIV Activity of Novel Isoindolinedione Analogues as Potent Non-nucleoside Reverse Transcriptase Inhibitors. *Chemical Biology & Drug Design*, *87*(2), 200–212. doi:10.1111/cbdd.12620 PMID:26212217

Kumari, G., Modi, M., Gupta, S. K., & Singh, R. K. (2011). Rhodium (II) acetate-catalyzed stereoselective synthesis, SAR and anti-HIV activity of novel oxindoles bearing cyclopropane ring. *European Journal of Medicinal Chemistry*, *46*(4), 1181–1188. doi:10.1016/j.ejmech.2011.01.037 PMID:21339030

Mourez, T., Simon, F., & Plantier, J.-C. (2013, July 01). Non-M Variants of Human Immunodeficiency Virus Type 1. *Clinical Microbiology Reviews*, *26*(3), 448–461. doi:10.1128/CMR.00012-13

NACO-Annual Report 2015-2016. (n.d.). Retrieved from http://naco.gov.in/ 202015-16_NACO.pdf

Patani, G. A., & LaVoie, E. J. (1996, January). Bioisosterism: A Rational Approach in Drug Design. *Chemical Reviews*, *96*(8), 3147–3176. doi:10.1021/cr950066q

Pauwels, R., Andries, K., Debyser, Z., Kukla, M. J., Schols, D., Breslin, H. J., ... De Clercq, E. (1994, December 01). New tetrahydroimidazo [4, 5, 1-jk][1, 4]-benzodiazepin-2 (1H)-one and-thione derivatives are potent inhibitors of human immunodeficiency virus type 1 replication and are synergistic with 2′, 3′-dideoxynucleoside analogs. *Antimicrobial Agents and Chemotherapy*, *38*(12), 2863–2870. doi:10.1128/AAC.38.12.2863

Ray, A. S. (2005). Intracellular interactions between nucleos(t)ide inhibitors of HIV reverse transcriptase. *AIDS Reviews*, *7*, 113–125. PubMed

Regina, G. L., Coluccia, A., & Silvestri, R. (2010). Looking for an active conformation of the future HIV type-1 non-nucleoside reverse transcriptase inhibitors. *Antiviral Chemistry & Chemotherapy*, *20*(6), 213–237. PubMed

Reust, C. E. (2011). Common adverse effects of antiretroviral therapy for HIV disease. *American Family Physician*, *83*(12).

Sharp, P. M., & Hahn, B. H. (2011). Origins of HIV and the AIDS pandemic. *Cold Spring Harbor Perspectives in Medicine*, *1*(1), a006841.

Silverman, R. B., & Holladay, M. W. (2014). *The organic chemistry of drug design and drug action*. Academic Press.

Singh, A., Yadav, D., Yadav, M., Dhamanage, A., Kulkarni, S., & Singh, R. K. (2015, March). Modeling, Synthesis and Biological Evaluation of N-Heteroaryl Compounds as Reverse Transcriptase Inhibitors Against HIV-1. *Chemical Biology & Drug Design*, *85*(3), 336–347. doi:10.1111/cbdd.12397 PMID:25055732

Singh, A., Yadav, M., Srivastava, R., Singh, N., Kaur, R., Gupta, S. K., & Singh, R. K. (2016, December). Design and anti-HIV activity of arylsulphonamides as non-nucleoside reverse transcriptase inhibitors. *Medicinal Chemistry Research*, *25*(12), 2842–2859. doi:10.100700044-016-1707-7

Singh, R. K., Miazga, A., Dąbrowska, A., Lipniacki, A., Piasek, A., Kulikowski, T., & Shugar, D. (2014). Myristoylated Derivatives of 2, 3-didehydro-2, 3-dideoxythymidine (stavudine) bi-Functional Prodrugs with Potent Anti-HIV-1 Activity and Low Cytotoxicity. *Antiviral Chemistry & Chemotherapy*, *23*(6), 231–235.

Singh, R. K., Yadav, D., Rai, D., Kumari, G., Pannecouque, C., & De Clercq, E. (2010, September). Synthesis, structure–activity relationship and antiviral activity of 3′-N, N-dimethylamino-2′, 3′-dideoxythymidine and its prodrugs. *European Journal of Medicinal Chemistry*, *45*(9), 3787–3793. doi:10.1016/j.ejmech.2010.05.028 PMID:20538384

Singh, U., & Singh, R. K. (2011). Implication for rational drug design. *Retrovirology*, *8*(S2), 82. doi:10.1186/1742-4690-8-S2-P82

Singh, U. P., Bhat, H. R., Verma, A., Kumawat, M. K., Kaur, R., Gupta, S. K., & Singh, R. K. (2013). discovery of a novel class of non-nucleoside reverse transcriptase inhibitors (NNRTIs) against HIV-1 and their antibacterial properties. *RSC Advances*, *3*(38), 17335–17348. doi:10.1039/c3ra41604f

Sluis-Cremer, N., & Tachedjian, G. (2008). Mechanisms of inhibition of HIV replication by non-nucleoside reverse transcriptase inhibitors. *Virus Research*, *134*(1), 147–156. PubMed

Tebit, D.M., & Arts, E.J. (2011). Tracking a century of global expansion and evolution of HIV to drive understanding and to combat disease. *Lancet Infect Dis*, *11*, 45–56. (10)70186-910.1016/S1473-3099

UNAIDS Fact Sheet. (2016). *Global statistics-2015*. Retrieved from http://www.unaids.org/sites/ default/

Veugelers, P. J., & Schechter, M. T. (1997). The incubation period of HIV infection. *Current Opinion in Infectious Diseases*, *10*(1), 7–11. doi:10.1097/00001432-199702000-00003

Wagner, J. R., Lee, C. T., Durrant, J. D., Malmstrom, R. D., Feher, V. A., & Amaro, R. E. (2016, June 8). Emerging computational methods for the rational discovery of allosteric drugs. *Chemical Reviews*, *116*(11), 6370–6390. doi:10.1021/acs.chemrev.5b00631 PMID:27074285

Yadav, M., Singh, A., & Singh, R. K. (2017). Docking studies on novel bisphenylbenzimidazoles (BPBIs) as non-nucleoside inhibitors of HIV-1 reverse transcriptase. *Indian Journal of Chemistry B*, *56*(7), 714-723.

Zhou, Z., Lin, X., & Madura, J. D. (2006). HIV-1 RT nonnucleoside inhibitors and their interaction with RT for antiviral drug development. *Infectious Disorders-Drug Targets*, *6*(4), 391-413.

Chapter 6
The Use of Liposomes in Enzymes and Drug Design:
Liposomes Drug Delivery System

Mahmoud Balbaa
Alexandria University, Egypt

Doaa Awad
Alexandria University, Egypt

ABSTRACT

Liposomes are phospholipid vesicles that share many of membranes properties. Liposomes can be easily prepared in a range of sizes. They are able to improve the unfavorable properties of many free drugs such as increasing the amount of drugs delivered to various diseased sites in addition to decreasing the drug toxicities. Encapsulation of enzymes and food ingredients, as well as antioxidants in liposomes also received a lot of awareness. Moreover, an increase for drugs delivered to various diseased tissues was achieved by encapsulating drugs in the liposomes. The topics of encapsulation of enzymes and food ingredients as well as antioxidants in liposomes were highly investigated.

INTRODUCTION

Liposomes are lipid bilayers enclosing aqueous solution in their core. Liposomes can either be passively or actively targeted to diseased tissues whose abnormal vasculature enables the liposomal accumulation. Inserting hydrophilic biomolecules such as polyethylene glycol within the liposomal bilayers was found to increase their stability in blood circulation and consequently raise their treatment efficacy. Liposomes vary in their sizes and can incorporate antioxidants, nutrients, enzymes or pharmaceutical drugs. Through encapsulation of various medications in liposomes, a decrease in their nonspecific toxicities and an increase in the amount delivered to the required location can be achieved. Medications encapsulated inside highly stable liposomal formulations are sequestered and require to be released so that the diseased sites receive appropriate doses of these medications. The main objective of this chapter is to elucidate the role of liposomes in enhancing the effectiveness of therapeutic agents including enzymes. The importance of liposomes in improving the limitations of drugs will also be clarified.

DOI: 10.4018/978-1-5225-5237-6.ch006

BACKGROUND

Liposomes are phospholipid vesicles first described by Bangham and his coworkers who showed that aqueous suspensions of phospholipids shared many of the structural and functional properties of membranes (Bangham, Standish, & Watkins, 1965). Liposomes can be prepared by dehydrating phospholipids dissolved in organic solvent. A thin phospholipid film is then obtained producing multilamellar vesicles upon its rehydration. Extrusion is a popular method of unilamellar liposomes preparation where several passes of multilamellar liposomal suspension through small polycarbonate filter membranes yields particles having a diameter near the pore size of the filter used. Consequently, the extrusion method enables the preparation of liposomes in a range of sizes (Hope, Bally, Webb, & Cullis, 1985).

Liposomes were recommended as drug delivery systems to improve the unfavorable properties of many free drugs such as the low stability, poor solubility, rapid metabolism or inappropriate biodistribution. The amphiphilic property of phospholipids enables the liposomes to carry either hydrophilic drugs in their aqueous core or hydrophobic drugs in the lipid bilayer (Leonetti, Scarsella, Semple, Molinari, D'Angelo, Stoppacciaro, Biroccio, & Zupi, 2004). Blood can be distributed to all tissues via the vasculature, which is affected and changed during the disease process. It was evidenced that inflammation results in an early phase of vasodilation followed by increased vessel permeability. Moreover, many of the drug delivery approaches in cancer therapy take advantage of the unique pathophysiology of tumor vasculature. That is, the enhanced permeability of tumor blood vessels and the decreased rate of clearance caused by the lack of functional lymphatic vessels resulted in the increased accumulation of macromolecules in tumors after intravenous administration (Claesson-Welsh, 2015).

1. Steric Stabilization of Liposomes

Most liposomes failed to stay in blood circulation for a few hours, owing to the opsonic activity of the plasma components and to the liposome removal by the cells of the mononuclear-phagocytic systems (Blume & Cevc, 1990). A change in liposome composition by incorporating a negatively charged glycolipid, such as monosialoganglioside produced a steric surface barrier suppressing the elimination of liposomes from the blood by the phagocytes (Gabizon & Papahadjopoulos, 1988). Stealth liposomes containing 10% distearoylphosphatidylethanolamine-conjugated poly-(ethylene glycol5000) (DSPE-PEG5000) stayed in blood and avoided liver and spleen for many hours. It was also assumed that PEG attracted water to the liposomes surface, presenting a barrier to the adherence of protein opsonins. The hydrophilic barrier also retarded disintegration of the liposomes through exchange and/or transfer of liposomal phospholipids to high density lipoproteins (Allen, 1998).

2. Targeted Delivery of Liposomes

A lot of work has been done on liposome targeting in order to achieve tissue or cell specific liposome uptake (Allen & Cullis, 2013). Specific accumulation of liposomal substances in the diseased sites was achieved by attaching of targeting agents to the liposomes. Epidermal growth factor was used as a tumor seeking agent because of its overexpression in many tumor cells (Chaidarun, Eggo, Sheppard, & Stewart, 1994). On the other hand, *in vivo-* tissue distribution of folate-targeted liposomes injected intravenously into mice bearing folate receptor-overexpressing tumors indicated the binding of those liposomes to the folate receptor in the tumor cell (Gabizon, Horowitz, Goren, Tzemach, Shmeeda, & Zalipsky, 2003). As

Eliaz and Szoka had incorporated multiple copies of the low molecular weight oligomer of hyaluronic acid into liposomes' bilayer, they found that the modified liposomes could bind to and be taken up by, cells with a high density of CD44 receptors that binds to hyaluronic acid. The results demonstrated that hyaluronic acid facilitated the recognition of liposomes by CD44-high-expressing melanoma cells in culture, whereas the control cells, which had a low CD44 density, similar to the CD44 density found on many normal cells, showed little uptake (Eliaz & Szoka, 2001). Moreover, the growth factor receptor $p185^{HER2}$, encoded by HER2 protooncogene, represented an attractive target antigen for cancer immunotherapies. Anti-$p185^{HER2}$ immunoliposomes were developed as a tumor-targeting vehicle. Park and coworkers concluded that anti-$p185^{HER2}$ immunoliposomes might be promising therapeutic vehicles in which drugs could be targeted to $p185^{HER2}$-overexpressing tumors (Park, Hong, Carter, Asgari, Guo, Keller, Wirth, Shalaby, Kotts, Wood, Papahadjopoulos, & Benz, 1995).

3. Liposome-Cell Interactions

Different mechanisms for the liposome uptake have been known. The occurrence of one of them depends on the size, charge and composition of liposomes as well as the cell type (Torchilin & Weissig, 2003). In the model of contact release, the contact of liposome and cell membrane leads to a higher permeability of the liposomal membrane, so that the content is released and can diffuse into the cytoplasm. Furthermore, the intermembrane transfer is an exchange of lipids between the liposomal and cell membranes that does not affect the membrane integrity (Sandra & Pagano, 1979). Additionally, the endocytosis is the major way for cellular uptake *in vivo* and is often receptor-mediated. Liposomes are taken up as sub-cellular vacuoles, termed as phagosomes or endosomes, originating by invagination of the plasma membrane. These vacuoles fuse with the lysosomes followed by a lysosomal digestion of the endosomal contents. The resulting fatty acids might be released from the cell or incorporated into the cell membrane. The liposomal content is then released into the lysosome where it leaks out or remain sequestered until exocytosis. On the other hand, fusion of the liposome with the cell membrane leads to a complete lipid intermixing and the release of the liposomal contents into the cytoplasm.

MAIN FOCUS OF THE CHAPTER

Liposomes were established to be a successful medication carrier. Although there were situations where the slow release of drugs out of liposomes was favorable (Charrois & Allen, 2004), the ability to control and produce a rapid release of a significant amount of drug directly at the diseased site would be extremely advantageous (Deshpande, Biswas, & Torchilin, 2013). Failure to release the drugs out of the liposomes leads to reduced therapeutic effect relative to the free drug (Meerum Terwogt, Groenewegen, Pluim, Maliepaard, Tibben, Huisman, ten Bokkel Huinink, Schot, Welbank, Voest, Beijnen, & Schellens, 2002). Therefore, the extent of content release from liposomes at the diseased site is an important factor for a successful therapy so that these sites can receive a proper dose of the drug (Deshpande, Biswas, & Torchilin, 2013). Temperature-sensitive materials and local hyperthermia were used to initiate a temperature-dependent change in the liposomal membrane in to enhance its permeability (Needham & Dewhirst, 2001). Besides, pH-sensitive liposomes were designed to facilitate liposome fusion with endosomes at low endosomal pHs (Guo, MacKay, & Szoka, 2003). However, experiments for improved drug release based on elevated levels of phospholipase A_2-catalyzed hydrolysis at the target tissue were performed (Davidsen, Jorgensen, Andresen, & Mouritsen, 2003). Moreover, the acoustic activity of air-containing liposomes enable them respond to ultrasound stimulation by releasing their contents (Huang & MacDonald, 2004).

It was confirmed that the boron cluster mercaptoundecahydrododecaborate ($Na_2B_{12}H_{11}SH$) and its perhalogenated derivatives triggered the release of liposomal contents in a dose-dependent manner (Gabel, Awad, Schaffran, Radovan, Daraban, Damian, Winterhalter, Karlsson, & Edwards, 2007; Awad, Bartok, Mostaghimi, Schrader, Sudumbrekar, Schaffran, Jenne, Eriksson, Winterhalter, Fritz, Edwards, & Gabel, 2015). The boron cluster has been already in clinical use for the treatment of glioblastoma with boron neutron capture therapy without having adverse reactions (Gabel, Foster, & Fairchild, 1987).

Cryo-transmission electron microscopy showed structural changes in 5 mM dipalmitoyl phosphatidyl choline (DPPC) liposomes induced by iodinated boron cluster (BI). The observed morphological changes in liposomes were supposed to be responsible for the release of their contents. These changes were obvious at lipid to cluster molar ratios equal 10:1 (Figure 1, top right). Several liposomes changed to opened disk-like structures. At higher concentrations of BI (16 mM), large bilayer sheets and needle-shaped structures were noticed (Figure 1, bottom left). Altering the lipid to cluster molar ratios to 1:10 led to the

Figure 1. Cryo-transmission electron microscope images of dipalmitoyl phosphatidyl choline liposomes in the absence (Top left) and presence of 0.5 mM (Top right), 16 mM (Bottom left), 50 mM (Bottom right) of iodinated boron cluster (BI). Arrows with a single arrowhead indicate open structures and disks. Arrows with triple arrowheads indicate closed liposomes. Scale bar in all pictures equals 100 nm (Awad, Bartok, Mostaghimi, Schrader, Sudumbrekar, Schaffran, Jenne, Eriksson, Winterhalter, Fritz, Edwards, & Gabel, 2015).

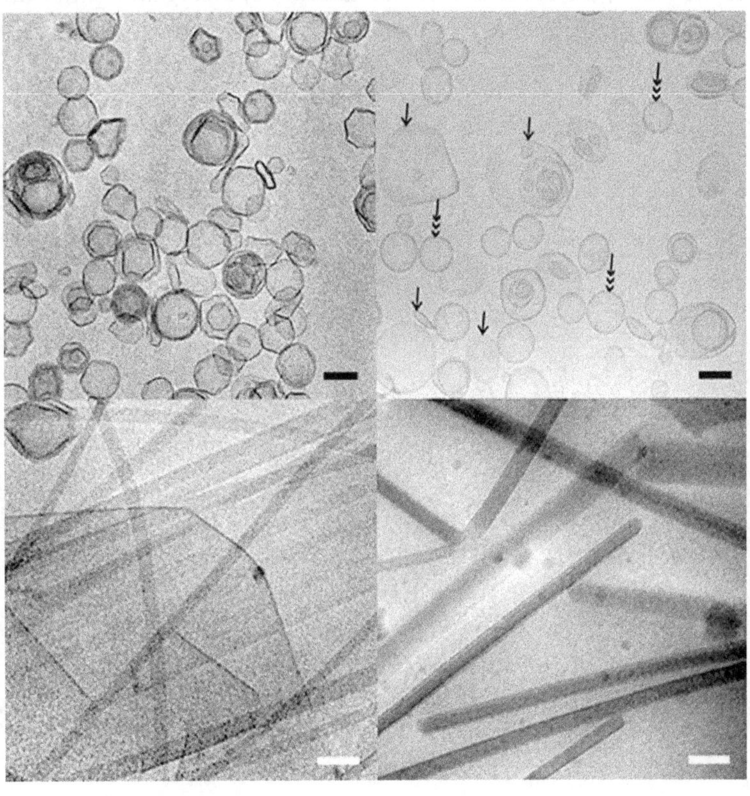

dominance of needle-shaped structures. The extent of release was determined by leakage experiments in which the liposomes were passively loaded with self-quenching concentration of carboxyfluorescein (100 mM). The dequenching of carboxyfluorescein's fluorescence indicates its liberation from DPPC liposomes. The increase of the fluorescence was measured and then normalized to 100% fluorescence after the addition of Triton X-100 at 0.05% final concentration for lysis of all liposomes. It was verified that BI is the most effective cluster in inducing leakage, whereas the chlorinated boron cluster (BCl) was demonstrated to be more effective than the brominated boron cluster (BBr). Surprisingly, intermediate BI concentrations around 0.1 mM led to a higher degree of leakage than a concentration of 0.5 mM (Figure 2). Leakage induction was explained by the binding of anions to phosphatidylcholine membranes that appears to follow Hofmeister series.

The release of the liposomal contents most probably takes place via pores formation either by the clusters or by complete destruction of the liposomes. Pores in lipid membranes can be either hydrophobic or hydrophilic (Figure 3).

In the hydrophobic pores, the clusters might coat the walls such that the hydrophobic lipid tails are not exposed to water, whereas in the hydrophilic pores, the cluster is incorporated between the phospholipids' head groups preventing their close contact (*Awad, Bartok, Mostaghimi, Schrader, Sudumbrekar, Schaffran, Jenne, Eriksson, Winterhalter, Fritz, Edwards, & Gabel, 2015*).

Figure 2. Carboxyfluorescein leakage from DSPC liposomes at 37°C induced by chlorinated boron cluster (BCl) (top left), brominated boron cluster (BBr) (top right), and iodinated boron cluster (BI) (bottom left for smaller concentrations, bottom right for higher concentrations). Concentrations of clusters in mM are given next to the end of the curve before the addition of detergent (Awad, Bartok, Mostaghimi, Schrader, Sudumbrekar, Schaffran, Jenne, Eriksson, Winterhalter, Fritz, Edwards, & Gabel, 2015).

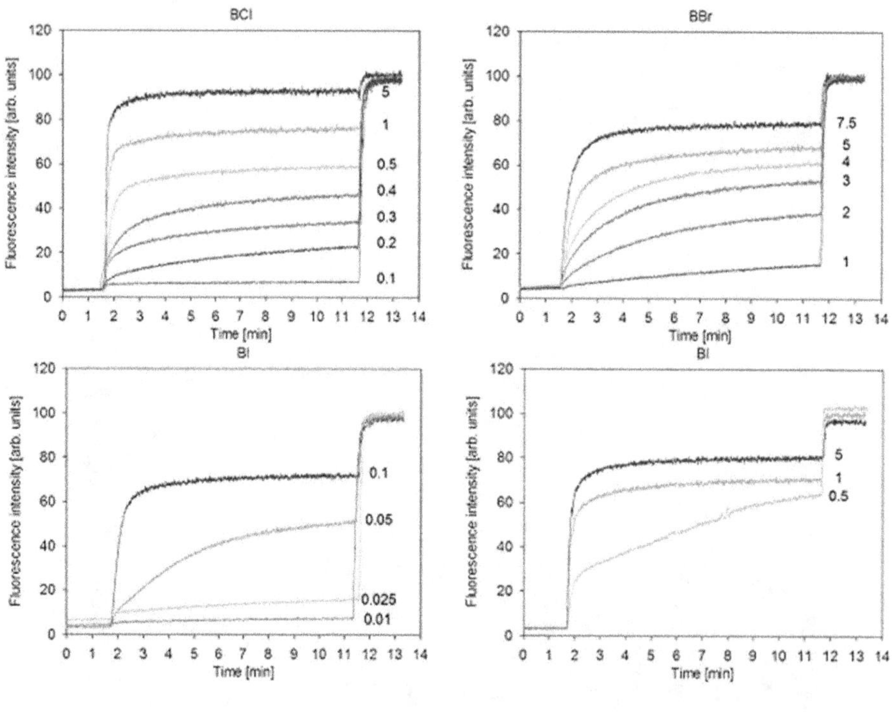

Figure 3. Models of pores with hydrophobic (top) and hydrophilic (bottom) walls, and the possible ways that the boron clusters can stabilize the holes. For the hydrophilic pores, the headgroups of lipids with exposure of hydrophobic chains to water are shown in a different color. Note that the relative sizes of the clusters and lipids are only approximate (Awad, Bartok, Mostaghimi, Schrader, Sudumbrekar, Schaffran, Jenne, Eriksson, Winterhalter, Fritz, Edwards, & Gabel, 2015).

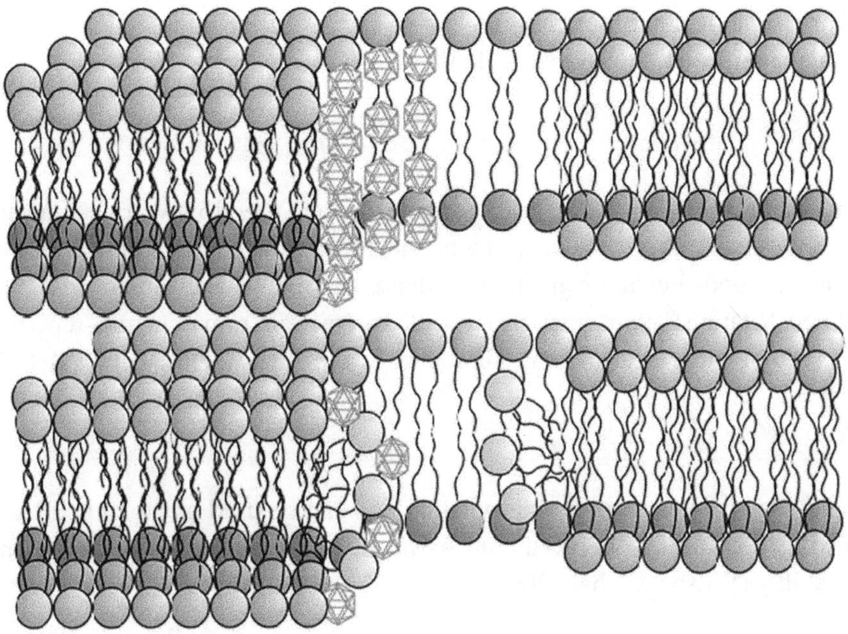

1. Liposome-Encapsulated Nutrients

One of the applications of liposomes is the delivery of some dietary and nutritional supplements. A liposome can be loaded with hydrophobic molecules in the phospholipid bilayer and/or hydrophilic molecules in the aqueous core (Liu, Ye, Liu, Liu, & Singh, 2013). Encapsulation of nutrients is a desirable approach in the food industry.

Liposomes are capable of enrichment of cheese with vitamins and minerals since various nutrients do not remain in food for a sufficient time. Prolonging the shelf life of cheese products can be attained by liposomes through limiting the interaction of nutrient with the external environment. On the other hand, encapsulation of cheese-ripening enzymes such as protease, peptidase or lipase in liposomes was found to speed up cheese ripening, which involves proteolysis and lipolysis reactions that give the desired smell, taste and texture of different types of cheese in addition to lowering the cost of cheese storage. The sustained release of encapsulated enzymes improved the enzyme distribution in the curd and reduced the enzyme loss in whey (Kheadr, Vuillemard, & El-Deeb, 2003; Mohammadi, Mahmoudzade, Atefi, Khosravi-Darani, & Mozafari, 2015).

2. Liposome-Encapsulated Antioxidants

Antioxidant therapy has drawbacks such as the poor solubility of antioxidants, their inability to cross membrane barriers, their rapid clearance from cells in addition to their tendency to be easily oxidized. The research is interested in the development of drug-delivery systems resulting in the selective delivery of antioxidants to tissues in sufficient concentrations to treat oxidant-induced tissue injuries. Improvement of the pharmacological and pharmacokinetic properties of the antioxidants can be achieved via encapsulating them in liposomes (Suntres, 2011).

a. Vitamin C

Liposomal formulation containing vitamin C was prepared and used to investigate the efficiency of its oral delivery. It was shown that oral delivery of vitamin C encapsulated in liposomes produced vitamin C concentrations in the blood circulation greater than those determined when un-encapsulated oral vitamin C was administered. Vitamin C-liposomes also provided protection from ischemia–reperfusion-mediated oxidative stress (Davis, Paris, Beals, Binns, Giordano, Scalzo, Schweder, Blair, & Bell, 2016).

b. Vitamin A

The hydrophobic vitamin A is an important vitamin and can be used as an antioxidant and as a dye. It was evidenced that encapsulation of vitamin A was effective against its oxidation and accordingly improving its stability (Wilson & Shah, 2007).

c. Vitamin E

It was also evidenced that α-Tocopherol, the major component of vitamin E, can be incorporated in the liposomal membrane leading to its stabilization. CoQ10 was used in cosmetics because of its ability to reduce wrinkle depth. Liposomal formulations containing α-Tocopherol and CoQ10 was prepared for the topical application and found to raise the CoQ10 content in rats' skin (Lee & Tsai, 2010).

d. Vitamin D3

It was implied that vitamin D3 properties against oxidation could be protected through its insertion in the hydrophobic lipid bilayer of nanoliposomes. The authors confirmed the stability of the aqueous dispersion of liposomal vitamin D3 and its ability to be used in beverage fortification. The absence of interactions between vitamin D3 and liposome's constituents was also established (Mohammadi, Ghanbarzadeh, & Hamishehkar, 2014).

e. Curcumin

Integration of curcumin, a hydrophobic polyphenolic compound, in the liposomal membranes elevated the curcumin's bioavailability and antioxidant capacity (Takahashi, Uechi, Takara, Asikin, & Wada, 2009). It was revealed that liposomal curcumin inhibited pancreatic carcinoma growth as well as angiogenesis (Mach, Mathew, Mosley, Kurzrock, & Smith, 2009).

f. Resveratrol

Resveratrol, a polyphenol found in grapes, is involved the cardiovascular protection due to its ability to inhibit the oxidation of low-density lipoproteins and inhibiting either platelet aggregation or proatherogenic eicosanoids production by human platelets and neutrophils (Brittes, Lúcio, Nunes, Lima, & Reis, 2010). Liposomal formulation containing resveratrol was established to be more effective in the improvement of cell injury than free resveratrol because of the low water solubility and stability of resveratrol (Kristl, Teskač, Caddeo, Abramović, & Šentjurc, 2009).

g. Astaxanthin

Astaxanthin, is a lipid-soluble carotenoid pigment with powerful antioxidant and antinflammatory properties. Astaxanthin's delivery in liposomes resulted in a reduction in its oxidation and accordingly an increase in its stability (Peng, Chang, Peng, & Chyau, 2010).

h. Glutathione (GSH)

It was found that dietary glutathione (GSH) was subjected to hydrolysis by intestinal and hepatic gamma-glutamyltransferase (Witschi, Reddy, Stofer, & Lauterburg, 1992), while a research showed that liposomal GSH was stable and efficient in treating mice having atherosclerosis via a reduction in oxidative stress (Rosenblat, Volkova, Coleman, & Aviram, 2007). Additionally, intratracheally administered GSH-liposomes improved lung injuries in animals due to the extended pulmonary retention of GSH-liposomes (Suntres & Shek, 1994). Likewise, liposomal formulations containing alpha/gamma-tocopherol alone or with N-acetylcysteine directly administered to the rats' trachea attenuated the fibrotic effects of lung injury (McClintock, Hoesel, Das, Till, Neff, Kunkel, Smith, & Ward, 2006).

3. Liposomes as Carriers for Enzymes

The limitations of the use of enzyme usage in clinical practice are due to their poor stability as well as their immunogenicity problems. Solutions to these problems are provided by liposomes as a delivery system for enzymes. It was proved that liposomal formulation containing uricase enzyme preserved uricase activity. The enables the catalysis of uric acid oxidation to allantoin, a soluble product readily excreted in urine. The enzymatic activity of uricase-containing liposomes was determined *in vitro* and then compared with the activity of the same amount of free uricase solution and it was evidenced that liposomal uricase was more effective than free uricase. Furthermore, the *in vivo* treatment using liposomal uricase was also found to reduce elevated serum uric acid concentrations when examined in a hyperuricemia rat model (Tan, Wang, Yang, Chen, Xiong, Zhang, Liu, Zhao, & Zhang, 2010). Furthermore, it was proved that Intravenous administration of liposomes containing superoxide dismutase (SOD) facilitated the delivery of SOD into the brain in both the infarct and noninfarcted subcortical area leading to significantly elevated levels of SOD activity and protection against neural injury (Imaizumi, Woolworth, Kinouchi, Chen, Fishman, & Chan, 1990). Further studies indicated that pegylated- catalase (PEG-catalase) and PEG-SOD liposomes were successfully used in the treatment of arthritis and lung injury in animals (Corvo, Boerman, Oyen, Bloois, Cruz, Crommelin, & Storm, 1999; McClintock, Hoesel, Das, Till, Neff, Kunkel, Smith, & Ward, 2006). It is noteworthy that inhaled nondrug-containing liposomes in healthy nonsmoking volunteers showed that liposome inhalation is well tolerated without side effects indicated that they are promising therapeutic agents (Thomas, Myers, Wichert, Schreier, & Gonzalez-Rothi, 1991).

4. Liposomes as Boron Delivery Vehicles for Boron Neutron Capture Therapy

For a successful boron neutron capture therapy, accumulation of ^{10}B in tumor to a larger extent than in the surrounding healthy tissue was necessary in order to destroy tumor cells while minimizing damage to healthy tissues (Barth, 1992). A major effort was directed toward the synthesis of boron-containing compounds representing boron carriers for this therapy. Dodecaborate cluster lipids are amphiphilic compounds that mimic the phospholipids and can accordingly be incorporated in liposomes. They are composed of the hydrophilic boron cluster in combination with a hydrophobic part that causes a different pharmacological effect than pure ionized cluster. They are also able to interact with the lipophilic part of the cell membrane leading to its disturbances (Justus, Awad, Hohnholt, Schaffran, Edwards, Karlsson, Damian, & Gabel, 2007; Schaffran, Jiang, Bergmann, Küstermann, Süss, Schubert, Wagner, Awad, & Gabel, 2014).

CONCLUSION

Bioavailability of various medications can be improved using liposomes due to their ability to deliver hydrophobic drugs to the site of disease. Liposomes are also able to reduce adverse effects of drugs via reducing their toxicity. Moreover, antioxidants-containing liposomes are valuable tools for protection against oxidative stress owing to the enhanced antioxidant properties of the liposomal formulation. Disruption of the liposomal bilayer membrane in order to trigger the release of their contents provides a highly advantageous approach in the drug delivery field. The oral administration and intravenous injection of various liposomal formulations, especially the targeted ones, are useful drug delivery approaches because of their ability to be successfully delivered to different types of cells.

REFERENCES

Allen, T. M. (1998). Liposomal drug formulations. Rationale for development and what we can expect for the future. *Drugs*, *56*(5), 747–756. doi:10.2165/00003495-199856050-00001 PMID:9829150

Allen, T. M., & Cullis, P. R. (2013). Liposomal drug delivery systems: From concept to clinical applications. *Advanced Drug Delivery Reviews*, *65*(1), 36–48. doi:10.1016/j.addr.2012.09.037 PMID:23036225

Awad, D., Bartok, M., Mostaghimi, F., Schrader, I., Sudumbrekar, N., Schaffran, T., ... Gabel, D. (2015). Halogenated dodecaborate clusters as agents to trigger release of liposomal contents. *ChemPlusChem*, *80*(4), 656–664. doi:10.1002/cplu.201402286

Bangham, A. D., Standish, M. M., & Watkins, J. C. (1965). Diffusion of univalent ions across the lamellae of swollen phospholipids. *Journal of Molecular Biology*, *13*(1), 238–252. doi:10.1016/S0022-2836(65)80093-6 PMID:5859039

Barth, R. F., Soloway, A. H., Fairchild, R. G., & Brugger, R. M. (1992). Boron neutron capture therapy for cancer. Realities and prospects. *Cancer*, *70*(12), 2995–3007. doi: PMID:1451084

Blume, G., & Cevc, G. (1990). Liposomes for the sustained drug release in vivo. *Biochimica et Biophysica Acta*, *1029*(1), 91–97. doi:10.1016/0005-2736(90)90440-Y PMID:2223816

Brittes, J., Lúcio, M., Nunes, C., Lima, J. L. F. C., & Reis, S. (2010). Effects of resveratrol on membrane biophysical properties: Relevance for its pharmacological effects. *Chemistry and Physics of Lipids*, *163*(8), 747–754. doi:10.1016/j.chemphyslip.2010.07.004 PMID:20691168

Chaidarun, S. S., Eggo, M. C., Sheppard, M. C., & Stewart, P. M. (1994). Expression of epidermal growth factor (EGF), its receptor, and related oncoprotein (erbB-2) in human pituitary tumors and response to EGF in vitro. *Endocrinology*, *135*(5), 2012–2021. doi:10.1210/endo.135.5.7956924 PMID:7956924

Charrois, G. J., & Allen, T. M. (2004). Drug release rate influences the pharmacokinetics, biodistribution, therapeutic activity, and toxicity of pegylated liposomal doxorubicin formulations in murine breast cancer. *Biochimica et Biophysica Acta*, *1663*(1-2), 167–177. doi:10.1016/j.bbamem.2004.03.006 PMID:15157619

Claesson-Welsh, L. (2015). Vascular permeability-the essentials. *Upsala Journal of Medical Sciences*, *120*(3), 135–143. doi:10.3109/03009734.2015.1064501 PMID:26220421

Corvo, M. L., Boerman, O. C., Oyen, W. J. G., Bloois, L. V., Cruz, M. E. M., Crommelin, D. J. A., & Storm, G. (1999). Intravenous administration of superoxide dismutase entrapped in long circulating liposomes. *In vivo* fate in a rat model of adjuvant arthritis. *Biochimica et Biophysica Acta*, *1419*(2), 325–334. doi:10.1016/S0005-2736(99)00081-4 PMID:10407083

Davidsen, J., Jorgensen, K., Andresen, T. L., & Mouritsen, O. G. (2003). Secreted phospholipase A_2 as a new enzymatic trigger mechanism for localised liposomal drug release and absorption in diseased tissue. *Biochimica et Biophysica Acta*, *1609*(1), 95–101. doi:10.1016/S0005-2736(02)00659-4 PMID:12507763

Davis, J. L., Paris, H. L., Beals, J. W., Binns, S. E., Giordano, G. R., Scalzo, R. L., ... Bell, C. (2016). Liposomal-encapsulated Ascorbic Acid: Influence on Vitamin C Bioavailability and Capacity to Protect Against Ischemia–Reperfusion Injury. *Nutrition and Metabolic Insights*, *9*, 25–30. doi:10.4137/NMI.S39764 PMID:27375360

Deshpande, P. P., Biswas, S., & Torchilin, V. P. (2013). Current trends in the use of liposomes for tumor targeting. *Nanomedicine (London)*, *8*(9), 1509–1528. doi:10.2217/nnm.13.118 PMID:23914966

Eliaz, R. E., & Szoka, F. C. Jr. (2001). Liposome-encapsulated doxorubicin targeted to CD44: A strategy to kill CD44-overexpressing tumor cells. *Cancer Research*, *61*(6), 2592–2601. PMID:11289136

Gabel, D., Awad, D., Schaffran, T., Radovan, D., Daraban, D., Damian, L., ... Edwards, K. (2007). The Anionic Boron Cluster (B(12)H(11)SH)(2-) as a means to trigger release of liposome contents. *ChemMedChem*, *2*(1), 51–53. doi:10.1002/cmdc.200600227 PMID:17154431

Gabel, D., Foster, S., & Fairchild, R. G. (1987). The Monte Carlo simulation of the biological effect of the 10B(n, alpha)7Li reaction in cells and tissue and its implication for boron neutron capture therapy. *Radiation Research*, *111*(1), 14–25. doi:10.2307/3577018 PMID:3602351

Gabizon, A., Horowitz, A. T., Goren, D., Tzemach, D., Shmeeda, H., & Zalipsky, S. (2003). In vivo fate of folate-targeted polyethylene-glycol liposomes in tumor-bearing mice. *Clinical Cancer Research, 9*(17), 6551–6559. PMID:14695160

Gabizon, A., & Papahadjopoulos, D. (1988). Liposome formulations with prolonged circulation time in blood and enhanced uptake by tumors. *Proceedings of the National Academy of Sciences of the United States of America, 85*(18), 6949–6953. doi:10.1073/pnas.85.18.6949 PMID:3413128

Guo, X., MacKay, J. A., & Szoka, F. C. Jr. (2003). Mechanism of pH-triggered collapse of phosphatidylethanolamine liposomes stabilized by an ortho ester polyethyleneglycol lipid. *Biophysical Journal, 84*(3), 1784–1795. doi:10.1016/S0006-3495(03)74986-8 PMID:12609880

Hope, M. J., Bally, M. B., Webb, G., & Cullis, P. R. (1985). Production of large unimellar vesicles by a rapid extrusion procedure. Characterization of size distribution, trapped volume and ability to maintain a membrane potential. *Biochimica et Biophysica Acta, 812*(1), 55–65. doi:10.1016/0005-2736(85)90521-8 PMID:23008845

Huang, S. L., & MacDonald, R. C. (2004). Acoustically active liposomes for drug encapsulation and ultrasound-triggered release. *Biochimica et Biophysica Acta, 1665*(1-2), 134–141. doi:10.1016/j.bbamem.2004.07.003 PMID:15471579

Imaizumi, S., Woolworth, V., Kinouchi, H., Chen, S. F., Fishman, R. A., & Chan, P. H. (1990). Liposome-entrapped superoxide dismutase ameliorates infarct volume in focal cerebral ischaemia. *Acta Neurochirurgica, 51*, 236–238. PMID:2089905

Justus, E., Awad, D., Hohnholt, M., Schaffran, T., Edwards, K., Karlsson, G., ... Gabel, D. (2007). Synthesis, liposomal preparation, and in vitro toxicity of two novel dodecaborate cluster lipids for boron neutron capture therapy. *Bioconjugate Chemistry, 18*(4), 1287–1293. doi:10.1021/bc070040t PMID:17569498

Kheadr, E. E., Vuillemard, J. C., & El-Deeb, S. A. (2003). Impact of liposome-encapsulated enzyme cocktails on Cheddar cheese ripening. *Food Research International, 36*(3), 241–252. doi:10.1016/S0963-9969(02)00166-7

Kristl, J., Teskač, K., Caddeo, C., Abramović, Z., & Šentjurc, M. (2009). Improvements of cellular stress response on resveratrol in liposomes. *European Journal of Pharmaceutics and Biopharmaceutics, 73*(2), 253–259. doi:10.1016/j.ejpb.2009.06.006 PMID:19527785

Lee, W. C., & Tsai, T. H. (2010). Preparation and characterization of liposomal coenzyme Q10 for *in vivo* topical application. *International Journal of Pharmaceutics, 395*(1-2), 78–83. doi:10.1016/j.ijpharm.2010.05.006 PMID:20635514

Leonetti, C., Scarsella, M., Semple, S. C., Molinari, A., D'Angelo, C., Stoppacciaro, A., ... Zupi, G. (2004). *In vivo* administration of liposomal vincristine sensitizes drug-resistant human solid tumors. *International Journal of Cancer, 110*(5), 767–774. doi:10.1002/ijc.20174 PMID:15146568

Liu, W., Ye, A., Liu, W., Liu, C., & Singh, H. (2013). Liposomes as food ingredients and nutraceutical delivery systems. *Agro Food Industry Hi-Tech, 24*(2), 68–71.

Mach, C. M., Mathew, L., Mosley, S. A., Kurzrock, R., & Smith, J. A. (2009). Determination of minimum effective dose and optimal dosing schedule for liposomal curcumin in a xenograft human pancreatic cancer model. *Anticancer Research, 29*(6), 1895–1899. PMID:19528445

McClintock, S. D., Hoesel, L. M., Das, S. K., Till, G. O., Neff, T., Kunkel, R. G., ... Ward, P. A. (2006). Attenuation of half sulfur mustard gas-induced acute lung injury in rats. *Journal of Applied Toxicology, 26*(2), 126–131. doi:10.1002/jat.1115 PMID:16252256

Meerum Terwogt, J. M., Groenewegen, G., Pluim, D., Maliepaard, M., Tibben, M. M., Huisman, A., ... Schellens, J. M. (2002). Phase I and pharmacokinetic study of SPI-77, a liposomal encapsulated dosage form of cisplatin. *Cancer Chemotherapy and Pharmacology, 49*(3), 201–210. doi:10.1007002800100371 PMID:11935212

Mohammadi, M., Ghanbarzadeh, B., & Hamishehkar, H. (2014). Formulation of nanoliposomal vitamin D3 for potential application in beverage fortification. *Advanced Pharmaceutical Bulletin, 4*(2), 569–575. PMID:25671191

Mohammadi, R., Mahmoudzade, M., Atefi, M., Khosravi-Darani, K., & Mozafari, M. R. (2015). Applications of nanoliposomes in cheese technology. *International Journal of Dairy Technology, 68*(1), 11–23. doi:10.1111/1471-0307.12174

Needham, D., & Dewhirst, M. W. (2001). The development and testing of a new temperature-sensitive drug delivery system for the treatment of solid tumors. *Advanced Drug Delivery Reviews, 53*(3), 285–305. doi:10.1016/S0169-409X(01)00233-2 PMID:11744173

Park, J. W., Hong, K., Carter, P., Asgari, H., Guo, L. Y., Keller, G. A., ... Benz, C. C. (1995). Development of anti-p185HER2 immunoliposomes for cancer therapy. *Proceedings of the National Academy of Sciences of the United States of America, 92*(5), 1327–1331. doi:10.1073/pnas.92.5.1327 PMID:7877976

Peng, C. H., Chang, C. H., Peng, R. Y., & Chyau, C. C. (2010). Improved membrane transport of astaxanthine by liposomal encapsulation. *European Journal of Pharmaceutics and Biopharmaceutics, 75*(2), 154–161. doi:10.1016/j.ejpb.2010.03.004 PMID:20226858

Rosenblat, M., Volkova, N., Coleman, R., & Aviram, M. (2007). Anti-oxidant and anti-atherogenic properties of liposomal glutathione: Studies *in vitro*, and in the atherosclerotic apolipoprotein E-deficient mice. *Atherosclerosis, 195*(2), e61–e68. doi:10.1016/j.atherosclerosis.2007.05.012 PMID:17588583

Sandra, A., & Pagano, R. E. (1979). Liposome-cell interactions. Studies of lipid transfer using isotopically asymmetric vesicles. *The Journal of Biological Chemistry, 254*(7), 2244–2249. PMID:429283

Schaffran, T., Jiang, N., Bergmann, M., Küstermann, E., Süss, R., Schubert, R., ... Gabel, D. (2014). Hemorrhage in mouse tumors induced by dodecaborate cluster lipids intended for boron neutron capture therapy. *International Journal of Nanomedicine, 9*, 3583–3590. PMID:25114527

Suntres, Z. E. (2011). Liposomal Antioxidants for Protection against Oxidant-Induced Damage. *Journal of Toxicology, 2011*, 1–16. doi:10.1155/2011/152474 PMID:21876690

Suntres, Z. E., & Shek, P. N. (1994). Incorporation of α-tocopherol in liposomes promotes the retention of liposome-encapsulated glutathione in the rat lung. *The Journal of Pharmacy and Pharmacology*, *46*(1), 23–28. doi:10.1111/j.2042-7158.1994.tb03714.x PMID:8201524

Takahashi, M., Uechi, S., Takara, K., Asikin, Y., & Wada, K. (2009). Evaluation of an oral carrier system in rats: Bioavailability and antioxidant properties of liposome-encapsulated curcumin. *Journal of Agricultural and Food Chemistry*, *57*(19), 9141–9146. doi:10.1021/jf9013923 PMID:19757811

Tan, Q. Y., Wang, N., Yang, H., Chen, L., Xiong, H. R., Zhang, L. K., ... Zhang, J. Q. (2010). Preparation and characterization of lipid vesicles containing uricase. *Drug Delivery*, *17*(1), 28–37. doi:10.3109/10717540903508953 PMID:22747073

Thomas, D. A., Myers, M. A., Wichert, B., Schreier, H., & Gonzalez-Rothi, R. J. (1991). Acute effects of liposome aerosol inhalation on pulmonary function in healthy human volunteers. *Chest*, *99*(5), 1268–1270. doi:10.1378/chest.99.5.1268 PMID:2019191

Torchilin, V., & Weissig, V. (2003). *Liposomes: a practical approach.* Oxford, UK: Oxford University Press.

Wilson, N., & Shah, N. P. (2007). Microencapsulation of Vitamins. *ASEAN Food Journal*, *14*(1), 1–14.

Witschi, A., Reddy, S., Stofer, B., & Lauterburg, B. H. (1992). The systemic availability of oral glutathione. *European Journal of Clinical Pharmacology*, *43*(6), 667–669. doi:10.1007/BF02284971 PMID:1362956

Chapter 7
Enzymatic Research Having Pharmaceutical Significance

Ishan H. Raval
Central Salt and Marine Chemicals Research Institute, India

Arvind Kumar Singh Chandel
Central Salt and Marine Chemicals Research Institute, India

ABSTRACT

The enzymes' biocatalysts act by lowering the activation energy without getting consumed in the reaction. The immense number of enzymes acts as a correctly matched orchestra to ensure that enormously complex life mechanisms and processes occur in a right direction. Sufficient quantity and accurate function of enzymes results in proper functional maintenance of body. The enzymes play a major role in the diagnosis, curing, biochemical investigation, and monitoring of many dreaded diseases of the century. The development of recombinant DNA technology had a significant impression on production levels of enzymes. Around 50% of the enzyme market is covered by recombinant enzymes. Because of development in molecular biology tools, several pharmaceutically enzymes have been identified and are being actively used in the pharmaceutical industry either for diagnostic or treatment. Information on this topic is very insufficient, and thus, the present chapter is an attempt to compile information on the sources, properties and applications of important therapeutic enzymes.

INTRODUCTION

In a few decades back the life expectancy of humans has increased from 45 years to 77 years globally because of significant advances in the medicine and pharmacy (Kola & Landis, 2004). The pharmaceutical industry is one of the fastest growing industries globally producing substantial revenue around the world (Kola & Landis, 2004). For the growth and development of any pharmaceutical industry, it must create unique products. Recombinant or therapeutically necessary enzymes come under the class of exclusive products which generate high revenue for the manufacturer. Pharmaceutical companies usually

DOI: 10.4018/978-1-5225-5237-6.ch007

invest money for very long time so to find essential factors for the diseases. Hence, the development of the novel drug targets and the lead compound is essential for humankind.

Drug targets can be of many kinds including receptors, transcription factors, growth regulators, cytokines, hormones, and enzymes. Enzymes are biocatalysts they catalyze most of the biological reactions by lowering the activation energy without interfering with initial and final energy states of the reaction. They are most significant drug targets as they play an essential role in metabolic pathways which are the culprit of the metabolic anomalies like cancer and neurodegenerative diseases and metabolic disorders.

There are so many enzymes that play a part in the cancer GTPases are a group of enzymes that play a crucial role as they involved in Ras-MAP kinase pathway (Bourne, Sanders, & McCormick, 1990, 1991; Hall, 1990). The GTPases activate Ras cancerous cells require a high amount of activation of this pathway for the rapid proliferation and survival of the cells thus, enzyme modifying the Ras can be used for the treatment of cancer. Previously described studies suggest the modification of C-terminal activating enzymes of Ras to be used as drug targets for cancerous cells (Gibbs, 1991). Apart from GTPase Farnesyl protein transferase also plays a key role in cancer progression by poly isoprenylation of the group to CAXX residues on the Ras (Goodman, Judd, Farnsworth, Powers, Gelb, Glomset, & Tamanoi, 1990; Schaber, O'hara, Garsky, Mosser, Bergstrom, Moores, ... Gibbs, 1990). Phosphorylation of proteins is essential. Initially, it was thought that phosphorylation was important only for the carbohydrate metabolism (Cohen & Frame, 2001). However studies of the role of phosphorylation started very early in the 80s, and researchers found that phosphorylation was an important factor for the cancer progression, one of the viral factor or Rous sarcoma virus (V-src) was found to be phosphorylation factor (Castagna, Takai, Kaibuchi, Sano, Kikkawa, & Nishizuka, 1982; Collett & Erikson, 1978). In 1980s Hiroyoshi discovered that some calmodulin binding inhibitors were found to inhibit the phosphorylation. A new generation of inhibitors found which inhibited phosphorylation by binding with CAMP and cGMP rather than binding with the calmodulin proteins. (Hidaka, Inagaki, Kawamoto, & Sasaki, 1984). Cyclooxygenases discovered in 1976. Further studies on them revealed that cyclooxygenase 2 was inducible by the proinflammatory cytokines. Both the cyclooxygenases (COX 1 & COX 2) play a role in the synthesis of prostaglandins. Nimesulide is an excellent example of COX 2 inhibitor which is used as the nonsteroidal anti-inflammatory drug (Vane, Bakhle, & Botting, 1998).

Carbohydrate metabolism is essential for any living organisms because it produces sources of energy for the survival reproduction and development of all the organisms. Parasites like Trypanosoma can evolve very rapidly making them tough to target by the immune system, so researchers decided to target the carbohydrate metabolism of the parasites to stop the infection (Opperdoes & Michels, 2001). For example, the glucose transporter of parasites govern more than 50% of metabolism of these organisms by targeting this receptor carbohydrate metabolism can be easily regulated (Bakker, Walsh, Ter Kuile, Mensonides, Michels, Opperdoes, & Westerhoff, 1999). Glucose transporter like GLUT 2 is known to play the significant role in the diabetic conditions in humans and drugs which target inhibiting the glucose transporter-like Canagliflozinare used as a treatment of Diabetes (Sarnoski-Brocavich & Hilas, 2013). NADP oxidases (NOX) are a class of enzymes essential for the phagocytic activities of macrophages, but earlier, only oxidases were believed to be responsible for the phagocytic activities. However, in the twenty-first-century, the scientist discovered the class of enzymes that playing a role in the phagocytic processes from there on the NOX became so important in the study of the immune system. There are many isoforms like Nox1, Nox2, Nox3, Nox4, Nox5, Duox1, and Duox2 which are found in humans and other isoforms are found in eukaryotes (Bedard & Krause, 2007; Krause, Lambeth, & Krönke, 2012; Lambeth, 2004). In short, the enzymes will play a crucial role in all the metabolic processes and

Table 1. Enzymes, their function and possible clinical significance in various conditions.

Name of Enzyme	Function	Associated Condition
Aldehyde Dehydrogenase	Oxidation	Alcohol flu reaction
Monoamine Oxidases (MAOs)	Oxidation of monoamines	Depression, Parkinson's disease
Cyclooxygenases (COXs)	Prostaglandin synthesis	Inflammation neurodegenerative disease.
Aromatase	Estrogen production	Breast cancer
Lipoxygenases	Deoxygenation of polyunsaturated fatty acid	Asthma, Crohn's disease
Thyroidal Peroxidase	Oxidation of iodide to iodine	Auto immune thyroid disease
Iodothyronine-5′Deiodinase	Oxidation of thyroid precursors	Hypothyroidism
Inosine-5′Monophosphate Dehydrogenase	Purine biosynthesis	Auto immune conditions
HMG-CoA Reductase	Production of cholesterol and isoprenoid	Cholesterol deposition conditions
5α-Testosterone Reductase	Metabolism of testosterone	Prostate enlargement baldness
Dihydrofolate Reductase (Bacterial Origin)	Folic acid synthesis	Bacterial infection
Dihydrofolate Reductase (Human Origin)	Tetrahydrofolate synthesis	Megaloblastic anaemia
Enoyl Reductase (Mycobacterial)	Fatty acid synthesis	Tuberculosis
Squalene Epoxidase (Fungal Origin)	Oxidation of squalene	Jock itch, Common fungal infection
Δ14 Reductase (Fungal Origin)	Ergosterol metabolism	Fungal infection on toe and nails
Xanthine Oxidase	Hypoxanthine oxidation	Usually increased during influenza
4-Hydroxyphenylpyruvate Dioxygenase	Tyrosine catabolism	Type III tyrosinemia
Ribonucleoside Diphosphate Reductase	Synthesis of deoxyribonucleotides from ribonucleotides	sickle-cell disease, chronic myelogenous leukaemia, cervical cancer, and polycythemia vera.
Protein Kinase C	Regulation of metabolic pathways	Plays a role in cancer progression
Peptidyl Transferase (Bacterial Origin)	Protein synthesis	Bacterial infections
Catecholamine-O-Methyltransferase	Inactivation of catecholamine neuro transmitters	Parkinson's disease
RNA Polymerase (Bacterial Origin)	Transcription of bacterial genes	Bacterial infections
Reverse Transcriptase	Reverse transcription of viral genes	AIDS
DNA Polymerases (Viral Origin)	Replication of viral DNA	Herpes, chickenpox etc.
GABA Transaminase	Succinate semialdehyde synthesis	Epilepsy
Tyrosine Kinases	Regulation of gene expression and other metabolic pathways	Cancer
Aspartyl Proteases (Retro Viral Origin)	Synthesis of functional retroviral proteins	AIDS
Serine Proteases (Bacterial Origin)	Bacterial cell wall synthesis	Bacterial infections.
Bacterial Lactamases	Degradation of β-lactam containing antibiotics	Bacterial infections.
Human Coagulation Factor	Blood clotting	Haemophilia

continued on following page

Table 1. Continued

Name of Enzyme	Function	Associated Condition
Human Angiotensin Converting Enzyme	Angiotensin activation	Heart attack
Human Carboxypeptidase A (Zn)	Activation of peptides	Kidney stones caused by cysteine and copper metabolic disorders
Human Enkephalinase	Degradation of enkaphaline opioids peptides	Diahhoreha
Acetylcholinesterase	Transmission of signals in neurones by breaking down acetyl choline to produce neurotransmitters	Glaucoma, paralysis, Alzheimer's disease etc.
Histone Deacetylases	Deacetylation of histone	Alzheimer's disease, Parkinson's disease
Viral Neuraminidase	Viral replication	Influenza
Histidine Decarboxylase	Decarboxylation of histidine to produce histamine	Cough common influenza
DNA Gyrases (Bacterial Origin)	Relaxes supercoiling	Bacterial infections
Mammalian Target of Rapamycin	Regulation of cell growth cell proliferation and various signalling pathways	Cancer, Tuberculosis etc.
Thymidylate Synthase	Catalyse the conversion of dUMP to dTMP	Cancer
Carbonic Anhydrase	Formation of bicarbonate	Glaucoma, altitude sickness

anomalies in these enzymes can lead to the various condition in metabolic processes and disease related to the function of these enzymes are stated in Table 1.

Apart from being used as drug targets these enzymes also used as the therapy itself, for example, the Adagen (pegadamase bovine), was utilized for the treatment of SCID, this represents the first successful application of an enzyme therapy for a congenital disease. The enzyme ADA cleaves the additional adenosine present in the blood circulation of these patients and decreases the toxicity to the immune system of the raised adenosine levels. The achievement of the treatment depends upon the alteration of ADA with PEG. PEG increases the half-life of the enzyme (initially less than 30 min) and decreases the chance of immunological reactions because of the bovine origin of the drug. The effective entrapment of native ADA in carrier erythrocytes also improves the half-life of the enzyme significantly.

Ceredase (alglucerase injection) for the treatment of Gaucher disease, a lysosomal storage disease (LSD), this was the first enzyme replacement therapy in which an exogenous enzyme targeted to its right compartment within the body. The effort to replace the missing glucocerebrosidase in Gaucher patients was started by Brady and colleagues, utilizing altered placental glucocerebrosidase (Ceredase1). The recombinant DNA technology afterward allowed the more efficient production of a glucocerebrosidase, Ceredase (imiglucerase), which approved in 1994. This medical success has covered the way for other enzyme therapies, in mainly those for other LSDs. An additional LSD that has attracted the interest of pharmaceutical companies is Fabry's disease, fat (glycolipid) storage disorder caused by a deficiency in a-galactosidase (Vellard, 2003). Table 2 indicates the use of enzymes in the treatment of various diseases.

TYROSINE KINASES

Tyrosine kinases are the class of enzymes involved in the regulation of many metabolic pathways. Tyrosine kinases transfer phosphate group on various molecules making the signaling pathways work flawlessly. A malfunction in the activity of this enzyme leads to cancer making it a significant drug target. Various inhibitors are found which can target these enzymes thus, providing potential treatment of deadly disease (Paul, 2004). Tyrosine kinases are the class of enzyme which is subclass of transferase and sub-subclass of kinases. Tyrosine kinases individually transfer the phosphate group to the tyrosine amino acid. The eukaryotic cell is a complex system and requires various modification in the proteins and tyrosine kinases plays key role in the post-translational modification of different proteins. Those proteins are essential for

Figure 1. Illustrates the area of enzyme application as therapeutics

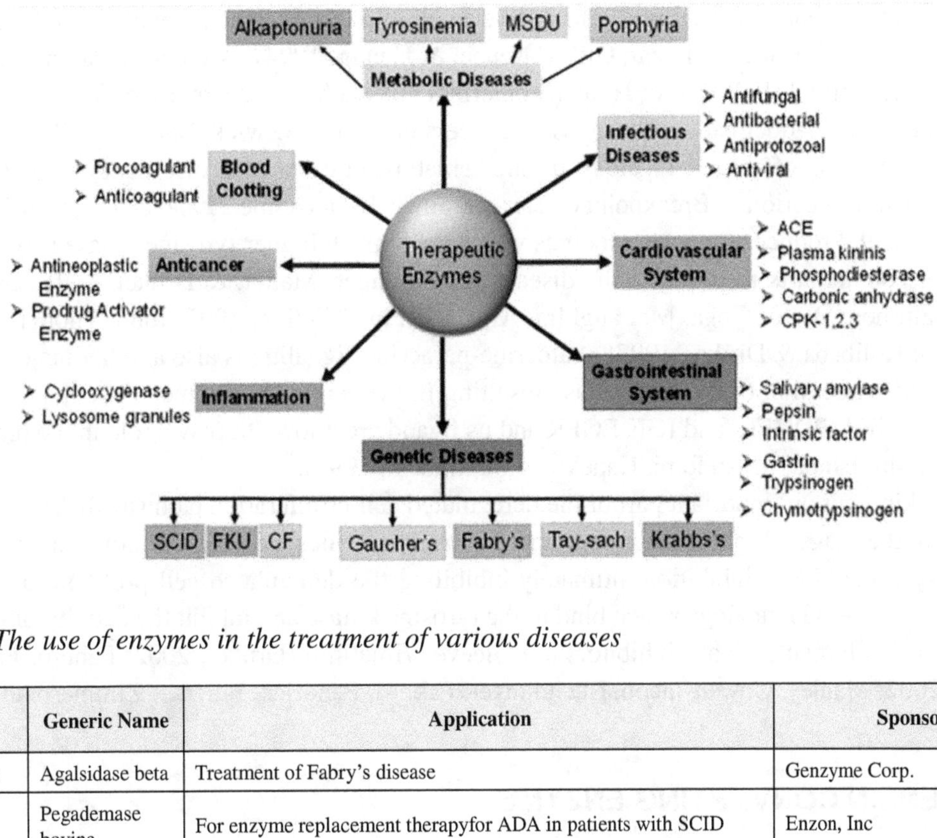

Table 2. The use of enzymes in the treatment of various diseases

Trade Name	Generic Name	Application	Sponsor
Fabrazyme	Agalsidase beta	Treatment of Fabry's disease	Genzyme Corp.
Adagen	Pegademase bovine	For enzyme replacement therapyfor ADA in patients with SCID	Enzon, Inc
Ceredase	Alglucerase injection	For replacement therapy in patients with Gaucher's disease type I	Genzyme Corporation
Cerezyme	Imiglucerase	Replacement therapy in patients with types I,II, and III Gaucher's disease	Genzyme Corporation
Pulmozyme1	Dornase a	To reduce mucous viscosity and enable the clearance of airway secretions in patients with CF	Genentech, Inc.
Oncaspar	Pegaspargase	Treatment of acute lymphocytic leukemia	Enzon, Inc
Aldurazyme	Laronidase	Treatment of patients with MPS I	BioMarin Pharmaceutical, Inc

the proper metabolism, migration, growth, and differentiation tyrosine kinase regulates the proliferation of the cells by getting involved in the apoptotic pathways (Hunter, 2000; Schlessinger, 2000).

Tyrosine kinases play key role in the neoplastic transformation of a cell as they regulate the cellular processes such as apoptosis. However, they generally regulate the deregulated cell proliferation, but the mutation in them can activate the unregulated cell proliferation (Blume-Jensen & Hunter, 2001). Tyrosine kinases are generally activated by the receptor-ligand binding resulting in the conformational change in the receptor. Ligand-receptor binding can be 1 ligand: 2 receptors (growth hormone and its receptor) or 2 ligands: 2 receptors (vascular endothelial growth receptor and vascular endothelial growth factor). Studies have suggested that binding of receptors to its ligands increases the concentration of receptor tyrosine kinases around the intracellular domains of the receptors. Increased concentration of tyrosine kinases around receptor activates the pathways. Interactive studies have suggested that tyrosine kinases transfer the ATP to tyrosine with the help of Mg2+ mediated mechanism (Paul, 2004). One of the main causes of cancer is the mutation in the tyrosine kinase-coupled receptors, for example, epidermal growth factor variant III (EGFR v III) can lead to glioblastoma by the deregulation of the tyrosine kinase (Nishikawa, Ji, Harmon, Lazar, Gill, Cavenee, & Huang, 1994). A point mutation in fibroblast growth factor receptor 3 (FGFR 3) or Human epidermal growth factor receptor 2 or 3 (EGFR 2&3) are important factors in bladder cancer and cervical cancers in humans (Zwick, Bange, & Ullrich, 2002).

Ninety-five percent of chronic myelodysplastic hematopoietic stem cell disorder syndrome happens because of the translocation of Breakpoint cluster region of chromosome 22 and c-ABL tyrosine kinase of chromosome 9. Produces the gene products which have much higher tyrosine kinase activities than their regular counterparts resulting in the disease (Buchdunger, Matter, & Druker, 2001; Daley, Van Etten, & Baltimore, 1990; Goga, McLaughlin, Afar, Saffran, & Witte, 1995; John, Thomas, Muft, & Padua, 2004; Kolibaba & Druker, 1997). Autocrine-paracrine signaling is also another major cause for the continuous expression of tyrosine kinase resulting in the carcinomas some of the major receptors involved are EGFR, PDGFR, and IGF. EGFR and its ligand are known to play a role in the lung cancer as well (Tateishi, Ishida, Mitsudomi, Kaneko, & Sugimachi, 1990).

Tyrosine kinase is an important part of the deregulated cell proliferation pathway makes it an excellent target for the drugs inhibiting the carcinoma. Several molecules have been discovered which shows promising tyrosine kinase inhibition ultimately inhibiting the deregulated cell proliferation. Most of these molecules are ATP analogs which bind to the tyrosine kinase and inhibit the activity of the same. Examples of this tyrosine kinase inhibitors are Gleevec /Imatinib (Druker, 2002; Fabbro, Parkinson, & Matter, 2002; Manley, Cowan-Jacob, Buchdunger, Fabbro, Fendrich, Furet,... Zimmermann, 2002)

ANGIOTENSIN CONVERTING ENZYME

Angiotensin converting enzyme (ACE) was first discovered by Skeggs and colleagues, and this enzyme can convert and I to ang II (called hypertensin I and hypertensin II at that time). This enzyme was termed "converting enzyme". ACE is a dipeptidyl carboxypeptidase that is present in two isoforms. ACE 1 also known as somatic ACE. Somatic ACE is expressed in many tissues and cell types including the kidneys, intestine, cardiovascular system, liver, uterus, adrenal gland, etc. although testicular ACE only can be found in sperm cells. Both ACE isoforms possess a short cytoplasmic fragment and hydrophobic transmembrane domain a. Somatic ACE has two homologous domains with two catalytic sites and Zn2+ binding regions. While testicular ACE only has one only one catalytic site and extracellular domain. The

Figure 2. The X-Ray crystal structure of the active EGFR kinase domain. Insulin-like growth receptor (IGFR) co-expression also observed in the development of breast carcinoma (Reeve, Morgan, Schwander, & Bleehen, 1993). An elevated level of phosphorylation of IGF-I was also found in the breast cancer (Resnik, Reichart, Huey, Webster, & Seely, 1998).

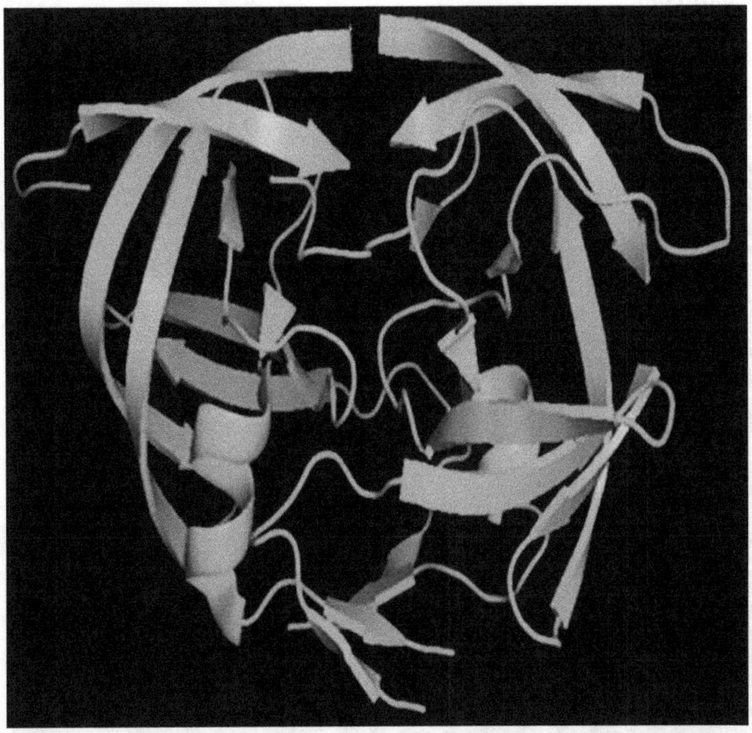

distribution of ACE 2 is more limited compared to ACE and has been identified in the human heart, testis, and kidney. In the vascular system, ACE occurs circulating in the blood and bound to the membrane of various cell types, including endothelial cells of blood vessels. ACE is situated on the luminal side of the endothelium with the C terminal domain to the plasma membrane. The blood circulating ACE initiates mainly from endothelial cells of blood vessels and has been released into the blood after the proteolytic cleavage of the C terminal domain. Due to the high vascularization of the ACE, the lungs are the major site for production of circulating ACE. Also being a vital enzyme for the transformation of and I to Ang II and degradation of bradykinin, ACE acts on many other natural substrates, including neurotensin, enkephalin, and substance P32 (Costerousse, Allegrini, Huang, Bounhik, & Alhenc-Gelas, 1994; Schunkert, Ingelfinger, Hirsch, Pinto, Remme, Jacob, & Dzau, 1993).

ACE is a type of zinc metalloenzyme; here zinc ion is vital for its activity by peptide hydrolysis. Consequently, ACE can be inhibited by metal-chelating agents. The E 384 residue of ACE has a dual function. First, it activates water as a nucleophile and second acid to cleave the C-N bond. The role of the chloride ion is very complex and is highly argued. The anion activation by chloride is a distinguishing feature of ACE. It experimentally determined that the activation of hydrolysis by chloride is fully dependent on the substrate. Although it increases hydrolysis rates, e.g. Hip-His-Leu, it inhibits hydrolysis of other substrates like Hip-Ala-Pro. Under the physiological conditions, ACE enzyme reaches about 60% of its maximal activity toward angiotensin I even though it reaches its full activity toward bradykinin. It is

Figure 3. These inhibitors bind irreversibly to the tyrosine kinase and prevent its activity ultimately blocking the metabolic pathway, In this case, de-regulated cell proliferation. Another example is Iressa which inhibits the EGF receptor tyrosine kinase (Fukuoka, Yano, Giaccone, Tamura, Nakagawa, & Douillard, 2003). There are so many tyrosine kinase blockers being under investigation that will be for the betterment of humankind.

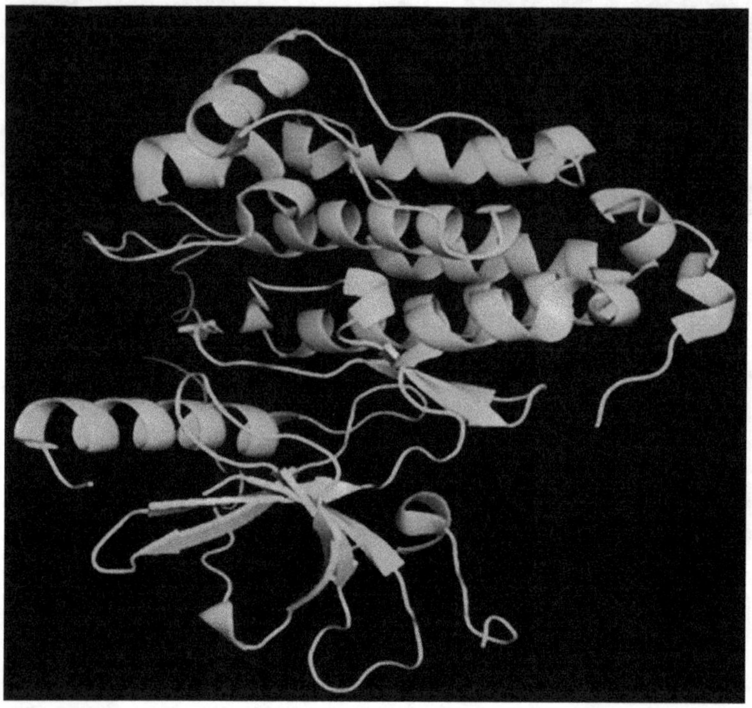

consequently assumed that the function of the anion activation in ACE delivers high substrate specificity (Costerousse, Allegrini, Huang, Bounhik, & Alhenc-Gelas, 1994; Masuyer, Schwager, Sturrock, Isaac, & Acharya, 2012). The regulation of ACE by large inter-individual variation of blood circulating ACE. Around 20-50% of the deviation can be accounted for by the ACE I/D polymorphism. However, serum ACE activity seems to be relatively stable when measured in the same individual at different occasions. ACEi changes the activity and expression of ACE in vitro and vivo. The straight effect of ACEi treatment is a reduced serum ACE activity (Schunkert, Ingelfinger, Hirsch, Pinto, Remme, Jacob, & Dzau, 1993).

These are the some ACE inhibitor and their brand name respectively, Ramipril (Altace/Prilace/Triatec/Tritace), Enalapril (Vasotec/Renitec /EnalaprilProfarma) Quinapril(Accupril),Ramipril(Altace/Prilace/Triatec/Tritace),Lisinopril(Listril/Lopril/Zestril,)Perindopril (Coversyl/Aceon/Perindo)Benazepril (Lotensin),Imidapril (Tanatril). At the beginning, the ACE inhibitors were approved for the treatment of hypertension and can be used the combination or alone in with another antihypertensive medications. After that, they were found beneficial for other cardiovascular and kidney diseases including Kidney complications of diabetes mellitus (diabetic nephropathy) acute myocardial infarction (heart attack) Cardiac failure (left ventricular systolic dysfunction).

Figure 4. Elucidates the functional involvement of angiotensin-converting enzyme in the human body

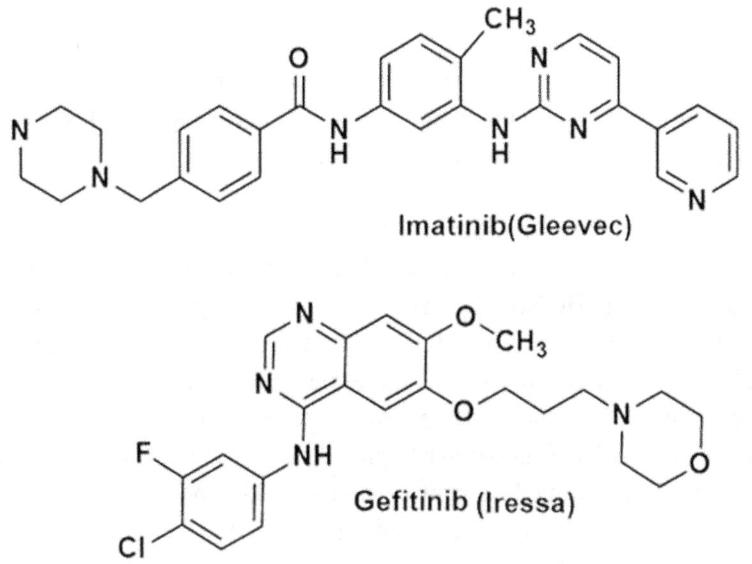

CYCLOOXYGENASES

Cyclooxygenases are the class of enzymes that synthesize the prostaglandins by oxidation of arachidonic acid. The major, studied isoforms of cyclooxygenases are COX 1 and COX 2. COX 2 is inducible by various signals from the surrounding cells for example during inflammation COX 2 expression is induced which ultimately induces the pain in the affected area. Cyclooxygenases are enzymes that are synthesized in many locations around the body such as stomach, kidney, uterine, platelets. Their function in stomach includes the production of cytoprotective prostaglandins these are mainly synthesized by COX 1 however small amount of COX 2 expression is also observed (Kargman, Charleson, Cartwright, Frank, Riendeau, Mancini, ... O'Neill, 1996). Cyclooxygenases are most important in the case of damaged renal function because they synthesize prostaglandins which regulate the blood flow in the in the damaged kidneys (Harris, McKanna, Akai, Jacobson, Dubois, & Breyer, 1994). In uterine cyclooxygenases induce the contraction by the production of prostaglandins and they also help in the induction of labor during childbirth. Apart from these organs cyclooxygenases play an important function in the central nervous system. Induction of fever and pain throughout the body is regulated by the cyclooxygenases COX 1, and Cox 2. There are so many molecules available in the market they inhibit the activity of COX 1 and COX 2 selectively or non-selectively. Examples of non-steroidal anti-inflammatory drugs (NSAIDs) include as Nimesulide and Aspirin are a good representative of NSAIDs Nimesulide is selective COX 2 inhibitor whereas Aspirin is non-selective COX 2 inhibitor both the drugs can be used as anti-inflammatory and antipyretic.

NADPH OXIDASES (NOX)

NADPH oxidases (NOX) are a group of enzymes associated with membranes of mesodermal cells. NADPH oxidases (NOX) mostly found in the phagocytes and B-lymphocytes. NOX produces free superoxide radicals by oxidation that help in the phagocytosis of the pathogens. This superoxide contributes to creating other Reactive oxygen species (ROS) which in turn assist in the phagocytosis. However, these ROS generated for phagocytosis damages surrounding cells also that's why tight regulation of NOX is required for the proper immunological function of phagocytes. Researchers discovered that enzyme comprises five components namely p40 PHOX (PHOX for PHagocyteOXidase), p47PHOX, p67PHOX, p22PHOXand gp91PHOX.In the resting cell(not doing phagocytosis), three of these five components p40PHOX, p47PHOXand p67PHOX exist in the cytosol as a complex and the other two elements p22PHOXand gp91PHOX are located in the membranes (Babior, 1999).

NOX play key role in the alcohol-induced hepatic diseases (Kono, Rusyn, Yin, Gäbele, Yamashina, & Dikalova, 2000) apart from that NOX are known to play the key role in the Parkinson's disease as well as coronary diseases (Inoue, Kawashima, Kanazawa, Yamada, Akita, & Yokoyama, 1998; Wu, Teismann, Tieu, Vila, Jackson-Lewis, Ischiropoulos, & Przedborski, 2003). There are some natural flavonoids which inhibit the activity of NOX, but the discovery of inhibitors of the NOX is an emerging field and can lead to the discovery of potent inhibitors which can help in various conditions.

ACETYLCHOLINESTERASE

Acetylcholinesterase is an enzyme which hydrolyses acetylcholine and other choline esters. It plays an important role in the production of neurotransmitters. It is found at the neuro-muscular junction of cholinergic neurons and also found in between neurons. Acetylcholinesterase has a very high turnover number it converts approximately 25000 acetylcholine molecules per second (Quinn, 1987). There are some pesticides which inhibit the Acetylcholinesterase such as Diazinon resulting in the paralysis of muscles or accumulation of the acetylcholine between neurons resulting in the choline plaques. Choline

Figure 5. Represents the structures of Aspirin and Nimesulide

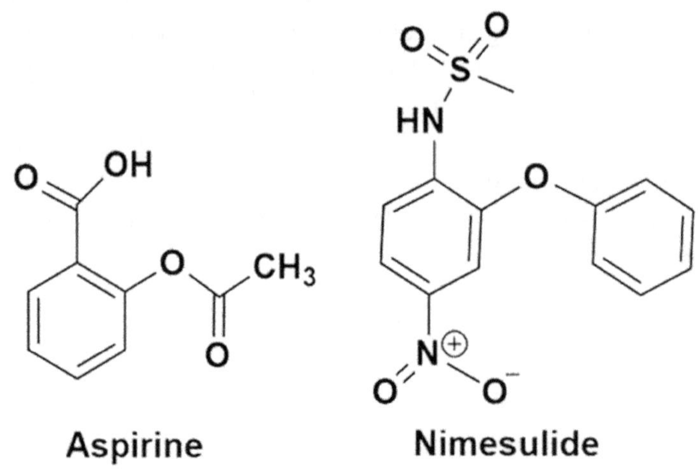

plaques are generally observed in the Alzheimer's disease. There are some inhibitors in the market which are used for treatment of Alzheimer's disease as well as depression for example Donepezil, which is marketed under the trade name Aricept is used for the treatment of Alzheimer's.

AMYLASE

Amylases are the class of enzymes which convert the complex starch into simple sugars like glucose and fructose. There are mainly three isoforms of this enzymes α-amylase, β-amylase, and γ-amylase. α-amylase hydrolyze the starch from the random location along the chain. β-amylase catalyze hydrolysis from the second α 1-4 linkage on the non-reducing end of the sugar. γ-amylase catalyze the hydrolysis from α 1-6 linkage of the starch molecule. Application of the amylase varies like they are widely used in juice production starch industries food processing industry diagnostic applications (Gurung, Ray, Bose, & Rai, 2013; Mohan, 2000). Apart from these amylases are also used in diabetes and cancer research as well for diagnostic purposes (Nielsen & Borchert, 2000).

GLUCOSE OXIDASE

Glucose oxidase is an enzyme which oxidizes glucose to δ-gluconolactone and hydrogen peroxide. The enzyme is mostly produced as the defense mechanism by some fungi and insect to get rid of bacteria in the presence of glucose and oxygen. It requires Flavin adenine dinucleotide (FAD) as co-factor to convert the D-glucose to δ-gluconolactone and peroxide. Application of this enzyme in the pharmaceutical industry is to measure the serum glucose level of the patients it can work for colorimetric assays as well as bio-sensors which represent the blood glucose level in digital glucometer and helps keep track of blood sugar for diabetic patients (Cass, Davis, Francis, Hill, Aston, Higgins, … Turner, 1984). Dipeptidyl peptidase 4 is also another enzyme which can be targeted as the treatment of diabetes research has suggested that targeting this enzyme has less side effect than conventional antidiabetic drugs (Parmar, Bhinchar, Bhatia, Chordia, Raval, Chauhan, … Kumar, 2014).

ASPARTYL PROTEASES

Aspartyl proteases are found in both prokaryotes eukaryotes and viruses. However presently will focus only of viral and bacterial proteases as they play a significant role in the pathogenesis of the viruses and bacteria. These enzymes contain an aspartyl molecule grouped with water, and they hydrolyze the protein. Aspartyl proteases are part of the viral gag or pol genes. For example, HIV-1 protease it cleaves essential proteins of gag and pol genes to make the functional virus particles. These virus particles will again infect new cells and cause AIDS (Acquired Immuno Deficiency Syndrome) this make it an excellent drug target for the treatment of HIV infection.

There are so many molecules available in the market like Saquinavir, Ritonavir (Kempf, Norbeck, Sham, & Zhao, 1996), Indinavir (Vacca, Dorsey, Guare, Holloway, Hungate, & Levin, 1995), etc. which effectively inhibit the aspartyl proteases of HIV. However, HIV is a very rapidly mutating retrovirus making them ineffective with the time new molecules must be discovered to prevent virus infection.

Figure 6. Represents the crystal structure of HIV-1 protease

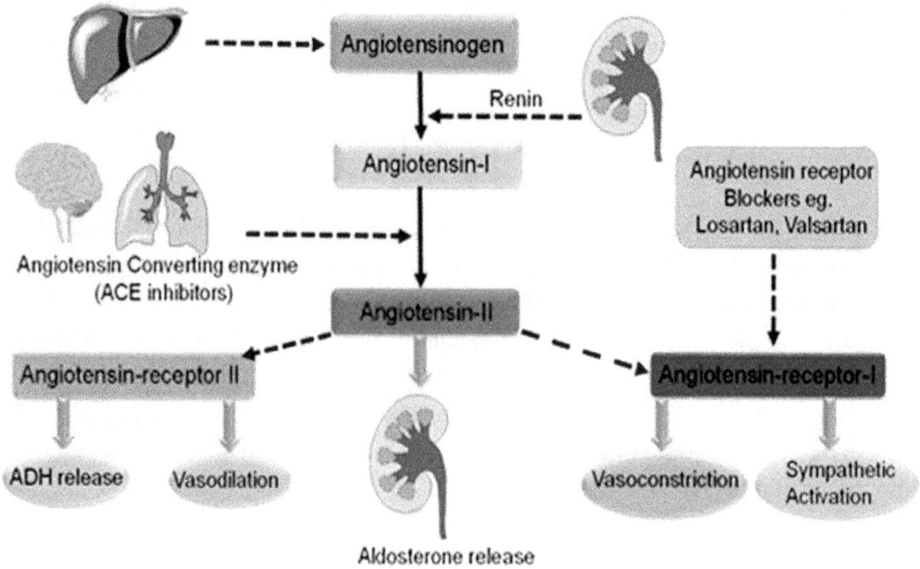

TRANSAMINASES

There are many transaminases, available in the market but here only two important transaminases are discussed Aspartate transaminase (AST) and Alanine transaminase (ALT). In 1954 Arthur Karmen and colleagues discovered these enzymes from the serum of human blood (Karmen, Wróblewski, & LaDue, 1955). Aspartate transaminase is an enzyme which transfers amino group between aspartate and glutamate reversibly using pyridoxal phosphate as co-factor. Aspartate transaminase is found in liver, Brain, heart and skeletal muscles however it mainly indicates the health of the liver. Alanine transaminase also requires pyridoxal phosphate as co-factor for the transfer of an amino group from L-Alanine to α-ketoglutarate. Alanine transaminase can be found in the plasma of all tissues, but mostly it's located in the liver. The ratio of AST/ALT is used for the assessment of liver health for example during the Hepatitis virus infection these enzyme activities increase than their normal level indicating hepatic inflammation.

RECOMBINANT ENZYMES

Recombinant DNA technology has significantly evolved over the period. Recombinant DNA technology has let the cloning of important enzymes into cell factories which produce the high level of cloned proteins in less time and with high efficiency. There are so many humanized proteins produced using this technology some examples are stated below. Recombinant human erythropoietin has been widely used for the treatment of anemia as well as damaged kidneys however it's very important for the normal function of almost all tissues as it helps in generation of red blood cells (Coleman & Brines, 2004).

Human insulin is another great example of recombinant proteins. Initially, it was produced using the cultured mammalian cells later it was cloned into a bacterium and was produced using either two-chain variant or as proinsulin, and it has provided great relief to the diabetic patients worldwide (Ladisch & Kohlmann, 1992). Tumor necrosis factor-α (TNFα) is essential protein because it controls the proliferation of transformed cells and can help in controlling the proliferation of cancerous cells (Sugarman, Aggarwal, Hass, Figari, Palladino, & Shepard, 1985). Recombinant thyroid stimulating hormone also has attracted attention as it can be used as a supplement after removal of thyroid in case of carcinomas however the use of recombinant thyroid hormone instead of Radioiodine remnant ablation is still controversial (Robbins, Larson, Sinha, Shaha, Divgi, Pentlow, … Tuttle, 2002).

Ornithine Decarboxylase

Ornithine decarboxylase is an enzyme that the rate-limiting for the polyamine biosynthesis pathway in the various organisms. Which acts a vital role in the synthesis of the polyamine such as trypanothione, T (SH) 2, it is the primarily reduced thiol which is responsible for the modulation of the various immune response. Also involved in the pathogenesis in visceral leishmaniasis and another parasitic disease. This enzyme mainly act through up-regulated IL-10 production decreased IL-12, nitric oxide production and *IFN*-γ production from CD4+ T cells Bimal et al. proved that approximately 77 kDa and examined its effects on the immunological responses in peripheral blood mononuclear cells of human visceral leishmaniasis cases. For these studies, the a-specific enzyme inhibitor of difluoromethylornithine (DFMO) tested as an inhibitor and used in parallel in all experiments. The recombinant Ornithine decarboxylase identified as having a direct correlation with parasite growth and significantly increased the number of promastigotes as well as axenic amastigotes after 96 h of culture. And also proved DFMO which can reduce the activity of ornithine decarboxylase of L. *donovani*, eliminating the parasite-induced immune suppression and inducing collateral host protective responses in visceral leishmaniasis.

CONCLUSION

Enzymes are the most important catalysts of mammalian systems any malfunction in their functions can lead to various disease such as cancer, diabetes, etc. However, enzymes are also essential drug targets that can be used for the treatment of deadly disease like AIDS, cancer, hypertension, etc. Not only they are excellent drug targets, but also, they can itself act as the therapy and some of them like insulin, TNFα, etc. are being used for the treatment of disorders.

ACKNOWLEDGMENT

IR is thankful for financial support from DST-INSPIRE fellowship. AK is thankful for financial support from CSIR. Arvind Kumar Singh Chandel served in the role of corresponding author during the work on this chapter.

REFERENCES

Babior, B. M. (1999). NADPH oxidase: An update. *Blood*, *93*(5), 1464–1476. PMID:10029572

Bakker, B. M., Michels, P. A. M., Opperdoes, F. R., & Westerhoff, H. V. (1999). What controls glycolysis in bloodstream form Trypanosoma brucei? *The Journal of Biological Chemistry*, *274*(21), 14551–14559. doi:10.1074/jbc.274.21.14551 PMID:10329645

Bakker, B. M., Walsh, M. C., Ter Kuile, B. H., Mensonides, F. I. C., Michels, P. A. M., Opperdoes, F. R., & Westerhoff, H. V. (1999). Contribution of glucose transport to the control of the glycolytic flux in Trypanosoma brucei. *Proceedings of the National Academy of Sciences of the United States of America*, *96*(18), 10098–10103. doi:10.1073/pnas.96.18.10098 PMID:10468568

Bedard, K., & Krause, K.-H. (2007). The NOX family of ROS-generating NADPH oxidases: Physiology and pathophysiology. *Physiological Reviews*, *87*(1), 245–313. doi:10.1152/physrev.00044.2005 PMID:17237347

Blume-Jensen, P., & Hunter, T. (2001). Oncogenic kinase signalling. *Nature*, *411*(, 6835), 355–365. doi:10.1038/35077225 PMID:11357143

Bourne, H. R., Sanders, D. A., & McCormick, F. (1990). The GTPase superfamily: A conserved switch for diverse cell functions. *Nature*, *348*(6297), 125–132. doi:10.1038/348125a0 PMID:2122258

Bourne, H. R., Sanders, D. A., & McCormick, F. (1991). The GTPase superfamily: Conserved structure and molecular mechanism. *Nature*, *349*(6305), 117–127. doi:10.1038/349117a0 PMID:1898771

Buchdunger, E., Matter, A., & Druker, B. J. (2001). Bcr-Abl inhibition as a modality of CML therapeutics. *Biochimica et Biophysica Acta (BBA)-. Revue Canadienne*, *1551*(1), M11–M18.

Cass, A. E. G., Davis, G., Francis, G. D., Hill, H. A. O., Aston, W. J., Higgins, I. J., ... Turner, A. P. F. (1984). Ferrocene-mediated enzyme electrode for amperometric determination of glucose. *Analytical Chemistry*, *56*(4), 667–671. doi:10.1021/ac00268a018 PMID:6721151

Castagna, M., Takai, Y., Kaibuchi, K., Sano, K., Kikkawa, U., & Nishizuka, Y. (1982). Direct activation of calcium-activated, phospholipid-dependent protein kinase by tumor-promoting phorbol esters. *The Journal of Biological Chemistry*, *257*(13), 7847–7851. PMID:7085651

Cohen, P., & Frame, S. (2001). The renaissance of GSK3. *Nature Reviews. Molecular Cell Biology*, *2*(10), 769–776. doi:10.1038/35096075 PMID:11584304

Coleman, T., & Brines, M. (2004). Science review: recombinant human erythropoietin in critical illness: a role beyond anemia? *Critical Care (London, England)*, *8*(5), 337–341. doi:10.1186/cc2897 PMID:15469595

Collett, M. S., & Erikson, R. L. (1978). Protein kinase activity associated with the avian sarcoma virus src gene product. *Proceedings of the National Academy of Sciences of the United States of America*, *75*(4), 2021–2024. doi:10.1073/pnas.75.4.2021 PMID:205879

Costerousse, O., Allegrini, J., Huang, H., Bounhik, J., & Alhenc-Gelas, F. (1994). Regulation of ACE gene expression and plasma levels during rat postnatal development. *American Journal of Physiology. Endocrinology and Metabolism*, *267*(5), E745–E753. doi:10.1152/ajpendo.1994.267.5.E745 PMID:7977726

Daley, G. Q., Van Etten, R. A., & Baltimore, D. (1990). Induction of Chronic Myelogenous Leukemia in Mice by the P210 ^bcr/abl Gene of the Philadelphia Chromosome. *Science*, *247*(4944), 824–830. doi:10.1126cience.2406902 PMID:2406902

Druker, B. J. (2002). STI571 (Gleevec™) as a paradigm for cancer therapy. *Trends in Molecular Medicine*, *8*(4), S14–S18. doi:10.1016/S1471-4914(02)02305-5 PMID:11927282

Fabbro, D., Parkinson, D., & Matter, A. (2002). Protein tyrosine kinase inhibitors: New treatment modalities? *Current Opinion in Pharmacology*, *2*(4), 374–381. doi:10.1016/S1471-4892(02)00179-0 PMID:12127869

Fukuoka, M., Yano, S., Giaccone, G., Tamura, T., Nakagawa, K., Douillard, J.-Y., ... Baselga, J. (2003). Multi-institutional randomized phase II trial of gefitinib for previously treated patients with advanced non--small-cell lung cancer. *Journal of Clinical Oncology*, *21*(12), 2237–2246. doi:10.1200/JCO.2003.10.038 PMID:12748244

Gibbs, J. B. (1991). Ras C-terminal processing enzymes-New drug targets? *Cell*, *65*(1), 1–4. doi:10.1016/0092-8674(91)90352-Y PMID:2013090

Goga, A., McLaughlin, J., Afar, D. E. H., Saffran, D. C., & Witte, O. N. (1995). Alternative signals to RAS for hematopoietic transformation by the BCR-ABL oncogene. *Cell*, *82*(6), 981–988. doi:10.1016/0092-8674(95)90277-5 PMID:7553858

Goodman, L. E., Judd, S. R., Farnsworth, C. C., Powers, S., Gelb, M. H., Glomset, J. A., & Tamanoi, F. (1990). Mutants of Saccharomyces cerevisiae defective in the farnesylation of Ras proteins. *Proceedings of the National Academy of Sciences of the United States of America*, *87*(24), 9665–9669. doi:10.1073/pnas.87.24.9665 PMID:2124698

Gurung, N., Ray, S., Bose, S., & Rai, V. (2013). A broader view: Microbial enzymes and their relevance in industries, medicine, and beyond. *BioMed Research International*. PMID:24106701

Hall, A. (1990). The cellular functions of small GTP-binding proteins. *Science*, *249*(4969), 635–640. doi:10.1126cience.2116664 PMID:2116664

Harris, R. C., McKanna, J. A., Akai, Y., Jacobson, H. R., Dubois, R. N., & Breyer, M. D. (1994). Cyclooxygenase-2 is associated with the macula densa of rat kidney and increases with salt restriction. *The Journal of Clinical Investigation*, *94*(6), 2504–2510. doi:10.1172/JCI117620 PMID:7989609

Hidaka, H., Inagaki, M., Kawamoto, S., & Sasaki, Y. (1984). Isoquinolinesulfonamides, novel and potent inhibitors of cyclic nucleotide-dependent protein kinase and protein kinase C. *Biochemistry*, *23*(21), 5036–5041. doi:10.1021/bi00316a032 PMID:6238627

Hunter, T. (2000). Signaling—2000 and beyond. *Cell*, *100*(1), 113–127. doi:10.1016/S0092-8674(00)81688-8 PMID:10647936

Inoue, N., Kawashima, S., Kanazawa, K., Yamada, S., Akita, H., & Yokoyama, M. (1998). Polymorphism of the NADH/NADPH oxidase p22 phox gene in patients with coronary artery disease. *Circulation*, *97*(2), 135–137. doi:10.1161/01.CIR.97.2.135 PMID:9445163

John, A. M., Thomas, N. S. B., Mufti, G. J., & Padua, R. A. (2004). Targeted therapies in myeloid leukemia. *Seminars in Cancer Biology*, *14*(1), 41–62. doi:10.1016/j.semcancer.2003.11.006 PMID:14757535

Kargman, S., Charleson, S., Cartwright, M., Frank, J., Riendeau, D., Mancini, J., ... O'Neill, G. (1996). Characterization of prostaglandin G/H synthase 1 and 2 in rat, dog, monkey, and human gastrointestinal tracts. *Gastroenterology*, *111*(2), 445–454. doi:10.1053/gast.1996.v111.pm8690211 PMID:8690211

Karmen, A., Wróblewski, F., & LaDue, J. S. (1955). Transaminase activity in human blood. *The Journal of Clinical Investigation*, *34*(1), 126–133. doi:10.1172/JCI103055 PMID:13221663

Kempf, D. J., Norbeck, D. W., Sham, H. L., & Zhao, C. (1996). *Retroviral protease inhibiting compounds*. Google Patents.

Kola, I., & Landis, J. (2004). Can the pharmaceutical industry reduce attrition rates? *Nature Reviews. Drug Discovery*, *3*(August), 1–5. doi:10.1038/nrd1470 PMID:15286737

Kolibaba, K. S., & Druker, B. J. (1997). Protein tyrosine kinases and cancer. *Biochimica et Biophysica Acta (BBA)-. Revue Canadienne*, *1333*(3), F217–F248.

Kono, H., Rusyn, I., Yin, M., Gäbele, E., Yamashina, S., Dikalova, A., ... Thurman, R. G. (2000). NADPH oxidase--derived free radicals are key oxidants in alcohol-induced liver disease. *The Journal of Clinical Investigation*, *106*(7), 867–872. doi:10.1172/JCI9020 PMID:11018074

Krause, K. H., Lambeth, D., & Krönke, M. (2012). NOX enzymes as drug targets. *Cellular and Molecular Life Sciences*, *69*(14), 2279–2282. doi:10.100700018-012-1006-5 PMID:22585058

Ladisch, M., & Kohlmann, K. (1992). Recombinant human insulin. *Biotechnology Progress*, *8*(6), 469–478. doi:10.1021/bp00018a001 PMID:1369033

Lambeth, J. D. (2004). NOX enzymes and the biology of reactive oxygen. *Nature Reviews. Immunology*, *4*(3), 181–189. doi:10.1038/nri1312 PMID:15039755

Manley, P. W., Cowan-Jacob, S. W., Buchdunger, E., Fabbro, D., Fendrich, G., Furet, P., ... Zimmermann, J. (2002). Imatinib: A selective tyrosine kinase inhibitor. *European Journal of Cancer*, *38*, S19–S27. doi:10.1016/S0959-8049(02)80599-8 PMID:12528769

Masuyer, G., Schwager, S. L. U., Sturrock, E. D., Isaac, R. E., & Acharya, K. R. (2012). Molecular recognition and regulation of human angiotensin-I converting enzyme (ACE) activity by natural inhibitory peptides. *Scientific Reports*, *2*(1), 717. doi:10.1038rep00717 PMID:23056909

Mohan, R. (2000). *Advances in microbial amylases*. Biotechnology and Applied.

Nielsen, J. E., & Borchert, T. V. (2000). Protein engineering of bacterial \$α\$-amylases. *Biochimica et Biophysica Acta (BBA)-. Protein Structure and Molecular Enzymology*, *1543*(2), 253–274. doi:10.1016/S0167-4838(00)00240-5

Nishikawa, R., Ji, X.-D., Harmon, R. C., Lazar, C. S., Gill, G. N., Cavenee, W. K., & Huang, H. J. (1994). A mutant epidermal growth factor receptor common in human glioma confers enhanced tumorigenicity. *Proceedings of the National Academy of Sciences of the United States of America, 91*(16), 7727–7731. doi:10.1073/pnas.91.16.7727 PMID:8052651

Opperdoes, F. R., & Michels, P. A. (2001). Enzymes of carbohydrate metabolism as potential drug targets. *International Journal for Parasitology, 31*(5–6), 482–490. doi:10.1016/S0020-7519(01)00155-2 PMID:11334933

Parmar, H. S., Bhinchar, M. K., Bhatia, M., Chordia, N., Raval, I., Chauhan, D. S., ... Kumar, A. (2014). Study on gluco-regulatory potential of glimepiride sulfonamide using in silico, in vitro and in vivo approaches. *Current Pharmaceutical Design, 20*(32), 5212–7. Retrieved from http://www.ncbi.nlm.nih.gov/pubmed/24641189

Paul, M. K., & Mukhopadhyay, A. K. (2004). Tyrosine kinase – Role and significance in Cancer. *International Journal of Medical Sciences, 1*(283), 101–115. doi:10.7150/ijms.1.101 PMID:15912202

Quinn, D. M. (1987). Acetylcholinesterase: Enzyme structure, reaction dynamics, and virtual transition states. *Chemical Reviews, 87*(5), 955–979. doi:10.1021/cr00081a005

Reeve, J. G., Morgan, J., Schwander, J., & Bleehen, N. M. (1993). Role for membrane and secreted insulin-like growth factor-binding protein-2 in the regulation of insulin-like growth factor action in lung tumors. *Cancer Research, 53*(19), 4680–4685. PMID:7691401

Resnik, J. L., Reichart, D. B., Huey, K., Webster, N. J. G., & Seely, B. L. (1998). Elevated insulin-like growth factor I receptor autophosphorylation and kinase activity in human breast cancer. *Cancer Research, 58*(6), 1159–1164. PMID:9515800

Robbins, R. J., Larson, S. M., Sinha, N., Shaha, A., Divgi, C., Pentlow, K. S., ... Tuttle, R. M. (2002). A retrospective review of the effectiveness of recombinant human TSH as a preparation for radioiodine thyroid remnant ablation. *Journal of Nuclear Medicine : Official Publication, Society of Nuclear Medicine, 43*(11), 1482–1488. PMID:12411552

Sarnoski-Brocavich, S., & Hilas, O. (2013). Canagliflozin (invokana), a novel oral agent for type-2 diabetes. *P&T, 38*(11), 656. PMID:24391386

Schaber, M. D., O'hara, M. B., Garsky, V. M., Mosser, S. C., Bergstrom, J. D., Moores, S. L., ... Gibbs, J. B. (1990). Polyisoprenylation of Ras in vitro by a farnesyl-protein transferase. *The Journal of Biological Chemistry, 265*(25), 14701–14704. PMID:2203759

Schlessinger, J. (2000). Cell signaling by receptor tyrosine kinases. *Cell, 103*(2), 211–225. doi:10.1016/S0092-8674(00)00114-8 PMID:11057895

Schunkert, H., Ingelfinger, J. R., Hirsch, A. T., Pinto, Y., Remme, W. J., Jacob, H., & Dzau, V. J. (1993). Feedback regulation of angiotensin converting enzyme activity and mRNA levels by angiotensin II. *Circulation Research, 72*(2), 312–318. doi:10.1161/01.RES.72.2.312 PMID:8380358

Sugarman, B. J., Aggarwal, B. B., Hass, P. E., Figari, I. S., Palladino, M. A., & Shepard, H. M. (1985). Recombinant human tumor necrosis factor-alpha: Effects on proliferation of normal and transformed cells in vitro. *Science, 230*(4728), 943–945. doi:10.1126cience.3933111 PMID:3933111

Tateishi, M., Ishida, T., Mitsudomi, T., Kaneko, S., & Sugimachi, K. (1990). Immunohistochemical evidence of autocrine growth factors in adenocarcinoma of the human lung. *Cancer Research, 50*(21), 7077–7080. PMID:2208175

Vacca, J. P., Dorsey, B. D., Guare, J. P., Holloway, M. K., Hungate, R. W., & Levin, R. B. (1995). *HIV protease inhibitors useful for the treatment of AIDS*. Google Patents.

Vane, J. R., Bakhle, Y. S., & Botting, R. M. (1998). Cyclooxygenases 1 and 2. *Annual Review of Pharmacology and Toxicology, 38*(1), 97–120. doi:10.1146/annurev.pharmtox.38.1.97 PMID:9597150

Vellard, M. (2003). The enzyme as drug: Application of enzymes as pharmaceuticals. *Current Opinion in Biotechnology, 14*(4), 444–450. doi:10.1016/S0958-1669(03)00092-2 PMID:12943856

Wu, D.-C., Teismann, P., Tieu, K., Vila, M., Jackson-Lewis, V., Ischiropoulos, H., & Przedborski, S. (2003). NADPH oxidase mediates oxidative stress in the 1-methyl-4-phenyl-1, 2, 3, 6-tetrahydropyridine model of Parkinson's disease. *Proceedings of the National Academy of Sciences of the United States of America, 100*(10), 6145–6150. doi:10.1073/pnas.0937239100 PMID:12721370

Yadav, A., Amit, A., Chaudhary, R., Chandel, A. S., Mahantesh, V., Suman, S. S., & Pandey, K. (2015). Leishmania donovani: Impairment of the cellular immune response against recombinant ornithine decarboxylase protein as a possible evasion strategy of Leishmania in visceral leishmaniasis. *International Journal for Parasitology, 45*(1), 33–42. doi:10.1016/j.ijpara.2014.08.013 PMID:25449949

Zwick, E., Bange, J., & Ullrich, A. (2002). Receptor tyrosine kinases as targets for anticancer drugs. *Trends in Molecular Medicine, 8*(1), 17–23. doi:10.1016/S1471-4914(01)02217-1 PMID:11796262

Chapter 8
Enzyme-Triggered Hydrogels for Pharmaceutical and Food Applications

Lakshmishri Upadrasta
Techno India, India

Vijay Kumar Garlapati
Jaypee University of Information Technology, India

Nafisa Lakdawala
Techno India, India

Rintu Banerjee
IIT Kharagpur, India

ABSTRACT

Enzyme-mediated polymeric hydrogels are drawing considerable attention in pharmaceutical and food sectors owing to their superior biocompatibility and process controllability under physiological conditions. Enzymes play a significant role in polymeric hydrogel formation through different mechanisms. Oxidases (e.g., horseradish peroxidase and tyrosinase) have demonstrated to drive the crosslinking of gel precursors by oxidizing the phenolic or acrylic moieties to free radicals. Transferases and hydrolases catalyze elongation of biopolymer chains which gradually self-assemble into hydrogels. Still more certain enzymes also participate in hydrogel formation by releasing gelation factors. Enhancement of the desired properties of certain hydrogels through the interior and exterior post-modifications has also been demonstrated by certain enzymes. Hence, in this chapter, the authors explore the different mechanisms of enzyme-mediated hydrogels preparations and its fabrication towards pharmaceutical and food sectors along with the discussion of recent trends and further prospects.

INTRODUCTION

Hydrophilic gels that are networks of polymer chains in which water is the dispersion medium (Jonker,

DOI: 10.4018/978-1-5225-5237-6.ch008

Borrmann, van Eck, van Delft, Löwik, & van Hest, 2015). Elastic networks with interstitial spaces contain as much as 90-99% w/w water. Hydrophilic groups attached to the polymeric backbone confer the potency to absorb water to the hydrogels while the property of resistance to dissolution arises from cross-links between network chains. Many materials, both naturally occurring and synthetic, fit the definition of hydrogels (Jonker, Borrmann, van Eck, van Delft, Löwik, & van Hest, 2015; Singh, Topuz, Hahn, Albrecht, & Groll, 2013).

Ideal Functional Features of a Hydrogel

- Maximum absorption/equilibrium swelling capacity in saline.
- Optimal desired rate of absorption (preferred particle size and porosity) depending on the application requirement.
- Maximum absorbency under load (AUL).
- Minimum soluble content and residual monomer.
- Least price.
- Maximum durability and stability in the swelling environment and during the storage.
- Phenomenal biodegradability without formation of toxic species following the degradation.
- pH-neutrality after swelling in water.
- Non toxic, Colorlessness, odorlessness etc.
- Photo stability.
- Re-wetting capability (if required) the hydrogel has to be able to give back the imbibed solution or to maintain it; depending on the application requirement (e.g., in agricultural or hygienic applications).

Optimization of the hydrogel production reaction variables must be executed to achieve an appropriate balance between the properties since a hydrogel with all the mentioned attributes is practically not feasible.

Figure 1. Representation of hydrogel

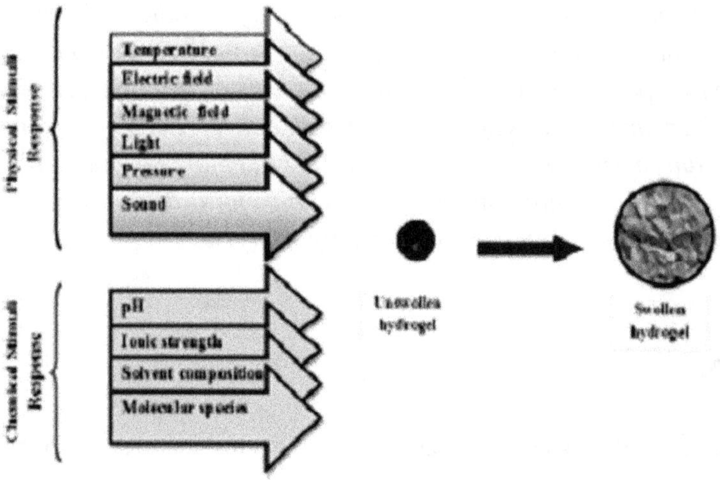

Methods of Preparation of Hydrogel

An Impetus for hydrogel preparation is gaining impetus for the exhaustive employment of hydrogel products in industrial and environmental sectors. This may be envisaged by:

- Chemical polymerization.
- Physical self-assembly.
- Reactions mediated by enzymes.
- Assembling of naturally occurring building blocks.

More often, these building blocks are a variety of macromolecules, including cross-linked polymers, entangled fibrillar networks (Zeng, She, Peng, Wei, & He, 2016; Phuong, Ho, Nguyen, Khoa, Quyen, & Lee, 2015), or colloidal assemblies (Moriyama, Minamihata, Wakabayashi, Goto, &Kamiya, 2014, Domeradzka, Werten, de Wolf, & de Vries, 2016; LaBella, Waykole, & Queen, 1968). In recent times, small amphiphilic molecules have emerged as a new class of hydrogelators, forming supramolecular or molecular hydrogels (Izydorczyk, Biliaderis, & Bushuk, 1990; Moore, Martinez-Munoz, & Hoseney, 1990; Takei, Sugihara, Ijima, & Kawakami, 2011).

CONVENTIONAL METHODS OF HYDROGEL PREPARATION

Hydrogels are synthesized in a number of 'chemical ways which may be:

- **One-Step Procedures:** Like polymerization and parallel cross-linking of multifunctional monomers
- **Multiple Step Procedures:** Involving the synthesis of polymer molecules having reactive groups and their subsequent cross-linking, possibly also by reacting polymers with suitable cross-linking agents (Teixeira, Feijen, van Blitterswijk, Dijkstra, & Karperien, 2012).

Compositionally, hydrogels are polymer networks possessing hydrophilic properties. Hence, they are prepared using only hydrophilic monomers and occasionally hydrophobic monomers to regulate the properties for specific applications. Synthetic polymers or natural polymers maybe employed for hydrogel preparation. In brief, any technique which can be used to create a cross-linked polymer can be used to produce a hydrogel (Trivić, Leskovac, Zeremski, Vrvić, & Winston, 2002, Su, Tang, He, Li, Wang, & Liao, 2014). Commonly used method of hydrogel preparation includes Copolymerization/cross-linking free-radical polymerizations by reacting hydrophilic monomers with multifunctional cross-linkers. The process of cross-linking to form hydrogels may consist of either of the following ways:

- Linking polymer chains via a chemical reaction.
- Using ionizing radiation to generate main-chain free radicals which can recombine as cross-link junctions.
- Through the physical interactions such as entanglements, electrostatics, and crystallite formation.

Bulk, solution, and suspension polymerization are the techniques while Monomer i.e polar ones more often, initiator, and cross-linker are the three integral parts that maybe used for hydrogel preparation. Diluents such as water or other aqueous solutions are used to control the heat of polymerization and final hydrogels properties.

Biological Method of Hydrogel Synthesis

Green formation technology that allows in situ hydrogelation under physiological conditions is currently in practice for the production of hydrogels used explicitly in medical and food applications. The desired properties of the resultant hydrogels include tunable gel properties, injectable properties, appropriate degradability, acceptable cytocompatibility, and anti-infection capacity, etc (Zeng, She, Peng, Wei, & He, 2016; Phuong, Ho, Nguyen, Khoa, Quyen, & Lee, 2015; Singh, Topuz, Hahn, Albrecht, & Groll, 2013).

In situ production of the functional hydrogels with high biosafety and biocompatibility is envisaged by enzymatic approach. These processes are advantageous because of their:

1. High specificity which reduces the generation of unwanted structures,
2. Mildoperable conditions, and
3. Feasible process control.

A large number of enzymes have been tested for hydrogel fabrication. Commonly used monomers for hydrogel fabrication are either natural polymers (e.g. dextran) or synthetic polymers (e.g. PEG) or small molecules. Selection of these molecules is based on the objective properties. Cross linkers employed are more often tyramine and its derivatives, whose amines offer ready coupling with the carboxyl monomers through EDC/NHS reactions.

Enzyme Mediated Hydrogel Preparation

Numerous enzyme mediated reactions contribute to hydrogel formation. Oxidation of phenolic moieties is demonstrated by enzymes like: HRP, TYR and LAC that primarily drive the C-C coupling of gel precursors. Also HRP/GOX/AcAc, GOX/Fe_{2+} and LAC/$O_2$are capable of initiating free radical polymerization. Elongation of polysaccharide chains which gradually self-assemble into hydrogels is catalyzed by transferases and hydrolases such as protease, lipase etc.

Undermentioned shown in the Table 1 is the list of enzymes used in hydrogel preparation:

Hydrogel formation by releasing gelation factors such as carboxylic acids, ammonia, etc may be executed by certain enzymatic reactions. Desired properties of hydrogels is made possible by enzyme mediated Post-modification processes such as interfacial coating, network development, mineralization.

Oxidases-Induced Crosslinking for Hydrogelation

1. **Horseradish Peroxidase** (HRP, EC1.11.1.7): Catalyzes the oxidation and polymerization of aromatic compounds in the presence of H_2O_2.HRP is one of the most commonly used enzymes for biosynthesis of hydrogels. Typicalhy drogelation triggered by HRP is illustrated in Figure 2.

Table 1. Enzymes in hydrogel preparation

Enzyme	Reaction	Role in Hydrogel Fabrication
Horse radish peroxidase Takei et al., (2011; 2013); Wang et al., (2015); Chen et al., (2002); Chen et al., (2002); Brown et al., (2016); Wang et al.,(2016)	Oxidation and polymerization of aromatic compound in the presence of hydrogen peroxide	i)Oxidative coupling of phenolic moieties in presence of hydrogen peroxide or thiol; ii)initiating radical polymerization in presence of acetylacetone.
Glucose oxidase Trivic et al., (2002); Wilson,R and Turner a.,(1992); Zavada et al., (2014): Johnson et al., (2009); Iwata et al.,(1991); Zeng et al., (2016)	Oxidation of beta glucose to gluconic acid with release of hydrogen peroxide from oxygen reduction	i)Supply hydrogen peroxide; ii)generate hydroxyl radicals with ferrous ion; iii)reduce specific substrates to be radical initiators; iv)interfacial hydrogel coating; v)release calcium ion for alginate gelation.
Laccase Rocasalbas et al.,(2013);, Azarikiaet al., (2015); Berlanga et al., (2009); Bollag,M ., (1992)	Oxidation of diphenol to free radicals for subsequent non enzymatic coupling	i)Crosslinking through C-C coupling; ii)generate ferric from ferrous ion for alginate gelation.
Protease Ferreeira et al.,(2002, 2005); Ramos et al., (2006); Carvalho et al., (2007)	Hydrolysis of peptide bonds in a polypeptide chain	i) Produce precursors via transesterification of acrylate to polysaccharides; ii) post modification of the hydrogel networks; iii) one pot transesterification and crosslinking of polysaccharides.
Lipase Das et al., (2015); Chronopoulou et al., (2010); Qin et al., (2010); Ferreira et al., (2002)	Hydrolysis of lipids(fats)to produce glycerol and fatty acids	i)Produce precursors via transesterification of acrylate to polysaccharides; ii) one pot transesterification and crosslinking of polysaccharides; iii) produce precursors via esterification.
Phospholyrase Takata et al., (2015); Kadokawa J and Kaneko Y., (2013); Izawa et al., (2009)	Addition of a phosphate group from an inorganic phosphate	Gradual chain growth via transesterification and consequent formation via self assembly.
Phosphatase Li et al., (2010); Wang et al., (2011)	Cleavage of phosphate groups from substrates	i)Hydrogelation of precursors to desired hydrogelators which from hydrogel via self assembly; ii) sol gel transition.
Kinase Suk et al., (1997); Yang et al., (2006); Martin et al., (1991)	Transfering a phosphate group from ATP to an acceptor	i)Phosphorylation of fibrinogen for thrombin induced gelation; ii) sol gel transition.
Transglutaminase Yung et al., (2007); Yin et al., (2012),Sanborn et al., (2002); Hu et al., (2003)	Formation of an isopeptide bond between the gamma carboxyamide group of glutamine and the amino group of lysine	Generating a permanent network of polypeptides.
Urease Chenite et al., (2006); Jee et al., (2016)	Hydrolysis of urea into ammonia and carbon dioxide	i)In situ generation of base for hydrogelation; ii)calcification of hydrogels.

Figure 2. Hydrogelation triggered by HRP

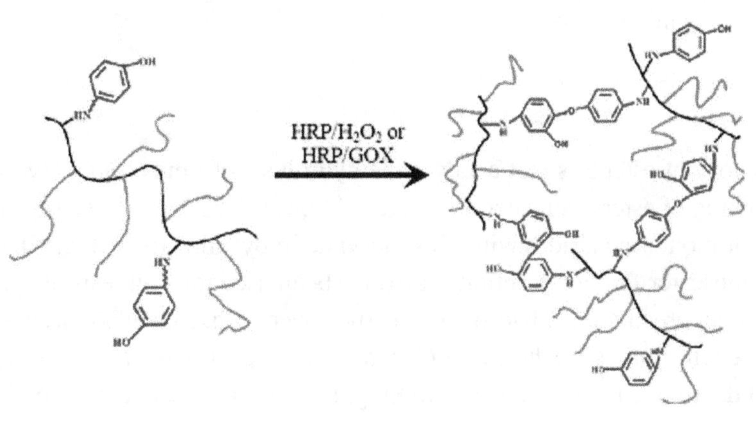

Gelation initiated by HRP/$H_2O_2$are very fast, usually within seconds or minutes. Dosage of enzyme and hydrogen peroxide can be tuned for controlling the physical properties of hydrogel such as swelling, porosity, mechanical strength.

2. **Glucose Oxidases (GOX)**: Are employed to gradually generate H_2O_2 and thereby form a mild initiation system with HRP, which mitigates the detrimental effect of excess H_2O_2, GOX also functions under physiological temperature and pH ranges. Replacement of exogeneous $H_2O_{2\,w}$ith enzymatic supply provide the scope of improvement of the mechanical properties of resultant hydrogel.
3. **Another Typical System:** using HRP/H_2O_2 can initiate radical polymerization by introducing acetylacetone (AcAc) to form an AcAc-HRP-H_2O_2 ternary system. HRP catalyzes H2O2 reacting with the enol form of ACAC to generate carbon-centered AcAcradicals. The radicals could then initiate the polymerization of a precursor (e.g.PEGMA).H2O2can also be yielded from the catalysis of GOX. The hydrogel in forms of layer or microsphere can be fabricated to be applied in colorimetric glucose detection using this methodology.
4. **Auto-Oxidation of Thiol Substrates for Generation of HRP-Catalyzed**: Can be obtained in presence of O_2, H_2O_2.A hydrogel with redox-sensitive degradation behavior in reductive cytosol-like environments may be obtained in this way. This simple gelation system, requires a basic pH (pH 8.5) and prolonged gelation time (>110 min).

Enzymatic Chain Elongation for Hydrogel Formation

Protease

Protease is an enzyme that catalyzes the hydrolysis of peptide bonds in a polypeptide chain. Certain proteases like *Bacillus subtili* sprotease (EC 3.4.21.62), are used to activate polysaccharides for hydrogel fabrication. Different strategies of the enzymes to obtain dextran-based hydrogels include:

- A one-pot transesterification and crosslinking of dextran with divinyladipate (DVA) in neat dimethylsulfoxide (DMSO) wherein both Proleather FG-F and two lipases yield>58% conversion of the dextran and finally result in the formation of a gel.
- Crosslinking of enzymatically activated dextran ester using enzymatic means, e.g. the addition of tetramethylethylenediamine (TEMED) and ammoniumpersulfate (APS).
    a. One-pot enzymatic reactions are time intensive for hydrogelation, probably due to the use of neat polar organic solvents.

Lipase

Lipase is a subcategory of esterases that catalyse the hydrolysis of lipids (fats). Hydrogels by polymerization and crosslinking of esters using certain bacterial lipases (EC 3.1.1.3).These enzymes are used in transesterification of monosaccharides with vinyl acrylate in pyridine to generate the 6-acryloyl esters. Lipases mediated transesterification reactions have also been used for synthesis of hyaluronic acid vinyl esters: a novel hydrogel precursor.3D hydrogel constructs were subsequently fabricated with addition of a photo initiator. Certain lipases produced by *Candida rugosa* and *Pseudomonas cepacian* could also inspire the one-pot dextranactivation and crosslinking. Lower monomer conversion is envisaged by the two lipases though they follow a similar reaction pathway as protease. Lipase-mediated esterification is gaining popularity for synthesizing precursors or hydrogelators.

Under catalysis of lipase, graft copolymerization of acrylic acid onto gum tragacanth is executed. Swelled hydrogel with excellent water holding and urea releasing capacities is obtained when glutaraldehyde is previously mixed in the polymerization system.

Peptide chain growth is also mediated by Lipase. Bond formation between 9-fluorenylmethoxyc arbonylphenylalanine(Fmoc–Phe) and the dipeptide diphenylalanine using lipase as catalyst. Lipase-catalyzed inclusion of p-hydroxybenzylalcohol to peptidebolaamphiphiles formed an activated diester building block that self-assembled to produce nanofibrillarthixotropichydrogels. After sophisticated post-modification steps, thixotropic nanofibrillar hydrogel maybe obtained that could serves as a scaffold for stem-cell proliferation.

Phosphorylase, Kinase and Phosphatase

Phosphotransferase catalyzes the addition of a phosphate group to an acceptor. These group of enzymes include:

- Phosphorylase,
- Kinase, and
- Phosphatases.

The three enzymes support hydrogel formation by acting in different mechanisms: Phosphorylases from different sources have been employed for the chain elongation of carbohydrates. Atypical synthesis of new α-(1→4)-D-glucosidic link ages using phosphorylase-catalyzed glycosylation involves glucose 1-phosphate (Glc-1-P), glycogen and the enzyme, e.g. glucanphosphorylase (EC 2.4.1.4). Glc-1-P acts as a glycosyl donor, leading to gradual chain growth and consequent formation of hydrogels.

pH-responsive amphoteric glycogen hydrogels are prepared by Phosphorylase-catalyzed successive reactions. These reactions use a thermostable α-glucanphosphorylase to synthesize an amphotericglycogen first; and then on incubation of the subsequent amphoteric glycogen with Glc-1-P and phosphorylase yield hydrogel. Subsequently the hydrogel is solublized in base due to the dissociation of amylose double helical conformation and by adjusting the pH to 9, amylosechains reformed double helices, which served as the crosslinking spots for re-hydrogelation.

Phosphatases transform a substrate-derived precursor to a hydrogelator that subsequently forms three-dimensional fibrous networks by a self-assembly process.

Kinases have also demonstrated role in hydrogelationas has been reported by several authors. Thrombininucedgelation of fibrinogen phosphorylated by kinases, synergistic stimulation offibrinogen gelation by casein kinase II in the presence of polycationic compounds and regulation of a supramolecular hydrogelby an interesting kinase/phosphatase switch are certain cases of kinase-mediated hydrogel formation.

Transglutaminase

Transglutaminase (EC 2.3.2.13) catalyze the formation of an isopeptide bond between the γ-carboxyamide group of glutamine and the ε-amino group of lysine. These group of enzymes are capable of generating the requisite permanent network of polypeptides for a tissue engineering scaffold. Gelatin, rationally synthesized peptides, peptide-PEG conjugates, casein-polysaccharide hybrids, etc.. are the probable precursors used for this process. The gelation time is determined by precursor functionalities, initial enzyme loadings and substrate kinetics. For drug releasing nd tissue engineering applications, microbial transglutaminase is the enzyme of choice as it maintains a high activity level over a broad range of working conditions (50% at 37°C, max. at 50°C; 90% pH 5 ~ 8).

The Release of Gelation Factors from Enzyme Catalysis

In Situ Generation of Metal Ions

Divalent and multivalent metal ions crosslink the carboxylate groups on polysaccharide chains, forming insoluble network junctions. A cascade reaction initiated with GOX is used to design /fabricate calcium alginate hydrogels. Glucose is firstly oxidized to gluconic acid which subsequently reacts with calcium carbonate to release calcium ions. The resulting alginate hydrogels demonstrate extensive applications as matrices for drug release or cell immobilization. Ferroxidase activity of laccase oxidizes Fe2+ into Fe3+ which binds to alginate polyanions. Monodisperse hydrogel nanoparticles are subsequently formed as a result of cross-linking between alginate chains via the newly formed trivalent cations. The growth of hydrogel nanoparticles is tunable by controlling the rate of enzymatic reaction (Liu et al., 2012, Bocharova et al., 2015).

In Situ Generation of Base

Crosslinking rate is determined by the concentration of bases in low-pH-induced gelation processes. Homogeneous intra networks construction is possible by Gradual in situ generation of soluble base. Urease, hydrolyzing urea into ammonium and carbon dioxide, allows in situ pH modulation with minimized gradients. Gelation time maybe regulated by moderating the sufficient enzyme and substrate participation (Chenite, Gori, Shive, Desrosiers, & Buschmann, 2006; Jee, Bánsági, Taylor, & Pojman, 2016).

Urease/urea are capable of triggering the base catalyzed Michael addition of a soluble trithiol to PEGDA. Polymerization of the two starting chemicals occurs via autocatalytic reactions at base pH.A time gap of frontal polymerization and gelation occur until the pH switches to 7 by urease catalysis. Subsequent degradation of the accumulated bases of hydrogels, enable convenient tuning of the gelation rate and the gel lifetime by changing the concentration of enzyme or substrate.

Post-Modification of the Hydrogels through Enzymes

Interfacial Coating

Formation of a multi-layered hydrogel is significant in manufacturing smart microdevices for controlled drug delivery. Post-modification of hydrogel substrates with interfacial gel coatings is one the effective ways to obtain desired characteristics, e.g., physical strength, water permeability, size exclusion, cell type sensitivity. Sequential multilayer coatings could be obtained by reswelling the layered hydrogel with glucose following with exposure to the precursor media. There exists a positive correlation between the layer thickness and the major biocatalysis parameters including glucose content, Fe^{2+} concentration and reaction time etc (Johnson, DeForest, Pendurti, Anseth, & Bowman, 2010; Shenoy & Bowman, 2012).

Re-Assembly of Networks

In addition to the case of enzymes-switched sol-gel transition, many studies on the development of hydrogel networks by protease have been reported. Protease-assisted photolithography technology has been employed to modify inert hydrogels with arbitrary large-area patterns and functional sites. The sequential steps include:

In these cases, original PEG hydrogel film contained a bisacrylated peptide crosslinker with an amino acid sequence digestible to protease. The peptide was caged by a photolabile moiety is used to cage the peptide which could be decaged under UV light at a patterned area.

Protease triggers the proteolysis of peptide following the patterns, resulting in fresh nucleophilic amine groups for further functionalization . The cleavage of native peptide network by protease could lead to the transition of chemical crosslinking to physical crosslinking.

In the protease-modified hydrogel, the newly formed secondary, physically cross-linked network had an increase in storage modulus than the original one. The enzymatic hydrogel development strategy can also be applied in the preparation of biomedical materials. Encapsulated collagenase and collagen microspheres are preceded with cells into the photo -crosslinked alginate hydrogels. Collagenase degraded collagen microspheres enlarged the space for cell spreading and migration. The scaffold properties are tunably by changing the amount of collagenase and microcarriers (Guz & Tang, 2010, Ayub & Kofinas 2015; Zhong, Sun, Wei, & Zhu, 2014).

Calcification of Hydrogels

Mineralization of hydrogels with calcium salts (or simplified as calcification) is an important step in biomimetic synthesis of bone-like composites and nacre-like materials. Enzyme-directed calcification has been recently developed to overcome blockage of diffusion pathways inside hydrogel matrices. Alkaline phosphatase (ALP) releases inorganic phosphate groups from hydrolysis of organic phosphoesters resulting in local deposition of carbonated apatite. Gradual enzymatic harvesting of phosphate groups minimized the unwanted precipitates during the mineral crystal growth in hydrogel (Gkioni, Leeuwenburgh, Douglas, Mikos, & Jansen, 2010; Rauner, Buenger, Schuller, & Tiller, 2015; Munro, Green, Dangerfield, & McGrath, 2011; Spoerke, Anthony, & Stupp, 2009)

APPLICATIONS OF HYDROGELS

Hydrogels have numerous applications in pharmaceutical, medical, food and industrial engineering fields. Hydrogel materials are increasingly studied for applications in the biological area for a number of reasons:

- Hydrogels provide suitable semiwet, three-dimensional environments for molecular-level biological interaction;
- Many hydrogels provide inert surfaces that prevent nonspecific adsorption of proteins, a property known as antifouling;
- Biological molecules can be covalently incorporated into hydrogel structures using a range of well-established chemistries;

- Hydrogel mechanical properties are highly tunable, for example elasticity can be tailored by modifying cross-link densities; and
- Specific responsive Hydrogels can be designed to change properties (e.g. swelling/collapse or solution-to-gel transitions) to externally applied triggers, such as temperature, ionic strength, solvent polarity, electric/magnetic field, light, or small (bio)molecules.

The volume of hydrogels can be easily changed due to a small change in the solvent, pH, ionic strength and temperature. The main applications revolving around drug delivery, pharmaceutical and biomedical industries (Ullah, Othman, Javed, Ahmad, & Akil, 2015).

Biomedical and Pharmaceutical Applications

The mimic nature of hydrogels as human organs as a response to environmental conditions (pH, temperature, ionic strength and solvent) makes hydrogels are the choice to use which medical implants, diagnostic devices to artificial muscles, stabilization of bone implants, decreasing thrombosis and prosthetic muscles or organs. The usage of hydrogels in urinary catheters prevents the bacterial colonization on the surface with providing a slippery surface to enhance the biocompatibility. Hydrogels can also be used in contraction of muscles as a response to different physical stimuli such as electric stimuli which have applications in the development of artificial muscles ((Ullah, Othman, Javed, Ahmad, & Akil, 2015).

In Contact Lens

Apart from usage in soft contact lenses, hydrogels also have applications in drug delivery systems to the eye. Having the bursting problem with conventional hydrogels, researchers developed "biomimetic hydrogels" and succeeded to load relevant amounts of therapeutic dosages of H1-antihistamines (Hu, X., Lingyun, H., Wang, H., Yang, X., Zhang, G., & Wang, 2011; Venkatesh, Sizemore, & Byrne, 2007). Addition of cyclodextrins to hydrogels also imparts enhanced swelling ratio along with the tensile strength (Xu, Li, & Sun, 2010).

In Wound Dressings

Hydrogels have also applicability in dry wound care as debriding agents, moist dressings, and paste compounds (Murphy & Evans, 2012). The hydrogels 'moisture donor' effect will help in autolytic debridement, enhanced collagenase production (Stashak, 2004).

The expansion of crosslinked polymer chains helps in absorbing and retaining contaminant exudates within the gel mass which results in separation of detritus, odor molecules and bacteria in the liquid. Due to the high water content, cooling and hydrating effect, hydrogels helps in oxygen transmission to the wounds and serve as emergency burns treatment (Osti & Osti, 2004). The inclusion of hydrogels in wound dressing along with other materials helps in forming composite structures which are suitable for many types of wounds (Shah, Kydonieus, Jamshidi, Decker, & Chang, 1996).

In Tissue Engineering

In tissue engineering, hydrogels can be applied as space-filling agents (for bulkiness purpose to prevent adhesion), delivery vehicles for drug moieties and as stimuli to ensure the development of a required tissue. The hydrogel scaffolds serve as drug delivery applications for the promotion of angiogenesis and also have applications in engineering the different body tissues such as cartilage, bone, and smooth muscle (Drury & Mooney 2003). Biocompatibility is the main property of hydrogels which makes the hydrogels for an indispensable role in "Tissue Engineering" discipline (Chen, Blevins, Park, & Park, 2000). The hydrogels composed of natural polypeptides such as alginate, collagen, agarose, chitosan, gelatin are suitable for tissue engineering applications (Hunt, Chen, van Veen, & Bryan, 2014; Jabbari, Yaszemski, & Currier, 2006).

In Drug Delivery Systems

Controlled drug delivery is the main feature of hydrogels to survive in the pharmaceutical industry. PVA based hydrogels are used to deliver the drugs to liver, intestine, brain, blood and different types of tumors. The composition and cross-linking of PVA plays an important role in the drug releasing time (Kim & Lee, 1992). The hydrophilic and hydrophobic block copolymers of Diacrylates of PEG and PCL imparts negative thermosensitive swelling behavior to PEG-PCL based hydrogels. The biocompatibility, elasticity and functionality features make PEG-PCL based hydrogels suitable for different pharmaceutical applications. PHEMA-co-DMAEMA based hydrogels are the choice for delivering the drugs based on the Zero-order kinetics and controlled release (Peppas 1997). Temperature-responsive hydrogels based on polyacrylamide derivatives are efficient with the rhythmic release of drugs which shrinks at well-known temperatures (Ramanan, Chellamuthu, Tang, & Nguyen, 2006).

Food Applications

Hydrogel particles are a part of different calorie food formulations due to their flexibility in changing the texture, appearance, flavour characteristics (Chung, Degner, Decker, & McClements, 2013). The well-studied applications of hydrogels in the food industry are mainly in packaging sector to develop eco-friendly packaging solutions by using polymer backbone. The direct usage of polymers in the packaging solutions are not cost-efficient, won't have high tensile strength, and won'thaswater- resistant properties (Gorrasi, Bugatti, & Vittoria, 2012). These drawbacks can be easily overcome by usage of hydrogels with desirable properties. The hydrogels formed from stiff and rigid linear polysaccharides and proteinaceous material are well-suited for food packaging applications. (Schmitt & Turgeon, 2011). Different protein-polysaccharide combinations have been recently reported for food packaging applications (Table 2). The promising features of hydrogel-based packaging include moisture-resistant, anti-bacterial shelf-life enhancement and oxidation resistance.

Table 2. Application of Protein- polysaccharide based hydrogels in food packaging (Chung, Degner, Decker, & McClements, 2013)

Application	Protein-Polysaccharide Base	Reference
Food Wrapping	Beta-lactoglobulin - pectin	Scrinis and Lyons, 2010
Food Packaging	Starch-cellulose	Miller et al., (2009)
Antimicrobial Packaging	Chitosan-gelatin	Gómez-Estaca et al., (2010)
Moisture Senstitve Packaging	Methyl cellulose-Whey protein	Baldwin et al., (2011)
Inner Package	Zein - Starch	Leroy et al., (2012)
Food Covering	Gelatin-pectin	Gómez-Guillén et al., (2011)

RECENT ADVANCEMENTS

The recent progress in hydrogel delivery systems is an improvement of nutraceutical bioavailability. As a part of this hydrogels are prepared by mixing the hydrophilic nutraceuticals with biopolymer solution and hydrophobic nutraceuticals are trapped in lipid droplets. Selection of suitable biopolymers, cross-linking agents, and adopting methods play a role in fabricating the hydrogel beads. The release of nutraceuticals can also alter by taking care of bead composition, structure and charge which is the utmost important for utilization of nutraceuticals in functional foods (McClements 2017). Protein hydrogels from silk fibroin and elastinis also one advancement in tissue engineering field which has superior biological compatibility and can be easily degraded by proteolytic enzymes (Silva, Fabry, &Boccaccini, 2014). These protein hydrogels are successfully utilized for articular cartilage repairs (Parkes, Myant, Dini, & Cann, 2015). Elastin-based protein hydrogels are also have huge potential in tissue engineering aspects of elastic tissues like skin, lung and vasculature (Annabi, Mithieux, Camci-Unal, Dokmeci, Weiss, & Khademhosseini, 2013). Hydrogels have become an attractive model to study the engineering of cell microenvironment in vitro. As a part of this, hydrogels will be developed with the functions of magnetic response, electrical conductivity and photo response. The developed hydrogels can be employed to engineer the 3D cellular microenvironment, to spatially control the bioactivity, mechanical properties and to study the contraction of cardiomyocytes and nerve regeneration (Zhang, Liu, Li, & Dong, 2015).

FUTURE PROSPECTS

There are several exciting developments in prospect for the future of enzymatic hydrogelation.

Synergistic actions of dual enzymes are promising in constructing multifunctional hydrogel materials (e.g. HRP/MTG bi-enzymatic crosslinking approach) although the process control is still challenging. It requires intensive studies on finding the optimal mutual conditions for enzymes with different properties.

Biomimetic enzyme nano-complexes in particular magnetic nanoparticles specifically functionalized are attracting in this field due to the feasibility of rationial tuning of the catalyst features. Their biosafety and biocompatibility should be extensively evaluated for practical use in biomedicine and health care.

Novel green media are investigated for supporting enzyme activity in the polymerization. The candidates include micro emulsions, ionic liquids, and deep eutectic solvents. These solvents may offer more advantages than expected. For instance, ionic liquids that have excellent dissolution capacities are desired for homogeneous synthesis hydrogels from recalcitrant natural polymers (e.g. lignin, cellulose).

Nowadays, Nanotechnology is an integral part of many research domains. There is a need to miniaturize these hydrogels towards nanofabrication for enhanced durability, mechanical properties and biocompatibility. Coming to the food industry, hydrogels have to address the needs of food sector such as satiety control, prebiotics encapsulation, controlling of texture for healthier food formulations and in targeted delivery of specific parts of the digestive tract. The engineering aspects of the microenvironment of hydrogels have to be addressed towards precise temporal control of cell microenvironment with multiple functionalities. Steps have to be adopted towards the development of "Green Hydrogels" by replacing the existing carcinogenic solvents with green solvents. Further, new approaches should be investigated in tissue engineering discipline to under the mechanism behind the stimulation of blood vessels by the polymers.

REFERENCES

Annabi, N., Mithieux, S. M., Camci-Unal, G., Dokmeci, M. R., Weiss, A. S., & Khademhosseini, A. (2013). Elastomeric recombinant protein-based biomaterials. *Biochemical Engineering Journal, 77*, 110–118. doi:10.1016/j.bej.2013.05.006 PMID:23935392

Ayyub, O. B., & Kofinas, P. (2015). Enzyme induced stiffening of nanoparticle–hydrogel composites with structural color. *ACS Nano, 9*(8), 8004–8011. doi:10.1021/acsnano.5b01514 PMID:26196060

Azarikia, F., Wu, B. C., Abbasi, S., & McClements, D. J. (2015). Stabilization of biopolymer microgels formed by electrostatic complexation: Influence of enzyme (laccase) cross-linking on pH, thermal, and mechanical stability. *Food Research International, 78*, 18–26. doi:10.1016/j.foodres.2015.11.013 PMID:28433280

Baldwin, E. A., Hagenmaier, R., & Bai, J. (Eds.). (2011). *Edible coatings and films to improve food quality*. CRC Press.

Berlanga-Reyes, C. M., Carvajal-Millán, E., Juvera, G. C., Rascón-Chu, A., Marquez-Escalante, J. A., & Martínez-López, A. L. (2009). Laccase induced maize bran arabinoxylan gels. *Food Science and Biotechnology, 18*(4), 1027–1029.

Bocharova, V., Sharp, D., Jones, A., Cheng, S., Griffin, P. J., Agapov, A. L., ... Sokolov, A. P. (2015). Enzyme induced formation of monodisperse hydrogel nanoparticles tunable in size. *Chemistry of Materials, 27*(7), 2557–2565. doi:10.1021/acs.chemmater.5b00187

Bollag, J. M. (1992). Enzymes catalyzing oxidative coupling reactions of pollutants. *Metal Ions in Biological Systems, 28*, 205–217.

Brown, J.E., Partlow, B.P., Berman, A.M., House, M.D., & Kaplan, D.L. (2016). Injectable silk-based biomaterials for cervical tissue augmentation: An in vitro study. *American Journal of Obstetrics and Gynecology, 214*(1), 118. e1- e9.

Carvalho, J., Gonçalves, C., Gil, A. M., & Gama, F. M. (2007). Production and characterization of a new dextrin based hydrogel. *European Polymer Journal, 43*(7), 3050–3059. doi:10.1016/j.eurpolymj.2007.02.046

Chen, J., Blevins, W. E., Park, H., & Park, K. (2000). Gastric retention properties of super porous hydrogel composites. *Journal of Controlled Release*, *64*(1-3), 39–51. doi:10.1016/S0168-3659(99)00139-X PMID:10640644

Chen, T., Embree, H. D., Wu, L. Q., & Payne, G. F. (2002). In vitro protein–polysaccharide conjugation: Tyrosinasecatalyzed conjugation of gelatin and chitosan. *Biopolymers*, *64*(6), 292–302. doi:10.1002/bip.10196 PMID:12124847

Chenite, A., Gori, S., Shive, M., Desrosiers, E., & Buschmann, M. (2006). Monolithic gelation of chitosan solutions via enzymatic hydrolysis of urea. *Carbohydrate Polymers*, *64*(3), 419–424. doi:10.1016/j.carbpol.2005.12.010

Chronopoulou, L., Lorenzoni, S., Masci, G., Dentini, M., Togna, A. R., Togna, G., ... Palocci, C. (2010). Lipase-supported synthesis of peptidic hydrogels. *Soft Matter*, *6*(11), 2525–2532. doi:10.1039/c001658f

Chung, C., Degner, B., Decker, E. A., & McClements, D. J. (2013). Oil –filled hydrogel particles for reduced for reduced-fat food applications: Fabrication, characterization, and properties. *Innovative Food Science & Emerging Technologies*, *20*, 324–334. doi:10.1016/j.ifset.2013.08.006

Das, A. K., Maity, I., Parmar, H. S., McDonald, T. O., & Konda, M. (2015). Lipase-catalyzed dissipative self-assembly of a thixotropic peptide bolaamphiphile hydrogel for human umbilical cord stem-cell proliferation. *Biomacromolecules*, *16*(4), 1157–1168. doi:10.1021/bm501835v PMID:25714334

Domeradzka, N. E., Werten, M. W., de Wolf, F. A., & de Vries, R. (2016). Protein cross-linking tools for the construction of nanomaterials. *Current Opinion in Biotechnology*, *39*, 61–67. doi:10.1016/j.copbio.2016.01.003 PMID:26871735

Drury, J. L., & Mooney, D. J. (2003). Hydrogels for tissue engineering: Scaffold design variables and applications. *Biomaterials*, *24*(24), 4337–4351. doi:10.1016/S0142-9612(03)00340-5 PMID:12922147

Ferreira, L., Carvalho, R., Gil, M. H., & Dordick, J. S. (2002). Enzymatic synthesis of inulin-containing hydrogels. *Biomacromolecules*, *3*(2), 333–341. doi:10.1021/bm010150h PMID:11888320

Ferreira, L., Gil, M. H., Cabrita, A. M., & Dordick, J. S. (2005). Biocatalytic synthesis of highly ordered degradable dextran-based hydrogels. *Biomaterials*, *26*(23), 4707–4716. doi:10.1016/j.biomaterials.2004.11.051 PMID:15763250

Ferreira, L., Gil, M. H., & Dordick, J. S. (2002). Enzymatic synthesis of dextran-containing hydrogels. *Biomaterials*, *23*(19), 3957–3967. doi:10.1016/S0142-9612(02)00132-1 PMID:12162328

Gkioni, K., Leeuwenburgh, S. C., Douglas, T. E., Mikos, A. G., & Jansen, J. A. (2010). Mineralization of hydrogels for bone regeneration. *Tissue Engineering. Part B, Reviews*, *16*(6), 577–585. doi:10.1089/ten.teb.2010.0462 PMID:20735319

Gómez-Estaca, J., de Lacey, A. L., López-Caballero, M., Gómez-Guillén, M., & Montero, P. (2010). Biodegradable gelatin–chitosan films incorporated with essential oils as antimicrobial agents for fish preservation. *Food Microbiology*, *27*(7), 889–896. doi:10.1016/j.fm.2010.05.012 PMID:20688230

Gómez-Guillén, M., Giménez, B., López-Caballero, M., & Montero, M. (2011). Functional and bioactive properties of collagen and gelatin from alternative sources: A review. *Food Hydrocolloids*, *25*(8), 1813–1827. doi:10.1016/j.foodhyd.2011.02.007

Gorrasi, G., Bugatti, V., & Vittoria, V. (2012). Pectins filled with LDH-antimicrobial molecules: Preparation, characterization and physical properties. *Carbohydrate Polymers*, *89*(1), 132–137. doi:10.1016/j.carbpol.2012.02.061 PMID:24750614

Gu, Z., & Tang, Y. (2010). Enzyme-assisted photolithography for spatial functionalization of hydrogels. *Lab on a Chip*, *10*(15), 1946–1951. doi:10.1039/c001335h PMID:20436969

Hu, B. H., & Messersmith, P. B. (2003). Rational design of transglutaminase substrate peptides for rapid enzymatic formation of hydrogels. *Journal of the American Chemical Society*, *125*(47), 14298–14299. doi:10.1021/ja038593b PMID:14624577

Hu, X., Lingyun, H., Wang, H., Yang, X., Zhang, G., Wang, G., & … . (2011). Hydrogel Contact Lens for Extended Delivery of Ophthalmic Drugs. *International Journal of Polymer Science*, *2011*, 1–9. doi:10.1155/2011/814163

Hunt, J. A., Chen, R., van Veen, T., & Bryan, N. (2014). Hydrogels for tissue engineering and regenerative medicine. *Journal of Materials Chemistry. B, Materials for Biology and Medicine*, *2*(33), 5319–5338. doi:10.1039/C4TB00775A

Iwata, H., Hata, Y., Matsuda, T., & Ikada, Y. (1991). Initiation of radical polymerization by glucose oxidase utilizing dissolved oxygen. *Journal of Polymer Science. Part A, Polymer Chemistry*, *29*(8), 1217–1218. doi:10.1002/pola.1991.080290818

Izawa, H., Nawaji, M., Kaneko, Y., & Kadokawa, J. (2009). Preparation of glycogenbased polysaccharide materials by phosphorylase catalyzed chain elongation of glycogen. *Macromolecular Bioscience*, *9*(11), 1098–1104. doi:10.1002/mabi.200900106 PMID:19637153

Izydorczyk, M., Biliaderis, C., & Bushuk, W. (1990). Oxidative gelation studies of water-soluble pentosans from wheat. *Journal of Cereal Science*, *11*(2), 153–169. doi:10.1016/S0733-5210(09)80117-3

Jabbari, E., Yaszemski, M. J., & Currier, B. L. (2006). *European Patent: 1 664 168 B1*. European Patent Office.

Jee, E., Bánsági, T. Jr, Taylor, A. F., & Pojman, J. A. (2016). Temporal control of gelation and polymerization fronts driven by an autocatalytic enzyme reaction. *Angewandte Chemie International Edition*, *55*(6), 2127–2131. doi:10.1002/anie.201510604 PMID:26732469

Johnson, L. M., DeForest, C. A., Pendurti, A., Anseth, K. S., & Bowman, C. N. (2010). Formation of three-dimensional hydrogel multilayers using enzyme-mediated redox chain initiation. *ACS Applied Materials & Interfaces*, *2*(7), 1963–1972. doi:10.1021/am100275n PMID:20586415

Johnson, L. M., Fairbanks, B. D., Anseth, K. S., & Bowman, C. N. (2009). Enzyme-mediated redox initiation for hydrogel generation and cellular encapsulation. *Biomacromolecules*, *10*(11), 3114–3121. doi:10.1021/bm900846m PMID:19821604

Jonker, A. M., Borrmann, A., van Eck, E. R., van Delft, F. L., Löwik, D. W., & van Hest, J. (2015). A fast and activatable crosslinking strategy for hydrogel formation. *Advanced Materials*, *27*(7), 1235–1240. doi:10.1002/adma.201404448 PMID:25535032

Kadokawa, J. I., & Kaneko, Y. (2013). Phosphorylase-catalyzed enzymatic polymerization. In *Advances in the Engineering of Polysaccharide Materials: by Phosphorylase-Catalyzed Enzymatic Chain-Elongation* (p. 43). CRC Press.

Kim, C. J., & Lee, P. I. (1992). Composite poly (vinyl alcohol) beads for controlled drug delivery. *Pharmaceutical Research*, *9*(1), 10–16. doi:10.1023/A:1018963223484 PMID:1589392

LaBella, F., Waykole, P., & Queen, G. (1968). Formation of insoluble gels and dityrosine by the action of peroxidase on soluble collagens. *Biochemical and Biophysical Research Communications*, *30*(4), 333–338. doi:10.1016/0006-291X(68)90746-8 PMID:5637039

Leroy, E., Jacquet, P., Coativy, G., Reguerre, A., & Lourdin, D. (2012). Compatibilization of starch–zein melt processed blends by an ionic liquid used as plasticizer. *Carbohydrate Polymers*, *89*(3), 955–963. doi:10.1016/j.carbpol.2012.04.044 PMID:24750886

Li, X., Gao, Y., Kuang, Y., & Xu, B. (2010). Enzymatic formation of a photo responsive supramolecular hydrogel. *Chemical Communications*, *46*(29), 5364–5366. doi:10.1039/c0cc00163e PMID:20582376

Liu, Y., Javvaji, V., Raghavan, S. R., Bentley, W. E., & Payne, G. F. (2012). Glucose oxidase-mediated gelation: A simple test to detect glucose in food products. *Journal of Agricultural and Food Chemistry*, *60*(36), 8963–8967. doi:10.1021/jf301376b PMID:22906038

Martin, B. D., Ampofo, S. A., Linhardt, R. J., & Dordick, J. S. (1992). Biocatalytic synthesis of sugar-containing polyacrylate-based hydrogels. *Macromolecules*, *25*(26), 7081–7085. doi:10.1021/ma00052a001

Martin, S. C., Forsberg, P. O., & Eriksson, S. D. (1991). The effects of in vitro phosphorylation and dephosphorylation on the thrombin-induced gelation and plasmin degradation of fibrinogen. *Thrombosis Research*, *61*(3), 243–252. doi:10.1016/0049-3848(91)90100-B PMID:1827546

McClements, D. J. (2017). Recent progress in hydrogel delivery systems for improving nutraceutical-bioavilability. *Food Hydrocolloids*, *68*, 238–245. doi:10.1016/j.foodhyd.2016.05.037

Moore, A., Martinez-Munoz, I., & Hoseney, R. (1990). Factors affecting the oxidative gelation of wheat water-solubles. *Cereal Chemistry*, *67*(1), 81–84.

Moriyama, K., Minamihata, K., Wakabayashi, R., Goto, M., & Kamiya, N. (2014). Enzymatic preparation of a redox-responsive hydrogel for encapsulating and releasing living cells. *Chemical Communications*, *50*(44), 5895–5898. doi:10.1039/C3CC49766F PMID:24761434

Müller, C. M., Laurindo, J. B., & Yamashita, F. (2009). Effect of cellulose fibers addition on the mechanical properties and water vapor barrier of starch-based films. *Food Hydrocolloids*, *23*(5), 1328–1333. doi:10.1016/j.foodhyd.2008.09.002

Munro, N. H., Green, D. W., Dangerfield, A., & McGrath, K. M. (2011). Biomimetic mineralisation of polymeric scaffolds using a combined soaking and Kitano approach. *Dalton Transactions (Cambridge, England)*, *40*(36), 9259–9268. doi:10.1039/c1dt11056j PMID:21829841

Murphy, P. S., & Evans, G. R. D. (2012). Advances in Wound Healing: A Review of Current Wound Healing Products. *Plastic Surgery International*, *2*, 1–8. doi:10.1155/2012/190436 PMID:22567251

Osti, E., & Osti, F. (2004). Treatment of cutaneous burns with burnishield (Hydrogel) and a semi-permeable adhesive film. *Annals of Burns and Fire Disasters*, *3*, 137–141.

Parkes, M., Myant, C., Dini, D., & Cann, P. (2015). Tribology-optimised silk protein hydrogels for articular cartilage repair. *Tribology International*, *89*, 9–18. doi:10.1016/j.triboint.2014.11.024

Peppas, N. A. (1997). Hydrogels and drug delivery. *Current Opinion in Colloid & Interface Science*, *2*(5), 531–537. doi:10.1016/S1359-0294(97)80103-3

Phuong, N. T., Ho, V. A., Nguyen, D. H., Khoa, N. C., Quyen, T. N., & Lee, Y. (2015). Enzyme-mediated fabrication of an oxidized chitosan hydrogel as a tissue sealant. *Journal of Bioactive and Compatible Polymers*, *30*(4), 1–12. doi:10.1177/0883911515578760

Ramanan, R. M. K., Chellamuthu, P., Tang, L., & Nguyen, K. T. (2006). Development of a temperature-sensitive composite hydrogel for drug delivery applications. *Biotechnology Progress*, *22*(1), 118–125. doi:10.1021/bp0501367 PMID:16454501

Ramos, R., Carvalho, V., & Gama, M. (2006). Novel hydrogel obtained by chitosan and dextrin-VA co-polymerization. *Biotechnology Letters*, *28*(16), 127912–127984. doi:10.100710529-006-9087-6 PMID:16802098

Rauner, N., Buenger, L., Schuller, S., & Tiller, J. C. (2015). Post polymerization of urease induced calcified, polymer hydrogels. *Macromolecular Rapid Communications*, *36*(2), 224–230. doi:10.1002/marc.201400426 PMID:25284027

Rocasalbas, G., Francesko, A., Touriño, S., Fernández-Francos, X., Guebitz, G. M., & Tzanov, T. (2013). Laccase-assisted formation of bioactive chitosan/gelatin hydrogel stabilized with plant polyphenols. *Carbohydrate Polymers*, *92*(2), 989–996. doi:10.1016/j.carbpol.2012.10.045 PMID:23399119

Sanborn, T. J., Messersmith, P. B., & Barron, A. E. (2002). In situ crosslinking of a biomimetic peptide-PEG hydrogel via thermally triggered activation of factor XIII. *Biomaterials*, *23*(13), 2703–2710. doi:10.1016/S0142-9612(02)00002-9 PMID:12059019

Schmitt, C., & Turgeon, S. L. (2011). Protein/polysaccharide complexes and coacervates in food systems. *Advances in Colloid and Interface Science*, *167*(1-2), 63–70. doi:10.1016/j.cis.2010.10.001 PMID:21056401

Scrinis, G., & Lyons, K. (2010). *Nanotechnology and the techno-corporate agri-food paradigm, Food security, nutrition and sustainability*. Earthscan Publisher.

Shah, K. R., Kydonieus, A., Jamshidi, K., Decker, S. C., & Chang, T. (1996). *US Patent 5,527,271*. US Patent Office.

Shenoy, R., & Bowman, C. N. (2012). Kinetics of interfacial radical polymerization initiated by a glucose-oxidase mediated redox system. *Biomaterials*, *33*(29), 6909–6914. doi:10.1016/j.biomaterials.2012.06.014 PMID:22795542

Silva, R., Fabry, B., & Boccaccini, A. R. (2014). Fibrous protein-based hydrogels for cell encapsulation. *Biomaterials*, *35*(25), 6727–6738. doi:10.1016/j.biomaterials.2014.04.078 PMID:24836951

Singh, S., Topuz, F., Hahn, K., Albrecht, K., & Groll, J. (2013). Embedding of active proteins and living cells in redoxsensitive hydrogels and nanogels through enzymatic crosslinking. *Angewandte Chemie International Edition*, *52*(10), 3000–3003. doi:10.1002/anie.201206266 PMID:23386357

Sperinde, J. J., & Griffith, L. G. (2000). Control and prediction of gelation kinetics in enzymatically cross-linked poly (ethylene glycol) hydrogels. *Macromolecules*, *33*(15), 5476–5480. doi:10.1021/ma000459d

Spoerke, E. D., Anthony, S. G., & Stupp, S. I. (2009). Enzyme directed templating of artificial bone mineral. *Advanced Materials*, *21*(4), 425–430. doi:10.1002/adma.200802242 PMID:22068437

Stashak, T. S., Farstvedt, E., & Othic, A. (2004). Update on wound dressings: Indications and best use. *Clinical Techniques in Equine Practice*, *3*(2), 148–163. doi:10.1053/j.ctep.2004.08.006

Su, T., Tang, Z., He, H., Li, W., Wang, X., Liao, C., ... Wang, Q. (2014). Glucose oxidase triggers gelation of N-hydroxyimide–heparin conjugates to form enzyme-responsive hydrogels for cell-specific drug delivery. *Chemical Science (Cambridge)*, *5*(11), 4204–4209. doi:10.1039/C4SC01603C

Suk, K., Lee, J. Y., & Kim, S. H. (1997). Synergistic stimulation of fibrinogen gelation by casein kinase II and polycationic compounds. *IUBMB Life*, *42*(3), 487–495. doi:10.1080/15216549700202891 PMID:9247706

Takata, Y., Yamamoto, K., & Kadokawa, J. (2015). Preparation of pHresponsive amphoteric glycogen hydrogels by αglucanphosphorylasecatalyzed successive enzymatic reactions. *Macromolecular Chemistry and Physics*, *216*(13), 1415–1420. doi:10.1002/macp.201500151

Takei, T., Sugihara, K., Ijima, H., & Kawakami, K. (2011). In situ gellable sugar beet pectin via enzyme-catalyzed coupling reaction of feruloyl groups for biomedical applications. *Journal of Bioscience and Bioengineering*, *112*(5), 491–494. doi:10.1016/j.jbiosc.2011.07.006 PMID:21807557

Takei, T., Sugihara, K., Yoshida, M., & Kawakami, K. (2013). Injectable and biodegradable sugar beet pectin/gelatin hydrogels for biomedical applications. *Journal of Biomaterials Science. Polymer Edition*, *24*(11), 1333–1342. doi:10.1080/09205063.2012.757727 PMID:23796034

Teixeira, L. S. M., Feijen, J., van Blitterswijk, C. A., Dijkstra, P. J., & Karperien, M. (2012). Enzyme-catalyzed crosslinkable hydrogels: Emerging strategies for tissue engineering. *Biomaterials*, *33*(5), 1281–1290. doi:10.1016/j.biomaterials.2011.10.067 PMID:22118821

Trivić, S., Leskovac, V., Zeremski, J., Vrvić, M., & Winston, G. W. (2002). Bioorganic mechanisms of the formation of free radicals catalyzed by glucose oxidase. *Bioorganic Chemistry*, *30*(2), 95–106. doi:10.1006/bioo.2001.1229 PMID:12020134

Ullah, F., Othman, M. B. H., Javed, F., Ahmad, Z., & Akil, H. M. (2015). Classification, processing and applications of hydrogels: A review. *Materials Science and Engineering C*, *57*(1), 414–433. doi:10.1016/j.msec.2015.07.053 PMID:26354282

Venkatesh, S., Sizemore, S. P., & Byrne, M. E. (2007). Biomimetic hydrogels for enhanced loading and extended release of ocular therapeutics. *Biomaterials*, *28*(4), 717–724. doi:10.1016/j.biomaterials.2006.09.007 PMID:17007923

Wang, H., Yang, C., Wang, L., Kong, D., Zhang, Y., & Yang, Z. (2011). Self-assembled nanospheres as a novel delivery system for taxol: A molecular hydrogel with nanosphere morphology. *Chemical Communications*, *47*(15), 4439–4441. doi:10.1039/c1cc10506j PMID:21387062

Wang, R., Xu, D. I., Liang, L., Xu, T. T., Liu, W., Ouyang, P. K., ... Xu, H. (2016). Enzymatically crosslinked epsilon-poly-L-lysine hydrogels with inherent antibacterial properties for wound infection prevention. *RSC Advances*, *6*(11), 8620–8627. doi:10.1039/C5RA15616E

Wang, Z., Chen, W., Cui, Z., He, K., & Yu, W. (2015). Studies on photoyellowing of silk fibroin and alteration of its tyrosine content. *Journal of the Textile Institute*, *2015*, 1–7.

Wilson, R., & Turner, A. (1992). Glucose oxidase: An ideal enzyme. *Biosensors & Bioelectronics*, *7*(3), 165–185. doi:10.1016/0956-5663(92)87013-F

Xu, J., Li, X., & Sun, F. (2010). Cyclodextrin-containing hydrogels for contact lenses as a platform for drug incorporation and release. *Acta Biomaterialia*, *6*(2), 486–493. doi:10.1016/j.actbio.2009.07.021 PMID:19619677

Yang, Z., Liang, G., Wang, L., & Xu, B. (2006). Using a kinase/phosphatase switch to regulate a supramolecular hydrogel and forming the supramolecular hydrogel in vivo. *Journal of the American Chemical Society*, *128*(9), 3038–3043. doi:10.1021/ja057412y PMID:16506785

Yin, W., Su, R., Qi, W., & He, Z. (2012). A casein-polysaccharide hybrid hydrogel cross-linked by transglutaminase for drug delivery. *Journal of Materials Science*, *47*(4), 2045–2055. doi:10.100710853-011-6005-7

Yung, C., Wu, L., Tullman, J., Payne, G., Bentley, W., & Barbari, T. (2007). Transglutaminasecross-linked gelatin as a tissue engineering scaffold. *Journal of Biomedical Materials Research. Part A*, *83*(4), 1039–1046. doi:10.1002/jbm.a.31431 PMID:17584898

Zavada, S., McHardy, N., & Scott, T. (2014). Oxygen-mediated enzymatic polymerization of thiol–ene hydrogels. *Journal of Materials Chemistry. B, Materials for Biology and Medicine*, *2*(17), 2598–2605. doi:10.1039/C3TB21794A PMID:24995128

Zeng, Z., She, Y., Peng, Z., Wei, J., & He, X. (2016). Enzyme-mediated in situ formation of pH-sensitive nanogels for proteins delivery. *RSC Advances*, *6*(10), 8032–8042. doi:10.1039/C5RA25133H

Zhang, X., Liu, M., Li, Y., Dong, Y., Pingguan-Murphy, B., Lu, T. J., & Xu, F. (2015). Engineering cell microenvironment using novel functional hydrogels. *European Polymer Journal*, *72*, 590–601. doi:10.1016/j.eurpolymj.2015.03.019

Zhong, M., Sun, J., Wei, D., Zhu, Y., Guo, L., Wei, Q., ... Zhang, X. (2014). Establishing a cell-affinitive interface and spreading space in a 3D hydrogel by introduction of microcarriers and an enzyme. *Journal of Materials Chemistry. B, Materials for Biology and Medicine*, *2*(38), 6601–6610. doi:10.1039/C4TB00887A

Chapter 9
Laccase Catalysis:
A Green Approach in Bioactive Compound Synthesis

Helina Patel
Natubhai V. Patel College of Pure and Applied Sciences, India

Akshaya Gupte
Natubhai V. Patel College of Pure and Applied Sciences, India

ABSTRACT

The search for cost-effective and environmental benign practices for the production of bioactive compounds has gained considerable attention since last decade, due to increasing demand of eco-friendly processes. Many industries have started adopting routes for the development of green chemistry by employing enzymatic approaches to overcome the limitations of physico-chemical methods and environmental concerns. Laccase is one such enzyme which has gained considerable attention in recent years as a biocatalyst in organic synthesis. Laccases possess versatile biochemical properties and the reactions catalyzed by laccase require only molecular oxygen with concomitant release of water as a byproduct. They have been widely used for reactions such as dimerization, polymerization, coupling, and grafting reactions and for antibiotic modifications. This chapter summarizes the advances that have been made in developing technologies based on laccase mediated reactions in the field of medicine, agriculture, food, and pharmaceuticals.

INTRODUCTION

The development of bioprocesses for producing bioactive compounds has received superior attention in recent years due to their enormous applications of these compounds in food, pharmaceutical and chemical industries. The bioactive compounds are present in small quantities in plants and food products (Kris-Etherton, Hecker, Bonanome, Coval, Binkoski, Hilpert, … & Etherton, 2002), sponges (Müller, Grebenjuk, Le Pennec, Schröder, Brümmer, Hentschel, ... & Breter, 2004), bacteria and fungi (Debbab, Aly, Lin, & Proksch, 2010). They are supposed to be beneficial and nutritionally rich, thus making

DOI: 10.4018/978-1-5225-5237-6.ch009

them a key functional component in nutraceuticals. The market value of these compounds is likely to increase by 7.03% over a period of 5 years (2013-2018). Commonly observed bioactive compounds are mainly secondary metabolites which include antibiotics, phenolic compounds, food grade pigments, growth factors, mycotoxins and alkaloids (Hölker, Höfer, & Lenz, 2004). In last few years, bioactive compounds and their method of synthesis have been intensively studied for the benefits they provide to human health such as anti-oxidant, anti-allergic, anti-inflammatory and anti-mutagenic properties (Ham, Kim, Moon, Chung, Cui, Han, ... & Choe, 2009; Parvathy, Negi, & Srinivas, 2009), decrease in frequency of degenerative diseases like diabetes and cancer (Kim, Shin, & Jang, 2009; Martins, Aguilar, Garza-Rodriguez, Mussatto, & Teixeira, 2011), and reduction in threat of cardiovascular diseases (Jiménez, Serrano, Tabernero, Arranz, Díaz-Rubio, García-Diz, ...& Saura-Calixto, 2008). Apart from their use in pharmaceutical sector, they are also being employed in food industry for functional food (nutraceuticals) production, in chemical industries, agrochemicals, geo-medicine, cosmetics and nanobioscience (Guaadaoui, Benaicha, Elmajdoub, Bellaoui, & Hamal, 2014).

There are several methods that are being employed for the production and recovery of bioactive compounds some of which include, use of microwave and ultrasound assisted techniques, fermentation techniques, solid liquid extraction in heat-reflux systems using organic solvents, use of supercritical fluids and, high pressure practices (Markom, Hasan, Daud, Singh, & Jahim, 2007; Martins, Aguilar, Garza-Rodriguez, Mussatto, & Teixeira, 2010). However, these physico-chemical processes are low yielding, energy demanding and often lead to generation of excessive amounts of waste showing adverse effect on the environment (Kudanga, Nyanhongo, Guebitz, & Burton, 2017). The 12 Principles of Green Chemistry framed by Anastas and Warner (1998) guides the chemists in scheming safer and environmentally benign chemical compounds and processes. Many of these principles are based on the elimination and reduction of waste generation. The standard metrics such as E-factor are being used for assessing the waste generation, it measures kilogram of waste produced per kilogram of product formed, thereby highlighting the inefficiencies of any process (Sheldon, 1992). It is reported that the amount of waste generated per kilogram of any fine chemical or pharmaceutical product synthesis was 5-100 times higher than the desired product (Li & Trost, 2008). To overcome the waste generation associated with any process environmental concerns have created an awareness in encouraging the use of green methods in pharmaceutical industries (Constable, Dunn, Hayler, Humphrey, Leazer Jr, Linderman, ... & Zaks, 2007; Kudanga, Nyanhongo, Guebitz, & Burton, 2017), leading to the development of newer, eco-friendly and cost effective practices for bioactive compound synthesis.

In chemical processes, catalysis is an important foundational pillar of green chemistry (Anastas & Warner, 2001). The use of catalysts in chemical transformations significantly reduces the amount of chemicals required and also the amount of waste generated during a process, thus fulfilling both environmental and economic objectives (Cannatelli & Ragauskas, 2017). Biocatalysts, or enzymes, are gaining a notable attention in organic synthesis, as they are very active in aqueous solvents at room temperature, non-toxic, and biodegradable. They can be easily obtained from the biological systems and involve less process steps during a reaction. Unlike classical chemical catalysts, enzymes are typically highly selective, a quality which is important for the synthesis of therapeutically valuable compounds (Maugh, 1984). Thus, nowadays research has been focused on the use of microorganisms, isolated enzymes and other natural resources for the generation of fuels, valuable materials, and chemicals.

One such group of enzymes that have gained considerable widespread attention as biocatalyst in organic synthesis is laccase. The oxidative power of laccase has long been exploited for various applications such as delignification, bleaching of craft pulp, fiber modification, in textile industries, and in synthesis of fine chemicals. Over the last decade, many studies have been published reporting the enzyme production using lignocellulosic agricultural residue as a substrate (Patel & Gupte, 2016), their characteristics, various physico-chemical properties and their uses in various industrial applications (Morozova, Shumakovich, Gorbacheva, Shleev, & Yaropolov, 2007a; Senthivelan, Kanagaraj, & Panda, 2016). The usefulness of this enzyme can be further illustrated by their use in several biotechnological applications such as bioremediation, biosensor development, food, cosmetic and pharmaceutical industries (Kudanga, Burton, Nyanhongo, & Guebitz, 2011; Pezzella, Guarino, & Piscitelli, 2015). The present chapter will extensively provide the current knowledge on laccases regarding their source, structure, oxidative mechanism and their use as biocatalyst in production of biologically active compounds, mainly in synthesis or modification of anti-oxidants, antibiotics and other therapeutically important drugs.

LACCASES: DISTRIBUTION AND THEIR FUNCTION IN NATURAL ENVIRONMENT

Laccases (benzenediol: oxygen oxido-reductases, EC 1.10.3.2) are biologically important enzymes that belong to the family of multi-copper oxidases (Messerschmidt, 1997). These copper oxidases possess a distinct redox ability to oxidize diverse range of compounds (including phenols, diphenols, phenolic and alkyl amines, and methoxy substituted phenols) whilst concomitantly reducing molecular oxygen to water and thus are also considered as "green catalysts" for various industrial applications to reduce environmental pollution (Arora & Sharma, 2010; Kudanga, Burton, Nyanhongo, & Guebitz, 2012). Attributing to their wide substrate specificity while requiring only air as an oxidant and producing water as the by-product, laccases are considered as ideal oxidative catalysts for their use in variety of industrial applications.

Laccase was discovered as a component of the sap of the Japanese lacquer tree *Rhus vernicifera* by Yoshida (1883) and later after a decade, Bertrand (1894) isolated and purified the enzyme and reported its mechanism of action. Laccases have been predominantly isolated from plants and fungi however; enzymes are also being detected to a lesser extent amongst few species of bacteria, insects, and algae (Claus, 2004; Dwivedi, Singh, Pandey, & Kumar, 2011). The biological roles and physiological function of laccase is different in various organisms, fungal laccases aid in delignification, fungal morphogenesis, and fruiting body formation; in bacteria, laccases play a role in melanin pigment synthesis, and in production of brown spore pigment that primarily provides protection to the spore coat against hydrogen peroxide and UV radiation (Dwivedi, Singh, Pandey, & Kumar, 2011; Sharma, Goel, & Capalash, 2007; Tetsch, Bend, & Hölker, 2006). In plants, laccases are involved in lignin biosynthesis and cell wall formation (Mayer & Staples, 2002) and, in insects, laccases perform sclerotization process for epidermal cuticle synthesis and are supposed to play role in tanning process by catalyzing the reaction between structural proteins in insects and catechol derivatives (Sakurai & Kataoka, 2007; Kramer, Kanost, Hopkins, Jiang, Zhu, Xu, ...& Turecek, 2001). Laccase has also been isolated from soil algae where it is anticipated that they contribute to the turnover of soil organic matter (Otto, Schlosser, & Reisser, 2010).

LACCASE STRUCTURE AND CATALYTIC MECHANISM

The particulars of the three dimensional structure of an enzyme can be acquired from the purified crystals of an enzyme. Recently, a review has been published by Hakulinen and Rouvinen (2015) surveying the crystal structures of all the available laccases. Laccases are typically extra- or intracellular glycoproteins containing four copper atoms bound to three redox sites (Kunamneni, Camarero, García-Burgos, Plou, Ballesteros, & Alcalde, 2008). Fungal laccases are typically monomeric proteins however; several exhibit homodimeric structures, e.g., *Trametes villosa* (Yaver, Xu, Golightly, Brown, Brown, Rey, ...& Dalboge, 1996), *Pleurotus pulmonarius* (de Souza & Peralta, 2003) and oligomeric forms of laccases are also in existence. Laccases have an average molecular weight ranging between 50-130 kDa. The glycosylation rate is between 10-25% with a carbohydrate moiety comprises of galactose, mannose and *N*-acetyl glucosamine. The presence of this carbohydrate moiety imparts stability to the enzyme. (Morozova, Shumakovich, Gorbacheva, Shleev, & Yaropolov, 2007a). The carbohydrate moieties present in enzyme are connected to the polypeptide chain via *N*-linkages (Ko, Leem, & Choi, 2001), and these are believed to increase the thermostability of protein (Slomczynski, Nakas, & Tanenbaum, 1995), and provide structural protection against proteolytic degradation, and inactivation by free radicals (Yoshitake, Katayama, Nakamura, Iimura, Kawai, & Morohoshi, 1993).

The catalytic centre of a typical fungal laccase contain 4 copper atoms viz., type I copper (T1), type II copper (T2), and two type III copper (T3) based on their different spectroscopic characteristics (Solomon, Sundaram, & Machonkin, 1996). The T1 Cu site performs oxidation of substrate and is also believed to be involved in electron transfer, whereas, the reduction of oxygen to water takes place at the trinuclear cluster (T2 and T3 Cu sites) (Polak & Jarosz-Wilkolazka, 2012; Senthivelan, Kanagaraj, & Panda, 2016). The first step involved in the catalytic cycle of laccase is the mono electronic oxidation of substrate at T1 Cu site, where transfer of electron from substrate to T1 Cu site and formation of substrate radical occurs which is associated with the reduction of T1 copper. Subsequently, in later steps internal transfer of electron from T1 copper to T2/T3 cluster occurs followed by binding of oxygen to T2/T3 Cu centre and finally electron transfer from trinuclear T2/T3 copper atoms on oxygen and its reduction to water (Mikolasch & Schauer, 2009; Solomon, Augustine, & Yoon, 2008). The internal transfer of electron from the T1 Cu site to the trinuclear cluster take place either through-bond via highly conserved Cys-His motif, or through-space (Bertrand, Jolivalt, Briozzo, Caminade, Joly, Madzak, & Mougin, 2002). This electron transfer is highly affected by pH, where at high pH level it is identified that the redox potential (E^o)of the substrate will decrease considerably compared to E^o of laccase, permitting actively favorable electron transfer from substrate to T1 copper ion (Cannatelli & Ragauskas, 2017). At increased pH level, the hydroxide anion binds to the T2 and T3 copper ions, thus the internal electron transfer is inhibited, thereby obstructing the catalytic efficiency of an enzyme (Xu, 1997), and hence pH control is important during laccase catalyzed reactions.

The E^o plays a chief role in maintaining the overall kinetics during the substrate oxidation at T1 Cu site, the kinetic study reveals that difference in E^o between the substrate and T1 Cu determines the rate of electron transfer stating that this is the rate-limiting stage of the complete catalytic phase (Xu, 1996). Laccases are divided into two types, depending on the arrangement and properties of copper center, “low-redox potential” and “high-redox potential” laccase (Kunamneni, Camarero, García-Burgos, Plou, Ballesteros, & Alcalde, 2008). The low redox potential laccases are distributed usually in bacteria, insects, and plants (Jung, Xu, & Li, 2002) and high-redox potential laccases are found largely in basidiomycetes (Quaratino, Federici, Petruccioli, Fenice, & D’Annibale, 2007; Hernández-Luna, Gutiérrez-Soto, &

Salcedo-Martínez, 2008). The E^o of fungal laccase ranges between 0.44 and 0.8 V which is noticeably higher than laccase from plant and bacterial origin (Kersten, Kalyanaraman, Hammel, Reinhammar, & Kirk, 1990), and, relatively lower compared to ligninolytic peroxidases (>1.0 V). For that reason, the action of laccase is limited to the oxidation of phenolic lignin compound with less than 20% of lignin polymer content (Kawai, Umezawa, Shimada, Higuchi, Koide, Nishida, ...& Haraguchi, 1987) while, non-phenolic substrates with E^o exceeding 1.3 V cannot be oxidized directly by laccase (Canas & Camarero, 2010).

LACCASE-MEDIATOR SYSTEM (LMS)

The scope of laccase catalyzed oxidation of substrates with high E^o have been extended by the use of redox mediators in the laccase-mediator systems. These redox mediators are low molecular weight compounds that expand the catalytic ability of laccase to oxidize substrates with greater E^o than their own; as for example oxidation of recalcitrant compounds such as non-phenolic lignin units (Barreca, Fabbrini, Galli, Gentili, & Ljunggren, 2003), and oxidation mediated by poly aromatic hydrocarbons (Riva, 2006). These mediators gets readily oxidized by laccase and serves as diffusible electron transporters between active site of the enzyme and target substrate (Figure 1), allowing the oxidation of complex polymeric substrates that are unable to penetrate into the active site of the enzyme structure due to steric hindrances (Canas & Camarero, 2010). Once oxidized by laccase, and stabilized in more or less stable radicals, mediators diffuse far away from the enzyme pocket and enable the oxidation of bulky target substrate with high E^o in a mechanism different from that of enzymatic one (Cannatelli & Ragauskas, 2017).

Bourbonnais and Paice (1990) proved that the presence of mediator prolonged the catalytic activity of laccase towards non-phenolic compounds and established that laccase (from *Trametes versicolor*) along with its common non-phenolic substrate 2,2′-azino-bis (3-ethylbenzthiazoline-6-sulfonic acid) diammonium salt (ABTS) efficiently oxidized veratryl alcohol, a non-phenolic lignin-model compound. More than hundred different mediators have been described and LMS has been applied successfully for bleaching of kraft pulp, delignification, bioremediation of xenobiotic compounds (Morozova, Shumakovich, Shleev, & Yaropolov, 2007b), to perform chemical transformations (Wells, Teria, & Eve, 2006), in conversion of diols to lactones (Diaz-Rodriguez et al., 2012), for conversion of aromatic methyl groups and benzyl alcohols to benzaldehydes (Potthast, Rosena, Chen, & Gratzl, 1995; Potthast, Rosenau, Chen, & Gratzl, 1996), and for regeneration of coenzyme $NAD(P)^+$ (Pham, Hollmann, Kracher, Preims, Haltrich, & Ludwig, 2015).

Diverse range of compounds have been broadly studied as effective mediators of laccases and these mediator compounds can be either artificially or naturally derived. The artificial mediators of laccase include: ABTS, compounds containing N-OH group that includes: *N*-hydroxyphthalimide (NHPI), 1-hydroxybenzotriazole (HBT), violuric acid (VIO) and, *N*-acetyl- *N*-phenylhydroxylamine (NHA), the stable radical 2,2,6,6-tetramethylpiperidin-1-yloxyl (TEMPO) and *N*-heterocycles (e.g., promazine (PZ), 1-nitroso-naphthol-3,6-disulfonic acid (NNDS)) (Astolfi, Brandi, Galli, Gentili, Gerini, Greci, & Lanzalunga, 2005; Solís-Oba, Ugalde-Saldívar, González, & Viniegra-González, 2005) (Figure 2). Depending upon the chemical structure and redox potential of mediator, LMS function via different mechanisms: (1) ABTS follows electron transfer (ET) mechanism, and this mechanism is possible with low redox potential substrate; (2) *N*-hydroxy type mediators attack weak C-H bonds and oxidize the substrate via hydrogen atom transfer (HAT) pathway and; (3) stable radical TEMPO reacts via ionic mechanism (Figure 3) (Fabbrini, Galli, & Gentili, 2002; d'Acunzo, Baiocco, Fabbrini, Galli, & Gentili, 2002). In ET mechanism, the first step of oxidation by ABTS lead to the formation of cationic radicals ($ABTS^{\bullet+}$) which finally gets oxidized to dication ($ABTS^{2+}$) (Baiocco, Barreca, Fabbrini, Galli, & Gentili, 2003). However, the oxidation of *N*-hydroxy mediators by laccase produces a highly reactive nitroxyl radical (>N-O•), due to the enzymatic exclusion of an electron followed by release of a proton (Xu, Kulys, Duke, Li, Krikstopaitis, Deussen, ...& Schneider, 2000). Hence in HAT pathway, the enthalpic stability between the dissociated bond (C-H) in a substrate and bond (NO-H) forming in a mediator is a driving force for the mechanism (Cantarella, Galli, & Gentili, 2003). Whereas, the case with the stable nitroxyl radical TEMPO is dissimilar as it performs a non-radical-ionic oxidation mechanism with laccase, forming an oxoammonium ion in which the oxidation to $>N=O^+$ occurs at E^o similar to that of a high redox fungal laccase (Fabbrini, Galli, & Gentili, 2002).

Sustainable practices that uses artificial mediators is however not a feasible method due to elevated commercial costs of mediators, their own toxicity and ability to generate toxic derivatives. Certain mediators also cause inactivation of laccase even under mild reaction conditions (Kurniawati & Nicell, 2007) and, the use of polymerized mediators can lead to the formation of dark precipitates causing residual coloration of wastewater (Ryan, Leukes, & Burton, 2007). Owing to these consequences, substitution with natural mediators could be less toxic and economically advantageous. Many compounds secreted during natural degradation of lignin by white-rot fungi (acetosyringone, acetovanillone, *p*-coumaric acid, syringaldehyde, vanillin, veratryl alcohol, ferulic acid, sinapic acid) (Figure 2) and some other natural metabolites including phenols, 4-hydroxybenzyl alcohol, aniline, and 4-hydroxybenzoic acid produced by fungi can serve as effective natural mediators of laccase (Barneto, Aracri, Andreu & Vidal, 2012; Johannes & Majcherczyk, 2000). However, in certain laccase mediated reactions the nitroxy radicals (formed during oxidation of N-OH compounds) were found to be more stable than phenoxy radicals (formed during oxidation of phenols) (Xu, Kulys, Duke, Li, Krikstopaitis, Deussen, ...& Schneider, 2000).

The mediating ability of redox mediators of natural origin in lignin biodegradation was first evidenced by Eggert et al. (1996). The particular study described laccase catalyzed oxidation of synthetic lignin and non-phenolic substrate in presence of a mediator, 3-Hydroxyanthranilic acid (3-HAA). 3-HAA is a secondary metabolite synthesized by the ligninolytic fungi *Pycnoporus cinnabarinus*. Though, the use of 3-HAA was questioned in later studies as the oxidative coupling of this mediator produces cinnabarinic acid, which is unable to mediate further oxidation of non-phenolic compounds (Li, Horanyi, Collins, Phillips, & Eriksson, 2001). Various mediators of natural origin were studied using recalcitrant dye as a substrate wherein, syringaldehyde and acetosyringone were found to be efficient as compared to the synthetic ones (NNDS and HBT) (Camarero, Ibarra, Martínez, & Martínez, 2005). Degradation of benzo[a]pyrene, pyrene and anthracene using laccase from *Pycnoporus cinnabarinus* with *p*-coumaric acid as mediator have also been reported by Canas and Camarero (2010). Moreover, the presence of redox mediator drives laccase towards the oxidation of non-phenolic compounds, mainly benzyl alcohol groups to their corresponding aldehyde and ketone. This biotransformation was useful in the oxidation of adlerol to its corresponding ketone, adlerone (Barreca, Fabbrini, Galli, Gentili, & Ljunggren, 2003), in oxidation of benzylic and aliphatic alcohols (Fabbrini, Galli, Gentili, & Macchitella, 2001), and allylic and propargyl alcohols (Barilli, Belinghieri, Passarella, Lesma, Riva, Silvani, & Danieli,, 2004) to their corresponding aldehydes and ketones, respectively.

LACCASE FOR SYNTHESIS OF BIOACTIVE COMPOUNDS

Organic synthesis of the chemicals undergoes numerous shortcomings, including the high cost of chemicals, cumbersome multi-step reactions and toxicity of reagents. Thus, in recent decades, laccase has become an attractive catalyst in the field of synthetic chemistry due to their high value of oxidation potential and, their capability to react with extensive range of substrates. Additionally, they are also able to catalyze bond formation reactions even under mild reaction conditions and, therefore are attractive among organic chemists for developing green routes in synthetic chemistry. In organic synthesis laccases are mainly employed for oxidation, bond formation, coupling and polymerization reactions (Mikolasch, Hammer, Jonas, Popowski, Stielow, & Schauer, 2002). The product range of laccase can be widened by performing coupling reaction by linking laccase substrate with a variable reaction partner to form new heteromolecular hybrid molecules (Mikolasch & Schauer, 2009).

Phenolic Compounds as Bioactive Materials

The versatility of laccase has long been exemplified for production of several compounds of pharmaceutical industry mainly, in synthesis of bioactive compounds and drugs like anti-cancer drugs, phenolic anti-oxidants, antibiotics and is also added in cosmetics to reduce their toxicity (Rocasalbas, Francesko, Touriño, Fernández-Francos, Guebitz, & Tzanov, 2013). Phenolic compounds are habitually found as secondary metabolites, secreted by plants and, are termed as the *first line in plant defense against infection* (Matern & Kneusel, 1988), because of the protection they provide to plants during stress and harsh environmental conditions (Bhattacharya, Sood, & Citovsky, 2010). Phenolic compounds hold both therapeutic and nutritional benefits and offer range of bioactivities such as anti-allergenic, anti-inflammatory, anti-microbial, anti-artherogenic, anti-thrombotic and anti-oxidant activities and are well known for providing protection against some cardiovascular diseases (Pasha, Saeed, Waqas, Anjum, & Arshad, 2013). Thereby, driving the researchers towards the synthesis of phenolic compounds employing laccase as a catalyst. All the reactions catalyzed by laccase listed in Table 1 and those discussed herein yielded products with diverse biological activities.

Laccase catalyzed oxidation of phenolic compounds leads to the formation of reactive radical intermediates, consequently, these radicals undergo radical-radical coupling of monomers (homomolecular coupling reaction) to form dimeric, oligomeric or polymeric products; radical cross coupling to form cross coupling products (heteromolecular coupling reaction); and *in situ* formation of *ortho-* and *para-*quinones from corresponding catechols and hydroquinones, respectively (Kudanga, Nyanhongo, Guebitz, & Burton, 2011) (Figure 4). The dimerization, oligomerization and polymerization capability of laccase have been widely applied for the synthesis of several anti-oxidant compounds. Burton and Davids (2012) utilized laccase from *Trametes pubescens* as a catalytic agent for the enhancement of anti-oxidant activity of hydroxytyrosol, a phenolic compound. Using hydroxytyrosol as a laccase substrate, dimers, oligomers and polymers of hydroxytyrosol were successfully obtained with superior anti-oxidant activities and, the dimer (consists of four hydroxyl groups) obtained had 87.6% radical scavenging capability as compared to hydroxytyrosol monomer (consists of two hydroxyl groups) with 33%. The product obtained is industrially useful as an anti-oxidant additive and also is a potential ingredient of skin care products and nutraceuticals. Researchers have also investigated the structural-activity relationship of anti-oxidants. The bioactivity of an anti-oxidant is dependent on the presence of active functional groups (hydroxyl, alkyl, or aniline) attached to its aromatic ring and, on the number of active groups attached to the antioxidant i.e., more the active groups present, more bioactive the anti-oxidant is (Bendary, Francis, Ali, Sarwat, & El Hady, 2013).

Rutin is poorly water-soluble flavonoid glycoside and is commonly found in market as a dietary supplement with notable anti-oxidant activity. Improvement in the properties of a natural phenolic compound rutin was effectively carried out using laccase. Kurisawa et al. (2003b) utilized *Myceliophthora* laccase to catalyze the synthesis of polymerized rutin, (poly(rutin)) with notably better water solubility and radical scavenging properties. Rutin was found to inhibit the protein disulfide isomerase, an enzyme accountable for blood clotting (thrombosis), when secreted fast from platelets and endothelial cells and, was discovered as an efficient anti-thrombotic agent (Jasuja, Passam, Kennedy, Kim, van Hessem, Lin, ... & Furie, 2012). A study has been published describing the use of laccase derived from *Rhus vernicifera* to catalyze domino oxidation of phenylpropanoids (Wan, Lu, Akiyama, Miyakoshi, & Du, 2007). The dimeric form of phenylpropanoid units are referred as lignan, which is obtained from the bark of *Eucommia ulmoides*, a traditional Chinese medicine. This traditional medicine is known for anti-hypertensive (Luo, Ma, Chen, Yao, Wan, Yang, & Hang, 2004; Greenway, Liu, Yu, & Gupta, 2011), anti-oxidant (Zhang, Su, & Zhang, 2013), anti-cancer (Li & Zhang, 2008) and antibiotic activity (JI & SU, 2008). The oxidation of phenylpropanoids catalyzed by *Rhus* laccase produced pinoresinol (8% and 23.5% yield using crude and purified *Rhus* laccase, respectively), a plant lignan which is reported to have strongest anti-inflammatory properties in human intestinal Caco-2 cells (During, Debouche, Raas, & Larondelle, 2012). Fini et al. (2008) also published a report on the chemo-preventive properties of pinoresinol rich olive oil that cause inactivation of ATM-p53 cascade in colon cancer cell lines. Another compound produced as a result of oxidation of phenylpropanoids was dehydrodiisoeugenol (24.5% and 25% yield using crude and purified *Rhus* laccase, respectively), which is commonly used for curing gastrointestinal disorders (Li & Yang, 2012), and also can be used as an anti-inflammatory or anti-oxidant agent (Murakami, Shoji, Hirata, Tanaka, Yokoe, & Fujisawa, 2005).

Laccase assisted oxidation reaction was also employed for the dimerization of natural phenolic compounds. Two dimers of ferulic acid, β-5 and β-β dimers were successfully obtained by oxidative dimerization of a natural phenolic monomer, ferulic acid in a laccase catalyzed reaction reported by Adelakun et al. (2012a). The β-5 dimer obtained showed higher anti-oxidant activity, while β-β dimer had lesser activity than ferulic acid monomer. The dimers can be served as an anti-oxidant additive for cosmetic and other pharmaceutical products. This improved anti-oxidant activity of β-5 dimer is due to the increased electron donating group on the compound and the carboxylic acid group associated with the neighboring unsaturated C-C double bond, which is responsible for providing extra sites to free radicals to attack (Srinivasan, Sudheer, & Menon, 2007). Moreover, it is also reported that the functional groups such as hydroxyl, alkyl, or aniline groups enhance the anti-oxidant activity of a compound (Bendary, Francis, Ali, Sarwat, & El Hady, 2013). Similarly, laccase from *T. pubescens* catalyzed the oxidative dimerization of 2,6-dimethoxyphenol (2,6-DMP), resulting in formation of a C-C linked dimer with 2-fold higher anti-oxidant activity compared to 2,6-DMP monomer (Adelakun, Kudanga, Green, le Roes-Hill, & Burton, 2012b) (Figure 5). This enhanced activity of 2,6-DMP dimer is attributed to the presence of more number of functional groups having ability to donate electron (Matsuura & Ohkatsu, 2000), reduced O-H bond dissociation energy, and increased radical stability (Sánchez-Moreno, Larrauri, & Saura-Calixto, 1998). The dimer attained has found potential use in the advancement of nutraceuticals and also as a component for cosmetic products.

Quinones are one of the important classes of natural and synthetic compounds with several beneficial effects. A number of studies have been conducted using laccase to generate reactive *ortho*- and *para-* quinone species and their further reaction with carbon, nitrogen and sulfur derived nucleophiles to produce biomaterials (Witayakran, 2008). A subgroup of quinones are 1,4-naphthoquinones and several derivatives of this compound have exhibited biological responses such as anti-allergic, anti-fungal, anti-bacterial, apoptosis, etc. and also have displayed anti-cancer activity (Wellington & Kolesnikova, 2012). Witayakran and Ragauskas (2007) reported a laccase catalyzed synthesis of *ortho-* and *para-* naphthoquinones via Diels-Alder reaction in an aqueous one-pot reaction. In a particular reaction, substituted hydroquinones and catechols were coupled with various dienes resulting in formation of 1,4-naphthoquinones and 1,2-naphthoquinones, respectively (Figure 6). In another study, Wellington and Kolesnikova (2012) exploited laccase from *Myceliophthora thermophila* for the synthesis of aminonaphthoquinones through nuclear monoamination of 1,4-naphthohydroquinone with primary aromatic amines by enabling C-N bond formation. The procedure resulted in formation of compounds that exhibited high potency against UACC62 (melanoma), TK10 (renal), and MCF (breast) cancer cell lines, and weak cytotoxicity on HeLa cell lines.

The synthetic approach for the production of benzofurans involve the use of sodium iodate or electrochemical method to carry out the oxidation of catechol in the presence of 1,3-dicarbonyl compounds via oxidative Michael addition mechanism (Nematollahi & Forooghi, 2002; Pei, Li, Bu, Gu, & Chan, 2006). However, number of other researchers have published the reports on synthesis of benzofuran products via laccase catalyzed oxidation-Michael addition reaction of catechols with cyclic, heterocyclic, and aliphatic 1,3-dicarbonyls (Wellington, Qwebani-Ogunleye, Kolesnikova, Brady, & de Koning, 2013; Kidwai, Jain, Sharma, & Kuhad, 2013a; Kidwai, Jain, Sharma, & Kuhad, 2013b). The synthesis of benzofurans was achieved by exploiting the oxidation potential of laccase to start an enzymatic cascade reaction between catechols and 1,3-dicarbonyl compounds (Figure 7). The reaction products obtained has a benzofuran core, a chief structural motif found in pharmaceutically active complexes having anti-microbial, anti-oxidant, and anti-HIV-1 properties (Rida, El-Hawash, Fahmy, Hazza, & El-Meligy, 2006).

A process has been developed for laccase catalyzed domino reactions between catechol and thiols to produce catechol thioethers. Laccase catalyzed the oxidation of a catechol to corresponding *o*-benzoquinone which subsequently undergoes 1,4-addition with a thiol by nucleophilic conjugate addition yielding catechol thioethers under milder reaction conditions (Abdel-Mohsen, Conrad, & Beifuss, 2014) (Figure 8). Catechol thioethers find application as powerful anti-oxidant and anti-bacterial agents (Adibi, Rashidi, Khodaei, Alizadeh, Majnooni, Pakravan, ...& Nematollahi, 2011). In a related study, laccase catalyzed reaction of substituted catechols with 2,3-dihydro-2-thioxopyrimidin-4(1*H*)-one yielded corresponding catechol thioethers and pyrimidobenzothiazoles, depending upon the substitution pattern of 2,3-dihydro-2-thioxopyrimidin-4(1*H*)-ones. The resulted catechol thioethers and pyrimidobenzothiazoles displayed potent cytotoxic activity when tested against HepG2 cell line (Abdel-Mohsen, Conrad, Harms, Nohr, & Beifuss, 2017) (Figure 8).

The potential of fungal laccase from *Myceliophthora thermophila* was investigated by Engelmann et al. (2015) for its ability to carry out the oxidative C-C couplings of phenolic compounds. The selective oxidative coupling of disubstituted phenols [2,6-DMP, 2,6-dimethoxyphenol (2,6-DMOP) and 2,6-diisopropylphenol (2,6-DIPP)], showed highly symmetric coupling products through laccase assisted oxidative dimerization reaction of disubstituted phenolic compounds. The oxidative coupling of 2,6-DIPP resulted into the formation of reduction product, 2,6-diisopropylphenol-diquinone(2,6-DIIP-DQ), a biphenol, commonly known antibacterial agent dipropofol (Ogata, Sato, Kunikane, Oka, Seki, Urano, ...& Endo, 2005; Ogata, Oka, Seki, Hoshi, Takatsu, Mashino, ...& Endo, 2007). The biphenol and quinoid units are found to be the structural unit for the synthesis of various pharmacologically important compounds.

Cannatelli and Ragauskas (2015a) employed a laccase assisted route to carry out the synthesis of 2,3-ethylenedithio-1,4-quinones by cross-coupling of 1,2-ethanedithiol with substituted hydroquinones (Figure 9). In a particular reaction, laccase first catalyzed the oxidation of hydroquinone to their corresponding 1,4-quinone derivative, which further undergoes nucleophilic addition by 1,2-ethanedithiol followed by more oxidation and addition steps, subsequently forming respective 2,3-ethylenedithio-1,4-quinone compounds. The synthesized cross coupling products of 1,2-ethanedithiol with hydroquinone contained a substructure of 2,3-ethylenedithio-1,4-quinone. The presence of this substructure is observed in compounds that have displayed pharmacological activity, for example the anti-proliferative 3',4'-(ethylenedithio)avarone and the cancerostatic agent dithianon (Božić, Novaković, Gašić, Juranić, Stanojković, Tufegdžić, ...& Sladić, 2010). While the coupling products of 1,2-ethanedithiol with naphthoquinone exhibited anti-bacterial activity (Tandon, Yadav, Singh, Vaish, Chaturvedi, & Shukla, 2005), as well as inhibited tumor metastasis (Bresnick, 2013). In a related study, laccase from *Trametes villosa* was exploited for the α-arylation of benzoylacetonitrile with several substituted hydroquinones, to yield benzylic nitriles (Cannatelli & Ragauskas (2015b) (Figure 10). Benzylic nitriles serves as a key component in manufacturing of several bioactive materials and drugs such as anti-helmintic and analgesics drugs (Fleming, Yao, Ravikumar, Funk, & Shook, 2010; Kermanshai, McCarry, Rosenfeld, Summers, Weretilnyk, & Sorger, 2001).

Alkaloids and Antibiotics as Bioactive Materials

Apart from the synthesis of phenolic bioactive compounds that are catalyzed by laccases, they are also being employed for the synthesis of biomaterials other than phenolic origin such as alkaloids, antibiotics, and other bioactive polymers. Many researchers have exploited the oxidation potential of laccase in production and modification of alkaloids and their derivatives. Sagui et al. (2009) reported the production of anhydrovinblastine, an anti-neoplastic bisindole alkaloid, by a laccase mediated coupling of catharanthine and vindoline. The reaction product was obtained with 56% yield which is higher than the chemical methods and enzyme cocktail biocatalysis procedures used so far. Anhydrovinblastine finds application in synthesis of anti-cancer and anti-tumor drugs (van der Heijden, Jacobs, Snoeijer, Hallard, & Verpoorte, 2004).

Another example of a laccase aided synthesis of biomaterials is the oxidation of ergot alkaloids to form ergoline derivatives. Ergot alkaloids are therapeutically important class of bioactive compounds and thus spurred research is being carried out in their enzymatic synthesis. During ancient era, ergot alkaloids were known for its poisonous nature, however, this image of ergot alkaloids has been changed after many findings, and they are now considered as highly valuable source of pharmaceuticals. Ergot alkaloids are being used as therapeutically important drug in treatment of syndromes like Parkinson's, hypertension, uterine antonia, migraines, etc., (Rutschmann, Stadler, Berde, & Schild, 1978). Chirivi and co-workers for the first time successfully showed the addition of hydroxyl group to the C-4 position of ergoline ring in the absence of chemical mediator (TEMPO) using a reaction catalyzed by *Trametes versicolor* laccase. During the reaction instead of expected oxidation to be occurred at CH_2OH group of trans-dihydrolysergol, hydroxylation reaction occurred at C-4 site, and the functionalization of ergot alkaloid at C-4 site has not been reported earlier by any researcher even by chemical means (Chirivì, Fontana, Monti, Ottolina, Riva, & Danieli, 2012).

Laccase ability to catalyze homo-coupling or hetero-coupling reactions is an opportunity for scientists in synthesizing new bioactive molecules, or in modification of the existing ones. One such important class of bioactive compounds that has been researched widely for its anti-microbial action against wide range of microorganisms is antibiotics. However, the challenge confronted in using antibiotics is the resistance power that is developed by the microorganisms, as a result of which antibiotics loses its effectiveness. This has been the universal problem being met with ESKAPE (*Enterococcus faecium, Staphylococcus aureus, Klebsiella pneumoniae, Acinetobacter baumannii, Pseudomonas aeruginosa,* and *Enterobacter* sp.) pathogens, known for their ability to develop new mechanisms to "escape" the strength of antibiotics (Lewis, 2013). According to the World Health Organization survey, the cases for multi-drug resistant tuberculosis are increasing drastically and that 480,000 cases were reported in the year 2013. The increasing drug resistance developed by microorganisms and decrease in the effect of antibiotics has urged the chemists in synthesizing a new antibiotics or modifying the existing one with higher potency (Aminov, 2010).

A number of reports have been published on the synthesis and modification of antibiotics by a laccase assisted reactions. The enzyme catalyzed production of antibiotics occurs mainly via phenolic oxidation (Agematu, Kominato, Shibamoto, Yoshioka, Nishida, Okamoto, ...& Murao, 1993a), phenolic oxidative coupling (Anyanwutaku, Petroski, & Rosazza, 1994), and oxidation coupled with nuclear amination (Mikolasch, A., Niedermeyer, T. H. J., Lalk, M., Witt, S., Seefeldt, S., Hammer, E., ...& Lindequist, 2006; 2007; Mikolasch, Wurster, Lalk, Witt, Seefeldt, Hammer, ...& Lindequist, 2008a). Agematu et al. (1993b) was the first to report the modification of antibiotic via laccase catalyzed oxidative dimerization of penicillin X however; the dimers obtained as a result of reaction did not display effective anti-microbial action. The succeeding studies have thus focused onto the production of antibiotics by heteromolecular coupling of antibiotic to obtain a product with improved activity. As a result of this, Anyanwutaku et al. (1994) carried out the phenolic oxidative coupling of hydroquinone and mitramycin, resulting in formation of heteromolecular dimers. The dimeric antibiotic has displayed lower activity as compared to substrates. Furthermore, Mikolasch and co-workers (2006) have successfully synthesized eight novel penicillin by coupling 2,5-dihydroxybenzoic acid derivatives to ampicillin or amoxicillin, with 98% yield obtained within 3h. The derivatives obtained showed remarkable bioactivity, inhibited the growth of gram positive bacteria and protected mice against lethal *Staphylococcus aureus* infection (ATCC 6538 and 3841). However, derivatives failed to display activity higher than ampicillin or amoxicillin.

In a subsequent study, Mikolasch et al. (2007) have synthesized derivatives of penicillin and cephalosporin by oxidation reaction coupled with nuclear amination resulting in formation of C-N coupled aminoquinonoid heterodimers (Figure 11). The heterodimers efficiently inhibited the growth of gram-positive bacterial strains including vancomycin-resistant Enterococci and methicillin resistant *Staphylococcus aureus*, and also protected mice infected with *Staphylococcus aureus*. Employing laccase catalyzed oxidation coupled with nuclear amination, novel derivatives of antibiotics were synthesized by cross-coupling N-analogues corollosporines, and β-lactam antibiotics (cefadroxil, amoxicillin, and ampicillin) with 2,5-dihydroxybenzoic acid derivatives and methylcatechols, respectively, resulting in formation of heterodimers with improved bioactivity (Mikolasch, Wurster, Lalk, Witt, Seefeldt, Hammer, ...& Lindequist, 2008a,b). Osiadacz and co-workers successfully carried out the dimerization of 4-methyl-3-hydroxyanthranilic acid to 2-amino-4,6- dimethyl-3-phenoxazinone-1,9-carboxylic acid (actinocin) using laccase from *Trametes versicolor* (Figure 12). Actinocin is a phenoxazinone chromophore found in actinomycin, a polypeptide antibiotic, and also has proven its potential as an effective anti-cancer drug, by blocking transcription of tumor cell DNA (Osiadacz, Al-Adhami, Bajraszewska, Fischer, & Peczyñska-Czoch, 1999). More examples of antibiotics synthesized via laccase assisted reaction are listed in Table 1.

Natural Polymers as Biomaterials

Laccase exhibits excellent proficiency in synthesizing polymeric structures and in functionalization of polymers with various phenolic compounds by means of grafting reactions. The usefulness of laccase in detoxification and diminishing the environmental pollutants by catalyzing polymerization reactions have been reported earlier in many studies (Uchida, Fukuda, Miyamoto, Kawabata, Suzuki, & Uwajima, 2001; Schultz, Jonas, Hammer, & Schauer, 2001; Ruttimann-Johnson & Lamar, 1996). However, recent researches have been focused on the grafting of laccase oxidized phenolic compounds onto natural polymers mainly chitosan, a linear polysaccharide (Aljawish, Chevalot, Jasniewski, Revol-Junelles, Scher, & Muniglia, 2014a; Božič, Gorgieva, & Kokol, 2012a). Chitosan is a biopolymer, naturally obtained from the hard outer skeleton of shrimp and crab shells and finds application in wine making to prevent spoilage and as a fining agent (Elmacı, Gülgör, Tokatlı, Erten, İşci, & Özçelik, 2015), in medicine as an anti-bacterial agent (Raafat, Von Bargen, Haas, & Sahl, 2008), and in food industry to improve the quality and shelf life of foods (No, Meyers, Prinyawiwatkul, & Xu, 2007). However, due to poor solubility and anti-oxidant activity their use is still limited. As stated by Bendary et al. (2013) the bioactivity of an anti-oxidant is dependent on the number of active functional groups attached to the aromatic ring, therefore, the poor antioxidant capacity of chitosan is attributed to the presence of less number of hydroxyl group attached onto the biopolymer.

To enhance the anti-oxidant power of chitosan, Bozic and co-workers have performed series of grafting reactions of laccase-oxidized phenolics (caffeic acid, gallic acid, tannic acid, and quercetin) onto chitosan backbone forming respective chitosan derivatives (Božič, Štrancar, & Kokol, 2013; Božič, Gorgieva, & Kokol, 2012a; Božič, Gorgieva, & Kokol, 2012b). Chitosan grafted with all the phenolics displayed better ABTS radical scavenging activity however; the highest was for quercetin and tannic acid-functionalized chitosan. Furthermore, the caffeic acid and gallic acid-functionalized chitosan derivatives displayed significant anti-bacterial activity against *Escherichia coli* and *Listeria monocytogenes* as compared to natural chitosan and thus, can be used in pharmaceutical and cosmetic products. In a related study, Aljawish et al. (2014a; 2014b) functionalized chitosan with ferulic acid and ethyl ferulate employing *Myceliophthora thermophila* laccase. The reaction resulted in the formation of respective phenolic-functionalized derivatives with enhanced ABTS radical scavenging ability than natural biopolymer. Similarly, chitosan/gelatin hydrogels cross-linked with plant phenolic extract using laccase were resistant to degradation by lysozyme, and also inhibited the growth of bacteria, thereby, finding application as chronic wound healing bioactive hydrogels (Rocasalbas, Francesko, Touriño, Fernández-Francos, Guebitz, & Tzanov, 2013).

CONCLUSION AND FUTURE RESEARCH DIRECTIONS

The potential of laccase as green biocatalysts for the synthesis of bioactive compounds is well documented. The broad substrate spectrum of laccase has permitted the access to variety of biologically active molecules that are not accessible by standard procedures. The applications reviewed may offer a green alternative to the present harsh physico-chemical methods which are costly, less specific and are environmentally hostile. More novel applications of laccase will continue to discover with the time. But, the major drawback that confronts the scientists is, high concentrations of organic solvents that are frequently being used during reactions to overcome the problem of radical proliferation and polymerization reactions, but the same solvents also inactivate the enzymes. In regard to this concern, developments in field of protein engineering have paved the way for improvement in stability and specificity of the enzyme and tolerance to organic solvent and ionic liquids, thus, making laccase a promising "green tool" for industrial processes. Moreover, future research should not only highlight the synthetic properties of enzyme, but also reaction engineering to optimize the synthesis which consequently will increase the yield of value added products. So in future it could facilitate the commercialization of laccase mediated processes, bringing it from bench to industrial level.

ACKNOWLEDGMENT

The authors are thankful to the Department of Biotechnology (DBT sanction no. BT/PR5859/PID/6/696/2012) and Ministry of Science and Technology, New Delhi, for their financial support. The authors also wish to acknowledge SICART, V.V. Nagar, Gujarat, for providing the necessary instrumentation facilities.

REFERENCES

Abdel-Mohsen, H. T., Conrad, J., & Beifuss, U. (2014). Laccase-catalyzed synthesis of catechol thioethers by reaction of catechols with thiols using air as an oxidant. *Green Chemistry*, *16*(1), 90–95. doi:10.1039/C3GC41968A

Abdel-Mohsen, H. T., Conrad, J., Harms, K., Nohr, D., & Beifuss, U. (2017). Laccase-catalyzed green synthesis and cytotoxic activity of novel pyrimidobenzothiazoles and catechol thioethers. *RSC Advances*, *7*(28), 17427–17441. doi:10.1039/C6RA28102H

Adelakun, O. E., Kudanga, T., Green, I. R., le Roes-Hill, M., & Burton, S. G. (2012b). Enzymatic modification of 2, 6-dimethoxyphenol for the synthesis of dimers with high antioxidant capacity. *Process Biochemistry*, *47*(12), 1926–1932. doi:10.1016/j.procbio.2012.06.027

Adelakun, O. E., Kudanga, T., Parker, A., Green, I. R., le Roes-Hill, M., & Burton, S. G. (2012a). Laccase-catalyzed dimerization of ferulic acid amplifies antioxidant activity. *Journal of Molecular Catalysis. B, Enzymatic*, *74*(1-2), 29–35. doi:10.1016/j.molcatb.2011.08.010

Adibi, H., Rashidi, A., Khodaei, M. M., Alizadeh, A., Majnooni, M. B., Pakravan, N., ... Nematollahi, D. (2011). Catecholthioether derivatives: Preliminary study of in-vitro antimicrobial and antioxidant activities. *Chemical & Pharmaceutical Bulletin*, *59*(9), 1149–1152. doi:10.1248/cpb.59.1149 PMID:21881259

Agematu, H., Kominato, K., Shibamoto, N., Yoshioka, T., Nishida, H., Okamoto, R., ... Murao, S. (1993a). Transformation of 7-(4-hydroxyphenylacetamido) cephalosporanic acid into a new cephalosporin antibiotic, 7-[1-oxaspiro (2.5) octa-6-oxo-4, 7-diene-2-carboxamido] cephalosporanic acid, by laccase. *Bioscience, Biotechnology, and Biochemistry*, *57*(8), 1387–1388. doi:10.1271/bbb.57.1387

Agematu, H., Tsuchida, T., Kominato, K., Shibamoto, N., Yoshioka, T., Nishida, H., ... Murao, S. (1993b). Enzymatic dimerization of penicillin X. *The Journal of Antibiotics*, *46*(1), 141–148. doi:10.7164/antibiotics.46.141 PMID:8436547

Aljawish, A., Chevalot, I., Jasniewski, J., Paris, C., Scher, J., & Muniglia, L. (2014b). Laccase-catalysed oxidation of ferulic acid and ethyl ferulate in aqueous medium: A green procedure for the synthesis of new compounds. *Food Chemistry*, *145*, 1046–1054. doi:10.1016/j.foodchem.2013.07.119 PMID:24128582

Aljawish, A., Chevalot, I., Jasniewski, J., Revol-Junelles, A. M., Scher, J., & Muniglia, L. (2014a). Laccase-catalysed functionalisation of chitosan by ferulic acid and ethyl ferulate: Evaluation of physicochemical and biofunctional properties. *Food Chemistry*, *161*, 279–287. doi:10.1016/j.foodchem.2014.03.076 PMID:24837951

Aminov, R. I. (2010). A brief history of the antibiotic era: Lessons learned and challenges for the future. *Frontiers in Microbiology*, *1*, 134. doi:10.3389/fmicb.2010.00134 PMID:21687759

Anastas, P. T., Kirchhoff, M. M., & Williamson, T. C. (2001). Catalysis as a foundational pillar of green chemistry. *Applied Catalysis A, General*, *221*(1-2), 3–13. doi:10.1016/S0926-860X(01)00793-1

Anastas, P. T., & Warner, J. C. (1998). Principles of green chemistry. *Green chemistry. Theory into Practice*, 29–56.

Anyanwutaku, I. O., Petroski, R. J., & Rosazza, J. P. (1994). Oxidative coupling of mithramycin and hydroquinone catalyzed by copper oxidases and benzoquinone.Implications for the mechanism of action of aureolic acid antibiotics. *Bioorganic & Medicinal Chemistry*, *2*(6), 543–551. doi:10.1016/0968-0896(94)80025-1 PMID:8000877

Arora, D. S., & Sharma, R. K. (2010). Ligninolytic fungal laccases and their biotechnological applications. *Applied Biochemistry and Biotechnology*, *160*(6), 1760–1788. doi:10.100712010-009-8676-y PMID:19513857

Astolfi, P., Brandi, P., Galli, C., Gentili, P., Gerini, M. F., Greci, L., & Lanzalunga, O. (2005). New mediators for the enzyme laccase: Mechanistic features and selectivity in the oxidation of non-phenolic substrates. *New Journal of Chemistry*, *29*(10), 1308–1317. doi:10.1039/b507657a

Baiocco, P., Barreca, A. M., Fabbrini, M., Galli, C., & Gentili, P. (2003). Promoting laccase activity towards non-phenolic substrates: A mechanistic investigation with some laccase–mediator systems. *Organic & Biomolecular Chemistry*, *1*(1), 191–197. doi:10.1039/B208951C PMID:12929410

Barilli, A., Belinghieri, F., Passarella, D., Lesma, G., Riva, S., Silvani, A., & Danieli, B. (2004). Enzyme assisted enantioselective synthesis of the alkaloid (+)-aloperine. *Tetrahedron, Asymmetry*, *15*(18), 2921–2925. doi:10.1016/j.tetasy.2004.06.056

Barneto, A. G., Aracri, E., Andreu, G., & Vidal, T. (2012). Investigating the structure–effect relationships of various natural phenols used as laccase mediators in the biobleaching of kenaf and sisal pulps. *Bioresource Technology*, *112*, 327–335. doi:10.1016/j.biortech.2012.02.136 PMID:22437048

Barreca, A. M., Fabbrini, M., Galli, C., Gentili, P., & Ljunggren, S. (2003). Laccase/mediated oxidation of a lignin model for improved delignification procedures. *Journal of Molecular Catalysis. B, Enzymatic*, *26*(1-2), 105–110. doi:10.1016/j.molcatb.2003.08.001

Bendary, E., Francis, R. R., Ali, H. M. G., Sarwat, M. I., & El Hady, S. (2013). Antioxidant and structure–activity relationships (SARs) of some phenolic and anilines compounds. *Annals of Agricultural Science*, *58*(2), 173–181. doi:10.1016/j.aoas.2013.07.002

Bertrand, G. (1894). Sur le latex de l'arbre à laque. *Comptes Rendus Hebdomadaires des Séances de l'Académie des Sciences*, *118*, 1215–1218.

Bertrand, T., Jolivalt, C., Briozzo, P., Caminade, E., Joly, N., Madzak, C., & Mougin, C. (2002). Crystal structure of a four-copper laccase complexed with an arylamine: Insights into substrate recognition and correlation with kinetics. *Biochemistry*, *41*(23), 7325–7333. doi:10.1021/bi0201318 PMID:12044164

Bhattacharya, A., Sood, P., & Citovsky, V. (2010). The roles of plant phenolics in defence and communication during *Agrobacterium* and *Rhizobium* infection. *Molecular Plant Pathology*, *11*, 705–719. PMID:20696007

Bourbonnais, R., & Paice, M. G. (1990). Oxidation of non-phenolic substrates: An expanded role for laccase in lignin biodegradation. *FEBS Letters*, *267*(1), 99–102. doi:10.1016/0014-5793(90)80298-W PMID:2365094

Božič, M., Gorgieva, S., & Kokol, V. (2012a). Homogeneous and heterogeneous methods for laccase-mediated functionalization of chitosan by tannic acid and quercetin. *Carbohydrate Polymers*, *89*(3), 854–864. doi:10.1016/j.carbpol.2012.04.021 PMID:24750872

Božič, M., Gorgieva, S., & Kokol, V. (2012b). Laccase-mediated functionalization of chitosan by caffeic and gallic acids for modulating antioxidant and antimicrobial properties. *Carbohydrate Polymers*, *87*(4), 2388–2398. doi:10.1016/j.carbpol.2011.11.006

Božič, M., Štrancar, J., & Kokol, V. (2013). Laccase-initiated reaction between phenolic acids and chitosan. *Reactive & Functional Polymers*, *73*(10), 1377–1383. doi:10.1016/j.reactfunctpolym.2013.01.005

Božić, T., Novaković, I., Gašić, M. J., Juranić, Z., Stanojković, T., Tufegdžić, S., ... Sladić, D. (2010). Synthesis and biological activity of derivatives of the marine quinone avarone. *European Journal of Medicinal Chemistry*, *45*(3), 923–929. doi:10.1016/j.ejmech.2009.11.033 PMID:19995673

Bresnick, A. R. (2013). Chemical agents for the prevention of inhibition or tumor metastasis *U.S. Patent US20130090355 A1*.

Burton, S. G., & Davids, L. M. (2014). *U.S. Patent No. 8,663,608*. Washington, DC: U.S. Patent and Trademark Office.

Camarero, S., Ibarra, D., Martínez, M. J., & Martínez, Á. T. (2005). Lignin-derived compounds as efficient laccase mediators for decolorization of different types of recalcitrant dyes. *Applied and Environmental Microbiology*, *71*(4), 1775–1784. doi:10.1128/AEM.71.4.1775-1784.2005 PMID:15812000

Cañas, A. I., & Camarero, S. (2010). Laccases and their natural mediators: Biotechnological tools for sustainable eco-friendly processes. *Biotechnology Advances*, *28*(6), 694–705. doi:10.1016/j.biotechadv.2010.05.002 PMID:20471466

Cannatelli, M. D., & Ragauskas, A. J. (2015a). Laccase-catalyzed synthesis of 2, 3-ethylenedithio-1, 4-quinones. *Journal of Molecular Catalysis. B, Enzymatic*, *119*, 85–89. doi:10.1016/j.molcatb.2015.05.016

Cannatelli, M. D., & Ragauskas, A. J. (2015b). Laccase-catalyzed α-arylation of benzoylacetonitrile with substituted hydroquinones. *Chemical Engineering Research & Design*, *97*, 128–134. doi:10.1016/j.cherd.2014.08.021

Cannatelli, M. D., & Ragauskas, A. J. (2017). Two decades of laccases: Advancing sustainability in the chemical industry. *Chemical Record (New York, N.Y.)*, *17*(1), 122–140. doi:10.1002/tcr.201600033 PMID:27492131

Cantarella, G., Galli, C., & Gentili, P. (2003). Free radical versus electron-transfer routes of oxidation of hydrocarbons by laccase/mediator systems: Catalytic or stoichiometric procedures. *Journal of Molecular Catalysis. B, Enzymatic*, *22*(3-4), 135–144. doi:10.1016/S1381-1177(03)00014-6

Chakroun, H., Bouaziz, M., Yangui, T., Blibech, I., Dhouib, A., & Sayadi, S. (2013). Enzymatic transformation of tyrosol by *Trametes trogii* laccases: Identification of the product and study of its biological activities. *Journal of Molecular Catalysis. B, Enzymatic*, *87*, 11–17. doi:10.1016/j.molcatb.2012.10.006

Chirivì, C., Fontana, G., Monti, D., Ottolina, G., Riva, S., & Danieli, B. (2012). The quest for new mild and selective modifications of natural structures: Laccase-catalysed oxidation of ergot alkaloids leads to unexpected stereoselective C-4 hydroxylation. *Chemistry (Weinheim an der Bergstrasse, Germany)*, *18*(33), 10355–10361. doi:10.1002/chem.201201076 PMID:22777708

Chung, J. E., Kurisawa, M., Uyama, H., & Kobayashi, S. (2003). Enzymatic synthesis and antioxidant property of gelatin-catechin conjugates. *Biotechnology Letters*, *25*(23), 1993–1997. doi:10.1023/B:BILE.0000004391.27564.8e PMID:14719812

Claus, H. (2004). Laccases: Structure, reactions, distribution. *Micron (Oxford, England)*, *35*(1-2), 93–96. doi:10.1016/j.micron.2003.10.029 PMID:15036303

Constable, D. J., Dunn, P. J., Hayler, J. D., Humphrey, G. R., Leazer, J. L. Jr, Linderman, R. J., ... Zaks, A. (2007). Key green chemistry research areas—a perspective from pharmaceutical manufacturers. *Green Chemistry*, *9*(5), 411–420. doi:10.1039/B703488C

Constantin, M. A., Conrad, J., & Beifuss, U. (2012). Laccase-catalyzed oxidative phenolic coupling of vanillidene derivatives. *Green Chemistry*, *14*(9), 2375–2379. doi:10.1039/c2gc35848d

d'Acunzo, F., Baiocco, P., Fabbrini, M., Galli, C., & Gentili, P. (2002). A mechanistic survey of the oxidation of alcohols and ethers with the enzyme laccase and its mediation by TEMPO. *European Journal of Organic Chemistry*, *2002*(24), 4195–4201. doi:

de Souza, C. G. M., & Peralta, R. M. (2003). Purification and characterization of the main laccase produced by the white-rot fungus *Pleurotus pulmonarius* on wheat bran solid state medium. *Journal of Basic Microbiology*, *43*(4), 278–286. doi:10.1002/jobm.200390031 PMID:12872309

Debbab, A., Aly, A. H., Lin, W. H., & Proksch, P. (2010). Bioactive compounds from marine bacteria and fungi. *Microbial Biotechnology*, *3*(5), 544–563. doi:10.1111/j.1751-7915.2010.00179.x PMID:21255352

Díaz-Rodríguez, A., Lavandera, I., Kanbak-Aksu, S., Sheldon, R. A., Gotor, V., & Gotor-Fernández, V. (2012). From diols to lactones under aerobic conditions using a laccase/TEMPO catalytic system in aqueous medium. *Advanced Synthesis & Catalysis*, *354*(18), 3405–3408. doi:10.1002/adsc.201200892

During, A., Debouche, C., Raas, T., & Larondelle, Y. (2012). Among plant lignans, pinoresinol has the strongest antiinflammatory properties in human intestinal Caco-2 cells. *The Journal of Nutrition, 142*(10), 1798–1805. doi:10.3945/jn.112.162453 PMID:22955517

Dwivedi, U. N., Singh, P., Pandey, V. P., & Kumar, A. (2011). Structure–function relationship among bacterial, fungal and plant laccases. *Journal of Molecular Catalysis. B, Enzymatic, 68*(2), 117–128. doi:10.1016/j.molcatb.2010.11.002

Eggert, C. (1997). Laccase-catalyzed formation of cinnabarinic acid is responsible for antibacterial activity of *Pycnoporus cinnabarinus. Microbiological Research, 152*(3), 315–318. doi:10.1016/S0944-5013(97)80046-8 PMID:9352667

Eggert, C., Temp, U., Dean, J. F., & Eriksson, K. E. L. (1996). A fungal metabolite mediates degradation of non-phenolic lignin structures and synthetic lignin by laccase. *FEBS Letters, 391*(1-2), 144–148. doi:10.1016/0014-5793(96)00719-3 PMID:8706903

Elmacı, S. B., Gülgör, G., Tokatlı, M., Erten, H., İşci, A., & Özçelik, F. (2015). Effectiveness of chitosan against wine-related microorganisms. *Antonie van Leeuwenhoek, 107*(3), 675–686. doi:10.100710482-014-0362-6 PMID:25528342

Emirdağ-Öztürk, S., Hajdok, S., Conrad, J., & Beifuss, U. (2013). Laccase-catalyzed reaction of 3-tert-butyl-1H-pyrazol-5 (4H)-one with substituted catechols using air as an oxidant. *Tetrahedron, 69*(18), 3664–3668. doi:10.1016/j.tet.2013.03.023

Engelmann, C., Illner, S., & Kragl, U. (2015). Laccase initiated C-C couplings: Various techniques for reaction monitoring. *Process Biochemistry, 50*(10), 1591–1599. doi:10.1016/j.procbio.2015.05.011

Fabbrini, M., Galli, C., & Gentili, P. (2002). Comparing the catalytic efficiency of some mediators of laccase. *Journal of Molecular Catalysis. B, Enzymatic, 16*(5-6), 231–240. doi:10.1016/S1381-1177(01)00067-4

Fabbrini, M., Galli, C., Gentili, P., & Macchitella, D. (2001). An oxidation of alcohols by oxygen with the enzyme laccase and mediation by TEMPO. *Tetrahedron Letters, 42*(43), 7551–7553. doi:10.1016/S0040-4039(01)01463-0

Fini, L., Hotchkiss, E., Fogliano, V., Graziani, G., Romano, M., De Vol, E. B., ... Ricciardiello, L. (2007). Chemopreventive properties of pinoresinol-rich olive oil involve a selective activation of the ATM–p53 cascade in colon cancer cell lines. *Carcinogenesis, 29*(1), 139–146. doi:10.1093/carcin/bgm255 PMID:17999988

Fleming, F. F., Yao, L., Ravikumar, P. C., Funk, L., & Shook, B. C. (2010). Nitrile-containing pharmaceuticals: Efficacious roles of the nitrile pharmacophore. *Journal of Medicinal Chemistry, 53*(22), 7902–7917. doi:10.1021/jm100762r PMID:20804202

Gavezzotti, P., Vavříková, E., Valentová, K., Fronza, G., Kudanga, T., Kuzma, M., ... Křen, V. (2014). Enzymatic oxidative dimerization of silymarin flavonolignans. *Journal of Molecular Catalysis. B, Enzymatic, 109*, 24–30. doi:10.1016/j.molcatb.2014.07.012

Greenway, F., Liu, Z., Yu, Y., & Gupta, A. (2011). A clinical trial testing the safety and efficacy of a standardized *Eucommia ulmoides*oliver bark extract to treat hypertension. *Alternative Medicine Review*, *16*, 338–347. PMID:22214253

Guaadaoui, A., Benaicha, S., Elmajdoub, N., Bellaoui, M., & Hamal, A. (2014). What is a bioactive compound? A combined definition for a preliminary consensus. *International Journal of Nutrition and Food Sciences*, *3*, 174–179.

Hahn, V., Mikolasch, A., Manda, K., Gördes, D., Thurow, K., & Schauer, F. (2009b). Laccase-catalyzed carbon–nitrogen bond formation: Coupling and derivatization of unprotected L-phenylalanine with different para-hydroquinones. *Amino Acids*, *37*(2), 315–321. doi:10.100700726-008-0154-2 PMID:18695937

Hahn, V., Mikolasch, A., Wende, K., Bartrow, H., Lindequist, U., & Schauer, F. (2009a). Synthesis of model morpholine derivatives with biological activities by laccase-catalysed reactions. *Biotechnology and Applied Biochemistry*, *54*(4), 187–195. doi:10.1042/BA20090219 PMID:19788410

Hakulinen, N., & Rouvinen, J. (2015). Three-dimensional structures of laccases. *Cellular and Molecular Life Sciences*, *72*(5), 857–868. doi:10.100700018-014-1827-5 PMID:25586561

Ham, S. S., Kim, S. H., Moon, S. Y., Chung, M. J., Cui, C. B., Han, E. K., ... Choe, M. (2009). Antimutagenic effects of subfractions of Chaga mushroom (*Inonotus obliquus*) extract. *Mutation Research/ Genetic Toxicology and Environmental Mutagenesis*, *672*(1), 55–59. doi:10.1016/j.mrgentox.2008.10.002 PMID:18992843

Hernández-Luna, C. E., Gutiérrez-Soto, G., & Salcedo-Martínez, S. M. (2008). Screening for decolorizing basidiomycetes in Mexico. *World Journal of Microbiology & Biotechnology*, *24*(4), 465–473. doi:10.100711274-007-9495-3

Hölker, U., Höfer, M., & Lenz, J. (2004). Biotechnological advantages of laboratory-scale solid-state fermentation with fungi. *Applied Microbiology and Biotechnology*, *64*, 175–186. doi:10.100700253-003-1504-3 PMID:14963614

Jadhav, S. B., & Singhal, R. S. (2014). Laccase–gum Arabic conjugate for preparation of water-soluble oligomer of catechin with enhanced antioxidant activity. *Food Chemistry*, *150*, 9–16. doi:10.1016/j.foodchem.2013.10.127 PMID:24360412

Jasuja, R., Passam, F. H., Kennedy, D. R., Kim, S. H., van Hessem, L., Lin, L., ... Furie, B. C. (2012). Protein disulfide isomerase inhibitors constitute a new class of antithrombotic agents. *The Journal of Clinical Investigation*, *122*(6), 2104–2113. doi:10.1172/JCI61228 PMID:22565308

Ji, Z. P., & Su, Y. Q. (2008). Study on Antimicrobial Activities of Extracts from *Eucommia ulmoides* Oliv. Leaves. *Chemistry and Industry of Forest Products, 2*, 16.

Jiménez, J. P., Serrano, J., Tabernero, M., Arranz, S., Díaz-Rubio, M. E., García-Diz, L., ... Saura-Calixto, F. (2008). Effects of grape antioxidant dietary fiber in cardiovascular disease risk factors. *Nutrition (Burbank, Los Angeles County, Calif.)*, *24*(7-8), 646–653. doi:10.1016/j.nut.2008.03.012 PMID:18485668

Joana Gil-Chávez, G., Villa, J. A., Fernando Ayala-Zavala, J., Basilio Heredia, J., Sepulveda, D., Yahia, E. M., & González-Aguilar, G. A. (2013). Technologies for extraction and production of bioactive compounds to be used as nutraceuticals and food ingredients: An overview. *Comprehensive Reviews in Food Science and Food Safety*, *12*(1), 5–23. doi:10.1111/1541-4337.12005

Johannes, C., & Majcherczyk, A. (2000). Laccase activity tests and laccase inhibitors. *Journal of Biotechnology*, *78*(2), 193–199. doi:10.1016/S0168-1656(00)00208-X PMID:10725542

Jung, H., Xu, F., & Li, K. (2002). Purification and characterization of laccase from wood-degrading fungus *Trichophyton rubrum* LKY-7. *Enzyme and Microbial Technology*, *30*(2), 161–168. doi:10.1016/S0141-0229(01)00485-9

Kawai, S., Umezawa, T., Shimada, M., Higuchi, T., Koide, K., Nishida, T., ... Haraguchi, T. (1987). C-alpha-C-beta cleavage of phenolic beta-1 lignin substructure model compound by laccase of *Coriolus versicolor. Journal of the Japan Wood Research Society (Japan)*, *33*, 792–797.

Kermanshai, R., McCarry, B. E., Rosenfeld, J., Summers, P. S., Weretilnyk, E. A., & Sorger, G. J. (2001). Benzyl isothiocyanate is the chief or sole anthelmintic in papaya seed extracts. *Phytochemistry*, *57*(3), 427–435. doi:10.1016/S0031-9422(01)00077-2 PMID:11393524

Kersten, P. J., Kalyanaraman, B., Hammel, K. E., Reinhammar, B., & Kirk, T. K. (1990). Comparison of lignin peroxidase, horseradish peroxidase and laccase in the oxidation of methoxybenzenes. *The Biochemical Journal*, *268*(2), 475–480. doi:10.1042/bj2680475 PMID:2163614

Kidwai, M., Jain, A., Sharma, A., & Kuhad, R. C. (2013a). Laccase—a natural source for the synthesis of benzofuro [2, 3-c] pyrazolin-5-ones. *Catalysis Science & Technology*, *3*(1), 230–234. doi:10.1039/C2CY20452E

Kidwai, M., Jain, A., Sharma, A., & Kuhad, R. C. (2013b). Laccase-catalysed reaction between Meldrum's acid and catechols/hydroquinones–An investigation. *Comptes Rendus. Chimie*, *16*(8), 728–735. doi:10.1016/j.crci.2013.04.006

Kim, G. N., Shin, J. G., & Jang, H. D. (2009). Antioxidant and antidiabetic activity of Dangyuja (*Citrus grandis Osbeck*) extract treated with *Aspergillus saitoi. Food Chemistry*, *117*(1), 35–41. doi:10.1016/j.foodchem.2009.03.072

Ko, E. M., Leem, Y. E., & Choi, H. (2001). Purification and characterization of laccase isozymes from the white-rot basidiomycete *Ganoderma lucidum. Applied Microbiology and Biotechnology*, *57*, 98–102. doi:10.1007002530100727 PMID:11693941

Kramer, K. J., Kanost, M. R., Hopkins, T. L., Jiang, H., Zhu, Y. C., Xu, R., ... Turecek, F. (2001). Oxidative conjugation of catechols with proteins in insect skeletal systems. *Tetrahedron*, *57*(2), 385–392. doi:10.1016/S0040-4020(00)00949-2

Kris-Etherton, P. M., Hecker, K. D., Bonanome, A., Coval, S. M., Binkoski, A. E., Hilpert, K. F., ... Etherton, T. D. (2002). Bioactive compounds in foods: Their role in the prevention of cardiovascular disease and cancer. *The American Journal of Medicine*, *113*(9), 71–88. doi:10.1016/S0002-9343(01)00995-0 PMID:12566142

Kudanga, T., Burton, S., Nyanhongo, G. S., & Guebitz, G. M. (2012). Versatility of oxidoreductases in the remediation of environmental pollutants. *Frontiers in Bioscience (Elite Edition)*, *4*(3), 1127–1149. doi:10.2741/e446 PMID:22201941

Kudanga, T., Nemadziva, B., & Le Roes-Hill, M. (2017). Laccase catalysis for the synthesis of bioactive compounds. *Applied Microbiology and Biotechnology*, *101*(1), 13–33. doi:10.100700253-016-7987-5 PMID:27872999

Kudanga, T., Nyanhongo, G. S., Guebitz, G. M., & Burton, S. (2011). Potential applications of laccase-mediated coupling and grafting reactions: A review. *Enzyme and Microbial Technology*, *48*(3), 195–208. doi:10.1016/j.enzmictec.2010.11.007 PMID:22112901

Kunamneni, A., Camarero, S., García-Burgos, C., Plou, F. J., Ballesteros, A., & Alcalde, M. (2008). Engineering and applications of fungal laccases for organic synthesis. *Microbial Cell Factories*, *7*(1), 32. doi:10.1186/1475-2859-7-32 PMID:19019256

Kurisawa, M., Chung, J. E., Uyama, H., & Kobayashi, S. (2003a). Laccase-catalyzed synthesis and antioxidant property of poly (catechin). *Macromolecular Bioscience*, *3*(12), 758–764. doi:10.1002/mabi.200300038

Kurisawa, M., Chung, J. E., Uyama, H., & Kobayashi, S. (2003b). Enzymatic synthesis and antioxidant properties of poly (rutin). *Biomacromolecules*, *4*(5), 1394–1399. doi:10.1021/bm034136b PMID:12959611

Kurniawati, S., & Nicell, J. A. (2007). Efficacy of mediators for enhancing the laccase-catalyzed oxidation of aqueous phenol. *Enzyme and Microbial Technology*, *41*(3), 353–361. doi:10.1016/j.enzmictec.2007.03.003

Lewis, K. (2013). Platforms for antibiotic discovery. *Nature Reviews. Drug Discovery*, *12*(5), 371–387. doi:10.1038/nrd3975 PMID:23629505

Li, C. J., & Trost, B. M. (2008). Green chemistry for chemical synthesis. *Proceedings of the National Academy of Sciences of the United States of America*, *105*(36), 13197–13202. doi:10.1073/pnas.0804348105 PMID:18768813

Li, F., & Yang, X. W. (2012). Analysis of anti-inflammatory dehydrodiisoeugenol and metabolites excreted in rat feces and urine using HPLC-UV. *Biomedical Chromatography*, *26*(6), 703–707. doi:10.1002/bmc.1717 PMID:21932389

Li, K., Horanyi, P. S., Collins, R., Phillips, R. S., & Eriksson, K. E. L. (2001). Investigation of the role of 3-hydroxyanthranilic acid in the degradation of lignin by white-rot fungus *Pycnoporus cinnabarinus*. *Enzyme and Microbial Technology*, *28*(4-5), 301–307. doi:10.1016/S0141-0229(00)00332-X PMID:11240183

Li, X. J., & Zhang, H. Y. (2008). Western-medicine-validated anti-tumor agents and traditional Chinese medicine. *Trends in Molecular Medicine*, *14*(1), 1–2. doi:10.1016/j.molmed.2007.11.002 PMID:18054521

Luo, X., Ma, M., Chen, B., Yao, S., Wan, Z., Yang, D., & Hang, H. (2004). Analysis of Nine Bioactive Compounds in *Eucommia ulmoides* Oliv.and Their Preparation by HPLC-Photodiode Array Detection and Mass Spectrometry. *Journal of Liquid Chromatography & Related Technologies, 27*(1), 63–81. doi:10.1081/JLC-120027086

Manda, K. (2006). *Biotransformation von Aminosäuren, Peptiden und dihydroxylierten Aromaten durch Laccasen aus Pilzen zur Herstellung von Hybridmolekülen* (Unpublished doctoral dissertation). University of Greifswald, Germany.

Manda, K., Hammer, E., Mikolasch, A., Gördes, D., Thurow, K., & Schauer, F. (2006). Laccase-induced derivatization of unprotected amino acid L-tryptophan by coupling with p-hydroquinone 2, 5-dihydroxy-N-(2-hydroxyethyl)-benzamide. *Amino Acids, 31*(4), 409–419. doi:10.100700726-005-0276-8 PMID:16583315

Markom, M., Hasan, M., Daud, W. R. W., Singh, H., & Jahim, J. M. (2007). Extraction of hydrolysable tannins from *Phyllanthus niruri*Linn.: Effects of solvents and extraction methods. *Separation and Purification Technology, 52*(3), 487–496. doi:10.1016/j.seppur.2006.06.003

Martins, S., Aguilar, C. N., Garza-Rodriguez, I. D. L., Mussatto, S. I., & Teixeira, J. A. (2010). Kinetic study of nordihydroguaiaretic acid recovery from *Larrea tridentata* by microwave-assisted extraction. *Journal of Chemical Technology and Biotechnology (Oxford, Oxfordshire), 85*(8), 1142–1147. doi:10.1002/jctb.2412

Martins, S., Mussatto, S. I., Martínez-Avila, G., Montañez-Saenz, J., Aguilar, C. N., & Teixeira, J. A. (2011). Bioactive phenolic compounds: Production and extraction by solid-state fermentation. A review. *Biotechnology Advances, 29*(3), 365–373. doi:10.1016/j.biotechadv.2011.01.008 PMID:21291993

Matern, U., & Kneusel, R. E. (1988). Phenolic compounds in plant disease resistance. *Phytoparasitica, 16*(2), 153–170. doi:10.1007/BF02980469

Matsuura, T., & Ohkatsu, Y. (2000). Phenolic antioxidants: Effect of o-benzyl substituents. *Polymer Degradation & Stability, 70*(1), 59–63. doi:10.1016/S0141-3910(00)00088-4

Maugh, T. H. (1984). Semisynthetic enzymes are new catalysts; enzymes may be the ultimate catalysts, especially if their specificities can be changed by chemical modification. *Science, 223*(4632), 154–157. doi:10.1126cience.223.4632.154 PMID:17733792

Mayer, A. M., & Staples, R. C. (2002). Laccase: New functions for an old enzyme. *Phytochemistry, 60*(6), 551–565. doi:10.1016/S0031-9422(02)00171-1 PMID:12126701

Messerschmidt, A. (Ed.). (1997). *Multi-copper oxidases*. Singapore: World Scientific. doi:10.1142/3146

Mikolasch, A., Hammer, E., Jonas, U., Popowski, K., Stielow, A., & Schauer, F. (2002). Synthesis of 3-(3, 4-dihydroxyphenyl)-propionic acid derivatives by N-coupling of amines using laccase. *Tetrahedron, 58*(38), 7589–7593. doi:10.1016/S0040-4020(02)00872-4

Mikolasch, A., Hessel, S., Salazar, M. G., Neumann, H., Manda, K., Gōrdes, D., ... Beller, M. (2008b). Synthesis of new N-analogous corollosporine derivatives with antibacterial activity by laccase-catalyzed amination. *Chemical & Pharmaceutical Bulletin, 56*(6), 781–786. doi:10.1248/cpb.56.781 PMID:18520080

Mikolasch, A., Manda, K., Schlüter, R., Lalk, M., Witt, S., Seefeldt, S., ... Lindequist, U. (2012). Comparative analyses of laccase-catalyzed amination reactions for production of novel β-lactam antibiotics. *Biotechnology and Applied Biochemistry*, *59*(4), 295–306. doi:10.1002/bab.1026 PMID:23586863

Mikolasch, A., Niedermeyer, T. H. J., Lalk, M., Witt, S., Seefeldt, S., Hammer, E., ... Lindequist, U. (2006). Novel penicillins synthesized by biotransformation using laccase from *Trametes* species. *Chemical & Pharmaceutical Bulletin*, *54*(5), 632–638. doi:10.1248/cpb.54.632 PMID:16651757

Mikolasch, A., Niedermeyer, T. H. J., Lalk, M., Witt, S., Seefeldt, S., Hammer, E., ... Lindequist, U. (2007). Novel cephalosporins synthesized by amination of 2, 5-dihydroxybenzoic acid derivatives using fungal laccases II. *Chemical & Pharmaceutical Bulletin*, *55*(3), 412–416. doi:10.1248/cpb.55.412 PMID:17329882

Mikolasch, A., & Schauer, F. (2009). Fungal laccases as tools for the synthesis of new hybrid molecules and biomaterials. *Applied Microbiology and Biotechnology*, *82*(4), 605–624. doi:10.100700253-009-1869-z PMID:19183983

Mikolasch, A., Wurster, M., Lalk, M., Witt, S., Seefeldt, S., Hammer, E., ... Lindequist, U. (2008a). Novel β-lactam antibiotics synthesized by amination of catechols using fungal laccase. *Chemical & Pharmaceutical Bulletin*, *56*(7), 902–907. doi:10.1248/cpb.56.902 PMID:18591799

Morozova, O. V., Shumakovich, G. P., Gorbacheva, M. A., Shleev, S. V., & Yaropolov, A. I. (2007a). "Blue" laccases. *Biochemistry (Moscow)*, *72*(10), 1136–1150. doi:10.1134/S0006297907100112 PMID:18021071

Morozova, O. V., Shumakovich, G. P., Shleev, S. V., & Yaropolov, Y. I. (2007b). Laccase-mediator systems and their applications: A review. *Applied Biochemistry and Microbiology*, *43*(5), 523–535. doi:10.1134/S0003683807050055 PMID:18038679

Müller, W. E. G., Grebenjuk, V. A., Le Pennec, G., Schröder, H. C., Brümmer, F., Hentschel, U., ... Breter, H. J. (2004). Sustainable production of bioactive compounds by sponges—cell culture and gene cluster approach: A review. *Marine Biotechnology (New York, N.Y.)*, *6*, 105–117. doi:10.100710126-002-0098-6 PMID:15085406

Murakami, Y., Shoji, M., Hirata, A., Tanaka, S., Yokoe, I., & Fujisawa, S. (2005). Dehydrodiisoeugenol, an isoeugenol dimer, inhibits lipopolysaccharide-stimulated nuclear factor kappa B activation and cyclooxygenase-2 expression in macrophages. *Archives of Biochemistry and Biophysics*, *434*(2), 326–332. doi:10.1016/j.abb.2004.11.013 PMID:15639233

Nematollahi, D., & Forooghi, Z. (2002). Electrochemical oxidation of catechols in the presence of 4-hydroxy-6-methyl-2-pyrone. *Tetrahedron*, *58*(24), 4949–4953. doi:10.1016/S0040-4020(02)00422-2

Nicotra, S., Cramarossa, M. R., Mucci, A., Pagnoni, U. M., Riva, S., & Forti, L. (2004a). Biotransformation of resveratrol: Synthesis of trans-dehydrodimers catalyzed by laccases from *Myceliophtora thermophila* and from *Trametes pubescens*. *Tetrahedron*, *60*(3), 595–600. doi:10.1016/j.tet.2003.10.117

Nicotra, S., Intra, A., Ottolina, G., Riva, S., & Danieli, B. (2004b). Laccase-mediated oxidation of the steroid hormone 17β-estradiol in organic solvents. *Tetrahedron, Asymmetry*, *15*(18), 2927–2931. doi:10.1016/j.tetasy.2004.06.034

No, H. K., Meyers, S. P., Prinyawiwatkul, W., & Xu, Z. (2007). Applications of chitosan for improvement of quality and shelf life of foods: A review. *Journal of Food Science*, *72*(5), 87–100. doi:10.1111/j.1750-3841.2007.00383.x PMID:17995743

Ogata, M., Oka, K., Seki, M., Hoshi, M., Takatsu, H., Mashino, T., ... Endo, T. (2007). Solubilization of dipropofol, an antibacterial agent, using saccharide and ascorbic acid. *Biological & Pharmaceutical Bulletin*, *30*(8), 1565–1568. doi:10.1248/bpb.30.1565 PMID:17666822

Ogata, M., Sato, K. T., Kunikane, T., Oka, K., Seki, M., Urano, S., ... Endo, T. (2005). Antibacterial activity of dipropofol and related compounds. *Biological & Pharmaceutical Bulletin*, *28*(6), 1120–1122. doi:10.1248/bpb.28.1120 PMID:15930760

Osiadacz, J., Al-Adhami, A. J., Bajraszewska, D., Fischer, P., & Peczyñska-Czoch, W. (1999). On the use of *Trametes versicolor* laccase for the conversion of 4-methyl-3-hydroxyanthranilic acid to actinocin chromophore. *Journal of Biotechnology*, *72*(1-2), 141–149. doi:10.1016/S0168-1656(99)00100-5

Otto, B., Schlosser, D., & Reisser, W. (2010). First description of a laccase-like enzyme in soil algae. *Archives of Microbiology*, *192*(9), 759–768. doi:10.100700203-010-0603-7 PMID:20623267

Parvathy, K. S., Negi, P. S., & Srinivas, P. (2009). Antioxidant, antimutagenic and antibacterial activities of curcumin-β-diglucoside. *Food Chemistry*, *115*(1), 265–271. doi:10.1016/j.foodchem.2008.12.036

Pasha, I., Saeed, F., Waqas, K., Anjum, F. M., & Arshad, M. U. (2013). Nutraceutical and functional scenario of wheat straw. *Critical Reviews in Food Science and Nutrition*, *53*(3), 287–295. doi:10.1080/10408398.2010.528080 PMID:23216000

Patel, H., & Gupte, A. (2016). Optimization of different culture conditions for enhanced laccase production and its purification from *Tricholoma giganteum* AGHP. *Bioresources and Bioprocessing*, *3*(1), 11. doi:10.118640643-016-0088-6

Pei, L. X., Li, Y. M., Bu, X. Z., Gu, L. Q., & Chan, A. S. (2006). One-pot synthesis of 5, 6-dihydroxylated benzo [b] furan derivatives. *Tetrahedron Letters*, *47*(15), 2615–2618. doi:10.1016/j.tetlet.2006.02.055

Pezzella, C., Guarino, L., & Piscitelli, A. (2015). How to enjoy laccases. *Cellular and Molecular Life Sciences*, *72*(5), 923–940. doi:10.100700018-014-1823-9 PMID:25577278

Pham, N. H., Hollmann, F., Kracher, D., Preims, M., Haltrich, D., & Ludwig, R. (2015). Engineering an enzymatic regeneration system for NAD(P)H oxidation. *Journal of Molecular Catalysis. B, Enzymatic*, *120*, 38–46. doi:10.1016/j.molcatb.2015.06.011

Pietruszka, J., & Wang, C. (2012). Laccase-catalysed α-arylation of cyclic β-dicarbonyl compounds. *Green Chemistry*, *14*(9), 2402–2409. doi:10.1039/c2gc35662g

Poeckel, D., Niedermeyer, T. H., Pham, H. T., Mikolasch, A., Mundt, S., Lindequist, U., ... Werz, O. (2006). Inhibition of human 5-lipoxygenase and anti-neoplastic effects by 2-amino-1, 4-benzoquinones. *Medicinal Chemistry (Shariqah, United Arab Emirates)*, *2*(6), 591–595. doi:10.2174/157340641060206059l PMID:17105440

Polak, J., & Jarosz-Wilkolazka, A. (2012). Fungal laccases as green catalysts for dye synthesis. *Process Biochemistry*, *47*(9), 1295–1307. doi:10.1016/j.procbio.2012.05.006

Potthast, A., Rosenau, T., Chen, C. L., & Gratzl, J. S. (1995). Selective enzymic oxidation of aromatic methyl groups to aldehydes. *The Journal of Organic Chemistry*, *60*(14), 4320–4321. doi:10.1021/jo00119a006

Potthast, A., Rosenau, T., Chen, C. L., & Gratzl, J. S. (1996). A novel method for conversion of benzyl alcohols to benzaldehydes by laccase-catalysed oxidation. *Journal of Molecular Catalysis A Chemical*, *108*(1), 5–9. doi:10.1016/1381-1169(95)00251-0

Quaratino, D., Federici, F., Petruccioli, M., Fenice, M., & D'Annibale, A. (2007). Production, purification and partial characterisation of a novel laccase from the white-rot fungus *Panus tigrinus* CBS 577.79. *Antonie van Leeuwenhoek*, *91*(1), 57–69. doi:10.100710482-006-9096-4 PMID:17013550

Raafat, D., Von Bargen, K., Haas, A., & Sahl, H. G. (2008). Insights into the mode of action of chitosan as an antibacterial compound. *Applied and Environmental Microbiology*, *74*(12), 3764–3773. doi:10.1128/AEM.00453-08 PMID:18456858

Rida, S. M., El-Hawash, S. A., Fahmy, H. T., Hazza, A. A., & El-Meligy, M. M. (2006). Synthesis and in vitro evaluation of some novel benzofuran derivatives as potential anti-HIV-1, anticancer, and antimicrobial agents. *Archives of Pharmacal Research*, *29*(1), 16–25. doi:10.1007/BF02977463 PMID:16491838

Riva, S. (2006). Laccases: Blue enzymes for green chemistry. *Trends in Biotechnology*, *24*(5), 219–226. doi:10.1016/j.tibtech.2006.03.006 PMID:16574262

Rocasalbas, G., Francesko, A., Touriño, S., Fernández-Francos, X., Guebitz, G. M., & Tzanov, T. (2013). Laccase-assisted formation of bioactive chitosan/gelatin hydrogel stabilized with plant polyphenols. *Carbohydrate Polymers*, *92*(2), 989–996. doi:10.1016/j.carbpol.2012.10.045 PMID:23399119

Rutschmann, J., Stadler, P. A., Berde, B., & Schild, H. O. (1978). Ergot alkaloids and related compounds. *Handbook of Experimental Pharmacology*, *49*, 61–62.

Rüttimann-Johnson, C. A. R. M. E. N., & Lamar, R. T. (1996). Polymerization of pentachlorophenol and ferulic acid by fungal extracellular lignin-degrading enzymes. *Applied and Environmental Microbiology*, *62*, 3890–3893. PMID:8967777

Ryan, D., Leukes, W., & Burton, S. (2007). Improving the bioremediation of phenolic wastewaters by *Trametes versicolor. Bioresource Technology*, *98*(3), 579–587. doi:10.1016/j.biortech.2006.02.001 PMID:16545562

Sagui, F., Chirivì, C., Fontana, G., Nicotra, S., Passarella, D., Riva, S., & Danieli, B. (2009). Laccase-catalyzed coupling of catharanthine and vindoline: An efficient approach to the bisindole alkaloid anhydrovinblastine. *Tetrahedron*, *65*(1), 312–317. doi:10.1016/j.tet.2008.10.064

Sakurai, T., & Kataoka, K. (2007). Structure and function of type I copper in multicopper oxidases. *Cellular and Molecular Life Sciences*, *64*(19-20), 2642–2656. doi:10.100700018-007-7183-y PMID:17639274

Sánchez-Moreno, C., Larrauri, J. A., & Saura-Calixto, F. (1998). A procedure to measure the antiradical efficiency of polyphenols. *Journal of the Science of Food and Agriculture*, *76*(2), 270–276. doi:

Schultz, A., Jonas, U., Hammer, E., & Schauer, F. (2001). Dehalogenation of chlorinated hydroxybiphenyls by fungal laccase. *Applied and Environmental Microbiology*, *67*(9), 4377–4381. doi:10.1128/AEM.67.9.4377-4381.2001 PMID:11526052

Senthivelan, T., Kanagaraj, J., & Panda, R. (2016). Recent trends in fungal laccase for various industrial applications: An eco-friendly approach-A review. *Biotechnology and Bioprocess Engineering; BBE*, *21*(1), 19–38. doi:10.100712257-015-0278-7

Sharma, P., Goel, R., & Capalash, N. (2007). Bacterial laccases. *World Journal of Microbiology & Biotechnology*, *23*(6), 823–832. doi:10.100711274-006-9305-3

Sheldon, R. A. (1992). Organic synthesis-past, present and future. *Chemistry & Industry*, *23*, 903–906.

Slomczynski, D., Nakas, J., & Tanenbaum, S. W. (1995). Production and Characterization of Laccase from *Botrytis cinerea* 61-34. *Applied and Environmental Microbiology*, *61*, 907–912. PMID:16534974

Solís-Oba, M., Ugalde-Saldívar, V. M., González, I., & Viniegra-González, G. (2005). An electrochemical–spectrophotometrical study of the oxidized forms of the mediator 2, 2′-azino-bis-(3-ethylbenzothiazoline-6-sulfonic acid) produced by immobilized laccase. *Journal of Electroanalytical Chemistry*, *579*(1), 59–66. doi:10.1016/j.jelechem.2005.01.025

Solomon, E. I., Augustine, A. J., & Yoon, J. (2008). O_2 Reduction to H_2O by the multicopper oxidases. *Dalton Transactions (Cambridge, England)*, *30*(30), 3921–3932. doi:10.1039/b800799c PMID:18648693

Solomon, E. I., Sundaram, U. M., & Machonkin, T. E. (1996). Multicopper oxidases and oxygenases. *Chemical Reviews*, *96*(7), 2563–2606. doi:10.1021/cr950046o PMID:11848837

Srinivasan, M., Sudheer, A. R., & Menon, V. P. (2007). Ferulic acid: Therapeutic potential through its antioxidant property. *Journal of Clinical Biochemistry and Nutrition*, *40*(2), 92–100. doi:10.3164/jcbn.40.92 PMID:18188410

Tandon, V. K., Yadav, D. B., Singh, R. V., Vaish, M., Chaturvedi, A. K., & Shukla, P. K. (2005). Synthesis and biological evaluation of novel 1, 4-naphthoquinone derivatives as antibacterial and antiviral agents. *Bioorganic & Medicinal Chemistry Letters*, *15*(14), 3463–3466. doi:10.1016/j.bmcl.2005.04.075 PMID:15950468

Tetsch, L., Bend, J., & Hölker, U. (2006). Molecular and enzymatic characterisation of extra-and intracellular laccases from the acidophilic ascomycete *Hortaea acidophila*. *Antonie van Leeuwenhoek*, *90*(2), 183–194. doi:10.100710482-006-9064-z PMID:16871425

Uchida, H., Fukuda, T., Miyamoto, H., Kawabata, T., Suzuki, M., & Uwajima, T. (2001). Polymerization of Bisphenol A by purified laccase from *Trametes villosa*. *Biochemical and Biophysical Research Communications*, *287*(2), 355–358. doi:10.1006/bbrc.2001.5593 PMID:11554734

van der Heijden, R., Jacobs, D. I., Snoeijer, W., Hallard, D., & Verpoorte, R. (2004). The Catharanthus alkaloids: Pharmacognosy and biotechnology. *Current Medicinal Chemistry*, *11*(5), 607–628. doi:10.2174/0929867043455846 PMID:15032608

Wan, Y., Lu, R., Akiyama, K., Miyakoshi, T., & Du, Y. (2007). Enzymatic synthesis of bioactive compounds by *Rhus* laccase from Chinese *Rhus vernicifera. Science in China. Series B, Chemistry*, *50*(2), 179–182. doi:10.100711426-007-0022-z

Wellington, K. W., & Kolesnikova, N. I. (2012). A laccase-catalysed one-pot synthesis of aminonaphthoquinones and their anticancer activity. *Bioorganic & Medicinal Chemistry*, *20*(14), 4472–4481. doi:10.1016/j.bmc.2012.05.028 PMID:22682920

Wellington, K. W., Qwebani-Ogunleye, T., Kolesnikova, N. I., Brady, D., & de Koning, C. B. (2013). One-pot laccase-catalysed synthesis of 5, 6-dihydroxylated benzo [b] furans and catechol derivatives, and their anticancer activity. *Archiv der Pharmazie*, *346*(4), 266–277. doi:10.1002/ardp.201200413 PMID:23447437

Wells, A., Teria, M., & Eve, T. (2006). Green oxidations with laccase–mediator systems. *Biochemical Society Transactions*, *34*(2), 304–308. doi:10.1042/BST0340304 PMID:16545100

Witayakran, S. (2008). *Laccase in organic synthesis and its applications* (Doctoral dissertation). Georgia Institute of Technology.

Witayakran, S., & Ragauskas, A. J. (2007). One-pot synthesis of 1, 4-naphthoquinones and related structures with laccase. *Green Chemistry*, *9*(5), 475–480. doi:10.1039/b606686k

Xu, F. (1996). Oxidation of phenols, anilines, and benzenethiols by fungal laccases: Correlation between activity and redox potentials as well as halide inhibition. *Biochemistry*, *35*(23), 7608–7614. doi:10.1021/bi952971a PMID:8652543

Xu, F. (1997). Effects of redox potential and hydroxide inhibition on the pH activity profile of fungal laccases. *The Journal of Biological Chemistry*, *272*(2), 924–928. doi:10.1074/jbc.272.2.924 PMID:8995383

Xu, F., Kulys, J. J., Duke, K., Li, K., Krikstopaitis, K., Deussen, H. J. W., ... Schneider, P. (2000). Redox chemistry in laccase-catalyzed oxidation of N-hydroxy compounds. *Applied and Environmental Microbiology*, *66*(5), 2052–2056. doi:10.1128/AEM.66.5.2052-2056.2000 PMID:10788380

Yaver, D. S., Xu, F., Golightly, E. J., Brown, K. M., Brown, S. H., Rey, M. W., ... Dalboge, H. (1996). Purification, characterization, molecular cloning, and expression of two laccase genes from the white rot basidiomycete *Trametes villosa. Applied and Environmental Microbiology*, *62*, 834–841. PMID:8975613

Yoshida, H. (1883). LXIII.—chemistry of lacquer (Urushi). Part I. communication from the chemical society of Tokyo. *Journal of the Chemical Society Transactions*, *43*(0), 472–486. doi:10.1039/CT8834300472

Yoshitake, A., Katayama, Y., Nakamura, M., Iimura, Y., Kawai, S., & Morohoshi, N. (1993). N-linked carbohydrate chains protect laccase III from proteolysis in *Coriolus versicolor. Microbiology*, *139*, 179–185.

Zhang, Q., Su, Y., & Zhang, J. (2013). Seasonal difference in antioxidant capacity and active compounds contents of *Eucommia ulmoides* Oliver leaf. *Molecules (Basel, Switzerland)*, *18*(12), 1857–1868. doi:10.3390/molecules18021857 PMID:23377129

KEY TERMS AND DEFINITIONS

Bioactive Molecules: A compounds or a molecule that shows beneficial effects on living organisms and promotes good health (e.g., nutraceuticals, therapeutic agents, antioxidants, etc.).

Biocatalyst: Any biologically available source or compounds that is useful to speed up any chemical reaction.

Biotransformation: The process of transformation or alteration of any chemical compound or a drug carried out by living system (e.g., enzyme, microorganism, etc.).

Catalyst: A substance which increases the rate of reaction without undergoing any structural alteration.

Enzyme: A compound produced or secreted by any living organism which acts as a catalyst in biochemical reactions.

Green Chemistry: The invention or technology focused on the development or designing of chemical compounds without using or liberating any hazardous substances.

Sustainability: An ability to preserve or develop the economic balance, environment, etc. without the depletion of natural resources.

APPENDIX

Figure 1. Schematic representation of laccase catalyzed oxidation of substrate in absence and in presence of mediator

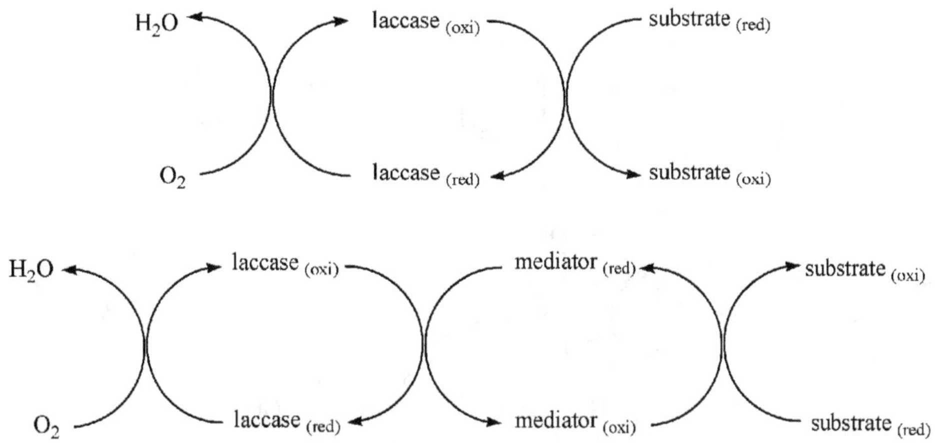

Figure 2. Chemical structures of phenolic and non-phenolic compounds classified as (a) artificial and (b) natural laccase mediators

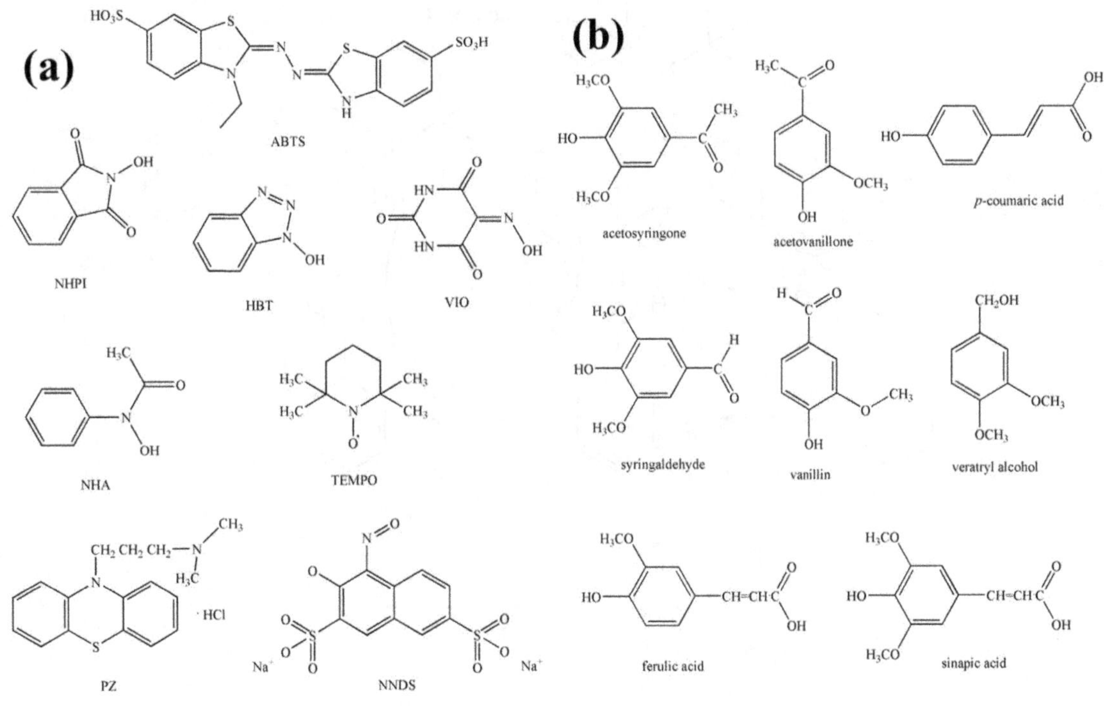

Figure 3. Different reaction mechanisms followed by LMS depending on the nature of substrate i) Electron transfer, ii) Hydrogen atom transfer, and iii) Ionic oxidation mechanism (Adapted from Cannatelli & Ragauskas, 2017)

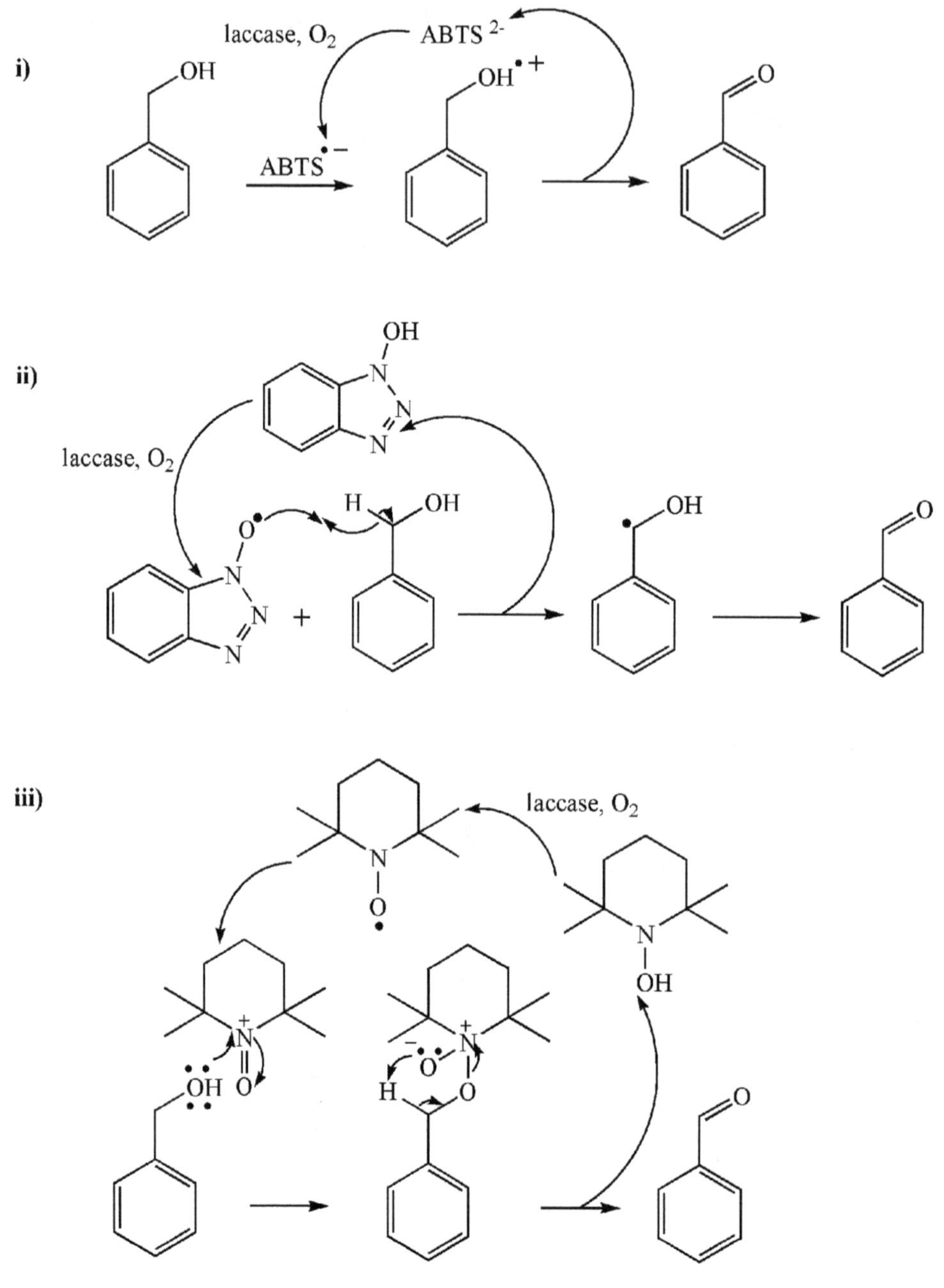

Figure 4. Laccase catalyzed formation of i) ortho- and ii) para- quinones from catechols and hydroquinones, respectively, adapted from Cannatelli & Ragauskas, 2017

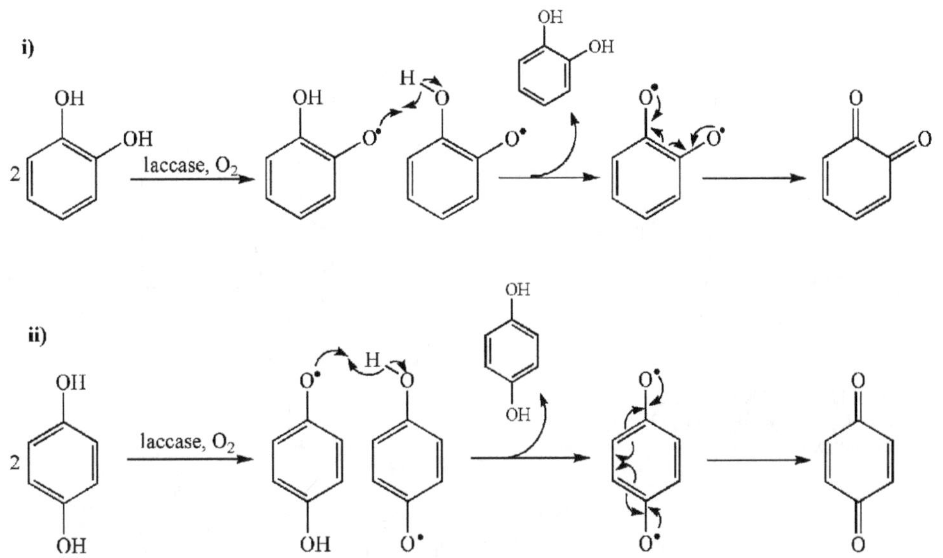

Figure 5. A reaction mechanism proposed for the synthesis of C-C linked homomolecular dimer of 2,6-DMP (Adelakun et al., 2012b)

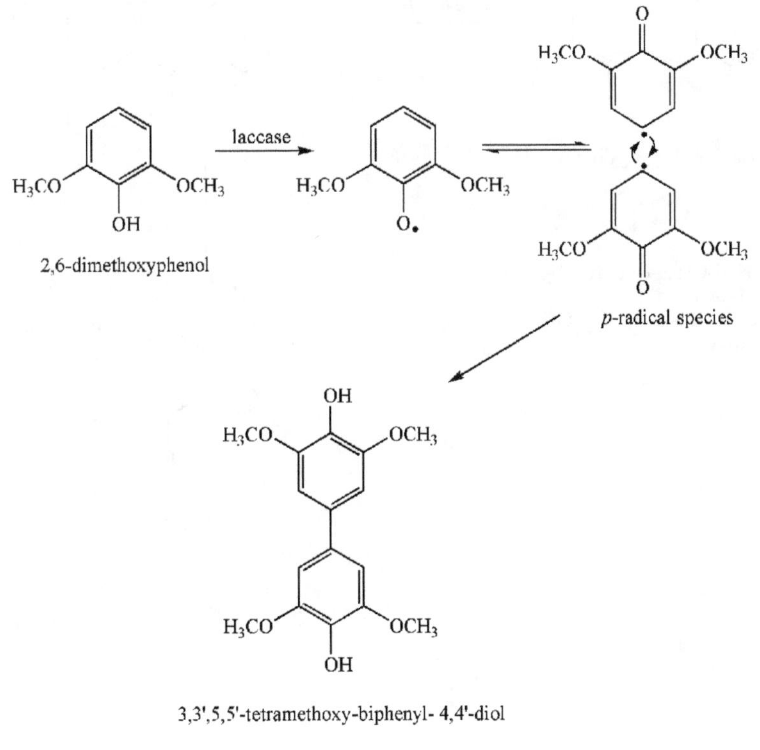

Figure 6. Laccase catalyzed reaction of i) hydroquinone and ii) catechol with substituted dienes resulting into formation of 1,4-naphthoquinones and 1,2-naphthoquinone derivatives, respectively (Witayakran & Ragauskas, 2007)

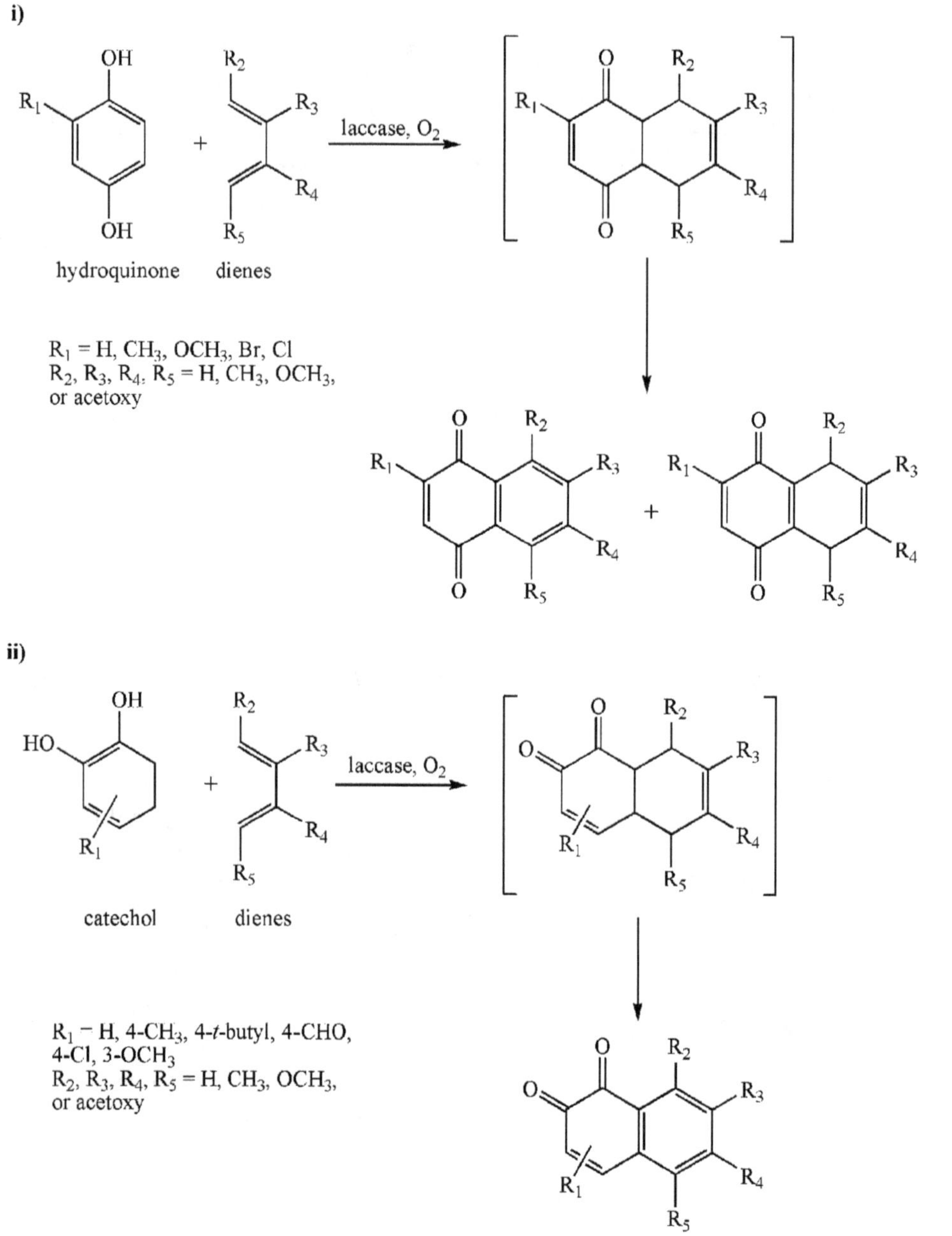

Figure 7. Synthesis of benzofuran products via laccase catalyzed oxidation-Michael addition reaction between substituted catechols and 1,3-dicarbonyl compounds (Wellington et al., 2013)

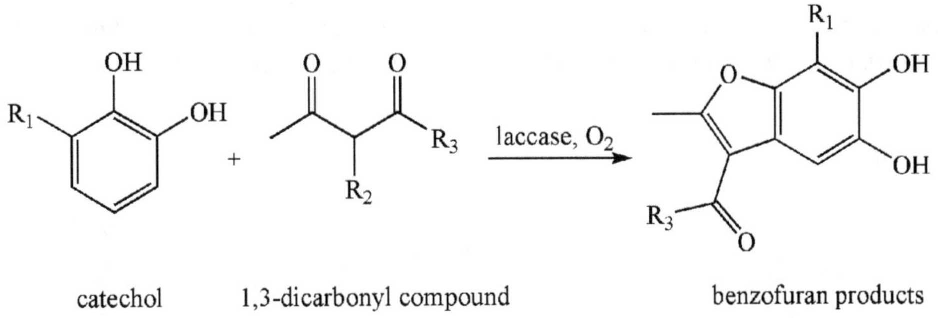

R_1= H, CH_3
R_2= H, Cl
R_3= CH_3, OCH_2CH_3

*Figure 8. Laccase catalyzed coupling of **a)** catechol(1) with 2-mercaptobenzoxazole(X=O), and 2-mercaptobenzothiazole(X=S)(2) to produce catechol thioethers(3) (Adapted from Abdel-Mohsen et al., 2014); **b)**(i) substituted catechols(1) and 2,3-dihydro-2-thioxopyrimidin-4(1H)-ones(2) to produce catechol thioethers(3); (ii) catechol(1) and substituted 2,3-dihydro-2-thioxopyrimidin-4(1H)-ones(2) to produce pyrimidobenzothiazoles(4) (Adapted from Abdel-Mohsen et al., 2017)*

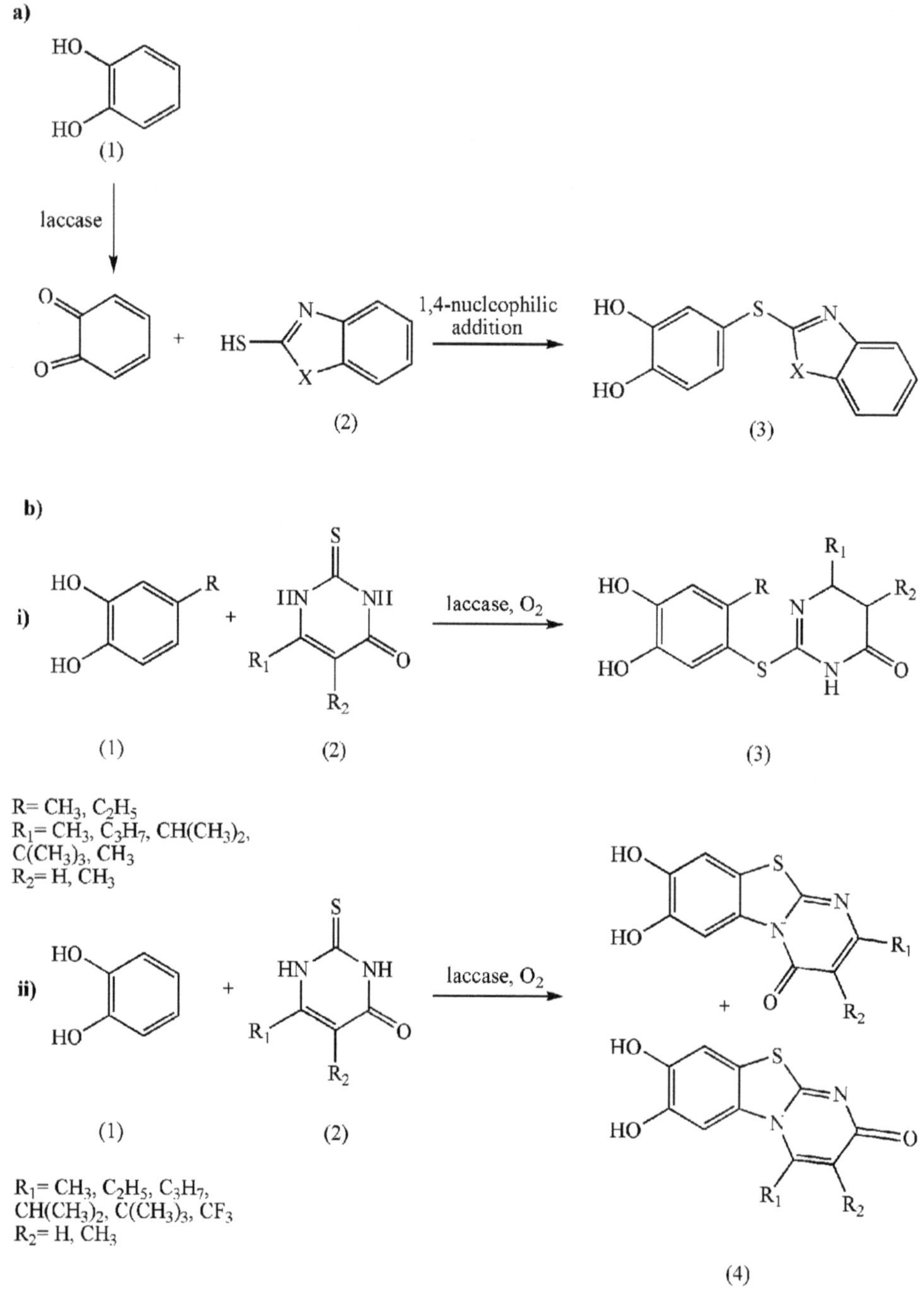

Figure 9. Schematic representation of a proposed reaction mechanism for laccase catalyzed cross-coupling of 1,2-ethanedithiol(2) with substituted hydroquinones(1) to yield 2,3-ethylenedithio-1,4-quinones(3) (Cannatelli & Ragauskas, 2015a)

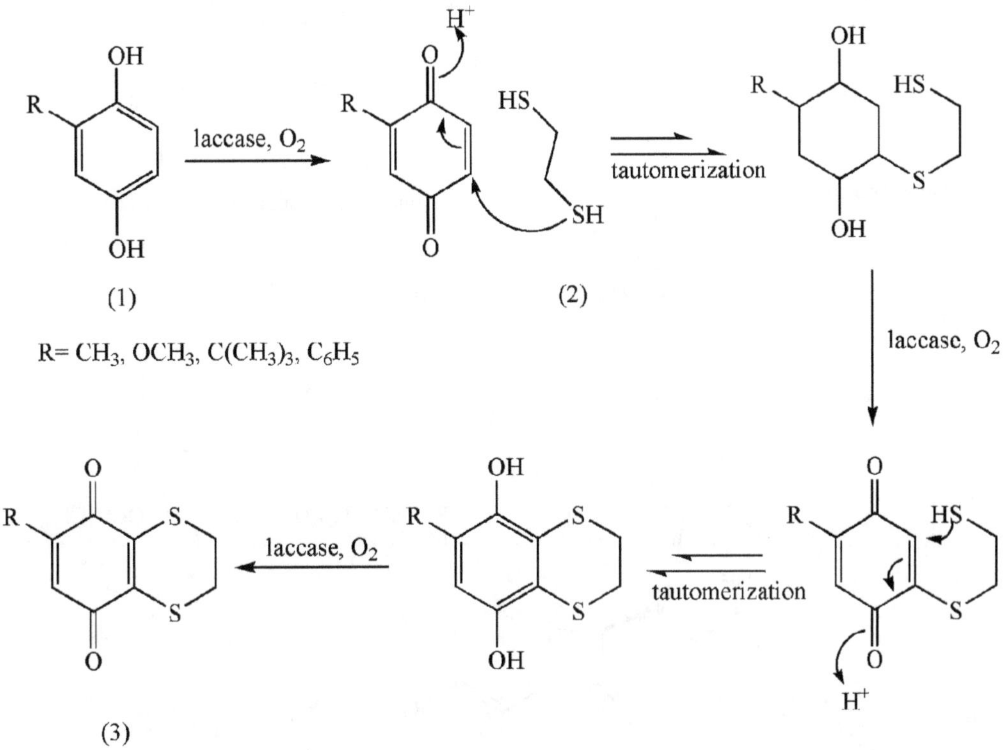

Figure 10. Laccase catalyzed synthesis of benzylic nitriles via cross-coupling of substituted hydroquinones with benzoylacetonitrile (Cannatelli & Ragauskas, 2015b)

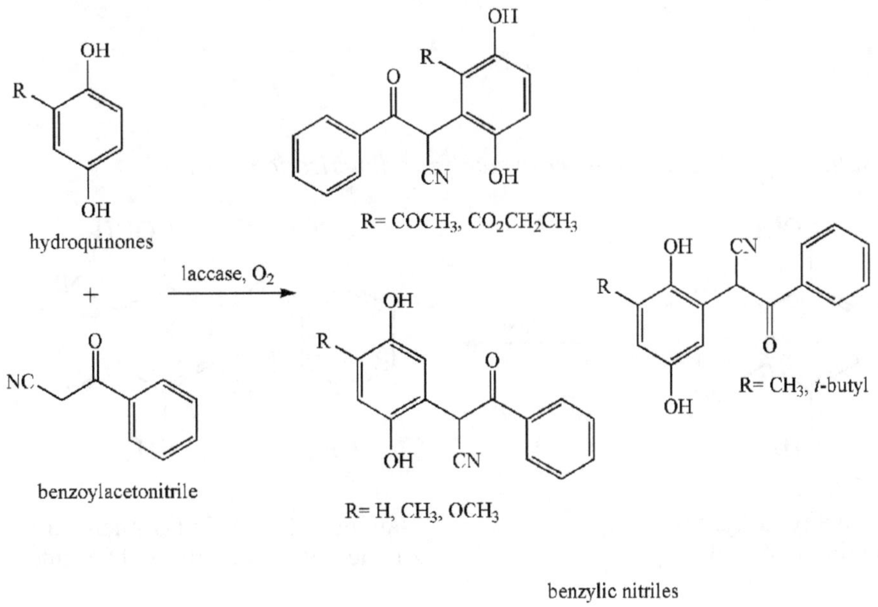

Figure 11. The synthesis of heterodimer derivatives of aminocephalosporins via laccase catalyzed oxidation coupled with nuclear amination reaction (Mikolasch et al., 2007)

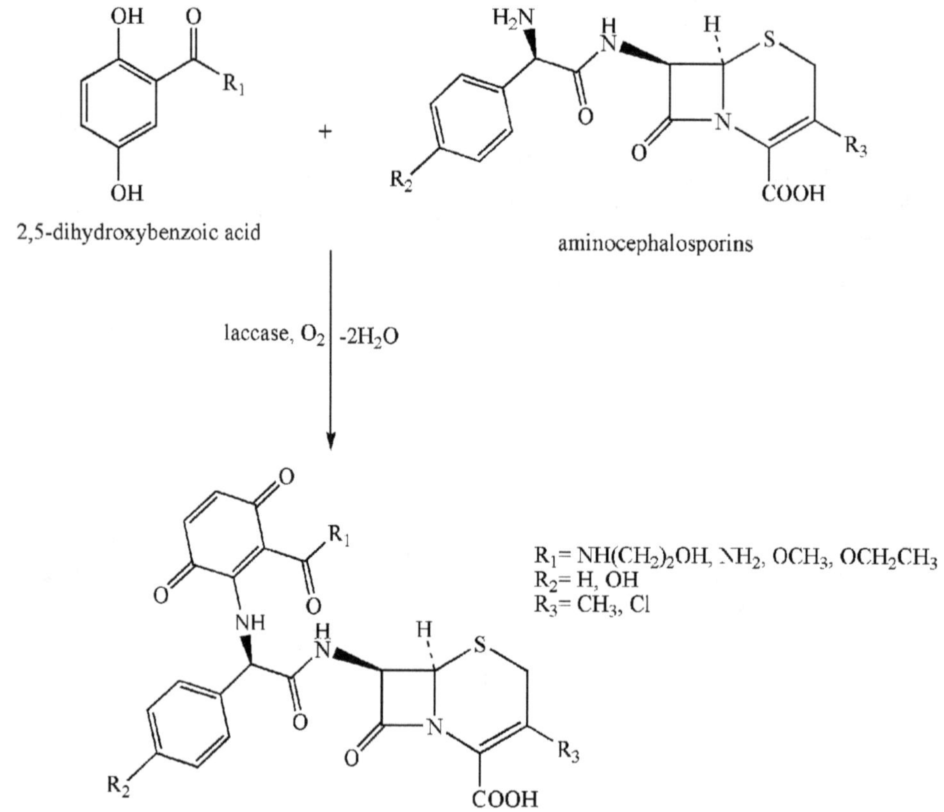

Figure 12. Synthesis of actinocin via laccase catalyzed dimerization reaction

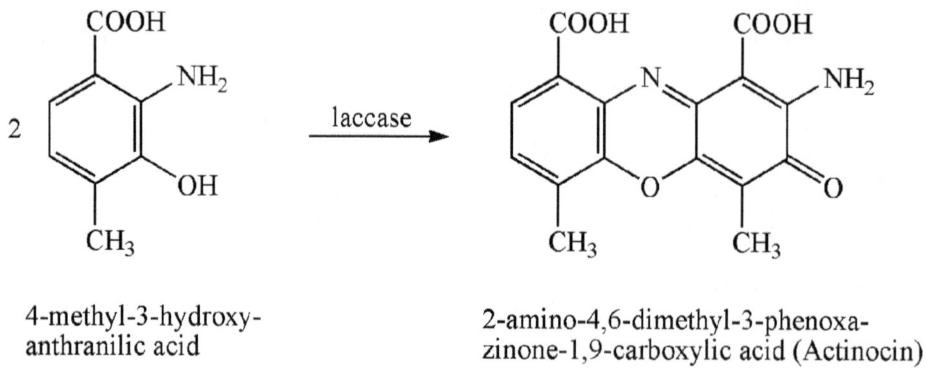

Chapter 10
Laccase-Mediated Treatment of Pharmaceutical Wastes

Hamid Forootanfar
Kerman University of Medical Sciences, Iran

Shokouh Arjmand
Kerman University of Medical Sciences, Iran

Mina Behzadi
Kerman University of Medical Sciences, Iran

Mohammad Ali Faramarzi
Tehran University of Medical Sciences, Iran

ABSTRACT

Laccases are versatile multi-copper enzymes belonging to the superfamily of oxidase enzymes, which have been known since the nineteenth century. Recent discoveries have refined investigators' views of the potential of laccase as a magic tool for remarkable biotechnological purposes. A literature review of the capabilities of laccases, their assorted substrates, and their molecular mechanism of action now indicates the emergence of a new direction for laccase application as part of an arsenal in the fight against the contamination of water supplies by a number of frequently prescribed medications. This chapter provides a critical review of the literature and reveals the pivotal role of laccases in the elimination and detoxification of pharmaceutical contaminants in aquatic environments and wastewaters.

1. INTRODUCTION

Since the first reports of enzyme discovery in the 18th century, applications of these valuable biologically derived macromolecules have been widely developed for many biotechnological and industrial procedures (Copeland, 2000). Enzymes are non-toxic, biodegradable, and environmental friendly catalysts that speed up chemical reactions with great specificity and at mild conditions of low pressure and temperature, as well as near neutral pH (Copeland, 2000; Yagi, 2006). Consequently, these magic tools make the

DOI: 10.4018/978-1-5225-5237-6.ch010

performance of many otherwise impossible and difficult reactions (especially enantio- and site-specific reactions) faster, less demanding, and more economical (Copeland, 2000; Yagi, 2006).

Based on their mode of action, enzymes are divided into several categories that include oxidoreductases, transferases, hydrolases, lyases, isomerases, and ligases (Purich & Allison, 2002). Laccases (benzenediol:oxygen oxidoreductase, EC 1.10.3.2), which are among the oldest and most extensively studied group of oxidoreductases, have shown multipotential utilities that range from industrial and environmental purposes to synthetic and diagnostic applications (Jafari, Rezaei, Rezaie, Dilmaghani, Khoshayand, & Faramarzi, 2017; Shraddha, Shekher, Sehgal, Kamthania, & Kumar, 2011; Yang, Li, Ng, Deng, Lin, & Ye, 2017). Laccases are generally monomeric, dimeric, and/or tetrameric glycoproteins with miscellaneous substrate ranges that can be even further expanded by the use of either synthetic or natural redox mediators (Mogharabi & Faramarzi, 2014; Shraddha, Shekher, Sehgal, Kamthania, & Kumar, 2011). These copper-containing oxidases are mainly produced by white-rot basidiomycetes, where they play physiological roles in the degradation of lignin biopolymers for natural recycling (Forootanfar & Faramarzi, 2015; Rezaei, Shahverdi, & Faramarzi, 2017; Rezaie, Rezaei, Jafari, Forootanfar, Khoshayand, & Faramarzi, 2017; Yang, Li, Ng, Deng, Lin, & Ye, 2017). Their wide spectrum of substrate specificities has drawn attention and made laccases the first choice in numerous exploitations, including dye decolorization, biodegradation and bioremediation processes, the separation of lignin in paper industries, and the removal of phenolic compounds from food (Asadgol, Forootanfar, Rezaei, Mahvi, & Faramarzi, 2014; Ashrafi, Nasseri, Alimohammadi, Mahvi, & Faramarzi, 2015; Forootanfar, Rezaei, Zeinvand-Lorestani, Tahmasbi, Mogharabi, Ameri, & Faramarzi, 2016; Mirzadeh, Khezri, Rezaei, Forootanfar, Mahvi, & Faramarzi Mirzadeh, 2014; Rezaei, Tahmasbi, Mogharabi, Firuzyar, Ameri, Khoshayand, & Faramarz, 2015; Yang, Li, Ng, Deng, Lin, & Ye, 2017). Another aim of many investigations has been the use of laccases and/or laccase producing microorganisms for the elimination of pharmaceuticals and endocrine disrupting chemicals (EDCs) present as hazardous materials in wastewater (Ashrafi, Nasseri, Alimohammadi, Mahvi, & Faramarzi, 2015; Becker, Rodriguez-Mozaz, Insa, Schoevaart, Barceló, de Cazes, … Wagner, 2017; Macellaro, Pezzella, Cicatiello, Sannia, & Piscitelli, 2014; Rahmani, Faramarzi, Mahvi, Gholami, Esrafili, Forootanfar, & Farzadkia, 2015; Tahmasbi, Khoshayand, Bozorgi-Koushalshahi, Heidary, Ghazi-Khansari, & Faramarzi, 2016; Yousefi-Ahmadipour, Bozorgi-Koshalshahi, Mogharabi, Amini, Ghazi-Khansari, & Faramarzi, 2016; Zeinvand-Lorestani, Sabzevari, Setayesh, Amini, Nili-Ahmadabadi, & Faramarzi, 2015). This type of enzyme-based bioremediation can overcome problems encountered when removing pollutants with physiochemical procedures, as these can be time consuming and costly and can generate metabolites with higher toxicity than the original pollutants (Azimi, Nafissi-Varcheh, Mogharabi, Faramarzi, & Aboofazeli, 2016).

This chapter considers all the above-mentioned points about the advantages of enzymatic treatment of wastewater, and takes an environmental point of view to represent how laccases are presently being employed to assist in the bioremediation, biotransformation, and removal of pharmaceutical wastes. A further goal was to provide a state-of-the-art review of the different categories of pharmaceuticals being biodegraded by laccases and to present a simplified scheme regarding their mechanisms of action. This chapter presents how far we have come and what else can be done to improve our current knowledge and to optimize laccase-mediated treatments of pharmaceutical wastes, as well as to broaden the categories of these compounds.

2. BACKGROUND

The occurrence of pharmaceuticals in the environment and the possible devastating effects they may leave, especially in water supplies, is now becoming an emerging concern, particularly when considering that these compounds may find a way into drinking water resources and ecosystems (Khetan & Collins, 2007). The lack of analysis protocols and standardized detection methods for all pharmaceuticals and the limitations for use of high cost techniques and analytical instruments for detecting them in wastewater and other water resources create an even worse situation (Brausch, Connors, Brooks, & Rand, 2012; Castiglioni, Thomas, Kasprzyk-Hordern, Vandam, & Griffiths, 2014; Contado, 2015). Contamination by some frequently prescribed drugs, such as antibiotics, analgesics, anti-depressants, contraceptives, and anti-cancer agents, also should not be overlooked, and the fears of acute and long-term exposure to these chemicals now loom large (Gros, Petrović, Ginebreda, & Barceló, 2010; Ikehata, Jodeiri Naghashkar, & Gamal El-Din, 2006; Pomati, Orlandi, Clerici, Luciani, & Zuccato, 2008; Vasquez, Lambrianides, Schneider, Kümmerer, & Fatta-Kassinos, 2014).

In this regard, several lines of evidence have explained the hormone-like effects of some chemical structures and pharmaceuticals and have identified them as endocrine disrupting chemicals (EDCs) as they mimic the functions of endogenous hormones in the body and display both agonistic and antagonistic abilities (Cabana, Jones, & Agathos, 2007; Gadipelly & Pérez-González, 2014; Gasser, Ammann, Shahgaldian, & Corvini, 2014). The outcome of this type of exposure is a drastic disturbance in hormone synthesis, secretion, transformation, binding, and elimination, or simply disruption of hormonal dynamics and kinetics (Auriol, Filali-Meknassi, Tyagi, Adams, & Surampalli, 2006; Sumpter, 1998). Other studies have also determined the importance of "cocktail effects," where the simultaneous exposure to a combination of different chemicals occurs and a biomedical interaction may result (Margot, Maillard, Rossi, Barry, & Holliger, 2013; Watts, 2008). The transformation and metabolism of wastewaters by living organisms might also convert polluting pharmaceuticals into metabolites of greater toxicity (Ritter, Solomon, Sibley, Hall, Keen, Mattu, & Linton, 2002). One factor to bear in mind is that these agents are designed to be biologically and pharmacologically active, which differentiates these pharmaceuticals from other chemical pollutants. Moreover, they are designed in a way to have maximum durability within the body, and especially to show resistance to acid hydrolysis, while some are turned into active metabolites that make their disposal even more difficult (Bajaj, 2002; Eisenman, Mues, Weber, Frases, Chaskes, Gerfen, & Casadevall, 2007).

These considerations, taken together, have prompted a search for an ideal treatment protocol to resolve the problem of pharmaceuticals as pollutants. Numerous physicochemical methods, including electrochemical techniques, coagulation, adsorption, flocculation, ion-exchange, reverse osmosis, oxidation, TiO_2 photo-catalysis, and ozonation, are undergoing development for dealing with pharmaceutical contaminants (Bevilaqua, Cammarota, Freire, & Sant'Anna Jr., 2002; Dan, Yang, Dai, Chen, Wang, & Tao, 2013; Hela, Brandtner, Widek, & Schuh, 2003; Kasprzyk-Hordern, Dinsdale, & Guwy, 2009; Sutar & Rathod, 2015; L. Yang, Yu, & Ray, 2008). However, some constraints, particularly the high cost and time-consuming nature of these methods, as well as probable higher toxicity of the produced metabolites arising from them, have lent support to the idea that biological treatments, either by microbes and/or by their derived enzymes, could represent a more promising solution (Bevilaqua, Cammarota, Freire, & Sant'Anna Jr., 2002; Dan, Yang, Dai, Chen, Wang, & Tao, 2013; Hela, Brandtner, Widek, & Schuh, 2003; Kasprzyk-Hordern, Dinsdale, & Guwy, 2009; Sutar & Rathod, 2015; Yang, Yu, & Ray, 2008).

Enzymatic treatment approaches have included the use of peroxidases, tyrosinases, hydrolases, P450 cytochrome monooxidases, and lyases. Among the most efficient candidate enzymes are the laccases (benzenediol: oxygen oxidoreductase, EC 1.10.3.2), versatile copper-containing oxidases found mainly in the white-rot basidiomycetes (Faramarzi & Forootanfar, 2011; Forootanfar & Faramarzi, 2015; Forootanfar, Faramarzi, Shahverdi, & Yazdi, 2011; Heidary, Khoobi, Ghasemi, Habibi, & Faramarzi, 2014). Laccases have been increasingly applied in the removal of environmentally hazardous chemicals since the first report of their occurrence in resin ducts of the lacquer tree *Rhus vernicifera* (Aghaie-Khouzani, Forootanfar, Moshfegh, Khoshayand, & Faramarzi, 2012; Cañas & Camarero, 2010; Claus, 2003; Dehghanifard, Jonidi Jafari, Rezaei Kalantary, Mahvi, Faramarzi, & Esrafili, 2013; Riva, 2006; Sadighi & Faramarzi, 2013). The laccase enzymes are widely distributed in nature, and have been reported in higher plants, yeast, fungal pathogens, algal strains, lichen species, bacteria, and even insects (Bugg, Ahmad, Hardiman, & Rahmanpour, 2011; Claus, 2003; Diaconu, Litescu, & Radu, 2010; Dittmer, Suderman, Jiang, Zhu, Gorman, Kramer, & Kanost, 2004; Rezaei, Shahverdi, & Faramarzi, 2014; Strong & Claus, 2011; Zhao & Kwan, 1999). Laccases are believed to have evolved phylogenetically from the azurine protein (a prokaryote-derived protein) produced by *Thiobacillus ferrooxidans*. Evolutionary investigations have provided estimates that duplication-divergence events of an ancestral gene played a main role in development of novel families of laccase genes and in the production of laccase isoforms (Forootanfar, Movahednia, Yaghmaei, Tabatabaei-Sameni, Rastegar, Sadighi, & Faramarzi, 2012).

Unlike the case for peroxidases, laccase activity can occur in the absence of toxic H_2O_2. Furthermore, as the only remaining by-product of laccase-involving reactions is water (Figure 1a), it is considered an "eco-friendly" enzyme capable of utilizing a broad range of substrates consisting of both phenolic and non-phenolic compounds (Forootanfar, Moezzi, Aghaie-Khozani, Mahmoudjanlou, Ameri, Niknejad, & Faramarzi, 2012; Mogharabi, Nassiri-Koopaei, Bozorgi-Koushalshahi, Nafissi-Varcheh, Bagherzadeh, & Faramarzi, 2012). Although attention has mainly focused on how to put laccases to good use in removing pharmaceuticals from the environment, many questions still remain regarding which pharmaceutical categories could be possible targets for this enzyme. This chapter attempts to provide a critical review of the literature on pharmaceuticals currently being removed by different laccases and to spotlight the pivotal role of laccases in this type of bioremediation.

Figure 1.

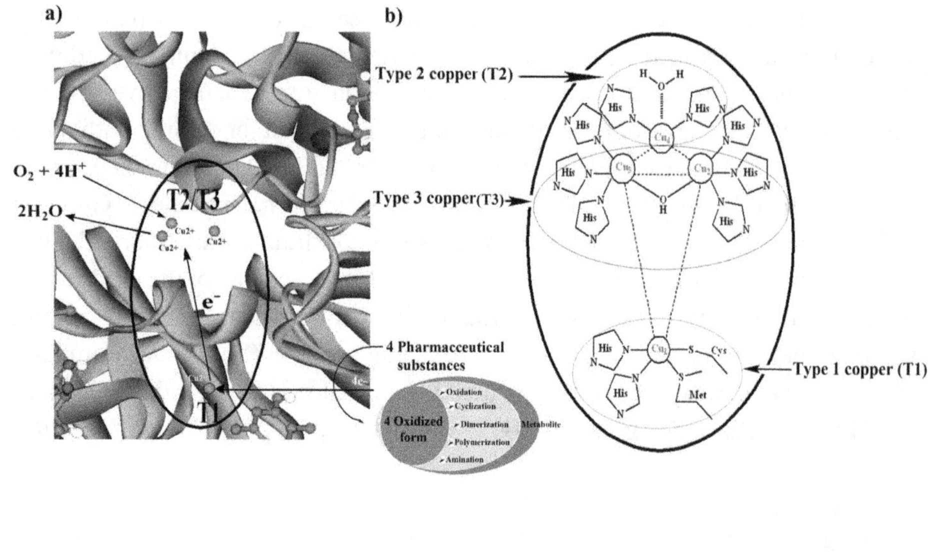

3. HOW DO LACCASES WORK?

Although the exact mechanism of electron transfer is not yet fully elucidated, to gain a thorough insight on how laccases work, we need to take a closer look at their active site, which is highly conserved among both bacteria and fungi (Lee, Kim, & Cho, 2015; Prasad, Vindal, Narayana, Ramakrishna, Kunal, & Srinivas, 2012; Schaetzle, Barrière, & Baronian, 2008). The active site of laccases usually comprises one binuclear and two mononuclear copper centers (designated T1, T2, and T3, respectively), which are classified based on their tendency to capture electrons and on their paramagnetic electron resonance characteristics (Bertrand, Jolivalt, Briozzo, Caminade, Joly, Madzak, & Mougin, 2002; Claus, 2004; Ramírez, Mano, Andreu, Ruzgas, Heller, Gorton, & Shleev, 2008; Sakurai & Kataoka, 2007; Shleev, Tkac, Christenson, Ruzgas, Yaropolov, Whittaker, & Gorton, 2005; Shleev, Morozova, Nikitina, Gorshina, Rusinova, Serezhenkov, & Yaropolov, 2004; Strong & Claus, 2011) (Figure 1a). The T1 center is surrounded by one cysteine and two histidine residues, and its high redox potential makes it a likely initiator of the catalytic cycle by consuming four substrates and producing four electrons as a consequence (Claus, 2004; Pita, Gutierrez-Sanchez, Olea, Velez, Garcia-Diego, Shleev, & De Lacey, 2011; Shleev, Morozova, Nikitina, Gorshina, Rusinova, Serezhenkov, & Yaropolov, 2004; Strong & Claus, 2011) (Figure 1b). Electrons are then transferred to the trinuclear cluster formed by both mononuclear T2 and binuclear T3, which have two and six histidine residues, respectively, as ligands (Claus, 2004; Lawton, Sayavedra-Soto, Arp, & Rosenzweig, 2009; Sergey Shleev; Strong & Claus, 2011) (Figure 1a and Figure 1b). This is the site where dioxygen binding and reduction occurs to generate two water molecules and terminate the electron transfer cycle (Bento, Silva, Chen, Martins, Lindley, & Soares, 2010; Claus, 2004; Enguita, Marçal, Martins, Grenha, Henriques, Lindley, & Carrondo, 2004; Hakulinen, Kiiskinen, Kruus, Saloheimo, Paananen, Koivula, & Rouvinen, 2002) (Figure 1a). Finally, after completion of these events, the generated oxidized free radicals can undergo non-enzymatic and spontaneous reactions, such as oxidative coupling of monomers, degradation of polymers, and ring cleavage of aromatic compounds (Claus, 2004; Strong & Claus, 2011). All four copper ions described above are in the 2^+ oxidation state in the resting form of enzyme (Figure 1b).

Atypical "Yellow" and "White" laccases have been frequently reported, which differ from "Blue" laccases by the absence of the characteristic peak at 610 nm and by the possession of only one copper ion, together with an additional transition metal atom, like iron, manganese, or zinc at the active site (Forootanfar & Faramarzi, 2015). *Gaeumannomyces graminis* (a phytopathogenic ascomycete) and *Agaricus bisporus* D621 (from the basidiomycete family) are among the yellow laccase producers (Forootanfar & Faramarzi, 2015). Min et al. (2001) described the purification of a white laccase from *Phellinus ribis*; this laccase possesses one Cu, two Zn, and one Mn in its active site, instead of the usual four copper atoms. Substrate molecules with suitable redox potential can directly interact with the laccase active site, but problems arise when laccase is faced with compounds that are either bulky or have much higher ionization potentials than laccase does and are accordingly incapable of penetrating into the active site or igniting the reaction (Kunamneni, Plou, Ballesteros, & Alcalde, 2008; Riva, 2006). Here, nature can be mimicked by the use of either natural [such as vanillin, methyl vanillate, acetovanillone, syringaldehyde (4-hydroxy-3,5-dimethoxybenzaldehyde, SYR), acetosyringone] or synthetic laccase mediators (like ABTS and HBT) (Forootanfar & Faramarzi, 2015) (Figure 2). This means that the oxidized radical forms of mediators can act as bridges between these targets and laccases to allow the reaction to proceed (Johannes & Majcherczyk, 2000; Couto, Sanromán, & Gübitz, 2005). In other words, the redox mediators can broaden the range of laccase substrates due to the shuttling role of these

Figure 2.

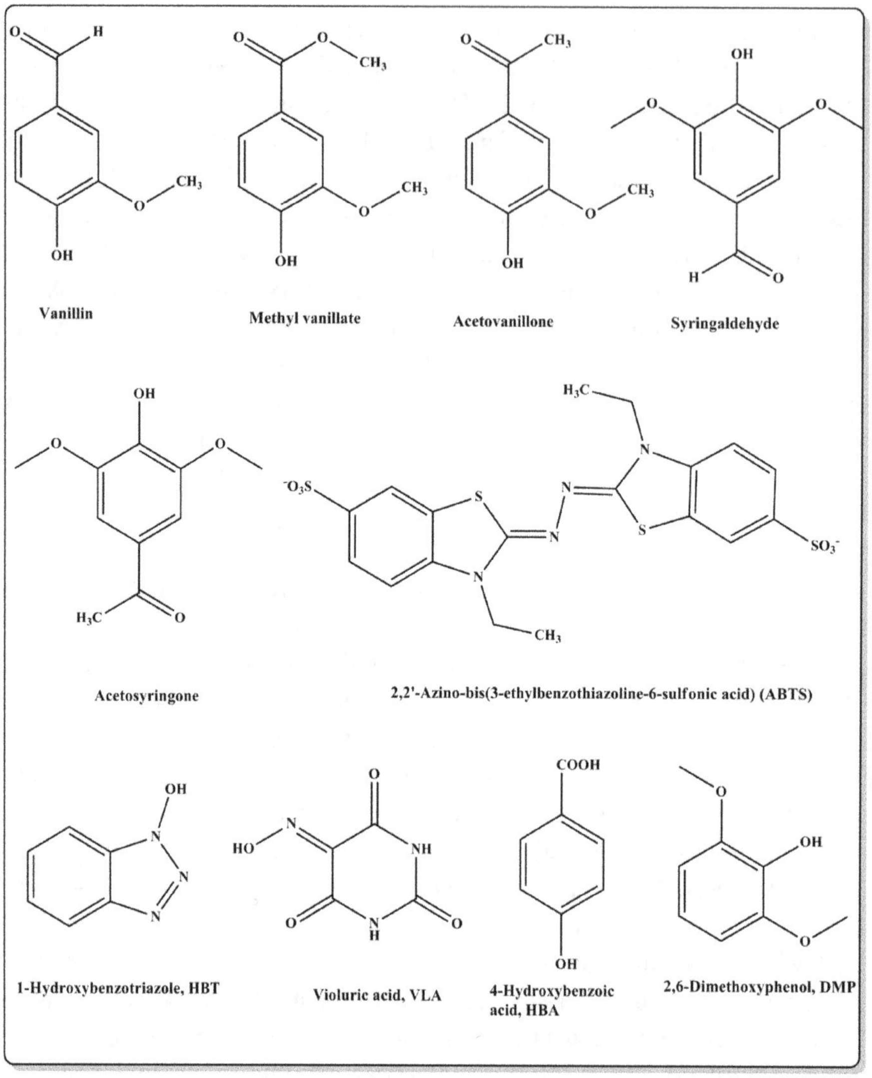

mediators between the enzyme and non-specific laccase substrates (Forootanfar & Faramarzi, 2015; Mogharabi & Faramarzi, 2014). However, previous reports indicate that yellow laccases are capable of catalyzing the oxidation of nonphenolic aromatic structures without the aid of these types of mediators (Forootanfar & Faramarzi, 2015).

Depending on the potential of the T1 site at the active site, the laccases can generally be categorized into three groups: 1) "low-redox potential" (about 430 mV), which include most of the bacterial-derived laccases; 2) "medium-redox potential" (470-710 mV), which include the laccases of *Trichoderma harzianum* WL1 and *Coprinus cinereus*; and 3) "high-redox potential," which include most of the basidiomycete-derived laccases with potentials of 780 mV (Forootanfar & Faramarzi, 2015). Note that the reduction potentials are lower for bacterial laccases than for fungal ones, which makes limitations in the choice of substrates for bacterial laccases not beyond the realm of possibility (Kanbi, Antonyuk,

Hough, Hall, Dodd, & Hasnain, 2002; Santhanam, Vivanco, Decker, & Reardon, 2011; Xu, Berka, Wahleithner, Nelson, Shuster, Brown, & Solomon, 1998). However, the unique properties of bacterial-derived laccases, such as their stability under the extreme conditions (high temperatures, alkaline pH, and high salt concentrations) have introduced them as attractive biocatalysts in the field of environmental bioremediation, despite their unsuitable redox potentials (Forootanfar & Faramarzi, 2015).

Laccases attack a wide variety of substrates and can extend beyond the compounds that are not its substrates when assisted by a mediator such as ABTS (Mayer, 2002; Sharma, Goel, & Capalash, 2007). The laccase substrates usually overlap with the substrates of other oxidizing enzymes, particularly tyrosinases, so substrate specification studies should be performed after purification steps (Forootanfar & Faramarzi, 2015). Those enzymes showing oxidizing ability for a specific substrate, like syringaldazine (4-hydroxy-3,5-dimethoxybenzaldehyde azine), but are unable to oxidize tyrosine, can be considered to be laccases (Forootanfar & Faramarzi, 2015). About 100 natural and artificial substrates have now been identified for different laccases and this number is destined to increase (Betancor, Johnson, & Luckarift, 2013; Mathioni, Patel, Riddick, Sweigard, Czymmek, Caplan, & Donofrio, 2013). These substrates have an astonishing range, from polycyclic aromatic hydrocarbons (PAH) and certain lipids to simple inorganic ions such as potassium triiodide (Forootanfar & Faramarzi, 2015; Forootanfar, Faramarzi, Shahverdi, & Yazdi, 2011). Monophenols, polyamines, aminophenols, aryl diamines, hydroxyindoles, aliphatic amines, benzenethiols, methoxy substituted phenols, and lignins are other examples of laccases substrates (Riva, 2006; Strong & Claus, 2011). This vast substrate variation is ascribed to the different redox potentials of various laccase enzymes (Alcalde, Bulter, Zumárraga, García-Arellano, Mencía, Plou, & Ballesteros, 2005; Ayala, Roman, & Vazquez-Duhalt, 2007; Chiacchierini, 2004; Tavares, Cristóvão, Gamelas, Loureiro, Boaventura, & Macedo, 2009). Electrophilic functional groups, particularly at the *ortho* position of phenolic rings, drastically decrease the ability of compounds to accommodate, whereas nucleophilic functional group substitution positively enhances a substrate's affinity for the enzyme (Majeau, Brar, & Tyagi, 2010; Pozdnyakova, Rodakiewicz-Nowak, & Turkovskaya, 2004).

4. LACCASES AND PHARMACEUTICAL WASTES

4.1. Antimicrobials

Antimicrobials are extensively used for both human and veterinary therapy and are among the most frequently detected pharmaceutical pollutants in wastewaters and the environment (D'Abrosca, Fiorentino, Izzo, Cefarelli, Pascarella, Uzzo, & Monaco, 2008; Llorca, Rodríguez-Mozaz, Couillerot, Panigoni, de Gunzburg, Bayer, … Barceló, 2014; Sarmah, Meyer, & Boxall, 2006). Antimicrobials are even able to act on non-targeted organisms, but what makes antibiotics pollution most worrisome is the promotion of resistance in bacteria to commonly used antimicrobials (Kim & Aga, 2007; Weng, Liu, & Lai, 2013). As these multi drug resistant bacteria continue to divide, they can spread their obtained antibiotic-resistance genes via horizontal gene transfer to other species, thereby paving the way for treatment failures for most infectious diseases (Dzidic & Bedeković, 2003). However, the currently available techniques are limited in their ability to remove these xenobiotics from water supplies (Chang, Hu, Wang, & Shao, 2008; Guo, Zhu, & Li, 2014). These complications have attracted the interest of researchers to bring laccase into play to deal with sulfonamides (SAs), penicillins (PCs), tetracyclines (TCs), trimethoprim (TMP), and fluoroquinolones (FQs).

Typical laccase mediator systems (LMS) have applied HBT, ABTS, violuric acid (VLA), etc. to attain significant removal of antimicrobial agents. For example, a study conducted by Weng et al. (2013) reported the removal of two SAs, sulfadimethoxine (SDM) and sulfamonomethoxine (SMM) (Figure 3a), which are highly consumed veterinary medications, by using a novel laccase from *Perenniporia* sp. strain TFRI 707 coupled with VLA, ABTS, SYR, and 4-hydroxybenzoic acid (4-HBA) as mediators. Limited reduction of SAs was observed in the presence of laccase alone at all tested temperatures, whereas LMS considerably degraded both SDM and SMM, with the fastest oxidation rate observed using 1 mM VLA (pH 4, temperature of 40–60 °C), 2 mM ABTS (pH 4, 50–60 °C), and 2 mM SYR (pH 6, 50 °C) (Weng, Liu, & Lai, 2013). However, HBA (2 mM) resulted in maximum removal at pH 4, and at 30 °C and 60

Figure 3.

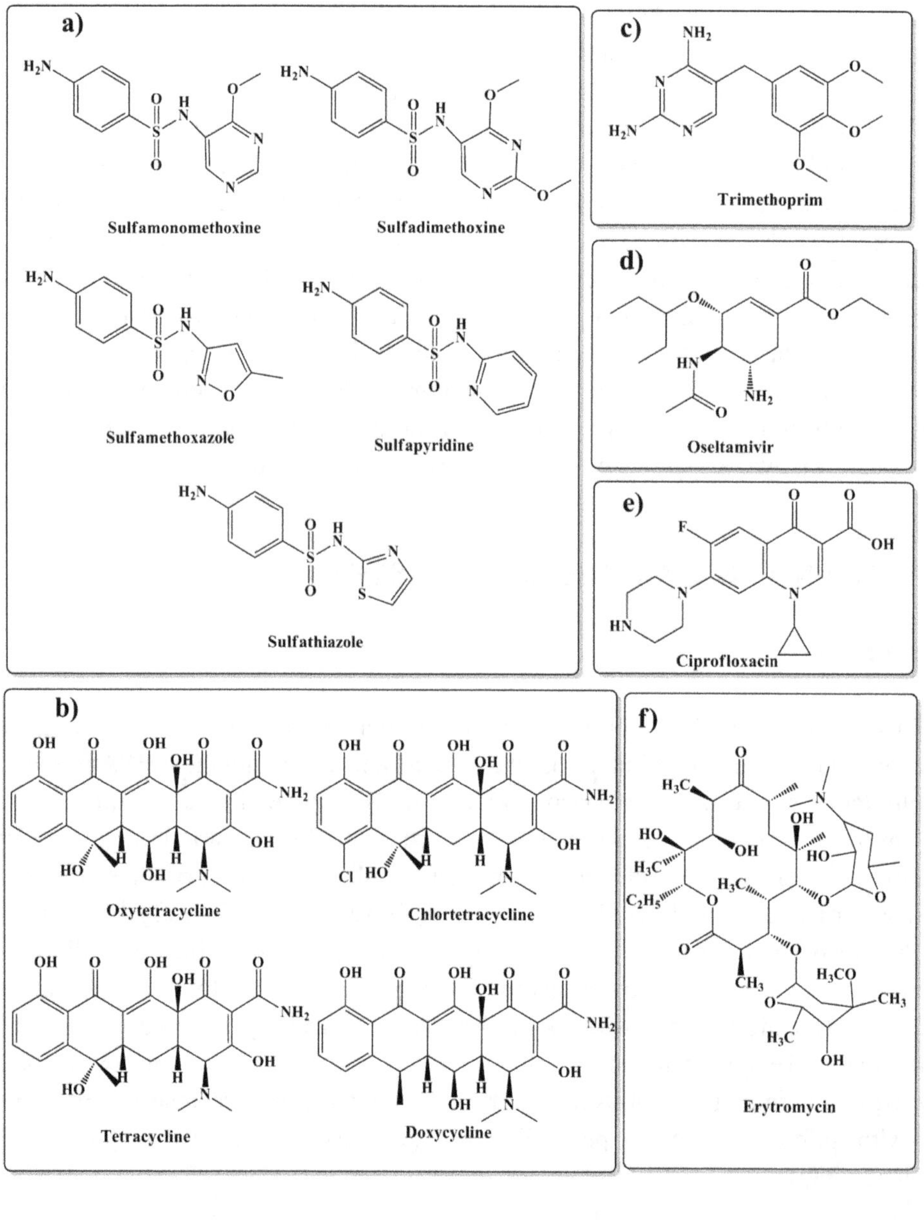

°C for SDM and SMM, respectively (Weng, Liu, & Lai, 2013). VLA and HBA were concluded to have lower toxicity (as assessed using the bioluminescent bacterium *Vibrio fischeri*) of SA solutions when compared to other mediators, and VLA was suggested as the best mediator to work with owing to its high removal rate, low temperature requirement, and low energy cost (Weng, Liu, & Lai, 2013). However, in some cases, the relative toxicity of the laccase-treated sample was enhanced, especially when LMS was applied. For example, Becker et al. (2017) ascribed the higher toxicity of laccase-treated antibiotics to the formation of quinonoid structures when a laccase-SYR system was applied. Guo et al. (2014) have previously indicated that an initial concentration of laccase (6 IU) assisted by 1 mM VLA, ABTS, and SYR, can favorably transform SDM and SMM, whereas no satisfactory results were achieved for other mediators, including HBA, 4-cyanophenol (CP), and 4-hydroxyacetophenone (HAP).

A recent study utilized *Phanerochaete chrysosporium*, a white-rot laccase producing fungus, for sulfamethoxazole (SMX) (Figure 3a) bioremediation (Guo, Zhu, & Li, 2014). SMX can inhibit the growth of the fungus in a concentration independent manner at concentrations varying from 10–30 mg/L, indicating a strong tolerance of *P. chrysosporium* to SMX (Guo, Zhu, & Li, 2014). In a defined nitrogen-limited liquid medium containing Cu^{2+} as a laccase inducer, the assessed degradation of an initial 10 mg/L of SMX was 53% removal after 24 h and 74% at the end of the experiment (Guo, Zhu, & Li, 2014). In vitro analysis of a crude enzyme also confirmed increasing laccase activity results in faster SMX degradation (Guo, Zhu, & Li, 2014). In many cases, the combination of laccases with other non-enzymatic processes further improves the removal efficiency. For instance, co-culture of *Alcaligenes faecalis* (a non-laccase secreting bacteria) with the laccase producing fungal strain, *P. sanguineus* or its purified laccase increased SMX removal (X. Li, Xu, Cheng, & Yuan, 2016) (Table 1).

Another white-rot fungus, *Trametes versicolor*, has demonstrated encouraging results for removal of sulfapyridine (SPY) and sulfathiazole (STZ) (Figure 3a) from wastewaters (García-Galán, Rodríguez-Rodríguez, Vicent, Caminal, Díaz-Cruz, & Barceló, 2011). The data were indicative of entire elimination of SPY and 88% removal of STZ in fungal cultures, underlining the longer incubation time of 96 h required for removal of STZ when compared to SPY (García-Galán, Rodríguez-Rodríguez, Vicent, Caminal, Díaz-Cruz, & Barceló, 2011). The level of degradation increased significantly to more than 75% following application of a purified laccase or laccase mediators [HBT, acetosyringone, and ABTS] (García-Galán, Rodríguez-Rodríguez, Vicent, Caminal, Díaz-Cruz, & Barceló, 2011). Eight and five intermediate compounds were substantially identified by the use of ultra-performance liquid chromatography-quadrupole time of flight mass spectrometry (UPLC-QqTOF-MS), and SPY laccase-aided degradation was confirmed to produce formaldehyde and other related hydroxylated products and desulfonated moieties (García-Galán, Rodríguez-Rodríguez, Vicent, Caminal, Díaz-Cruz, & Barceló, 2011) (Table 1).

Immobilization techniques also have advantages, such as increasing the stability and production yield, as well as the reusability feature, so many studies have focused on the potential application of immobilized laccases for pharmaceutical pollutant removal. For example, Rahmani et al. (2015) have provided evidence that immobilization of laccase (obtained from *T. versicolor*) on porous silica beads allowed a 66 and 77% detoxification of SMZ and STZ, respectively. Immobilization also made the blue enzyme more stable at broader ranges of both pH and temperature when compared to the free laccase (Rahmani, Faramarzi, Mahvi, Gholami, Esrafili, Forootanfar, & Farzadkia, 2015). The immobilized laccase retained 63.3% and 82.6% of its initial activity towards SMZ and STZ, respectively, after 10 application runs (Rahmani, Faramarzi, Mahvi, Gholami, Esrafili, Forootanfar, & Farzadkia, 2015). When these SAs were subjected to treatment with HBT-aided laccase at high temperatures, the enzymatic removal was increased, so that the maximum enzymatic degradation of both SAs were seen at 50 °C and pH 5 (Rahmani, Faramarzi, Mahvi, Gholami, Esrafili, Forootanfar, & Farzadkia, 2015). These workers also reported a decrease in the microtoxicity of laccase-treated SA solutions and suggested that this biocatalyst system can be an efficient armamentarium against SAs (Rahmani, Faramarzi, Mahvi, Gholami, Esrafili, Forootanfar, & Farzadkia, 2015).

Concomitantly, three other studies attempted to effect degradation of TCs through various methods that had one parameter in common: laccases (Table 1). In this regard, Migliore et al. (2012) demonstrated that mycelia of *Pleurotus ostreatus*, a ligninolytic fungus, could absorb oxytetracycline (OTC) (Figure 3b) for degradation by laccase, as confirmed by mass spectroscopy analyses. The fungus not only remained viable upon exposure to large amounts of the drug, but also almost cleared the initial concentration of OTC (50 µg/L during the 14-day period of the experiment (Migliore, Fiori, Spadoni, & Galli, 2012) (Table 1). The tentative transformation product was considered to be 2-acetyl-2-decarboxyamidoxytetracycline. Notably, neither abiotic and extracellular laccase were able to degrade OTC (Migliore, Fiori, Spadoni, & Galli, 2012). A more comprehensive study investigated the fate of tetracycline (TC), doxycycline (DC), chlortetracycline (CTC), and OTC (Figure 3b) after treatment with laccase produced by *T. versicolor* in the presence of HBT (Suda, Hata, Kawai, Okamura, & Nishida, 2012). Laccase alone could not eliminate these four TCs, but when coupled with HBT in a LMS, it completely removed CTC and DC together after 15 min, while TC and OTC were removed after 1 h at a rate even greater than that obtained with manganese peroxidase (MnP) (Suda, Hata, Kawai, Okamura, & Nishida, 2012). A complete reversal of growth inhibition of *E. coli* and *Bacillus subtilis* as well as the green alga *Pseudokirchneriella subcapatita* indicated the potential of this LMS to eradicate TC ecotoxicity (Suda, Hata, Kawai, Okamura, & Nishida, 2012). A more recent study used an on-line turbulent flow system coupled to a hybrid linear ion trap–high resolution mass spectrometer to detect transformation products of a 100 µg/mL TC solution treated with laccase alone (Llorca, Rodríguez-Mozaz, Couillerot, Panigoni, de Gunzburg, Bayer, … Barceló, 2014). The laccase could detoxify 78% of TC after 18 h and produced three different metabolites designated TP618, TP396, and TP431 with a degradation rate of nearly 70% ((Llorca, Rodríguez-Mozaz, Couillerot, Panigoni, de Gunzburg, Bayer, … Barceló, 2014). Oxidation, dihydroxylation, and bi-demethylation of rings A and C were postulated as the major reactions involved(Llorca, Rodríguez-Mozaz, Couillerot, Panigoni, de Gunzburg, Bayer, … Barceló, 2014).

In addition to these antimicrobial agents (SAs and TCs), the penicillins (PCs) and fluoroquinolones (FQs) have also been subjected to removal by laccase and LMS (Table 1). Becker et al. (2017) described a potential application of the immobilized laccase of *T. versicolor* and laccase-SYR for removal of twelve SAs, six PCs, ten FQs, and four TCs. Of these structures, the FQs were the most recalcitrant compounds, followed by PCs, TCs, and SAs. The authors ascribed the higher biodegradability of SAs and PCs to the strong electron donation characteristic of the amine group and phenol ring, respectively, which are absent in FQs (Becker, Rodriguez-Mozaz, Insa, Schoevaart, Barceló, de Cazes, … Wagner, 2017). Similar results were observed in the case of trimethoprim (Figure 3c) (67% removal using laccase-SYR system), which possesses three methoxy groups and two amine groups in its structure. By contrast, substitutions with electronegative moieties decreased the oxidation by laccases ((Becker, Rodriguez-Mozaz, Insa, Schoevaart, Barceló, de Cazes, … Wagner, 2017).

Flanking these findings, another study used a bioplastic formulation entrapping propagules of *P. chrysosporium* for the removal of oseltamivir (Figure 3d), the first commercially developed oral neuraminidase inhibitor, and antibiotics such as ciprofloxacin (Figure 3e), SMX, and erythromycin (Figure 3f) (Accinelli, Saccà, Batisson, Fick, Mencarelli, & Grabic, 2010) (Table 1). Oseltamivir was identified as a recalcitrant compound, as only low amounts of it were eliminated after 10 days of incubation whereas more than 80% of the erythromycin was removed after 5 days (Accinelli, Saccà, Batisson, Fick, Mencarelli, & Grabic, 2010). The highest removal percentage was observed for ciprofloxacin (60% for effluent water and 80% for an activated sludge-mixed liquor) (Accinelli, Saccà, Batisson, Fick, Mencarelli, & Grabic, 2010). Further microbial analyses have shown that a bioplastic matrix would markedly sustain the growth of an applied fungus, which could open a new avenue for application to other types of pharmaceutical contaminants (Accinelli, Saccà, Batisson, Fick, Mencarelli, & Grabic, 2010).

4.2. Lipid Regulating Agents

Previous work in a fluidized bed reactor assessed the capability of *T. versicolor* pellets to remove environmentally relevant concentrations of clofibric acid (CLOF) (Figure 4a), a lipid regulator commonly found in wastewaters (Cruz-Morató, Jelić, Perez, Petrović, Barceló, Marco-Urrea, & Vicent, 2013). In the first hour, 33% of CLOF was eliminated and its complete detoxification occurred within four days of the experiment and reached a steady state by day 12 (Cruz-Morató, Jelić, Perez, Petrović, Barceló, Marco-Urrea, & Vicent, 2013). Hydroxy-CLOF was identified by QqTOF-MS as the main transformation product of this catalytic reaction and its concentration remained constant in the medium (Cruz-Morató, Jelić, Perez, Petrović, Barceló, Marco-Urrea, & Vicent, 2013). The only drawback encountered in this series of experiments was the greater toxicity observed for the final treated effluent when compared to the initial feed, which could be ascribed to the presence of hydroxyl-CLOF (Cruz-Morató, Jelić, Perez, Petrović, Barceló, Marco-Urrea, & Vicent, 2013) (Table 1).

Figure 4.

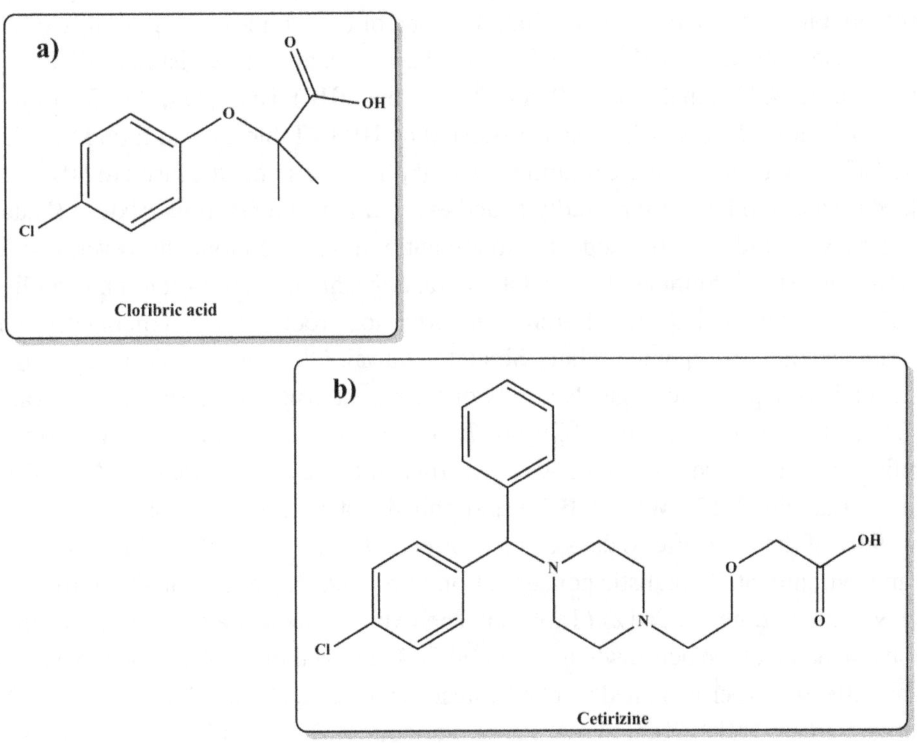

4.3. Antihistamines

A literature review revealed only one available study pointing out the enzymatic activity of laccase toward a frequently used antihistamine, cetirizine hydrochloride (Figure 4b), under the influence of ultrasound irradiation to enhance the degradation rate of cetirizine (Sutar & Rathod, 2015). Under optimized conditions, 91% of the cetirizine was degraded. Among the intermediates, three prominent ones were identified using a Thermo Electron Corporation LC/MS and data analysis (Sutar & Rathod, 2015). As a final point, the authors depicted the degradation pathway for cetirizine, with further elucidation of its probable mechanism (Sutar & Rathod, 2015) (Table 1).

4.4. Central Nervous System Drugs

Central nervous system (CNS) agents are highly prescribed medicines due to their diverse therapeutic applications. They can act as antidepressants, antiepileptics, local anesthetics, antipsychotics, muscle relaxants, analgesics, and sedative-hypnotics (Cooper, Siewicki, & Phillips, 2008). The special impacts of these agents on living organisms and humans are still a matter of debate, and more studies are needed to fill the gaps in our understanding (Palmer & Alavijeh, 2012). Most studies have focused on carbamazepine (CBZ) (Figure 5) as a recalcitrant pollutant in wastewaters, as it is highly resistant to degradation by physicochemical and conventional biological processes (Carballa, Omil, Lema, Llompart, García-Jares, Rodríguez, & Ternes, 2004; Clara, Strenn, & Kreuzinger, 2004; Joss, Keller, Alder, Göbel, McArdell, Ternes, & Siegrist, 2005; Miao & Metcalfe, 2003; Radjenovic, Petrovic, & Barceló, 2007; Suárez, Carballa, Omil, & Lema, 2008; Zhang, Geissen, & Gal, 2008) (Table 1). Some methods have tackled the resistance of CBZ, but the production of hazardous and toxic by-products have limited the application of these methods (Negrón-Encarnación & Arce, 2007). At present, the only microorganism showing a potential for use in CBZ bioremediation is one of the white-rot fungi family (Golan-Rozen, Chefetz, Ben-Ari, Geva, & Hadar, 2011; Hata, Shintate, Kawai, Okamura, & Nishida, 2010; Marco-Urrea, Pérez-Trujillo, Vicent, & Caminal, 2009; Zhang & Geissen, 2010). Hata et al. (2010) studied repeated treatments of CBZ with a *T. versicolor* laccase assisted by HBT (Table 1), and found that 22% of CBZ was eliminated after 24 h when a single treatment was applied. This rate increased to 60% removal after 48 h when laccase and HBT were repeatedly added every 8 h to the reaction mixture (Hata, Shintate, Kawai, Okamura, & Nishida, 2010). Repeated treatment with laccase alone, however, resulted in only 39% CBZ elimination (Hata, Shintate, Kawai, Okamura, & Nishida, 2010) (Table 1). In addition, HPLC analyses of CBZ residues revealed four different transformation products; two of them were characterized as 9(10H) acridone and CBZ-epoxide (Hata, Shintate, Kawai, Okamura, & Nishida, 2010) (Table 1). The latter one retains the pharmacological activity of the parent drug (Hata, Shintate, Kawai, Okamura, & Nishida, 2010). In another study, the ability of *T. versicolor* to decimate CBZ in Erlenmeyer flasks and an air pulsed fluidized bed bioreactor was investigated (Jelic, Cruz-Morató, Marco-Urrea, Sarrà, Perez, Vicent, & Barcelo, 2012). When CBZ was supplied at a concentration of 9 mg/ml, *T. versicolor* could remove 94% of the CBZ after 6 days of incubation, whereas only 61% of the CBZ was degraded in 7 days at an environmentally realistic concentration of 50 μg/mL (Jelic, Cruz-Morató, Marco-Urrea, Sarrà, Perez, Vicent, & Barcelo, 2012) (Table 1). The experiment was also carried out in both batch and continuous mode in a bioreactor, where 96% and 54% removal of CBZ was achieved, respectively, within two days after the experiment had reached a steady state (Jelic, Cruz-Morató, Marco-Urrea, Sarrà, Perez, Vicent, & Barcelo, 2012) (Table 1). Acridine, acridone, CBZ-epoxide, and dihydroxycarbamaze-

Figure 5.

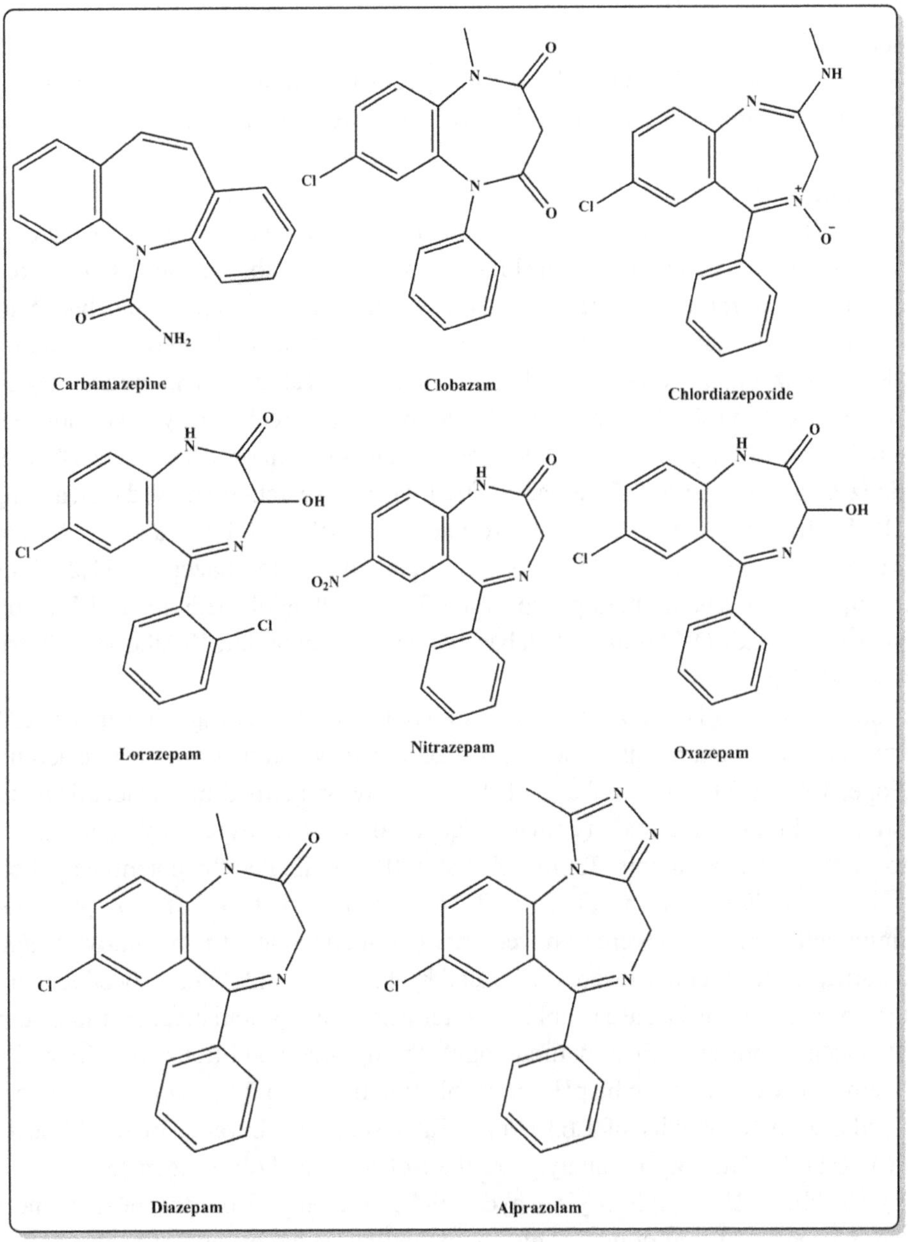

pine were identified as unique biodegraded metabolites of CBZ, and the results of acute toxicity tests confirmed their low toxicity in both batch and continuous operation mode, with 24% and 77% values for EC_{50} indicating a greater toxicity of CBZ than its metabolites (Jelic, Cruz-Morató, Marco-Urrea, Sarrà, Perez, Vicent, & Barcelo, 2012) (Table 1).

In agreement with these findings, the growth of another laccase-producing fungus, *P. chrysosporium*, was sustained with a specially designed plate bioreactor that used a 10 pore per inch polyether foam to detoxify CBZ under non-sterile conditions in both batch and continuous mode (Zhang & Geißen, 2012).

Unlike the previous study by Zhang and Geissen. (2010) that achieved limited degradation of CBZ using MnP and lignin peroxidase (LiP), a marked improvement (60–80%) was attained. The continuous bioreactor operation led to stable elimination of CBZ for a long period (about 100 days) (Zhang & Geissen, 2010; Zhang & Geißen, 2012). Overall, and in agreement with a literature survey, the presence of the strong electron-attracting amide group in the CBZ structure may be responsible for its recalcitrance against enzymatic treatment via laccases.

Disposal of some benzodiazepines in aquatic solutions was examined using LMS by Ostadhadi et al. (Ostadhadi-Dehkordi, Tabatabaei-Sameni, Forootanfar, Kolahdouz, Ghazi-Khansari, & Faramarzi, 2012). Clobazam, chlordiazepoxide, and lorazepam (Figure 5) were intractably resistant to degradation by the purified laccase of *Paraconiothyrium variabile*, whereas nitrazepam, oxazepam, and diazepam (Figure 5) underwent 27.3, 18.7, and 18.6% decomposition, respectively (Ostadhadi-Dehkordi, Tabatabaei-Sameni, Forootanfar, Kolahdouz, Ghazi-Khansari, & Faramarzi, 2012) (Table 1). The highest degradation was reported for alprazolam (45.6%) (Figure 5), and higher removal percentages were established by either enhancement of laccase activity or use of a mediator (Ostadhadi-Dehkordi, Tabatabaei-Sameni, Forootanfar, Kolahdouz, Ghazi-Khansari, & Faramarzi, 2012). These researchers showed that, among the tested mediators (ABTS, HBT, VLA, and 2,6-dimethoxyphenol [DMP]), HBT has a better performance and enhanced the degradation percentage to 73% for nitrazepam, 61.4% for diazepam, 71.2% for oxazepam, and 88.1% for alprazolam. The authors proposed HBT as a potential mediator in LMS for removing benzodiazepines (Ostadhadi-Dehkordi, Tabatabaei-Sameni, Forootanfar, Kolahdouz, Ghazi-Khansari, & Faramarzi, 2012) (Table 1).

Tricyclic antidepressants (TCAs) are another group of CNS-related drugs that are prescribed for the treatment of mental depression but also show anti-inflammatory and neuroprotective activity (Gruber, Hudson, & Pope, 1996) (Table 1). Leakage of these widely prescribed pharmaceuticals into sewage and surface water and their acute toxicity toward aquatic organisms, especially algae, have persuaded researchers to combat these pollutants. Tahmasbi et al. (2016) evaluated the potential application of the purified laccase of *P. varibile* for removal of five TCAs and found that 67% of clomipramine (Figure 6) and 82% of imipramine (Figure 6) were removed after 6 h of enzymatic treatment, while amitriptyline, doxepin, and nortriptyline (Figure 6) were eliminated by 11%, 6%, and 23%, respectively after 72 h incubation in the presence of this laccase (Table 1). After performing a statistical optimization procedure (using Box-Behnken response surface methodology), the optimal conditions for elimination of imipramine (99% removal) were found to be pH 4.9, incubation time of 5.7 h, imipramine concentration of 0.12 μg/mL, and enzymatic activity of 1.63 U/mL (Tahmasbi, Khoshayand, Bozorgi-koushalshahi, & Heidary, 2016) (Table1). The spectral analyses of the isolated metabolite suggested 3-(2,7-dihydroxy-10,11-dihydro-5*H*-dibenzo[*b*,*f*]azepin-5-yl)-1-hydroxy-*N*,*N*-dimethyl-3-oxopropan-1-amine oxide as the laccase-treated metabolite of imipramine. This compound had a lesser cytotoxic effect on the Caco-2 cell line when compared to the parent molecule (imipramine).

4.5. Sexual Hormone Analogs

As previously stated, the presence of steroid estrogens in the aquatic environment is becoming an excruciating concern (Auriol, Filali-Meknassi, & Tyagi, 2006; Sumpter, 1998). These hormones are able to mimic what hormones are doing in the body and act as EDCs, so the traces left in wastewaters may alter physiological functions in both humans and biota and could lead to genotoxic damage, altered metabolism, nephrotoxicity, and reproductive failure (Auriol, Filali-Meknassi, Tyagi, & Adams, 2007; Auriol,

Figure 6.

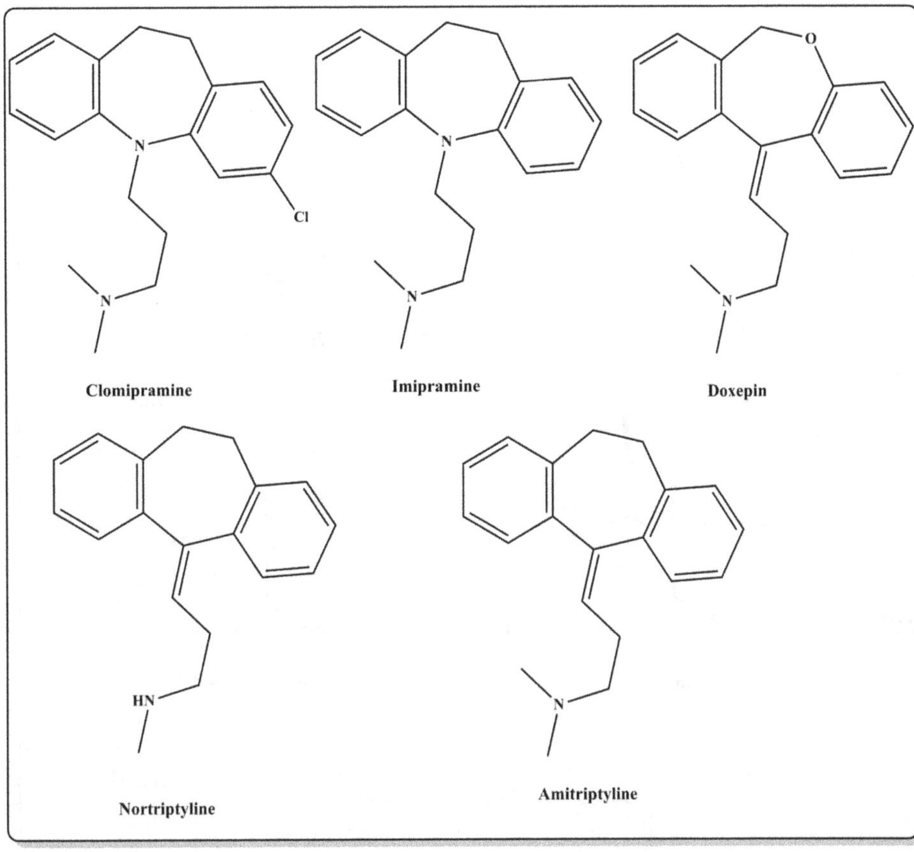

Filali-Meknassi, & Tyagi, 2006; Cabana, Jones, & Agathos, 2007; Choi, Yoo, & Lee, 2004; Colborn, vom Saal, & Soto, 1993; Filby, Neuparth, Thorpe, Owen, Galloway, & Tyler, 2007; Gagné, Blaise, & André, 2006; Jobling, 2002; Liney, Hagger, Tyler, Depledge, Galloway, & Jobling, 2005). The hazardous effects of these estrogenic compounds, whether synthetic or natural, their incomplete removal during conventional treatment of municipal wastewaters, and their potential to disrupt the endocrine system even at trace concentrations, have directed research toward the application of laccases for detoxification of water supplies as an alternative to the conventional time-consuming and high cost methods, which usually produce toxic byproducts (Auriol, Filali-Meknassi, Adams, & Tyagi, 2006; Auriol, Filali-Meknassi, Tyagi, & Adams, 2007; Nicell, Wagner, & Sakurai, 2003). Accordingly, Auriol et al. (2007) demonstrated that a 20 U/mL activity of *T. versicolor* laccase can completely and pH dependently remove estriol (E3), 17ß-estradiol (E2), and estrone (E1) (Figure 7), which are natural estrogen candidates, and a synthetic estrogen, 17α-ethenylestradiol (EE2) (Figure 7), from municipal and synthetic wastewaters (Table 1). The enzyme kinetics profile was matched by the Michaelis-Menten equation with a similar approximate affinity toward each of the studied estrogens under experimental conditions at pH 7 and temperature of 25±1°C (Auriol, Filali-Meknassi, Adams, & Tyagi, 2006; Auriol, Filali-Meknassi, Tyagi, & Adams, 2007). Moreover, the use of HBT as a mediator for this catalytic reaction enhanced the overall efficiency of the laccase catalytic system (Auriol, Filali-Meknassi, Adams, & Tyagi, 2006; Auriol, Filali-Meknassi,

Figure 7.

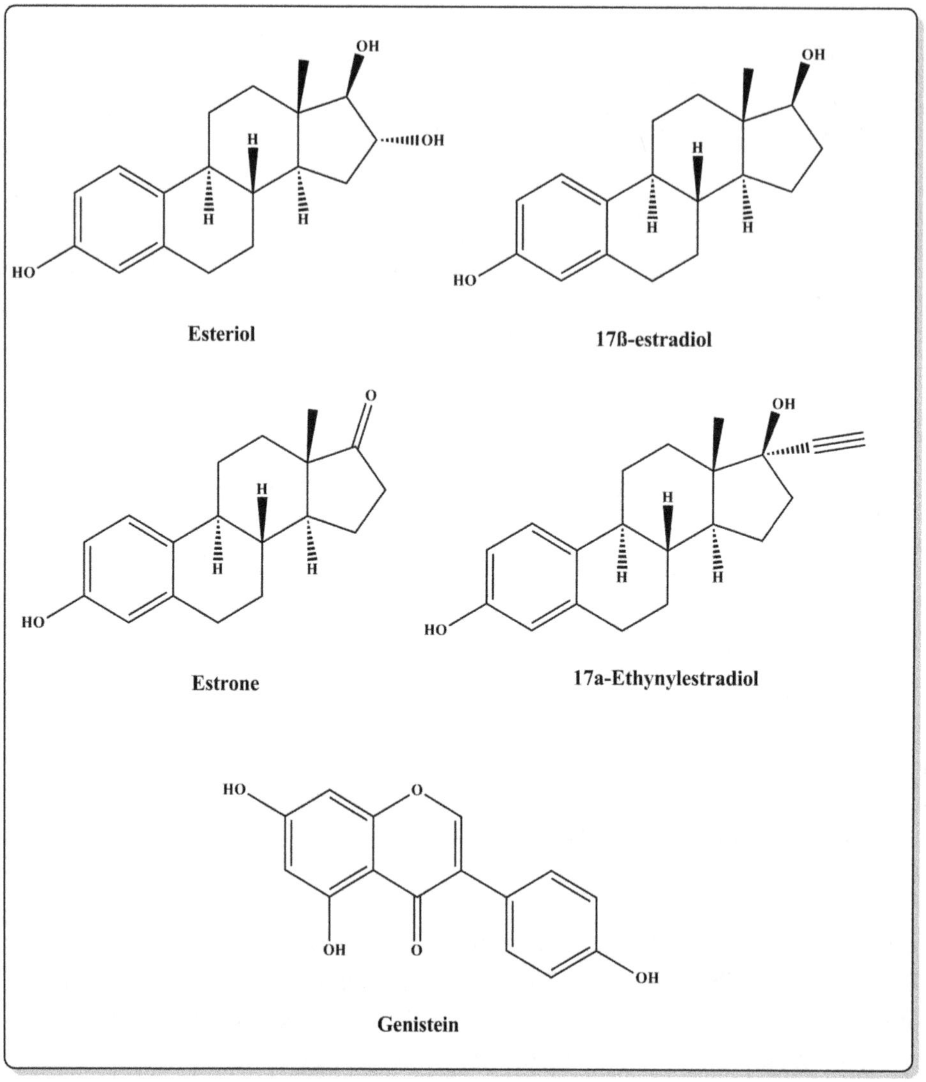

Tyagi, & Adams, 2007). Blanquez et al. (2008) confirmed the technical feasibility of biodegradation of E2 and EE2 in a continuous bioreactor with a suspended biomass of *T. versicolor* after 26 days at a hydraulic retention time of 120 h (Table 1). This white-rot fungus degraded more than 97% of these estrogenic compounds after 24 h in batch mode at volumetric removal rates of 0.43 and 0.44 mg/l/h; these rates decreased to 0.16 mg/mL and 0.09 mg/L/h in continuous mode for E2 and EE2, respectively (Blánquez & Guieysse, 2008). Of note, no MnP or LiP activity was observed and laccase was confirmed to contribute to this enzymatic biodegradation (Blánquez & Guieysse, 2008). Production of laccase was evident in *P. chrysosporium* and *Ganoderma lucidum* and they were also shown to have the capacity to remove steroid estrogens with the same kinetics as *T. versicolor* (Blánquez & Guieysse, 2008) (Table 1).

Along these lines, several other studies aimed at degradation of estrogens by different approaches. The commercial purified laccase obtained from *Myceliophthora thermophila* can degrade E1, E2, and EE2 immediately at all pH values with no marked effects on the degradation process (Lloret, Eibes, Lú-Chau, Moreira, Feijoo, & Lema, 2010a) (Table 1). However, the optimal pH was 4 and the reaction proceeded well even at relatively low concentrations of laccase (Lloret, Eibes, Lú-Chau, Moreira, Feijoo, & Lema, 2010a). SYR and HBT considerably augmented the laccase catalytic reaction, though E2 and EE2 were completely removed even in the absence of any mediator after only 24 h of laccase treatment. This finding indicated that laccase application for bioremediation at the industrial scale could use the natural mediator SYR to reduce the costs even more (Lloret, Eibes, Lú-Chau, Moreira, Feijoo, & Lema, 2010a).

Lloret et al. (2012) took further steps to reduce the costs of this treatment strategy by investigating the impacts of pH, free laccase activity, aeration or oxygenation, and estrogen feeding rate on laccase catalytic behavior. Under the best determined conditions, where oxygen was supplied and initial laccase activity was decreased, an oxidation of 94.1% for E1 and 95.5% of E2 was obtained (Lloret, Eibes, Feijoo, Moreira, & Lema, 2012b) (Table 1). Continuous removal of E1 and E2 by an enzymatic membrane reactor has also been examined and resulted in 95% elimination of E1 and nearly complete removal of E2 (Lloret, Eibes, Lú-Chau, Moreira, Feijoo, & Lema, 2012b). As outlined in other studies, the residual products of degradation exhibited negligible estrogenic activity (Auriol, Filali-Meknassi, Adams, Tyagi, Noguerol, & Piña, 2008; Lloret, Eibes, Lú-Chau, Moreira, Feijoo, & Lema, 2012b).

In parallel, in an attempt to improve the reaction yield and enzyme stability, the process was carried out in a fluidized bed reactor with immobilized laccase with promising results (Lloret, Eibes, Feijoo, Moreira, & Lema, 2012a). Another study by Tamagawa et al. (2005) reported the removal of about 93% of genistein (Figure 7), an isoflavone with several biological effects, by the use of the white-rot fungus *Phanerocaete sordida* YK-624 (Tamagawa, Hirai, Kawai, & Nishida, 2005) (Table 1). Ligninolytic activity during the degradation process and HPLC data indicated complete removal of genistein and diminished estrogenic activity after 4 h by MnP, laccase, or a laccase-HBT system (Kar, Tandon, & Saha, 2002; Si & Liu, 2007; Tamagawa, Hirai, Kawai, & Nishida, 2005; Wang, Waltenberger, Pferschy-Wenzig, Blunder, Liu, Malainer, & Atanasov, 2014) (Table 1). Taken together, the results confirm the applicability of laccases as a tool for large-scale detoxification of effluents in the aquatic environment from pharmaceuticals that act as EDCs. Notably, EDCs cover a wide range of chemical substances and numerous studies have examined laccase application for the detoxification of these chemicals in the environment but are beyond the scope of this review. We have only focused on pharmaceuticals or endogenous hormones amenable to elimination by laccase.

4.6. Non-Steroidal Anti-Inflammatory Drugs (NSAIDs)

Although they are found only in trace amounts in wastewaters, NSAIDs are another category of pharmaceuticals that are receiving substantial attention due to their extensive use among human and animals, as well as their possible cytotoxicity for mankind and animals (Domaradzka, Guzik, & Wojcieszyńska, 2015; Heberer, 2002; Heberer, Reddersen, & Mechlinski, 2002; Ternes, 1998). Naproxen, diclofenac, ibuprofen, ketoprofen, and mefenamic acid (MFA) (Figure 8) are among the most commonly prescribed NSAIDs that have been the focus of researchers. In this regard, naproxen at concentrations of 10 mg/L and 55 μg/L was almost completely eliminated (95%) after 6 h incubation in the presence of laccase from *T. versicolor*. Degradation was also significantly improved when a mediator like HBT accompanied the purified *T. versicolor* laccase (Marco-urrea, Pérez-trujillo, Blánquez, Vicent, & Caminal, 2010). Structural elucidation studies using HPLC-DAD-MS and NMR identified three main intermediates of naproxen, including 2-(6-hydroxynaphthalen-2-yl) propionic acid, 1-(6-metoxynaphthalen-2-yl) ethanone, and 6-O-

Figure 8.

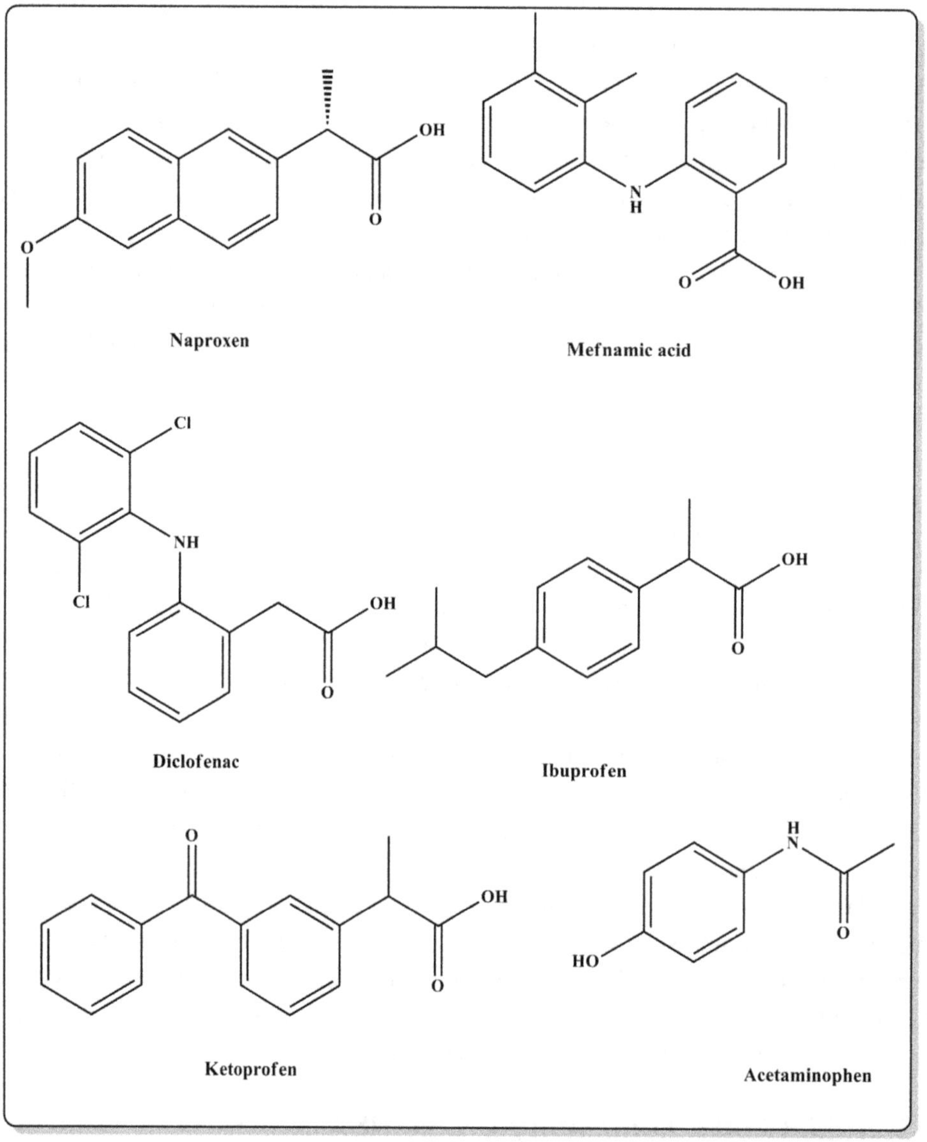

desmethylnaproxen. Of these, the last was formed as the result of cytochrome 450 monooxygenase action (Table 1). A Microtox® test confirmed the nontoxicity of the medium, which reinforced application of a LMS for removal of naproxen (Marco-Urrea, Pérez-Trujillo, Cruz-Morató, Caminal, & Vicent, 2010).

Another study monitored the elimination of naproxen in the presence of *T. versicolor* as a diagnostic tool and indicator of fungal activity for bioremediation of sewage sludge and solid state matrices (Rodríguez-Rodríguez, Marco-Urrea, & Caminal, 2010b). Naproxen could be totally degraded in sludge solid cultures within 72 h with 38% bulking materials while 47% could be removed from 25% bio slurry cultures within 24 h of the experiment (Rodríguez-Rodríguez, Marco-Urrea, & Caminal, 2010a). In the case of ibuprofen, a complete removal of this NSAID was seen using four fungal strains, including *T. versicolor,* after a 7-day incubation period (Marco-Urrea, Pérez-Trujillo, Vicent, & Caminal, 2009) (Table

1). Notably, the degradation of ibuprofen occurred rapidly immediately after its addition and diminished after 3 h of incubation (Marco-Urrea, Pérez-Trujillo, Vicent, & Caminal, 2009). In addition, 1,2-dihydroxy ibuprofen was characterized as the major metabolite (Marco-Urrea, Pérez-Trujillo, Vicent, & Caminal, 2009). Despite this extensive depletion of ibuprofen, in vitro studies of laccase-mediator systems have indicated no significant drug transformation. These findings suggested that different enzymatic systems or intracellular pathways might have a crucial role in degradation of this pharmaceutical ((Marco-Urrea, Pérez-Trujillo, Vicent, & Caminal, 2009).

To clean up water sources containing ketoprofen, a polycyclic NSAID, a defined liquid medium containing concentrations of 10 mg/L and 40 µg/L ketoprofen was used; the higher concentration resembles the real wastewater concentration of ketoprofen (Marco-Urrea, Pérez-Trujillo, Cruz-Morató, Caminal, & Vicent, 2010b). Within 24 h of treatment with *T. versicolor,* no trace of the 10 mg/L ketoprofen was detected and almost full removal of the environmental concentration of ketoprofen occurred after 3 h of incubation (Marco-Urrea, Pérez-Trujillo, Cruz-Morató, Caminal, & Vicent, 2010b). Three major intermediates were reported and identified by HPLC-DAD-MS and NMR (Marco-Urrea, Pérez-Trujillo, Cruz-Morató, Caminal, & Vicent, 2010b). The extracellular activity of laccase (using ABTS as a substrate) was detected, but no oxidized product was observed following exposure to ketoprofen, which suggested no role for laccase in the degradation process, so the involvement of laccase, at least in the first step of degradation, was ruled out (Marco-Urrea, Pérez-Trujillo, Cruz-Morató, Caminal, & Vicent, 2010b).

Some studies have also been carried out in an effort to biodegrade diclofenac, another widely distributed NSAID in water resources. For example, Nguyen et al. (2014) utilized a novel enzymatic membrane reactor to enhance both the reusability and efficacy of a diclofenac biotransformation process. Periodic re-injection of laccase obtained from *Aspergillus oryzae* was impeded by an ultrafiltration membrane which permitted continuous removal of diclofenac. Its removal percentage was increased from 60 to 80% by adding SYR to the reactions as a mediator (Nguyen, Hai, Price, Leusch, Roddick, Mcadam, & Nghiem, 2014). Additionally, covalently immobilization of laccase on a membrane consisting of polyvinyl alcohol (PVA)/ chitosan (CS)/ multi-walled carbon nanotubes (MWNTs) afforded pH and thermal stability, and increased the diclofenac removal efficiency. An initial concentration of diclofenac (12.5 mg/L) was promptly diminished within 2 h of the experiment and no amount of diclofenac was detected by 6 h (Nguyen, Hai, Price, Leusch, Roddick, Mcadam, & Nghiem, 2014). The PVA/CS/MWNT membrane was assumed to act as a bridge for electron shuttling between the enzyme and diclofenac and this improved the removal efficiency (Nguyen, Hai, Price, Leusch, Roddick, Mcadam, & Nghiem, 2014). Due to the high porosity of the PVA/CS/MWNT, the surface area was increased and the amount of loaded enzyme was therefore augmented (R. Xu, Tang, Zhou, Li, & Zhang, 2015). Furthermore, a 62.7% transformation efficacy for diclofenac over seven cycles was reported, indicating the significant reusability of the enzyme (Xu, Tang, Zhou, Li, & Zhang, 2015).

In accordance with these studies, Marco-Urrea et al. (2010) have provided evidence for fast and almost complete degradation of diclofenac by applying *T. vericolor* pellets in a defined liquid medium. A high and environmentally relevant concentration of diclofenac (10 mg/L, 45 µg/L, respectively) was biodegraded by 94% after only 1 h. They also showed that 5-hydroxy diclofenac and 4'-hydroxy diclofenac were the main intermediates (Marco-Urrea, Pérez-Trujillo, Cruz-Morató, Caminal, & Vicent, 2010a). However, results obtained from the metabolite structure studies suggested that laccase may not be the key enzyme for diclofenac detoxification, as most of the diclofenac was transformed before the enzyme activity had reached its peak (Marco-Urrea, Pérez-Trujillo, Cruz-Morató, Caminal, & Vicent, 2010a). A Microtox® test indicated the removal of diclofenac as well as its produced metabolites within 24 h of treatment, in parallel with a decrease in ecotoxicity that could open a new avenue of research into further investigation of the large scale application of this method (Marco-Urrea, Pérez-Trujillo, Cruz-Morató, Caminal, & Vicent, 2010a).

Another study assessed a fungal membrane bioreactor inoculated with *T. versicolor* that operated under non-sterile conditions (Yang, Hai, Nghiem, Nguyen, Roddick, & Price, 2013). Biodegradation was proposed as the principle approach for degradation of diclofenac. However, due to the non-sterile conditions, which influenced enzyme activity, the correlation of laccase activity and compound removal percentages was ambiguous (Yang, Hai, Nghiem, Nguyen, Roddick, & Price, 2013). Sathishkumar et al. (2012) employed an immobilized laccase for detoxification of diclofenac and showed better tolerability to pH and temperature differences. The biocatalyst, a laccase-poly(lactic-co-glycolic acid) (PLGA) nanofiber, showed an 18% decrease in activity when compared to its free counterpart, but a complete diclofenac degradation was achieved in three reuse cycles (Sathishkumar, Chae, Unnithan, Palvannan, Kim, Lee, & Oh, 2012). Application of SYR showed promising results by maintaining enzyme functionality due to the absence of interfering products, such as oligomerized species, and by extending enzyme reusability (Sathishkumar, Chae, Unnithan, Palvannan, Kim, Lee, & Oh, 2012).

By contrast, Lloret et al. (2013) indicated that laccase from *T. versicolor* can biotransform both diclofenac and naproxen. Detoxification of 70–94% naproxen took place under acidic conditions after a 24 h treatment in the presence of HBT, while diclofenac was entirely detoxified and became undetectable in an acidic environment at pH 4 within 30 min to 4 h of experiment, regardless of the presence or absence of mediators such as ABTS, HBT, and SYR (Lloret, Eibes, Moreira, Feijoo, & Lema, 2013) (Table 1). The major intermediates were structurally elucidated by GC-MS as decarboxylated moieties, which were suggested to be less toxic than the parent compound (Lloret, Eibes, Moreira, Feijoo, & Lema, 2013). In agreement with this study, Lloret et al. (2010) reported that the purified laccase of *Myceliophthora thermophile* was able to completely remove diclofenac in the presence of SYR, VA, and HBT, with VA being the most efficient mediator ((Lloret, Eibes, Lú-Chau, Moreira, Feijoo, & Lema, 2010).

(Table 1). The *M. thermophile* laccase gave a marked elimination of diclofenac in the absence of all applied mediators (Lloret, Eibes, Feijoo, Moreira, & Lema, 2012b). By contrast, partial degradation of naproxen was observed with the free laccase, and this was improved by bringing HBT and violuric acid into play (50% and 30%, respectively) (Lloret, Eibes, Lú-Chau, Moreira, Feijoo, & Lema, 2010).

The removal of MFA was observed with the use of the white-rot fungus *Phanerochaete sordida* YK-624 and the results indicated an effective depletion of MFA (Hata, Kawai, Okamura, & Nishida, 2010). The initial amounts were decreased by 60 and 90% after 3 and 6 days, respectively, with notable and almost complete elimination of the parent compound, based on the results of an acute lethal toxicity test (Hata, Kawai, Okamura, & Nishida, 2010). Four hydroxylated metabolites were identified and the metabolite structures implied a cytochrome metabolic pathway, but further investigations are required to confirm the role of laccase activity in combination with other enzymes (Hata, Kawai, Okamura, & Nishida, 2010).

A significant depletion of micropollutants, including MFA, was observed using the bacterial laccase from *Streptomyces cyaneus* (Margot, Bennati-Granier, Maillard, Blánquez, Barry, & Holliger, 2013). The commercial laccase from *T. versicolor* was also utilized to compare the removal efficiency and stability of the derived enzyme under various conditions (Margot, Bennati-Granier, Maillard, Blánquez, Barry, & Holliger, 2013). While the fungal laccase showed superior properties, the performance of the bacterial laccase in simulated municipal wastewater was acceptable (Margot, Bennati-Granier, Maillard, Blánquez, Barry, & Holliger, 2013). Additional studies performed by Margot et al. (2013) revealed that the proper conditions for enzyme functionality depended on the compound structure (Margot, Bennati-Granier,

Maillard, Blánquez, Barry, & Holliger, 2013). They evaluated four criteria, including pH, reaction time, temperature, and laccase activity to determine the removal efficiency of MFA, and they determined that pH was the most effective factor (Margot, Bennati-Granier, Maillard, Blánquez, Barry, & Holliger, 2013). They achieved 90% removal of MFA using 730 U/L laccase at the optimal pH and 25 °C within about 2 h (Margot, Bennati-Granier, Maillard, Blánquez, Barry, & Holliger, 2013). A reasonable percentage of 85% and above in real wastewater condition supports laccase as a suitable tool for overcoming the water micropollutant crisis (Margot, Bennati-Granier, Maillard, Blánquez, Barry, & Holliger, 2013).

Acetaminophen (Figure 8) is not pharmacologically categorized as an NSAID, but its antipyretic and pain reliever characteristics placed it in a similar group. The extensive over-the-counter and prescribed consumption of acetaminophen has established it as one of the most widely used human medicines (J. Li, Ye, & Gan, 2014). Thus, finding it among the most dominant pharmaceutically active compounds (PhAs) detected in surface and ground waters, sewage effluents, and drinking water of several countries around the world is not surprising (Li, Ye, & Gan, 2014; Ozgun, Basak, & Cinar, n.d.; Sim, Lee, & Oh, 2010; Yang, Yu, & Ray, 2008). The outcome of several recent studies based on laccase use as a way to overcome this critical threat is therefore encouraging.

Combined cross-linked enzyme aggregates (Combi-CLEA) are multi-purpose biocatalysts that have been employed for the detoxification of pharmaceuticals in several studies (Touahar, Haroune, Ba, Bellenger, & Cabana, 2014) (Table 1). Their increased conversion yield, improved stability, and maintained activity under extreme conditions are among the reasons to consider transforming many selected compounds under conditions resembling what is observed in real wastewaters (Touahar, Haroune, Ba, Bellenger, & Cabana, 2014). For instance, Touahar et al. (2014) studied the elimination of several pharmaceuticals, including acetaminophen, using three cross-linked oxidative enzymes, including the laccase from *T. versicolor*. The removal of acetaminophen by free laccase was 95%, and the results for compound degradation through Combi-CLEA were significant and led to complete drug removal (Touahar, Haroune, Ba, Bellenger, & Cabana, 2014) (Table 1). Assessment of PhAC elimination in a municipal wastewater showed 25% removal of acetaminophen using a Combi-CLEA laccase mediated system; this was the lowest transformation when compared to all the tested compounds (Touahar, Haroune, Ba, Bellenger, & Cabana, 2014).

Ba et al. (2014) investigated the degradation of acetaminophen using a Combi-CLEA system, taking into consideration the advantages of this system over the free enzyme. They applied a cross-linked *T. versicolor* laccase and mushroom tyrosinase using chitosan and N-(3-dimethylaminopropyl)-N′-ethylcarbodiimide hydrochloride (EDAC) to make aggregates (Ba, Haroune, Cruz-Morató, Jacquet, Touahar, Bellenger, & Cabana, 2014). Assessment of a real wastewater sample showed a more than 80% to nearly complete drug removal from municipal wastewater and more than 90% compound degradation in hospital wastewater (Ba, Haroune, Cruz-Morató, Jacquet, Touahar, Bellenger, & Cabana, 2014). The production of less toxic materials and the ease of their separation when compared to other costly detoxification methods for acetaminophen are convincing reasons for further investigational attempts (Ba, Haroune, Cruz-Morató, Jacquet, Touahar, Bellenger, & Cabana, 2014).

Another promising outcome for elimination of aromatic pharmaceutical compounds was shown by the development of a novel hybrid bioreactor (HBR) (Ba, Jones, & Cabana, 2014). A CLEA-Lac (including *T. versicolor* laccase and a polysulfone hollow fiber microfilter membrane) was employed for elimination of acetaminophen, MFA, and CBZ (Ba, Jones, & Cabana, 2014). The microfilter alone caused the removal of the three drugs (from 50 to 90% for acetaminophen), but a higher drug removal (99% for acetaminophen) was observed due to the synergistic action of the microfilter and CLEA-lac during operation in an aqueous solution (Ba, Jones, & Cabana, 2014). The removal procedure under continuous operation with wastewater showed near complete elimination of acetaminophen and MFA within 24 h of treatment (Ba, Jones, & Cabana, 2014).

5. FUTURE RESEARCH DIRECTIONS

As final points, the results currently indicate that laccase could be a magic tool to combat pharmaceutical pollution and could be employed for other drug categories. Data are being presented for new approaches for increasing catalytic function of laccases, but further studies are necessary to seek other widespread pharmaceutical pollutants that are suitable for elimination by laccases, and similar approaches should be explored to find a promising strain of laccase-producing microorganisms that can detoxify marked amounts of xenobiotics either without mediators or with trace amounts. Considering the laccase catalytic behavior, some produced metabolites may have their own toxicity, and this is still a challenging problem. The performance of toxicity tests should be essential to validate the nontoxicity of this eco-friendly method. The development of techniques for the immediate determination of related toxicity or reduction of this toxicity is therefore another future trend in this area. The comparisons and analyses of the structural activity relationships (SAR) of pharmaceuticals detoxified with laccases can potentially uncover more chemicals that are also suitable candidates for elimination by this enzyme. Extended research may also aim to find different fast-acting and highly potent mediators to accelerate the process of bioremediation and make it more cost effective as well as eco-friendly.

6. CONCLUSION

Pharmaceuticals pass into water supplies, and as they are designed to have pharmacological and biological activities throughout the body, they may pose a considerable threat to aquatic organisms and to humans relying on safe drinking water. The evidence is compelling that the use of biological methods is highly effective for elimination of recalcitrant pharmaceuticals. Among these, the 'Green' ones are particularly attractive, especially when covering a broad range of substrates. However, a new era of research has only recently arrived in terms of laccase utilization to remove pharmaceuticals from wastewaters. The conceptual framework is supported by evidence indicating that laccase is capable of detoxifying almost all tested drugs and its capability is improved when it is accompanied by a mediator.

To recap, laccase catalytic behavior is thought to involve four substrates and electrons that ultimately oxidize the parent molecule and produce two water molecules and some metabolites. Determining whether these metabolites have their own toxicity is still a challenging problem, and toxicity tests are essential to validate the nontoxicity of this eco-friendly remediation method. Given the complexity of this issue, research emphasis should focus on the concentrations, kind of mediators, fungal strains, and the maximum or optimum conditions used in previous studies and then compare these in tandem.

REFERENCES

Accinelli, C., Saccà, M. L., Batisson, I., Fick, J., Mencarelli, M., & Grabic, R. (2010). Removal of oseltamivir (Tamiflu) and other selected pharmaceuticals from wastewater using a granular bioplastic formulation entrapping propagules of *Phanerochaete chrysosporium*. *Chemosphere*, *81*(3), 436–443. doi:10.1016/j.chemosphere.2010.06.074 PMID:20673959

Aghaie-Khouzani, M., Forootanfar, H., Moshfegh, M., Khoshayand, M. R., & Faramarzi, M. A. (2012). Decolorization of some synthetic dyes using optimized culture broth of laccase producing ascomycete *Paraconiothyrium variabile*. *Biochemical Engineering Journal*, *60*, 9–15. doi:10.1016/j.bej.2011.09.002

Alcalde, M., Bulter, T., Zumárraga, M., García-Arellano, H., Mencía, M., Plou, F. J., & Ballesteros, A. (2005). Screening mutant libraries of fungal laccases in the presence of organic solvents. *Journal of Biomolecular Screening*, *10*(6), 624–631. doi:10.1177/1087057105277058 PMID:16103414

Asadgol, Z., Forootanfar, H., Rezaei, S., Mahvi, A. H., & Faramarzi, M. A. (2014). Removal of phenol and bisphenol-A catalyzed by laccase in aqueous solution. *Journal of Environmental Health Science & Engineering*, *12*(1), 93. doi:10.1186/2052-336X-12-93 PMID:25031840

Ashrafi, S. D., Nasseri, S., Alimohammadi, M., Mahvi, A. H., & Faramarzi, M. A. (2015). Optimization of the enzymatic elimination of flumequine by laccase-mediated system using response surface methodology. *Desalination and Water Treatment*, *57*(31), 14478–14487. doi:10.1080/19443994.2015.1063462

Auriol, M., Filali-Meknassi, Y., Adams, C. D., & Tyagi, R. D. (2006). Natural and synthetic hormone removal using the horseradish peroxidase enzyme: Temperature and pH effects. *Water Research*, *40*(15), 2847–2856. doi:10.1016/j.watres.2006.05.032 PMID:16849026

Auriol, M., Filali-Meknassi, Y., Adams, C. D., Tyagi, R. D., Noguerol, T.-N., & Piña, B. (2008). Removal of estrogenic activity of natural and synthetic hormones from a municipal wastewater: Efficiency of horseradish peroxidase and laccase from *Trametes versicolor*. *Chemosphere*, *70*(3), 445–452. doi:10.1016/j.chemosphere.2007.06.064 PMID:17897698

Auriol, M., Filali-Meknassi, Y., Tyagi, R. D., & Adams, C. D. (2007). Laccase-catalyzed conversion of natural and synthetic hormones from a municipal wastewater. *Water Research*, *41*(15), 3281–3288. doi:10.1016/j.watres.2007.05.008 PMID:17585984

Auriol, M., Filali-Meknassi, Y., Tyagi, R. D., Adams, C. D., & Surampalli, R. Y. (2006). Endocrine disrupting compounds removal from wastewater, a new challenge. *Process Biochemistry*, *41*(3), 525–539. doi:10.1016/j.procbio.2005.09.017

Ayala, M., Roman, R., & Vazquez-Duhalt, R. (2007). A catalytic approach to estimate the redox potential of heme-peroxidases. *Biochemical and Biophysical Research Communications*, *357*(3), 804–808. doi:10.1016/j.bbrc.2007.04.020 PMID:17442271

Azimi, M., Nafissi-Varcheh, N., Mogharabi, M., Faramarzi, M. A., & Aboofazeli, R. (2016). Study of laccase activity and stability in the presence of ionic and non-ionic surfactants and the bioconversion of indole in laccase-TX-100 system. *Journal of Molecular Catalysis. B, Enzymatic*, *126*, 69–75. doi:10.1016/j.molcatb.2016.02.001

Ba, S., Haroune, L., Cruz-Morató, C., Jacquet, C., Touahar, I. E., Bellenger, J.-P., & Cabana, H. (2014). Synthesis and characterization of combined cross-linked laccase and tyrosinase aggregates transforming acetaminophen as a model phenolic compound in wastewaters. *The Science of the Total Environment*, *487*, 748–755. doi:10.1016/j.scitotenv.2013.10.004 PMID:24867811

Ba, S., Jones, J. P., & Cabana, H. (2014). Hybrid bioreactor (HBR) of hollow fiber microfilter membrane and cross-linked laccase aggregates eliminate aromatic pharmaceuticals in wastewaters. *Journal of Hazardous Materials*, *280*, 662–670. doi:10.1016/j.jhazmat.2014.08.062 PMID:25218263

Bajaj, P. (2002). Finishing of textile materials. *Journal of Applied Polymer Science*, *83*(3), 631–659. doi:10.1002/app.2262

Becker, D., Rodriguez-Mozaz, S., Insa, S., Schoevaart, R., Barceló, D., de Cazes, M., ... Wagner, M. (2017). Removal of endocrine disrupting chemicals in wastewater by enzymatic treatment with fungal laccases. *Organic Process Research & Development*, *21*(4), 480–491. doi:10.1021/acs.oprd.6b00361

Bento, I., Silva, C. S., Chen, Z., Martins, L. O., Lindley, P. F., & Soares, C. M. (2010). Mechanisms underlying dioxygen reduction in laccases. Structural and modelling studies focusing on proton transfer. *BMC Structural Biology*, *10*(1), 28. doi:10.1186/1472-6807-10-28 PMID:20822511

Bertrand, T., Jolivalt, C., Briozzo, P., Caminade, E., Joly, N., Madzak, C., & Mougin, C. (2002). Crystal structure of a four-copper laccase complexed with an arylamine: Insights into substrate recognition and correlation with kinetics. *Biochemistry*, *41*(23), 7325–7333. doi:10.1021/bi0201318 PMID:12044164

Betancor, L., Johnson, G. R., & Luckarift, H. R. (2013). Stabilized laccases as heterogeneous bioelectrocatalysts. *ChemCatChem*, *5*(1), 46–60. doi:10.1002/cctc.201200611

Bevilaqua, J. V., Cammarota, M. C., Freire, D. M. G., & Sant'Anna, G. L. Jr. (2002). Phenol removal through combined biological and enzymatic treatments. *Brazilian Journal of Chemical Engineering*, *19*(2), 151–158. doi:10.1590/S0104-66322002000200010

Blánquez, P., & Guieysse, B. (2008). Continuous biodegradation of 17β-estradiol and 17α-ethynylestradiol by *Trametes versicolor*. *Journal of Hazardous Materials*, *150*(2), 459–462. doi:10.1016/j.jhazmat.2007.09.085 PMID:17996366

Brausch, J. M., Connors, K. A., Brooks, B. W., & Rand, G. M. (2012). Human pharmaceuticals in the aquatic environment: A review of recent toxicological studies and considerations for toxicity testing. *Reviews of Environmental Contamination and Toxicology*, *218*, 1–99. doi:10.1007/978-1-4614-3137-4_1 PMID:22488604

Bugg, T. D. H., Ahmad, M., Hardiman, E. M., & Rahmanpour, R. (2011). Pathways for degradation of lignin in bacteria and fungi. *Natural Product Reports*, *28*(12), 1883–1896. doi:10.1039/c1np00042j PMID:21918777

Cabana, H., Jones, J. P., & Agathos, S. N. (2007). Elimination of endocrine disrupting chemicals using white rot fungi and their lignin modifying enzymes: A review. *Engineering in Life Sciences*, *7*(5), 429–456. doi:10.1002/elsc.200700017

Cañas, A. I., & Camarero, S. (2010). Laccases and their natural mediators: Biotechnological tools for sustainable eco-friendly processes. *Biotechnology Advances*, *28*(6), 694–705. doi:10.1016/j.biotechadv.2010.05.002 PMID:20471466

Carballa, M., Omil, F., Lema, J. M., Llompart, M., García-Jares, C., Rodríguez, I., & Ternes, T. (2004). Behavior of pharmaceuticals, cosmetics and hormones in a sewage treatment plant. *Water Research*, *38*(12), 2918–2926. doi:10.1016/j.watres.2004.03.029 PMID:15223286

Castiglioni, S., Thomas, K. V., Kasprzyk-Hordern, B., Vandam, L., & Griffiths, P. (2014). Testing wastewater to detect illicit drugs: State of the art, potential and research needs. *The Science of the Total Environment*, *487*, 613–620. doi:10.1016/j.scitotenv.2013.10.034 PMID:24963530

Chang, H., Hu, J., Wang, L., & Shao, B. (2008). Occurrence of sulfonamide antibiotics in sewage treatment plants. *Chinese Science Bulletin*, *53*(4), 514–520. doi:10.100711434-008-0123-x

Chiacchierini, E., Restuccia, D., & Vinci, G. (2004). Bioremediation of food industry effluents: Recent applications of free and immobilised polyphenoloxidases. *Food Science & Technology International*, *10*(6), 373–382. doi:10.1177/1082013204049388

Choi, S. M., Yoo, S. D., & Lee, B. M. (2004). Toxicological characteristics of endocrine-disrupting chemicals: Developmental toxicity, carcinogenicity, and mutagenicity. *Journal of Toxicology and Environmental Health, Part B. Critical Review*, *7*, 1–24. PMID:14681080

Clara, M., Strenn, B., & Kreuzinger, N. (2004). Carbamazepine as a possible anthropogenic marker in the aquatic environment: Investigations on the behaviour of carbamazepine in wastewater treatment and during groundwater infiltration. *Water Research*, *38*(4), 947–954. doi:10.1016/j.watres.2003.10.058 PMID:14769414

Claus, H. (2003). Laccases and their occurrence in prokaryotes. *Archives of Microbiology*, *179*(3), 145–150. doi:10.100700203-002-0510-7 PMID:12610719

Claus, H. (2004). Laccases: Structure, reactions, distribution. *Micron (Oxford, England)*, *35*(1-2), 93–96. doi:10.1016/j.micron.2003.10.029 PMID:15036303

Colborn, T., vom Saal, F. S., & Soto, A. M. (1993). Developmental effects of endocrine-disrupting chemicals in wildlife and humans. *Environmental Health Perspectives*, *101*(5), 378–384. doi:10.1289/ehp.93101378 PMID:8080506

Contado, C. (2015). Nanomaterials in consumer products: A challenging analytical problem. *Frontiers in Chemistry*, *3*, 48. doi:10.3389/fchem.2015.00048 PMID:26301216

Cooper, E. R., Siewicki, T. C., & Phillips, K. (2008). Preliminary risk assessment database and risk ranking of pharmaceuticals in the environment. *The Science of the Total Environment*, *398*(1-3), 26–33. doi:10.1016/j.scitotenv.2008.02.061 PMID:18440598

Copeland, R. A. (2000). *Enzymes : a practical introduction to structure, mechanism, and data analysis*. Wiley. doi:10.1002/0471220639

Cruz-Morató, C., Jelić, A., Perez, S., Petrović, M., Barceló, D., Marco-Urrea, E., & Vicent, T. (2013). Continuous treatment of clofibric acid by *Trametes versicolor* in a fluidized bed bioreactor: Identification of transformation products and toxicity assessment. *Biochemical Engineering Journal, 75*, 79–85. doi:10.1016/j.bej.2013.03.020

D'Abrosca, B., Fiorentino, A., Izzo, A., Cefarelli, G., Pascarella, M. T., Uzzo, P., & Monaco, P. (2008). Phytotoxicity evaluation of five pharmaceutical pollutants detected in surface water on germination and growth of cultivated and spontaneous plants. *Journal of Environmental Science and Health. Part A, Toxic/ Hazardous Substances & Environmental Engineering, 43*(3), 285–294. doi:10.1080/10934520701792803 PMID:18205060

Dan, A., Yang, Y., Dai, Y. N., Chen, C. X., Wang, S. Y., & Tao, R. (2013). Removal and factors influencing removal of sulfonamides and trimethoprim from domestic sewage in constructed wetlands. *Bioresource Technology, 146*, 363–370. doi:10.1016/j.biortech.2013.07.050 PMID:23954243

Dehghanifard, E., Jonidi Jafari, A., Rezaei Kalantary, R., Mahvi, A. H., Faramarzi, M. A., & Esrafili, A. (2013). Biodegradation of 2,4-dinitrophenol with laccase immobilized on nano-porous silica beads. *Iranian Journal of Environmental Health Sciences & Engineering, 10*(1), 25. doi:10.1186/1735-2746-10-25 PMID:23547870

Diaconu, M., Litescu, S. C., & Radu, G. L. (2010). Laccase–MWCNT–chitosan biosensor—A new tool for total polyphenolic content evaluation from in vitro cultivated plants. *Sensors and Actuators. B, Chemical, 145*(2), 800–806. doi:10.1016/j.snb.2010.01.064

Dittmer, N. T., Suderman, R. J., Jiang, H., Zhu, Y.-C., Gorman, M. J., Kramer, K. J., & Kanost, M. R. (2004). Characterization of cDNAs encoding putative laccase-like multicopper oxidases and developmental expression in the tobacco hornworm, Manduca sexta, and the malaria mosquito, Anopheles gambiae. *Insect Biochemistry and Molecular Biology, 34*(1), 29–41. doi:10.1016/j.ibmb.2003.08.003 PMID:14723895

Domaradzka, D., Guzik, U., & Wojcieszyńska, D. (2015). Biodegradation and biotransformation of polycyclic non-steroidal anti-inflammatory drugs. *Reviews in Environmental Science and Biotechnology, 14*(2), 229–239. doi:10.100711157-015-9364-8

Dzidic, S., & Bedeković, V. (2003). Horizontal gene transfer-emerging multidrug resistance in hospital bacteria. *Acta Pharmacologica Sinica, 24*, 519–526. PMID:12791177

Eisenman, H. C., Mues, M., Weber, S. E., Frases, S., Chaskes, S., Gerfen, G., & Casadevall, A. (2007). *Cryptococcus neoformans* laccase catalyses melanin synthesis from both D- and L-DOPA. *Microbiology, 153*(12), 3954–3962. doi:10.1099/mic.0.2007/011049-0 PMID:18048910

Enguita, F. J., Marçal, D., Martins, L. O., Grenha, R., Henriques, A. O., Lindley, P. F., & Carrondo, M. A. (2004). Substrate and dioxygen binding to the endospore coat laccase from *Bacillus subtilis. The Journal of Biological Chemistry, 279*(22), 23472–23476. doi:10.1074/jbc.M314000200 PMID:14764581

Faramarzi, M. A., & Forootanfar, H. (2011). Biosynthesis and characterization of gold nanoparticles produced by laccase from *Paraconiothyrium variabile. Colloids and Surfaces. B, Biointerfaces, 87*(1), 23–27. doi:10.1016/j.colsurfb.2011.04.022 PMID:21616647

Filby, A. L., Neuparth, T., Thorpe, K. L., Owen, R., Galloway, T. S., & Tyler, C. R. (2007). Health impacts of estrogens in the environment, considering complex mixture effects. *Environmental Health Perspectives*, *115*(12), 1704–1710. doi:10.1289/ehp.10443 PMID:18087587

Forootanfar, H., & Faramarzi, M. A. (2015). Insights into laccase producing organisms, fermentation states, purification strategies, and biotechnological applications. *Biotechnology Progress*, *31*(6), 1443–1463. doi:10.1002/btpr.2173 PMID:26399693

Forootanfar, H., Faramarzi, M. A., Shahverdi, A. R., & Yazdi, M. T. (2011). Purification and biochemical characterization of extracellular laccase from the ascomycete *Paraconiothyrium variabile*. *Bioresource Technology*, *102*(2), 1808–1814. doi:10.1016/j.biortech.2010.09.043 PMID:20933400

Forootanfar, H., Moezzi, A., Aghaie-Khozani, M., Mahmoudjanlou, Y., Ameri, A., Niknejad, F., & Faramarzi, M. A. (2012). Synthetic dye decolorization by three sources of fungal laccase. *Iranian Journal of Environmental Health Sciences & Engineering*, *9*(1), 27. doi:10.1186/1735-2746-9-27 PMID:23369690

Forootanfar, H., Movahednia, M. M., Yaghmaei, S., Tabatabaei-Sameni, M., Rastegar, H., Sadighi, A., & Faramarzi, M. A. (2012). Removal of chlorophenolic derivatives by soil isolated ascomycete of *Paraconiothyrium variabile* and studying the role of its extracellular laccase. *Journal of Hazardous Materials*, *209–210*, 199–203. doi:10.1016/j.jhazmat.2012.01.012 PMID:22277342

Forootanfar, H., Rezaei, S., Zeinvand-Lorestani, H., Tahmasbi, H., Mogharabi, M., Ameri, A., & Faramarzi, M. A. (2016). Studies on the laccase-mediated decolorization, kinetic, and microtoxicity of some synthetic azo dyes. *Journal of Environmental Health Science & Engineering*, *14*(1), 7. doi:10.118640201-016-0248-9 PMID:27182441

Gadipelly, C., Pérez-González, A., Yadav, G. D., Ortiz, I., Ibáñez, R., Rathod, V. K., & Marathe, K. V. (2014). Pharmaceutical industry waste water: Review of the technologies for water treatment and reuse. *Industrial & Engineering Chemistry Research*, *53*(29), 11571–11592. doi:10.1021/ie501210j

Gagné, F., Blaise, C., & André, C. (2006). Occurrence of pharmaceutical products in a municipal effluent and toxicity to rainbow trout (Oncorhynchus mykiss) hepatocytes. *Ecotoxicology and Environmental Safety*, *64*(3), 329–336. doi:10.1016/j.ecoenv.2005.04.004 PMID:15923035

García-Galán, M. J., Rodríguez-Rodríguez, C. E., Vicent, T., Caminal, G., Díaz-Cruz, M. S., & Barceló, D. (2011). Biodegradation of sulfamethazine by *Trametes versicolor*: Removal from sewage sludge and identification of intermediate products by UPLC-QqTOF-MS. *The Science of the Total Environment*, *409*(24), 5505–5512. doi:10.1016/j.scitotenv.2011.08.022 PMID:21944200

Gasser, C. A., Ammann, E. M., Shahgaldian, P., & Corvini, P. F.-X. (2014). Laccases to take on the challenge of emerging organic contaminants in wastewater. *Applied Microbiology and Biotechnology*, *98*(24), 9931–9952. doi:10.100700253-014-6177-6 PMID:25359481

Golan-Rozen, N., Chefetz, B., Ben-Ari, J., Geva, J., & Hadar, Y. (2011). Transformation of the recalcitrant pharmaceutical compound carbamazepine by *Pleurotus ostreatus*: Role of cytochrome P450 monooxygenase and manganese peroxidase. *Environmental Science & Technology*, *45*(16), 6800–6805. doi:10.1021/es200298t PMID:21744850

Gros, M., Petrović, M., Ginebreda, A., & Barceló, D. (2010). Removal of pharmaceuticals during wastewater treatment and environmental risk assessment using hazard indexes. *Environment International*, *36*(1), 15–26. doi:10.1016/j.envint.2009.09.002 PMID:19819553

Gruber, A. J., Hudson, J. I., & Pope, H. G. Jr. (1996). The management of treatment-resistant depression in disorders on the interface of psychiatry and medicine. *The Psychiatric Clinics of North America*, *19*(2), 351–369. doi:10.1016/S0193-953X(05)70292-6 PMID:8827194

Guo, X. L., Zhu, Z. W., & Li, H. L. (2014). Biodegradation of sulfamethoxazole by *Phanerochaete chrysosporium*. *Journal of Molecular Liquids*, *198*, 169–172. doi:10.1016/j.molliq.2014.06.017

Hakulinen, N., Kiiskinen, L.-L., Kruus, K., Saloheimo, M., Paananen, A., Koivula, A., & Rouvinen, J. (2002). Crystal structure of a laccase from *Melanocarpus albomyces* with an intact trinuclear copper site. *Nature Structural Biology*, *9*, 601–605. PMID:12118243

Hata, T., Kawai, S., Okamura, H., & Nishida, T. (2010). Removal of diclofenac and mefenamic acid by the white rot fungus *Phanerochaete sordida* YK-624 and identification of their metabolites after fungal transformation. *Biodegradation*, *21*(5), 681–689. doi:10.100710532-010-9334-3 PMID:20127144

Hata, T., Shintate, H., Kawai, S., Okamura, H., & Nishida, T. (2010). Elimination of carbamazepine by repeated treatment with laccase in the presence of 1-hydroxybenzotriazole. *Journal of Hazardous Materials*, *181*(1-3), 1175–1178. doi:10.1016/j.jhazmat.2010.05.103 PMID:20619797

Heberer, T. (2002). Tracking persistent pharmaceutical residues from municipal sewage to drinking water. *Journal of Hydrology (Amsterdam)*, *266*(3-4), 175–189. doi:10.1016/S0022-1694(02)00165-8

Heberer, T., Reddersen, K., & Mechlinski, A. (2002). From municipal sewage to drinking water: fate and removal of pharmaceutical residues in the aquatic environment in urban areas. *Water Science and Technology: A Journal of the International Association on Water Pollution Research, 46*, 81–88.

Heidary, M., Khoobi, M., Ghasemi, S., Habibi, Z., & Faramarzi, M. A. (2014). Synthesis of quinazolinones from alcohols *via* laccase-mediated tandem oxidation. *Advanced Synthesis & Catalysis*, *356*(8), 1789–1794. doi:10.1002/adsc.201400103

Hela, W., Brandtner, M., Widek, R., & Schuh, R. (2003). Determination of sulfonamides in animal tissues using cation exchange reversed phase sorbent for sample cleanup and HPLC-DAD for detection. *Food Chemistry*, *83*(4), 601–608. doi:10.1016/S0308-8146(03)00254-1

Ikehata, K., Jodeiri Naghashkar, N., & Gamal El-Din, M. (2006). Degradation of aqueous pharmaceuticals by ozonation and advanced oxidation processes: A review. *Ozone Science and Engineering*, *28*(6), 353–414. doi:10.1080/01919510600985937

Jafari, N., Rezaei, S., Rezaie, R., Dilmaghani, H., Khoshayand, M. R., & Faramarzi, M. A. (2017). Improved production and characterization of a highly stable laccase from the halophilic bacterium *Chromohalobacter salexigens* for the efficient delignification of almond shell bio-waste. *International Journal of Biological Macromolecules*, *105*, 489–498. doi:10.1016/j.ijbiomac.2017.07.055 PMID:28709895

Jelic, A., Cruz-Morató, C., Marco-Urrea, E., Sarrà, M., Perez, S., Vicent, T., & Barcelo, D. (2012). Degradation of carbamazepine by *Trametes versicolor* in an air pulsed fluidized bed bioreactor and identification of intermediates. *Water Research*, *46*(4), 955–964. doi:10.1016/j.watres.2011.11.063 PMID:22178304

Jobling, S., Beresford, N., Nolan, M., Rodgers-Gray, T., Brighty, G. C., Sumpter, J. P., & Tyler, C. R. (2002). Altered sexual maturation and gamete production in wild roach (*Rutilus rutilus*) living in rivers that receive treated sewage effluents. *Biology of Reproduction*, *66*(2), 272–281. doi:10.1095/biolreprod66.2.272 PMID:11804939

Johannes, C., & Majcherczyk, A. (2000). Natural mediators in the oxidation of polycyclic aromatic hydrocarbons by laccase mediator systems. *Applied and Environmental Microbiology*, *66*(2), 524–528. doi:10.1128/AEM.66.2.524-528.2000 PMID:10653713

Joss, A., Keller, E., Alder, A. C., Göbel, A., McArdell, C. S., Ternes, T., & Siegrist, H. (2005). Removal of pharmaceuticals and fragrances in biological wastewater treatment. *Water Research*, *39*(14), 3139–3152. doi:10.1016/j.watres.2005.05.031 PMID:16043210

Kanbi, L. D., Antonyuk, S., Hough, M. A., Hall, J. F., Dodd, F. E., & Hasnain, S. S. (2002). Crystal structures of the Met148Leu and Ser86Asp mutants of Rusticyanin from *Thiobacillus ferrooxidans*: Insights into the structural relationship with the cupredoxins and the multi copper proteins. *Journal of Molecular Biology*, *320*(2), 263–275. doi:10.1016/S0022-2836(02)00443-6 PMID:12079384

Kar, P. K., Tandon, V., & Saha, N. (2002). Anthelmintic efficacy of *Flemingia vestita*: Genistein-induced effect on the activity of nitric oxide synthase and nitric oxide in the trematode parasite, Fasciolopsis buski. *Parasitology International*, *51*(3), 249–257. doi:10.1016/S1383-5769(02)00032-6 PMID:12243779

Kasprzyk-Hordern, B., Dinsdale, R. M., & Guwy, A. J. (2009). The removal of pharmaceuticals, personal care products, endocrine disruptors and illicit drugs during wastewater treatment and its impact on the quality of receiving waters. *Water Research*, *43*(2), 363–380. doi:10.1016/j.watres.2008.10.047 PMID:19022470

Khetan, S. K., & Collins, T. J. (2007). Human pharmaceuticals in the aquatic environment: A challenge to Green Chemistry. *Chemical Reviews*, *107*(6), 2319–2364. doi:10.1021/cr020441w PMID:17530905

Kim, S., & Aga, D. S. (2007). Potential ecological and human health impacts of antibiotics and antibiotic-resistant bacteria from wastewater treatment plants. *Journal of Toxicology and Environmental Health. Part B, Critical Reviews*, *10*(8), 559–573. doi:10.1080/15287390600975137 PMID:18049923

Kunamneni, A., Plou, F. J., Ballesteros, A., & Alcalde, M. (2008). Laccases and their applications: A patent review. *Recent Patents on Biotechnology*, *2*(1), 10–24. doi:10.2174/187220808783330965 PMID:19075849

Lawton, T. J., Sayavedra-Soto, L. A., Arp, D. J., & Rosenzweig, A. C. (2009). Crystal structure of a two-domain multicopper oxidase: Implications for the evolution of multicopper blue proteins. *The Journal of Biological Chemistry*, *284*(15), 10174–10180. doi:10.1074/jbc.M900179200 PMID:19224923

Lee, Y.-Y., Kim, T. G., & Cho, K.-S. (2015). Isolation and characterization of a novel electricity-producing yeast, *Candida* sp. IR11. *Bioresource Technology*, *192*, 556–563. doi:10.1016/j.biortech.2015.06.038 PMID:26092068

Li, J., Ye, Q., & Gan, J. (2014). Degradation and transformation products of acetaminophen in soil. *Water Research*, *49*, 44–52. doi:10.1016/j.watres.2013.11.008 PMID:24316789

Li, X., Xu, Q.-M., Cheng, J.-S., & Yuan, Y.-J. (2016). Improving the bioremoval of sulfamethoxazole and alleviating cytotoxicity of its biotransformation by laccase producing system under coculture of *Pycnoporus sanguineus* and *Alcaligenes faecalis*. *Bioresource Technology*, *220*, 333–340. doi:10.1016/j.biortech.2016.08.088 PMID:27591519

Liney, K. E., Hagger, J. A., Tyler, C. R., Depledge, M. H., Galloway, T. S., & Jobling, S. (2005). Health effects in fish of long-term exposure to effluents from wastewater treatment works. *Environmental Health Perspectives*, *114*(S-1), 81–89. doi:10.1289/ehp.8058 PMID:16818251

Llorca, M., Rodríguez-Mozaz, S., Couillerot, O., Panigoni, K., de Gunzburg, J., Bayer, S., ... Barceló, D. (2014). Identification of new transformation products during enzymatic treatment of tetracycline and erythromycin antibiotics at laboratory scale by an on-line turbulent flow liquid-chromatography coupled to a high resolution mass spectrometer LTQ-Orbitrap. *Chemosphere*, *119C*, 90–98. doi:10.1016/j.chemosphere.2014.05.072 PMID:24972175

Lloret, L., Eibes, G., Feijoo, G., Moreira, M. T., & Lema, J. M. (2012a). Continuous operation of a fluidized bed reactor for the removal of estrogens by immobilized laccase on Eupergit supports. *Journal of Biotechnology*, *162*(4), 404–406. doi:10.1016/j.jbiotec.2012.04.007 PMID:22546624

Lloret, L., Eibes, G., Feijoo, G., Moreira, M. T., & Lema, J. M. (2012b). Degradation of estrogens by laccase from *Myceliophthora thermophila* in fed-batch and enzymatic membrane reactors. *Journal of Hazardous Materials*, *213–214*, 175–183. doi:10.1016/j.jhazmat.2012.01.082 PMID:22342900

Lloret, L., Eibes, G., Lú-Chau, T., Moreira, M. T., Feijoo, G., & Lema, J. M. (2010). Laccase-catalyzed degradation of anti-inflammatories and estrogens. *Biochemical Engineering Journal*, *51*(3), 124–131. doi:10.1016/j.bej.2010.06.005

Lloret, L., Eibes, G., Lú-Chau, T. A., Moreira, M. T., Feijoo, G., & Lema, J. M. (2010). Laccase-catalyzed degradation of anti-inflammatories and estrogens. *Biochemical Engineering Journal*, *51*(3), 124–131. doi:10.1016/j.bej.2010.06.005

Lloret, L., Eibes, G., Moreira, M. T. T., Feijoo, G., & Lema, J. M. M. (2013). On the use of a high-redox potential laccase as an alternative for the transformation of non-steroidal anti-inflammatory drugs (NSAIDs). *Journal of Molecular Catalysis. B, Enzymatic*, *97*, 233–242. doi:10.1016/j.molcatb.2013.08.021

Macellaro, G., Pezzella, C., Cicatiello, P., Sannia, G., & Piscitelli, A. (2014). Fungal laccases degradation of endocrine disrupting compounds. *BioMed Research International*, *2014*, 614038. doi:10.1155/2014/614038 PMID:24829908

Majeau, J. A., Brar, S. K., & Tyagi, R. D. (2010). Laccases for removal of recalcitrant and emerging pollutants. *Bioresource Technology*, *101*(7), 2331–2350. doi:10.1016/j.biortech.2009.10.087 PMID:19948398

Marco-urrea, E., Pérez-trujillo, M., Blánquez, P., Vicent, T., & Caminal, G. (2010). Biodegradation of the analgesic naproxen by *Trametes versicolor* and identification of intermediates using HPLC-DAD-MS and NMR. *Bioresource Technology*, *101*(7), 2159–2166. doi:10.1016/j.biortech.2009.11.019 PMID:20004093

Marco-Urrea, E., Pérez-Trujillo, M., Cruz-Morató, C., Caminal, G., & Vicent, T. (2010a). Degradation of the drug sodium diclofenac by *Trametes versicolor* pellets and identification of some intermediates by NMR. *Journal of Hazardous Materials*, *176*(1-3), 836–842. doi:10.1016/j.jhazmat.2009.11.112 PMID:20031320

Marco-Urrea, E., Pérez-Trujillo, M., Cruz-Morató, C., Caminal, G., & Vicent, T. (2010b). White-rot fungus-mediated degradation of the analgesic ketoprofen and identification of intermediates by HPLC–DAD–MS and NMR. *Chemosphere*, *78*(4), 474–481. doi:10.1016/j.chemosphere.2009.10.009 PMID:19913277

Marco-Urrea, E., Pérez-Trujillo, M., Vicent, T., & Caminal, G. (2009). Ability of white-rot fungi to remove selected pharmaceuticals and identification of degradation products of ibuprofen by *Trametes versicolor*. *Chemosphere*, *74*(6), 765–772. doi:10.1016/j.chemosphere.2008.10.040 PMID:19062071

Margot, J., Bennati-Granier, C., Maillard, J., Blánquez, P., Barry, D., & Holliger, C. (2013). Bacterial versus fungal laccase: Potential for micropollutant degradation. *AMB Express*, *3*(1), 63. doi:10.1186/2191-0855-3-63 PMID:24152339

Margot, J., Maillard, J., Rossi, L., Barry, D. A., & Holliger, C. (2013). Influence of treatment conditions on the oxidation of micropollutants by *Trametes versicolor* laccase. *New Biotechnology*, *30*(6), 803–813. doi:10.1016/j.nbt.2013.06.004 PMID:23831273

Mathioni, S. M., Patel, N., Riddick, B., Sweigard, J. A., Czymmek, K. J., Caplan, J. L., & Donofrio, N. M. (2013). Transcriptomics of the rice blast fungus *Magnaporthe oryzae* in response to the bacterial antagonist Lysobacter enzymogenes reveals candidate fungal defense response genes. *PLoS One*, *8*(10), e76487. doi:10.1371/journal.pone.0076487 PMID:24098512

Mayer, A. (2002). Laccase: New functions for an old enzyme. *Phytochemistry*, *60*(6), 551–565. doi:10.1016/S0031-9422(02)00171-1 PMID:12126701

Miao, X.-S., & Metcalfe, C. D. (2003). Determination of carbamazepine and its metabolites in aqueous samples using liquid chromatography-electrospray tandem mass spectrometry. *Analytical Chemistry*, *75*(15), 3731–3738. doi:10.1021/ac030082k PMID:14572037

Migliore, L., Fiori, M., Spadoni, A., & Galli, E. (2012). Biodegradation of oxytetracycline by *Pleurotus ostreatus* mycelium: A mycoremediation technique. *Journal of Hazardous Materials*, *215–216*, 227–232. doi:10.1016/j.jhazmat.2012.02.056 PMID:22436341

Min, K.-L., Kim, Y.-H., Kim, Y. W., Jung, H. S., & Hah, Y. C. (2001). Characterization of a novel laccase produced by the wood-rotting fungus *Phellinus ribis*. *Archives of Biochemistry and Biophysics*, *392*(2), 279–286. doi:10.1006/abbi.2001.2459 PMID:11488603

Mirzadeh, S.-S., Khezri, S.-M., Rezaei, S., Forootanfar, H., Mahvi, A. H., & Faramarzi, M. A. (2014). Decolorization of two synthetic dyes using the purified laccase of *Paraconiothyrium variabile* immobilized on porous silica beads. *Journal of Environmental Health Science & Engineering*, *12*(1), 6. doi:10.1186/2052-336X-12-6 PMID:24393474

Mogharabi, M., & Faramarzi, M. A. (2014). Laccase and laccase-mediated systems in the synthesis of organic compounds. *Advanced Synthesis & Catalysis*, *356*(5), 897–927. doi:10.1002/adsc.201300960

Mogharabi, M., Nassiri-Koopaei, N., Bozorgi-Koushalshahi, M., Nafissi-Varcheh, N., Bagherzadeh, G., & Faramarzi, M. A. (2012). Immobilization of laccase in alginate-gelatin mixed gel and decolorization of synthetic dyes. *Bioinorganic Chemistry and Applications*, *2012*, 1–6. doi:10.1155/2012/823830 PMID:22899898

Negr6n-Encarnaci6n, I., & Arce, R. (2007). Light-induced transformations of aza-aromatic pollutants adsorbed on models of atmospheric particulate matter: Acridine and 9(10-H) acridone. *Atmospheric Environment*, *41*(32), 6771–6783. doi:10.1016/j.atmosenv.2007.04.062 PMID:18836521

Nguyen, L. N., Hai, F. I., Price, W. E., Leusch, F. D. L., Roddick, F., Mcadam, E. J., & Nghiem, L. D. (2014). Continuous biotransformation of bisphenol A and diclofenac by laccase in an enzymatic membrane reactor. *International Biodeterioration & Biodegradation*, *95*, 25–32. doi:10.1016/j.ibiod.2014.05.017

Nicell, J. A., Wagner, M., & Sakurai, A. (2003). Peroxidase-catalyzed oxidation of phenols in industrial wastewaters., 93–124.

Nugroho Prasetyo, E., Kudanga, T., Østergaard, L., Rencoret, J., Gutiérrez, A., del Río, J. C., & Guebitz, G. M. (2010). Polymerization of lignosulfonates by the laccase-HBT (1-hydroxybenzotriazole) system improves dispersibility. *Bioresource Technology*, *101*(14), 5054–5062. doi:10.1016/j.biortech.2010.01.048 PMID:20176477

Ostadhadi-Dehkordi, S., Tabatabaei-Sameni, M., Forootanfar, H., Kolahdouz, S., Ghazi-Khansari, M., & Faramarzi, M. A. (2012). Degradation of some benzodiazepines by a laccase-mediated system in aqueous solution. *Bioresource Technology*, *125*, 344–347. doi:10.1016/j.biortech.2012.09.039 PMID:23069616

Ozgun, D., Basak, S., & Cinar, O. (2014). The inhibition effect of three different pharmaceuticals in domestic wastewater. *World Journal of Environmental Research*, *4*, 1–6.

Palmer, A. M., & Alavijeh, M. S. (2012). Translational CNS medicines research. *Drug Discovery Today*, *17*(19-20), 1068–1078. doi:10.1016/j.drudis.2012.05.001 PMID:22580061

Pita, M., Gutierrez-Sanchez, C., Olea, D., Velez, M., Garcia-Diego, C., Shleev, S., & De Lacey, A. L. (2011). High Redox Potential Cathode Based on Laccase Covalently Attached to Gold Electrode. *The Journal of Physical Chemistry C*, *115*(27), 13420–13428. doi:10.1021/jp203643h

Pomati, F., Orlandi, C., Clerici, M., Luciani, F., & Zuccato, E. (2008). Effects and interactions in an environmentally relevant mixture of pharmaceuticals. *Toxicological Sciences : An Official Journal of the Society of Toxicology*, *102*(1), 129–137. doi:10.1093/toxsci/kfm291 PMID:18048368

Pozdnyakova, N., Rodakiewicz-Nowak, J., & Turkovskaya, O. (2004). Catalytic properties of yellow laccase from *Pleurotus ostreatus* D1. *Journal of Molecular Catalysis. B, Enzymatic*, *30*(1), 19–24. doi:10.1016/j.molcatb.2004.03.005

Prasad, N. K., Vindal, V., Narayana, S. L., Ramakrishna, V., Kunal, S. P., & Srinivas, M. (2012). In silico analysis of *Pycnoporus cinnabarinus* laccase active site with toxic industrial dyes. *Journal of Molecular Modeling*, *18*(5), 2013–2019. doi:10.100700894-011-1215-0 PMID:21877154

Purich, D. L., & Allison, R. D. (2002). *The enzyme reference : a comprehensive guidebook to enzyme nomenclature, reactions, and methods*. Academic Press. doi:10.1016/B978-012568041-7/50002-2

Radjenovic, J., Petrovic, M., & Barceló, D. (2007). Analysis of pharmaceuticals in wastewater and removal using a membrane bioreactor. *Analytical and Bioanalytical Chemistry*, *387*(4), 1365–1377. doi:10.100700216-006-0883-6 PMID:17115140

Rahmani, K., Faramarzi, M. A., Mahvi, A. H., Gholami, M., Esrafili, A., Forootanfar, H., & Farzadkia, M. (2015). Elimination and detoxification of sulfathiazole and sulfamethoxazole assisted by laccase immobilized on porous silica beads. *International Biodeterioration & Biodegradation*, *97*, 107–114. doi:10.1016/j.ibiod.2014.10.018

Ramírez, P., Mano, N., Andreu, R., Ruzgas, T., Heller, A., Gorton, L., & Shleev, S. (2008). Direct electron transfer from graphite and functionalized gold electrodes to T1 and T2/T3 copper centers of bilirubin oxidase. *Biochimica et Biophysica Acta*, *1777*(10), 1364–1369. doi:10.1016/j.bbabio.2008.06.010 PMID:18639515

Rezaei, S., Shahverdi, A. R., & Faramarzi, M. A. (2014). An extremely halophilic laccase from the saline water isolate *Chromohalobacter* sp. *Molecular Biology Research Communications*, *3*, 260.

Rezaei, S., Shahverdi, A. R., & Faramarzi, M. A. (2017). Isolation, one-step affinity purification, and characterization of a polyextremotolerant laccase from the halophilic bacterium *Aquisalibacillus elongatus* and its application in the delignification of sugar beet pulp. *Bioresource Technology*, *230*, 67–75. doi:10.1016/j.biortech.2017.01.036 PMID:28161622

Rezaei, S., Tahmasbi, H., Mogharabi, M., Ameri, A., Forootanfar, H., Khoshayand, M. R., & Faramarzi, M. A. (2015). Laccase-catalyzed decolorization and detoxification of Acid Blue 92: Statistical optimization, microtoxicity, kinetics, and energetics. *Journal of Environmental Health Science & Engineering*, *13*(1), 31. doi:10.118640201-015-0183-1 PMID:25908991

Rezaei, S., Tahmasbi, H., Mogharabi, M., Firuzyar, S., Ameri, A., Khoshayand, M. R., & Faramarzi, M. A. (2015). Efficient decolorization and detoxification of reactive orange 7 using laccase isolated from *Paraconiothyrium variabile*, kinetics and energetics. *Journal of the Taiwan Institute of Chemical Engineers*, *56*, 113–121. doi:10.1016/j.jtice.2015.04.008

Rezaie, R., Rezaei, S., Jafari, N., Forootanfar, H., Khoshayand, M. R., & Faramarzi, M. A. (2017). Delignification and detoxification of peanut shell bio-waste using an extremely halophilic laccase from an Aquisalibacillus elongatus isolate. *Extremophiles*, *21*(6), 993–1004. doi:10.100700792-017-0958-7 PMID:28871494

Ritter, L., Solomon, K., Sibley, P., Hall, K., Keen, P., Mattu, G., & Linton, B. (2002). Sources, pathways, and relative risks of contaminants in surface water and groundwater: A perspective prepared for the Walkerton inquiry. *Journal of Toxicology and Environmental Health. Part A.*, *65*, 1–142. doi:10.1080/152873902753338572 PMID:11809004

Riva, S. (2006). Laccases: Blue enzymes for green chemistry. *Trends in Biotechnology*, *24*(5), 219–226. doi:10.1016/j.tibtech.2006.03.006 PMID:16574262

Rodríguez Couto, S., Sanromán, M., & Gübitz, G. M. (2005). Influence of redox mediators and metal ions on synthetic acid dye decolourization by crude laccase from *Trametes hirsuta. Chemosphere, 58*(4), 417–422. doi:10.1016/j.chemosphere.2004.09.033 PMID:15620733

Rodríguez-Rodríguez, C. E., Marco-Urrea, E., & Caminal, G. (2010a). Degradation of naproxen and carbamazepine in spiked sludge by slurry and solid-phase *Trametes versicolor* systems. *Bioresource Technology, 101*(7), 2259–2266. doi:10.1016/j.biortech.2009.11.089 PMID:20031398

Rodríguez-Rodríguez, C. E., Marco-Urrea, E., & Caminal, G. (2010b). Naproxen degradation test to monitor *Trametes versicolor* activity in solid-state bioremediation processes, *179*, 1152–1155.

Sadighi, A., & Faramarzi, M. A. (2013). Congo red decolorization by immobilized laccase through chitosan nanoparticles on the glass beads. *Journal of the Taiwan Institute of Chemical Engineers, 44*(2), 156–162. doi:10.1016/j.jtice.2012.09.012

Sakurai, T., & Kataoka, K. (2007). Structure and function of type I copper in multicopper oxidases. *Cellular and Molecular Life Sciences, 64*(19-20), 2642–2656. doi:10.100700018-007-7183-y PMID:17639274

Santhanam, N., Vivanco, J. M., Decker, S. R., & Reardon, K. F. (2011). Expression of industrially relevant laccases: Prokaryotic style. *Trends in Biotechnology, 29*(10), 480–489. doi:10.1016/j.tibtech.2011.04.005 PMID:21640417

Sarmah, A. K., Meyer, M. T., & Boxall, A. B. A. (2006). A global perspective on the use, sales, exposure pathways, occurrence, fate and effects of veterinary antibiotics (VAs) in the environment. *Chemosphere, 65*(5), 725–759. doi:10.1016/j.chemosphere.2006.03.026 PMID:16677683

Sathishkumar, P., Chae, J.-C. C., Unnithan, A. R., Palvannan, T., Kim, H. Y., Lee, K.-J. J., & Oh, B.-T. T. (2012). Laccase-poly(lactic-co-glycolic acid) (PLGA) nanofiber: Highly stable, reusable, and efficacious for the transformation of diclofenac. *Enzyme and Microbial Technology, 51*(2), 113–118. doi:10.1016/j.enzmictec.2012.05.001 PMID:22664196

Schaetzle, O., Barrière, F., & Baronian, K. (2008). Bacteria and yeasts as catalysts in microbial fuel cells: Electron transfer from micro-organisms to electrodes for green electricity. *Energy & Environmental Science, 1*(6), 607–620. doi:10.1039/b810642h

Sharma, P., Goel, R., & Capalash, N. (2007). Bacterial laccases. *World Journal of Microbiology & Biotechnology, 23*(6), 823–832. doi:10.100711274-006-9305-3

Shleev, S., Christenson, A., Serezhenkov, V., Burbaev, D., Yaropolov, A., Gorton, L., & Ruzgas, T. (2005). Electrochemical redox transformations of T1 and T2 copper sites in native Trametes hirsuta laccase at gold electrode. *The Biochemical Journal, 385*(3), 745–754. doi:10.1042/BJ20041015 PMID:15453829

Shleev, S., Tkac, J., Christenson, A., Ruzgas, T., Yaropolov, A. I., Whittaker, J. W., & Gorton, L. (2005). Direct electron transfer between copper-containing proteins and electrodes. *Biosensors & Bioelectronics, 20*(12), 2517–2554. doi:10.1016/j.bios.2004.10.003 PMID:15854824

Shleev, S. V., Morozova, O. V., Nikitina, O. V., Gorshina, E. S., Rusinova, T. V., Serezhenkov, V. A., & Yaropolov, A. I. (2004). Comparison of physico-chemical characteristics of four laccases from different basidiomycetes. *Biochimie, 86*(9-10), 693–703. doi:10.1016/j.biochi.2004.08.005 PMID:15556280

Shraddha, S. (2011). R., Sehgal, S., Kamthania, M., & Kumar, A. (2011). Laccase: Microbial sources, production, purification, and potential biotechnological applications. *Enzyme Research*, •••, 217861. PMID:21755038

Si, H., Liu, D., Si, H., & Liu, D. (2007). Phytochemical genistein in the regulation of vascular function: New insights. *Current Medicinal Chemistry*, *14*(24), 2581–2589. doi:10.2174/092986707782023325 PMID:17979711

Sim, W.-J., Lee, J.-W., & Oh, J.-E. (2010). Occurrence and fate of pharmaceuticals in wastewater treatment plants and rivers in Korea. *Environmental Pollution*, *158*(5), 1938–1947. doi:10.1016/j.envpol.2009.10.036 PMID:19913963

Strong, P. J., & Claus, H. (2011). Laccase: A Review of Its Past and Its Future in Bioremediation. *Critical Reviews in Environmental Science and Technology*, *41*(4), 373–434. doi:10.1080/10643380902945706

Suárez, S., Carballa, M., Omil, F., & Lema, J. M. (2008). How are pharmaceutical and personal care products (PPCPs) removed from urban wastewaters? *Reviews in Environmental Science and Biotechnology*, *7*(2), 125–138. doi:10.100711157-008-9130-2

Suda, T., Hata, T., Kawai, S., Okamura, H., & Nishida, T. (2012). Treatment of tetracycline antibiotics by laccase in the presence of 1-hydroxybenzotriazole. *Bioresource Technology*, *103*(1), 498–501. doi:10.1016/j.biortech.2011.10.041 PMID:22071243

Sumpter, J. P. (1998). Xenoendorine disrupters--environmental impacts. *Toxicology Letters*, *102–103*, 337–342. doi:10.1016/S0378-4274(98)00328-2 PMID:10022275

Sutar, R. S., & Rathod, V. K. (2015). Ultrasonics sonochemistry ultrasound assisted enzyme catalyzed degradation of Cetirizine dihydrochloride. *Ultrasonics Sonochemistry*, *24*, 80–86. doi:10.1016/j.ultsonch.2014.10.016 PMID:25515235

Tahmasbi, H., Khoshayand, M. R., Bozorgi-Koushalshahi, M., Heidary, M., Ghazi-Khansari, M., & Faramarzi, M. A. (2016). Biocatalytic conversion and detoxification of imipramine by the laccase-mediated system. *International Biodeterioration & Biodegradation*, *108*, 1–8. doi:10.1016/j.ibiod.2015.11.029

Tamagawa, Y., Hirai, H., Kawai, S., & Nishida, T. (2005). Removal of estrogenic activity of endocrine-disrupting genistein by ligninolytic enzymes from white rot fungi. *FEMS Microbiology Letters*, *244*(1), 93–98. doi:10.1016/j.femsle.2005.01.023 PMID:15727826

Tavares, A. P. M., Cristóvão, R. O., Gamelas, J. A. F., Loureiro, J. M., Boaventura, R. A. R., & Macedo, E. A. (2009). Sequential decolourization of reactive textile dyes by laccase mediator system. *Journal of Chemical Technology and Biotechnology (Oxford, Oxfordshire)*, *84*(3), 442–446. doi:10.1002/jctb.2060

Ternes, T. A. (1998). Occurrence of drugs in German sewage treatment plants and rivers1Dedicated to Professor Dr. Klaus Haberer on the occasion of his 70th birthday. *Water Research*, *32*(11), 3245–3260. doi:10.1016/S0043-1354(98)00099-2

Touahar, I. E., Haroune, L., Ba, S., Bellenger, J.-P. P., & Cabana, H. (2014). Characterization of combined cross-linked enzyme aggregates from laccase, versatile peroxidase and glucose oxidase, and their utilization for the elimination of pharmaceuticals. *The Science of the Total Environment*, *481*, 90–99. doi:10.1016/j.scitotenv.2014.01.132 PMID:24589758

Vasquez, M. I., Lambrianides, A., Schneider, M., Kümmerer, K., & Fatta-Kassinos, D. (2014). Environmental side effects of pharmaceutical cocktails: What we know and what we should know. *Journal of Hazardous Materials*, *279*, 169–189. doi:10.1016/j.jhazmat.2014.06.069 PMID:25061892

Wang, L., Waltenberger, B., Pferschy-Wenzig, E.-M., Blunder, M., Liu, X., Malainer, C., & Atanasov, A. G. (2014). Natural product agonists of peroxisome proliferator-activated receptor gamma (PPARγ): A review. *Biochemical Pharmacology*, *92*(1), 73–89. doi:10.1016/j.bcp.2014.07.018 PMID:25083916

Watts, G. (2008). How clean is your water? *BMJ (Clinical Research Ed.)*, *337*(7661), a237. doi:10.1136/bmj.a237 PMID:18595920

Weng, S.-S., Liu, S.-M., & Lai, H.-T. (2013). Application parameters of laccase-mediator systems for treatment of sulfonamide antibiotics. *Bioresource Technology*, *141*, 152–159. doi:10.1016/j.biortech.2013.02.093 PMID:23561949

Xu, F., Berka, R. M., Wahleithner, J. A., Nelson, B. A., Shuster, J. R., Brown, S. H., & Solomon, E. I. (1998). Site-directed mutations in fungal laccase: Effect on redox potential, activity and pH profile. *The Biochemical Journal*, *334*(1), 63–70. doi:10.1042/bj3340063 PMID:9693103

Xu, R., Tang, R., Zhou, Q., Li, F., & Zhang, B. (2015). Enhancement of catalytic activity of immobilized laccase for diclofenac biodegradation by carbon nanotubes. *Chemical Engineering Journal*, *262*, 88–95. doi:10.1016/j.cej.2014.09.072

Yagi, K. (2006). *Mechanism of Enzyme Action-an Approach Through the Study of Slow Reactions* (pp. 1–36). John Wiley & Sons, Inc.

Yang, J., Li, W., Ng, T. B., Deng, X., Lin, J., & Ye, X. (2017). Laccases: Production, expression regulation, and applications in pharmaceutical biodegradation. *Frontiers in Microbiology*, *8*, 832. doi:10.3389/fmicb.2017.00832 PMID:28559880

Yang, L., Yu, L. E., & Ray, M. B. (2008). Degradation of paracetamol in aqueous solutions by TiO_2 photocatalysis. *Water Research*, *42*(13), 3480–3488. doi:10.1016/j.watres.2008.04.023 PMID:18519147

Yang, S., Hai, F. I., Nghiem, L. D., Nguyen, L. N., Roddick, F., & Price, W. E. (2013). Removal of bisphenol A and diclofenac by a novel fungal membrane bioreactor operated under non-sterile conditions. *International Biodeterioration & Biodegradation*, *85*, 483–490. doi:10.1016/j.ibiod.2013.03.012

Yousefi-Ahmadipour, A., Bozorgi-Koshalshahi, M., Mogharabi, M., Amini, M., Ghazi-Khansari, M., & Faramarzi, M. A. (2016). Laccase-catalyzed treatment of ketoconazole, identification of biotransformed metabolites, determination of kinetic parameters, and evaluation of micro-toxicity. *Journal of Molecular Catalysis. B, Enzymatic*, *133*, 77–84. doi:10.1016/j.molcatb.2016.07.015

Zeinvand-Lorestani, H., Sabzevari, O., Setayesh, N., Amini, M., Nili-Ahmadabadi, A., & Faramarzi, M. A. (2015). Comparative study of in vitro prooxidative properties and genotoxicity induced by aflatoxin B1 and its laccase-mediated detoxification products. *Chemosphere*, *135*, 1–6. doi:10.1016/j.chemosphere.2015.03.036 PMID:25876029

Zhang, Y., & Geissen, S.-U. (2010). In vitro degradation of carbamazepine and diclofenac by crude lignin peroxidase. *Journal of Hazardous Materials*, *176*(1-3), 1089–1092. doi:10.1016/j.jhazmat.2009.10.133 PMID:19945218

Zhang, Y., & Geißen, S. U. (2012). Elimination of carbamazepine in a non-sterile fungal bioreactor. *Bioresource Technology*, *112*, 221–227. doi:10.1016/j.biortech.2012.02.073 PMID:22409982

Zhang, Y., Geissen, S.-U., & Gal, C. (2008). Carbamazepine and diclofenac: Removal in wastewater treatment plants and occurrence in water bodies. *Chemosphere*, *73*(8), 1151–1161. doi:10.1016/j.chemosphere.2008.07.086 PMID:18793791

Zhao, J., & Kwan, H. S. (1999). Characterization, molecular cloning, and differential expression analysis of laccase genes from the edible mushroom *Lentinula edodes*. *Applied and Environmental Microbiology*, *65*, 4908–4913. PMID:10543802

ADDITIONAL READING

Barrios-Estrada, C., de Jesús Rostro-Alanis, M., Muñoz-Gutiérrez, B. D., Iqbal, H. M. N., Kannan, S., & Parra-Saldívar, R. (2018). Emergent contaminants: Endocrine disruptors and their laccase-assisted degradation – A review. *The Science of the Total Environment*, *612*, 1516–1531. doi:10.1016/j.scitotenv.2017.09.013 PMID:28915546

Canas, A. I., & Camarero, S. (2010). Laccases and their natural mediators: Biotechnological tools for sustainable eco-friendly processes. *Biotechnology Advances*, *28*(6), 694–705. doi:10.1016/j.biotechadv.2010.05.002 PMID:20471466

Chandra, R., & Chowdhary, P. (2015). Properties of bacterial laccases and their application for bioremediation of industrial wastes. *Environmental Science. Processes & Impacts*, *17*(2), 326–342. doi:10.1039/C4EM00627E PMID:25590782

Forootanfar, H., & Faramarzi, M. A. (2015). Insights into laccase producing organisms, fermentation states, purification strategies, and biotechnological applications. *Biotechnology Progress*, *31*(6), 1443–1463. doi:10.1002/btpr.2173 PMID:26399693

Giardina, P., & Sannia, G. (2015). Laccases: Old enzymes with a promising future. *Cellular and Molecular Life Sciences*, *72*(5), 855–856. doi:10.100700018-014-1821-y PMID:25575563

Mate, D. M., & Alcalde, M. (2017). Laccase: A multi-purpose biocatalyst at the forefront of biotechnology. *Microbial Biotechnology*, *10*(6), 1457–1467. doi:10.1111/1751-7915.12422 PMID:27696775

Mogharabi, M., & Faramarzi, M. A. (2014). Laccase and laccase-mediated systems in the synthesis of organic compounds. *Advanced Synthesis & Catalysis*, *356*(5), 897–927. doi:10.1002/adsc.201300960

Senthivelan, T., Kanagaraj, J., & Panda, R. C. (2016). Recent trends in fungal laccase for various industrial applications: An eco-friendly approach - A review. *Biotechnology and Bioprocess Engineering; BBE, 21*(1), 19–38. doi:10.100712257-015-0278-7

Strong, P. J., & Claus, H. (2011). Laccase: A review of its past and its future in bioremediation. *Critical Reviews in Environmental Science and Technology, 41*(4), 373–434. doi:10.1080/10643380902945706

Yang, J., Li, W., Ng, T. B., Deng, X., Lin, J., & Ye, X. (2017). Laccases: Production, expression regulation, and applications in pharmaceutical biodegradation. *Frontiers in Microbiology, 8*, 1–24. PMID:28559880

KEY TERMS AND DEFINITIONS

Biodegradation: Disintegration of materials by bacteria, fungi, or other biological means.

Bioremediation: A process used to treat contaminated media, including water, soil, and subsurface materials, by altering environmental conditions to stimulate the growth of microorganisms and degrade the target pollutants.

Biotransformation: The chemical modification (or modifications) made to a chemical compound by a biological organism.

Cross-Linked Enzyme Aggregate: An immobilized enzyme prepared via crosslinking of the physical enzyme aggregates with a bifunctional cross-linker.

Detoxification: The physiological or medicinal removal of toxic substances from a living organism, including the human body, carried out mainly by the liver.

Endocrine Disrupting Chemicals: Compounds that can interfere with endocrine (or hormone) systems at certain doses. These disruptions can cause cancerous tumors, birth defects, and other developmental disorders.

Nonsteroidal Anti-Inflammatory Drugs (NSAIDs): Drug classes that reduce pain, decrease fever, and in higher doses, decrease inflammation.

Xenobiotic: A chemical substance found within an organism that is not naturally produced by or expected to be present within that organism. It can also cover substances that are present in much higher concentrations than are usual.

APPENDIX

Table 1. The use of laccase in elimination of pharmaceutical substances.

Pharmaceutical Category	Pharmaceutical Substance	Source of Laccase	Enzyme Mediator	Removal Percentage	Metabolite Toxicity	Metabolite(s) Detection Method	Reference
Antimicrobials							
Sulfonamides	Sulfadimethoxine Sulfamonomethoxine	*Perenniporia* strain TFRI 707	VLA[a], ABTS[b], SYR[c], and HBA[d]	100% after 30 min with ABTS; 100% after 15 min with VLA; >95% after 60 min with SYR	VLA and HBA resulted in lower toxicity; ABTS and SYR increased the toxicity	HPLC	(Weng, Liu, & Lai, 2013)
	Sulfamethoxazole	*Pycnoporus sanguineus and Alcaligenes faecalis*	-	Nearly complete removal	ND[e]	UPLC-QqTOF -MS	(Li, Xu, Cheng, & Yuan, 2016)
Tetracyclines	Tetracycline	*Trametes versicolor*	-	78%	ND[e]	HLITHR-MS[f]	(Llorca et al., 2014)
	Oxytetracycline	*Pleurotus ostreatus* strain SMR684	-	Almost completely removed	ND[e]	MS analysis	(Migliore, Fiori, Spadoni, & Galli, 2012)
Quinolone	Ciprofloxacine	*Phanerochaete chrysosporium*	-	Nearly complete removal	ND[e]	-	(Accinelli et al., 2010)
Macrolides	Erythromycine	*P. chrysosporium*	-	More than 80%	-	-	(Accinelli et al., 2010)
Lipid regulating agents							
	Clofibric acid	*T. versicolor*	-	33%	Higher toxicity than its parent compound	-	(Cruz-Morató et al., 2013)
Antihistamines							
	Cetirizine	*Aspergillus oryzae*	-	91%	ND[e]	LC/ MS	(Sutar & Rathod, 2014)
CNS-related agents							
Anticonvulsant	Carbamazepine	*P. chrysosporium* ME-446	HBT[g]	22%	ND[e]	HPLC	(Hata, Shintate, Kawai, Okamura, & Nishida, 2010)
	Carbamazepine	*T. versicolor*	-	94% after 6 days of incubation	Lower toxicity than its parent compound	-	(Jelic et al., 2012)
Benzodiazepines	Clobazam Chlordiazepoxide Lorazepam Nitrazepam Oxazepam Diazepam Alperazolam	*P. variabile* *P. variabile* *P. variabile* *P. variabile* *P. variabile* *P. variabile* *P. variabile*	ABTS, HBT, VA[h], and DMP[i]	Resistant Resistant Resistant 27.3% 18.7% 18.6% 45.6%	ND[e] ND[e] ND[e] ND[e] ND[e] ND[e] ND[e]	- - - - - - -	(Ostadhadi-Dehkordi et al., 2012)
Tricyclic antidepressants	Clomipramine Imipramine Amitriptyline Doxepin Nortriptyline	*P. variabile* *P. variabile* *P. variabile* *P. variabile* *P. variabile*	-	67% 82% 11% 6% 23%	ND[e] ND[e] ND[e] ND[e] ND[e]	-	(Tahmasbi et al., 2016)
NSAIDs							
	Naproxen	*T. versicolor*	HBT	Completely removed	-	HPLC-DAD-MS	(Marco-Urrea, Pérez-Trujillo, Blánquez, Vicent, & Caminal, 2010)

continued on following page

Table 1. Continued

Pharmaceutical Category	Pharmaceutical Substance	Source of Laccase	Enzyme Mediator	Removal Percentage	Metabolite Toxicity	Metabolite(s) Detection Method	Reference
	Naproxen	*Myceliophthora thermophila*	HBT and VLA	30%	ND[e]	-	(Lloret et al., 2010)
	Ibuprofen	*T. versicolor*	-	Completely depleted after 3 h	Higher toxicity than Ibuprofen	GC-NMR	(Marco-Urrea, Pérez-Trujillo, Vicent, & Caminal, 2009)
	Diclofenac	*T. versicolor*	ABTS, HBT, and SYR	100%	-	-	(Lloret, Eibes, Moreira, Feijoo, & Lema, 2013)
	Diclofenac	*Myceliophthora thermophila*	SYR, VLA, and Vanillin	-	ND[e]	-	(Lloret et al., 2010)
	Acetaminophen	Combined cross-linked enzyme aggregates of laccase (*T. versicolor*), versatile peroxidase (*Bjerkandera adusta*) and glucose oxidase (*Aspergillus niger*)	-	More than 80%	ND[e]	UPLC/MS	(Touahar, Haroune, Ba, Bellenger, & Cabana, 2014)
Sexual hormone analogs							
Estrogens	Estrone 17β-Estradiol Estriol 17α-Ethinylestradiol	*T. versicolor* *T. versicolor* *T. versicolor* *T. versicolor*	HBT	100% removal in the presence of HBT (100 μM)	ND[e]	LC-MS	(Auriol, Filali-Meknassi, Tyagi, & Adams, 2007)
	17β-Estradiol 17α-Ethinylestradiol	*T. versicolor* *T. versicolor*	-	More than of 97%	ND[e]	HPLC	(Blánquez & Guieysse, 2008)
	Estrone 17β-Estradiol 17α-Ethinylestradiol	*M. thermophila* *M. thermophila* *M. thermophila*	SYR and HBT	- 100% 100%	ND[e] ND[e] ND[e]	HPLC	(Lloret et al., 2010)
	Estrone 17β-Estradiol	*M. thermophila*	-	95% 100%	ND[e] ND[e]	GC–MS	(Lloret, Eibes, Feijoo, Moreira, & Lema, 2012)
	Genistein	*Phanerochaete sordid* YK-624	HBT	Completely removed	ND[e]	HPLC	(Tamagawa, Hirai, Kawai, & Nishida, 2005)

[a]Violuric acid; [b] 2,2'-Azino-bis(3-ethylbenzothiazoline-6-sulphonic acid); [c] Syringaldehyde; [d] 4-Hydroxybenzoic acid; [e] Not Determined; [f] Hybrid linear ion trap- high resolution mass spectroscopy; [g] Hydroxybenzotriazole; [h] Vanillic acid; [i] 2,6-Dimethoxyphenol

Chapter 11
Laccase From White Rot Fungi Having Significant Role in Food, Pharma, and Other Industries

Ankita Kushwaha
University of Allahabad, India

Shweta Maurya
University of Allahabad, India

Ravi Kant. Pathak
University of Allahabad, India

Sonam Agarwal
University of Allahabad, India

Pankaj Kumar Chaurasia
MNNIT Allahabad, India & and LS College Muzaffarpur, India

M. P. Singh
University of Allahabad, India

ABSTRACT

Laccases (E.C. 1.10.3.2 benzenediol: oxygen oxidoreductase) are an interesting group of N glycosylated multicopper blue oxidase enzymes and the widely studied enzyme having a broad range of substrate specificity of both phenolic and non-phenolic compounds. They are widely found in fungi, bacteria plant, insects, and in lichen. They catalyze the oxidation of various phenolic and non-phenolic compounds, with the concomitant reduction of molecular oxygen to water. They could increase productivity, efficiency, and quality of products without a costly investment. This chapter depicts the applications of laccase enzyme from white rot fungi, having various industrial (such as textile dye bleaching, paper and pulp bleaching, food includes the baking, it also utilized in fruit juice industry to improve the quality and stabilization of some perishable products having plant oils), pharmaceutical (as it has potential for the synthesis of several useful drugs such anticancerous, antioxidants, synthesis of hormone derivatives because of their high value of oxidation potential) significance.

DOI: 10.4018/978-1-5225-5237-6.ch011

INTRODUCTION

The oxidation reactions are essential in several industries. The conventional method of oxidation reactions system have disadvantages like they are non-specific in nature and in which some undesirable side-reactions take place. They also uses environmentally hazardous chemicals. The biological enzymatic oxidation reaction system acts as biodegradable catalyst. They are specific in nature and enzymatic reactions are carried out in mild environmental conditions, acting as biodegradable catalyst. Therefore, laccases (benzenediol: oxygen oxidoreductase; EC 1.10.3.2) have potential for the above mentioned purposes (Couto, Toca, & Herrera, 2006). They oxidize polyphenols, methoxy-substituted phenols, diamines and a considerable range of other compounds except tyrosine (as do the tyrosinases) (Thurston, 1994).They are extensively distributed in fungi. They occur mostly in ascomycetes, deuteromycetes and basidiomycetes. According to mode of action of wood decaying fungi, they are categorized into soft rot fungi, brown rot and white rot. Soft rot belongs to ascomycetes group whereas brown and white rot belongs to basidiomycetes. White rot fungi, degrade lignocellulosic substrate which leaves a white rot residue during the degradation of lignin, hence known as white rot fungi (Kushwaha, Agarwal, Gupta, Maurya, Chaurasia, Singh, & Singh, 2017). In addition to fungi, plants and bacteria, the presence of laccases have also been reported in wasp venom as well as in insects. More or less, all white rot fungi are the laccase producers such as (except for *Phanerochaete chrysosporium)Pleurotus ostreatus, Coriolus sanguineus, Trametes hirsuta, Trametes vercicolour, Trametes villosa, Coriolopsis polyzona, Phlebia radiate, Podospora anserine, Lentinus tigrinus, Pleurotus eryngii, Fomes durrismus, Pleurotus sajor caju, Trametes trogii,* etc. Some of the examples of laccase producing bacteria are *Azospirrullum lipoferum* (first laccase producing bacteria), *Streptomyces*, *Anabaena azollae* and *Actinobacteria* . They are thermally more stable than fungal laccases. They are also stable at high pH and high concentrations of chloride and copper ions (Chaurasia, Bharati, Sharma, Singh, Yadav, & Yadava, 2015). They are dimeric or tetrameric glycoprotein, which contains four copper atoms per monomer distributed in three redox sites (Gianfreda, Xu, & Bollag, 1999). Laccase was first described by Yoshida. He extracted it from the exudates of the Japanese lacquer tree *Rhus vernicifera* (Yoshida 1883). In 1985, Bertrand discovered their characteristic as a metal containing oxidase whereas in the year 1896, both Bertrand and Laborde observed the presence of laccase in fungi for the first time (Desai & Nityanand 2011). Extracellular fungal laccases are extracellular proteins which has the molecular size of approximately 60–70 kDa and with acidic isoelectric point around pH 4.0(Baldrian 2006). These enzymes are more likely stable in extracellular role because they are often produced as highly glycosylated derivatives in which the carbohydrate moieties increase their hydrophilicity. They are generally glycosylated, with an extent of glycosylation ranging between 10 and 25% and only in a few cases higher than 30% (Shleev, Morozova, & Nikitina, 2004; De Souza, & Peralta, 2003). These promising features may contribute to the high stability of the enzyme (Dur´an, Rosa, D'Annibale, & Gianfreda, 2002). Most of the fungal laccases are monomeric proteins. However, few of them exhibit a homodimeric structure, in which the enzyme are being composed of two identical subunits with molecular weight typical for monomeric laccases, e.g., in *Phellinus ribis* (Min, Kim, Kim, Jung, & Hah, 2001), *Pleurotus pulmonarius* (De Souza, & Peralta, 2003), *Trametes villosa* (Yaver, Xu, Golightly, Brown, Brown, & Rey, 1996) and the mycorrhizal fungus *Cantharellus cibarius* (Palmieri, Giardina, Bianco, Scaloni, Capassoi, & Sannia, 1997).The electrons are removed from the reducing substrate molecules and transferred to oxygen in order to form water without the step of hydrogen peroxide formation (Ducros, Brzozowski, Wilson, Brown, Stergard, Schneider, Yaver, Pedersen, & Davies, 1998). Laccases have a wide substrate range which can serve industrial purposes.

The simple requirements of laccase catalysis (presence of substrate and O_2), its apparent stability and lack of inhibition (as has been observed with H_2O_2 for peroxidase), make this enzyme both suitable and attractive for nutitional, industrial and pharmaceutical applications. In addition, laccase can oxidize a wide range of organic and inorganic substrates, including mono, di, polyphenols, aminophenols, methoxyphenols as well as metal complexes which are the major reason for their attractiveness for dozens of biotechnological applications (Upadhayay, Shrivastava, & Agrawal, 2016).

Mechanism of Laccase Action

Laccase only attacks the phenolic subunits of lignin which leads to aryl–alkyl cleavage, Cα oxidation and Cα–Cβ cleavage.The catalysis of laccase involves the following steps:

1. Reduction of the type 1 copper by reducing substrate.
2. Internal electron transfer from the type 1 to the type 2 and type 3 copper.
3. Reduction of oxygen to water at the types 2 and 3 copper site (Ferraroni, Myasoedova, Schmatchenko, Leontievsky, Golovleva, & Scozzafava, 2007).

Fungal laccases are very stable and have very high catalytic efficiency. Therefore acts as synthetic biocatalyst. After the development of different high redox-potential mediators such as ABTS (2,2' [Azino-bis-(3-ethylbonzthiazoline-6-sulphonic acid) diammonium salt]), HBT (1-hydroxybenzotriazole) etc, which have tendency to enhance the catalytic substrate range of laccases. By the help of these mediators, synthetic and industrial applications of laccases have been enhanced (Chaurasia, Bharati, & Sunil Kumar Singh, 2013; Chaurasia, Bharati, Sharma, Singh, Yadav, & Yadava, 2015). It has frequently observed that laccase has various applications in industries like food industry paper and pulp, textile, cosmetic as well as pharmaceutical (Pezzella, Guarino, & Piscitelli, 2005).

Food Industry

There are wide applications of Laccase enzyme in food industry such as wine and beer stabilization, fruit juice processing, baking, improving of food sensory parameter, sugar beet pectin gelation and determination of certain compounds in beverages. Laccase substrate such as phenols, unsaturated fatty acid, carbohydrate and thiol containing proteins which are essential for food and beverages and their modification leads to the quality improvement new functionalities and cost reduction. (Osma, Herrera, & Couto, 2010; Minussi, Pastore, & Duran, 2002; Kirk, Borchert, & Fuglsang, 2002). Laccase enzyme can be used as additives in food and beverage processing that modify and enhance the appearance of food and beverage.

Wine Stabilization

Wine stabilization is the one of the important use of laccase in food industry (Minussi, Pastore, & Duran, 2002). The chemical composition of must and wine are ethanol, organic acids (responsible for aroma), salts, and phenolic compounds (responsible for colour and taste) present in different kinds of wines (Brenna & Bianchi, 1994). In wine, many groups of phenolic compounds are found such as cinnamic acid derivatives and catechins.But anthocyanidins are found in rose and red wines. Malvidin (anthocyanidin monoglucoside) is the major colour source compound found in grapes (Gibson, 1997). The phenolic compounds composition and percentage in wine depends on various factors such as grape variations processing (Cheynier, Arellano, Souquet, & Moutounet, 1997; Macheix, Sapis, & Fleuriet, 1991). The fresh wine should remain constant at least first year of storage. Polyphenols (derived of the coumaric acid derivatives, flavans and anthocyanidins) in wine play an important role in oxidative reactions which are catalysed by iron, copper, enzymes, aldehydes, amino acids and proteins that causes turbidity,colour intensification and aroma and flavour alterations. Therefore, the above oxidative process is known as madeirization (Zamorani, Spettoli, Lante, Crapisi, & Pasini, 1993).

Cantarelli et al (1986) reported that mutant laccase from *Polyporus versicolor* (pH around 2.7) effectively removed the phenolic (70% catechin) as well as polyphenolic(90% anthocyanidins) compounds from wine. The treatment of laccase action is also seen in black grape juice.It removed approximately 50% of the total polyphenols. The efficacy of laccase is demonstrated by must treated and untreated with laccase and thus it is confirmed by HPLC analysis. Therefore, laccase from *Polyporus* may be used as a fining agent with the addition of silica solution or a thermal treatment. Futher the ultrafiltration process is required for the removal of the oxidized products.

Laccase from *Trametes versicolor* showed that removal of phenolic and polyphenolic compounds from wines. In a model solution it removed 90% of ferulic acid as well as 34% of phenolic compounds (Minussi, Pastore, & Duran, 1998).

Zamorani et al (1989) compared the different types of enzymatic extracts such as anthocyanase, laccase, phenolase and tannase in the production of wine from Pinot grapes. The result showed that laccase was more effective in polyphenol removal than other enzymatic extracts. The results obtained with Trebbiano wine processed by the mentioned sequence (laccase and silica solution) were compared with the data obtained by clarification (caseinate+active and carbon+bentonite) in presence and absence of sulphur dioxide. The organoleptic analyses confirmed that wines treated with laccase and silica solution or sulphur dioxide showed better organoleptic madeirization-resistant features. The use of the laccase was shown to be highly effective, preferable or practically identical to those obtained by the traditional processing.

Maier et al (1990) assess the Riesling wines prepared with oxidation of must and laccase treatment or without the oxidation of must. The results showed that wines made by laccase treatment were the best and suggesting that a stable and high quality wine can be made with little or no added SO_2.

Cantarelli and Giovanelli (1991) used 5–20 U/ml of added laccase to the must, which was then filtered with and without polyvinylpolypyrrolidone (PVPP) or clarified with bentonite or gelatine and silica solution, or sulphited. The results demonstrated that the enzymatic treatment (laccase) offers an opportunity for future research.

Another disadvantage of wine storage is the characteristic cork off-flavour that comes in wine due to the contact with cork stoppers, which is correlated to phenols naturally present in cork and or produced by microbial degradation. The use of laccase for the treatment of cork stoppers for wine bottles has been patented as an effective method to reduce the cork taint by phenols oxidation (Conrad, Sponholz, & Berker, 2000). Laccase has been commercialized for patented in the name of Suberyme® (Kushwaha, Agarwal, Gupta, Maurya, Chaurasia, Singh, & Singh, 2017) brand name Suberase (Novozymes, Denmark)

Beer Stabilization

The unrelenting problem in brewing industry is the hazes develop at the time of storage (MacMurrough, Madigan, Kelly, & O'Rourke, 1999). Long term storage of beer depends on many factors such as temperature, oxygen etc. Haze formation in beer is due to the formation of protein precipitation which is stimulated by the naturally occurring proanthocyanidins polyphenols (Mathiasen, 1995). This type of haze formation called the chill-haze which appears on cooling and disappear at room temperature or above. The chemistry behind this is the formation of hydrogen bonding or hydrophobic interactions with the proline residue of protein. Soon after nucleophilic substitution of phenolic rings(protein) by sulphdryl groups which leads to the formation of permanent haze. This permanent haze does not removed even at high temperature. The removal of polyphenols are traditionally done by treatment with polyvinylpolypyrrolidone(PVPP). The insoluble PVPP is also a problem in the wastewater due to its low biodegradability. Therefore, another traditional method for treating the above problem is by laccase.

According to Mathiasen (1995), laccase may remove any excess oxygen at the end of the process thereby storage life of beer is also enhanced as well as laccase also removed polyphenols which is present in the beer. This polyphenol complexes formed may be removed by filtration. The potential of laccase utilization in beer stabilization is also confirmed by Rossi et al (1988) and Giovanelli (1989).A commercialised product named "Flavourstar", manufactured by Novozymes which is marketed for using in brewing beer to prevent the formation of off-flavour compounds (e.g.,trans-2-nonenal) by scavenging the oxygen (FAO,2004).

Fruit Juice Processing

The important constituent of naturally occurring beverages are phenolic compounds and their oxidative products. They contribute colour and taste to the natural beverages. Fruit juices especially apple and grape juice, where excessive oxidation of phenolics take place which in turn is detrimental to the oranoleptic quality of the fruit juice. Several authors have proposed the use of laccases as stabilizing agents, due to their ability to oxidize most of phenols in fruit juices. Contradictory results have been published. Sammartino et al (1998) demonstrated that enzymatically treated apple juice was less stable than the one stabilized by conventional methods (SO_2 added as metabisulfite, polyvinylpolypyrrolidone (PVPP), bentonite) with the use of free and immobilised laccase. They showed that the enzymatically treated juice was less stable than the one conventionally treated.

Laccase treatment increased susceptibility to browning during storage (Giovanelli & Ravasini, 1993; Gokmen, 1998). Many researchers evaluated effective phenol removal followed by filtration step and observed that it could improve fruit juice (Minussi 2002; Ritter, Maier, Schoepplein, & Dietrich, 1992; Cantarelli & Giovanelli,1990).

According to the Neifar et al (2011) laccase treated pomegranate juice by RSM process with ultrafiltration process (for controlling the haze formation and browning of the pomegranate juice) resulted in 30% increase in clarity of juice in comparison to untreated juice. Reduction in polyphenol content was also seen in treated juice with laccase. (laccase concentration 5U/mL; incubation time 300 min; incubation temperature 20°C)

Gassara-Chatti (2013) reported significant reduction of polyphenolic content and improved clarity of fruit juice using ligninolytic enzymes from *Phanerochaete chrysosporium* encapsulated in polyacrylamide hydrogel for clarification of mixed juice of berry and pomegranate.

Artik et al (2004) observed the significant decrease in polyphenolic content by around 70% and clarity amelioration and stability of sour cherry juice when treated with laccase enzyme (heating to 50ºC for 6 hours and filtering through 20 kDa membrane after 1 hour of oxidation).

Baking

In the process of making bread, an additive is being added to the dough, in order to make improvement in volume,texture, flavour and freshness of bread. Laccase was first reported by Labat et al (2000). Now laccase are important in baking because of their ability to cross link biopolymers. In baking, use of laccase result in better strength, stability and reduced stickiness which results in enhanced machinability of the dough. The improved machinability is of particular importance for doughs that are processed industrially. An increased volume and an improved crumb structure and softness of the dough and or baked products were observed (Labat, Morel, & Rouau, 2000; Selinheimo, Kruus, Buchert, Hopia, & Autio, 2006)

The viscoelastic properties of wheat flour dough depend primarily on the dough's containing protein prolamins .The glutenins provide elasticity whereas gliadin provides viscosity and extensibility in a dough system. Arabinoxylans(AX) are present in low content in wheat flour which help in determining the dough-handling properties and bread quality (Labat, Morel, & Rouau, 2000). Laccase catalyses gelation by dimerization of feruloylated esters in feruloylated arabinoxylans(FA) (Espinoza, Morel, & Rouau, 1998; Espinoza, Morel, Surget, Asther, Moukha, Sigoillot, & Rouau, 1999; Espinoza, & Rouau, 1999).

According to Si (1994) reported that when a laccase enzyme is enzyme added to dough, it may exert an oxidizing effect on the dough constituents and thereby serve to improve the strength of gluten structures in dough or baked products.

Renzetti et al (2010) showed that a commercial preparation of laccase enzyme significantly improved the bread-making performances of oat flour and the textural quality of oat bread by increasing specific volume and lowering crumb hardness and chewiness. The improved bread-making performances could be related to the increased softness, deformability and elasticity of oat batters with laccase supplementation.

Selinheimo et al (2006) showed that a laccase from the white-rot fungus *Trametes hirsuta* increased the maximum resistance of dough and decreased the dough extensibility in both flour and gluten doughs.

Therefore, from the above observations it was concluded that the effect of laccase was mainly due to the cross-linking of the esterified ferulic acid (FA) on the arabinoxylan (AX) fraction of dough resulting in a strong AX network.

Improvement of Food Sensory Parameter

The exterior odour and aspect of food product influences the consumer's perception about the quality of food product. Laccases have been used to improve food sensory parameters by acting on processes affecting physicochemical deterioration of food products, or enhancing their flavour and taste properties.

Takemori et al (1992) used crude laccase from *Coriolus versicolor* to improve the flavour and taste of cacao nib and its products. Bitterness and other unpleasant tastes were removed by the laccase treatment, and the chocolate manufactured from the cacao mass tasted better than the untreated one. Reduction of tannin content in cocoa pod husk after laccase treatment has been found to improve its nutritive value when used as ingredient for animal feeds (Mensah, Adamafio, Kwarteng, & Rodrigues, 2012).

Bouwens et al. (1997) reported that the color of tea based products could be improved when treated with laccase from a *Pleurotus* species. The oxidation step converted the colorless catechins which is present in the leaves to a complex mixture of yellow and orange to dark brown substances such as the flavins and the arubigins. In the same manner, chopped olives in an olive-water mixture were treated with laccase from *Trametes villosa*. Therefore, the bitterness was significantly reduced while the color turned darker compared to the untreated one. (Novo Nordisk A/S, 1995).

According to Petersen et al (1996), laccase enzyme catalyzed the deoxygenation of oil or product comprising an oil. Oils, especially vegetable oils (e.g., soybean oil), are present in many food items such as salads dressings, French mayonnaise, and other sauces. Soybean oil contains a large amount of linoleic and linolenic acids that can react with dissolved oxygen in the product producing undesirable volatile compounds. They also reported that the laccase catalyzed deoxygenation reaction could prevent the oxidation of linoleic and linolenic compounds. Laccase is characterized by a group of blue multi-copper enzyme. It degrades both phenolic and non-phenolic compounds with the concomitant four electron reduction of molecular oxygen to water. It acts on both o- and p-quinones. Therefore, dissolved oxygen could be removed in order to maintain the flavour quality of oil. Other food products such as juices, soups, concentrates, puree, pastes, and sauces etc. can be deoxygenated by the treated with laccase enzyme.

Sugar Beet Pectin Gelation

Sugar beet pectin is a food ingredient with specific functional properties. Laccase have been applied to enzymatic gelation of sugar beet pectin .It may form gels by an oxidative cross-linking of ferulic acid (Norsker, Jensen, & Adler-Nissen, 2000). Such kinds of gels are of particular interest in food industry due to their thermo-irreversibility, allowing a foodstuff to be heated without losing its gel structure.

Peroxidase or laccase can perform the oxidation gelation. Norsker et al (2000) evaluated the gelling effect of laccases and peroxidise in food industry. Peroxidase requires the addition of hydrogen peroxidise while laccase using oxygen. The obtained gel is thermo-irreversible in nature. Thus, it makes more interesting for food industry as the product can be heated without losing its gel structure.

The enzymatic gelation was studied in three food products such as luncheon meat (chopped heat-treated meat emulsion), black currant juice and milk with added sugar beet pectin (SBP). Gelation were occurred in all the food products. However, in two products such as in milk and black currant juice some unwanted side effects were obtained while in luncheon meat no side effects were found. Hence, it is more advantageous to add laccase to food products.

Kuuva et al (2003) concluded that by using laccases as cross-linking agents together with calcium, the ratio of covalent and electrostatic cross-links of sugar beet pectin gels can be varied and it can be possible to tailor different types of gel structures.

Littoz and McClements (2008) showed that laccase could be used to covalently cross-link beet pectin molecules adsorbed to the surfaces of protein-coated lipid droplets at pH 4.5, hence suggesting that emulsions should be enhanced and efficient so that their performance could be prepared using a biomimetic approach that utilised laccase enzyme to cross-link adsorbed biopolymers.

Laccase-mediated cross-linking of food proteins has also been used in order to develop added-value products in the food industry (Jund & Wicker, 2012; Mattinen, Kruus, Bucher, Nielsen, Andersen, & Steffensen, 2005) as well as to influence protein digestibility and reduce their allergenicity (Stanic, Monogioudi, Dilek, Radosavljevic, Markovic, Vuckovic, Raija, Mattinen, Buchert, & Cirkovic, 2010; Tantoush, Stanic, Stojadinovic, Ognjenovic, Mihajlovic, Markovic, & Cirkovic, 2011). Beta-Lactoglobulin (BLG)contain approximately 10–15% of total milk proteins which is an important nutrient of dairy products. At the same time, one of the major milk allergens causing serious health risk in cow's milk-allergic patients (Kontopidis, 2004). Laccase from *Trametes versicolor* has been used to cross-link BLG phenolic groups in the presence of a sour cherry extract as a natural source of phenolic mediators (Tantoush, Stanic, Stojadinovic, Ognjenovic, Mihajlovic, Markovic, & Cirkovic, 2011). Enzymatic processing by cross-linking may be used to tailor BLG properties, by improving its safety and facilitating its digestibility while conserving beneficial health effects such as the radical-scavenging activity of peptides released during enzymatic digestion.

Compounds Detection in Beverages

Laccase based different biosensors have been developed in order to measure polyphenols in food products such as wine, beer and tea. Ghindilis et al (1992) first verified the potential of a biosensor based on immobilized laccase for the detection of tannin in tea of different brands. Afterwards many reports about the development of laccase based biosensor have been published .

Fusco et al (2010) showed the laccase (from different species of *Trametes*) based biosensor were immobilized on carbon nanotubes screen printed electrodes using polyazetidine prepolymer (PAP) for the determination of polyphenol index in wines. They also report that biosensor performance mainly depended upon the source of laccase enzyme.. Thus, values obtained by using *T. hirsuta* laccase were close to those determined by Folin-Ciocalteu method whereas polyphenol index measured with *T. versicolor* laccase was discordant to that found with the reference assay. Many data have been reported that laccase based biosensor performance is mainly dependent on the laccase source used.

Ibarra-Escutia et al (2010) optimised and developed an amperometric laccase (*T.versicolor*) based biosensor for phenolic compounds monitoring in tea infusion. The biosensor developed showed an excellent stability and exhibited good performance in terms of response time,sensitivity, operational stability, and manufacturing processes, simplicity and can be used for accurate determination of the phenolic content without any pre-treatment of the sample. Disposable biosensor based on platinum nanoparticles reduced grapheme oxide laccase biocomposite for the determination of total polyphenolic content in tea infusions (Eremia, Vasilescu, Radoi, Litescu, & Radu, 2013).

Prasetyo et al (2010) studied the use of tetramethoxyazobismethylene quinone (TMAMQ) for measuring the antioxidant activity of a wide range of structurally diverse molecules present in food and humans. TMAMQ was generated by the oxidation of syringaldazine with laccases and therefore, they are used for the detection of antioxidant activity present in different food products.

Montereali et al (2010) showed the detection of polyphenols in musts and wines from Imola (Italy) through an amperometric biosensor based on the tyrosinase and laccase from *Trametes versicolor*. Both enzymes were immobilised on graphite screen-printed electrodes modified with ferrocene.However, the presence of SO_2 clearly inhibited the enzymatic activity and thus, the measurements on musts and wines recently bottled were seriously affected.

Texile Industry

Laccase have been wide application in textile industry which includes the fibre bleaching and dyeing and improve textile properties. Natural pigments, cellulosic material and other non-cellulosic material removal before fibre dyeing and finishing one mandatory step involved in textile process known as bleaching. Laccase have been used as natural bleaching agent (fiber preserving) alternative to hydrogen peroxide which is commonly used as chemical bleaching agent (Pereira, Bastos, Tzanov, Cavaco-Paulo, & Guebitz, 2005; Tzanov, Basto, Gu¨bitz, & Cavaco-Paulo,2003).

Ren and Diller (2007) studied the oxidioreductases for the modification of linen fibers. Glucose oxidase has been successfully applied for bleaching of cellulosic materials such as cotton, while laccases, lignin and manganese peroxidases have been found useful for delignification processes of pulp fibers. Glucose oxidase, laccase and combinations of both enzymes were applied to unbleached linen. When the combination of laccase and glucose oxidase enzymes are used for bleaching of linen fabric showed higher efficacy than the control either one of the enzymes applied alone.

Abou et al (2010) studied the ecofriendly laccase hydrogen peroxide or ultrasound assisted bleaching of linen fabrics and its influence in dyeing efficiency. Enchanced bleaching efficiency occurred by the combination of laccase and ultrasound treatment has been also verified on linen fabrics. In this study, the time/dye uptake isotherms were also enhanced after combined bleaching treatment

Tian et al (2012) evaluates the bleaching efficiency of the hydrogen peroxide bleaching process combined with laccase-mediated system pre-treatment (LMS-HPBP) in the treatment of scoured cotton fabric. In LMS-HPBP processes that are more environment friendly than the conventional hydrogen peroxide bleaching process (CHPBP): (i) bleaching with lower dosage of hydrogen peroxide; (ii) bleaching at reduced temperature; (iii) bleaching for shortened time. Whiteness, retained tensile strength and *K*/*S* values of cotton fabric samples treated by above three processes were similar to or higher than those by CHPBP. X-ray diffraction (XRD) analysis also demonstrated that the three processes rendered fabric of both lower crystallinity and bigger crystallite size than those by CHPBP. They developed the "green" short-flow process to treat cotton fabric in an energy-saving process.

In 1996, Novozyme (Novo Nordisk,Denmark) was launched, a new industrial application of laccase enzyme in denim finishing called DeniLite®.It was the first industrial laccase and bleaching enzyme acting with the help of a mediator molecule. In 2001, the company Zytex (Zytex Pvt. Ltd., Mumbai, India) developed Zylite, a formulation based on LMS capable of degrading indigo in a very specific way.

Montazer and Maryan (2010) verified that the combination of laccases with cellulases in denim processing helped to improve the lightness and decrease staining on garment.

Pazarlıog˘lu et al (2005) reported the effective application of laccase treatment for stone washing effects of denim fabric without using a mediator.

Functional Modification of Textile Fibers

Modification of protein fibers, such as wool and cotton allows improving their functional properties. Water repellence is associated to self-cleaning properties, and antioxidant and antimicrobial features are required in finishing design.

Wool fibres have been modified with nordihydroguaiaretic acid (NDGA) in order to improve their performance. This water insoluble bi-functional phenolic compound has been grafted on wool through a laccase enzyme catalyzed reaction in an aqueous–ethanol mixture. The physical and mechanical properties of wool fabrics such as shrink resistance, crease recovery and tensile strength were improved by the grafting of NDGA were improved by the grafting of NDGA. In addition, the NDGA imparted to the textile material strong antioxidant activity and UV protection due to the formation of inter/intramolecular bridges and cross-links among wool protein (Hossain, Gonza´lez, Juan, & Tzanov, 2010).

An enzymatic method using laccases (*Trametes* sp.) for grafting the water insoluble phenolic compound lauryl gallate on wool fabric was developed. The enzymatic coating of wool with lauryl gallate provided a multifunctional textile material with antioxidant, antibacterial and water repellent properties (Hossain, Gonza´lez, Lozano, & Tzanov, 2009). A patent application about the use of *Trametes versicolor* laccase to increase the shrink resistance of wool has been published by Kim et al (2007) and Yoon (1998) and also investigated by Lantto et al (2004).

Textile Dyeing

Laccase-mediated coupling and grafting reactions have been used for colouring of various textiles (Schroder, Aichernig, Gu¨bitz, & Kokol, 2007; Hadzhiyska, Calafell, Gibert, Daga`, & Tzanov, 2006; Blanco, Gonza´lez, Monmany, & Tzanov, 2009).

Hadzhiyska et al (2006) studied the cellulose textile was dyed in situ by a polymeric dye generated by oxidative coupling of 2,5-diaminobenzenesulfonic acid (DABSA, colourless) and 1-hydroxyphenol (catechol) with laccase-treated fibers, resulting into 70% of dye fixation. However, it has been reported that all fabrics showed high wash fastness but low light and friction resistance.

Schroeder et al (2007) used a response surface methodology (RSM) approach to determine the best conditions for laccase-induced coating of flax fibers and fabrics by a *Trametes hirsuta* laccase. The evaluation was based on the obtained coloration and color depth. A screening was carried out with different phenols for their potential as monomers for enzyme-catalyzed polymerization resulting in a coating with antibacterial performance.While all the methoxyphenols showed different coloration with weak fastness properties, bacterial growth of *Bacillus subtilis* and *Staphylococcus aureus* was reduced significantly using ferulic acid and hydroquinone.

Kim et al (2007 and 2008) studied the use of the natural flavonoids present in the cotton as anchors to attach other phenolic compounds to the surface of fiber. Recently, trends of biological and environmentally friendly concepts are emerging which replace the chemical treatments of fabrics. They developed the colouration of flax fabrics by the use of enzymatic(Laccase from *Trametes hirsuta*)oxidation of flavonoids (morin, quercetin). Later they polymerized them for better coloured polymeric solution which further could be applied to the colouration of flax fabrics. The physical and chemical factors such as temperature, reaction time, presence of NaCl or the use of bleached or unbleached fabrics were studied in order to increase the color of the fabrics and the color fastness. The increase of temperature, the presence of salt and the use of unbleached fabrics allowed the final colour to be improved. Colourized flax fabrics with oxidized quercetin solution showed a colour fixation two times higher than the fabrics colorized with oxidized morin. Finally, the polymerization of flavonoids and their binding to fibers were verified using Fourier transform infrared spectroscopy (FT-IR). The above results confirmed that the process was environmentally friendly so it would be used for the coloration of flax fabrics. A similar technique could also be extended to the treatment of other types of fabrics in textile processes.

Tzanov et al (2003 b) reported the potential of laccase in dyeing of wool. They used the enzymatic system laccase, 2,5-diaminobenzenesulfonic acid (dye precursor) and catechol and resorcinol(dye modifier) for wool dyeing. For enzymatic wool dyeing they performed as batch wise process and at desirable temperature and pH where the enzyme activity was maximum. They have shown that by varying the concentration of the modifiers a nd the time of laccase treatment better results were obtained in comparison to the conventional process.

Suparno et al (2007) studied colour in leather to exploit the chemistry of creating the natural skin and hair pigments, melanins. Substrates such as 1,8-dihydroxynaphthalene and other dihydroxynaphthalenes can be oxidised and polymerised to create pigment products by reactions catalysed by laccase(*Coriolus versicolor*). They not only create colour but if the reactions are conducted in the presence of collagen, the pigments are bound to the collagen by covalent bonds, which confer a tanning effect.

Paper and Pulp Industry

The industrial preparation of paper requires separation either by degradation and removal of lignin (chemical pulping) or by physically tearing the fibre apart (mechanical pulping). Separation of wood fibres from each other and then processing them into sheets leads to the formation of paper from wood (Madhavi & Lele 2009).

Taspinar and Kolankaya (1998) studied the possibility of chlorine removal from chlorine bleached Kraft pulp by using laccase produced from white-rot fungus, *Trametes* (*Coriolus*) *versicolor*. Pulp bleaching is currently achieved by treating pulps with chlorine-based chemicals. The highly bleached and brightened pulp for production of white paper. The discharged water of this treatment process are very rich with respect to which results in the formation chlorinated organic compounds that could be toxic and mutagenic. Kuhad et al (1997) suggested the replacement of conventional method of delignification/bleaching because it was chlorine-based which causes pollution. Therefore, pre-treatments of wood pulp with ligninolytic enzymes might provide milder and cleaner strategies of delignification that are also respectful of the integrity of cellulose.

Call and Mücke (1997) studied the development of LMS delignification technologies for kraft pulps. In addition, laccase is more readily available and easier to manipulate than both lignin peroxidise (LiP) and manganese-dependent peroxidase (MnP). The LMS has already found practical applications such as the Lignozym®-process.

The determination of delignification/bio bleaching were studied on wood pulps and little is known about the efficiency of the LMS on non-wood pulps including those used for manufacturing specialty papers. Camarero et al (2004) reported the efficient bleaching of non-wood high-quality paper pulp using laccase-mediator system (LMS). High-quality flax pulp was bleached by using a laccase-mediator system in a totally chlorine free sequence. Three fungal laccases (from *Pycnoporus cinnabarinus, Trametes versicolor and Pleurotus eryngii*) and two mediators, 2,2 -azinobis(3-ethylbenzothiazoline-6- sulfonic acid) and 1-hydroxybenzotriazole (HBT), were compared. Laccases from *P. cinnabarinus and T. versicolor* were used in the presence of HBT gave the best results such as high brightness and low lignin content (kappa number). The former laccase also resulted in the best preservation of cellulose and the largest removal of residual lignin as revealed by analytical pyrolysis, and was selected for subsequent TCF bleaching.

Lund et al (2002) investigated the aromatic polymer lignin can be modified through promotion of oxidative coupling between phenolic groups on lignin and various phenols. The reaction is initiated by an oxidation of both components, e.g., by using the oxidoreductases laccase (*Trametes villosa*). They studied the incorporation of water-soluble phenols into kraft lignin, using laccase as catalyst. Markers such as several phenols with carboxylic or sulfonic acid groups were used for the incorporation. When the phenol guaiacol sulfonate was incorporated into kraft lignin, the lignin became water-soluble at pH 2.4 and a low ionic strength due to the introduction of sulfonic acid groups. The content of sulfonic acid groups in the product was 0.5–0.6 mmol/g lignin. A lower amount of 4-hydroxyphenylacetic acid was incorporated under similar condition.

Chandra et al (2002), reported the physical properties of high-yield kraft and fibers were treated at high consistency (20%) with laccase and syringic, vanillic, or 4-hydroxybenzoic acid. Treatment with laccase and 4-hydroxybenzoic acid resulted in a 20-point increase in kappa number and a 100% increase in bulk acid groups. The treated and untreated pulp revealed that the laccase-grafted fibers had a two-fold enrichment in acid groups strongly suggesting a laccase-facilitated coupling of 4-hydroxybenzoic acid to the fiber surface.

Laccase based biobleaching process offers an environment friendly way to improve pulp and paper production. Therefore, laccases from white rot fungi can be used for the treatment of effluents from pulp mills or from other industries containing chlorolignins or phenolic compounds.

Nanobiotechnology and Biosensors

During the past two decades, bio electrochemistry has received increased attention. Progress on bio electrochemistry has been integrated into analytical applications, e.g.in biosensors working as detectors in clinical and environmental analysis (Haghighi, Gorton, Ruzgas, & Jönsson 2003). A biosensor is a device that detects, transmits and records information regarding a physiological or biochemical change. Laccase-containing biosensors for detecting O_2, glucose, aromatic amines, phenolic compounds and a wide variety of reducing substrates have been developed (Kushwaha, Agarwal, Gupta, Maurya, Chaurasia, Singh, & Singh, 2017). They also investigated the characterization of graphite electrodes treated with laccase enzyme from *Trametes versicolor.* Therefore, use for bioelectrochemical monitoring of phenolic compounds in flow injection analysis.

Laccase also studied in biosensors in order to detect the various phenolic, oxygen and azides. Because it does not involve the addition cofactors in electron transfer reactions. Biosensors are specific, sensitive, reliable, portable and simple in operation. Therefore, laccase can be applied as a biosensor.

Bauer et al (1999) studied the new enzyme sensors for morphine and codeine based on morphine dehydrogenase and laccase (*Coriolus hirsutus*). They quantitatively analyzed the presence of codeine or morphine in drugs. Likewise catecholamines(Lisdat, Wollenberger, Makower, Hortnagl, Pfeiffer, & Scheller, 1997; Leite, Lupetti, Fatibello-Filho, Vieira, & de Barbosa, 2003; Ferry & Leech, 2005), plant flavonoids (Jarosz-Wilkołazka, Ruzgas, & Gorton, 2004) and also for electroimmunoassay (Kuznetsov, Shumakovich, Koroleva, & Yaropolov, 2001) have been developed.

Hammond and Whiteside (1995) investigated the formation of polymer microstructures by selective deposition of polyion multilayers using patterned Self-Assembled Monolayers (SAMS) as a template. They demonstrated by patterning of regions of the surface with SAMs of a given functionality, they direct the deposition of polymer multilayer films to form polymeric microstructures. It includes the patterning of polyions onto glass surfaces using trichlorosilanes patterned with pCP.12. They also studied the polymerldye systems and systems containing electrochromic polymers as the basis for new diffractive sensor systems. Applications for organic thin films and organic- inorganic hybrids include optical waveguides, sensors, LEDs, and other electro-optical or photoresponsive systems.

Chen et al (1998) reported the microcontact printing of self-assembled monolayers (SAMs) of alkanethiolates on gold. They manufactured substrates that contained micrometer-scale islands of extracellular matrix (ECM) separated by non-adhesive regions such that the pattern of islands determined the distribution and position of bovine and human endothelial cells.Therefore, micropatterning technology is essential for construction of biosurface devices and also for the investigation of the fundamental biology of cell-ECM interactions.

Sigal et al (1998) studied the effect of surface wettability on the nonspecific adsorption of proteins and detergents to self-assembled monolayers (SAMs) of alkanethiolates on gold. Regarding laccases, the immobilisation has an important influence on the biosensor sensitivity (Freire, Durán, & Kubota, 2001).

Roy et al (2005) reported that Cross-linked enzyme crystals (CLECs) from *Trametes versicolor* was crystallized, cross-linked and lyophilized with cyclodextrin. The laccase treated CLEC was found to be highly active towards phenols and non phenols such as 2-amino phenol, guaiacol, catechol, pyrogallol, catechin and ABTS respectively.. The CLEC laccase was embedded in 30% polyvinylpropylidone (PVP) gel and mounted into an electrode to make the biosensor.

Cabrita et al (2005) have studied immobilised laccase from *Coriolus versicolor* on N-Hydroxysuccinimide-terminated self-assembled monolayers on gold surface. The reactivity of the terminated N-hydroxysuccinimide (NHS) towards amine functionalities has been tested for the covalent attachment of Dopamine This process could be useful for the further development of biosensors. In addition, Ferry and Leech (2005) investigated an enzyme electrode based on the co-immobilisation of an osmium redox polymer and a laccase from *T. versicolor* on glassy carbon electrodes and applied to ultrasensitive amperometric detection of the catecholamine neurotransmitters dopamine, epinephrine and Norepinephrine.

Laccase from *Pleurotus ostreatus* could be immobilised on the cathode of biofuel cells that could provide power; for example for small transmitter systems (Chen, Barton, Binyamin, Gao, Zhang, & Kim, 2001; Calabrese, Pickard, Vazquez-Duhalt, & Heller, 2002). Biofuel cells are extremely attractive from an environmental point of view because electrical energy is generated without combusting fuel thus providing a cleaner source of energy.

Cosmetic Industry

An increasing interest has been focused on laccase application in the cosmetic indusrty. Laccase can act as an oxidizing agent, substituting H_2O_2 that is why they are less irritant (Fu, Nyanhongo, Gubitz, Cavaco, Paulo, & Kim, 2012; Lavanya, Dhankar, Chhikara, & Sheoran, 2014). Laccase acts as a biocatalyst in the hair dye synthesis, deodorants, toothpaste, mouthwash, detergent, soaps. Hydrogen peroxide and phenylenediamines are the two chemicals used in hair dyes. These chemical hair dyes are irritant, allergic as well as carcinogenic in nature (Chen, Chen, Chern, Hsu, Huang, Chung, & Chye, 2006; Huang, Hung, Kang, Chen, & Chai, 2007) and caused severe hair damage (Araujo, Fernandes, Cavaco-Paulo, & Gomes, 2011; Takada, Nakamura, Matsuo, Inoue, Someya, & Shimogaki, 2003). Recently, cosmetic and dermatological preparations containing proteins for skin lightening have also been developed (Golz-Berner, Walze, Zastrow, & Doucet, 2004).

Pharmaceutical Industry

Laccases have the ability to synthesize products which may be highly valuable from pharmaceutical point of view. There are two major roles firstly they are directly involved in the synthesis of new medicinally valuable compounds via different methodologies or application for the bioremediation of by products or hazardous organic pollutants released from the pharmaceutical industries (Chaurasia, Bharati, & Sarma, 2016).

El-Fakharany et al (2010) studied, the potential activity of laccase from *P. ostreatus* against hepatitis C virus(HCV) infectivity in PBMCs and HepG2 cells. They investigated that *P. ostrestus* laccase has been purified to homogeneity. The optimum temperature and pH of this laccase were 60ºC and pH 4.0, respectively. It blocked HCV entry into both HepG2 and PBMCs cells in vitro.

Role of Laccases in Anticancer Activity

The first chemical has been prepared from 4-methyl-3-hydroxyanthranilic acid into actinocin. This compound has anticancer potential and functions as blocking the transcription of DNA from the tumor cell (Burton, 2003). Vinblastine (cytostatic) is also anticancer drugs, which is useful for the treatment of leukemia. Plant like *Catharanthus roseus* naturally produces vinblastine in small amount. Katarantine and vindoline are the precursors of this compound. Laccase is used to convert katarantine and vindoline precursors into vinblastine. Around 40% conversion of these molecules into the final product has been obtained using laccase (Yaropolov, Skorobogatko, Vartanov & Varfolomeyev, 1994). Mitomycins belongs pharmaceutical active compounds, which have antibiotic and cytotoxic activity. Laccase-mediated synthesis of 5-alkylamino- and 2,5-bis(alkylamino)-[1,4]-benzoquinones, showing structural similarity to natural mitomycins. These also serve as chemotherapeutics in the curing of several kinds of cancer like solid carcinomas (Herter, Michalik, Mikolasch, Schmidt, Wohlgemuth, Bornscheuer, & Schauer, 2013).

Role of Laccases in Antiviral Activity

Some researcher investigated that laccase enzymes are used in preparation of antiviral medicine. Example of such drugs is 3-(3, 4-dihydroxyphenyl)-propionic acid. Aromatic and aliphatic amines can be converted into it using laccase based oxidation. The derivatives have the antiviral natural activity and can be used for pharmacological purposes. They studied the novel dimers of the biologically active phenolic compound totarol which were synthesized using the phenol oxidase enzyme laccase, obtained from *Trametes pusbescens*, in organic solvent medium. Two dimeric products, linked either by carbon-carbon or bycarbon-oxygen bonds, were isolated and characterized. The effect of changes in various parameters such as solvent, temperature, pH and buffer concentration on the conversion of totarol bylaccase was investigated. The nature of the organic solvent, in particular, was found to affect the nature and the ratio of the products obtained (Ncanana, Baratto, Roncaglia, Riva, & Burton, 2007).

Role of Laccases Enhancing Antioxidant Activity

Catechins have the antioxidant ability to scavenge free radicals. These catechins consist of small units of tannins andwhich are important antioxidants occurs in tea, herbs and vegetables. By scavenging of free radicals catechins are important in preventing several diseases including cancer, inflammatory and cardiovascular diseases. They have less antioxidant activity. Therefore these properties can be increased by using laccase and has resulted in the conversion of catechins in several products with enhanced antioxidant ability (Kurisawa, Chung, Uyama, & Kobayashi, 2003).

Role of Laccase in Anti-Microbial Activity

Laccase plays an important role in antimicrobial activity, because it has electrochemical mode of action to penetrate in cell wall of organisms. Due to penetration essential metabolites are leaked and cell function physically disrupt. Various reports shows that white rot fungi have anti-microbial property like *Pleuotus ostreatus* against *E. coli* and *C. albicans* respectively (Othman, Elshafei1, Hassan1, Haroun, Elsayed1, & Farrag, 2014). Novel Cephalosporins were synthesized by amination of 2,5-dihydroxybenzoic acid derivatives using fungal laccases from *Trametes* sp. or Myceliophthora thermophila. These have inhibited the growth of several Gram positive bacteria by performing agar diffusion assay, among them were methicillin-resistant *S. aureus* and vancomycin-resistant *Enterococci*. The results shown that the laccase are used for the synthesis of potential new antibiotics (Mikolasch, Niedermeyer, Lalk, Witt, Seefeldt, Hammer, Schauer, Gesell, Hessel, Jülich, & Lindequist, 2006). Cyclosporin, a naturally occurring cyclic peptidefound in the fungus *Tolypocladium inflatum*. It possesses immunosuppressive activity and used for curing organ transplantation rejection. These are also used in other therapeutic areas, such as psoriasis, asthma and other autoimmune diseases (Yang, Kevin, Molino, Haydar, Yeyu, Bois, Maeng, Hemenway, Rich, Khmelnitsky, Friedrich, Peace, & Michels, 2009). Laccase from *Trametes* species can be applicable for the synthesis of novel penicillins by heteromolecular coupling of two different compounds in order to characterize the products of the reaction and to analyze the biological activity of the novel penicillins. These are inhibited growth of several gram positive bacterial strains against methicillin–resistant *Staphyllococcus aureus* and Vanomycin-resistant *Entercocci* (Mikolasch, Niedermeyer, Lalk, Witt, Seefeldt, Hammer, Schauer, Gesell, Hessel, Jülich, & Lindequist, 2006). Biocatalytical oxidation of iodide with together production of elemental iodine used to kill a large number of bacteria,fungi and viruses (Ihssen, Schubert, Meyer, & Richter, 2014). Wood rotting basidiomycetes fungi *Pycnosporous cinnabarinus* gives antibacterial activity against a number of bacterial strains. The maximum zone of inhibinition was obtained Gram positive bacteria of the genus *Streptococcus* (Claudia, 1997).

Role of Laccase in Anti HIV

Hu et al (2011) reported laccase from *Agrocybe cylindracea* with HIV-1 reverse transcriptase inhibitor activity and antiproliferative activity against HepG2 and MCF7cells was isolated.

Wanga and Ng (2004) reported that the laccase from *Tricholoma giganteum* inhibited the retroviral reverse transcriptase is within the range of potencies exhibited by the aforementioned mushroom proteins and other natural products. It inhibited HIV-1 reverse transcriptase with an IC50 of 2.2 lM.

Wong et al (2010) studied the potentially exploitable activities of Laccase from *Pleurotus cornucopiae*. It inhibited proliferation of murine leukemia cell line L1210 and human hepatoma cell line HepG2, and reduced the activity of HIV-1 reverse transcriptase with an IC50 of 22 μM. There was neither mitogenic activity toward mouse splenocytes seen nor hemagglutinating/hemolytic activity toward rabbit erythrocytes.

Another laccase has been shown capable of fighting aceruloplasminemia (a medical condition of lacking ceruloplasmin, a multi-Cu serum oxidase whose ferroxidase activity regulates iron homeostasis) (Harris, Davis-Kaplan, Gitlin, & Kaplan, 2004).

CONCLUSION

This book chapter illustrates that laccase from white rot fungi is a promising enzyme with a great potential application in several industries such as food, textile, paper and pulp,nanobiotechnology and biosensor,cosmetic as well as pharmaceutical industry. By the use of this enzyme one could improve productivity, efficiency and quality of products without high investment costs and has the advantage of being a mild technology. However, the limitation is to produce large volumes of highly active enzyme at an affordable cost for large scale applications. This can be accomplished in the heterlogous host by recombinant DNA technology. To obtain robust, active, stable, less expensive enzymes, protein engineering or chemical means can be used.

REFERENCES

Abou-Okeil, A., El-Shafie, A., & El Zawahry, M. M. (2010). Ecofriendly laccase–hydrogen peroxide/ultrasound-assisted bleaching of linen fabrics and its influence on dyeing efficiency. *Ultrasonics Sonochemistry*, *17*(2), 383–390. doi:10.1016/j.ultsonch.2009.08.007 PMID:19748814

Araujo, R., Fernandes, M., Cavaco-Paulo, A., & Gomes, A. (2011). Biology of human hair: Know your hair to control it. *Advances in Biochemical Engineering/Biotechnology*, *13*, 121–143. PMID:21072698

Artik, N., Karhan, M., & Aydar, G. (2004). Effects of polyphenoloxidase (LACCASE) application on clarity stability of sour cherry juice. *Journal of Food Technology*, *2*(4), 237–243.

Baldrian, P. (2006). Fungal laccases-occurrence and properties. *FEMS Microbiology Reviews*, *30*(2), 215–242. doi:10.1111/j.1574-4976.2005.00010.x PMID:16472305

Bauer, C. G., Kuhn, A., Gajovic, N., Skorobogatko, O., Holt, P. J., Bruce, N. C., ... Scheller, F. W. (1999). New enzyme sensors for morphine and codeine based on morphine dehydrogenase and laccase. *Fresenius' Journal of Analytical Chemistry*, *364*(1-2), 179–183. doi:10.1007002160051320

Bertrand, G. (1985). Sur la laccase et sur le pouvoir oxydant de cette diastase. *Comp Rendus L'Acad Sci.*, *120*, 266–269.

Blanco, D. C., Gonza'lez, D. M., Monmany, D. M., & Tzanov, T. (2009). Dyeing properties, synthesis, isolation and characterization of an in situ generated phenolic pigment, covalently bound to cotton. *Enzyme and Microbial Technology*, *44*(6-7), 380–385. doi:10.1016/j.enzmictec.2009.02.003

Bouwens, E. M., Trivedi, K., Vliet, C. V., & Winkel, C. (1997). *Method of enhancing color in a tea based foodstuff European patent application*. EP 760213 A1.

Bouwens, E.M., Trivedi, K., Vliet, C.V., & Winkel, C. (1999). *Method of enhancing color in a tea based foodstuff.* US5879730 A.

Brenna, O., & Bianchi, E. (1994). Immobilized laccase for phenolic removal in must and wine. *Biotechnology Letters*, *16*(1), 35–40. doi:10.1007/BF01022620

Burton, S. (2003). Laccases and phenol oxidases in organic synthesis. *Current Organic Chemistry*, *7*(13), 1317–1331. doi:10.2174/1385272033486477

Cabrita, J. F., Abrantes, L. M., & Viana, A. S. (2005). N-Hydroxysuccinimide-terminated self-assembled monolayers on gold for biomolecules immobilisation. *Electrochimica Acta*, *50*(10), 2117–2124. doi:10.1016/j.electacta.2004.09.019

Calabrese, B. S., Pickard, M., Vazquez-Duhalt, R., & Heller, A. (2002). Electroreduction of O_2 to water at 0.6 V (NHE) at pH 7 on the 'wired' *Pleurotus ostreatus* Laccase Cathode. *Biosensors & Bioelectronics*, *17*(11-12), 1071–1074. doi:10.1016/S0956-5663(02)00100-8 PMID:12392957

Call, H. P., & Mücke, I. (1997). History, overview and applications of mediated lignolytic systems, especially laccase-mediator systems (LignozymR process). *Journal of Biotechnology*, *53*(2-3), 163–202. doi:10.1016/S0168-1656(97)01683-0

Camarero, S., Garcia, O., Vidal, T., Colom, J., del Río, J. C., Gutiérrez, A., ... Martínez, Á. T. (2004). Efficient bleaching of non-wood high-quality paper pulp using laccase-mediator system. *Enzyme and Microbial Technology*, *35*(2-3), 113–120. doi:10.1016/j.enzmictec.2003.10.019

Cantarelli, C. (1986). Trattamenti enzimatici sui costituenti fenolici dei mosti come prevenzione della maderizzazione. *Vini d'Italia.*, *3*, 87–98.

Cantarelli, C., & Giovanelli, G. (1990). Stabilization of pome and grape juice against phenolic deterioration by enzymic treatments. *Inst Fruchtsaft Union Wiss Tech Komm.*, *21*, 35–57.

Cantarelli, C., & Giovanelli, G. (1991). La stabilizzazione dei vini bianchi mediante ossidazione enzimatica. *Vini d'Italia.*, *33*, 21–32.

Chandra, R. P., & Ragauskas, A. J. (2002). Evaluating laccase-facilitated coupling of phenolic acids to high-yield kraft pulps. *Enzyme and Microbial Technology*, *30*(7), 855–861. doi:10.1016/S0141-0229(02)00020-0

Chaurasia, , P. K., Bharati, S. L., & Singh, S. K. (2013). Comparative Studies on the Blue and Yellow Laccases. *Research in Plant Sciences.*, *1*(2), 32–37.

Chaurasia, P. K., Bharati, S. L., & Sarma, C. (2016). *Laccases in Pharmaceutical Chemistry: A Comprehensive Appraisal*. Academic Press.

Chaurasia, P. K., Bharati, S. L., Sharma, M., Singh, S. K., Yadav, S. S., & Yadava, S. (2015). Fungal Laccases and their Biotechnological Significances in the Current Perspective. *RE:view*, *19*(19), 1916–1934.

Chen, C. S., Mrkisch, M., Huang, S., Whistesides, G. M., & Ingber, D. E. (1998). Micropatterned surfaces for control of cell shape, position, and function. *Biotechnology Progress*, *14*(3), 356–363. doi:10.1021/bp980031m PMID:9622515

Chen, S. C., Chen, C. H., Chern, C. L., Hsu, L. S., Huang, Y. C., Chung, K. T., & Chye, S. M. (2006). p-Phenylenediamine induces p53-mediated apoptosis in Mardin-Darby canine kidney cells. *Toxicology In Vitro*, *20*(6), 801–807. doi:10.1016/j.tiv.2005.12.003 PMID:16469477

Chen, T., Barton, S. C., Binyamin, G., Gao, Z., Zhang, Y., Kim, H. H., & Heller, A. (2001). A miniature biofuel cell. *Journal of the American Chemical Society*, *123*(35), 8630–8631. doi:10.1021/ja0163164 PMID:11525685

Cheynier, V., Arellano, H., Souquet, J. M., & Moutounet, M. (1997). Estimation of the oxidative changes in phenolic compounds of Carignane during winemaking. *American Journal of Enology and Viticulture*, *48*, 225–228.

Claudia, E. (1997). Laccase-catalyzed formation of cinnabarinic acid is responsible for antibacterial activity of Pycnoporus cinnabarinus. *Microbiological Research*, *152*(3), 315–318. doi:10.1016/S0944-5013(97)80046-8 PMID:9352667

Conrad, L. S., Sponholz, W. R., & Berker, O. (2000). *Treatment of cork with a phenol oxidizing enzyme*. US Patent 6152966.

Couto, S. G., Toca, J. L., & Herrera, J. L. (2006). Industrial and biotechnological applications of laccases: A review. *Biotechnology Advances*, *24*(5), 500–513. doi:10.1016/j.biotechadv.2006.04.003 PMID:16716556

De Souza, C. G. M., & Peralta, R. M. (2003). Purification and characterization of the main laccase produced by the white rot fungus *Pleurotus pulmonarius* on wheat bran solid state Medium. *Journal of Basic Microbiology, 43*(4), 278–286. doi:10.1002/jobm.200390031 PMID:12872309

Desai, S. S., & Nityanand, C. (2011). Microbial laccases and their applications:a review. *Asian J Biotechnol, 3*(2), 98–124. doi:10.3923/ajbkr.2011.98.124

Ducros, V., Brzozowski, A. M., Wilson, K. S., Brown, S. H., Stergard, P., Schneider, P., ... Davies, G. J. (1998). Crystal structure of the type-2 Cu depleted laccase from *Coprinus cinereus* at 2.2A° resolution. *Nature Structural Biology, 5*(4), 310–316. doi:10.1038/nsb0498-310 PMID:9546223

Dur´an, Rosa, D'Annibale, & Gianfreda. (2002). Applications of laccases and tyrosinases (phenoloxidases) immobilized on different supports: a review. *Enzyme and Microbial Technology, 31*(7), 907–931.

El-Fakharany, E. M., Haroun, B. M., Ng, T. B., & Redwan, E. R. (2010). Oyster mushroom laccase inhibits hepatitis C virus entry into Peripheral blood cells and hepatoma cells. *Protein and Peptide Letters, 17*, 1031–1039. doi:10.2174/092986610791498948 PMID:20156183

Eremia, S. A., Vasilescu, I., Radoi, A., Litescu, S. C., & Radu, G. L. (2013). Disposable biosensor based on platinum nanoparticles-reduced graphene oxide-laccase biocomposite for the determination of total polyphenolic content. *Talanta, 110*, 164–170. doi:10.1016/j.talanta.2013.02.029 PMID:23618190

Espinoza, F. M. C., Morel, M. H., & Rouau, X. (1998). Effect of lysine, tyrosine, cysteine and glutathione on the oxidative crosslinking of feruloylated arabinoxylans by a fungal laccase. *Journal of Agricultural and Food Chemistry, 46*, 2589.

Espinoza, F. M. C., Morel, M. H., Surget, A., Asther, M., Moukha, S., Sigoillot, J. C., & Rouau, X. (1999). Attempt to crosslinkferuloylated arabinoxylans and proteins with a fungal laccase. *Food Hydrocolloids, 13*(1), 65–71. doi:10.1016/S0268-005X(98)00072-1

Espinoza, F. M. C., & Rouau, X. (1999). Effect of cysteinyl caffeic acid, caffeic acid, and l-dopa on the oxidative crosslinking of feruloylated arabinoxylans by a fungal laccase. *Journal of Agricultural and Food Chemistry, 47*(2), 497–503. doi:10.1021/jf980723q PMID:10563923

FAO. (2004). *Laccase from* Myceliophthora thermophila *expressed in* Aspergillus oryzae. Chemical and Technical Assessment (CTA). FAO.

Ferraroni, M., Myasoedova, N. M., Schmatchenko, V., Leontievsky, A. A., Golovleva, L. A., Scozzafava, A., & Briganti, F. (2007). Crystal structure of a blue laccase from Lentinus tigrinus: Evidences for intermediates in the molecular oxygen reductive splitting by multicopper oxidases. *BMC Structural Biology, 7*(1), 60. doi:10.1186/1472-6807-7-60 PMID:17897461

Ferry, Y., & Leech, D. (2005). Amperometric detection of catecholamine neurotransmitters using electrocatalytic substrate recycling at a laccase electrode. *Electroanalysis, 17*(2), 2113–2119. doi:10.1002/elan.200403069

Freire, R. S., Durán, N., & Kubota, L. T. (2001). Effects of fungal laccase immobilization procedures for the development of a biosensor for phenol compounds. *Talanta, 54*(4), 681–686. doi:10.1016/S0039-9140(01)00318-6 PMID:18968290

Fu, J., Nyanhongo, G. S., Gubitz, G. M., Cavaco-Paulo, A., & Kim, S. (2012). Enzymatic colouration with laccase and peroxidases: Recent progress. *Biocatalysis and Biotransformation*, *30*(1), 125–140. do i:10.3109/10242422.2012.649563

Fusco, D. (2010). Laccase based biosensor for the determination of polyphenol index in wine. *Talanta*, *81*(1-2), 235–240. doi:10.1016/j.talanta.2009.11.063 PMID:20188914

Gassara-Chatti, F., Brar, S. K., Ajila, C. M., Verma, M., Tyagi, R. D., & Valero, J. R. (2013). Encapsulation of ligninolytic enzymes and its application in clarification of juice. *Food Chemistry*, *137*(1-4), 18–24. doi:10.1016/j.foodchem.2012.09.083 PMID:23199985

Ghindilis, A. L., Gavrilova, V. P., & Yaropolov, A. I. (1992). Laccase based biosensor for determination of polyphenols: Determination of catechols in tea. *Biosensors & Bioelectronics*, *7*(2), 127–131. doi:10.1016/0956-5663(92)90017-H PMID:1637524

Gianfreda, L., Xu, F., & Bollag, J. M. (1999). Laccases: A useful group of oxidoreductive enzymes. *Bioremediation Journal*, *3*(1), 125. doi:10.1080/10889869991219163

Gibson, J. F. (1997). Chemistry in the vineyards. *Education in Chemistry, 34*, 123–126.

Giovanelli, G. (1989). Enzymic treatment of malt polyphenols for beer stabilization. *Industrie delle Bevande*, *18*, 497–502.

Giovanelli, G., & Ravasini, G. (1993). Apple juice stabilization by combined enzyme- membrane filtration process. *Lebensmittel-Wissenschaft + Technologie*, *26*(1), 1–7. doi:10.1006/fstl.1993.1001

Gokmen, V., Borneman, Z., & Nijhuis, H. H. (1998). Improved ultrafiltration for color reduction and stabilization of apple juice. *Journal of Food Science*, *63*(3), 504–507. doi:10.1111/j.1365-2621.1998.tb15773.x

Golz-Berner, K., Walze, B., Zastrow, L., & Doucet, O. (2004). *Cosmetic or dermatological preparation with skin-lightening proteins*. WO2004017931.

Hadzhiyska, H., Calafell, M., Gibert, J. M., Daga', J. M., & Tzanov, T. (2006). Laccase-assisted dyeing of cotton. *Biotechnology Letters*, *28*(10), 755–759. doi:10.100710529-006-9043-5 PMID:16791731

Haghighi, B., Gorton, L., Ruzgas, T., & Jönsson, L. J. (2003). Characterization of graphite electrodes modified with laccase from Trametes versicolor and their use for bioelectrochemical monitoring of phenolic compounds in flow injection analysis. *Analytica Chimica Acta*, *487*(1), 314. doi:10.1016/S0003-2670(03)00077-1

Hammond, P. T., & Whitesides, G. M. (1995). Formation of polymer microstructures by selective deposition of poly ion multilayers using patterned self assembled self assembled monolayers as a template. *Macromolecules*, *28*(22), 7569–7571. doi:10.1021/ma00126a040

Harris, Z. L., Davis-Kaplan, S. R., Gitlin, J. D., & Kaplan, J. (2004). A fungal multicopper oxidase restores iron homeostasis in aceruloplasminemia. *Blood*, *103*(12), 4672–4673. doi:10.1182/blood-2003-11-4060 PMID:14739215

Herter, S., Michalik, D., Mikolasch, A., Schmidt, M., Wohlgemuth, R., Bornscheuer, U., & Schauer, F. (2013). Laccase-mediated synthesis of 2-methoxy-3-methyl-5-(alkylamino)- and 3-methyl-2,5-is(alkylamino)-[1,4]-benzoquinones. *Journal of Molecular Catalysis. B, Enzymatic*, *90*, 91–97. doi:10.1016/j.molcatb.2013.01.020

Hossain, Kh. M. G., Gonza'lez, M. D., Juan, A. R., & Tzanov, T. (2010). Enzyme-mediated coupling of a bi-functional phenolic compound onto wool to enhance its physical, mechanical and functional properties. *Enzyme and Microbial Technology*, *46*(3-4), 326–330. doi:10.1016/j.enzmictec.2009.12.008

Hossain, Kh. M. G., Gonza'lez, M. D., Lozano, G. R., & Tzanov, T. (2009). Multifunctional modification of wool using an enzymatic process in aqueous-organic media. *Journal of Biotechnology*, *141*(1-2), 58–63. doi:10.1016/j.jbiotec.2009.02.011 PMID:19428731

Hu, D. D., Zhang, R. Y., Zhang, G. Q., Wang, H. X., & Ng, T. B. (2011). A laccase with antiproliferative activity against tumor cells from an edible mushroom, white common Agrocybe cylindracea. *Phytomedicine*, *18*(5), 374–379. doi:10.1016/j.phymed.2010.07.004 PMID:20739163

Huang, Y. C., Hung, W. C., Kang, W. Y., Chen, W. T., & Chai, C. Y. (2007). p-Phenylenediamine induced DNA damage in SV-40 immortalized human uroepithelial cells and expression of mutant p53 and COX-2 proteins. *Toxicology Letters*, *170*(2), 116–123. doi:10.1016/j.toxlet.2007.02.011 PMID:17403587

Ibarra-Escutia, P., Gomez, J. J., Calas-Blanchard, C., Marty, J. L., & Ramirez-Silva, M. T. (2010). Amperometric biosensor based on a high resolution photopolymer deposited onto a screen-printed electrodefor phenolic compounds monitoring in tea infusions. *Talanta*, *81*(4-5), 1636–1642. doi:10.1016/j.talanta.2010.03.017 PMID:20441951

Ihssen, J., Schubert, M., Meyer, L. T., & Richter, M. (2014). Laccase Catalyzed Synthesis of Iodinated Phenolic Compounds with Antifungal Activity. *PLoS One*, *9*(3), 89924. doi:10.1371/journal.pone.0089924 PMID:24594755

Jarosz-Wilkołazka, A., Ruzgas, T., & Gorton, L. (2004). Use of laccase-modified electrode for amperometric detection of plant flavonoids. *Enzyme and Microbial Technology*, *35*(2-3), 238–241. doi:10.1016/j.enzmictec.2004.04.016

Jung, J., & Wicker, L. (2012). Laccase mediated conjugation of heat treated beta-lactoglobulin and sugar beet pectin. *Carbohydrate Polymers*, *89*(4), 1244–1249. doi:10.1016/j.carbpol.2012.04.028 PMID:24750938

Kim, S. Y., Lopez, C., Guebitz, G. M., & Cavaco-Paulo, A. (2008). Biological coloration of flax fabrics with flavonoids using laccase from *Trametes hirsuta. Engineering in Life Sciences*, *8*(3), 324–330. doi:10.1002/elsc.200700061

Kim, S. Y., Moldes, D., & Cavaco-Paulo, A. (2007). Laccase for enzymatic coloration of unbleached cotton. *Enzyme and Microbial Technology*, *40*(7), 1788–1793. doi:10.1016/j.enzmictec.2007.01.002

Kirk, O., Borchert, T. V., & Fuglsang, C. C. (2002). Industrial enzyme applications. *Current Opinion in Biotechnology*, *13*(4), 345–351. doi:10.1016/S0958-1669(02)00328-2 PMID:12323357

Kontopidis, G., Holt, C., & Sawyer, L. (2004). Beta-lactoglobulin:binding properties, structure, and function. *Journal of Dairy Science*, *87*(4), 785–796. doi:10.3168/jds.S0022-0302(04)73222-1 PMID:15259212

Kuhad, R. C., Singh, A., & Eriksson, K. E. L. (1997). Microorganisms and enzymes involved in the degradation of plant fiber cell wall. In K. E. L. Eriksson (Ed.), *Biotechnology in the Pulp and paper industry. Advances in biochemical engineering biotechnology*. Berlin: Springer Verlag. doi:10.1007/BFb0102072

Kurisawa, M., Chung, J. E., Uyama, H., & Kobayashi, S. (2003). Laccase-catalyzed synthesis and antioxidant property of poly(catechin). *Macromolecular Bioscience*, *3*(12), 758–764. doi:10.1002/mabi.200300038

Kushwaha, A., Agarwal, S., Gupta, K. K., Maurya, S., Chaurasia, P. K., Singh, A. K., & Singh, M. P. (2017). Laccase Enzyme from white rot fungi: An overview and its Application. In Incredible world of Biotechnology. Nova Science Publishers Inc.

Kuuva, T., Lantto, R., Reinikainen, T., Buchert, J., & Autio, K. (2003). Rheological properties of laccase-induced sugar beet pectin gels. *Food Hydrocolloids*, *17*(5), 679–684. doi:10.1016/S0268-005X(03)00034-1

Kuznetsov, B. A., Shumakovich, G. P., Koroleva, O. V., & Yaropolov, A. I. (2001). On applicability of laccase as label in the mediated and mediatorless electro immunoassay: Effect of distance on the direct electron transfer between laccase and electrode. *Biosensors & Bioelectronics*, *16*(1-2), 73–84. doi:10.1016/S0956-5663(00)00135-4 PMID:11261856

Labat, E., Morel, M. H., & Rouau, X. (2000). Wheat gluten phenolic acids: Occurrence and fate upon mixing. *Journal of Agricultural and Food Chemistry*, *48*(12), 6280–6283. doi:10.1021/jf0006271 PMID:11312800

Labat, E., Morel, M. H., & Rouau, X. (2000). Effects of laccase and ferulic acid on wheat flour doughs. *Cereal Chemistry*, *77*(6), 823–828. doi:10.1094/CCHEM.2000.77.6.823

Lantto, R., Scha¨nberg, C., & Buchert, J. (2004). Effects of laccase mediator combinations on wool. *Textile Research Journal*, *74*(8), 713–717. doi:10.1177/004051750407400809

Lavanya, C., Dhankar, R., Chhikara, S., & Sheoran, S. (2014). Degradation of toxic dyes: A review. *International Journal of Current Microbiology and Applied Sciences*, *3*(6), 189–199.

Leite, O. D., Lupetti, K. O., Fatibello-Filho, O., Vieira, I. C., & de Barbosa, A. M. (2003). Synergic effect studies of the bi-enzymatic system laccaseperoxidase in a voltammetric biosensor for catecholamines. *Talanta*, *59*(5), 889–896. doi:10.1016/S0039-9140(02)00650-1 PMID:18968977

Lisdat, F., Wollenberger, U., Makower, A., Hortnagl, H., Pfeiffer, D., & Scheller, F. W. (1997). Catecholamine detection using enzymatic amplification. *Biosensors & Bioelectronics*, *12*(12), 1199–1211. doi:10.1016/S0956-5663(97)00098-5 PMID:9474769

Littoz, F., & McClements, D. J. (2008). Bio-mimetic approach to improving emulsion stability: Cross-linking adsorbed beet pectin layers using laccase. *Food Hydrocolloids*, *22*(7), 1203–1211. doi:10.1016/j.foodhyd.2007.06.009

Lund, M., & Ragauskas, A. J. (2001). Enzymatic modification of kraft lignin through oxidative coupling with water-soluble phenols. *Applied Microbiology and Biotechnology*, *55*(6), 699–703. doi:10.1007002530000561 PMID:11525617

Macheix, J. J., Sapis, J. C., Fleuriet, A., & Lee, C. Y. (1991). Phenolic compounds and polyphenoloxidases in relation to browning in grapes and wines. *Critical Reviews in Food Science and Nutrition*, *30*(4), 441–486. doi:10.1080/10408399109527552 PMID:1910524

MacMurrough, I., Madigan, D., Kelly, R., & O'Rourke, T. (1999). Haze formation. Shelf-life prediction for lager beer. *Food Technology*, *53*, 58–62.

Madhavi, V., & Lele, S. S. (2009). Laccase: Properties and applications. *Bioresource Technology*, *4*(4), 1694–1717.

Maier, G., Mayer, P., & Dietrich, H. (1990). Application of a polyphenoloxidase to stabilization of apple juices. *Deutsche Lebensmittel-Rundschau*, *86*, 137–142.

Mathiasen, T.E. (1995). *Laccase and beer storage. PCT international application*. WO 9521240 A2.

Mattinen, M. L., Kruus, K., Buchert, J., Nielsen, J. H., Andersen, H. J., & Steffensen, C. L. (2005). Laccase-catalysed polymerization of tyrosine-containing peptides. *The FEBS Journal*, *272*(14), 3640–3650. doi:10.1111/j.1742-4658.2005.04786.x PMID:16008563

Mensah, C. A., Adamafio, N. A., Kwarteng, A. K., & Rodrigues, F. K. (2012). Reduced tannin content of laccase-treated cocoa (Theobroma cacao) pod husk. *Int J Biol Chem.*, *6*(1), 31–36. doi:10.3923/ijbc.2012.31.36

Mikolasch, A., Niedermeyer, T. H., Lalk, M., Witt, S., Seefeldt, S., Hammer, E., ... Lindequist, U. (2006). Novel penicillins synthesized by biotransformation using laccase from *Trametes spec. Chemical & Pharmaceutical Bulletin*, *54*(5), 632–638. doi:10.1248/cpb.54.632 PMID:16651757

Min, K. L., Kim, Y. H., Kim, Y. W., Jung, H. S., & Hah, Y. C. (2001). Characterization of a novel laccase produced by the wood-rotting fungus *Phellinus ribis*. *Archives of Biochemistry and Biophysics*, *392*(2), 279–286. doi:10.1006/abbi.2001.2459 PMID:11488603

Minussi, R. C., Pastore, G. M., & Duran, N. (1998). Phenol removal and stabilization of red wine in the presence of laccase from *Trametes versicolor*. In *Brazilian Meeting on the Chemistry of Food and Beverages*, Sa˜o Carlos, Brazil.

Minussi, R.C., Pastore, G.M., & Duran, N. (2002). Potential applications of laccase in the food industry. *Trends in Food Science and Technology, 13*(6-7), 205–216.

Minussi, R. C., Pereira, D. G., Pastore, G. M., & Duran, N. (1998).Enzymatic oxidation of ferulic acid by laccase. *IV National Congress of Biotechnology, Talca-Chile, Resume'n.*

Montazer, M., & Maryan, S. A. (2010). Influences of different enzymatic treatment on denim garment. *Applied Biochemistry and Biotechnology*, *160*(7), 2114–2128. doi:10.100712010-009-8727-4 PMID:19728171

Montereali, M. R., Seta, L. D., Vastarella, W., & Pilloton, R. (2010). A disposable Laccase-Tyrosinase based biosensor for amperometric detection of phenolic compounds in must andWine. *Journal of Molecular Catalysis. B, Enzymatic*, *64*(3-4), 189–194. doi:10.1016/j.molcatb.2009.07.014

Ncanana, S., Baratto, L., Roncaglia, L., Riva, S., & Burton, S. G. (2007). Laccase mediated oxidation of totarol. *Advanced Synthesis & Catalysis*, *349*(8-9), 1507–1513. doi:10.1002/adsc.200700005

Neifar, M., Ghorbel, E. R., Kamoun, A., Baklouti, S., Mokni, A., Jaouani, A., & Chaabouni, E. S. (2011). Effective clarification of pomegranate juice using laccase treatment optimized by response surface methodology followed by ultrafiltration. *Journal of Food Process Engineering*, *34*(4), 1199–1219. doi:10.1111/j.1745-4530.2009.00523.x

Norsker, M., Jensen, M., & Adler-Nissen, J. (2000). Enzymatic gelation of sugar beet pectin in food products. *Food Hydrocolloids*, *14*(3), 237–243. doi:10.1016/S0268-005X(00)00016-3

Osma, J. F., Herrera, J. L. T., & Couto, S. R. (2010). Uses of laccases in the food industry. *Enzyme Research*, *2010*, 918761. doi:10.4061/2010/918761 PMID:21048873

Othman, A. M., Elshafei, A., Hassan, M., Haroun, B., Elsayed, M., & Farrag, A. (2014). Purification, Biochemical Characterization and Applications of *Pleurotus ostreatus* ARC280. Laccase. *British Microbiology Research Journal*, *4*(12), 1418–1439. doi:10.9734/BMRJ/2014/11218

Palmieri, G., Giardina, P., Bianco, C., Scaloni, A., Capassoi, A., & Sannia, G. (1997). A novel white laccase from *Pleurotus ostreatus*. *The Journal of Biological Chemistry*, *272*(50), 31301–31307. doi:10.1074/jbc.272.50.31301 PMID:9395457

Pazarlıog˘lu, N. K., Sariişik, M., & Telefoncu, A. (2005). Laccase: Production by *Trametes versicolor* and application to denim washing. *Process Biochemistry (Barking, London, England)*, *40*(5), 1673–1678. doi:10.1016/j.procbio.2004.06.052

Pereira, L., Bastos, C., Tzanov, T., Cavaco-Paulo, A., & Guebitz, G. M. (2005). Environmentally friendly bleaching of cotton using laccases. *Environmental Chemistry Letters*, *3*(2), 66–69. doi:10.100710311-005-0004-3

Petersen, B. R., Mathiasen, T. E., Peelen, B., & Andersen, H. (1996). Use of laccase for deoxygenation of oil-containing product such as salad dressing. *PCT Int. Appl. WO*, *9635768*, A1.

Pezzella, C., Guarino, L., & Piscitelli, A.A. (2015). How to enjoy laccases Cell. *Mol. Life Sci.*

Prasetyo, E. N., Kudanga, T. W., Steiner, W., Murkovic, M., Nyanhongo, G. S., & Guebitz, G. M. (2010). Laccase-generated tetramethoxyazobismethylene quinone (TMAMQ) as a tool for antioxidant activity measurement. *Food Chemistry*, *118*(2), 437–444. doi:10.1016/j.foodchem.2009.04.102

Ren, X., & Diller, B. G. (2007). Oxidoreductases for modification of linen fibers. *Colloid Surface A*, *299*(1-3), 15–21. doi:10.1016/j.colsurfa.2006.11.011

Renzetti, S., Courtin, C. M., Delcour, J. A., & Arendt, E. K. (2010). Oxidative and proteolytic enzyme preparations as promising improvers for oat bread formulations: Rheological, biochemical and microstructural background. *Food Chemistry*, *119*(4), 1465–1473. doi:10.1016/j.foodchem.2009.09.028

Ritter, G., Maier, G., Schoepplein, E., & Dietrich, H. (1992). The application of polyphenoloxidase in the processing of apple juice. *Bulletin de Liaison-Groupe Polyphenols.*, *16*, 209–212.

Rossi, M., Giovanelli, G., Cantarelli, C., & Brenna, O. (1988). Effects of laccase and other enzymes on barley wort phenolics as a possible treatment to prevent haze in beer. *Bulletin de Liaison-Groupe Polyphenols.*, *14*, 85–88.

Roy, J. J., Abraham, T. E., Abhijith, K. S., Kumar, P. V. S., & Thakur, M. S. (2005). Biosensor for the determination of phenols based on Cross-Linked Enzyme Crystals (CLEC) of laccase. *Biosensors & Bioelectronics*, *21*(1), 206–211. doi:10.1016/j.bios.2004.08.024 PMID:15967371

Sammartino, M., Piacquadio, P., De Stefano, G., & Sciancalepore, V. (1998). Apple juice stabilization by conventional and innovative methods. *Industrie delle Bevande*, *27*, 367–369.

Schroder, M., Aichernig, N., Gu¨bitz, G. M., & Kokol, V. (2007). Enzymatic coating of lignocellulosic surfaces with polyphenols. *Biotechnology Journal*, *2*(3), 334–341. doi:10.1002/biot.200600209 PMID:17260331

Selinheimo, E., Kruus, K., Buchert, J., Hopia, A., & Autio, K. (2006). Effects of laccase, xylanase and their combination on the rheological properties of wheat doughs. *Journal of Cereal Science*, *43*(2), 152–159. doi:10.1016/j.jcs.2005.08.007

Shleev, S. V., Morozova, O.V., & Nikitina O.V. (2004). Comparison of physico-chemical characteristics of four laccases from different basidiomycetes. *Biochimie, 86*(9-10), 693–703.

Si, J. Q. (1994). *Use of laccase in baking industry*. International patent application PCT/DK94/00232.

Si, J.Q. (1994). *Use of laccase in baking*. PCT Int. Appl. WO 9428728A1.

Sigal, G. B., Mrksich, M., & Whitesides, G. M. (1998). Effect of surface wett ability on the adsorption of proteins of proteins and detergents. *Journal of the American Chemical Society*, *120*(14), 3464–3473. doi:10.1021/ja970819l

Stanic, D., Monogioudi, E., Dilek, E., Radosavljevic, J., Markovic, A. M., Vuckovic, O., ... Cirkovic, T. (2010). Digestibility and allergenicity assessment of enzymatically crosslinked beta-casein. *Molecular Nutrition & Food Research*, *54*(9), 1273–1284. doi:10.1002/mnfr.200900184 PMID:20205191

Suparno, O., Covington, A. D., & Evans, C. S. (2007). Application of diphenols for dyeing. *J Soc Leath Tech Ch.*, *91*, 139–141.

Takada, K., Nakamura, A., Matsuo, N., Inoue, A., Someya, K., & And Shimogaki, H. (2003). Influence of oxidative and/or reductive treatment on human hair (I): Analysis of hair-damage after oxidative and/or reductive treatment. *Journal of Oleo Science*, *52*(10), 541–548. doi:10.5650/jos.52.541

Takemori, T., Ito, Y., Ito, M., & Yoshama, M. (1992). Flavor and taste improvement of cacao nib by enzymatic treatment. *Japan Kokai Tokkyo Koho JP*, *04126037*, A2.

Tantoush, Z., Stanic, D., Stojadinovic, M., Ognjenovic, J., Mihajlovic, L., Markovic, A. M., & Cirkovic, T. (2011). Digestibility and allergenicity of b-lactoglobulin following laccase-mediated cross-linking in the presence of sour cherry phenolics. *Food Chemistry*, *125*(1), 84–91. doi:10.1016/j.foodchem.2010.08.040

Taspinar, A., & Kolankaya, N. (1998). Optimization of enzymatic chlorine removal from kraft pulp. *Bulletin of Environmental Contamination and Toxicology*, *61*, 15–21. doi:10.1007001289900723 PMID:9657825

Thurston, C. F. (1994). The structure and function of fungal laccases. *Microbiology, 140*(1), 19–26. doi:10.1099/13500872-140-1-19

Tian, L., Branford-White, C., Wang, W., Nie, H., & Zhu, L. (2012). Laccase-mediated system pretreatment to enhance the effect of hydrogen peroxide bleaching of cotton fabric. *International Journal of Biological Macromolecules, 50*(3), 782–787. doi:10.1016/j.ijbiomac.2011.11.025 PMID:22155402

Tzanov, T., Basto, C., Gu¨bitz, G. M., & Cavaco-Paulo, A. (2003 a). Laccases to improve the whiteness in a conventional bleaching of Cotton. *Macromolecular Materials and Engineering, 288*(10), 807–810. doi:10.1002/mame.200300100

Tzanov, T., Silva, C., Zille, A., Oliveira, J., & Cavaco-Paulo, A. (2003 b). Effect of some process parameters in enzymatic dyeing of wool. *Applied Biochemistry and Biotechnology, 111*(1), 1–14. doi:10.1385/ABAB:111:1:1 PMID:14566065

Upadhayay, P., Shrivastava, R., & Agrawal, P.K. (2016).Bioprospecting and Biotechnological applications of fungal laccase. *Biotech.*

Wanga, H. X., & Ng, T. B. (2004). Purification of a novel low-molecular-mass laccase with HIV-1 reverse transcriptase inhibitory activity from the mushroom *Tricholoma giganteum. Biochemical and Biophysical Research Communications, 315*(2), 450–454. doi:10.1016/j.bbrc.2004.01.064 PMID:14766229

Wong, J. H., Ng, T. B., Jiang, Y., Liu, F., Sze, S. C., & Zhang, K. Y. (2010). Purification and characterization of a laccase with inhibitory activity toward HIV-1 reverse transcriptase and tumor cells from an edible mushroom (*Pleurotus cornucopiae*). *Protein and Peptide Letters, 17*(8), 1040–1047. doi:10.2174/092986610791498966 PMID:19807674

Yang, Z., Kevin, P., Molino, B. F., Haydar, S. N., Yeyu, C., Bois, F., ... Michels, P. C. (2009). Novel Oxidation of Cyclosporin A: Preparation of Cyclosporin Methyl Vinyl Ketone (Cs-MVK). *Synlett, 18*, 2935–2938.

Yaropolov, A. I., Skorobogatko, O. V., Vartanov, S. S., & Varfolomeyev, S. D. (1994). Laccase: Properties, catalytic mechanism and applicability. *Applied Biochemistry and Biotechnology, 49*(3), 257–280. doi:10.1007/BF02783061

Yaver, D. S., Xu, F., Golightly, E. J., Brown, K. M., Brown, S. H., & Rey, M. W. (1996). Purification, characterization, molecular cloning, and expression of two laccase genes from the white rot basidiomycete *Trametes villosa. Applied and Environmental Microbiology, 62*, 834–841. PMID:8975613

Yoon, M. Y. (1998). *Process for improved shrink resistance in wool.* WO Patent 98/27264.

Yoshida, H. (1883). Chemistry of lacquer (Urushi) part 1. *Journal of the Chemical Society, 43*(0), 472–486. doi:10.1039/CT8834300472

Zamorani, A. (1989). Enzymatic processing of musts and wines. In Biotechnology applications in beverage production. Elsevier Applied Science. doi:10.1007/978-94-009-1113-0_14

Zamorani, A., Spettoli, P., Lante, A., Crapisi, A., & Pasini, G. (1993). Immobilized laccase and tyrosinase: An approach for wine stabilization. *Italian Journal of Food Science, 4*, 409–414.

Chapter 12
The Potential Application of Peroxidase Enzyme for the Treatment of Industry Wastes

Sonam Agarwal
University of Allahabad, India

Krishna Kumar Gupta
University of Allahabad, India

Vivek Kumar Chaturvedi
University of Allahabad, India

Ankita Kushwaha
University of Allahabad, India

Pankaj Kumar Chaurasia
MNNIT Allahabad, India & and LS College Muzaffarpur, India

M. P. Singh
University of Allahabad, India

ABSTRACT

Environmental pollution is becoming one of the major threats around the world because of the release of toxic and hazardous substances from food, pharmaceutical, and other industries as well. These wastes are mainly dumped indiscriminately which ultimately reached water bodies, thereby affecting marine ecosystem. Therefore, effective effluent treatment is an important step which can help in conserving our water resources. White rot fungus (WRF) have been shown to degrade and mineralize a wide variety of wastes because of their nonspecific extracellular lignin mineralizing enzymes (LMEs). These enzymes are used for the decolorization of synthetic dyes. They help in the degradation of pesticides, polycyclic aromatic hydrocarbons (PAHs), and pharmaceuticals wastes like- anti-inflammatory, lipid regulatory, antiepileptic drugs, endocrine disrupting chemicals, etc. They also help in degrading the food waste and convert them into useful products which can be used as food, feed, fodder; some of these wastes are lignocellulosic waste, viticulture waste, olive mill waste, molasses waste, etc.

DOI: 10.4018/978-1-5225-5237-6.ch012

INTRODUCTION

Wastes from different industries like food industry, pharma industry, paper and mill industry, dye industry, etc. are hazardous in nature and their proper treatment is necessary. These wastes after conventional treatment are mainly dumped into water bodies or spilled on the land. The organic micropollutants in the effluent of waste water treatment plants (WWTPs) still include toxic substances which usually settle on the banks of the river or are present on the soil (Marco-Urrea, Pérez-Trujillo, Vicent, & Caminal, 2009). Generally, conventional biological treatment technologies do not remove pollutants effectively. These pollutants can adversely affect key biotransformation process of other pollutants, like denitrification, nitrogen fixation, degradation of organic compounds, etc (Heberer, 2002; Rooklidge, 2004). Therefore, an efficient technique for their total destruction is very necessary, and one such process is bioprocess involving micro-organisms and their enzymes as biocatalysts. This technique is cost effective with attractive properties, such as low energy requirements, easy process, broad range specificity and operational over wide range of pH, temperature and ionic strengths (Torres, 2003). White rot fungi (WRF) and their lignin modifying enzymes (LMEs) have been used extensively for the treatment of these toxic compounds. They help in the bioremediation and biodegradation of many toxic compounds like dye stuffs, PAHs, pesticides, antibiotics, phenols, polychlorinated biphenyls (PCBs), anti-depressants, endocrine disrupting chemicals, agricultural waste, etc (Torres, 2003; Mayer, 2002; Reddy, 1995; Pointing, 2001). WRF includes basidiomycetes and litter-decomposing fungi which are capable of extensive aerobic lignin polymerization and mineralization. They produce extracellular enzymes which help in the degradation of wide range of xenobiotics because of their low substrate specificity. Peroxidases are widely distributed in nature. They (EC 1.11.1.7) are heme proteins and contain iron (III) protoporphyrin IX (ferriprotoporphyrin IX) as the prosthetic group. Their molecular weight ranges from 30,000 to150,000 Da. These are a group of oxidoreductases that catalysethe reduction of peroxides, such as hydrogen peroxide and the oxidation of a variety of organic and inorganic compounds. They secrete low molecular weight mediators that enlarge the spectrum of compounds they are able to oxidize. The LMEs of WRF are lignin peroxidases (LiP, E.C. 1.11.1.14), manganese- dependent peroxidases (MnP, E.C. 1.11.1.13), versatile peoxidases (VP, E.C. 1.11.1.16) and laccases (Lac, E.C. 1.10.3.2). These enzymes are produced during WRF secondary metabolism since lignin oxidation provides no net energy to the fungus (Cabana, 2007).

LiP, MnP and VP belong to the heme peroxidases family and for the oxidation of substrates, they require hydrogen peroxide (Wong, 2009). LiP is a monomeric hemoprotein with heme group in its active centre. It was first discovered in the extracellular medium of *Phanerochaete chrysosporium.* Its molecular weight and pI range from 38 to 43 kDa and 3.3 to 4.7 pI, respectively (Perez, Munoz-Dorado, De la Rubia, & Martinez, 2002). Fe (III) of LiP is pentacoordinated to the four heme tetrapyrrole nitrogen and to a histidineresidue. LiPs are oxidized by H_2O_2 to give a two electron-oxidized intermediate (Compound I) in which the iron is present as Fe (IV) and a free radical resides on the tetrapyrrole ring (or on a nearby amino acid). Compound I then oxidizes a donor substrate by one electron, yielding a substrate-free radical andCompound II, in which the iron is still present as Fe (IV) but no radical is present on the tetrapyrrole. Compound II then oxidizes a second molecule of donor substrate, giving another substrate-free radical and the resting state of the peroxidase (Hammel & Cullen, 2008). In presence of hydrogen peroxide, LiP oxidizes substrates containing methoxybenzene groups such as veratryl alcohol generating cation radicals involved in the carbon-carbon and ether bonds cleavage in lignin model compounds (Tien, Kirk, Morgan, Maytield, Kuwahara, & Gold, 1988).

MnP is glycosylated protein, with molecular mass ranging from 45 to 60 kDa. Mn (II) is oxidized to Mn (III) by MnP. This product Mn (III) is stabilized by chelating Mn (II) with organic acid chelators. Mn (III) is a strong oxidant which oxidizes phenolic compounds. For non-phenolic compounds, MnP oxidizes it in the presence of Mn (II) by the peroxidation of unsaturated lipids (Perez, Munoz-Dorado, De la Rubia, & Martinez, 2002). MnP, in the presence of hydrogen peroxide and Mn (II) is able to oxidize a wide range of organic compounds, by converting Mn (II) into Mn (III), which from complexes with α-hydroxy acids and oxidizes lignin- like substrates (Glenn, Lakshmi, & Gold, 1986).

VP exhibits both the property of LiP and MnP which naturally degrades lignin. In the absence of exogenous H_2O_2 and in the presence of Mn (II), VP oxidizes hydroquinone. The oxidation of hydroquinone in the presence of Mn (II) is important in the degradation of wood because ligninolytic enzymes are too large to penetrate into non- modified wood cell walls (Perez, Munoz-Dorado, De la Rubia, & Martinez, 2002). VP oxidizes both phenolic and non-phenolic compounds. This book chapter focus on the remediation and degradation potential of LiP, MnP and VP, whether alone or in combination, on different types of wastes which are toxic to environment if left untreated.

Pharmaceutical Industry

Many pharmacologically active substances are found in the aqueous environment, this include pharmaceuticals and personal care products and endocrine disrupting chemicals (Lienert, Güdel, & Escher, 2007). Pharmaceuticals and their metabolites can be found in the surface water, ground water and drinking water. Apart from this, many antibiotics are found in water samples like hospital wastewater, municipal wastewater, effluent of wastewater treatment plant, antibiotics industry wastewater, livestock farm mud and wastewater, surface water, underground water and drinking water (Wen, Jia, & Li, 2009). It has been found that low dose man-made chemicals are released into environment and they have potential to affect the endocrine system of living organisms present in ecosystems (Tyler, Jobling, & Sumpter, 1998). Some of the unprescribed drugs are used and after its intake, they are excreted with urine and feces either as active substance or metabolites (Marco-Urrea, Pérez-Trujillo, Vicent, & Caminal, 2009). Some of the pharmaceutical products, their nature and their remediation through peroxidases are given below:

- **Anti-Inflammatory Drugs:** Diclofenac (DFC) is taken or applied to reduce inflammation and as an analgesic in reducing pain in certain conditions. Naproxen is used to treat pain or inflammation caused by conditions such as arthritis, ankylosing spondylitis, tendinitis, bursitis, gout, or menstrual cramps (Rossi, 2013). Indometacin is used to reduce fever, pain, stiffness, and swelling from inflammation (Brayfield, 2014). Mefenamic acid is used to treat mild to moderate pain, including menstrual pain, and sometimes to prevent migraines associated with menstruation (Pringsheim, 2008). Ketoprofen is prescribed for arthritis-related inflammatory pains or severe toothaches that result in the inflammation of the gums and treatment of nerve pain such as sciatica, postherpetic neuralgia and referred pain for radiculopathy.
- **Antipyretic:** Acetaminophen is used to treat pain and fever.
- **Antibiotics:** Tetracycline, an antibiotic is used to treat a number of infections. This includes acne, cholera, brucellosis, plague, malaria, and syphilis. Oxytetracylcine (OTC), also an antibiotic, is used to treat acne, chest infection, eye infection, genital infection, pneumonia and bronchitis. Trimethoprim (TMP) is an antibiotic used mainly in the treatment of bladder infections, middle ear infections and travelers' diarrhea. Ciprofloxacin is an antibiotic used to treat number of bacte-

rial infections; this includes bone and joint infections, intra-abdominal infections, certain type of infectious diarrhea, respiratory tract infections, skin infections, typhoid fever, and urinary tract infections. Sulfamethoxazole (SMZ) is an antibiotic used for bacterial infections such as urinary tract infections, bronchitis, and prostatitis and is effective against both gram negative and positive bacteria.

- **Anti-Depressants:** Citalopram is used off-label to treat anxiety, panic disorder, dysthymia, premenstrual dysphoric disorder, body dysmorphic disorder and obsessive–compulsive disorder. Fluoxetine, is used for the treatment of major depressive disorder, obsessive–compulsive disorder (OCD), bulimia nervosa, panic disorder, and premenstrual dysphoric disorder.
- **Other Drugs:** Fenofibrate is used to reduce cholesterol levels in people at risk of cardiovascular disease; it reduces both low-density lipoprotein (LDL) and very low density lipoprotein (VLDL) levels, as well as increases high-density lipoprotein (HDL) levels and reduces triglyceride levels (Yang & Keating, 2009). Bezafibrate is a fibrate drug used as a lipid-lowering agent to treat hyperlipidaemia. It helps to lower LDL cholesterol and triglyceride in the blood, and increase HDL. Diazepam is used to treat anxiety, alcohol withdrawal syndrome, benzodiazepine withdrawal syndrome, muscle spasms, seizures, trouble sleeping, and restless legs syndrome (Calcaterra & Barrow, 2014). Carbamazepine (CBZ) is an anticonvulsant and mood- stabilizing drug used in the treatment of epilepsy, bipolar disorder and trigeminal neuralgia (Ceron-Litvoc, Soares, Geddes, Litvoc, & De Lima, 2009).

Remediation of the Drugs

Zhang and Geißen (2012) reported that LiP from *P. chrysosporium* degraded diclofenac (DFC) at pH3.0–4.5 (24 ppm H_2O_2) and at 3–24 ppm H_2O_2 (pH 4.0) and the degradation efficiency of carbamazepine (CBZ) was mostly below 10%. However, the gradual addition of H_2O_2 slightly elevated the degradation efficiency of CBZ to 15%, but the addition of veratryl alcohol and the high temperature (30°C) did not enhance the degradation of CBZ and negative effects were not found on the degradation of DFC.

Wen et al (2009) observed that lignin peroxidase prepared from *Phanerochaete chrysosporium* tetracycline (TC) and oxytetracycline (OTC) were removed rapidly by LiP in the reaction system. After 5 min about 95% of the initial TC and OTC were converted and after 30 min, only traces of TC and OTC (<0.25 mg L1) were detectable. In the control tests by using either heat-inactivated LiP or H_2O_2 alone, TC and OTC concentration did not show any decrease throughout the entire experiment (90 min), which proves that the removal of TC and OTC was enzymatic. The degradation of TC and OTC by LiP was pH and temperature dependent and was enhanced by increasing the concentration of veratryl alcohol (VA) and initial H_2O_2. The optimized degradation conditions were determined as pH 4.2, 37°C, 2 mM VA, 0.4 mM H_2O_2.

Touahar et al (2014) reported that combined effect of VP from *Bjerkandera adusta* and glucose oxidase (GOD) from *Aspergillus niger* was responsible for the removal of large number of pharmaceuticals. VP mediated treatments (VP with Mn (II) and H_2O_2, and VPGOD with Mn (II) and glucose) helped in the complete or high percentage of removal of pharmaceuticals like acetaminophen, naproxen, diclofenac, indometacin and mefenamic acid. VP-GOD pair also removed fenofibrate, ketoprofen and bezafibrate by 43%, 44% and 59%, respectively. However, this pair decreased contents of some pharmaceutical products in very low percentage like caffeine (16%), carbamazepine (20%), diazepam(18%), trimethoprim(14%) and ciprofloxacin (5%) only. They also reported the effect of VP alone on these products and the results obtained are carbamazepine (54%), ketoprofen (54%), trimethoprim (42%) and ciprofloxacin (28%). The limitation of this combination was the required presence of manganese sulfate and hydrogen peroxide, therefore they used the combination of laccase (from *Trametes versicolor*), VP and GOD. This combination gave better results for some products which were earlier not removed by VP alone or VP-GOD pair. For acetaminophen, indometacin and diclofenac near complete removal was achieved by the combination of the three enzymes and caffeine by 41%, diazepam by 32% and ciprofloxacin by 60%. Significant decrease by more than 20% was observed for ketoprofen, naproxen and mefenamic acid using the combination of three enzymes.

Eibes et al (2011) reported that versatile peroxidase (VP) from the ligninolytic fungus *Bjerkandera adusta* was responsible for the oxidation of pharmaceuticals belonging to medicinal classes like antibiotic, anti-inflammatory, anti-depressants, tranquilizer and endocrine disrupting chemicals. The degradation of sulfamethoxazole (antibiotic) after treatment with VP was found to be around 80% after 7 h with an average degradation rate of 0.24 mgl-1h-1. For citalopram (CTL) and fluoxetine (FLX) (anti-depressants) the degradation achieved was 18% and 10%, respectively, after enzymatic treatment. For diclofenac (DCF) and naproxen (NPX) (anti-inflammatories), the degradation was 100% and 55%, respectively with 200 Ul-1 of VP. A higher initial activity of VP at 1,000 Ul-1 led to 80% removal of NPX in 7 h. For DCF, low level of VP was checked at 10 Ul-1 VP and 1M H_2O_2 and total degradation of DCF occurred after only 25 minutes.

Pleurotus sp. P1, *Pleurotus ostreatus* and an unidentified basidiomycete, named as BNI helped in 47% removal of carbamazepine (CBZ) after 28 days of incubation, with 42% attributable to biodegradation. BNI was grown in the presence of CBZ and enzyme activity for LiP and MnP were evaluated. The maximum activity of LiP reached on 14th day with 663 U L-1 activity, while for MnP, it was detected maximum on 21st day with 628 U L-1 activity (Santosa, Grossmana, Sartorattob, Ponezib, & Durrant, 2012).

Endocrine Disrupting Chemicals (EDCs)

A large number of data has shown the hormone like effects of anthropogenic chemicals in the environment. These hormones when released into environment causes hormonal disruptions in the populations present in that ecosystem. These EDCs act as hormone agonist or antagonist and disrupt the synthesis of endogenous hormones or hormone receptors. These EDCs disturb the synthesis, secretion, transport, binding, action and elimination of the endogenous hormones which are responsible for maintaining homeostasis, reproduction, development and integrity in living organisms and their progeny (Cabana, 2007).

Eibes et al (2011) reported the enzymatic oxidations of E1, E2 (natural estrogens) and EE2 (17α-Ethynylestradiol, synthetic estrogen). Different experimental conditions were analyzed for E2 and EE2: (i) VP activity between 10 and 50 U l-1 and (ii) H_2O_2 ranging from 1 to 5 lM min-1and 1mM malonate, which remained constant in all experiments. In 15 minutes or less, complete degradation of the three compounds was obtained. Average degradation rates ranged from 12 mg l-1 h-1 (when 1 lM H_2O_2 min-1 was used) to 35 mg l-1 h-1 (E2 with 55 U l-1 VP and 5 lM H_2O_2 min-1). Both variables, H_2O_2 and VP concentration, had a positive effect on the oxidation in the evaluated range. Increasing their concentration led to the highest kinetic coefficients. In this way, almost complete degradation would be probably achieved immediately by using higher VP concentrations. Nevertheless, this system can be considered highly efficient, since almost complete degradation of all estrogens was obtained after 5–10 min under minimal requirement conditions. However, Santos et al made use of three WRF, *Pleurotus* sp. P1, *Pleurotus ostreatus* and an unidentified basidiomycete, named as BNI, for the oxidation of EE2. All the three strains were capable of degrading EE2. *Pleurotus* sp. P1 was grown in the presence of EE2 and the enzyme activity was evaluated in every 2 days over a period of 6 days. The maximum activity of MnP of 5122 UL-1 was found after 6 days of incubation and high levels of MnP was detected on 4th and 6th day (Santosa, Grossmana, Sartorattob, Ponezib, & Durrant, 2012).

Food Industry

In the last decades, large-scale food industries have increased over the worldwide. As food industries are increasing, their processing leads to the loss or wastage of food. Food waste is a global problem. It is defined as any uneaten food or food preparation residues from residences or commercial establishments. The amount of food waste is expressed in terms of billions of dollars. Each year food worth around billions of dollars is wasted. The food waste ends at landfill sites, where it converted into methane, carbon dioxide and other greenhouse gasses by anaerobic digestion, which leads to climate changes. Food waste pollution affects our habitat and each of us by causing serious health problems by polluted water. On average, in US more than 70 million cases of food-borne illnesses occur in every year, resulting in approximately 5,000 deaths per year (Melikoglu, 2013; Beede & Bloom 1995; Heller & Keoleian, 2003; Kantor, 1997). White rot fungus grow on wide variety of lignocellulosic wastes, upon which the enzymes secreted by mushrooms act which helps in digesting the surrounding waste and which in turn provides mushroom nourishment, nutrients and organic matter (Agarwal, Kushwaha, Verma, & Singh, 2017). Enzymes are specific biological catalyst in bioremediation processes and are made in order to protect the environment from the damage caused by food waste. Cellulose and lignin is major constituent of plant biomass as food waste. Each polymer is degraded by a variety of enzymes (laccase, LiP and MnP) produced by white rot fungus (Perez, Munoz-Dorado, De la Rubia, & Martinez, 2002; Baldrian, 2005). Crude lignin peroxidases, although believed to catalyse the partial lignin digestion in food waste (Hammel & Moen, 1991).

It is investigated that *Pleurotus* species that contains enzymatic complex system, includes peroxidases which is used for the biodegradation of viticulture wastes. These residues through solid state fermentation by *Pleurotus* species is used to produce food for human consumption (Nchez, Ysunza, Garciäa, & Esqueda, 2002). During olive oil production, huge amount of by-product is released as Olive mill wastewaters (OMW) in Mediterranean area. The dark liquid coloured waste, contains water (83–92%), minerals (1–2%) and organic substances(4–16%), such as phenols, sugars, organic acids, polyalcohols, colloids, tannins and lipids (Chiacchierini, Restuccia, & Vinci, 2004). OMW were shown to increase phytotoxicity thus rendering biodegradability very difficult and non-usable (Ramos-Cormenzana, Monteoliva-Sanchez, & Lopez, 1995). Some researcher using manganese peroxidase (MnP) from the white-rot fungus *Panus tigrinus*, studied over OMW and observed a decreasing content of polyphenols in the food waste (Fenice, 2003).

Vahabzadeh et al (2004) reported the decolorization of molasses wastewater (MWW) from an ethanolic fermentation plant by *Phanerochaete chrysosporium.* By incubating MWW with the spores (2.5 – 106/ml), about 75% of the decolorization takes place on the 5th day of the incubation. The colour removal was related to the activity of peroxidase enzyme of *Phanerochaete chrysosporium,* with LiP activity of 185 U/l and MnP activity with 25 U/l. Some operating variables were also used to see their effect on decolorization as well as the enzyme activity. MnP was highest with enzyme activity 25 U/l, when Mn (II) was added to the diluted MWW.

Songulashvili et al (2006) reported that *Phellinus robustus* produced manganese peroxidase (MnP) (1,000–11,300 U l–1) during the fermentation of seven food wastes (Corn bran, Chicken feathers, Wheat bran, Kiwi fruits, Banana peels, Mandarin peels and Ethanol production residue).

Industry Waste

Large amount of toxic organo-pollutants are produced, which contaminate soil, water, ground water and air. Different types of wastes from industry are produced such as waste water, phenols, hydrocarbons, insecticides, paints, dyes and xenobiotics. White rot fungi have ability to mineralize lignin by using extracellular enzymes. Basidiomycetes fungi *Phanerochaete chrysosporium* secretes two heme peroxidases i.e. lignin peroxidase (EC 1.11.1.7) and manganese peroxidase (EC 1.11.1.7) (Gold & Alic, 1993; Hammel, 1993). White-rot fungi (WRP) produce oxidases enzymes which are extracellular in nature with many applications. These enzymes are involved in the degradation of lignocellulosic substrates and various xenobiotic compounds and toxic dyes (Wesenberg, 2003). Researchers studied that these enzymes had a prospective biotechnical applications in biopulping, biobleaching, and soil bioremediation (Aust, 1990). WRP have capacities to produce polymeric products by removing xenobiotic substances and make them useful for bioremediation. However, studies on *P. chrysosporium* are not suitable for large-scale production system for LiP and MnP proteins but researchers have been attempt to optimize LiP and MnP production (Feijoo, Dosoretz, & Lema, 1995).

Removal of Phenolic Contaminant and Related Compounds

Nowadays, industrial pollution has been a leading factor causing the degradation of the environment around us, affecting the water quality by which the human health related issues arises. The conventional processes for phenol removal are divided into three categories: 1.) biological; 2.) physical; and 3.) chemical processes. Biological treatment is used for removal of phenolic compounds due to the more availability of microorganisms which degrade organic compounds (Marrot, Barrios-Martinez, Moulin, & Roche, 2006). A pathway for degradation of aromatic compounds is to dehydroxylate the benzene ring to form catechol derivatives and after that ortho/meta oxidative cleavage takes place to open the aromatic ring (Melo, Kholi, Patwardhan, & D'Souza, 2005). Moreover, biological treatment has several limitations like prolonged start-up for microbial acclimatization (Firozjaee, Najafpour, Asgari, & Khavarpour, 2012) sluggish rates of microbial degradation because of the pollutant toxicity (Steiert & Crawford, 1985) and the inhibition of microbial growth due to high concentrations of phenolic compounds (Gernjak, Krutzler, Glaser, Malato, Caceres, Bauer, & Fernandez-Alba, 2003).

Physical process for removal of phenolic contaminants is based upon the adsorption onto the surface of activated carbon. It is one of the most widely used and efficient techniques for treating high concentrations and low volumes of phenolic water waste (Tan, Ahmad,& Hameed, 2009). A number of attempts have been done from agricultural by-products such as palm seed coat (Rengaraj, Moon, Sivabalan, Arabindoo, & Murugesan, 2002), rice husk ash (Kermani, Pourmoghaddas, Bina, & Khazaei, 2006), bagasse ash and wood charcoal (Mukherjee, Sunil Kumar, Misra, & Fan, 2007) and oil palm shell (Mukherjee, Sunil Kumar, Misra, & Fan, 2007) with hopeful results. By the help of chemical process, phenol can also be removed from the water wastes. The ozone treatment of phenol-containing water has been widely studied with high removal rate and ozone mass transfer (Chedeville, Debacq, Almanza, & Porte, 2007; Chedeville, Debacq, & Porte, 2009).

Decolorization of Synthetic Dye

A large amount of dyes are produced yearly with a multiplicity of color and chemical structure. These dyes are used for different purpose such as in textile dyeing, paper printing, colour photography, and in petroleum products (Kaur, Kumar, Garg, & Kaur, 2015). When these synthetic dyes are released into industrial effluents, they cause environmental pollution (Ong, Keng, Lee, Ha, & Hung, 2011; Singh & Srivastava, 2016). Many synthetic dyes are not easy to remove by normal waste water treatment method. Currently available methods, such as reverse osmosis, chemical oxidation, and adsorption, have some disadvantages, like they are less efficient, limited applicability and high cost rate. Therefore, biodegradation of wastewater effluents containing dye solution is now growing in the degradation of dyes, since this method is less expensive. Researcher studied the decolorization of different azo dyes by *Phanerochaete chrysosporium* RP 78 under optimized conditions (Ghasemi, Tabandeh, Bambai, & Rao, 2010). An edible fungi *Pleurotus ostreatus* produced an extracellular peroxidase to decolorize remazol brilliant blue and other different groups including triarylm ethane, heterocyclic azo and polymeric dyes. Decolorization capacity of Bromophenol blue was more in comparison to methylene blue and toluidine blue O (Shin & Kim, 1997). Lignin peroxidase (LiP) from the white rot fungus has the ability to decolourize DBGLL dye by their asymmetric cleavage (Umesh, Vishal, Amar, & Sanjay, 2008). In textile industries, indigo dye is commonly used for cellulosic cotton fibres and during textile processing many chemical effluents containing this dye is discharged and thereby causing serious environmental problems. Comparative study using many basidiomycete fungi was used to analyze the decolourization of indigo dye, and it was observed that *Phellinus gilus* decolorized by 100%, *Pleurotus sajor-caju* by 94%, *Pycnoporus sanguineus*by 91% and *Phanerochaete chrysosporium* by 75% (Doralice & Regina, 2001). Another researcher analyzed the decolorization of many synthetic dyes, such as Amido black, congo red, trypan blue, methyl green and remazol brilliant blue R (RBB), by white rot fungus *P. pulmonarius* (Tychanowicz, Zilly, de Souza, & Peralta, 2001). Manganese peroxidase (MnP) and manganese-independent peroxidase (MIP) produced from the white rot fungus (WRF) *Pleurotus ostreatus* decolorized sulfonphthalein (SP) dyes (Shrivastava, Christian, & Vyas, 2005). Edible mushroom *Pleurotus ostreatus* produced an extracellular peroxidase that can decolorize Bromophenol blue (98%), while methylene blue and toluidine blue O were decolorized by 10% (Bansal and Kanwar, 2013). White-rot fungus *Phanerochaete chrysosporium* decolorize methylene blue (MB) dye up to 14-40% by lignin peroxidase (LiP) enzyme in sludge sewage treatment plant (STP) (Alama, Mansora, & Jalal, 2009). Ligninolytic enzyme manganese peroxidase (MnP) degrades azo dyes up to 90% (Mielgo, Lopez, Moreira, Feijoo, & Lema, 2003).

Degradation of Polychlorinated Biphenyl (PCB) Pesticides

PCBs are a family of 209 compounds called as congeners and is commercially produced as Aroclorsvia chlorination of biphenyl (Aroclor, 1956; Voogt & Brinkman, 1989). PCBs are a component of crude oil, creosote, and coal (Harayama, 1997). Peroxidases extracted from some fungal species have the potential to transform several pesticides into harmless types. Peroxidases and phenol oxidases react with PAH's and transform them to less toxic as well as easy to degrade. PAHs are oxidized by peroxidases such as lignin (Weber, Gaus, & Tysklind, 2008) and manganese peroxidase (Harford-Cross, Carmichael, Allan, England, Rouch, & Wong, 2000). *Phanerochaete chyrosporium* having peroxidase activity is shown to detoxify herbicides such as more toxic atrazine to less toxic compounds and would certainly find many potential applications in agriculture (Rob, Ball, Tuncer, & Wilson, 1997).

Removal of Endocrine Disruptive Chemicals (EDC)

EDCs are a group of compounds, which are able to act as agonists or antagonists of hormones because of their chemical structure. They inhibit the synthesis, secretion, transport, binding, action and elimination of the endogenous hormones (Cabana, Jones, & Agathos, 2007). The presence of EDCs in the aquatic systems showed a major environmental issue (Ahel, Giger, & Koch, 1994; Staples, Dorn, Klecka, O'Block, & Harris, 1998). The occurrence and distribution of these phenolic EDCs have been present in drinking water (Wang, Ying, Chen, Zhang, Zhao, Lai, Chen, & Tao, 2012; Wang, Ying, Zhao, Liu, Yang, Zhou, Tao, & Su, 2011). 10 U/mL of manganese peroxidase from *Pleurotus ostreatus* eradicated 0.4mM bisphenol in 1 h (Huang & Weber 2005). Peroxidases are very valuable in degradation of different environmental pollutants such as chloroanilines and polycyclic aromatic hydrocarbons (Renner, 1980).

White rot fungus *Phanerochaete sordid* produces manganese peroxidase (MnP), which decreases the level of endocrine-disrupting genistein containing media up to 93% (Tamagawa, Hirai, Kawai, & Nishida, 2015). Another researcher also investigated that LiP of *Phanerochaete sordid* YK-624 decreases level of five bisphenol A (BPA) after a 24-h treatment (Wang, Ying, Chen, Zhang, Zhao, Lai, Chen, & Tao, 2012). Bisphenol A is used as a common EDC in plastics and epoxy resins production. Many oxidoreductases present in white rot fungus have potential to remediate EDC such as manganese peroxidase, lignin peroxidase and laccases (Hain & Qayyum, 2013).

Degradation of Chlorinated Alkanes, Alkenes and Insecticide

Tri Chlorinated Ethanol is mineralized by *P. chrysosporium* cultures which are grown aerobically (Marco-Urrea, Pérez-Trujillo, Vicent, & Caminal, 2009). Lignin peroxidase (LiP) of *P. chrysosporium* in the presence of tertiary alcohol, H_2O_2, and EDTA plays an essential role in the biodegradation of the corresponding reduced chlorinated radicals (Gomez-Toribio, Garcıa-Martın, Martınez, Martınez, & Guillen, 2009). DDT (1,1,1-trichloro-2,2-bis(4-chlorophenyl)(ethane) was one of the first synthetic pesticides to gain wide acceptance. It has been used for pest controlling agriculture and forestry worldwide since the 1940s. DDT and its metabolites are highly ecotoxic to higher organisms. Therefore, it is important to develop effective methods of remediation (Agarwal & Singh, 2015). DDT is a chlorinated dioxin which are mineralized by the white rot fungi *P. chrysosporium*, *P. ostreatus*, *T. versicolor* and *Phellinus weirii* (Purnomo, Mori, Kamei, Nishii, & Kondo, 2010).

CONCLUSION

The present book chapter shows that the peroxidase enzymes of white rot fungus play an important role in the degradation of toxic compounds, which are discharged from different chemical and agricultural industries, which ultimately is found on the river or on soil, which in turn causes environmental issues along with severe human health issues. The peroxidase enzymes sometimes alone or along with other operating variables help degrade these wastes. Several factors help in the degradation like temperature, pH, substrate, ionic strength, which also decide the rate of mechanism of the substrates and enzymes. These enzymes degrade high molecular structured wastes in to low molecular weight compounds which can be easily be degraded, without harming our environment. Therefore, focus should be on the use of biological process for the removal of toxic compounds, because this method is cost effective and is environment friendly.

REFERENCES

Agarwal, S., Kushwaha, A., Verma, V., & Singh, M. P. (2017). Nutritional attributes of *Pleurotus* mushroom. In V. V. Singh & A. K. Singh (Eds.), *Incredible World of Biotechnology, M.P* (pp. 13–24). New York, USA: Nova Science Publishers.

Agarwal, S., & Singh, M. P. (2015). Oyster Mushroom: An Effective Tool of Bioremediation and Sustainable Development. In Insight. *International Journal of Sciences*, *2*, 14–25.

Ahel, M., Giger, W., & Koch, M. (1994). Behaviour of alkylphenolpolyethoxylate surfactants in the aquatic environment I Occurrence and transformation in sewage treatment. *Water Research*, *28*(5), 1131–1142. doi:10.1016/0043-1354(94)90200-3

Alama, M. Z., Mansora, M. F., & Jalal, K. C. A. (2009). Optimization of decolorization of methylene blue by lignin peroxidase enzyme produced from sewage sludge with *Phanerocheate chrysosporium*. *Journal of Hazardous Materials*, *162*(2-3), 708–715. doi:10.1016/j.jhazmat.2008.05.085 PMID:18599210

Aroclor. (1956). Advertisement by Monsanto Chemical Company. *Chem. Eng.*, 386.

Aust, S. D. (1990). Degradation of environmental pollutants by *Phanerochaete chrysosporium*. *Microbial Ecology*, *20*(1), 197–209. doi:10.1007/BF02543877 PMID:24193974

Baldrian, P., Valášková, V., Merhautová, V., & Gabriel, J. (2005). Degradation of lignocellulose by *Pleurotus ostreatus* in the presence of copper, manganese, lead and zinc. *Research in Microbiology*, *156*(5-6), 670–676. doi:10.1016/j.resmic.2005.03.007 PMID:15921894

Bansal, N., & Kanwar, S. S. (2013). Peroxidase(s) in Environment Protection. *The Scientific World Journal*, *2013*, 1–9. doi:10.1155/2013/714639 PMID:24453894

Beede, D. N., & Bloom, D. E. (1995). The Economics of Municipal Solid Waste. *The World Bank Research Observer*, *10*(2), 113–150. doi:10.1093/wbro/10.2.113

Brayfield, A. (Ed.). (2014). *"Indometacin". Martindale: The Complete Drug Reference*. London, UK: Pharmaceutical Press.

Cabana, H., Jones, J. P., & Agathos, S. N. (2007). Elimination of Endocrine Disrupting Chemicals using White Rot Fungi and their Lignin Modifying Enzymes: A Review. *Engineering in Life Sciences*, *7*(5), 429–456. doi:10.1002/elsc.200700017

Calcaterra, N. E., & Barrow, J. C. (2014). Classics in chemical neuroscience: Diazepam (valium). *ACS Chemical Neuroscience*, *5*(4), 253–260. doi:10.1021/cn5000056 PMID:24552479

Ceron-Litvoc, D., Soares, B. G., Geddes, J., Litvoc, J., & De Lima, M. S. (2009). Comparison of carbamazepine and lithium in treatment of bipolar disorder, a systematic review of randomized controlled trials. *Human Psychopharmacology*, *24*(1), 19–28. doi:10.1002/hup.990 PMID:19053079

Chedeville, O., Debacq, M., Almanza, M. F., & Porte, C. (2007). Use of an ejector for phenol containing water treatment by ozonation. *Separation and Purification Technology*, *57*(2), 201–208. doi:10.1016/j.seppur.2007.04.004

Chedeville, O., Debacq, M., & Porte, C. (2009). Removal of phenolic compounds present in olive mill wastewaters by ozonation. *Desalination*, *249*(2), 865–869. doi:10.1016/j.desal.2009.04.014

Chiacchierini, E., Restuccia, D., & Vinci, G. (2004). Bioremediation of Food Industry Effluents: Recent Applications of Free and Immobilised Polyphenoloxidases. *Food Science & Technology International*, *10*(6), 373–382. doi:10.1177/1082013204049388

Doralice, S. L. B., & Regina, T. R. M. (2001). Decolorization of textile indigo dye by ligninolytic fungi. *Journal of Biotechnology*, *89*(2-3), 141–145. doi:10.1016/S0168-1656(01)00304-2 PMID:11500207

Eibes, G., Debernardi, G., Feijoo, G., Moreira, M. T., & Lema, J. M. (2011). Oxidation of pharmaceutically active compounds by a ligninolytic fungal peroxidase. *Biodegradation*, *22*(3), 539–550. doi:10.100710532-010-9426-0 PMID:20972884

Feijoo, G., Dosoretz, C., & Lema, J. M. (1995). Production of lignin-peroxidase by *Phanerochaete chrysosporium* in a packed bed bioreactor operated in semicontinuous mode. *Journal of Biotechnology*, *42*(3), 247–253. doi:10.1016/0168-1656(95)00085-5

Fenice, M., Sermanni, G. G., Federici, F., & D'Annibale, A. (2003). Submerged and solid-state production of laccase and Mn-peroxidase by *Panustigrinus* on olive mill wastewater-based media. *Journal of Biotechnology*, *100*(1), 77–85. doi:10.1016/S0168-1656(02)00241-9 PMID:12413788

Firozjaee, T. T., Najafpour, G. D., Asgari, A., & Khavarpour, M. (2013). Biological treatment of phenolicwastewater in an anaerobiccontinuous stirred tank reactor. *Chemical Industry & Chemical Engineering Quarterly*, *19*(2), 173–179. doi:10.2298/CICEQ120216052F

Gernjak, W., Krutzler, T., Glaser, A., Malato, S., Caceres, T., Bauer, R., & Fernandez-Alba, A. R. (2003). Photo-Fenton treatment of water containing natural phenolic pollutants. *Chemosphere*, *50*(1), 71–78. doi:10.1016/S0045-6535(02)00403-4 PMID:12656231

Ghasemi, F., Tabandeh, F., Bambai, B., & Rao, K. R. S. (2010). Decolorization of different azo dyes by *Phanerochaete chrysosporium* RP78 under optimal condition. *International Journal of Environmental Science and Technology*, *7*(3), 457–464. doi:10.1007/BF03326155

Glenn, J. K., Lakshmi, A., & Gold, M. H. (1986). Mn(II) oxidation is the principal function of the extracellular Mn-Peroxidase from *Phanerochaete chrysosporium*. *Archives of Biochemistry and Biophysics*, *251*(2), 688–696. doi:10.1016/0003-9861(86)90378-4 PMID:3800395

Gold, M. H., & Alic, M. (1993). Molecular biology of the lignin-degrading Basidiomycete *Phanerochaete chrysosporium*. *Microbiological Reviews*, *57*, 605–622. PMID:8246842

Gomez-Toribio, V., Garcıa-Martın, A. B., Martınez, M. J., Martınez, A. T., & Guillen, F. (2009). Enhancing the production of hydroxyl radicals by *Pleurotus eryngii* via quinone redox cycling for pollutant removal. *Applied and Environmental Microbiology*, *75*(12), 3954–3962. doi:10.1128/AEM.02138-08 PMID:19376890

Hain, Q., & Qayyum, S. (2013). Biological and enzymatic treatment of bisphenolA and other endocrine disrupting compounds: A review. *Critical Reviews in Biotechnology*, *33*(3), 260–292. doi:10.3109/07388551.2012.694409 PMID:22712546

Hammel, K. E., & Cullen, D. (2008). Role of fungal peroxidases in biological ligninolysis. *Current Opinion in Plant Biology, 11*(3), 349–355. doi:10.1016/j.pbi.2008.02.003 PMID:18359268

Hammel, K. E., Jensen, K. A. Jr, Mozuch, M. D., Landucci, L. L., Tien, M., & Pease, E. A. (1993). Ligninolysis by a purified lignin peroxidase. *The Journal of Biological Chemistry, 268*, 12274–12281. PMID:8509364

Hammel, K. E., & Moen, M. A. (1991). Depolymerization of a synthetic lignin in vitro by lignin peroxidise. *Enzyme and Microbial Technology*, 13.

Harayama, S. (1997). Polycyclic aromatic hydrocarbon bioremediation design. *Current Opinion in Biotechnology, 8*(3), 268–273. doi:10.1016/S0958-1669(97)80002-X PMID:9206005

Harford-Cross, C. F., Carmichael, A. B., Allan, F. K., England, P. A., Rouch, D. A., & Wong, L. (2000). Protein engineering of cytochrome P458 (cam) (CYP101) for the oxidation of polycyclic aromatic hydrocarbons. *Protein Engineering, 13*(2), 121–128. doi:10.1093/protein/13.2.121 PMID:10708651

Heberer, T. (2002). Occurrence, fate and removal of pharmaceutical residues in the aquatic environment: A review of recent research data. *Toxicology Letters, 131*(1-2), 5–17. doi:10.1016/S0378-4274(02)00041-3 PMID:11988354

Heller, M. C., & Keoleian, G. A. (2003). Assessing the sustainability of the US food system: A life cycle perspective. *Agricultural Systems, 76*(3), 1007–1041. doi:10.1016/S0308-521X(02)00027-6

Huang, Q., & Weber, W. J. (2005). Transformation and removal of bisphenol A from aqueous phase via peroxidase-mediated oxidative coupling reactions: Efficacy, products, and pathways. *Environmental Science & Technology, 39*(16), 6029–6036. doi:10.1021/es050036x PMID:16173560

Kantor, L.S., Lipton, K., Manchester, A. and Oliveira V. (1997). Estimating and Addressing America's Food Losses. *Food Review*, 1-12.

Kaur, B., Kumar, B., Garg, N., & Kaur, N. (2015). Statistical optimization of conditions for decolorization of synthetic dyes by *Cordyceps militaris* MTCC 3936 Using RSM. *BioMed Research International, 2015*, 1–17. doi:10.1155/2015/536745 PMID:25722980

Kermani, M., Pourmoghaddas, H., Bina, B., & Khazaei, Z. (2006). Removal of phenol from aqueous solutions by rice husk ash and activated carbon. *Pakistan Journal of Biological Sciences, 9*(10), 1905–1910. doi:10.3923/pjbs.2006.1905.1910

Lienert, J., Güdel, K., & Escher, B. I. (2007). Screening method for ecotoxicological hazard assessment of 42 pharmaceuticals considering human metabolism and excretory routes. *Environmental Science & Technology, 41*(12), 4471–4478. doi:10.1021/es0627693 PMID:17626454

Marco-Urrea, E., Pérez-Trujillo, M., Vicent, T., & Caminal, G. (2009). Ability of white-rot fungi to remove selected pharmaceuticals and identification of degradation products of ibuprofen by *Trametes versicolor. Chemosphere, 74*(6), 765–772. doi:10.1016/j.chemosphere.2008.10.040 PMID:19062071

Marrot, B., Barrios-Martinez, A., Moulin, P., & Roche, N. (2006). Biodegradation of high phenol concentration by activated sludge in an immersed membrane bioreactor. *Biochemical Engineering Journal, 30*(2), 174–183. doi:10.1016/j.bej.2006.03.006

Mayer, A. M., & Staples, R. C. (2002). Laccase: New functions for an old enzyme. *Phytochemistry*, *60*(6), 551–565. doi:10.1016/S0031-9422(02)00171-1 PMID:12126701

Melikoglu, M., Lin, C. S. K., & Webb, C. (2013). Analysing global food waste problem: Pinpointing the facts and estimating the energy content. *Cent. Eur. J. Eng.*, *3*(2), 157–164.

Melo, J. S., Kholi, S., Patwardhan, A. W., & D'Souza, S. F. (2005). Effect of oxygen transfer limitations in phenol biodegradation. *Process Biochemistry*, *40*(2), 625–628. doi:10.1016/j.procbio.2004.01.049

Mielgo, I. C., Lopez, C., Moreira, M. T., Feijoo, G., & Lema, J. M. (2003). Oxidative Degradation of Azo Dyes by Manganese Peroxidase under Optimized Condition. *Biotechnology Progress*, *19*(2), 325–331. doi:10.1021/bp020136w PMID:12675567

Mukherjee, S., Sunil Kumar, S., Misra, A. K., & Fan, M. (2007). Removal of phenols from water environment by activated carbon,bagasse ash and wood charcoal. *Chemical Engineering Journal*, *129*(1-3), 133–142. doi:10.1016/j.cej.2006.10.030

Nchez, A. S., Ysunza, F., Garciäa, M. J. B., & Esqueda, M. (2002). Biodegradation of Viticulture Wastes by *Pleurotus*: A Source of Microbial and Human Food and Its Potential Use in Animal Feeding. *Journal of Agricultural and Food Chemistry*, 50. PMID:11958618

Ong, S. T., Keng, P. S., Lee, W. N., Ha, S. T., & Hung, W. T. (2011). Dye waste treatment. *Water (Basel)*, *3*(1), 157–176. doi:10.3390/w3010157

Perez, J., Munoz-Dorado, J., De la Rubia, T., & Martinez, J. (2002). Biodegradation and biological treatments of cellulose, hemicellulose and lignin: An overview. *International Microbiology*, *5*(2), 53–63. doi:10.100710123-002-0062-3 PMID:12180781

Pointing, S. B. (2001). Feasibility of bioremediation by white-rot fungi. *Applied Microbiology and Biotechnology*, *57*(1-2), 20–33. doi:10.1007002530100745 PMID:11693920

Pringsheim, T., Davenport, W. J., & Dodick, D. (2008). Acute Treatment and Prevention of Menstrually Related Migraine Headache: Evidence-Based. *Revista de Neurología*, *70*(17), 1555–1563. doi:10.1212/01.wnl.0000310638.54698.36 PMID:18427072

Purnomo, A. S., Mori, T., Kamei, I., Nishii, T., & Kondo, R. (2010). Application of mushroom waste medium from *Pleurotus ostreatus* for bioremediation of DDT-contaminated soil. *International Biodeterioration & Biodegradation*, *64*(5), 397–402. doi:10.1016/j.ibiod.2010.04.007

Ramos-Cormenzana, A., Monteoliva-Sanchez, M. and Lopez, M.J. (1995). Bioremediation of Alpechin. *Intonational Biodeterioration & Biodegradation*, 249-268.

Reddy, C. A. (1995). The potential for white-rot fungi in the treatment of pollutants. *Current Opinion in Biotechnology*, *6*(3), 320–328. doi:10.1016/0958-1669(95)80054-9

Rengaraj, S., Moon, S. H., Sivabalan, R., Arabindoo, B., & Murugesan, V. (2002). Agricultural solid waste for the removal of organics: Adsorption of phenol from water and wastewater by palm seed coat activated carbon. *Waste Management (New York, N.Y.)*, *22*(5), 543–548. doi:10.1016/S0956-053X(01)00016-2 PMID:12092764

Renner, G. (1980). Metabolic studies on pentachloronitrobenzene (PCNB) in rats. *Xenobiotica*, *7*(8), 537–550. doi:10.3109/00498258009033788 PMID:7445521

Rob, A., Ball, A. S., Tuncer, M., & Wilson, M. T. (1997). Novel Applications of Peroxidase. *Journal of Chemical Education*, *74*(2), 212. doi:10.1021/ed074p212

Rooklidge, S. J. (2004). Environmental antimicrobial contamination from terraccumulation and diffuse pollution pathways. *The Science of the Total Environment*, *325*(1-3), 1–13. doi:10.1016/j.scitotenv.2003.11.007 PMID:15144773

Rossi, S. (Ed.). (2013). *Australian Medicines Handbook*. Adelaide: The Australian Medicines Handbook Unit Trust.

Santosa, I. J. S., Grossmana, M. J., Sartorattob, A., Ponezib, A. N., & Durrant, L. R. (2012). Degradation of the Recalcitrant Pharmaceuticals Carbamazepine and 17α-Ethinylestradiol by Ligninolytic Fungi. *Chemical Engineering Transactions*, *27*, 169–174.

Shin, J. S., & Kim, B. G. (1997). Kinetic resolution of *α*-methylbenzylamine with *ω*-transaminase screened fromsoil microorganisms: Application of a biphasic system to overcome product inhibition. B*iotechnology and Bioengineering*, *55(2)*, 348–335. doi: PMID:18636493

Shrivastava, R., Christian, V., & Vyas, B. R. M. (2005). Enzymatic decolorization of sulfonphthalein dyes. *Enzyme and Microbial Technology*, *36*(2-3), 333–337. doi:10.1016/j.enzmictec.2004.09.004

Singh, M. P., & Srivastava, A. K. (2016). Decolorization of Synthetic Textile Dye and Enzymes Production by Improved Strains of *Pleurotus* Species. *Cellular and Molecular Biology*, *62*, 145. doi:10.4172/1165-158X.1000128

Songulashvili, G., Elisashvili, V., Wasser, S., Nevo, E., & Hadar, Y. (2006). Laccase and manganese peroxidase activities of *Phellinus robustus* and *Ganoderma adspersum* grown on food industry wastes in submerged fermentation. *Biotechnology Letters*, *28*(18), 1425–1429. doi:10.100710529-006-9109-4 PMID:16823599

Staples, C. A., Dorn, P. B., Klecka, G. M., O'Block, S. T., & Harris, L. R. (1998). A review of the environmental fate, effects, and exposures ofbisphenol A. *Chemosphere*, *36*(10), 2149–2173. doi:10.1016/S0045-6535(97)10133-3 PMID:9566294

Steiert, J. G., & Crawford, R. L. (1985). Microbial degradation of chlorinated phenols. *Trends in Biotechnology*, *3*(12), 300–305. doi:10.1016/0167-7799(85)90032-0

Tamagawa, Y., Hirai, H., Kawai, S., & Nishida, T. (2005). Removal of estrogenic activity of endocrine-disrupting genistein byligninolytic enzymes from white rot fungi. *FEMS Microbiology Letters*, *244*(1), 93–98. doi:10.1016/j.femsle.2005.01.023 PMID:15727826

Tan, I. A. W., Ahmad, A. L., & Hameed, B. H. (2009). Fixed-bed adsorption performance of oil palm shell-based activated carbon for removal of 2,4,6-trichlorophenol. *Bioresource Technology*, *100*(3), 1494–1496. doi:10.1016/j.biortech.2008.08.017 PMID:18809316

Tien, M., Kirk, T. K., Morgan, M. A., Maytield, M. B., Kuwahara, M., & Gold, M. H. (1988). Lignin Peroxidase of *Phanerochaete chrysosporium*. *Methods in Enzymology*, *161*, 238–249. doi:10.1016/0076-6879(88)61025-1

Torres, E., Bustos-Jaimes, I., & Le Borgne, S. (2003). Potential use of oxidative enzymes for the detoxification of organic pollutants. *Applied Catalysis B: Environmental*, *46*(1), 1–15. doi:10.1016/S0926-3373(03)00228-5

Touahar, I. E., Haroune, L., Ba, S., Bellenger, J. P., & Cabana, H. (2014). Characterization of combined cross-linked enzyme aggregates from laccase, versatile peroxidase and glucose oxidase, and their utilization for the elimination of pharmaceuticals. *The Science of the Total Environment*, *481*, 90–99. doi:10.1016/j.scitotenv.2014.01.132 PMID:24589758

Tychanowicz, G. K., Zilly, A., de Souza, C. G. M., & Peralta, R. M. (2001). Decolourisation of industrial dyes by solid-state cultures of *Pleurotus pulmonarius*. *Process Biochemistry*, *39*(7), 855–859. doi:10.1016/S0032-9592(03)00194-8

Tyler, C. R., Jobling, S., & Sumpter, J. P. (1998). Endocrine disruption in wildlife: A critical review of the evidence. *Critical Reviews in Toxicology*, *28*(4), 319–361. doi:10.1080/10408449891344236 PMID:9711432

Umesh, U. J., Vishal, V. D., Amar, A. T., & Sanjay, P. G. (2008). Decolorization of Direct Blue GLL with enhanced lignin peroxidase enzyme production in *Comamonas* sp UVS. *Journal of Chemical Technology and Biotechnology (Oxford, Oxfordshire)*, *84*, 126–132.

Vahabzadeh, F., Mehranian, M., & Saatari, A. R. (2004). Color removal ability of *Phanerochaete chrysosporium* in relation to lignin peroxidase and manganese peroxidase produced in molasses wastewater. *World Journal of Microbiology & Biotechnology*, *20*(8), 859–864. doi:10.100711274-004-9005-9

Voogt, D., & Brinkman, P. (1989). U. A. T. Production, properties and usage of polychlorinated biphenyls in Halogenated Biphenyls, Terphenyls, Naphthalenes, Dibenzodioxins, and Related Products. *Science*, 3–45.

Wang, J., Majima, N., Hirai, H., & Kawagishi, H. (2012). Effective Removal of Endocrine-Disrupting Compounds by Lignin Peroxidase from the White-Rot Fungus *Phanerochaete sordida* YK-624. *Current Microbiology*, *64*(3), 300–303. doi:10.100700284-011-0067-2 PMID:22203416

Wang, L., Ying, G. G., Chen, F., Zhang, L. J., Zhao, J. L., Lai, H. J., ... Tao, R. (2012). Monitoring of selected estrogenic compounds and estrogenic activity in surface water and sediment of the Yellow River in China using combined chemical and biological tools. *Environmental Pollution*, *165*, 241–249. doi:10.1016/j.envpol.2011.10.005 PMID:22015335

Wang, L., Ying, G. G., Zhao, J. L., Liu, S., Yang, B., Zhou, L. J., ... Su, H. C. (2011). Assessing estrogenic activity in surface water and sediment of the Liao Riversystem in northeast China using combined chemical and biological tools. *Environmental Pollution*, *159*(1), 148–156. doi:10.1016/j.envpol.2010.09.017 PMID:20951481

Weber, R., Gaus, C., Tysklind, M., Johnston, P., Forter, M., Hollert, H., ... Zennegg, M. (2008). Dioxin- and POP contaminated sites-contemporary and future relevance and challenges: Overview on background aims and scope of the series. *Environmental Science and Pollution Research International*, *15*(5), 363–393. doi:10.100711356-008-0024-1 PMID:18597132

Wen, X., Jia, Y., & Li, J. (2009). Degradation of tetracycline and oxytetracycline by crude lignin peroxidase prepared from *Phanerochaete chrysosporium* – A white rot fungus. *Chemosphere*, *75*(8), 1003–1007. doi:10.1016/j.chemosphere.2009.01.052 PMID:19232429

Wesenberg, D., Kyriakides, I., & Agathos, S. N. (2003). White-rot fungi and their enzymes for the treatment of industrial dye effluents. *Biotechnology Advances*, *22*(2), 161–187. doi:10.1016/j.biotechadv.2003.08.011 PMID:14623049

Wong, D. W. S. (2009). Structure and action mechanism of ligninolytic enzymes. *Applied Biochemistry and Biotechnology*, *157*(2), 174–209. doi:10.100712010-008-8279-z PMID:18581264

Yang, L. P., & Keating, G. M. (2009). Fenofibric acid: In combination therapy in the treatment of mixed dyslipidemia. *American Journal of Cardiovascular Drugs*, *9*(6), 401–409. doi:10.2165/11203920-000000000-00000 PMID:19929038

Zhang, Y., & Geißen, S. U. (2012). Elimination of carbamazepine in a non-sterile fungal bioreactor. *Bioresource Technology*, *112*, 221–227. doi:10.1016/j.biortech.2012.02.073 PMID:22409982

Chapter 13
Biotechnology of Microbial Xylanase:
Overview

Hooi Ling Ho
UCSI University, Malaysia

ABSTRACT

Xylanases are inducible enzymes responsible for the complete hydrolysis of xylan into xylose. Both solid state fermentation (SsF) and submerged fermentation (SmF) are used in the production of xylanase. SsF has become a popular approach due to its economic value. In fact, higher biomass and lower protein breakdown are among the factors involved in determining the production of xylanases in SsF. Agricultural extracts which are abundantly available in the environment such as rice bran and wheat bran are commonly used as the potential carbon source in xylanases production. Xylanase is indeed one of the valuable enzymes which show immense potential in vast industrial applications. The demand for xylanase is increasing because of its prodigious utilization in pulp and paper, bakery, food and beverage, detergents, textile, and animal feed. Xylanase has therefore become one of the important commercial enzymes in recent years.

INTRODUCTION

Xylanases are inducible enzymes which responsible for the complete hydrolysis of xylan into simpler compounds, mainly xylose (Gupta & Kar, 2009). Xylanases are genetically single chain glycoproteins with molecular weight of 6 kDa to 80 kDa. Xylanases are active between pH 4.5 to 6.5 from 40°C to 60°C. Xylanases are produced by numerous numbers of different fungi.Various strains of filamentous fungi such as *Aspergillus niger*, *Aspergillus oryzae* and *Trichoderma spp*have been reported to be the potent producers of xylanases. However, xylanases production is typically restricted to *Aspergillus spp* and *Trichoderma spp* in the industrial scale (Dietmar, Bernd, Kulbe, Walter, & Silvia, 1996). Meanwhile, *Aspergillus spp* are normally selected and optimised for xylanases production. Apart from xylanases, *Aspergillus spp* also produce huge variety of extracellular enzymes including amylase, cellulase and protease (Pandey, Nigam, Soccol, Soccol, Singh, & Mohan, 2000). Xylanase shows tremendous potential in many industrial processes especially in textiles, leather, detergents and baking (Bhatnagar, & Imelda-Joseph, 2010). Both solid state fermentation (SsF) and submerged fermentation (SmF) are used in the production

DOI: 10.4018/978-1-5225-5237-6.ch013

of fungal xylanase. SsF has become a popular approach to produce xylanase due to its economical value that does not involved complicated technology. Viniegra-Gonzalez et al. (2003) studied the comparison between SsF and SmF in terms of enzymes production. In fact, they identified out that higher biomass and lower protein breakdown were among the factors involved in determining the production of enzymes in SsF. Besides that, the utilization of inexpensive agricultural extracts in SsF is environmentally sound because besides providing sufficient nutrients as carbon source, it reduces pollutions to the surroundings. Hence, SsF is more economical compared to SmF. Indeed, SsF is in fact, an attractive and economical method for xylanases production especially for fungal cultivations. SsF produces higher enzymes productivity using lower operation and capital cost (Lonsane, Ghildyal, Budiatman, & Ramakrishna, 1985). Malaysia with abundant natural rainforest will be of great advantage in xylanase production using SmF and SsF. Natural resources and agricultural extracts which are abundantly available in the environment such as rice bran, wheat bran, palm kernel cake and soybean hulls are used as potential carbon sources in xylanases production under SmF and SsF (Pang & Ibrahim 2005).

Nonetheless, only a few studies on the optimization of medium formulation for the maximum microbial xylanases production have been conducted. The application of agricultural wastes as the carbon source for the industrial xylanases production are scarce and not comprehensively studied compared and reported in SmF and SsF. Being the simple, non-toxic and cost-effective carbon source to yield xylanases, the replacement of xylan as the substrate with agricultural extracts in SmF and SsF is of great interest particularly in industrial production. Agricultural extracts are the good alternative carbon source due to their similarity in the polymers structure of xylan, as a result, these lignocellulose residualsare suitable to use as the prime carbon source for xylanases production. Besides that, the lack of precise information of the optimum growth conditions on the microbial xylanases production in SmF and SsF also lead to the vast studies over the past few years. There are several crucial fermentation parameters including carbon source, temperature, pH medium and agitation speed used to elucidate and optimize the production of xylanases in SmF and SsF. These parameters are determined by the types of microorganism that yield xylanases. Precisely, suitable parameters allow the proper proliferation of microorganisms to produce high concentration of xylanases. Carbon source provides prerequisite nutrients for growth of microorganisms in SmF and SsF. In order to enhance the xylanases production, cheap but effective carbon source such as agricultural extract is generally added to supply as the prime nutrient. SsF has become a known interest to produce xylanases because of its economical process of using agricultural extracts. Thus, lower cost production of xylanases is easily achieved. Therefore, proper optimization of fermentation parameters would able to produce the desirable xylanases at the maximum level. Additionally, optimum pH medium and temperature possess huge positive impact on the growth of microorganisms and xylanases production. Indeed, optimal pH and temperature in agitated culture show greater xylanases activity compared to non-agitated culture. Nonetheless, further increase of the optimum agitation speed would cause irregular morphology of microorganisms that ultimately lead to xylanases interference. Nevertheless, continuous research and development efforts are being given to SsF to be as compatible as SmF, making it more practicable especially for industrial production.

All in all, xylanase is one of the valuable enzymes which show immense potential in both biotechnological and industrial applications. Notably, filamentous fungi of *Aspergillus spp* have always been the preference choice because they produce higher activity of xylanase than other fungi, yeast and bacteria. In the present day, the demand for xylanase is increasing because of its prodigious utilization in vast industries. Recently, there has been much concern and immense importance on xylanase due to its broad potential industrial applications especially pulp and paper industry. One of the foremost industrial application of xylanase is involved in chlorine-free bleaching process in pulp and paper industry whereby xylanase is added into the pulp to degrade xylan found within the lignin residuals. As a result, it allows easy bleaching process of the pulp and hence, it increases the brightness of paper. Notably, xylanase plays an effective pivotal role in pulp and paper industry due to its simplicity and economical properties. Other applications of xylanase are in bakery, textile, food and beverage, detergents, animal feed industry and so forth. Increasing applications of xylanase in various industries have developed xylanase to become more valuable. In fact, recent market trends revealed that xylanase took a major chunk of share up to 20% of the world enzyme market along with cellulase and pectinase. The technology of utilising xylanase has reached the stage where this enzyme has become indispensable. Nonetheless, high production costs and low yields of xylanase hinder its further potentials. As a result, they remain as the paramount problems for industrial applications especially in large industrial scale. In order for commercial applications, xylanase should be produced rapidly in huge quantity using simple and inexpensive substrate. Therefore, investigations on the ability of xylanase-producing fungal strains utilising inexpensive substrate have been extensively carried out.

In fact, much effort has been done on the optimisation of fermentation medium and growth conditions to further reduce the production costs of xylanase. There always continuously endeavors in enhancing and developing xylanase into a profitable enzyme besides with lower costs of production using different types of agricultural extracts at the optimal growth conditions, it also involves the recovery and purification of xylanase. Furthermore, the characterization study of xylanase is equally important where the understanding of its optimum parameters is necessary for its applications in various industries. As a result, step-wise manner was generally applied for the characterization study of purified xylanase. Characterization study using different parameters including temperature and pH stability of xylanase, effects of temperature, pH, substrate specificity, substrate concentration, incubation time, metal ion, chelating agent, detergent and organic solvent on xylanase activity was commonly elucidated. In this chapter, the elucidation of xylanase production from *Aspergillus brasiliensis* using wheat bran as the sole carbon source under the optimum growth conditions in SmF was performed followed by purification using a two-step column chromatography and characterization of the purified xylanase using step-wise manner. In conclusion, this chapter summarised the economical utilisation and recycling of wheat bran in the production of xylanase by *A. brasiliensis* under SmF followed by purification using column chromatography and step-wise manner of characterisation study. All these aspects are reviewed in this chapter.

XYLAN SUBSTRATE FOR XYLANASE ACTIVITY

Xylan or known as wood gum is a type of heterogeneous polysaccharide found in the cell wall of many plant species (Wong, Tan, & Saddler, 1988). Xylan is consisted of complicated structures of hemicelluloses with the backbone of β-1,4-linked xylopyranose subunits and its substituent of O-acetyl, α-L-arabinofuranosyl, α-1,2-glucoronic or 4-O-methylglucoronic acid which are accounted for approximately 15% to 30% of the total dry weight in wood component of angiosperms (Subramaniyan & Prema, 2002). Hemicellulose is also called as heteropolysaccharides comprised of low molecular weight polymers that consisted of 80 to 200 units degree of polymerisation (Sun, Tomkinson, Wang, & Xiao, 2000). Thus, it is also collectively viewed as the second most abundant biomass component in plants besides cellulose (Rifaat, Nagieb, & Ahmedv, 2005). Hemicellulose has broad applications because of its non-toxic biodegradable features. Apart from that, hemicellulose is bonded to other components such as cellulose, proteins, lignin and phenolic compounds by hydrogen, covalent, ionic bond as well as hydrophobic interaction (Peng, Bian, Peng, Guan, Xu, & Sun, 2012). Xylan provides structural support to the plants besides polysaccharides storage in the seeds for germination purposes. Xylan of the plant cell wall is greatly differed depending on their origins and molecules that attached to the xylan backbone. The degradation of xylan occurs when β-1,4-xylanase cleaves the polysaccharide backbone of xylan followed by β-xylosidases hydrolyzes xylo-oligosaccharides to xylose (Haltrich, Nidetzky, Kulbe, Steiner, & Zupancic, 1996). Xylan is categorized into homoxylans, glucuronoxylans, arabinoglucuronoxylans, arabinoxylans, glucuronoarabinoxylans and heteroxylans depending on its degree of substitution and classes of side groups (Girio, Fonseca, Carvalheiro, Duarte, Marques, & Bogel-Lukasik, 2010). Homoxylans are typically found in seaweeds whereas glucuronoxylans are found in hardwoods and arabinoglucuronoxylans in softwoods. Hardwood xylan such as O-acetyl-4-O-methylglucuronoxylans has higher degree of polymerization with about 150 to 200 than softwood xylan, arabino-4-O-methylglucuronoxylans which only about 70 to 130 (Kulkarni, Shendye, & Rao, 1999). On the other hand, arabinoxylans, glucuronoarabinoxylans and heteroxylans are particularly found in cereals. In addition, cereal xylan consists of D-glucuronic acid (Kulkarni, Shendye, & Rao, 1999). The naturally occurring lignocellulosic plant biomass such as sawdust, sugarcane bagasse, barley, wheat bran and rice bran are mainly consisted of three groups of polymers, which are cellulose, hemicellulose and lignin. Cellulose and hemicellulose are sugar rich raw materials which often used in fermentation process. Hemicellulose is a linear and heterogenous mixture of five different sugars consists of D-xylose, D-mannose, D-glucose, D-galactose and L-arabinose. Hemicellulose has shorter chains with its amorphous not a crystalline structure making it easier to hydrolyze than cellulose (Mussatto & Teixeira 2010).

In general, xylan is a yellow, water soluble, and gummy polysaccharide which is found in most of the plant cell wall. Hydrolysis of xylan is an important step to liberate valuable products, mainly xylose. Xylose can be further process into ethanol. Hydrolysis of xylan can be done by two different methods which are chemical hydrolysis and enzyme hydrolysis. Chemical hydrolysis of xylan is typically used in many industries but results in number of toxic and undesirable substances that are being produced during the process causes the pollution to the environment. In order to get rid of the problem, enzyme hydrolysis of xylan such as xylanase to hydrolyse xylan into xylose which is environmental friendly is more preferable (Ninawe & Kuhad 2005). Xylitol, a five-carbon sugar alcohol is one of the major end products of xylan and is produced by fermentation of xylose. Xylitol has been widely used as a natural food sweetener, sugar substitute for diabetics and also as a dental caries reducer (Mussatto & Teixeira 2010). In the past, the conversion of xylan into xylose was conventionally carried out using acid hydrolysis process. Nowadays, many studies considered xylan as one of the new substrates for the production of biofuels, pharmaceuticals and solvents.

XYLANASE (ENDO- β-1,4-XYLANASE)

Xylanase (endo-β-1,4-xylanase) is an extracellular enzyme that comprised of β-xylosidase, arabinofuranosidase and acetylxylanase (Krisana, Rutchadaporn, Jarupan, Lily Sutipa, & Kanyawim, 2005). These enzymes are found to be synthesised by eukaryotes and prokaryotes. However, the main producers are amongst fungi followed by bacteria. In general, xylanase has been categorized as glycosyl hydrolase Family F10 and G11 based on the amino acid sequence similarity and three-dimensional structure analysis. Predominantly, fungal xylanase is belonged to Family F10 because greater parts of its members are endo-β-xylanase which is larger with higher molecular mass approximately 35 kDa than Family G11 xylanase which is only about 20 kDa, hence they have better capability in cleaving the glycosidic bonds (Torronen, Harkki, & Rouvinen, 1994). Torronen et al. (1994) reviewed that the 3D structures of Family G11 xylanase has the overall shape of a "right hand". A wide variety of fungi and bacteria produce extra-cellular xylanase which act on hemicellulosic material to liberate xylose. The main highlight of xylanase is its efficiency in the degradation of xylan into several xylose units by cleaving β-1,4-glycosidic bonds of xylan backbone in a random manner within the lignocellulosic materials (Haltrich, Nidetzky, Kulbe, Steiner, & Zupancic, 1996). Additionally for a complete degradation of xylan, β-xylosidase proceeds cooperatively by hydrolyzing the xylooligosaccharides into xylose. In this respect, the end product such as xylose is being utilized commercially in vast industries worldwide. Xylanase is the advantageous choice for the degradation of xylan because it has high specificity with slight reaction state, insignificant loss of substrate and side product output (Kulkarni, Shendye, & Rao, 1999). Studies have shown that the production of xylanase, an inducible enzyme is influenced by either pure xylan or substrates that contain generous amount of xylan.

XYLANASE ACTIVITY

Since xylanase is one of the most valuable enzymes in biotechnology applications, there are a lot of studies regarding xylanase activity. Xylanase activity is determined by measuring the reducing sugar released from xylan. In general, reducing sugar is the sugar that contains aldehyde group which is oxidized by oxidizing agent to become carboxylic acid. Xylanase activity is influenced by the incubation temperature in the range of 45 to 60°C (Coral, Arikan, Unaldi, & Guvenmez, 2002). Furthermore, xylanase activity is also depended on the types of substrates. Other types of buffer used to dissolve substrate include sodium acetate buffer, pH 5.0 (Rezende, de, Melo, Barbosa, Vasconcelos, & Endo, 2002), phosphate buffer, pH 5.0 (Abdel-Naby, Mohamed, & Kwon, 1992), citrate phosphate buffer, pH 6.0 (Szendefy, Szakacs, & Christopher, 2006) and potassium phosphate buffer, pH 6.0 (Singh, Tyagi, Datt, & Upadhyaya, 2009). In fact, there are several numbers of xylanase activity assays used for the determination of reducing sugars, each with their own definition of xylanase activity unit. These xylanase activity assays are differed in their assays procedure such as temperature, duration of incubation and substrate used. However, the principle is to quantify xylanase activity from the detection of reducing sugars released from the respective substrate. For example, one unit of xylanase activity is defined as the amount of xylanase required to release one micromole of xylose/mL of enzyme extract under the assay condition. Indeed, the detection of the reducing sugar xylose can be determined using several methods. One of the methods was from Khanna and Gauri (1993) that required the addition of sodium phosphate buffer, pH 7.2 with incubation temperature at 37°C and incubation in a boiling water bath for 15 minutes. In a journal published by Gessesse and Gashe (1997), xylanase activity was assayed with the aided of glycine-NaOH buffer, pH 9 and incubation at 50°C for 10 minutes followed by incubation for 5 minutes in boiling water. Another method was derived from Bailey et al. (1992) in which citrate buffer, pH 5.3 with incubation temperature of 30°C and incubation in a boiling water bath for 5 minutes. Bailey et al. (1992) reported that 3, 5-dinitrosalicylic (DNS) method has been widely used in most laboratories as the most common and effective method in determination of xylanase activity (Maciel, de Souza, Vandenberghe, Windson, Haminiuk, Fendrich, Bianca, da Silva, Brandalize, Pandey, & Soccol, 2008). DNS method uses xylose standard curve as a standard to determine the amount of reducing sugar released. In addition, Rochelle salt used in DNS method plays an important role in color stabilization besides preventing the reagent from dissolving oxygen. The higher the amount of xylose, the darker the color of the enzyme-xylose complex formed. As a result, more light will be absorbed. The amount of reducing sugar released was quantified using DNS reagent by Miller (1959). The DNS reagent was used to halt the reaction between xylan and xylanase to produce xylose. Another method to measure the reducing sugar is the Nelson and Somogyi method. It requires incubation in a boiling water bath for 15 minutes. The addition of Somogyi copper reagent is used to stop the reaction process by reducing copper from cupric to cuprous and conclusively to cuprous oxide. With the addition of Nelson arsenomolybdate reagent, it aids in the reduction of molybdenum acid to molybdenum blue (Breuil & Saddler, 1985). The measurement of the reducing sugar is conducted by spectrophotometric reading at 540 nm. Comparatively, DNS method is more common method to measure the reducing sugar of xylose compared to Somogyi and Nelson method because it is less hazardous, faster and more reliable (Haltrich, Nidetzky, Kulbe, Steiner, & Zupancic, 1996). DNS method was abundantly reported in most studies of xylanase activity assays to measure xylose according to (Singh, Tyagi, Datt, & Upadhyaya, 2009; Yang, Xu, Wang, & Yang, 2004; Kavya & Padmavathi 2009; Kaur, Dutt, & Tyagi, 2011; Murugan, Arnold, Pongiya, & Narayanan, 2011).

INDUSTRIAL APPLICATIONS OF XYLANASE

Xylanase has become a growing interest lately because of their promising applications in various industries. Indeed, in the present day, the demand for xylanase is rising because of its remarkable usages in vast industries. Among those important xylanase industrial applications are pulp and paper (Gupta & Kar 2009; Kulkarni et al. 1999; Shah & Madamwar 2005a, 2005b), baking (Gupta & Kar, 2009) and also food and beverage (Gupta & Kar 2009, Pang & Ibrahim, 2005, Kavya & Padmavathi, 2009). Xylanase is commercially applied to replace the use of harsh chemical such as chlorine which is known to be toxic and harmful to the environment during the bleaching process of kraft pulps in pulp and paper industry. Xylanase acts upon the surface layer of cellulose fibers to increase the permeability of the pulp for better chemical penetration through depolymerization process. Xylanase also acts within the inner fibers layer to aid in the bleaching process. As a result, xylanase pretreatment in pulp and paper industry has been reported to reduce the usage of chlorine. Utilization of xylanase is more economical and environmental friendly, besides reduces the treatment time of pulps that results in greater final brightness of products (Pang & Ibrahim, 2005, Wong, Tan, & Saddler, 1988). In the countries likes Western European and North America, their environmental regulations are strictly restricted the use of chlorine in the bleaching process in pulp and paper industry (Kulkarni, Shendye, & Rao, 1999). Indeed, mannanase, a type of glycanase that helps in depolymerisation of hemicelluloses backbone of wood, especially on softwood, is combined with xylanase in order to improve the bleaching process of kraft pulp (Kulkarni, Shendye, & Rao, 1999). Consequently, xylanase as a low-cost bio-bleaching agent gives benefit in both environmentally and economically. Livestock feed such as corns and wheat that contains high amount of non-starch polysaccharide causes problem in nutrients digestibility due to the present of arabinoxylans that act as anti-nutritional factor. Consequently, xylanase supplement is added to improve effectiveness in nutrients digestion and absorption. In food and beverage industry, xylitol a product of xylose possesses wide utilization in food, pharmaceutical, cosmetic and oral products. Xylitol is commonly produced as a sweetener for diabetics because it does not require insulin for its breakdown process. Additionally, xylitol also serves as sweetener to the glucose-6-phosphate dehydrogenase deficient community because it does not require glucose-6-phosphate dehydrogenase in its metabolism (Ylikahri, 1979).In addition, it demonstrates to provide good dental healthcare therefore xylitol is introduced into chewing gum, sweets, soft drinks, ice-cream and toothpaste. In general, xylitol is produced through two types of processes including chemical and microbiological. Chemical process involves the reduction of xylose from hemicellulose of xylan whereas another involves the hydrolysis of xylan into xylose followed by hydrogenation (Nigam & Singh, 1995a). In the end, xylitol undergoes purification and crystallisation process (Nigam & Singh, 1995b). Microbiological process involves the production of xylitol from xylose through hydrolysis of lignocellulose xylan-rich materials.

Apart from that, xylanase in desirable amount also has been used to further fasten the bread making process to improve bread crumbliness, shelf life and bread volume (Courtin & Delcour, 2001). In general, there are several of enzymes typically used in the baking process such as xylanase, protease, α-amylase, glucose oxidase, pentosanase and lipase. Enzymes used for baking are denatured to inactive protein during the baking process and thus they have to be thermo-labile enzymes. Hence, xylanase plays an important role in improving the water absorption in dough, increasing dough volume and decreasing dough firmness by changing the water insoluble hemicelluloses into soluble from. Moreover, xylanase also improves the gluten elasticity, bread crumb structure and volume by reacting to both soluble and insoluble pentosans in flour. Besides that, shelf life of bread is prolonged by the addition of xylanase as anti-staling agent in bread storage. Xylanase also help in improving the softening effect on dough. Furthermore, Jiang et al. (2005) and Romanowska et al. (2003) reviewed that xylanase from *A. niger* also aids in improving the loaf volume and bread quality. On the other hand, in brewing industry, the problem of viscous polysaccharides such as xylan in the final beer filtration process is solved by pretreatment with xylanase to break down the xylan. As a result, it increases the filtration rates and prevents the accumulation of unwanted materials during the process. In juices making industry, enzymes are typically used in order to maximize the production of 'clear' juices. Xylanase is commonly added with other cell wall degrading enzymes such as pectinase and cellulase in the clarification and filtration process to clarify the juice. In other words, xylanase plays a role in degrading pectins, starch and hemicellulose components to reduce turbidity and to increase the volume of juice during pulp processing steps (Dhiman, Garg, Sharma, Mahajan, & Methoxy, 2011). Additionally, other potential applications of xylanase are also involved in the bioremediation and bioconversion of agricultural, municipal and food wastes for the production of fermentable bioproducts and renewable biofuel such as bioethanol. Table 1 summarizes the major applications of xylanase in the industry.

PRODUCERS OF XYLANASE

Recently, considerable attention has been focused on the use of microorganisms in industrial xylanase production. Microorganisms such as bacteria, fungi, yeast and actinomycetes are among the producers of xylanase. They are found to be active at temperature ranging from 40°C to 60°C with the optimal pH medium in neutral scope for bacteria while slight acidic pH is the best condition for fungal xylanase. Nonetheless, there are microorganisms classified as thermophiles, alkaliphiles and acidophiles that are capable of producing xylanase. Thermophiles are microorganisms that are actively stable from 50°C to 80°C. Thermophilic fungi such as *Paecilomyces thermiophila* J18 yielded 18580 U/g of xylanase activity within 7 days of fermentation by utilising wheat straw (Yang, Yan, Jiang, Li, Tian, & Wang, 2006). On the other hand, *Thermomyces lanuginosus* (D_2W_3) produced 48000 U/g of xylanase activ-

Table 1. Major applications of xylanase in industry

Industry	Application	Reference
Pulp and paper	As a bio-bleaching agent to improve the pre-bleaching process of kraft pulp	Wong et al. (1988), Kantelinen et al. (1993); Dietmar et al. (1996); Kulkarni et al. (1999); Pang and Ibrahim, 2005; Shah and Madamwar, 2005a, 2005b; Gupta and Kar, 2009
Baking	Improve loaf volume, texture, stability and bread quality. Prolong the shelf life of bread	Bhat and Hazlewood, 2001; Jiang et al. (2005)
Food and beverage	Clarification of wine and juices. As alternative sweeteners and food additives	Dietmar et al. (1996); Pang and Ibrahim, 2005, Fang, 2007
Animal feed	Improve animal feed digestibility and increase feed efficiency	Dietmar et al. (1996), Shah and Madamwar 2005a, 2005b; Maciel et al. (2008)
Agriculture	Bioconversion of lignocelluloses into fuels and other chemicals	Gawande and Kamat 1999; Kavya and Padmavathi, 2009

ity using sorghum straw within 6 days of fermentation (Sonia, Chadha, & Saini, 2005). Other types of thermophiles include *Caldicellulosiruptor spp*, *Thermoascus spp* and *Thermotogaspp*. The founding of thermostable xylanase enhances the production in pulp paper industry. Indeed, *Aspergillus spp* (Haq, Tasneem, Raana, Khan, Mukhtar, & Javed, 2004) *Penicillium spp* (Fadel & Fouda, 1993), *Streptomyces spp* (Kansoh & Gammel, 2001), *Bacillus spp* (Rashid, 1999) and *Trichoderma spp* (Azin, Moravej, & Zareh, 2007; Seyis & Aksoz, 2005) have been mostcomprehensively investigated and manipulated in the production of xylanase.

Alkaliphiles on the other hand are mostly from *Bacillusspp*. Microorganisms that grouped into this category of alkaliphiles are survived well at pH above 8. A study by Kohli et al. (2001) revealed that *Thermoactinomyces thalophilis* subgroup C produced xylanase actively at pH 8.5, *Streptomyces spp* RCK-2010 at pH 8 (Kumar, Rishi, Bhuvnesh, Yogender, & Ramesh, 2012) and *Bacillus spp Sam-3* at pH 8 (Shah, Sidid, Ahmad, & Rele, 1999). Meanwhile, acidophiles is a term refers to microorganisms that are capable to flourish under pH 3. Acidophilic xylanase microorganisms are *Aureobasidium pululans* var. *Melanigenum* optimally produced xylanase at pH 2 (Ohta, Moriyama, Tanaka, Shige, & Akimoto, 2001) and *Penicillium occitanis* Pol6 at pH 3 (Driss, Brhiri, Elleuch, Bouly Stals, Miled, Blibech, Ghorbel, & Chaabouni, 2011). The alkaliphilic xylanase application is important in pulp and paper industry whereas for acidophilic xylanase, it is needed in food and feed industry. Kavya and Padmavathi (2009) reviewed that xylanase production is higher in optimised conditions which reached 12.65 U/mL when compared to under optimised conditions which only about 9.38 U/mL. Majority of xylanase achieved their optimum xylanase activity in the range of pH 5.0 to 6.0, which are slight acidic condition at temperature between 40°C to 70°C (Dutta, Sengupta, Sahoo, Ray, Bhattacharjee, & Ghosh, 2007). Meanwhile, Fang et al. (2007) stated that both xylanase production and cell growth by *Aspergillus carneus* reached their optimisation at 30°C. On the other hand, fungi such as *Aspergillus spp* and *Trichoderma spp* are the main producers of xylanase as they have shown to produce the highest level of xylanase compared to yeast and bacteria (Kulkarni, Shendye, & Rao, 1999). Other reason would be fungi species are able to excrete out xylanase. As a result, the extraction process of xylanase is conducted in ease. In fact, it is an economical way to produce this enzyme as fungi are cheaper to grow using widely available natural substrates rather than pure xylan.Table 2 illustrates the production of xylanase from various types of microorganisms reported in literature and Table 3 summarizes the characteristic of xylanases including molecular weight, optimum pH and temperature, Km, Vmax and pI values from different microorganisms.

Table 2. Production of xylanase by different types of microorganisms

Microorganism	Reference
Aspergillus niger	Pang and Ibrahim, 2005
Aspergillus niger	Zulfiqar et al. (2009)
Aspergillus niger	Kavya and Padmavathi, 2009
Aspergillus carneus	Fang et al. (2007)
Aspergillus foetidus	Shah and Madamwar, 2005a, 2005b
Bacillus spp	Gupta and Kar, 2009
Bacillus licheniformis	Archana and Satyanarayan, 1997
Paecilomyces themophila	Yang et al. (2006)
Penicillium chrysogenum	Jayant et al. (2011)
Streptomyces spp	Saurav and Kannabiran, 2010
Streptomyces cyaneus	Ninawe and Kuhad, 2005
Tricoderma viride	Juwaied et al. (2011)

Table 3. Characteristics of xylanases from different microorganisms

Species	Molecular Weight (kDa)	Optimum		K_m (mg/ml)	V_{max}	pI	References
		pH	Temperature (°C)				
Aspergillus Niger	13.5-14.0	5.5	45	-	-	9	Frederick et al. (1985)
Aspergillus Kawachii IFO 4308	26-35	2-5.5	50-60	-	-	3.5-6.7	Ito et al. (1992)
Aspergillus Nidulans	22-34	5.4	55	-	-	-	Fernandez-Epsinaret al. (1992)
Aspergillus Sydowii MG 49	30	5.5	60	-	-	-	Ghosh and Nanda, 1994
Aspergillus Fischeri Fxn1	31	6	60	4.88	5.88 µM/min/mg	-	Raj and Chandra, 1996
Aspergillus Sojae	32.7	5.0	60	-	-	3.50	Kimura et al. (1995)
	35.5	5.5	50	-	-	3.75	
Aspergillus Caespitosus	26.3	6.5	50	2.5	1679 µ/mg	-	Sandrim et al. (2005)
	27.0	7.0	55	3.9	113 µ/mg	-	
Aspergillus Ficuum AF-98	35	5	45	3.267	18.38 M/min/mg (beechwood xylan)	-	Lu et al. (2008)
	35	5	45	3.747	11.1 M/min/mg (birchwood xylan)	-	Lu et al. (2008)
Aspergillus Niger BCC14405	21	5	55	8.9	11100 U/mg	-	Asano et al. (2005)
Trichoderma Harzianum	20	5	50	0.58	0.106 µM/min/mg	-	Tan et al. (1985)
Trichoderma Reesei	20	5-5.5	45	3.0-6.8	-	9.0	Tenkanen et al. (1992)
	19	4-4.5	40	14.8-22.3	-	5.5	
Thermomyces Lanuginosus-SSBP	23.6	6.5	70-75	3.26	6300 µM/min/mg	3.8	Lin et al. (1999)
Thermomyces Lanuginosus DSM 5826	25.5	7	60-70	7.3	-	4.1	Cesar and Mrsa, 1996
Penicillium Purpurogenum	33	7.0	60	-	-	8.6	Belancic et al. (1995)
	23	3.5	50	-	-	5.9	
Fusarium Oxysporum	20.8	6	60	9.50	0.41 µM/min/mg	-	Christakopolous et al. (1996)
	23.5	6	55	8.45	0.37 µM/min/mg	-	

PRODUCTION OF XYLANASE VIA SUBMERGED FERMENTATION (SmF) AND SOLID STATE FERMENTATION (SsF)

Multiple biotechnological approaches including SmF and SsF are utilised for xylanase biosynthesis. Using SmF, microorganisms are cultured in a liquid medium containing required concentrations of nutrients for the optimisation of medium formulation. Similarly, growth conditions of incubation temperature and medium pH are easily maintained in this fermentation process. In general, nutrients composition in SmF are supplied in the medium in the form of cheaper and readily available complex materials such as undefined agricultural wastes including rice bran and rice straw (Wang, Yen, Shih, Chang, Chang, Wu, & Chai, 2003), corn cobs (Zulfiqar, Masood, Anjum, & Asgher, 2012), sugarcane bagasse, oat straw and wheat bran (Khandeparkar & Bhosle, 2007). SmF is now almost universal in the development of industrial enzymes in all other fermentation of industrial field. In the case of SmF, both of microorganism and substances are involved in submerged state in the liquid medium. Since the contents are in submerged state in the liquid medium, the modeling of the process is amenable and the transfer of heat and mass is more efficient. Methods for the design of fermentation equipment and for the evaluations of its performance are greatly improved by increased knowledge of factors affecting oxygen transfer in those systems requiring some degree of aeration (Arthur & Elmer, 1950). The production of commercially important enzymes in SmF has long been established.

Many reports have been studied on the xylanase production in the SmF and SsF (Sanghi, Garg, Kuhar, Kuhad, & Gupta, 2009; Irfan, Nadeem, Syed, & Baig, 2012; Kamble & Jadhav 2012; Jones, van, Dyk, & Pletschke, 2012). Currently, 80 to 90% of xylanase are produced in submerged culture because of the microbial biomass and the substrates are homogeneously distributed in a liquid medium. Besides that, Gaanappriya et al. (2011) also stated that SmF has a higher degree of intensification and higher level of automation. Due to the better understanding of scientific literature on the bacterial metabolism, characteristics and their response, it leads to the rising on the development of SmF for xylanase production. In order to develop a good fermentation, those parameters should be optimized and maintained according to the limitation of process, such as temperature, pH, substrate concentrations, size of inoculum, cultivation time and aeration. Sanghi et al. (2009) has proposed their investigation on optimizing the parameters on the production of xylanase from *B. subtilis* and its potential in bleaching of Kraft paper. Their work concluded that the optimization of fermentation conditions enhanced xylanase production by 1.5 fold as compared to under optimized conditions. In SmF of *B. subtilis* for xylanase production may be initiated by direct inoculation of spores to the sterile liquid medium. Similarly, the medium is incubated at a specified temperature and pH with constant aeration and agitation until the biomass product is reduced. Irfan et al. (2012) reported the selection of suitable medium also plays a vital role in xylanase production because it is a prerequisite to make the process cost effective. Besides that, in a study of Arthur and Elmer (1950), they stated that aeration taken in the general sense to mean the provision of an adequate oxygen supply is required in some of degree for all aerobic processes. When the transfer of oxygen supplied by aeration in SmF is treated as a series of rate processes, oxygen transfer equation for each step is developed. The physical absorption of oxygen is shown to be a function only of the design and operating characteristics of the equipment. Therefore, the advantage of using this technique for optimization process is the ease of handling various parameters whereby they can be monitored, periodic sampling of broth, and controlled if necessary by the addition of further nutrients. According to Kavya and Padmavathi (2009), low xylanase activity was obtained in non-agitated flasks, most probably due to oxygen or mass transfer limitations, while on agitation, high xylanase activity was produced, probably due to good oxygen supply. In addition, the volume used for fermentation also has a great impact on air supply, nutrient supply, growth of microorganism and production of enzyme (Mimura & Shinichi, 1999; Ivanova, Yankov, Kabaivanova, & Pashkkoulov, 2001). Therefore, in SmF, aeration and agitation are very important to ensure availability of oxygen, nutrients and other essential substances to the growing cells. Table 4 shows previous reports studied on the production of xylanase by different microorganisms in SmF.

SmF is the culture of microorganisms in liquefied medium as compared to SsF which involves the growth of microorganisms in the absence or near-absence of free moving water. SsF has been defined as the growth of microorganisms on moist substrates under controlled condition in the absence of free-flowing water (Shah & Madamwar, 2005b; Amare & Gashaw, 1999). SsF is an effective biotechnological tool which shows greater potential for the production of large variety of enzymes. In addition, SsF is performed using different types of microorganisms, such as fungi, yeast and bacteria in the production of several enzymes including xylanase. Large numbers of bacterial and fungal species are known to grow well on limited moist substrates in the absence or near-absence of free water. SsF process generally uses raw natural material as carbon and energy sources. In other words, SsF is a cost effective and economical process due to the use of simple growth and agro-industrial residues as the carbon sources for the production of enzyme. Agro-industrial residues or more commonly known as agricultural extracts from wheat bran, rice bran, soybean hulls and palm kernel cake are typically being considered as the best industrial substrates for SsF because they are abundantly available and low in cost (Pandey, Selvakumar, Soccol, & Nigam, 1999). In addition, SsF tends to generate value added products such as enzymes, flavors, organic acids, single cell protein and bioactive compounds from agro-industrial residues (Pang & Ibrahim, 2005; Mussatto & Teixeira, 2010). Indeed, SsF possesses several advantages over SmF in the production of enzymes including xylanase. SsF is a simple process. It produces higher fermentation productivity with lower contamination risks. It uses little amount of water which exclusively reduces the wastewater output. Besides that, it involves less expensive downstream processing steps with lower

Table 4. Previous reports studied on the production of xylanase by different microorganisms in submerged fermentation (SmF)

Microorganism	Xylanase Activity (U/mL)	Reference
Bacillus Circulans	2.20	Mullai and Rene, 2008
Bacillus Subtilis	408.90	Irfan et al. (2012)
Bacillus Subtilis	695.12	Nagar et al. (2010)
Bacillus Subtilis	128.00	Annamalai et al. (2009)
Bacillus Spp	19.10	Bocchini et al. (2005)
Streptomyces Spp	70.00	Nascimento et al. (2002)
Trichoderma Viride	2.45	Goyal et al. (2008)
Aspergillus Niger	44.10	Loera and Cordova, 2003
Aspergillus Niger	293.82	Bakri et al. (2008)
Mamillisphaeria Spp	242.70	Thanaporn et al. (2011)
Schizophyllum Commune	5.74	Haltrich et al. (1996)
Thermomyces Lanuginosus	2.70	Purkarthofer et al. (1993)

capital costs and energy demand. The usage of cheap and abundant carbon sources especially agricultural residuals for SmF and SsF that often results in a higher productivity of enzymes (Pandey, Nigam, Soccol, Soccol, Singh, & Mohan, 2000; Lonsane & Ramesh, 1990). In addition, SsF possesses lower risk of contamination because most of the contaminants including bacteria are not able to grow in the absence of free-flowing water (Lonsane & Ramesh, 1990). Nevertheless, there are several setbacks of SsF such as temperature, pH, water content and diffusivity of substance across barriers during the growth of microorganisms. Heat is generated during the fermentation process, when the availability of water is meager. In SsF, heat is not able to regulate which eventually builds up and causes the enzyme denaturation (Holker & Lenz, 2005). Table 5 shows previous reports studied on production of xylanase by different microorganisms in SsF.

MEDIUM FORMULATION FOR XYLANASE PRODUCTION

Medium composition is crucial to influence the xylanase activity. The production of microbial xylanase in SmF is strongly affected by the types of carbon source used. When Seyis and Aksoz (2005) studied the effect of different carbon sources on xylanase activity, sucrose showed higher xylanase biosynthesis by *Tetracladium marchalianum* compared to glucose in SmF. On the other hand, complex undefined agricultural extracts were also reported as more preferable carbon source for xylanase activity by several studies. Agricultural extracts or lignocellulosic biomass are important material resource and energy source which shows greater potential in lignocellulosic biotechnology application such as pulp and paper, food and feed, and bioethanol production (Kuhad, Kuhad, Kapoor, Sharma, & Singh, 2007). As usual, agricultural extract constituents of cellulose, hemicelluloses, and lignin. Agricultural extracts contain high concentration of xylan about 15 to 30% in hemicellulose which is good as an inducer for xylanase production in SsF (Svarachorn, 1999). In addition, agricultural extracts such as wheat bran, rice bran, palm kernel cake, sugarcane baggase, maize, barley husk, soybean hulls and sawdust are typically used as carbon sources in the xylanase production of SsF (Pang & Ibrahim, 2005; Mussatto & Teixeira, 2010; Kavya & Padmavathi, 2009; Svarachorn, 1999). All in all, lignocellulosic materials are utilized in culture medium formulation because they are plenteous in nature, low in cost and possess high level

Table 5. Previous reports studied on production of xylanase by different microorganisms in solid state fermentation (SsF)

Microorganism	Xylanase activity	Reference
Arthrobacter Spp	819.00 U/mL	Murugan et al. (2011)
Bacillus Subtilis	12.00 U/mL	Kim et al. (2012)
Bacillus Subtilis	8964.00 U/mL	Sanghi et al. (2009)
Bacillus Subtilis	5.19 U/mg	Heck et al. (2002)
Bacillus Spp	720.00 U/mL	Gessesse et al. (1999)
Bacillus Pumilus	142.00 U/mL	Kapilan and Arasaratnam, 2011
Aspergillus Niger	23.97 U/mL	Pang and Ibrahim, 2005
Aspergillus Niger	12.65 U/mL	Kavya and Padmavathi, 2009
Coprinopsis Cinerea	695.80 U/mL	Kaur et al. (2011)

of carbohydrate content which are suitable to generate fermentable sugars. In some studies, nitrogen supplement of organic and inorganic sources were added to further increase the growth of microorganisms and therefore enhanced the production of xylanase.

EFFECT OF DIFFERENT CARBON SOURCES OF AGRICULTURAL EXTRACTS ON XYLANASE PRODUCTION

The production of microbial xylanase is strongly affected by the types of carbon source used. A fairly large number of carbon containing sources have been reported in various studies. Many studies used different types of carbon sources for optimizing medium formulation for xylanase production by various microorganisms.

Wheat Bran

Endosperm cell wall of wheat bran or scientifically known as *Triticum aestivum* contains 75% of non-starch polysaccharide (Mares & Stone, 1973). Wheat bran is a by-product of wheat milling industry in which it is originated from the outer layer of the wheat kernel. In general, wheat bran is defined as the outer hard layer of the grain. Wheat bran has been vastly used as animal feed for improvement of digestibility as well as in bakery products. According to Bergmans et al. (1996), the vital components of wheat bran are consisted of 46% non-starch polysaccharides, 10to 20% starch, 15to 22% proteins and 4to 8% lignin. Furthermore, arabinoxylan is the primary constituent of the non-starch polysaccharides in hemicelluloses. Wheat bran is considered as one of the extensively utilized carbon source for the growth of *Aspergillus spp* in the production of xylanase. Indeed, wheat bran is identified as the optimum carbon source for the production of xylanase by *Aspergillus spp*. According to Gawande and Kamat (1999) and Kavya and Padmavathi (2009), wheat bran was observed to produce the optimum xylanase activity of 9.87 U/mL by *A. niger* which was isolated from garden soil samples in Banglore, India. They also observed that xylanase activity of 68.91 and 74.5 U/mL were produced by *A. terreus* 5 and *A. niger* 44, respectively. Wheat bran is abundantly available and highly nutritious agricultural by-product that commonly used in fermentation process. Moreover, wheat bran is an economical by-product which highly abounding with xylan, and thus, it possesses vital potential in xylanase production by microorganisms (Gawande & Kamat, 1999). In a study conducted by Kavya and Padmavathi (2009), they observed that wheat bran achieved the maximum xylanase activity of 9.87 U/mL in the xylanase production by *A. niger* in SsF. Much higher xylanase production of 74.5 U/mL was achieved using wheat bran by *A. niger* when compared to *A. terreus* (Subramaniyan & Prema, 2002). In addition, Xu et al. (2008) showed that xylanase production by *A. niger* XY-1 reached 14637 U/g dry substrate of wheat bran in shake flask under optimized condition.

Rice Bran

Rice (*Oryza sativa*) bran is the outer layer of rice grain. It is a by-product of rice milling industry in polishing process which contains high in dietary fibers. It takes up 10% of weight in unpolished rice. In a study conducted by Fang et al. (2007), rice bran contained high level of hemicellulose which was about 7.55 $\pm$ 0.49%. Hemicellulose of rice bran is a heteropolysaccharide which consists of a backbone xylan and sugars such as arabinose, galactose and glucose as side chains (Harada, Tanaka, Fukuda, Hashimoto, & Murata, 2005). According to Kuhad et al. (2007), they showed that xylan of rice bran contained high level of xylose which was about 46% xylose besides 44.9% arabinose. Hence, rice bran is an inexpensive and abundantly available carbon source which typically used in fermentation process (Kavya & Padmavathi, 2009; Begum & Alimon, 2011). A study was conducted by Kavya and Padmavathi (2009), who observed rice bran was a good substrate on xylanase production by *A. niger* via SsF. Rice bran has been mainly supplied for animal feed as well as rice oil production in Asian counterparts. It has many nutritive values as unsaturated fatty acids with antioxidant and anti-cancer properties. However, the consumption of rice bran in human diet is limited due to the exposure of agro-toxics and contamination during the storage of rice bran (Faccin, Viera, Miotto, Barreto, & Amante, 2009). On a global basis, over 600 million tons of rice was harvested annually due to its high amount of nutrition values (Nutracea, 2012). Abundant agricultural extracts with estimated 4.5 million tons was produced annually worldwide mainly in China (Wang, Yen, Shih, Chang, Chang, Wu, & Chai, 2003). Rice bran has been suggested as the crucial carbon source for biotechnological production of enzymes including xylanase production by many different microorganisms such as *Streptomyces actuosus* A-151 (Wang, Yen, Shih, Chang, Chang, Wu, & Chai, 2003), *Paenibacillusspp* strain HC1 (Harada, Tanaka, Fukuda, Hashimoto, & Murata, 2005) and*Arthrobacterspp* MTCC 5214 (Khandeparkar & Bhosle, 2007), respectively.

Sugarcane Bagasse

Sugarcane (*Saccharum afficinarum*) bagasse is a by-product of sugarcane stalks in the form of fibrous residues after the production of sugarcane juice and sugar. Cellulose and hemicellulose are the two main polysaccharides that constituted 70% of the sugarcane bagasse (Brienzo, Siqueira, & Milagres, 2009). Specifically, it contains 50% cellulose, 25% hemicellulose and 25% lignin (Pandey, Selvakumar, Soccol, & Nigam, 1999). This enormous amount of polysaccharides in sugarcane bagasse is useful for biofuels and alcohol manufacturing processes. Furthermore, it is being endeavored into animal feed, enzymes and organic acids production (Bocchini, Oliveira, Gomes, & Silva, 2005). By utilizing sugarcane bagasse, fungi such as *Penicillium janthinellum* that isolated from plant material found in a termite colony were able to produce xylanase activity of 98 U/mL (Milagres, Laci, & Prade,1993). Indeed, *A. niger* grown on Mendel and Weber media yielded 65.8 U/mL of xylanase compared to *A. niger* LPB 326 that produced 2099 IU/g of xylanase using 65% sugarcane bagasse and 35% soybean meal, respectively (Maciel, de Souza, Vandenberghe, Windson, Haminiuk, Fendrich, Bianca, da Silva, Brandalize, Pandey, & Soccol, 2008). Notably, *Thermoascus aurantiacus* ATCC 204492 also produced 1597 U/g of xylanase using sugarcane bagasse as the sole carbon source (Milagres, Santos, Piovan, & Roberto, 2004).

Oil Palm (Palm Kernel Cake)

Oil palm (*Elaeis guineensis*) is a tropical palm tree found abundant in Malaysia, Indonesia and Thailand. They are commercially cultivated for their edible palm oil which possesses characteristic as resistance to both oxidation and long term exposure to heat. Apart from that, palm oil has been utilised in cosmetic, detergents and biodiesel (Stichnothe & Schuchardt, 2011).The rapid growth of this palm oil agricultural industry has developed Malaysia as the world's largest and most leading exporter of palm oil with the account of 52% of total world oils and fats exports in the year of 2006 (Sumathi, Chai, & Mohamed, 2008). Oil palm industry produces a lot of agricultural by-products, mainly palm kernel cake. Besides that, this oil palm waste also has a notable profitable market worldwide in which it is being used as animal feed because of its content that comprised of 15 to 18% crude protein. Again, Malaysia is the largest palm kernel cake exporter mainly to various Europe countries (Saw, Janaun, & Subbarao, 2008). Due to its plentiful in the industry, palm kernel cake has been utilised as a potential substrate for the production of enzymes especially in SsF. It has been known that palm kernel cake was used in the cultivation of several fungi such as *A. niger* USM AI 1, *A. niger* II, *A. niger* F4, *Trichodermaspp* and *Phanerochaete chrysosporium* (Hong, Ibrahim, & Omar, 2011). Indeed, palm kernel cake has also been applied as a substrate is more economical approach in the xylanase production using SsF due to its low cost value and enormous amount as agricultural by-product from oil palm industry (Subramaniyan & Prema, 2002; Iluyemi, Hanafi, Radziah, & Kamarudin, 2006). In a study published by Pang and Ibrahim (2005), *A. niger* USM AI 1 yielded 23.97 U/g of xylanase using palm kernel cake as the sole substrate. It was reported by Dusterhoft et al. (1992) that palm kernel cake comprised of 6% xylan compound suitable for xylan production. Likewise, in a study carried out by Hakim (2006), palm kernel cake produced the highest xylanase activity by *A. niger* using both SsF and SmF. Additionally, Pang and Ibrahim (2005) revealed that palm kernel cake contained high protein content which was a good substrate for xylanase production. From their results, palm kernel cake was managed to achieve 9.5 U/g of xylanase production by *Aspergillusspp* through SsF. Furthermore, xylanase activity of 192.50 U/g was produced by *A. niger* FTCC 5003 in column bioreactor under optimized condition using palm kernel cake as the prime carbon source (Abdeshahian, Samat, & Wan, 2009).

Soybean

Soybean (*Glycine max*) hulls are the outer shell of soybean seed that are being removed during the production of soy oil, soybean meal, protein and flour. They are commercially distributed in pellet forms as ruminant feed mainly for horses and cows. Soybean hulls represent 8% of a single soybean seed which comprise about 86% complex carbohydrates (Gnanasambandam, Mathias, & Proctor, 1998). They are also comprised of 40 to 45% cellulose and 30 to 35% xylan on a dry basis (Brijwani, & Vadlani, 2010). High amount of xylan is believed to produce high level of xylanase activity. Gawande and Kamat (1999) proved that the production of cellulase-free xylanase by *A. terreus* strain 5 and *A. niger* strain 44 using soybean hulls as substrate yielded 6.2 and 8.7 U/mL, respectively. Likewise, Kavya and Padmavathi (2009) showed that soybean hulls possessed great potential as a substrate on the xylanase production using *A. niger* in SsF. In another study conducted by Brijwani and Vadlani (2010), they observed that soybean hulls were a good substrate in the production of cellulolytic and xylanolytic enzymes including endoglucanase, β-glucosidase and xylanase. As a result, soybean hulls are suitable for fungal growth and enzymes production due to their high cellulosic composition (Brijwani & Vadlani, 2010).

Barley

Barley (*Hordeum vulgare*) is one of the major world crops today and it is grown worldwide due to its tremendous potential in industrial applications including food and animal feed, malting and brewing. The barley husk or 'cover' of barley is commonly removed during the pearling process. Barley grains comprise of two significant non-starch polysaccharides of 1,3-1,4-β-glucans and arabinoxylan that located within the endosperm and grain cell walls (Debyser, W., Derdelinckx, G., & Delcour, 1997a; Debyser, W., Derdelinckx, G., & Delcour, 1997b). In this respect, barley husk contains approximately

30% hemicelluloses whereas xylan accounts for about 25.7g/100g of hemicelluloses in barley husk (Krawczyk, Persson, Andersson, & Jonsson, 2007). 1,3-1,-4-β-glucanase hydrolyses 1,3-1,4-β-glucans while xylanase that secreted by the aleurone layer of barley acts in the depolymerization of arabinoxylan within the cell wall during germination process. As a result, some transferable nutrient components were produced during degradation of cell wall. Hence, several studies have shown the usage of barley in the form of bran or straw for the fermentation of microorganisms to produce xylanase enzyme. In the latest study conducted by Soliman et al.2012, they remarked that barley husk as a substrate was able to show tremendous potential in xylanase production by both *A. niger* and *T. viride* which achieved the highest xylanase activity of 12.5 ± 0.13 U/g substrate and 11.0 ± 0.13 U/g substrate, respectively. In addition, an increased of xylanase activity by *A. niger* on barley husk from 12.5 ± 0.13 U/g substrate before optimization to 42.0 ± 0.22 U/g substrate after optimization (Soliman, Abdel-Dayem, & El-Tanash, 2012). Moreover, barley straw was being used by *Sclerotinia sclerotiorum* S2 to produce xylanase activity of 0.83 U/mL (Ellouze, Fattouch, Mestiri, Mohammed, & Mohammed, 2008).

Sawdust

Sawdust is a waste product of wood processing which is constituted of 45 to 50% cellulose and 23 to 30% lignin. In fact, sawdust with high composition of beechwood is a good substrate for the production of xylanase enzyme (Bailey, Biely, & Poutanen, 1992). Sawdust also plays a crucial role as carbon source in xylanase production. Xylanase production was found to be optimum with 117.0 U/mL by *Arthrobacter spp* MTCC 6915 (Murugan, Arnold, Pongiya, & Narayanan, 2011) and 1.35 U/mL by *Penicillium chrysogenum* PCL501 (Okafor, Emezue, Okochi, Onyegeme-okerenta, & Nwodo-Chinedu, 2007a), respectively. On the other hand, xylanase production was 0.65 and 1.26 U/mL by *A. niger* ANL 301 (Okafor, Okochi, Onyegeme-okerenta, & Nwodo-Chinedu, 2007b) and *Rhizopus oryzae* ATCC 9363 (Bakir, Yavascaoglu, Guvenc, & Ersayin, 2001), respectively.

Maize

Maize or corn is a cereal crop that grown all over the world with the worldwide production about 785 million tons according to the International Institute of Tropical Agriculture (IITA), 2009. In a study conducted by Goyal et al. (2008), maize straw was the best carbon source for xylanase production if compared with jowar straw and barseem straw due to the fact that maize contained high level of hemicellulose. In addition, Doner and Hicks (1997) also identified that purified maize bran contained high level of hemicelluloses, there was consisted of 67.5% of arabinoxylan, 22.5% of cellulose and 2.4% of protein. Table 6 shows the effect of different carbon sources of agricultural extracts on the xylanase production by different microorganisms.

CULTURE CONDITIONS FOR XYLANASE PRODUCTION

Culture growth conditions including incubation temperature, pH and agitation speed play an important role in leading to greater xylanase production by various microorganisms.

Table 6. Effect of different carbon sources of agricultural extracts on the xylanase production by different microorganisms.

Microorganism	Carbon Source	Xylanase Activity (U/mL)	Reference
Arthrobacter Spp	Sawdust	117.00	Muragan et al. (2011)
Bacillus Circulans	Bagasse	8.40	Bocchini et al. (2005)
Bacillus Pumilus	Paddy husk	224.20	Kapilan et al. (2011)
Bacillus Subtilis	Wheat bran	410.00	Sanghi et al. (2009)
Bacillus Spp	Oat spelt	52.00	Anuradha et al. (2007)
Aspergillus Niger	Maltose	17.80	Simoes et al. (2009)
Streptomyces Chartreusis	Corn cobs	334.34	Li et al. (2011)
Trichoderma Viride	Sorbitol	169.00	Simoes et al. (2009)

Effect of Incubation Temperatures on Xylanase Production

The incubation temperature is one of the important parameters to determine the performance of xylanase fermentation. Generally, fungi including *Aspergillus spp* were cultivated at the temperature ranging from 25°C to 45°C. Djekrif-Dakhmouche et al. (2006) pointed out that *A. niger* ATCC 16404 was best cultured at 30°C for adequate growth to enhance the optimum liberation of xylanase. Likewise, in another study carried out by Ali et al. 2002, they observed that the ideal growth temperature for *A. niger* GCBT7 was at 30°C and they also proved that 30°C was the optimum temperature for both enzyme activity and citric acid production in a stirred tank fermentor. Moreover, Maciel et al. (2008) observed that 30°C was the optimum temperature for xylanase activity by *A. niger* LPB 326. These results were similar to the report by Shah and Madamwar (2005a, 2005b) which stated that 30°C was the optimum temperature for xylanase production by *A. foetidus*. Likewise, *A. niger* B03 was best cultured at 29°C (Dobrev, Pishtiyski, Stanchev, & Mircheva, 2007), *A. niger* LPB 326 was optimally cultured at 30°C ± 1°C (Maciel et al. 2008), *A. niger* KKS was at 30°C (Kim, Kang, & Lee, 1997) and *A. niger* 44 was at 35°C (Gawande & Kamat, 1999), respectively. On the other hand, the ideal temperature for *A. niger* An-1.15 to acquire the optimum xylanase activity was observed to be at 28°C (Yuan, Wang, Huai, & Qian, 2005). The optimum growth temperature for *A. niger* USM A1I was at 28°C ± 3°C which was similar to the optimum temperature for xylanase production (Subramaniyan & Prema, 2002). Kavya and Padmavathi (2009) showed that *A. niger* grown well at 28°C with the xylanase production of 8.98 U/mL. They also remarked that the cultivation temperature possessed huge impact to both xylanase production and growth of *A. niger*. Notably, *A. sulphureus* that isolated from soil in Northern China showed the optimum temperature for xylanase production occurred at the range of 30 to 35°C (Lu, Li, & Wu, 2003), *A. foetidus* MTCC 4898 grown best at 30°C (Shah & Madamwar, 2005a, 2005b; Chapla, Divecha, Madamwar, & Shah, 2010), *A. niger* FGSCA733 at 25°C (Ncube, Howard, Abotsi, van, Rensburg, & Ncube, 2012), *A. nidulans* CECT 2544 at temperature ranging from 37°C to 42°C (Fernandez-Espinar, Ramon, Pinaga, & Valles, 1992), *A. carneus* M34 at 50°C (Fang, Changa, Hsieh, & Fang, 2007), *A. niger* 44 at 35°C (Gawande & Kamat, 1999), *A. niger* isolated by preliminary screening at 28°C, respectively(Kavya & Padmavathi, 2009).

Effect of Different Initial pH Medium on Xylanase Production

The pH medium has a huge influence on the performance of microbial xylanase activity where it plays a significant part in initiating the excretion of xylanase enzyme. *A. niger* is able to thrive in a slightly acidic environment of pH 6.5. In fact, majority of fungi are able to thrive at pH ranging from 5.0 to 8.0. Slightly acidic pH of 6.0 was observed by Maciel et al. (2008) as the most suitable pH medium for fungi to flourish during fermentation process. Djekrif-Dakhmouche et al. (2006) discovered that the best initial pH range for *A. niger* was noted to be at 5 or 6 while *A. niger* B03 was at pH ranging from 6.2 to 6.4 (Dobrev, Pishtiyski, Stanchev, & Mircheva, 2007).*A. fumigatus* AR1 was observed to produce the maximum xylanase activity at pH 6.0 to 6.5 (Anthony, Chandra, Rajendran, & Gunasekaran, 2003), *A. nidulans* CECT 2544 at pH 6.8 (Fernandez-Espinar, Ramon, Pinaga, & Valles, 1992) and *A. carneus* M34 at pH 6, respectively (Fang, Changa, Hsieh, & Fang, 2007). Additionally, low pH value inhibits the growth of other microorganisms particularly bacteria. Maciel et al. (2008) observed that *A. niger* LPB 326 achieved optimal growth at low pH level around pH 6.0. Meanwhile, according to Springer 2009, *A. ochraceus* grown well at pH level between pH 3 and 10 whereas slower growth was anticipated at pH 2.2. Some isolated *A. niger* is natural acidophilic fungi which grown well at pH 4.5 to 4.8 (Gifford, Lahey, & Reyn, 2006). In addition, Jayant et al. (2011) observed that *A. niger* in their study grown well at the slight acidic pH 5. Likewise, *Penicillium implicatum* and *Fusarium solani* produced the maximum xylanase activity of 58.80U/mL and 28.60 U/mL at pH medium 5.5 and 6, respectively.

Effect of Different Agitation Speeds on Xylanase Production

Agitation speed plays a very significant role in the fermentation process because it affects the dissolved oxygen and mass transfer of cultivation medium (Dietmar, Bernd, Kulbe, Walter, & Silvia, 1996; Purwanto, Ibrahim, & Sudrajat, 2009). Filamentous fungi require appropriate agitation speed for uniform distribution of the spores and prevention of the emergence of inactive clump of fungal growth. In a study established by Palma et al. (1996), a low xylanase activity was obtained due to shear stress occurred at higher agitation speed of 400 per minute during the growth of fungi *Penicillium janthinellum* that isolated from decaying wood. On the contrary, the most desirable agitation speed to produce high xylanase enzyme was achieved with a speed of 60 per minute in some studies. Another study carried out by Purwanto et al. (2009), the optimal growth of *A. niger* was occurred at agitation speed of 100 rpm. However, when the agitation speed increases to 200 rpm, it causes higher shear stress and mechanical forces disruption on filamentous fungi and thus results in the formation of small pellet size of *Aspergillus spp*. Meanwhile, a study by Fang et al. (2007) demonstrated that the optimum xylanase production of 22.2 U/mL by *A. carneus* M34 was observed at agitation speed of 111.9 rpm when utilizing oat spelt xylan as the carbon source. On the contrary, Haltrich et al. (1996) stated that the static growth condition was more suitable and preferable for fungi. His statement was supported by Djekrif-Dakhmouche et al. (2006) that during fermentation process of *A. niger* ATCC 16404, there was no significant results detected on the production of the desirable enzyme at agitation speed of 150rpm to 200 rpm. Nevertheless, Darah et al. (2011) remarked that the optimum agitation speed for the maximum growth of *A. niger* FETL FT3 of 3.75 g/L was observed at 200 rpm. There are number of different agitation speeds used by different researchers to obtain the maximum growth of *Aspergillus spp.* in fermentation process such as 100rpm,130rpm, 180rpm and 200 rpm to achieve their optimal xylanase production (Begum, & Alimon, 2011, Dai, Liu, Jin, & Jing, 2011, Ali, Haq, Qadeer, & Iqbal, 2002, Darah, Sumathi, Jain, & Lim, 2011, Subramaniyan & Prema, 2002).

CONCLUSION

The development of enzyme industries are emerging and growing rapidly in the recent decade. In fact, the market of xylanase is expanding rapidly due to its greater potential in industrial uses, particularly in the biotechnological applications. Xylanase possesses growing demand globally because of its vast applications in wide industries. Xylanase produced from different fungi are found to be more satisfactory in terms of its activity which is easily determined by DNS method. Agricultural extracts or lignocellulosic materials are commonly utilized as carbon source in medium formulation because they are plenteous in nature, low in cost with high level of carbohydrate content which are suitable to encourage production of xylanase by various microorganisms especially *Aspergillus spp* and *Bacillusspp*. Moreover, agricultural extracts are the important raw materials resource and energy source to enhance the growth of microorganisms. They are basically constituent of cellulose, hemicellulose and lignin. Indeed, they contain high concentration of xylan which consisted of hemicellulose is potent as an inducer for xylanase production. Xylan is generally derived from plant cell wall component. It is the second most abundant polysaccharides after cellulose. Xylan as a type of hemicelluloses is depolymerized through its hydrolyzing enzyme of xylanase. Hence, xylanases have created a specific demand in global market due to its significant role in various applications in the manufacturing of pulp and paper, baking, animal feeds and many more. As previously described, xylanase is derived from various sources which include plants, snails, insects, crustaceans and microorganisms. However, microbial origin has found to be the greatest producer of xylanase, plays an important role in the production of xylanase in industrial scale. In fact, *Aspergillusspp* has been found to be the greatest xylanase producer using lignoecellulosic materials especially those from agricultural crops and extracts. They are great sources of xylan which are practically derived from wheat bran, rice bran, palm kernel cake, sugarcane baggase, maize, barley husk, soybean hulls and sawdust. Thus, utilization of agricultural extracts is notably economical and environmentally sound. One of the main factors is because of their cost effectiveness and availability in current market. Furthermore, SmF was also chosen primarily as the mode of fermentation for production of xylanase since it is easily operated in controlling its parameters including high oxygen and nutrients supply. As a result, due to these advantages, SmF was chosen over SsF. Besides that, wider area with higher culture volume in SmF is provided to the fungi to grow with proper optimized medium composition under optimum growth conditions control. Growth conditions including incubation temperature, pH medium and agitation speed also play a crucial role in the optimization of growth of microorganisms and xylanase production. Nonetheless, different microorganism strains isolated from different origins around the world might have slightly differences in their growth conditions for xylanase production. In a nutshell, many works are needed to carry out to monitor the xylanase production, efficiency and stability for its industrial applications to take place.

REFERENCES

Abdel-Naby, M. A., & Kwon, D.Y. (1992). Production of xylanase and xylosidase by *Aspergillus niger* NRC 107. *Korean Journal of Applied Microbial Biotechnology*, *20*(5), 543–550.

Abdeshahian, P., Samat, N., & Wan, Y. W. M. (2009). Xylanase production by *Aspergillus niger* FTCC 5003 using palm kernel cake in fermentative bioprocess. *Pakistan Journal of Biological Sciences*, *12*(15), 1049–1055. doi:10.3923/pjbs.2009.1049.1055 PMID:19943460

Ali, S., Haq, I., Qadeer, M. A., & Iqbal, J. (2002). Production of citric acid by *Aspergillus niger* using cane molasses in a stirred fermentor. *Electronic Journal of Biotechnology*, *5*(3), 258–271. doi:10.2225/vol5-issue3-fulltext-3

Amare, G., & Gashaw, M. (1999). High-level xylanase production by an alkaliphilic *Bacillus* sp. by using solid-state fermentation. *Enzyme and Microbial Technology*, *25*(1-2), 68–72. doi:10.1016/S0141-0229(99)00006-X

Annamalai, N., Thavasi, R., Jayalakshmi, S., & Balasubramanian, T. (2009). Production and optimization of cellulase-free, alkali-stable xylanase by *Bacillus pumilus* SV-85S in submerged fermentation. *Indian Journal of Biotechnology*, *8*, 291–297.

Anthony, T. K., Chandra, R., Rajendran, A., & Gunasekaran, P. (2003). High molecular weight cellulase-free xylanase from alkali-tolerant *Aspergillus fumigatus* AR1. *Enzyme and Microbial Technology*, *32*(6), 647–654. doi:10.1016/S0141-0229(03)00050-4

Anuradha, P., Vijayalakshmi, K. N. D. P., & Sridevi, K. (2007). Production and properties of alkaline xylanase from *Bacillus* sp. isolated from sugarcane fields. *Current Science*, *92*, 1283–1286.

Archana, A., & Satyanarayan, T. (1997). Xylanase production by thermophilic *Bacillus licheniformis* A99 in solid state fermentation. *Enzyme and Microbial Technology*, *21*(1), 12–17. doi:10.1016/S0141-0229(96)00207-4

Arthur, W. H., & Elmer, L. (1950). Oxygen transfer in submerged fermentation. *Industrial & Engineering Chemistry Research*, *42*(9), 1792–1801. doi:10.1021/ie50489a031

Asano, K., Sriprang, R., Gobsuk, J., Eurwilaichitr, L., Tanapongpipat, S., & Kirtikara, K. (2005). Endo-1,4-β-xylanase B from *Aspergillusniger* BCC14405 isolated in Thailand: Purification, characterization and gene isolation. *Journal of Biochemistry and Molecular Biology*, *38*(1), 17–23. PMID:15715941

Azin, M., Moravej, R., & Zareh, D. (2007). Production of xylanase by *Trichoderma longibrachiatum* on a mixture of wheat bran and wheat straw: Optimization of culture condition by Taguchi method. *Enzyme and Microbial Technology*, *40*(4), 801–805. doi:10.1016/j.enzmictec.2006.06.013

Bailey, M. J., Biely, P., & Poutanen, K. (1992). Interlaboratory testing of methods for assay of xylanase activity. *Journal of Biotechnology*, *23*(3), 257–270. doi:10.1016/0168-1656(92)90074-J

Bakir, U., Yavascaoglu, S., Guvenc, F., & Ersayin, A. (2001). An endo-β-1,4-xylanase from *Rhizopus oryzae*: Production, partial purification and biochemical characterisation. *Enzyme and Microbial Technology*, *29*(6-7), 328–334. doi:10.1016/S0141-0229(01)00379-9

Bakri, Al-Jazairi, M., & Al-Kayat, G. (2008). Xylanase production by a newly isolated *Aspergillus niger* SS7 in submerged culture. *Polish Journal of Microbiology*, *57*(3), 249–251. PMID:19004246

Begum, M. F., & Alimon, A. R. (2011). Bioconversion and saccharification of some lignocellulosic wastes by *Aspergillus oryzae* ITCC-4857.01 for fermentable sugar production. *Electronic Journal of Biotechnology*, *14*(5), 1–9. doi:10.2225/vol14-issue5-fulltext-3

Belancic, A., Scarpa, J., Peirano, A., Diaz, R., Steiner, J., & Eyzayuirre, J. (1995). *Penicillium purpurogenum* produces several xylanases: Purification and properties of two of the enzymes. *Journal of Biotechnology*, *41*(1), 71–79. doi:10.1016/0168-1656(95)00057-W PMID:7640003

Bergmans, M. E. F., Beldman, G., Gruppen, H., & Voragen, A. G. J. (1996). Optimisation of the selective extraction of (glucurono) arabinoxylans from wheat bran: Use of barium and calcium hydroxide solution at elevated temperatures. *Journal of Cereal Science*, *23*(3), 235–245. doi:10.1006/jcrs.1996.0024

Bhat, M. K., & Hazlewood, G. P. (2001). Enzymology and other characteristics of cellulases and xylanases. In M. R. Bedford & G. G. Partridge (Eds.), Enzymes in Farm Animal Nutrition. Academic Press.

Bhatnagar, D., & Imelda-Joseph, P. R. R. (2010). Amylase and acid protease production by solid state fermentation using *Aspergillus niger* from mangrove swamp. *Indian Journal of Fisheries*, *57*(1), 45–51.

Bocchini, D. A., Oliveira, O. M. M. F., Gomes, E., & Silva, R. (2005). Use of sugarcane bagasse and grass hydrolysates as carbon sources for xylanase production by *Bacillus circulans* D1 in submerged fermentation. *Process Biochemistry*, *40*(12), 3653–3659. doi:10.1016/j.procbio.2005.02.021

Breuil, C., & Saddler, J. N. (1985). Comparison of the 3,5-dinitrosalicylic acid and Nelson-Somogyi methods of assaying for reducing sugars and determining cellulase activity. *Enzyme and Microbial Technology*, *7*(7), 327–332. doi:10.1016/0141-0229(85)90111-5

Brienzo, M., Siqueira, A. F., & Milagres, A. M. F. (2009). Search for optimum conditions of sugarcane bagasse hemicelluloses extraction. *Biochemical Engineering Journal*, *46*(2), 199–204. doi:10.1016/j.bej.2009.05.012

Brijwani, K., Oberoi, H. S., & Vadlani, P. V. (2010). Production of a cellulolytic enzyme system in mixed-culture solid-state fermentation of soybean hulls supplemented with wheat bran. *Process Biochemistry*, *45*(1), 120–128. doi:10.1016/j.procbio.2009.08.015

Brijwani, K., & Vadlani, P. V. (2010). *Solid state fermentation of soybean hulls for cellulolytic enzymes production. In Soybean: Application and Technology* (pp. 305–322). InTech.

Cesar, T., & Mrsa, V. (1996). Purification and properties of xylanase produced by *Thermomyces lanuginosus. Enzyme and Microbial Technology*, *19*(4), 289–296. doi:10.1016/0141-0229(95)00248-0

Chapla, D., Divecha, J., Madamwar, D., & Shah, A. (2010). Utilisation of agro-industrial waste for xylanase production by*Aspergillus foetidus*MTCC 4898 under solid state fermentation and its application in saccharification. *Biochemical Engineering Journal*, *49*(3), 361–369. doi:10.1016/j.bej.2010.01.012

Christakopoulos, P., Nerinckx, W., Kekos, D., Marcis, B., & Claeyssens, M. (1996). Purification and characterization of two low molecular mass alkaline xylanases from *Fusarium oxysporum* F3. *Journal of Biotechnology*, *51*(2), 181–189. doi:10.1016/0168-1656(96)01619-7 PMID:8987884

Coral, G., Arikan, B., Unaldi, M. N., & Guvenmez, H. K. (2002). Some properties of thermostable xylanase from an *Aspergillus niger* strain. *Annals of Microbiology*, *52*, 299–306.

Courtin, C. M., & Delcour, J. A. (2001). Relative activity of endoxylanases towards water-extractable and water-unextractable arabinoxylan. *Journal of Cereal Science*, *33*(3), 301–312. doi:10.1006/jcrs.2000.0354

Dai, X. J., Liu, M. Q., Jin, H. X., & Jing, M. Y. (2011). Optimisation of solid-state fermentation of *Aspergillus niger* JL-15 for xylanase production and xylo-oligosaccharides preparation. *Czech Journal of Food Sciences*, *29*(5), 557–567.

Darah, I., Sumathi, G., Jain, K., & Lim, S. H. (2011). Influence of agitation speed on tannase production and morphology of *Aspergillus niger* FETL FT3 in submerged fermentation. *Applied Biochemistry and Biotechnology*, *165*(7-8), 7–8. doi:10.100712010-011-9387-8 PMID:21947762

Debyser, W., Derdelinckx, G., & Delcour, J. A. (1997a). Arabinoxylans and arabinoxylan hydrolysing activities in barley malts and worts derived from them. *Journal of Cereal Science*, *26*(1), 67–74. doi:10.1006/jcrs.1996.0107

Debyser, W., Derdelinckx, G., & Delcour, J. A. (1997b). Arabinoxylan solubilization and inhibition of the barley malt xylanolytic system by wheat during brewing with wheat whole meal adjunct: Evidence for a new class of enzyme inhibitors. *Journal of the American Society of Brewing Chemists*, *55*, 153–156.

Dhiman, S. S., Garg, G., Sharma, J., & Mahajan, R. (2011). Characterisation of statistically produced xylanase for enrichment of fruit juice clarification process. *New Biotechnology*, *28*(6), 746–755. doi:10.1016/j.nbt.2010.11.004 PMID:21093618

Dietmar, H., Bernd, N., Kulbe, K. D., Walter, S., & Silvia, Z. (1996). Production of fungal xylanases. *Journal of Bioresource Technology*, *58*(2), 137–161. doi:10.1016/S0960-8524(96)00094-6

Djekrif-Dakhmouche, S., Gheribi-Aoulmi, S., Meraihi, Z., & Bennamoun, L. (2006). Application of a statistical design to the optimisation of culture medium for α-amylase production by *Aspergillus niger* ATCC 16404 grown on orange waste powder. *Journal of Food Engineering*, *73*(2), 190–197. doi:10.1016/j.jfoodeng.2005.01.021

Dobrev, G. T., Pishtiyski, I. G., Stanchev, V. S., & Mircheva, R. (2007). Optimisation of nutrient medium containing agricultural wastes for xylanase production by *Aspergillus niger* B03 using optimal composite experimental design. *Bioresource Technology*, *28*(14), 2671–2678. doi:10.1016/j.biortech.2006.09.022 PMID:17092711

Doner, L. W., & Hicks, K. B. (1997). Isolation of hemicelluloses from corn fiber by alkaline hydrogen peroxide extraction. *Cereal Chemistry*, *74*(2), 176–181. doi:10.1094/CCHEM.1997.74.2.176

Driss, D., Brhiri, F., Elleuch, L., Bouly, N., Stals, I., Miled, N., ... Chaabouni, S. E. (2011). Purification and properties of an extracellular acidophilic endo-1,4-β-xylanase, naturally deleted in the "thumb", from *Penicillium occitanis* Pol6. *Process Biochemistry*, *46*(6), 1299–1306. doi:10.1016/j.procbio.2011.02.022

Dusterhoft, E. M., Posthumus, M. A., & Voragen, A. G. J. (1992). Non-starch polysaccharides from sunflower (*Helianthus annuus*) meal and palm kernel (*Elaeis guineensis*) meal preparation of cell wall material and extraction of polysaccharide fractions. *Journal of the Science of Food and Agriculture, 59*, 151–160. doi:10.1002/jsfa.2740590204

Dutta, T., Sengupta, R., Sahoo, R., Ray, S. S., Bhattacharjee, A., & Ghosh, S. (2007). A novel cellulase free alkaliphilic xylanase from alkali tolerant *Penicillium citrinum*: Production, purification and characterization. *Letters in Applied Microbiology, 44*(2), 206–211. doi:10.1111/j.1472-765X.2006.02042.x PMID:17257262

Ellouze, O., Fattouch, S., Mestiri, F., Mohammed, R. A., & Mohammed, N. M. (2008). Optimisation of extracellular xylanase production by *Sclerotinia sclerotiorum* S2 using factorial design. *Indian Journal of Biochemistry & Biophysics, 45*, 404–409. PMID:19239127

Faccin, G. L., Viera, L. D. N., Miotto, L. A., Barreto, P. L. M., & Amante, E. R. (2009). Chemical, sensorial and rheological properties of a new organic rice bran beverage. *Rice Science, 16*(3), 226–234. doi:10.1016/S1672-6308(08)60083-9

Fadel, M., & Fouda, M. S. (1993). Physiological studies on xylanase production by *Penicillium fumiculosum* on some agricultural wastes. *Microbiology, 148*, 304–312. PMID:8368028

Fang, H. Y., Changa, S. M., Hsieh, M. C., & Fang, T. J. (2007). Production, optimization growth conditions and properties of the xylanase from *Aspergillus carneus* M34. *Journal of Molecular Catalysis. B, Enzymatic, 49*(1-4), 36–42. doi:10.1016/j.molcatb.2007.08.002

Fernandez-Espinar, M. T., Ramon, D., Pinaga, F., & Valles, S. (1992). Xylanase production by *Aspergillus nidulans*. *FEMS Microbiology, 91*, 91–96. doi:10.1111/j.1574-6968.1992.tb05190.x

Frederick, M. M., Kiang, C., Frederick, J. R., & Reilly, P. J. (1985). Purification and characterization of endo-xylanases from *Aspergillus niger*. I. Two isozymes active on xylan backbones near branch points. *Biotechnology and Bioengineering, 27*(4), 525–532. doi:10.1002/bit.260270420 PMID:18553703

Gaanappriya, M., Guhankumar, P., & Balakrishnan, V. (2011). Isolation of xylan degrading enzyme from *Trichoderma* spp. *Research in Plant Biology, 1*(3), 15–20.

Gawande, P. V., & Kamat, M. Y. (1999). Production of *Aspergillus* xylanase by lignocellulosic waste fermentation and its application. *Journal of Applied Microbiology, 87*(4), 511–519. doi:10.1046/j.1365-2672.1999.00843.x PMID:10583678

Gessesse, A., & Gashe, B. A. (1997). Production of alkaline xylanase by an alkaliphilic *Bacillus spp.* isolated from an alkaline soda lake. *Journal of Applied Microbiology, 83*(4), 402–406. doi:10.1046/j.1365-2672.1997.00242.x

Ghosh, M., & Nanda, G. (1994). Purification and some properties of xylanase from *Aspergillus sydowii* MG 49. *Applied and Environmental Microbiology, 60*, 4620–4623. PMID:7811101

Gifford, A. H., Lahey, T., & Reyn, C. F. V. (2006). Fatal hemoptysis from invasive *Aspergillus niger* in a patient with cavitary lung disease and *Mycobacterium avium complex* infection. *Medical Mycology*, *44*(6), 557–560. doi:10.1080/13693780600575635 PMID:16966174

Girio, F. M., Fonseca, C., Carvalheiro, F., Duarte, L. C., Marques, S., & Bogel-Lukasik, R. (2010). Hemicelluloses for fuel ethanol: A review. *Bioresource Technology*, *101*(13), 4775–4800. doi:10.1016/j.biortech.2010.01.088 PMID:20171088

Gnanasambandam, R., Mathias, M., & Proctor, A. (1998). Structure and performance of soy hull carbon adsorbents as affected by pyrolysis temperature. *Journal ofthe American Oil Chemists'. Society*, *75*(5), 615–621.

Goyal, M., Kalra, K. L., Sareen, V. K., & Soni, G. (2008). Xylanase production with xylan rich lignocellulosic wastes by a local soil isolate of *Trichoderma viride*. *Brazilian Journal of Microbiology*, *39*(3), 535–541. doi:10.1590/S1517-83822008000300025 PMID:24031262

Gupta, U., & Kar, R. (2009). Xylanase production by a thermo-tolerant *Bacillus* species under solid-state and submerged fermentation. *Brazilian Archives of Biology and Technology*, *52*(6), 1363–1371. doi:10.1590/S1516-89132009000600007

Hakim, A. B. R. (2006). *Study on the effect of substrates concentration in the production of xylanase from Aspergillus niger using palm kernel cake*. University College of Engineering and Technology.

Haltrich, D., Nidetzky, B., Kulbe, K. B., Steiner, W., & Zupancic, S. (1996). Production of fungal xylanases. *Bioresource Technology*, *58*(2), 137–161. doi:10.1016/S0960-8524(96)00094-6

Haq, I., Tasneem, M., Raana, K., Khan, A., Mukhtar, H., & Javed, M. (2004). Optimization of cultural conditions for the production of xylanase by chemically mutated strain of *A. niger* GCBCX-20. *International Journal of Agriculture and Biotechnology, 6,* 1115-1118.

Harada, K. M., Tanaka, K., Fukuda, Y., Hashimoto, W., & Murata, K. (2005). Degradation of rice bran hemicellulose by *Paenibacillus* sp. strain HC1: Gene cloning, characterization and function of β-D-glucosidase as an enzyme involved in degradation. *Archives of Microbiology*, *184*(4), 215–224. doi:10.100700203-005-0038-8 PMID:16205911

Heck, J. X., Hertz, P. F., & Ayub, M. A. Z. (2002). Cellulase and xylanase production by isolated amazon *Bacillus* strains using soybean industrial residue based solid-state cultivation. *Brazilian Journal of Microbiology*, *33*(3), 213–218. doi:10.1590/S1517-83822002000300005

Holker, U., & Lenz, J. (2005). Solid-state fermentation are there any biotechnological advantages? *Current Opinion in Microbiology*, *8*(3), 301–306. doi:10.1016/j.mib.2005.04.006 PMID:15939353

Hong, L. S., Ibrahim, D., & Omar, I. C. (2011). Lignocellulolytic materials-as a raw material for the production of fermentable sugars via solid state fermentation. *Asian Journal of Scientific Research*, *4*(1), 53–61. doi:10.3923/ajsr.2011.53.61

Iluyemi, F. B., Hanafi, M. M., Radziah, O., & Kamarudin, M. S. (2006). Fungal solid state culture of palm kernel cake. *Bioresource Technology*, *97*(3), 477–482. doi:10.1016/j.biortech.2005.03.005 PMID:16216731

Irfan, M., Nadeem, M., Syed, Q., & Baig, S. (2012). Effect of medium composition on xylanase production by *Bacillus subtilis* using various agricultural wastes. *American-Eurasian Journal of Agricultural and Environmental Sciences*, *12*(5), 561–565.

Ito, K., Ogasawara, H., Sugimoto, T., & Ishikawa, T. (1992). Purification and properties of acid stable xylanases from *Aspergillus kawachii*. *Bioscience, Biotechnology, and Biochemistry*, *56*(4), 547–550. doi:10.1271/bbb.56.547 PMID:27280644

Ivanova, V., Yankov, D., Kabaivanova, L., & Pashkkoulov, D. (2001). Simultaneous biosynthesis and purification of two extra cellular *Bacillus hydrolases* in aqueous two alpha amylase. *Journal of Biochemical Engineering*, *8*, 61–81.

Jayant, M., Rashmi, J., Shailendra, M., & Deepesh, Y. (2011). Production of cellulase by different co-culture of *Aspergillus niger* and *Penicillium chrysogenum* from waste paper, cotton waste and baggase. *Journal of Yeast and Fungal Research*, *2*(2), 24–27.

Jiang, Z., Li, X., Yang, S., Li, L., & Tan, S. (2005). Improvement of bread making quality of wheat flour by the hyperthermophilic xylanase B from *Thermotoga maritime*. *Food Research International*, *38*(1), 37–43. doi:10.1016/j.foodres.2004.07.007

Jones, S.M., van Dyk, J. S., & Pletschke, B.I. (2012). *Bacillus subtilis* SJ01 produces hemicellulose degrading multi-enzyme complexes. *BioResources*, *7*(1), 1294–1309.

Juwaied, A. A., Al-amiery, A. A. H., Abdumuniem, Z., & Anaam, U. (2011). Optimization of cellulase production by *Aspergillus niger* and *Tricoderma viride* using sugar cane waste. *Journal of Yeast and Fungal Research*, *2*(2), 19–23.

Kamble, R. D., & Jadhav, A. R. (2012). Isolation, purification and characterization of xylanase produced by a new species of *Bacillus* in solid state fermentation. *International Journal of Microbiology*. PMID:22315613

Kansoh, A. L., & Gammel, A. (2001). Xylanolytic activities of *Streptomyces sp.1*, taxonomy production, partial purification and utilization of agricultural wastes. *Acta Microbiologica et Immunologica Hungarica*, *48*(1), 39–52. doi:10.1556/AMicr.48.2001.1.5 PMID:11233699

Kantelinen, A., Hortling, B., Sundquist, J., Linko, M., & Viikari, L. (1993). Proposed mechanism of the enzymatic bleaching of kraft pulp with xylanases. *Holzforschung*, *47*(4), 318–324. doi:10.1515/hfsg.1993.47.4.318

Kapilan, R., & Arasaratnam, V. (2011). Paddy husk as support for solid state fermentation to produce xylanase from *Bacillus pumilus*. *Rice Science*, *18*(1), 36–45. doi:10.1016/S1672-6308(11)60006-1

Kaur, H., Dutt, D., & Tyagi, C. H. (2011). Production of novel alkali-thermo-tolerant cellulose-poor xylanases from *Coprinopsis cinerea* HK-1 NFCCI-2032. *BioResources*, *6*(2), 1376–1391.

Kavya, V., & Padmavathi, T. (2009). Optimization of growth conditions for xylanase production from *Aspergillus niger* in solid state fermentation. *Polish Journal of Microbiology*, *58*(2), 125–130. PMID:19824396

Khandeparkar, R., & Bhosle, N. B. (2007). Application of thermoalkalophilic xylanase from *Arthrobacter* sp. MTCC 5214 in biobleaching of kraft pulp. *Bioresource Technology, 98*(4), 897–903. doi:10.1016/j.biortech.2006.02.037 PMID:16617015

Khanna, S., & Gauri, S. (1993). Regulation, purification, and properties of xylanase from *Cellulomonas fimi. Enzyme and Microbial Technology, 15*(11), 990–995. doi:10.1016/0141-0229(93)90177-4

Kim, S. W., Kang, S. W., & Lee, J. S. (1997). Cellulase and xylanase production by *Aspergillus niger* KKS in various bioreactors. *Bioresource Technology, 59*(1), 63–67. doi:10.1016/S0960-8524(96)00127-7

Kim, Y. K., Lee, S. C., Cho, Y. Y., Oh, H. J., & Ko, Y. H. (2012). *Isolation of cellulolytic Bacillus subtilis strains from agricultural environments*. Microbiology.

Kimura, I., Sasahar, H., & Tajima, S. (1995). Purification and characterization of two xylanases and an arabinofuranosidase from *Aspergillus sojae. Journal of Fermentation and Bioengineering, 80*(4), 334–339. doi:10.1016/0922-338X(95)94200-B

Kohli, V., Nigam, P., Singh, D., & Chaudhary, K. (2001). Thermostable, alkalophilic and cellulose free xylanase production by *Thermoactinoyces thalophillus Subgroup C. Enzyme and Microbial Technology, 28*(7-8), 606–610. doi:10.1016/S0141-0229(01)00320-9 PMID:11339942

Krawczyk, H., Persson, T., Andersson, A., & Jonsson, A. S. (2007). Isolation of hemicelluloses from barley husks. *Food and Bioproducts Processing, 86*(1), 31–36. doi:10.1016/j.fbp.2007.10.018

Krisana, A., Rutchadaporn, S., Jarupan, G., Lily, E., Sutipa, T., & Kanyawim, K. (2005). Endo-1,4-xylanase B from *Aspergillus* cf. *niger* BCC14405 isolated in Thailand: Purification, characterisation and gene isolation. *Journal of Biochemistry and Molecular Biology, 38*(1), 17–23. PMID:15715941

Kuhad, R. C., Kuhad, S., Kapoor, M., Sharma, K. K., & Singh, A. (2007). Lignocellulolytic microorganisms, their enzymes and possible biotechnologies based on lignocellulolytic microorganisms and their enzymes. In R. C. Kuhad & A. Singh (Eds.), Lignocellulose Biotechnology: Future Prospects. Academic Press.

Kulkarni, N., Shendye, A., & Rao, M. (1999). Molecular and biotechnological aspects of xylanases. *FEMS Microbiology Reviews, 23*(4), 411–456. doi:10.1111/j.1574-6976.1999.tb00407.x PMID:10422261

Kumar, A., Rishi, G., Bhuvnesh, S., Yogender, P. K., & Ramesh, C. K. (2012). Xylanase production from an alkalophilic actinomycete isolate *Streptomyces* sp. RCK-2010, its characterization and application in saccharification of second generation biomass. *Journal of Molecular Catalysis. B, Enzymatic, 74*(3-4), 170–177. doi:10.1016/j.molcatb.2011.10.001

Li, X., Sun, B. Z., Lv, J., Li, Y. H. S. E., & Zhu, Y. (2011). Production and improved bleaching abilities of a thermostable xylanase from a newly isolated *Streptomyces chartreusis* strain. *African Journal of Biotechnology, 10*(64), 14132–14142. doi:10.5897/AJB10.2360

Lin, J., Ndlovu, L. M., Singh, S., & Pillay, B. (1999). Purification and biochemical characteristics of β-D-xylanase from a thermophilic fungus, *Thermomyces lanuginosus*-SSBP. *Biotechnology and Applied Biochemistry, 30*, 73–79. PMID:10467122

Loera, O., & Cordova, J. (2003). Improvement of xylanase production by a parasexual cross between *Aspergillus niger* strains. *Brazilian Archives of Biology and Technology*, *46*(2), 177–181. doi:10.1590/S1516-89132003000200006

Lonsane, B. K., Ghildyal, N. P., Budiatman, S., & Ramakrishna, S. V. (1985). Engineering aspects of solid state fermentation. *Enzyme and Microbial Technology*, *7*(6), 256–265. doi:10.1016/0141-0229(85)90083-3

Lonsane, B. K., & Ramesh, M. V. (1990). Production of bacterial thermostable α-amylase by solid state fermentation: A potential tool for achieving economy in enzyme production and starch hydrolysis. *Advances in Applied Microbiology*, *35*, 1–56. doi:10.1016/S0065-2164(08)70242-9 PMID:2205080

Lu, F. X., Lu, M., Lu, Z., Bie, X. M., Zhao, H. Z., & Wang, Y. (2008). Purification and characterization of xylanase from *Aspergillus ficuum* AF-98. *Bioresource Technology*, *99*(13), 5938–5941. doi:10.1016/j.biortech.2007.10.051 PMID:18068974

Lu, W., Li, D., & Wu, Y. (2003). Influence of water activity and temperature on xylanase biosynthesis in pilot-scale solid-state fermentation by *Aspergillus sulphureus*. *Enzyme and Microbial Technology*, *32*(2), 305–311. doi:10.1016/S0141-0229(02)00292-2

Maciel, G.M., de Souza, Vandenberghe, L. P., Windson, C., Haminiuk, I., Fendrich, R.C., … Soccol, C.R. (2008). Xylanase production by *Aspergillus niger* LPB 326 in solid state fermentation using statistical experimental design. *Food Technology and Biotechnology*, *46*(2), 183–189.

Mares, D. J., & Stone, B. A. (1973). Studies on wheat endosperm-II. Properties of wall components and studies on organisation in the wall. *Australian Journal of Biological Sciences*, *26*(4), 813–830. doi:10.1071/BI9730813

Milagres, A. M. F., Laci, L. S., & Prade, R. A. (1993). Characterisation of xylanase production by a local isolate of *Penicillium janthinellum*. *Enzyme and Microbial Technology*, *15*(3), 248–253. doi:10.1016/0141-0229(93)90145-R

Milagres, A. M. F., Santos, E., Piovan, T., & Roberto, I. C. (2004). Production of xylanase by *Thermoascus aurantiacus* from sugar cane bagasse in an aerated growth fermentor. *Process Biochemistry*, *39*(11), 1387–1391. doi:10.1016/S0032-9592(03)00272-3

Miller, G. L. (1959). Use of Dinitrosalicyclic acid reagent for determination of reducing sugar. *Analytical Chemistry*, *31*(3), 426–428. doi:10.1021/ac60147a030

Mimura, H., & Shinichi, N. (1999). Physiological characteristics of *Bacillus* sp. *Biocontrol Science and Technology*, *4*(2), 105–108. doi:10.4265/bio.4.105

Mullai, P., & Rene, E. R. (2008). A simple multi layered neural network model for predicting bacterial xylanase production by *Bacillus* sp. using *Avena sativa* as substrate. *Biotechnology (Faisalabad)*, *7*(3), 499–504. doi:10.3923/biotech.2008.499.504

Murugan, S., Arnold, D., Pongiya, U. D., & Narayanan, P. M. (2011). *Production of xylanase from Arthrobacter sp. MTCC 6915 using saw dust as substrate under solid state fermentation*. Enzyme Research.

Mussatto, S. I., & Teixeira, J. A. (2010). Lignocellulose as raw material in fermentation processes. In A. Méndez-Vilas (Ed.), *Current research, technology and education topics in applied microbiology and microbial biotechnology* (Vol. 2). Badajoz, Spain: Formatex Research Center.

Nagar, S., Mittal, A., & Gupta, V. K. (2012). A cost effective method for screening and isolation of xylan degrading bacteria using agro waste material. *Asian Journal of Biological Sciences, 5*(8), 384–394. doi:10.3923/ajbs.2012.384.394

Nascimento, R. P., Coelho, R. R. R., Marques, S., Alves, L., Girio, F. M., Bon, E. P. S., & Amaral-Collaco, M. T. (2002). Production and partial characterization of xylanase from *Streptomyces* sp. strain AMT-3 isolated from Brazilian cerrado soil. *Enzyme and Microbial Technology, 31*(4), 549–555. doi:10.1016/S0141-0229(02)00150-3

Ncube, T., Howard, R. L., Abotsi, E. K., van Rensburg, E. L. J., & Ncube, I. (2012). *Jatropha curcas* seed cake as substrate for production of xylanase and cellulase by *Aspergillus niger* FGSA733 in solid-state fermentation. *Industrial Crops and Products, 37*(1), 118–123. doi:10.1016/j.indcrop.2011.11.024

Nigam, P., & Singh, D. (1995a). Enzyme and microbial systems involved in starch processing. *Enzyme and Microbial Technology, 17*(9), 770–778. doi:10.1016/0141-0229(94)00003-A

Nigam, P., & Singh, D. (1995b). Processes for fermentative production of xylitol-a sugar substitute. *Process Biochemistry, 30*, 117–124.

Ninawe, S., & Kuhad, R. C. (2005). Use of xylan-rich cost effective agro-residues in the production of xylanase by *Streptomyces cyaneus* SN32. *Journal of Applied Microbiology, 99*(5), 1141–1148. doi:10.1111/j.1365-2672.2005.02711.x PMID:16238744

Nutracea. (2012). *Rice bran facts.* Retrieved from http://www.nutracea.com/RiceBranFacts

Ohta, K., Moriyama, S., Tanaka, H., Shige, T., & Akimoto, H. (2001). Purification and characterisation of an acidophilic xylanase from *Aureobasidium pullulans* var. *melanigenum* and sequence analysis of the encoding gene. *Journal of Bioscience and Bioengineering, 92*(3), 262–270. doi:10.1016/S1389-1723(01)80260-7 PMID:16233094

Okafor, U. A., Emezue, T. N., Okochi, V. I., Onyegeme-okerenta, B. M., & Nwodo-Chinedu, S. (2007a). Xylanase production by *Penicillium chrysogenum* PCL501 fermented on cellulosic wastes. *African Journal of Biochemistry Research, 1*(4), 48–53.

Okafor, U. A., Okochi, V. I., Onyegeme-okerenta, B. M., & Nwodo-Chinedu, S. (2007b). Xylanase production by *Aspergillus niger* ANL 301 using agro-wastes. *African Journal of Biotechnology, 6*(14), 1710–1714.

Palma, M. B., Milagres, A. M. F., Prata, A. M. R., & Manicilha, D. I. M. (1996). Influence of aeration and agitation on xylanase production. *Brazilian Journal of Biochemistry, 31*(2), 141–145.

Pandey, A., Nigam, P., Soccol, C. R., Soccol, V. T., Singh, D., & Mohan, R. (2000). Advances in microbial amylases. *Biotechnology and Applied Biochemistry, 31*(2), 135–152. doi:10.1042/BA19990073 PMID:10744959

Pandey, A., Selvakumar, P., Soccol, C. R., & Nigam, P. (1999). Solid state fermentation for the production of industrial enzymes. *Current Science*, *77*, 149–162.

Pang, P. K., & Ibrahim, C. O. (2005). Xylanase production by a local fungal isolate, *Aspergillus niger* USM AI 1 via solid state fermentation using palm kernel cake (PKC) as substrate. *Songklanakarin Journal of Science and Technology*, *27*(2), 325–336.

Peng, F., Bian, J., Peng, P., Guan, Y., Xu, F., & Sun, R. C. (2012). Fractional separation and structural features of hemicelluloses from sweet sorghum leaves. *BioResources*, *7*(4), 4744–4759. doi:10.15376/biores.7.4.4744-4759

Pitt, J. I., & Hocking, A. D. (2009). *Fungi and food spoilage*. Springer, US. doi:10.1007/978-0-387-92207-2

Purkarthofer, H., Lischnig, T., & Steiner, W. (1993). Thermostability of endo-β-xylanase from the thermophilic fungus *Thermomyces lanuginosus*. *Biotechnology Letters*, *15*(4), 411–414. doi:10.1007/BF00128286

Purwanto, L. A., Ibrahim, D., & Sudrajat, H. (2009). Effect of agitation speed on morphological changes in *Aspergillus niger* hyphae during production of tannase. *World Journal of Chemistry*, *4*(1), 34–38.

Raj, K. C., & Chandra, T. S. (1996). Purification and characterization of xylanase from alkalitolerant *Aspergillus fischeri* Fxn 1. *FEMS Microbiology Letters*, *145*(3), 457–461. doi:10.1111/j.1574-6968.1996.tb08616.x PMID:8978101

Rashid, A. N. (1999). Characterization of xylanase produced by *Bacillus pumilus* strain PJ19. *Journal of Microbiology and Biotechnology*, *9*(2), 157–162.

Rezende, M. I., Barbosa, A. M., Vasconcelos, A. F. D., & Endo, A. S. (2002). Xylanase production by *Trichoderma harzianum rifai* by solid state fermentation on sugarcane bagasse. *Brazilian Journal of Microbiology*, *33*(1), 67–72. doi:10.1590/S1517-83822002000100014

Rifaat, H. M., Nagieb, Z. A., & Ahmedv, Y. M. (2005). Production of xylanase by *Streptomyces* species and the bleaching effect on rice straw pulp. *Applied Ecology and Environmental Research*, *4*(1), 151–160. doi:10.15666/aeer/0401_151160

Romanowska, I., Polak, J., Janowska, K., & Bieleckim, S. (2003). The application of fungal endoxylanase in bread making. *Communications in Agricultural and Applied Biological Sciences*, *68*, 317–320. PMID:15296186

Sandrim, V. C., Rizzatti, A. C. S., Terenzi, H. F., Jorge, J. A., Milagres, A. M. F., & Polizeli, M. L. T. M. (2005). Purification and biochemical characterization of two xylanases produced by *Aspergillus caespitosus* and their potential for kraft pulp bleaching. *Process Biochemistry*, *40*(5), 1823–1828. doi:10.1016/j.procbio.2004.06.061

Sanghi, A., Garg, N., Kuhar, K., Kuhad, R. C., & Gupta, V. K. (2009). Enhanced production of cellulase-free xylanase by alkalophilic *Bacillus subtilis* ash and its application in biobleaching of kraft pulp. *BioResources*, *4*(3), 1109–1129.

Saurav, K., & Kannabiran, K. (2010). Diversity and optimization of process parameters for the growth of *Streptomyces* VITSVK9 spp. isolated from Bay of Begal, India. *Journal of Natural and Environmental Sciences, 1*(2), 56–65.

Saw, H. Y., Janaun, J., & Subbarao, D. (2008). Hydration properties of palm kernel cake. *Journal of Food Engineering, 89*(2), 227–231. doi:10.1016/j.jfoodeng.2008.04.027

Seyis, I., & Aksoz, N. (2005). Effect of carbon and nitrogen sources on xylanase production by *Trichoderma harzianum* 1073 D3. *International Biodeterioration & Biodegradation, 55*(2), 115–119. doi:10.1016/j.ibiod.2004.09.001

Shah, A. K., Sidid, S. S., Ahmad, A., & Rele, M. V. (1999). Treatment of bagasse pulp with cellulase-free xylanases from an alkalophilic *Bacillus spp.* Sam-3. *Bioresource Technology, 68*(2), 133–140. doi:10.1016/S0960-8524(98)00133-3

Shah, A. R., & Madamwar, D. (2005a). Xylanase production by a newly isolated *Aspergillus foetidus* strain and its characterisation. *Process Biochemistry, 40*(5), 1763–1771. doi:10.1016/j.procbio.2004.06.041

Shah, A. R., & Madamwar, D. (2005b). Xylanase production under solid-state fermentation and its characterization by an isolated strain of *Aspergillus foetidus* in India. *World Journal of Microbiology & Biotechnology, 21*(3), 233–243. doi:10.100711274-004-3622-1

Simoes, M. L. G., Tauk-Tornisielo, S. M., & Tapia, D. M. P. (2009). Screening of culture condition for xylanase production by filamentous fungi. *African Journal of Biotechnology, 8*(22), 6317–6326. doi:10.5897/AJB2009.000-9479

Singh, S., Tyagi, C. H., Datt, D., & Upadhyaya, J. S. (2009). Production of high level of cellulose-poor xylanases by wild strains of white-rot fungus *Coprinellus disseminatus* in solid-state fermentation. *New Biotechnology, 26*(3/4). PMID:19761879

Singh, S. K., Sczakas, G., Soccol, C. R., & Pandey, A. (2008). Production of enzymes by solid-state fermentation. In A. Pandey, C. R. Soccol, & C. Larroche (Eds.), *Current Development in Solid-state Fermentation* (pp. 183–204). New York, NY: Springer. doi:10.1007/978-0-387-75213-6_9

Soliman, H. M., Abdel-Dayem, A. S., & El-Tanash, A. B. (2012). Production of xylanase by *Aspergillus niger* and *Trichoderma viride* using some agriculture residues. *International Journal of Agricultural Research, 7*, 46–57. doi:10.3923/ijar.2012.46.57

Sonia, K. G., Chadha, B. S., & Saini, H. S. (2005). Sorghum straw for xylanase hyper-production by *Thermomyces lanuginosus* (D_2W_3) under solid-state fermentation. *Bioresource Technology, 96*(14), 1561–1569. doi:10.1016/j.biortech.2004.12.037 PMID:15978988

Stichnothe, H., & Schuchardt, F. (2011). Life cycle assessment of two palm oil production systems. *Biomass and Bioenergy, 35*(9), 3976–3984. doi:10.1016/j.biombioe.2011.06.001

Subramaniyan, S., & Prema, P. (2002). Biotechnology of microbial xylanases: Enzymology, molecular biology, and application. *Critical Reviews in Biotechnology, 22*(1), 33–64. doi:10.1080/07388550290789450 PMID:11958335

Sumathi, S., Chai, S. P., & Mohamed, A. R. (2008). Oil palm as a source of renewable energy in Malaysia. *Renewable & Sustainable Energy Reviews, 12*(9), 2404–2421. doi:10.1016/j.rser.2007.06.006

Sun, R. C., Tomkinson, J., Wang, Y. X., & Xiao, B. (2000). Physico-chemical and structural characterisation of hemicelluloses from wheat straw by an alkaline peroxide extraction. *Polymer, 41*(7), 2647–2656. doi:10.1016/S0032-3861(99)00436-X

Svarachorn, A. (1999). Production of fungal-xylanase using agricultural waste by solid state fermentation. *The Journal of Scientific Research Chulalongkorn University, 24*(1).

Szendefy, J., Szakacs, G., & Christopher, L. (2006). Potential of solid-state fermentation enzymes of *Aspergillus oryzae* in biobleaching of paper pulp. *Enzyme and Microbial Technology, 39*(6), 1354–1360. doi:10.1016/j.enzmictec.2006.06.016

Tan, L. U. L., Wong, K. K. Y., Yu, E. K. C., & Saddler, J. N. (1985). Purification and characterization of two D-xylanases from *Trichodermaharzianum. Enzyme and Microbial Technology, 7*(7), 425–430. doi:10.1016/0141-0229(85)90041-9

Tenkanen, H., Puls, J., & Poutanen, K. (1992). Two major xylanases of *Trichoderma reesei. Enzyme and Microbial Technology, 14*(7), 566–574. doi:10.1016/0141-0229(92)90128-B

Thanaporn, L., Parichart, K., Nakul, R., Phungjai, T., Saowanee, A., Paemanee, A., ... Eurwiilaichitr, L. (2011). Production of multi-fiber modifying enzyme from *Mamillisphaeria* sp. for refining of recycled paper pulp. *Bioscience, Biotechnology, and Biochemistry, 75*(12), 2297–2303. doi:10.1271/bbb.110431 PMID:22146707

Torronen, A., Harkki, A., & Rouvinen, J. (1994). Three dimensional structure of endo-1,4-β-xylanase II from *Trichoderma reesei*: Two conformational states in the active site. *The EMBO Journal, 13*(11), 2493–2501. PMID:8013449

Viniegra-Gonzalez, G., Favela-Torres, E., Aguilar, C. N., Romero-Gomez, S., Diaz-Godinez, G., & Augur, C. (2003). Advantages of fungal enzyme production in solid state over liquid fermentation system. *Biochemical Engineering Journal, 13*(2-3), 157–167. doi:10.1016/S1369-703X(02)00128-6

Wang, S. L., Yen, Y. H., Shih, I. L., Chang, A. C., Chang, W. T., Wu, W. C., & Chai, Y. D. (2003). Production of xylanases from rice bran by *Streptomyces actuosus* A-151. *Enzyme and Microbial Technology, 33*(7), 917–925. doi:10.1016/S0141-0229(03)00246-1

Wong, K. K. Y., Tan, L. U. L., & Saddler, J. N. (1988). Multiplicity of α-1,4 xylanase in microorganisms functions and applications. *Microbiological Reviews, 52*, 305–317. PMID:3141761

Xu, Y. M., Li, Y. L., Xu, S. C., Liu, Y., Wang, X., & Tang, J. W. (2008). Improvement of xylanase production by *Aspergillus niger* XY-1 using respond surface methodology for optimizing the medium composition. *Journal of Zhejiang University. Science, 9*(7), 558–566. doi:10.1631/jzus.B0820038 PMID:18600786

Yang, R., Xu, S., Wang, Z., & Yang, W. (2004). Aqueous extraction of corncob xylan and production of xylooligosacchrids. *Lebensmittel-Wissenschaft + Technologie, 38*(6), 677–682. doi:10.1016/j.lwt.2004.07.023

Yang, S. Q., Yan, Q. J., Jiang, Z. Q., Li, L. T., Tian, H. M., & Wang, Y. Z. (2006). High-level of xylanase production by the thermophilic *Paecilomyces thermophila* J18 on wheat straw in solid-state fermentation. *Bioresource Technology*, *97*(15), 1794–1800. doi:10.1016/j.biortech.2005.09.007 PMID:16230011

Ylikahri, R. (1979). Metabolic and nutritional aspects of xylitol. *Advances in Food Research*, *25*, 159–180. doi:10.1016/S0065-2628(08)60237-2 PMID:391001

Yuan, Q. P., Wang, J. D., Huai, Z., & Qian, Z. M. (2005). Effect of temperature shift on production of xylanase by *Aspergillus niger*. *Process Biochemistry*, *40*(10), 3255–3257. doi:10.1016/j.procbio.2005.03.020

Zulfiqar, A., Masood, S. B., Anjum, F. M., & Asgher, M. (2009). Effect of wheat bran concentration on xylanase biosynthesis by *Aspergillus niger*. *International Journal of Agriculture and Biology*, *11*, 571–576.

Chapter 14
Functional and Antioxidant Properties of Protein Hydrolysates From Ricinodendron Heudelotii(Bail.) Flours From Cameroon

Ngangoum Eric Serge
National Advanced School of Industrial Sciences, Cameroon

Mezajoug Kenfack Laurette Blandine
University of Ngaoundere, Cameroon

Sanjit Kumar
Vellore Institute of Technology, India

Mookambeswaran Vijayalakshmi
Vellore Institute of Technology, India

Tchiégang Clergé
University Institute of Technology, Cameroon

ABSTRACT

Ricinodendron heudelotii kernels were defatted and used as substrate to produce protein hydrolysatesusing papain (PHP), trypsin (PHT), proteases from Abrus precatorius (PHAp)and B. enneandra(PHBe). The degree of hydrolysis (DH), antioxidant (DPPH method), and functional properties of hydrolysates were performed. The DH value, whatever hydrolysis time, was highest with PHP. The water holding capacity decreased with the hydrolysis time from 21.50 ± 0.44% to 5.20 ± 0.07%. After 6h of hydrolysis, PHAp exhibited maximum solubility value (70.17 ± 2.15%) while PHBe had lower solubility value (18.43 ± 0.12%). The highest value of emulsifying activity index was found at pH 9 with 0.25% (w/v) hydrolysate concentration. Within the range of pH used (4-9), the best foam capacity and foam stability were exhibited by PHBe. PHP, PHAp, and PHT inhibited DPPH radical at 83.30 ± 0.46, 75.07 ± 0.15, and 56.78 ± 0.40%, respectively, at 6h of hydrolysis.

DOI: 10.4018/978-1-5225-5237-6.ch014

INTRODUCTION

Protein hydrolysates are a mixture of peptides and amino acids produced during enzymatic hydrolysis of proteins. They are recognized to exhibit certain functional properties that allow their use as a readily natural source of proteins for animal feed, human food and microorganism growth medium (Adjonu, Doran, Torley, & Agboola, 2013). Protein hydrolysates can act as functional foods through their antioxidant activity. Functional foods are known as foods that have a potentially positive biological effect on health a part of their basic traditional nutrition. Several authors have reported that protein hydrolysates from soy bean (Shih, Yang, & Kuo, 2002), mackerel fish (Wu, Chen, & Shiau, 2003), tilapia (Foh, Amadou, Foh, Kamara, & Xia, 2010) pink perch (Naqash & Nazeer, 2013), freshwater fish (Elavarasan & Shamasundar, 2015)and bovine plasma (Seo, Seo, & Yang, 2016) that possess antioxidant activity and can be used as natural antioxidant to substitute synthetic ones (Chi, Wang, Wang, Zhang, & Deng, 2015).Production of protein hydrolysates with good properties and useful for human nutrition requires enzymes whose availability is still a challenge due to their higher cost(Taha, Yamamah, Mohamed, Mohamed, & Wagdy, 2014).Therefore, protein hydrolysates production using local sources of proteases and substrates is a great asset for the population in developing countries.

Abrus precatorius and *Burnatia enneandra* are two edible domestic plants belonging to Fabaceae and Alismataceae family respectively. Proteolytic enzymes were identified and extracted from the leaves of *A. precatorius, B. enneandra* tubers (Mezajoug-Kenfack, Ngangoum, Tchiégang, & Linder, 2014; Ngangoum,Mezajoug Kenfack, Sanjit, Tchiégang& Vijayalakshmi, 2017). Their characteristics reveled that they can be used like papain, trypsin, alcalase and flavourzyme for protein hydrolysates production (Foh, Amadou, Foh, Kamara, & Xia, 2010; Martins, Costa, & Prentice-Hernández, 2010).

In Cameroon, some local kernels rich in proteins usually used by the population as spices, can serve as substrate to the hydrolysis. This is for example *Ricinodendron heudelotii* known as an endemic plant belonging to the Euphorbiaceae family. It is widely distributed in the equatorial forest of Madagascar, the Guinean Zone of West and Central Africa. Its kernels are consumed by the inhabitants for soup thickening (Mezajoug-Kenfack & Tchiégang, 2016). Research works carried out on description, physicochemical characterization and solubility of protein from *R. heudelotii* kernels, revealed that kernels contain more than 50% of lipids. Its defatted flour possess about 55% of crude proteins (Tchiégang, Kapseu, Ndjouenkeu, & Ngassoum, 1997). Proteins from defatted *R. heudelotii* kernels have good functional properties such as solubility and possess a good number of essential amino acids (Mezajoug-Kenfack, Arab-Tehrany, Tchiégang, & Linder, 2011). Nevertheless, to the best of our knowledge, no scientific work has been done regarding the production of protein hydrolysates from *R. heudelotii* kernel and their eventual antioxidant activity.

This work was undertaken to promote utilization of *R. heudelotii* proteins by producing protein hydrolysates using purified proteases from *A. precatorius* leavesand *B. enneandra* tubersas well as papain and trypsin. The specific goal was to determine and compare antioxidant and functional properties of protein hydrolysates from defatted *R. heudelotii* kernels flourprepared using papain, trypsin, purified proteases from *A. precatorius* leaves and*B. enneandra*tubers.

MATERIALS AND METHODS

Preparation of *R. Heudelotii* Defatted Kernels

R. heudelotii kernels were purchased from Akonolinga market, a local market in the Center Region of Cameroon. Cleaned kernels (1kg) were ground in a hammer mill for 3 minutes. The charge of grinding container was fixed at 2/3 (w/v) per round. Paste obtained was defatted through Soxhlet apparatus for 12h using hexane as solvent (AOAC, 1990). Finally, the meal was sieved with a 500μm mesh sieve and kept at 4°Cin an airtight container for further manipulations.

Chemical Composition of *R. Heudelotii* Defatted Kernels

Total carbohydrate, crude fiber and ash contents were performed in triplicate according to AOAC (1990). Nitrogen content of *R. heudelotii* defatted kernels was determined using micro-Kjeldahl techniqueaccording to AOAC (1980) method. Total protein content was evaluated by multiplying the nitrogen content by a factor of 6.25.

Production of Protein Hydrolysate from *R. Heudelotii* Defatted Kernels

Papain and trypsin were purchased from Sigma-Aldrich (USA) and Veronica Biotech (New Delhi, India) respectively. Prior to this work, local proteases used were purified from *A. precatorius* leaves (Ngangoum, Mezajoug Kenfack, Sanjit, Tchiégang, & Mookambeswaran, 2017) and *B. enneandra* tubers which are local endemic plants.Proteins were extracted from *R. heudelotii* defatted kernels using phosphate buffer (25mM, pH 8) and used as substrate to produce *R. heudelotii* protein hydrolysates. For this, enzymatic hydrolyses were performed with 4 enzymes namely: trypsin (phosphate buffer, 0.1 M; pH, 8.8; temperature, 37 °C),papain (phosphate buffer, 0.1 M; pH, 6.0; temperature, 37 °C) (Foh, Amadou, Foh, Kamara, & Xia, 2010; Martins, Costa, & Prentice-Hernández, 2010; Naqash & Nazeer, 2013; Wiriyaphan, Chitsomboon, & Yongsawadigul, 2012), purified protease from *A. precatorius*leaves (phosphate buffer, 50 mM; pH, 8.0; temperature, 40 °C) and purified protease from *B. enneandra*tubers (phosphate buffer, 50 mM; pH, 5.1; temperature, 40 °C) at enzyme /substrate ratio (10/100 v/v).

The substrate and each enzyme were mixed thoroughly and incubated at different times periods(30-360 min) at the optimal conditions of each protease (temperature, pH) with continuous stirring (Jamdar, Rajalakshmi, Pednekar, Juan, Yardi, & Sharma, 2010). The mixture was then heated in a boiling water bath at 100 °C for 10 min to inactivate the enzyme and centrifuged at 8000 rpm for 15 min. The supernatant was then collected and lyophilized (Christ Alpha 1-4 LSC). The lyophilized hydrolysates were packed in an airtight containers and stored at−20 °C until subjected to analyses.

Determination of the Degree of Hydrolysis

The degree of hydrolysis (DH)was determined according to the method of Alder-Nissen (1986). The samples were dissolved in 1% solution of sodium dodecyl sulfate. To the properly diluted sample (0.25 mL), 2 mL of sodium phosphate buffer (0.2 M, pH 8.2) and 2 mL of 0.01% trinitrobenzenesulfonic acid were added, followed by incubation in the dark for 60 min at 50°C. Before incubation, test tubes were covered with aluminum foil to avoid any acceleration of the reaction by light. The reaction was stopped by adding 4 ml of 0.1 N HCl and the absorbance was read at 420 nm. L-leucine solution was used as the standard and amino acid content was expressed in terms of L-leucine. Total amino acid content (L_{max}) of defatted flour was obtained after treated sample with HCl 6N at 100°C for 24h in an oven.

DH was calculated as follows:

$$DH = [(L_t - L_0) / (L_{max} - L_0)] * 100$$

Where $L_0$is the amount of amino acid released at time 0 h and L_t the amount of amino acid in the supernatant at time t.

Determination of Antioxidant Activity

The 1,1 diphenyl-2-picrylhydrazyl (DPPH) radical scavenging activity of proteinhydrolysates was done according to the method decribed by Intarasirisawat, Benjakul, Visessanguan and Wu (2012)with some modifications. The reaction mixture contained 100 µL of various concentration of proteinhydrolysates and mixed with 900µL of 0.1 mM DPPH inmethanol. The mixure was incubated for 30 min at room temperature in dark. After incubation, the decrease in the absorbance (A) was measuredat 517 nm. The blank tube contained methanol instead of protein hydrolysate. The butylated hydroxyanisole(BHA) 0.004mgmL^{-1} was used as positifcontrol whereas ascorbic acidhelped to draw the standard curve. The percentage of DPPH radical inhibition (I%) by the protein hydrolysates was calculated according to the following equation:

$$I(\%) = \frac{(A_{Blank} - A_{Sample})}{A_{Blank}} \times 100$$

Determination of Functional Properties

Solubility

The solubility of protein hydrolysatesfrom *R.heudelotii* was determined at different hydrolysistimesaccording to the method described by Naqash and Nazeer (2013) with a slight modification. Samples were dispersed in distilled water and the pH of the mixture was adjusted at 7 using HCl or NaOH. The reaction mixture having a concentration of 10 mg/mL was stirred for an hour and centrifuged at 10000 rpm for 15 min. Protein content in the supernatant was determined using Bradford method while the total protein in the sample was evaluated by Kjeldahl method. Protein solubility (S%) was calculated using the following equation:

$$S(\%) = \frac{\text{Protein content in supernatant}}{\text{Total protein content in sample } (\text{N x6.25})} \text{x100}$$

Water Holding Capacity

The water holding capacity (WHC) of *R. heudelotii* protein hydrolysates was evaluated according to the method described byOkezie, Akanbi, Otunola, and Adeyemi (2003) with some slight modifications. 50 mg of the sample (W_0) were dissolved in 5 mL distilled water and mixed for 2 min using an electric shaker. The mixture was allowed at room temperature(25°C) for 30 min and further, centrifuged at 12000 rpm for 15min. The pallet was collected, weighed (W_1) and dried in an oven for 5 h at 105°C, then reweighed(W_2). The WHC expressed in percentage was calculated as follows:

$$\mathrm{WHC}(\%) = \frac{\left(\mathrm{W}_2 - \mathrm{W}_0\right)}{\mathrm{W}_1} \times 100$$

Emulsifying Properties

The method described by Pearce and Kinsella (1978)with little modification was investigated to evaluate emulsifying activity index (EAI) and emulsion stability index (ESI). Emulsion was prepared by mixing 10 mL soya beans oil with 30 mL of protein hydrolysates solution (1%) and homogenized at 12000 rpm for 2min. A portion of 50µL of emulsion was taken from the bottom of the tube at 0 and 10 min respectively after homogenization and mixed thoroughly with 5mL sodium dodecyl sulphate (0.1%). Absorbance(A)of thediluted solution were measured at 500 nm and values was used to calculate EAI and ESI. The calculations were performed as follows:

i. $$EAI\left(m^2 / g\right) = \frac{2 \times 2.303 \times A_{500}}{0.25 \times protein \quad weight\ \left(g\right)}$$

ii. $$ESI\ \left(\mathrm{min}\right) = A_0 \times \Delta t / \Delta A$$

Where, $\Delta A = A_0 - A_{10}$ and $\Delta t = 10$ min

Foaming Properties

The foaming capacity (FC) and foam stability (FS) of protein hydrolysates from *R. heudelotii* were determined according to the method of Sathe and Salunkhe (1981). 10mL of protein hydrolysate solutions prepared at different concentrations in MilliQ water (0.1, 0.25, 0.5 and 0.75% w/v) were thoroughly homogenized in a 25 mL cylinder at 15000 rpm for 5min. The volume of the mixture was taken immediately after homogenization (V_T), then the cylinder was kept at room temperature (25°C) for 10 min and the volume later measured (V_t). FC and FS expressed in percentages were determined using the following formulas:

iii. $FC\ (\%) = \frac{V_T}{V_0} \times 100\ FS\ (\%) = \frac{Vt}{V_0} \times 100$

Where, V_0 is the initial volume before whipping.

RESULTS AND DISCUSSION

Proximate Composition of the Defatted Kernels and Degree of Hydrolysis

Proximate Composition

R. heudelotii kernels were defatted to remove oil contain in order to maximize protease actions during protein hydrolysates production. *R. heudelotii* kernel contains 52.34% total lipids. This values is similar to that of Tchiégang *et al.* (1997) who previously found that, the same kernel possessed more than 50% of total fat. The chemical composition of defatted *R. heudelotii* kernels flour with particle size less than 500µm is given in table1.From the results, the high ash content (10.42%) indicates that defatted *R. heudelotii* defatted kernels is an important source of mineral. The protein content (46.65%) was lower than that of defatted unsieved flour (55 – 62%) (Tchiégang, Kapseu, Ndjouenkeu, & Ngassoum, 1997) and higher compared to the value of 40.53%found by Mezajoug-Kenfack and Tchiégang (2016), with particle size between 400 - 500µm. With 46.65% of protein content, *R. heudelotii* defatted kernels flour is a good source of proteins compared to soy beans (40.3%) which is often used as substrate for protein hydrolysate production (Vollman, Fritz, Wagentristi, & Ruckenbauer, 2000).

Degree of Hydrolysis

The hydrolysis curves of proteins from*R. heudelotii* defatted kernels after 360 min of incubation with various proteases are presented in Figure 1. Initially, there is an exponential increase in DH with increased time of hydrolysis. The rate of enzymatic reaction reached its steady-state phase after 120 min for purified proteases from *A. precatorius* leaves, *B. enneandra* tubers and 240 min for papain and trypsin. The initial exponential increase of enzymatic hydrolysis rate indicated that maximum cleavage of peptide bonds occurred during this phase. The decrease of the reaction rate following by the steady-phase could be due to a slowering in the concentration of peptide bonds within the protein substrate available for enzymatic hydrolysis, the inhibition by the products or reduction of enzyme activity due to the denaturation (Intarasirisawat, Benjakul, Visessanguan, & Wu, 2012). Similar results in hydrolysis curves were reported for splendid squid (*Loligo formosana*) (Hamzeh, Benjakul, & Senphan, 2016) and yellow stripe trevally (*Selaroides leptolepis*) (Klompong, Benjakul, Kantachote, & Shahidi, 2007). As presented in Figure 1, papain shows the highest DH values (71.27%), followed by trypsin (54.74%) protease from *A. precatorius*leaves (38.91%) and protease from *B. enneandra* tubers(32.65%). This result indicates that papain hasthe highest efficiency while protease from *B. enneandra* tubers has the lowest in *R. heudelotii* defatted kernels flour protein hydrolysis. Indeed, alcalin proteases like papain, trypsin and protease from *A. precatorius* leavesreact faster on this type of subtrate (plant proteins) than neutral and acid protease like those of *B. enneandra* tubers (Klompong, Benjakul, Kantachote, & Shahidi, 2007).

Figure 1.

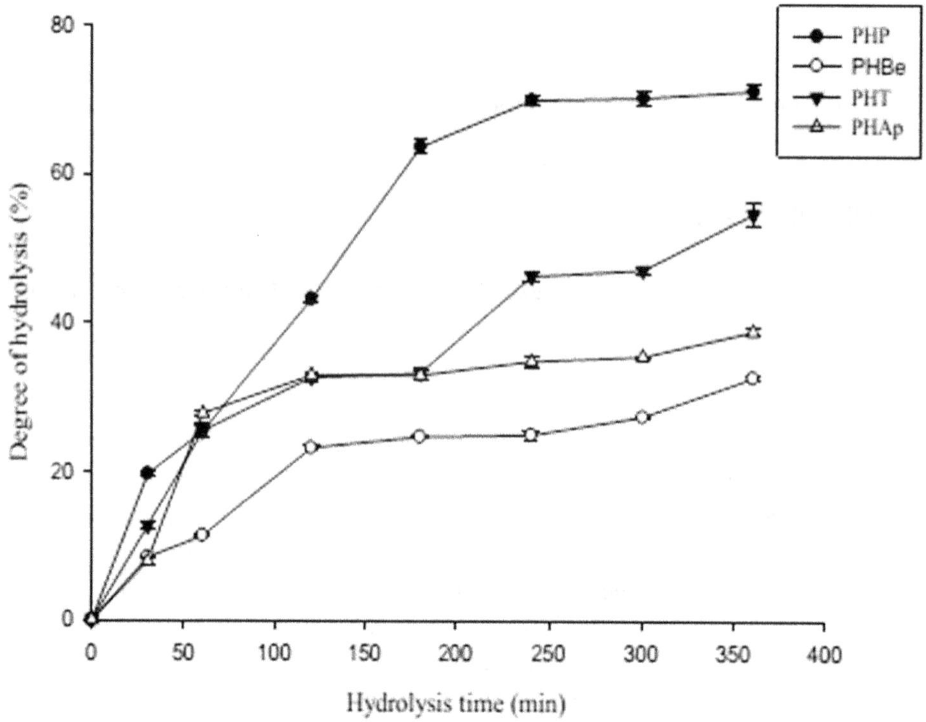

Antioxidant Activity

Protein hydrolysates generated from *R. heudelotii* defatted kernels flour using papain (PHP), trypsin (PHT), proteases from *A. precatorius* leaves(PHAp) and *B. enneandra* tubers (PHBe) were tested for their antioxidant activity with DPPH radical scavenging method. Apart from protein hydrolysates produced with protease from *B. enneandra* tubers, all protein hydrolysates prepared from *R. heudelotii* defatted kernels inhibited DPPH radical as shown in Figure 2. The antioxidant activity curve of PHP, PHT and PHAp increased with time compared to the activity from BHA which is a commercial antioxidant. The highest activity was obtained from PHP (83.03%) compared to PHAp (75.07%). The difference in the antioxidant activity can be justified by the proteolytic enzymes specificity. As a matter of fact, several authors have reported that the function of protein hydrolysates are governed by protein substrates, hydrolysis conditions and proteases used for hydrolysis (Foh, Amadou, Foh, Kamara, & Xia, 2010; Hamzeh, Benjakul, & Senphan, 2016; Jun, Park, Jung, & Kim, 2004; Klompong, Benjakul, Kantachote, & Shahidi, 2007).

The antioxidant activity increased with protein hydrolysate concentrations (Figure 2). The results revealed that protein hydrolysates from *R. heudelotii* defatted kernels prepared with papain, trypsin and protease from *A. precatorius* leaves have electron-donor peptides which can react with free unstable radicals. This reaction with free unstable radicals converted them to more stable products and consequently will enable to end the radical chain reaction.

Figure 2.

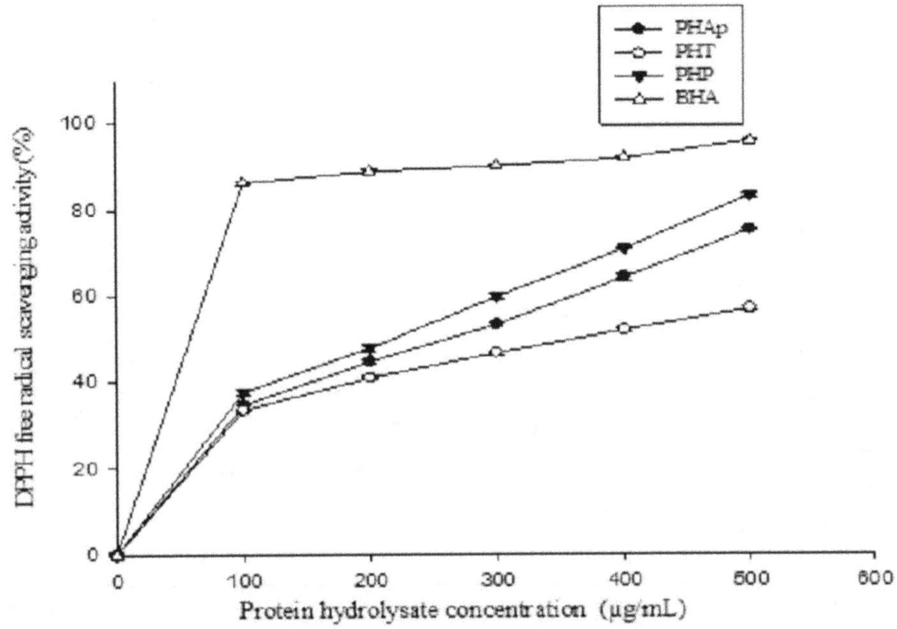

Functional Properties of Protein Hydrolysates From *R. Heudelotii* Defatted Kernels

Solubility

Protein hydrolysates (PH) were produced from*R. heudelotii* deffated kernelsusing papain (PHP), trypsin (PHT),protease from *A. precatorius* leaves (PHAp) and *B. enneadra* tubers (PHBe) as enzymes and the percentage of their solubilities were recorded. Figure 3 shows the trends of the solubilitycurves of PH generated with the proteases. Whatever the protease used, the solubility of PH increases with the hydrolysis time.The solubility of PHP and PHAp were higher than 55%, while those of PHT and PHBewere less than this value.The lowest solubility was obtained with PHBewhich alsoshows the lowest affinity of purified enzyme from *B. enneeandra* tuberswith the substrate. Resuts in Figure 1 indicated that the DH ofpapain is higher than that obtained with trypsin followed by that of proteases from*A. precatorius*-leavesand *B. enneandra*tubers.These results give an idea on the rate of protein hydrolysisof the substrate by the enzymes and can justify the difference of the solubility of PH produced.

Enzymatichydrolysisgenerates smaller peptides which have probably more polar groups such as $-NH_2$, -COOH, with the ability to form hydrogen bonds with water and thus increasessolubility (Gbogouri, Linder, Fanni, & Parmentier, 2004). As a consequence, *R. heudelotii* PH which containshigh quantity ofsmaller peptides (generated at the highest values of DH) were more soluble. This result is in line with some founding inthe literature which point out that the solubility of protein increase with the DH (Gbogouri, Linder, Fanni, & Parmentier, 2004; Klompong, Benjakul, Kantachote, & Shahidi, 2007).

Figure 3.

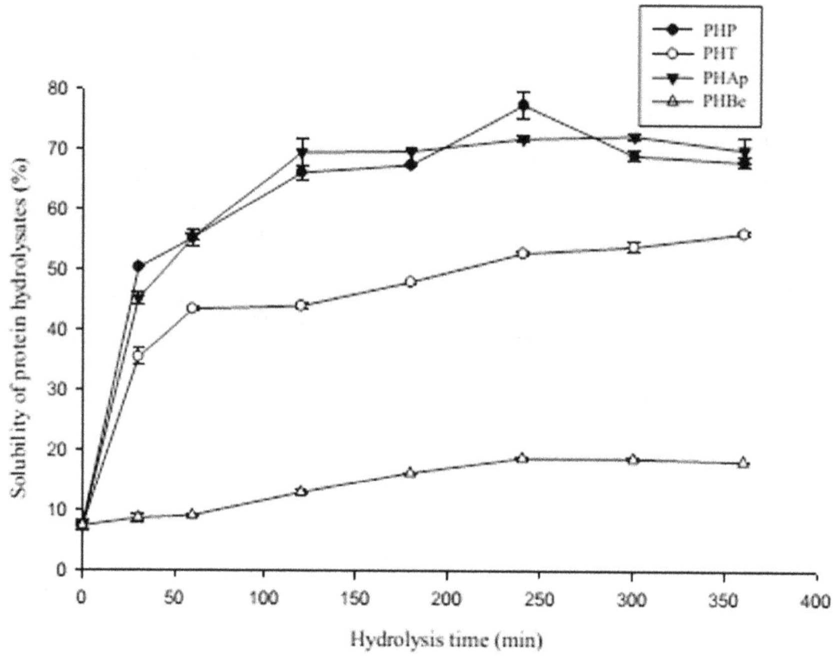

Water Holding Capacity

Water holding capacity (WHC) is the ability of a substance or a matrix to absorb water and retain it against a gravitational force within a protein hydrolysate matrix (Foh, Amadou, Foh, Kamara, & Xia, 2010). The WHC results of protein hydrolysates from *R. heudelotii* defatted kernels are shown in Figure 4. This figure easily brings out the decreasing trend of WHC with respect to the hydrolysis time. WHC

Figure 4.

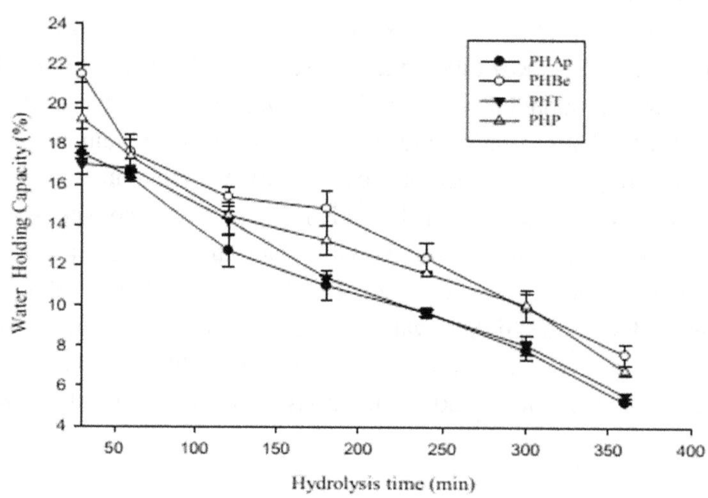

of protein hydrolysates produced with protease from *B. enneandra* tubers(21.50%) is higher than that generated with papain (19.20%), protease from *A. precatorius* leaves(17.50%) and trypsin (17%). This can be related to the difference in hydrolysis rate of the substrates by the enzymes. The decreasing trend of WHC of *R. heudelotii* hydrolysates could be due to hydrolytic process of the substrate by enzymes. Enzymatic hydrolysis causes possible disruption of the protein network which is responsible of WHC within the protein matrix (Meinlschmidt, Sussmann, Schweiggert-Weisz, & Eisner, 2016). This result corroborated that of Meinlschmidt *et al.* (2016). These authors pointed out a similar trend of WHC for soy protein isolates. Conversely, Balti, Bougatef, El-Hadj, Zekri, Barkia, and Nasri (2010) reported that WHC of cuttlefish protein hydrolysates increases from 2.5 to 5.5 mL/g with increasing degree of hydrolysis.

Emulsifying Property

The method described by Pearce and Kinsella with some modifications was performed to evaluate Emulsifying Activity Index (EAI), Emulsion Stability Index (ESI) at various concentrations and pH. The results are shown in Table 2. No matter the enzymes used, the highest EAI:62.85, 42.58, 41.07 and 74.39m^2/g for HPAp, HPBe, HPP and HPT respectively were obtained at pH9 with protein hydrolysates concentration of 0.25% (w/v). This high value of EAI at pH9 can be due to the maximum solubility of the peptides at that pH. At lower pH, peptides generated during enzymatic hydrolysis had lower amount of ionic charge and could not rapidly diffuse and adsorb at the interface (Klompong, Benjakul, Kantachote, & Shahidi, 2007). This result is also supported by that of Naqash and Nazeer (2013) who reported that EAI of pink perch (*Nemipterus japonicas*) increased as pH moved away from pH 4. Difference was found among EAI and ESI values from one protease to another and this can be attributed to enzyme specificity. This result corroborated those found by Klompong *et al.* (2007) who reported that the EAI and ESI vary when alcalase or flavourzyme was used as proteases for yellow stripe trevally (*Selaroides leptolepis*) protein hydrolysates preparation. Emulsions were more stable at theextreme pH values either 4 or 9. This can be due to the fact that at those pH, the ionization of proteins and protein hydrolysates is high and allows them to move faster in interface.

Foam Capacity and Stability

The effect of protein hydrolysate concentrations on the Foam Capacity (FC) and Foam Stability (FS) with respect to pHare presented in Figure 5. The FCand FS values increase from 4 to 93% and from 1 to 41.5%respectively with the increase of protein hydrolysates concentration whatever the protease used. Similar results were achieved with vitabosa (*Mucuna deeringiana*) and soybean (*Glycine max*) proteins(Acuña, González, & Aristizábal-Torres, 2012).However, a great difference was observed among the values from one protease to another. The highest FC (93%) and FS (41.5%) values were obtained with protein hydrolysates produced using protease from *B. enneandra* whilethose generated by trypsin showed the lowest FC (4%).The smallest value of FS (1%) was exhibited by protein hydrolysates from trypsin and protease from *A. precatorius* leaves. They were no explicit relationship between FC and FS of protein hydrolysates generated with papain, trypsin and protease from *A. precatorius*. The difference in FC and FS of *R. heudelotii* defatted kernels from one protease to another could result from the variations in protein hydrolysate contents and their solubility. Protein hydrolysates are amphiphilic molecules in nature and due to this, they tend to place themselves according to their solubility at the oil/water and air/water interface preventing foam coalescence and therefore stabilize them (Santiago, Maldonado-Valderrama,

Figure 5.

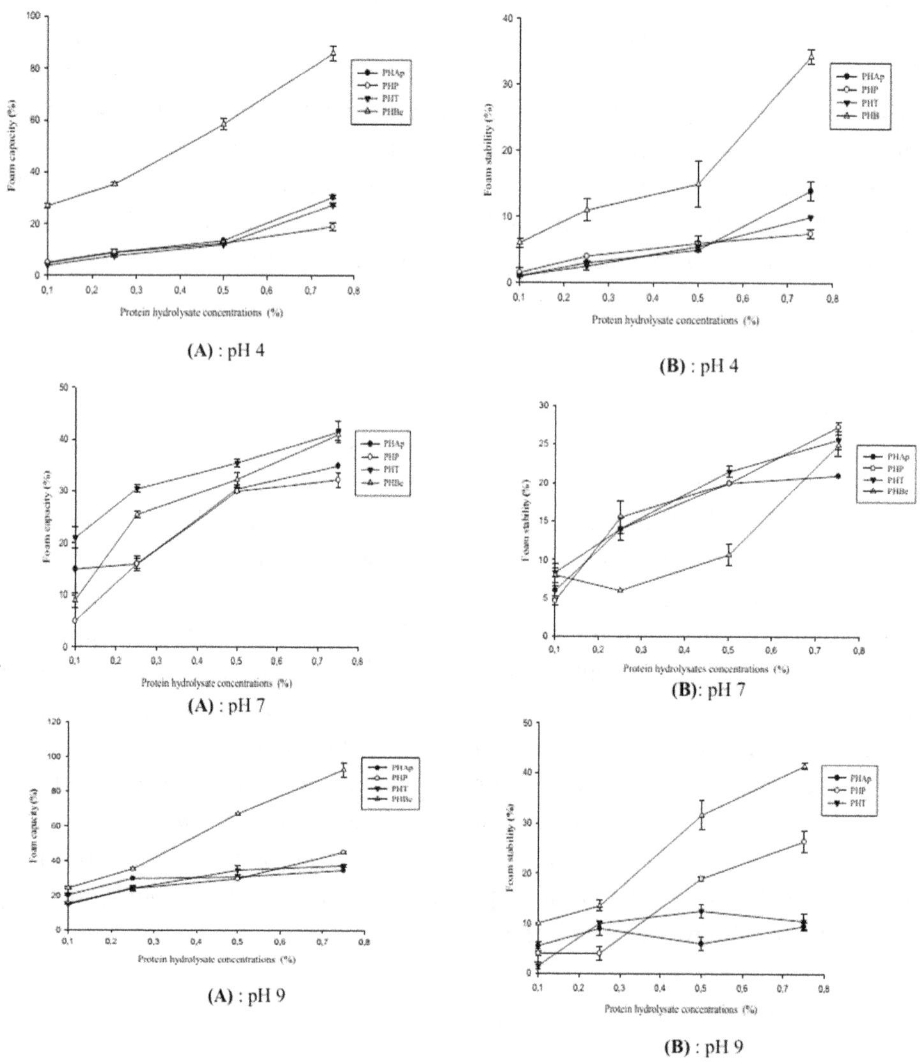

Martin-Molina, Haro-Pérez, Maldonado-Valderrama, Martin-Molina, & Gàlvez-Ruiz, 2008). FC and FS were also found to increase with increase in pH from 4. However, the best values were obtained at pH 7. According to Cheftel, Cuq, and Lorient (1985), FC and FS values are higherat the extreme pH values. At the extreme pH values, there exist more polar residues in proteins and protein hydrolysates structure that favor their solubility and can improve their movement and absorption in the interface.

CONCLUSION

Protein hydrolysates from *R. heudelotii* defatted kernels were prepared using purified proteases from local plants, *A. precatorius* leaves and *B. enneandra* tubers as well as papain and trypsin. These hydrolysates possessed functional properties which vary with enzymes used,time of hydrolysis and the protein hydrolysate concentrations. All protein hydrolysates prepared exhibited antioxidant activity at different protein hydrolysate concentrations, apart of protein hydrolysates generated with protease from *B. enneandra* tubers. Therefore *R. heudelotii* protein hydrolysates can be used in food processing as a natural additive either for their functional properties or antioxidant activities.

ACKNOWLEDGMENT

We are thoroughly thankful to the Department of Biotechnology Government of India (DBT), and TWAS(The Academy of Sciences for the developing World) for their intellectual and material contributions.

REFERENCES

Acuña, C. P. S., González, G. H. J., & Aristizábal-Torres, I. D. (2012). Physicochemical characteristics and functional properties of vitabosa (*Mucuna deeringiana*) and soybean (*Glycine max*). *Food Science and Technology (Campinas)*, *32*(1), 98–105. doi:10.1590/S0101-20612012005000007

Adjonu, R., Doran, G., Torley, P., & Agboola, S. (2013). Screening of whey protein isolate hydrolysates for their dual functionality: Influence of heat pre-treatment and enzyme specificity. *Food Chemistry*, *136*(3-4), 1435–1443. doi:10.1016/j.foodchem.2012.09.053 PMID:23194546

Alder-Nissen, J. (1986). *Enzymatic hydrolysis of food proteins*. London: Elsevier Applied Science.

AOAC. (1990). *Association of the Official Analytical Chemists* (K. Elrich, Ed.). Washington, DC: AOAC.

Balti, R., Bougatef, A., El-Hadj, A. N., Zekri, D., Barkia, A., & Nasri, M. (2010). Influence of degree of hydrolysis on functional properties and angiotensin I-converting enzyme inhibitory activity of protein hydrolysates from cuttlefish (*Sepia officinalis*) by-products. *Journal of the Science of Food and Agriculture*, *90*, 2006–2014. PMID:20583200

Cheftel, J. C., Cuq, J. L., & Lorient, D. (1985). *Protéines alimentaires: Biochimie - propriétés fonctionnelles, valeur nutritionnelle - Modifications chimiques*. Technique et Documentation, Lavoisier-Paris-France.

Chi, C. F., Wang, B., Wang, Y. M., Zhang, B., & Deng, S. G. (2015). Isolation and characterization of three antioxidant peptides from protein hydrolysate of bluefin leatherjacket (*Navodon septentrionalis*) heads. *Journal of Functional Foods*, *12*, 1–10. doi:10.1016/j.jff.2014.10.027

Elavarasan, K., & Shamasundar, B. A. (2015). Effect of oven drying and freeze drying on the antioxidant and functional properties of protein hydrolysates derived from freshwater fish (*Cirrhinus mrigala*) using papain enzyme. *Journal of Food Science and Technology*. doi:10.100713197-015-2084-9 PMID:27162411

Foh, M. B. K., Amadou, I., Foh, B. M., Kamara, M. K., & Xia, W. (2010). Functionality and Antioxidant Properties of Tilapia (*Oreochromis niloticus*) as Influenced by the degree of hydrolysis. *International Journal of Molecular Sciences*, *11*(12), 1851–1869. doi:10.3390/ijms11041851 PMID:20480046

Gbogouri, G. A., Linder, M., Fanni, J., & Parmentier, M. (2004). Influence of hydrolysis degree on the functional properties of salmon byproducts hydrolysates. *Journal of Food Science*, *69*(8), 615–622. doi:10.1111/j.1365-2621.2004.tb09909.x

Hamzeh, A., Benjakul, S., & Senphan, T. (2016). Comparative study on antioxidant activity of hydrolysates from splendid squid *(Loligo formosana)* gelatin and protein isolate prepared using protease from hepatopancreas of Pacific white shrimp *(Litopenaeus vannamei)*. *Journal of Food Science and Technology*, *53*(6), 3615–3623. doi:10.100713197-016-2348-z PMID:27777469

Intarasirisawat, R., Benjakul, S., Visessanguan, W., & Wu, J. (2012). Antioxidative and functional properties of protein hydrolysate from defatted skipjack (*Katsuwonous pelamis*) roe. *Food Chemistry*, *135*(4), 3039–3048. doi:10.1016/j.foodchem.2012.06.076 PMID:22980906

Jamdar, S. N., Rajalakshmi, V., Pednekar, M. D., Juan, F., Yardi, V., & Sharma, A. (2010). Influence of degree of hydrolysis on functional properties, antioxidant activity and ACE inhibitory activity of peanut protein hydrolysate. *Food Chemistry*, *121*(1), 178–184. doi:10.1016/j.foodchem.2009.12.027

Jun, S. Y., Park, P. J., Jung, W. K., & Kim, S. K. (2004). Purification and characterization of an antioxidative peptide from enzymatic hydrolysate of yellowfin sole (*Limanda aspera*) frame protein. *European Food Research and Technology*, *219*(1), 20–26. doi:10.100700217-004-0882-9

Klompong, V., Benjakul, S., Kantachote, D., & Shahidi, F. (2007). Antioxidative activity and functional properties of protein hydrolysate of yellow stripe trevally (*Selaroides leptolepis*) as influenced by the degree of hydrolysis and enzyme type. *Food Chemistry*, *102*(4), 1317–1327. doi:10.1016/j.foodchem.2006.07.016

Martins, V. G., Costa, J. A. V., & Prentice-Hernández, C. (2009). Hidrolisado protéico de pescado obtido por vias química enzimática a partir de corvina (*Micropogonias furnieri*). *Quimica Nova*, *32*(1), 61–66. doi:10.1590/S0100-40422009000100012

Meinlschmidt, P., Sussmann, D., Schweiggert-Weisz, U., & Eisner, P. (2016). Enzymatic treatment of soy protein isolates: Effects on the potential allergenicity, technofunctionality, and sensory properties. *Food Science & Nutrition*, *4*(1), 11–23. doi:10.1002/fsn3.253 PMID:26788306

Mezajoug-Kenfack, L. B., Arab-Tehrany, E., Tchiégang, C., & Linder, M. (2011). Compositional and nutritional studies of two defatted flours obtained from *Ricinodendron heudelotii* (Bail.) and *Tetracarpidium conophorum* (Müll. Arg.). *Food Global Science Books*, *5*(1), 52–57.

Mezajoug-Kenfack, L. B., Ngangoum, E. S., Tchiégang, C., & Linder, M. (2014). Proteolytic enzymes from *Abrus precatorius* Linn. *Burnatia enneandra* Micheli. and *Ziziphus mauritiana* Lam: Optimization of the extraction process by Response Surface Methodology. *The Journal of Bioprocess Technology*, *100*, 453–461.

Mezajoug-Kenfack, L. B., & Tchiégang, C. (2016). Physico chemical properties of defatted flours from *Ricinodendron heudelotii* (Bail.) and *Tetracarpidium conophorum* (Müll. Arg.) with respect to particle sizes. *Global Advanced Research Journal of Agricultural Science*, *5*(2), 94–102.

Naqash, S. Y., & Nazeer, R. A. (2013). Antioxidant and functional properties of protein hydrolysates from pink perch (*Nemipterus japonicus*) muscle. *Journal of Food Science and Technology*, *50*(5), 972–978. doi:10.100713197-011-0416-y PMID:24426005

Ngangoum, E. S., Mezajoug Kenfack, L. B., Sanjit, K., Tchiégang, C., & Mookambeswaran, V. M. (2017). Extraction, purification, and biochemical characterization of serine protease from leaves of Abrus precatorius. *Preparative Biochemistry & Biotechnology*, 1–9. doi:10.1080/2017/1373289 PMID:28857663

Okezie, U., Akanbi, C. T., Otunola, E. T., & Adeyemi, I. A. (2003). Effect of addition of ripe bananas on some physico-chemical properties of maize extract. *International Journal of Food Sciences and Nutrition*, *54*(6), 437–445. doi:10.1080/09637480220138106 PMID:14522689

Pearce, K. N., & Kinsella, J. E. (1978). Emulsifying properties of proteins: Evaluation of turbidimetric technique. *Journal of Agricultural and Food Chemistry*, *26*(3), 716–723. doi:10.1021/jf60217a041

Santiago, L. G., Maldonado-Valderrama, J., Martin-Molina, A., Haro-Pérez, C., Maldonado-Valderrama, J., Martin-Molina, A., & Gàlvez-Ruiz, M. J. (2008). Adsorption of soy protein isolate at air–water and oil–water interfaces. *Colloids and Surfaces. A, Physicochemical and Engineering Aspects*, *323*(1-3), 155–162. doi:10.1016/j.colsurfa.2007.11.001

Sathe, S. K., & Salunkhe, D. K. (1981). Functional properties of the Great Northern bean (*Phaseolus vulgaris* L.) protei:s: emulsion, foaming, viscosity and gelation properties. *Journal of Food Science*, *46*(1), 71–74. doi:10.1111/j.1365-2621.1981.tb14533.x

Seo, H.-W., Seo, J. K., & Yang, H. S. (2016). Supplementation of Pork Patties with Bovine Plasma Protein Hydrolysates Augments Antioxidant Properties and Improves Quality. *Korean J. Food Sci. An.*, *36*(2), 198–205. doi:10.5851/kosfa.2016.36.2.198 PMID:27194928

Shih, M. C., Yang, K. T., & Kuo, S. J. (2002). Quality and antioxidative activity of black soybean tofu as affected by Bean Cultivar. *Journal of Food Science*, *67*(2), 480–484. doi:10.1111/j.1365-2621.2002.tb10623.x

Taha, F. S., Yamamah, G. A., Mohamed, G. F., Mohamed, S. S., & Wagdy, S. M. (2014). Biological Activities of Sunflower Protein Hydrolysate. *World Applied Sciences Journal*, *30*(11), 462–1470.

Tchiégang, C., Kapseu, C., Ndjouenkeu, R., & Ngassoum, M. B. (1997). Amandes de *Ricinodendron heudelotii* (Bail): Matière première potentielle pour les industries agro-alimentaires tropicales. *Journal of Food Engineering*, *32*(1), 1–10. doi:10.1016/S0260-8774(97)00094-0

Vollman, J., Fritz, N. C., Wagentristi, H., & Ruckenbauer, P. (2000). Environmental and genetic variation of soybean seed protein content under Central European growing conditions. *Journal of the Science of Food and Agriculture*, *80*(9), 1300–1306. doi:

Wiriyaphan, C., Chitsomboon, B., & Yongsawadigul, J. (2012). Antioxidant activity of protein hydrolysates derived from threadfin bream surimi byproducts. *Food Chemistry*, *132*(1), 104–111. doi:10.1016/j.foodchem.2011.10.040 PMID:26434269

Wu, H. C., Chen, H. M., & Shiau, C. Y. (2003). Free amino acids and peptides as related to antioxidant properties in protein hydrolysates of mackerel (*Scomber austriasicus*). *Food Research International*, *36*(9-10), 949–957. doi:10.1016/S0963-9969(03)00104-2

APPENDIX

Values in this table are given as means ± standard deviation of three measurements.

Table 1. Proximate composition of R. heudelotiide fatted kernels

Chemical Parameters	Values in g/100gof Dry Weight
Ash	10.42 ± 0.72
Total proteins (N x 6.25)	46.65 ± 1.32
Carbohydrates	7.76 ± 0.86
Crude fibers	3.14 ± 0.46

Table 2. Emulsifying activity index (EAI) (m^2/g) and emulsifying stability index (ESI) (%) of protein hydrolysates from defatted R. heudelotii kernelsflour

	Protein Hydrolysates	pH 4		pH 7		pH 9	
	Concentration (%)	EAI	ESI	EAI	ESI	EAI	ESI
PHAp	0.1	7.48 ± 0.46^{c}	11.53 ± 0.85^{a}	4.36 ± 026^{b}	18.61 ± 1.02^{hg}	11.77 ± 0.48^{a}	**33.89 ± 0.47^{g}**
	0.25	17.33 ± 0.76^{g}	13.33 ± 0.16^{bc}	13.37 ± 0.73^{f}	16.81±0.38^{ef}	**62.85 ± 0.56^{l}**	27.41 ± 0.59^{f}
	0.5	6.26±0.08^{b}	15.74 ± 0.32^{d}	3.91 ± 0.23^{b}	31.24 ± 0.67^{k}	45.32 ±0.51^{k}	16.61± 1.80^{c}
	0.75	6.59 ± 0.98^{b}	12.84 ± 0.13^{ab}	4.03 ± 0.42^{b}	18.95 ±0.04^{hg}	17.96 ± 0.97^{c}	46.01 ± 1.79^{i}
PHBe	0.1	12.35 ± 0.22^{f}	14.53 ± 0.49^{cd}	14.17 ± 0.71^{f}	14.11±0.55^{cd}	18.99±0.46^{ce}	10.99±0.25^{a}
	0.25	37.21 ± 0.83^{j}	25.69 ±0.06^{g}	28.78 ± 0.71^{i}	11.23±0.23^{a}	**42.58 ± 0.78^{j}**	11.04±0.36^{a}
	0.5	18.73 ± 0.63^{h}	**27.11 ± 0.53^{h}**	20.30±0.81^{h}	12.43± 1.70^{ab}	41.47 ± 1.39^{ij}	13.46±0.06^{b}
	0.75	9.49±0.39^{e}	33.71 ± 0.91^{i}	12.10±0.62^{g}	12.49 ± 0.90^{ab}	21.32±0.43^{f}	16.35 ±0.58^{c}
PHP	0.1	8.91 ± 0.14^{de}	12.64 ±0.80^{ab}	17.85 ± 0.91^{g}	18.93 ± 1.27^{hg}	**41.07 ± 0.93^{i}**	10.54±0.43^{a}
	0.25	9.70 ±0.34^{e}	13.35 ± 1.02^{bc}	10.29 ± 0.35^{d}	25.33 ± 0.58^{j}	26.94 ± 0.27^{g}	14.05 ± 0.84^{b}
	0.5	4.30±0.53^{a}	18.93± 0.86^{e}	2.21 ± 0.11^{a}	18.74± 1.81^{hg}	15.85 ± 0.83^{b}	16.27 ± 0.56^{c}
	0.75	4.14±0.26^{a}	20.82±1.11^{f}	2.53 ± 0.26^{a}	22.55 ± 0.94^{i}	26.86±0.12^{g}	**27.10 ± 053^{f}**
PHT	0.1	16.58±0.46^{g}	**50.50 ± 1.07^{j}**	17.49 ± 0.79^{g}	17.49±0.51^{fg}	37.54±3.58^{h}	41.71 ±0.69^{h}
	0.25	31.09±0.23^{i}	20.95 ± 0.97^{f}	34.54 ± 0.66^{j}	20.08 ±0.86^{hi}	**74.39 ± 1.63^{m}**	20.37 ± 0.59^{e}
	0.5	8.53 ±0.59^{d}	19.25 ± 0.92^{e}	13.58 ± 0.94^{f}	12.78 ±0.78^{bc}	19.4±1.38^{ef}	18.36 ± 1.38^{d}
	0.75	6.72±0.60^{bc}	15.58 ± 0.53^{d}	8.42 ±0.16^{c}	15.48 ±0.23^{de}	20.65±1.74^{ef}	16.15 ± 0.18^{c}

Mean ± SD (n=3), Different small letter exponent of values in the same column indicate that there is no significant differences ($p<0.05$)

Chapter 15
Enzyme Use and Production in Industrial Biotechnology

Subir Kumar Nandy
Technical University of Denmark, Denmark

ABSTRACT

This chapter demonstrates the bioprocess strategies involved in the application and production of enzymes from an industrial view point. Moreover, bottlenecks in enzyme production and novel strategies to overcome the barriers are demonstrated here. Enzymes are produced from different sources of microorganisms and mostly all biological reactions happen due to the help of enzymes within a very short time. The different uses of enzymes are discussed in this chapter.

INTRODUCTION

A chemical reaction can be fastened by the catalyst. Enzymes are using as a biocatalyst in several biochemical reactions. Enzymes are also proteins that used to fast the reaction rate to form a product. Enzymes are available from variable sources such as from extreme environment or from microorganism from nature. Therefore, enzymes are the key factor to build and maintain the microorganisms such as bacteria or fungi and also all living organisms including human beings. Enzymes are so-called very small machines evolved millions of years to perform very specific biochemical tasks. Some enzymes have been designed by nature to build chemical compounds and others are either breaking them or modifying them over time. These breaking or modifying bond between molecules called reactions. Enzymes are used as a catalyst for the reaction so that the reaction is very fast (million times faster than normal reactions) and effective. As the population increases from last decades, researchers are working towards increasing the yield of feed/food, fruits and vegetables are in the increasing trend while a record of production of 1.74 billion tons in 2013 which is 9.4% more than in 2012 (WFO 2014). India is the second largest producer of fruits and vegetables with a global production share of 10% and 14% respectively (Ingale, Joshi, & Gupta, 2014). Therefore, total fruit and vegetables producing in all countries are limitless and their wastes are countless. Pollution and greenhouse gas effects are increased by the disposal of these fruit and vegetable wastes which is a matter of serious concern. Oelofse and Nahman, (2013) has shown an

DOI: 10.4018/978-1-5225-5237-6.ch015

amount of 4.14 t of CO2 equivalent per ton of waste food. Upscaling of microbial processing technology of these wastes into essential enzymes and organic acids can also be done industrially.

Enzymes are very specific with the substrate and once if enzymes have the right substrate to act then the process becomes very faster. This system has been described by a lock (enzyme) and key (substrate) model. In the old time, people use natural process by using hand to make yogurt without knowing the enzyme involved in the process. In this line, enzyme engineering was developed to use enzymes in industrial processes and also the production of enzymes. Not surprisingly, industries were among the first to recognize and use this most potential of enzymes in different fields. The first field was food industry where enzymes have a big impact such as beer, milk, cheese, bread, juice industry and later the process has been developed to other sectors such as detergents, the pharmaceutical industry. Today the global market for industrial enzymes is growing rapidly and is currently worth more than two billion euro per annum where the most potential sectors are detergent, starch processing, food and feed industries and the recovery of oil and gas. Finally, enzyme using is very efficient, most cost-effective, less time consuming or the fastest process in the industry, and moreover, it is regarded as green or environmentally friendly technology ever (Nandy, 2016).

Enzyme engineering has wide use in industries and the industrial enzyme markets are growing very fast. So, it will be important and attractive to summarize what the current market of the industrial enzyme is, how enzymes are produced, and what the difficulties are. In the review, this interesting topic started and described the application of enzymes, production and technology for enzymes.

DEVELOPMENT OF ENZYME APPLICATION

Enzyme applications were grown every day from last decades and new techniques have been discovered and use for application purposes on a daily basis. All industrial enzymes have their range of applications and these applications have been increased a lot while the cost of use reduced from 1975 onwards. After 8 years 49 enzymes applications in both analytical and industrial area were very well described by Godfrey. For example, in 1983, seven different enzyme applications were used for baking purpose but after 13 years, this number becomes 11. Surprisingly, 54 different application scopes were open by 1996 but at the same time, enzyme complexity becomes very high compared to the past (Godfrey, 1996). Therefore, different enzymes were applied for the same purpose by checking the efficacy for lowering the application cost. Novozymes is one of the biggest enzyme producers in the world today where thousands of enzymes were tested and few of them commercially attractive and viable due to their application cost and efficacy. Cowan, 2000 has been discussed on xylanases uses as animal feed which is reduced to 20-30% of the starting cost. Therefore, in this modern era, the development of production techniques tends to lower application cost for producing efficient enzymes industrially.

A process needs enzyme application and therefore a development of enzyme discovery required to carry out the process. On the other hand, any technology can be replaced by using enzyme process then either economy of the process is viable or the environmental sustainability benefits towards mankind or both have a positive approach together.

With the advent of new frontiers in biotechnology and bioprocess engineering, the spectrum of enzymes application has widened in many different fields, such as medicinal, food and additives, brewing, distilling, clinical and textiles industries. Enzymes in the bioprocessing are fast and effective technology to get the quick and efficient product. There are several industrial processes where enzymes use for commercial purposes. Proteases in face cream for hair removing, amylases in brewing to germinate barley, rennin in cheese production for coagulation are some specific example of enzyme used for commercial purposes. Enzymes are preferable in industries for various advantages. Enzymes always use as a catalyst in many industries in a small quantity in a reaction and make a pure product. Therefore, there is no requirement for extraction of by-product from the desired pure product. All the reactions with enzymes in general work in normal condition under low temperature, neutral pH and normal atmospheric pressure which saves time, cost and therefore, energy.

Biochemical reaction engineering principles are the main platform for using enzymes. Bioprocesses especially fermenters are the best possible process to use or produce enzymes. Enzymes can be used to get volumetric mass transfer coefficient (k_La) experimentally in any volume of the reactor to check the performance end efficiency of the process. Enzyme-catalyzed reaction is also related to the basic concept of transport phenomena. Enzyme kinetics is the example of the reaction of substrate and enzyme to form product under different condition (Villadsen, Nielsen, & Liden, 2011). Enzymes are working in the cytoplasm under a stable environment of the cell called intracellular and those secreted from the cell and work in the surrounding environment called extracellular. Extracellular enzymes are most robust because they have a wide range of pH and temperature and therefore much cheaper and therefore, the long shelf life for commercial use. All the enzymes used in the different sector are summarized in Table 1.

As discussed before, the industrial enzymes are used mostly in the food industry. First-time glucose is produced from starch in the 60s with the use of glucoamylase with a greater yield and higher purity

Table 1. Enzymes are using in different commercial applications.

Enzyme/s	Sector/Industry	Potential Market Value
Lipase, Protease, Cellulase, Amylase	Detergent (Novozymes)	$2.767 M by 2019 (http://www.marketsandmarkets.com/PressReleases/protein-hydrolysis-enzymes.asp)
Amylase, Glucoamylase, Glucose Isomerase	Processing Starch (Novozymes in Denmark, DuPont in U.S., Rouquette Freres in France etc.)	$2.238 M by 2018 (http://www.prnewswire.com/news-releases/alcohol-and-starchsugar-enzyme-market-by-type-carbohydrase-protease-and-lipase-by-application-industrial-specialty--geography---global-trends--forecasts-to-2018-256229371.html)
Protease, Phytase, Xylanase	For animal feed (Royal DSM NV)	$1.371 M by 2020 (http://www.marketsandmarkets.com/PressReleases/feed-enzyme.asp)
Xylanase	Pulp and Paper (Novozymes)	200 M DKK by 2024 (www.novozymes.com)
Arbinanase, Amylase	Processing of fruits or vegetables (Novozymes, DSM)	Food enzymes: $ 2.3 B by 2018 (http://www.iisc.ernet.in/currsci/jul10/articles22.htm)
Hydrolase	Oil and Gas (Novozymes)	$330 B by 2015 (PetroChina, China)
Chymosin	Dairy (Pfizer, Chr. Hansen)	Food enzymes: $ 2.3 B by 2018 (http://www.iisc.ernet.in/currsci/jul10/articles22.htm)
Urease, Pectinase	Wine (Carlsberg)	Feed enzymes: $ 1.37 M by 2020 (http://www.marketsandmarkets.com/PressReleases/feed-enzyme.asp)
Protease	Meat (Novozymes)	Feed enzymes: $ 1.37 M by 2020 (http://www.marketsandmarkets.com/PressReleases/feed-enzyme.asp)
Amylase	Textiles (Novozymes)	$2.7 M by 2019 (www.marketsandmarkets.com)
Amylase	Baking (Novozymes)	$695.1 M by 2019 (www.marketsandmarkets.com)
Beta-glucanase	Brewing (Carlsberg)	Feed enzymes: $ 1.37 M by 2020 (http://www.marketsandmarkets.com/PressReleases/feed-enzyme.asp)
Protease	Tanning (Novozymes)	Feed enzymes: $ 1.37 M by 2020 (http://www.marketsandmarkets.com/PressReleases/feed-enzyme.asp)

Source: Nandy, 2016.

with easier crystallization technology. This application replaced traditional acid hydrolysis method and the steam cost reduced by 30%, ash production by 50% and by-product by more than 90%. In 1973 immobilized glucose isomerase was invented (high fructose corn syrups from starch). Furthermore, amylase was also used to increase the bread, cake shelf life and beta-galactosidase in the production of cheese and other dairy products. Industrial enzymes have also a big impact on the beverage industry. In 1930s RohM and Haas launched the pectinase (Pectinol K) in the market for higher yield of juice products (apple juice). Nowadays detergent industry is the largest user of industrial enzymes (proteases, lipases, amylases, and cellulases). These enzymes are using based on their efficiency to remove dirt from the cloths quickly under a lower temperature which saves energy and prevents wastage of lot of chemical detergents over a longer time period under higher temperature. In 1960s research scientists found the great potential of using enzymes in the field of medical diagnostics, food and beverage analysis. Several test kits and reagents are produced from different industry for different analysis. Reagents for medical diagnostics have been developed and manufactured in a different industry (Roche, Trinity Biotech, and Merck). Dairy, food, and feed beverage analysis test kits and reagents are also developed and manufactured by different industry (Megazyme, Boehringer, and Merck). Several microorganisms have been selected to get different flavor compounds by enzymatic methods (Akacha & Gargouri, 2015). New advanced literature or studies have shown the current progress in the application of the enzyme engineering (listed in Table 2).

Table 2. Current progress in the application of enzyme engineering (Li, Yang, Yang, Zhu, & Wang, 2012) is divided into main categories and corresponding one known company is assigned with their applications

Industry Product/s	Enzyme	Known Company	Application/Benefits
Detergents	Proteinases	Novozymes, Genencor	Remove protein-based stains in fabrics into soluble amino acids.
	Lipases		Digest oils and fats like butter, sauces etc.
	Amylases		Remove resistant starch residues.
	Cellulases		Remove the fuzz and microfibers that gives the fabric a glossier appearance and soften the cotton.
Starch	Amylases, amyloglucosidases, and glucoamylases	Novozymes	Help to convert starch to glucose and other sugar syrups
	α-amylases		Increasing maltose and glucose content.
	β-amylases		Producing low molecular weight carbohydrates, for example, maltose.
	β-glucanases		Improving wort separation.
	Pillulanases		This secure maximum fermentability of the wort.
	Glucose isomerases		Help to convert glucose to fructose
Dairy	Lipases, chymosin, lysozymes, Rennin	Chr. Hansen	Cheese manufacturing
	Lactases, β-galactosidases		Used for breaking lactose to glucose and galactose in milk processing to avoid lactose intolerance.
Juice	Amylases, glucoamylases	AB Enzymes	Help to break starch into glucose.
	Laccase		Susceptibility to browning during storage increased.
	Naringinase and limoninases		Cause bitterness in citrus juices.
	Cellulases, hemicellulases		Maintain the texture and lower the viscosity.
	Pectinases		Increasing the juice production.

continued on following page

Table 2. Continued

Industry Product/s	Enzyme	Known Company	Application/Benefits
Textiles	Amylases	Genencor	Used to remove starch from woven fabrics
	Cellulases		Fuzz and microfibers remove from fabric to give glossier and smoother appearance.
	Laccases, glucose oxidases		Creating bleaching agent in whiteness.
	Pectinases		Reduce the outer cell layer to improve fiber extraction.
Cosmetic	Oxidases, peroxidases, polyphenol oxidases	DSM	Hair dyeing.
	Papain, bromelain, subtilisin		Peeling effects in skin care.
	Amyloglucosidase, glucose oxidases		Toothpaste and mouthwash.
	Transglutaminases		Hair waving.
Brewing	α-acetolactate-decarboxylases (ALDC)	Carlsberg	Fermentation time reduced where to prevent of diacetyl formation and make the beer test correct.
	Amylases, glucanases, proteinases		Make polysaccharides and proteins in the malt.
	Pentosanases, xylanases		Improving extraction and beer filtration process, hydrolyze pentosans of malt, barley, and wheat.
	Proteinases		Improve the yeast growth and reducing the clouding of beers.
	Amyloglucosidases		Help for low-calorie beer production ("light beer") and increasing the glucose content.
	β-glucanases		Used for improvement of filtration
Baking	Lipases	Novozymes	Gas cells stability improved in the dough.
	Oxidoreductases		Gluten strength is improved.
	α-amylases		Degrading starch (catalyzes) in flours and bread structure is controlled.
	β-xylanases		Using to control the handling and stability issues of dough and rising bread.
	Proteinases (Biscuit)		Reduces the protein in flour
Leather	Proteinases	Novozymes	Trypsin is used here to treat the leather with proteinase to make it more pliable. (process is known as bating)
Pulp and Paper	Mannanases	AB Enzymes	Increasing brightness by degrading residual glucomannan.
	Amylases		Cleaving the starch molecules which reduce viscosity for surface sizing in coatings.
	β-xylanases		Enhance the pulp-bleaching efficiency.
	Cellulases		Make flexible fibers and softness improved by cellulose hydrolyzing.
	Laccases		Improving brightness by bleaching.
	Lipases		Reduces or control pitch for the stickiness of paper in pulping processes.

Source: Nandy, 2016.

One of the best examples of enzyme process development is using the enzyme in the animal feed application. The main parameter for animal feed process is to check the digestibility issues of feed to animals. This digestibility depends on the variability factor of cereal and vegetable protein sources. There is a presence of beta-glucan in the endosperm wall of the grain (barley) and therefore, the enzyme is directly related to feeding strategy of animals. Brewing industries commonly use beta glucanases and therefore it will be easier to screen all developed enzymes to determine if these could be applied. In a similar fashion, xylanases were developed in wheat containing feeds and also for other enzyme activities for example phytase.

One enzyme is also not an all-rounder for all process because a specific enzyme has been developed for a specific reason and therefore for different conditions such as temperature or pH, different enzyme sources required for other activities. There are single or multi-activity enzymes products available for feed applications where more specific target product can be developed by using single activity products. These can be justified by "lock and key" hypothesis as described before. Wyss et. al., (1999) demonstrated the variation in catalytic properties of a range between phosphatases and phytases where substrate used in the initial characterization of an enzyme may not be the substrate against which it has the highest specific activity.

Andersen and Dalbøge, (1999) demonstrated expression cloning by which isolation of single enzyme activities coming from a screening process can be faster and more effective. All frontiers industries used these screening approaches where thousands of enzymes were screened in a robot and the ability to isolate single activities from this without having to make amino acid sequencing of purified single enzymes is of crucial importance in the development of new enzyme products.

ENZYME PRODUCTION

In the other hand, enzymes are produced from the microbes in the fermenter and the major producer of industrial enzymes is Novozymes (Villadsen, Nielsen, & Liden, 2011). One of the examples is that on-site enzyme productions in biorefineries have always some optimal strategy for lowering the enzyme conversion cost. There are known industrial enzyme producers such as *Aspergillus niger*, *Aspergillus oryzae*, *Tricoderma reesei* (Haltrich, Nidetzky, Kulbe, & Zupancic, 1997; Pel, de Winde, Archer, Dyer, Hofmann, & Schaap, 2007; Ward, 2012). The first enzyme produced industrially was an amylase in 1894 from a fungal microorganism which was used as a pharmaceutical product for the digestive disorder treatment (Crueger & Crueger, 1989). Bacillus species are considered to be the most important sources of amylase and have been used for enzyme production using solid state fermentation (SSF). SSF is used to produce glucoamylase using *A. niger* (Mamo & Gessesse, 1997; Fogerty & Kelly, 1990). Producing enzymes in bioprocess technology has been performed by SSF and liquid fermentation. But in both conditions, a strict rule is followed known as fermentation process such as sterile bioreactor to avoid contamination, sterile air to aerate, with a constant stirring speed for bacterial or fungi growth or maximize yield, maintain a constant pH and constant temperature for cooling effect to enzyme denaturing effect and finally nutrient requirements for energy and protein required for cell growth and maximize yield.

Therefore, enzymes that stable and work quickly are preferred most. Hot springs bacteria are most preferred due to this reason. The enzymes from hot springs are thermostable and can tolerate a wide range of pH variation and therefore use efficiently. Extracellular industrial enzymes are extracted from the microbes by centrifuging or filtering and then evaporate the water or using ultrafiltration method and finally dried for the customer. On the other hand, intracellular enzymes are much more expensive and very difficult to obtain.

Most enzymes are produced by submerges cultivation in fermentor/s. The industrial enzymes production process commonly follow the basic approach. The approach is described in short and summarized as follows:

- Enzyme selection,
- Production strain selection,
- Using genetic engineering, overproducing strain can be constructed,
- Medium optimization and process parameters conditions,
- Downstream process (recovery) optimization, and

- Production of stable and efficient (long shelf life) enzyme (Formulation).

In addition, after fermentation, the broth is stored in a tank and micro filtrate the broth. After this stage, the broth is sent to the homogenizer and then to centrifugation. It is also known that ultrafiltration can be used here. After the ultrafiltration, freeze drying is the important last step is used commonly. There are several other unit operations and processes in the formulation and the process stages can be changed according to the need (see Figure 1).

Batch, Fed-batch, and continuous cultivations are three major modes of operation for production of enzymes mainly from marine bacteria and fungi in the laboratory and pilot plant scales. Marine bacteria are used in batch mode under different processes such as submerged, immobilized and solid-state. But the continuous process is used only with submerged or with immobilized cells. Marine enzymes are unique in terms of stability and having novel properties because they found from the extreme environment. Apart from microorganisms such as bacteria and fungi, marine organisms like fishes, prawns, snakes, algae and other species were also studied to get the enzymes from the different environments like salt tolerance or cold adaptivity and other extreme conditions (Ghosh, Saha, Sana, & Mukherjee, 2005).

The industrial enzymes are produced from different organisms and their production processes are summarized in the following Table 3.

Scale-up and optimization of enzyme production processes at the pilot scales was also discussed based on 600-L fermenter where the economical process for enzyme production has been achieved (Junker et. al., 2001). Scale-up issues are very important for commercial purpose from laboratory scale fermenters. Process optimization for the scale-up required the medium optimization to determine the effect of various defined ingredients as well as the complex nitrogen sources on enzyme production. Other fermentation conditions such as inoculum transfer, agitation, the temperature for cultivation have also effect on optimized enzyme production for synthetic use.

QUALITY CONTROL (QC) OF ENZYMES

Performance of enzymes can be tested in lab scale based on the efficiency, stability, and robustness. Industrially, microorganisms are sorted based on high throughput screening (HTS) producing specific enzymes. HTS plays a crucial role in this phase where large numbers of enzymes are exposed to the particular target like as detergent. Few enzymes showing in terms of degree of confidence are processed further as lead enzymes for laboratory scale and finally one or two enzyme process to the industrial scale. The experiments are performed in the laboratories and the positive results are then optimized in

Figure 1. Block diagram to illustrate the enzyme production process
Source: Nandy, 2016.

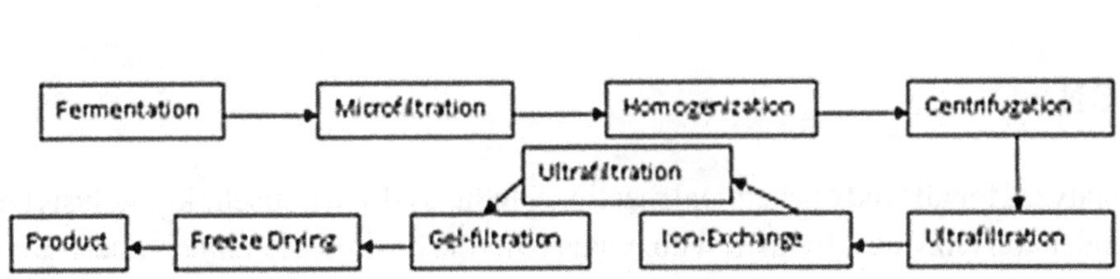

Table 3. The industrial enzymes production from different sources

Source	Enzymes	Microorganisms	Method/s
Fungal	Amylase	*Aspergillus oryzae, Trichoderma longibrachiatum*	Submerged cultivation (Batch, Fed-batch, and Continuous)
	Glucosidases	*Aspergillus flavus*	
	Proteases	*Aspergillus niger*	
	Pectinases	*Aspergillus niger*	
	Glucose oxidase	*Penicillium notatum*	
	Catalase	*Aspergillus niger*	
	α – Galactosidase	*Aspergillus niger*	
	Cellulase	*Trichoderma longibrachiatum*	
	β – Glucanase	*Aspergillus niger, Trichoderma longibrachiatum*	
	Lipase	*Aspergillus niger*	
	Phytase	*Aspergillus niger*	
	Xylanase	*Aspergillus niger, Trichoderma longibrachiatum*	
Bacterial	Amylases	*Bacillus subtilis, Bacillus amyloliquefaciens, Bacillus licheniformis*	
	β - Glucanase	*Bacillus subtilis*	
	Proteases	*Bacillus subtilis*	
	Penicillinase	*Bacillus subtilis*	
	Asparaginase	*Escherichia coli*	
	Pullulanase	*Bacillus acidopullulyticus*	
	Maltase	*Bacillus subtilis*	
	Mannanase	*Bacillus leutus*	
	Xylanase	*Bacillus subtilis*	
Yeast	Invertase	*Saccharomyces cerevisiae*	
	Lactase	*Saccharomyces fragilis*	

Sources: Gurung et. al., (2013).

terms of potency and selectivity. Physiology of the bacteria and the industrial enzyme produced from it should be very robust and stable for long shelf life for future use for the customers.

Life cycle assessment (LCA) is an analytical tool developed to facilitate the measurement of the relative environmental impacts by comparing two or more processes. LCA has been applied different processes in oil and fats industry under different enzymatic bioprocesses to compare the environmental impact. Mainly LCA is using to compare different CO2 emission savings from different processes and very efficient tool to describe (Thum & Oxenball, 2006).

CONCLUSION

Enzymes are catalyzed for all kind of chemical reactions. Industrial and household catalysis become more and more dependent on enzymes. Industrial enzymes with the desired activity can be obtained by optimizing bioprocess technology conditions and also by protein engineering. Enzyme engineering provides higher product quality, low manufacturing cost, low energy consumption and saves time. This drives the market growth to enhance cost efficiencies and productivity which grow the interest of consumers or customers. Specially, the development of scale-up of mono component enzymes production has resulted in a considerable widening of the range of applications in which enzymes can be used. These techniques are intrinsically less -resource demanding than the production of multi-component enzyme products. It is very crucial to understand the potential area in which the enzyme is to be used. Therefore, a collaboration between enzyme developer and the potential user in which both ends contribute such that the knowledge gathering from specialist/s can be easily used successfully and the process will be faster. This kind of joint ventures required to establish for validation purpose of new applications based on both existing and newly discovered enzyme activities. In this paper, several industrial and laboratory approaches have been summarized in terms of enzyme production, application, and uses. These approaches will help us in future to get an overall idea about enzymes and can be useful for further research and development.

REFERENCES

Akacha, N. B., & Gargouri, M. (2015). *Food and Bioproducts Processing*. Academic Press. doi:10.1016/j.fbp.2014.09.011

Andersen, L. D., & Dalbøge, H. D. (1999). Future Trend of Feed Enzymes. *Proceedings of the Roche-American Soybean Association Joint Seminar.*

Cowan, W. D. (2000). Review and Preview of The Development of Geographical Feed Enzyme Markets. *Proceedings of the 3rd European Symposium on Feed Enzymes.*

Crueger, W., & Crueger, A. (Eds.). (1989). Industrial Microbiology. Sinauer Association.

Fogarty, W. M., & Kelly, C. T. (1990). In W. M. Forgarty & C. T. Kelly (Eds.), *Microbial Enzymes and Biotechnology* (2nd ed.; pp. 71–132). London: Elsevier Science Publishers. doi:10.1007/978-94-009-0765-2_3

Ghosh, D., Saha, M., Sana, B., & Mukherjee, J. (2005). Article. *J. Marine Enzymes. Adv. Biochem. Eng. Biotechnol.*, *96*, 189–218.

Haltrich, D., Nidetzky, B., Kulbe, K. D., & Zupancic, S. (1997). Production of fungal xylanases. *Bioresource Technology*, *58*(2), 137–161. doi:10.1016/S0960-8524(96)00094-6

Ingale, S., Joshi, S. J., & Gupta, A. (2014). Production of Bioethanol using agricultural waste: Banana pseudo stem. *Brazilian Journal of Microbiology*, *45*(3), 885–892. doi:10.1590/S1517-83822014000300018 PMID:25477922

Junker, B., Moore, J., Sturr, M., & McLoughlin, K. (2001). Article. *Bioprocess and Biosystems Engineering*, *24*, 39–49. doi:10.1007004490100230

Li, S., Yang, X., Yang, S., Zhu, M., & Wang, X. (2012). Technology prospecting on Enzymes: Application, Marketing and Engineering. *Computational and Structural Biotechnology Journal*, *2*(3), e201209017. doi:10.5936/csbj.201209017 PMID:24688658

Mamo, G., & Gessesse, A. (1997). Article. *Biotechnology Techniques*, *11*(6), 447–450. doi:10.1023/A:1018437310837

Nandy, S. K. (2016). Bioprocess Technology Governs Enzyme use an production in Industrial Biotechnology: An Overview. *Enzyme Engg.*, *5*(01), 1. doi:10.4172/2329-6674.1000144

Oelofse, S. H. H., & Nahman, A. (2013). Estimating the magnitude of food waste generated in South Africa. *Waste Management & Research*, *31*(1), 80–86. doi:10.1177/0734242X12457117 PMID:22878934

Pel, H. J., de Winde, J. H., Archer, D. B., Dyer, P. S., Hofmann, G., Schaap, P. J., ... Stam, H. (2007). Genome sequencing and analysis of the versatile cell factory Aspergillus niger CBS 513.88. *Nature Biotechnology*, *25*(2), 221–231. doi:10.1038/nbt1282 PMID:17259976

Villadsen, J., Nielsen, J., & Liden, G. (2011). *Bioreaction engineering principles* (3rd ed.). Springer publishers. doi:10.1007/978-1-4419-9688-6

Ward, O. P. (2012). Production of recombinant proteins by filamentous fungi. *Biotechnology Advances*, *30*(5), 1119–1139. doi:10.1016/j.biotechadv.2011.09.012 PMID:21968147

World Farmers Organisation. (2014). Retrieved from http://www.wfo-oma.com/news/global-production-of-fruits-and-vegetables-increased-by-9-4.html

Wyss, M., Brugger, R., Kronenberger, A., Remy, R., Fimbel, R., Oesterhelt, G., & … van Loon, A.P.G.M. (1999). Biochemical Characterization of Fungal Phytases (myo-Inositol Hexakisphosphate Phosphohydrolases): Catalytic properties. *Applied and Environmental Microbiology*, *65*(2), 367–373. PMID:9925555

Chapter 16
Trends and Challenges in Enzymatic Bioengineering of Natural Products to Industrially Valuable Products:
Status and Future of Industrial Enzymology

Sujata Sinha
Indian Institute of Technology, India

ABSTRACT

Bioengineering of natural products to useful products are trending rapidly. These products are cost effective and eco-friendly and fitting into consumer demands of natural and organic. Wastes from industries, agriculture, fishery, dairy, etc. are being investigated for transformation to useful biomolecules for other industries like cosmetics, food supplements/preservation, dairy, etc. Biocatalytic transformation looks promising in the present scenario, but needs intensive research looking for novel enzymes/process and their optimization. Immobilization and scale up is also required for taking this process up to industrial level. Process improvement and downstream processing research for product purification is going on. Some of the areas which look promising are metagenomic screening of novel biocatalysts, gene cloning for overexpression and purification, etc. Bioreactor designing for scale up and simultaneous production and purification of desirable products are also being emphasised. Exploring biological activities after enzymatic reaction is one of the main areas of research nowadays.

INTRODUCTION

Demand for natural products is growing day by day as consumers are being more aware towards improved health and well- being.Even though these natural products have interesting properties which could be of great potential, their uses in various sectors are limited by poor solubility, lack of uniformity in their structure and function along with poor characterization etc. Their bioconversion to industrially valuable products using chemicals or using biological methods is of varying importance. Both these approaches

DOI: 10.4018/978-1-5225-5237-6.ch016

are used widely for hydrolysis of natural products but chemical methods are more common at industrial scale hydrolysis. However, nowadays chemical methods are being avoided due to obvious reasons of toxicity, poor ease of control, less mildness and environmental hazards. In themajority of cases, it leads to lower production yield due to theformation of by- products and lacks technology for large-scale production of industrially relevant products. Biocatalysts use is preferable over chemical methods for controlling reaction as itis easier and faster. Workingunder milder condition, better productyield due to lesser adverse chemical modification and higher specificity of biocatalysts (Kim & Rajapakse, 2005) gives them theedge over chemical methods. It has been reported thatenzymatically produced biomolecules show improved biological activities in some product categories, i.e. bioactive oligosaccharides and peptides and this route is eco-friendly. However, high cost and low availability of efficient enzymes discourage their uses for the industrial production of valuable products. This problem can be overcome by looking for more efficient novel biocatalysts, using these biocatalysts multiple times at commercial level for various batches of production and designing of a process to improve the overall efficiency which leads to yield improvement of desired products.

Lignocellulosic biomass (agro-industrial and agro wastes), fishery wastes (chitin/chitosan), food wastes (fat wastes/ wastes from cooking oil/fruit processing wastes) and dairy waste are mainly being explored using enzyme hydrolysis aspotential source of industrially valuable products like bioethanol, biodiesel, bioactive molecules (peptides/oligosaccharides), anti-oxidants, sugars etc.This chapter aims at summarising natural products being investigated for the industrially valuable product as a result of enzymatic hydrolysis. Trends in biochemical production of these molecules and their yield improvement in various ways have also been covered. Upcoming challenges in these areas of biomolecule production from natural products have been discussed in this chapter.

BACKGROUND

Studies on production of various industrially valuable products from cheap natural substrates using enzymes and their uses have been summarised in Table 1. Generation of bioactive peptides through enzymatic routes and strategies from food processing industry have already been studied in detail (Toldra, Reig, Aristoy, & Mora, 2017). Enzymatic or biological conversion of renewable lignocellulosic biomass to biofuel or other high-value product are considered to be having high potential. as lignin can be converted to different polymersusing laccase or laccase mediated system. Similarly, fat waste from cooking oil can be interesting substrates for production of biodiesel using microbial lipase. Fishery wastes or dairy wastes rich in protein with the help of protease enzyme can be used for generation of bioactive peptides. Phenolic antioxidants can also be generated using microbial enzymes from different wastes. A number of bioactive compounds like anti-oxidants (flavonols, flavonons, phenolic acids, Anthocyanidines, anthocyanins, carotenoids etc), antibacterial saponins, pectins, amino acid and proteins, lignin and xyloglucans, dietary fibers, glycosides and bioactive lipids have been reportedly present in fruit processing wastes (Banerjee, Singh, Vijayraghvan, MacFarlane, Ptti, & Arora, 2017). Although all these processes involve microbial enzymes these are not considered feasible at industrial scale in their native form. Enzyme engineering focusses on overexpression of enzyme proteins as well as other approaches for improvement in their specificity and stability to enhance the productyield. Very few immobilized enzymes have been used at commercial scale (DiCosimo, McAuliffe, Poulose, & Bohlmann, 2013) and advantageous forcontinuous production, enhanced stability and downstream processing.

Glucose isomerase immobilization for production of high fructose corn syrup (HFCS)and also enhanced production of glucose from cellulosic biomass for ethanol productionhave been done. Non-digestible fructooligosaccharides as prebiotics and hydroxymethyl furfural synthesis have also been possible using glucose isomerase enzyme in immobilized form.

Production of various industrial enzymes from agricultural wastes and their uses have already been summarised inthe literature (Bharathiraja, Suriya, Krishanan, Manivasagan, & Kim, 2017). Numerous studies of theenzymatic production of chitooligosaccharides(COS) from chitosan and their biological activities have already been reported (Kim & Rajapakse, 2005).

NATURAL PRODUCTS: POTENTIAL RAW MATERIAL FOR INDUSTRIES

Abundantly available natural polysaccharides like cellulose, chitin (1st and 2nd most abundant polysaccharides), chitosan (deacetylated form of chitin) are being explored for their uses in various sectors like food, textiles, agriculture, pharmaceuticals, cosmeticsetcLignocellulosic wastes refer to plant biomass and it is more than 60% of total biomass available and includes domestic wastes, food industry wastes, municipal wastes, grass, waste papers and agriculture residues etc (Bilal, Asgher, Iqbal, Hu, & Zhang, 2017). Some of the most commonly used lignocellulosic materials reported are wheat/rice straw, banana stalk, saw dust, soft/hard wood chips, cotton waste, corn cobs, cotton stalk and soybean stalk etc. By-products from agricultural and food processing industries are available in abundance and disposal of these wastes are creating problems related to environment and health.

Bioethanol in India is generally produced from molasses which is a by-product of sugar production. US, Brazil and China are other leading bioethanol producer where Corn, sugarcane, cassava and other stale food crops are used for this purpose. Wheat straw, an invasive weed *Prosopis juliflora* and another weed *Lantana camara*, water hyacinth, banana stems, cotton stem, sweet sorghum are being usedfor research for bioethanol technology (Jahnavi, Prashanthi, Sarvanthi, & Rao, 2017). Key components of lignocellulosic waste are cellulose and hemicellulose which are further hydrolyzed to monomers. Cellulose is crystalline and difficult to hydrolyze while amorphous nature of hemicellulose makes it easily degradable to sugars. Pre-treatment of cellulosic material is followed by hydrolysis which undergoes fermentation using microbial enzymes.

Converting citrus waste into value-added products through environmental friendly approaches using enzymesis one of the prime examples of such study (Sharma, Mahato, Cho, & Lee, 2017). Bioactive compounds from citrus wastes reportedly have therapeutic properties and show anti-cancer, anti-oxidant, anti-tumor and anti-inflammatory properties. They have also been shown to have anti-viral properties, anti-platelet aggregation and known to form protective enzyme in liver against genetic material damage in the cells. Industries related to citrus processing wastes generates a huge amount of waste (50% of wet fruit mass) which are of immensely valuable economically as it contains flavonoids, carotenoids, dietary fiber, sugar, polyphenol, ascorbic acid, essential oil and a considerable amount of some trace elements. High sugar content present in these wastes are also considered suitable for bioethanol production by removing inhibitory D-limonene compound.Physical methods like heat treatment, microwave assisted extraction, chemicals method using water and methanol extraction and enzymatic assisted extraction methods are used for getting value-added compounds from citrus wastes. However, enzymatic treatment of citrus pulp after centrifugation is considered as one of the most efficient methodsfor processing the pulp which shows high protein content and used for animal feed albeit lesser that that of generated from other agro- industrial waste products. Enzyme assisted extraction from citrus wastes have been reported for flavonoids and phenolic compounds (Li, Smith, & Hossain, 2006). Essential oils, alkaloids, flavonoids, coumarins, limonoids, carotenoids, phenolic acids and other value-added products (bioethanol, protease enzyme, oligosaccharides, citric acid, limonene, pectin, sugar) have been also obtained using microbial enzymes (cellulase, pectinase) and whole cells like *A niger, A terreus, Penicillium, Saccharomyces cerevisiae, Trichosporan penicillatum* etc.

Chitinous wastes contain mainly crustacean shell waste from crab, fish, lobster, krill, shrimp by-products (Thadathil &Velappan, 2014) etc which can be converted to high-value products. Chitin, a polysaccharide is present in these wastes which leads to chitosan production through deacetylation process and is obtained through deproteinization and demineralization process from the by-product wastes from fishery industries. Chitin and chitosan are known to have many useful biological activities which can be of great use to food, pharmaceuticals and cosmetics industry. It is estimated that around 10^{10}-10^{11} tons of chitin are produced every year by living organisms however, its uses in various sectors are limited by insolubility in water. Solubility is enhanced by deacetylation of chitin to change it to chitosan of various degree of deacetylation and oligosaccharides of chitin and chitosan. Chitosan is partially soluble in water and its useful properties are further enhanced by breaking it into smaller chains i.e. chitosan oligosaccharides. In some case, even chitin is hydrolyzed using chemical and enzymes to smaller oligosaccharides (Liang, Hsieh, & Wang, 2012). Hydrolysis of chitin/chitosan is achieved by chemical, physical and enzymatic methods. All these methods have their advantages and disadvantages but the enzymatic method is always preferable over other methods due to eco-friendly considerations, ease of reaction control and cost-effectiveness. Various bioprocess strategies using specific/non-specific enzymes in free and immobilized forms have been summarised in literature (Sinha, Chand, & Tripathi, 2016). By-products of protein industries like slaughterhouses, marine by-products, olive-mill, fishery waste industry etc and egg, soybean and peanut proteins controlledhydrolysis with enzymes are used for generation of bioactive peptides. These peptides are bioactive in nature and show health promoting activities like anti-oxidant activity and cholesterol reducing activity etc. Recently, plant peptides have also been reported which show antibiosis like lowering blood pressure and cholesterol, they also show anti-thrombosis, anti-oxidant and immunomodulatory activities (Maestri, Marmiroli, & Marmiroil, 2016).

BIOCATALYSTS FOR BIOENGINEERING: PAST, PRESENT, AND FUTURE

Biocatalysts,especially from microbial sources have been used widely for industrial production of various chemicals, pharmaceuticals, fermentation, agricultural and food production. Although, they are an excellent toolfor catalyzing industrial reaction they are not considered feasible for industrial scale for generation of high-value products. Genetic modification and overexpression of enzymatic proteins are considered useful for removing these limitations and this technique has also been used for improving enzyme characteristics like stability, efficiency, and specificity.

New Biocatalysts: Discovery and Screening Approaches

Various approaches for isolation, screening, and identification of enzyme producing microbial strains have been used. Ecological samples like, soil, water, compost, animal guts, plant parts where a high concentration of likely substrates can be present etc are used for thesource of microbes which are used for screening of potential enzymes (Sinha, Tripathi, & Chand, 2012).The first step for enzyme screening is to find the need for commercial enzymesandindustrial demand for the enzymes. Mainly two approaches are used for isolation of novel catalyst from environmental samples.Thetraditional approach involves growing in enrichment culture and screening of awide variety of microbes for thedesired activity. Enzyme and genes for selected proteins are isolated from selected microbes but this process leads to loss of alarge number of microbial diversity due to difficulties in isolation and enrichment of culture in pure form.

A more improved method of metagenomic screening has been devised over thetraditional method of screening, which involves exploitation and assessment of complete microbial genomic material present in nature. It is especially useful for samples like marine microbes which are considered as arich source of unique and diverse biocatalysts butonly less than 1% of microbes in any marine habitats can be cultivated in lab conditions (Uria & Zilda, 2016). DNA is directly extracted from environmental samples and gene is cloned, then the resulting libraries are screened for thetargeted gene for enzymes. However, it has been observed that positive clones in a screen can be enhanced by enrichment technique (Daniel, 2002). Metagenomic screening has been reportedly used for a number of enzymes like lipases, cellulases, amylases, chitinases, esterases, Lignases, xylanase and enzymes involved in biotin synthesis (Knietsch, 2003). However, total numbers of enzymes screened by this method are low as compared to traditional method.Metagenomic screening approaches includemainly two methods, thenucleotide sequence (PCR based) and enzyme activity based screening. Former method can access only homologs of the known gene in a wide variety of gene pool and less preferable over SIGEX (substrate induced gene expression screening)which is based on the known fact that catabolic gene expression can be induced by relevant substrates and metabolites of catabolic enzymes. However, this methodis also being criticised for lack of expression of theenzyme in active form in a cloning host even in presence of suitable substrates. It has also been discouraged in biocatalysts where transcriptional regulators are distantly situated to thedesired gene. In an effort to make SIGEX a high throughput screening this process has been combined with FACS (fluorescence activated cell sorter) and positive clones have been selected in liquid culture (Uchiyama, Abe, Ikemura, & Watanabe, 2005).

Immobilized Enzymes and CLEA (Cross-Linked Enzyme Aggregate)

Applications of catalyst in immobilizedform in various industries are well established (DiCosimo, McAuliffe, Poulose, & Bohlmann, 2013). Ease of product recovery, continuous production, reactor designing, improved efficiency/stability and reusability of enzymes are some of the advantages which outweigh the shortcomings associated (mass transfer limitations and loss of activity) with theenzyme in immobilized form. It is especially recommended where multienzyme reactions are involved in industrial processes and in such case co-immobilization of one or more enzymes can be achieved on a single matrix. Some of the successful example of industrial scale enzyme immobilization is glucose isomerase for production of HFCS,fructooligosaccharide and furfural production. Immobilized pectin lyase on Nylon-6 and use of immobilized pectinol-D on thecontinuous processing of citrus juice and debittering of juices by immobilized naringinase enzyme by entrapment in cryogels.

Nanomatrices such as nano- fibers and particles have been used as immobilization matrix in case of chitosanase enzyme for decreasing mass transfer limitation which is high in case of chitosan due to their highly viscous solution (Kuroiwa, Noguchi, Nakajima, Sato, Sukekuni, & Ichikawa, 2008; Sinha, Tripathi, & Chand, 2012). Controlling forenzymatic reaction at theappropriate time for production of larger oligomers are considered to be difficult, however, this problem has been partially solved with the help of immobilized chitosanase.

Chitosanase has been immobilized on various nanomatrices like nanoparticles and nanofibers (PAN-NFM) and has been used for multiple times in biochemical reaction for production of dimer to hexamer as a result of chitosan hydrolysis (Zeng & Zheng, 2002). Other biodegradable matrix used for this purpose are chitin, Agar or agarose gel particlesusing multipoint attachment and pentamers and hexamers were produced from chitosan hydrolysis (Kuroiwa, Ichikawa, Sato, Hiruta, Sato, & Mukataka, 2002) using this system. Controlling and stopping the hydrolytic reactions are easier, yield of larger COS are improved and mass transfer limitations are reduced when enzymes are used in immobilized form

COS was produced continuously using dual reactor system where the immobilized enzyme was packed in a column and was attached to ultrafiltration membrane reactor. Partially hydrolyzed chitosan was produced from a viscous solution of chitosan (DD 89%) and COS production took place in ultrafiltration membrane reactor (UFM)Reactor fouling of membrane reduced and drawbacks of bath reactor was removed where COS of smaller size is obtained as compared to this condition as hydrolysis reaction was difficult to stop and control and product inhibition was also averted.

Cellulase Enzyme Immobilization

Cellulose, an abundantly available renewable source of energy can be easily converted to glucose and then to bioethanol which can be used for clean and green energy. However, due to low specific activity, unstable nature of key enzyme, i.e. cellulase and lack of recovery and reuse of the same batch of the enzyme (Li, Henriksson, & Gellerstedt, 2007) make this process economicallyunviable. Immobilization has been suggested to improve stability, reusability, and reactor designing for efficient production in various research. It also helps in downstream processing during the production of sugars from cellulose and is helped by a favourable change in temperature and pH optima during the production process (Ikeda et al, 2015).Among various types of substrates (membranes, nanoparticles, gels etc) suggested for enzyme immobilization nanoparticles have been suggested and found to be best suited for water-insoluble substrates such as in case of cellulose (Ahmed, Chang, & Tsai, 2017).Cellulase enzyme loaded on MnO2 nanoparticle and was used for bioethanol production by SSF (solid state fermentation) and it was compared with free enzyme. Enzyme activity increased on immobilization on nanoparticle as the production of reducing sugar as well as bioethanol increased as compared to the enzyme in the free state (Cherian, Dharmendirakumar, & Baskar, 2015). It was also found to be more thermostable as compared to free form and showed repeated use and ethanol concentration increased upto 21.96 g/L. Two variants of cellulase enzyme C1 and C2 have been immobilized on nano-porous and porous silica using simple and cheap adsorption method which led to high hydrolysis yield using cellulose like environment and could be used for 4 to 5 cycles with more than 50% of activity remaining after that. It was effective for hydrolysis of not only model substrates but was effective against industrially produced lignocellulosic biomass (Ikeda, Parashar, Chae, & Bressler, 2015). Poly acrylic acid nanogels (150 nm diameter) fabricated via inverse phase micro-emulsion polymerization was enriched with carboxyl groups. Thermostability of enzyme improved over free enzyme and activity was more than free enzyme in acidic buffer. Immobilized enzymes could be recovered by centrifugation and used multiple times over free enzymes (Ahmed et al, 2017).Other studies related to cellulase immobilization are on magnetic nanoparticle (Fe3O4) using glutaraldehyde activation and MWCNT (multi walled carbon nano tubes) have been reported (Mubarak, Wong, Tan, Sahu, Abdullah, & Jayakumar, 2014).

INDUSTRIALLY RELEVANT VALUABLE PRODUCTS

Bioactive molecules (peptides, oligosaccharides) are known for their health-promoting activities. They help in reducing infection, show antioxidant properties, prebiotic properties, anti-microbial properties, anti-cancer properties ACE inhibitory activity, opioid activity, mineral binding, anti-appetizing, cyto-modulation, immunomodulation, antithrombotic activity etc and are of immense interest to industries and consumers today. Details of bioactive databases can be obtained from PEPBANK(http://pepbank.mgh.harvard.edu/), another available online tool for peptide sequence processing is BIOPEP, EROP-Moscow (http://erop.inbi.ras.ru/), Brainpeps (http://brainpeps.ugent.be/), PepBank, Peptide DB, Peptide Locator, SwePep, APD (antimicrobial database) which can be used after deciphering the sequence (Giacometti, & Tomljanovic, 2017; Dorpe, Bronsealer, Nielandt, Stalmens, Wynendaele, Audenaert, DeWiele, Burvenich, Peremans, Hsuchou, Tre, & Spiegeleer, 2012).Sugar, bioethanol and alternative fuel along with single cell protein are produced form renewable sources using cellulase enzyme from the fermentation process.

Oligosaccharides

Oligosaccharides (OS)are defined as polymers having 2-10 degree of polymerization are sometimes even higher numbers are joined together by glycosidic bonds (Nakakuki, 1993) which can be hydrolzsed either by acid or by enzymatically to produce monomer or sugar. Various types of OS are found naturally in fruits; vegetables, milk, honey etc. and play physiological and structural roles (Mussatto & Mancilha, 2007). Some of the oligosaccharides obtained from natural sources have been summarised in Table 2.

Originally developed as an artificial sweetenerare gaining importance due to their useful physiological and biochemical properties. Interesting biological functions of OS and their derivatives like low calorific value, noncariogenicity, prebiotic and other health-promoting properties (Gibson & Roberfroid, 1995) have increased their demand for food and nutraceutical industry. COS can also be used as bio preservatives, chemicals and cosmetics industries (Kim & Rajapakse, 2005) and some OS are also called non-digestible oligosaccharides (NDOs) as they are resistant to attack by digestive enzymes in human and animals, not absorbed by host and promote growth of beneficial bacteria in human colon (Mussatto & Manchilla, 2007). Functional food market was reported to be around 80 billion US dollars in 2008 (Vergari, Tibuzzi, & Basile, 2010) and OS considered to be having a major share in functional food market present which are present in food items as well as in free forms. Of 223 items listed by the Japanese government in 1991, more than 50% incorporated OS as functional components

legislated for ''Foods for Specified Health Use (FOSHU)which later changed to Foods with Health Claims (FHC) (Nakakuki, 2002). Examples of functional oligosaccharides are fructooligosaccahrides, maltooligosaccharides, gentiooligosaccharides, mannanoligosaccharides, xylooligosaccharides, cyclodextrin, agarooligosaccharides and COS etc. and can be found as natural components in milk, honey, fruits and vegetables such as onion, jerusalem, artichoke, chicory, leak, garlic, artichoke banana, rye, barley, bacon and salsify (Mussatto & Manchilla, 2007). Other examples are soybean, pulses, and leguminous seeds which contains galactosyl sucrose, raffinose, and stachyose respectively. Xylooligosaccharides (XOS) and galactose oligosaccharides are present in bamboo shoots and milk (particularly, colostrums) respectively either in free form and or as glycoconjugates (Voragen, 1998). Physiological properties like non-Digestibility, cariogenicity, bacteriostatic action, the selective proliferation of bifidobacteria, improvement of serum lipids and blood glucose, etc. (Nakakuki, 2002) make them desirable to be used as food ingredients. These are not utilized by mouth microflora, consequently, the production of acids and polyglucans doesn't take place (Crittenden & Plyne, 2002) and can be used as a low cariogenic sugar substitute in confectionary products, chewing gums etc. These are suitable for use by diabetic persons in sweets, the low calorie diet foods etc. (Crittenden & Plyne, 2002; Rivero-Urgell, 2001). Some OS can be utilized by the limited class of gut bacteria such as *Bifidobacteria* and *Lactobacillus* and beneficially affect the host health and are also considered as prebiotics, as they selectively stimulate the growth and metabolic activity of bacterial species beneficial for health (Crittenden & Plyne, 2002; Voragen, 1998). They have similar effect as those of dietary fibre (Mussatto & Manchilha, 2007) improve glucose control and metabolism of triglycerides (Delzenne & Roberfroid, 1994; Roberfroid & Slavin, 2000; Santos, 2002) are preferable over other dietary fibre because of their low sweetness, lower daily dose, pleasant texture/taste, stability, solubility in waterand their easy incorporation in food items (Totamatsu, 1994). Other properties which make them useful for incorporation in food items are sweetness, bitterness, hygroscopicity, water activity, inclusion capability etc. It is especially useful in food items where sucrose use is restricted due to high sweetness (Roberfroid & Slavin, 2000).

High molecular weight OS have high viscosity and leads to high body and mouthfeel and also being exploited in various food items. They also show have high moisture retaining capacity, prevention of excessive drying, low water activity and reduce microbial contamination (Crittenden & Plyne, 1996).

OS act as a specific substrate for enzymes, enzyme inhibitors, elicitors, etc,for example, Sulfated OS have been shown to inhibit hyaluronidases enzymes from bee venom, bovine testis and *Streptococcus agalactiae* (Salmen, Hoechstetter, Käsbauer, Paper, Bernhardt, & Buschauer, 2005; Nakakuki, 2002). A series of omega-epoxy alkyl glycosides of d-xylopyranose, xylobiose, and xylotriose were tested as potential active-site-directed inhibitors of xylanases from glycoside hydrolase families 10 and 11. A number of OS have been reported as substrates and inhibitors of glycosyltransferases and glycosidases enzymes (Huey, Ichikawa, & Shen, 1995).

Oligosaccharides: Industrial Applications

Bioactive oligosaccharides (BOS) have been used extensively in various industries like food, pharmaceuticals, cosmetics, animal feed, agrochemicals and their uses in various fields are increasing by each day. In the pharmaceutical industry, OS are ingredients of various pharmaceutical products used for treating constipation, hyperlipidemia, hypocholesterolemia, antihypertensive activities (Park, Je, & Kim, 2003) and immunostimulating properties (Remaud-Simeon, Willemot, Sarçabal, Montalk, & Monsan, 2000) etc. Cyclodextrins are known to reduce the undesirable effects of drug molecules in the various route of administration like oral, rectal, dermal and nasal (Singh, Sharma, & Banerjee, 2002; Del Valle, 2004). Lactulose for relieving constipation and portosystemic encephalopathy (Villamiel, Corzo, Foda, Montes, & Olano, 2002). The agarooligosaccharides with DP 2–4 are able to suppress the production of the pro-inflammatory cytokine TNF- α and the expression of iNOS (Enoki, Sagawa, Tominaga, Nishiyama, Koyama, Sakai, Yu, & Ikai, 2003). The oligosaccharides with DP 6–8 can elicit a physiological response in algae and could prevent death from endotoxin shock, as well as exhibit anti-cancer and anti-inflammatory effect (Chen et al 2004). Liv-Well, a product by Kunpoong Bio Co. Ltd. Korea is a COS has unique protective benefits like blood lipid adjustment, remarkable prevention of ethanol-induced liver injury and reduction of carbon tetrachloride toxicity.

OS are added as nutraceuticals in food products as they are from natural sources, anticariogenic in nature, are of lower calories and are known to have prebiotic properties. They are resistant to gastric acidity, hydrolysis by mammalian enzymes and gastrointestinal absorption, get fermented by the intestinal microflora and stimulate selectively the growth and/or activity of intestinal bacteria associated with health and wellbeing such as *Bifidobacteria* and *Lactobacilli*. Fructooligosaccharides are the most studied among OS and their effect on the growth of colon beneficial bacteria has been established (Gibson & Roberfroid, 1995; McBain & McFarlane, 1997). They are produced in large quantities in several countries and are added to several products such as biscuits, drinks, yogurts, breakfast cereals and sweeteners (Mizota, 1996). Lactulose, synthetic disaccharides are also known to stimulate the growth of prebiotics (Fadden & Owen, 1992). OS derived from cell wall polysaccharides like arabinogalactooligosaccharides made from soybeans, arabinooligosaccharides from sugar beet, arabinoxylooligosaccharidesfrom wheatand galacturonooligosaccharides were shown to have prebiotic potential (Rastall & Maitin, 2002). COS attracts greater interest in antimicrobial agents, antioxidants, and enhancers of the nutritional quality of food (Shahidi, Vidana, & Jeon, 1999). COS along with lysozyme enzyme synergistically have been reported to enhance the shelf life of meat products by the elimination of bacterial sp like *Pseudomonas* and *E coli* (Rao, Chander, & Sharma, 2008). COS have antibacterial, fungistatic properties and it increases the nutraceutical value without altering the organoleptic properties of food. It has potential to replace artificial preservatives (Kim & Rajapakse, 1995) in the food industry. A number of carbohydrate derivatives like mannan oligosaccharides, peptidoglycan and COS have also been reported to possess immunostimulating properties (Matsuo & Miyazono, 1993). OS can also be used as a bulking agent with artificial sweetener like aspartame or sucralose in sweet foods. It can alter the freezing temperature of frozen foods and can control the intensity of browning and it can lower water activity which leads to reduced microbial contamination (Totamatsu, 1994). Low Calorific value of OS makes it ideal for diabetics and can be used as low cariogenic sugar substitutes in products like confectionery, chewing gums, yogurts and drinks (Sako, Matsumoto, & Tanaka, 1999). Trehalose and gentiooligosaccharides has a bitter taste and is used for taste improvement of beverages, has effects on intestinal microflora, absorption of calcium and act as color stabilizing saccharides (Nakakuki, 2002). OS are immensely used in food industry especially desserts such as jellies, puddings, and sherbets; confectionary products such as candy, cookies, biscuits breakfast cereals, chocolates and sweets, bread and pastries, jams, meat and fish products (Voragen, 1998). They are also used for the improvement of texture and flavor of bread, pastries, pasta, jellies etc. Physiological and physico chemical property of OS varies with types of the mixture prepared during the preparation of food items so the most appropriate OS also varies with the food items (Crittenden & Plyne, 1996). For example, during bread preparation, Galactooligosaccharides (GOS) is added which is unbroken during fermentation and provide excellent taste and texture. It is added to infant food and given to the patient during hospitalized due to their susceptibility to change in intestinal microflora.

Oligosaccharides: Production and Synthesis

OS isobtained by extraction from natural sources, polysaccharides hydrolysis or can be synthesized from disaccharides using mechanical, chemical or enzymatic methods (Courtois, 2009). Xylooligosaccharides (XOS) can be prepared by controlled enzymatic hydrolysis of xylan (Moure, Gullon, Domı´nguez, & Parajo, 2006). Plants materials are used for extraction of raffinose oligosaccharides using water, methanol or ethanol extraction (Mussatto & Manchilla, 2007). However, for preparation of complex and highly pure OS, microbial enzymes like glucosyl and fructosyl transferase enzymes are found to be very effective. Potential of cell wall polysaccharides as the good source of OS has gained attention recently as these plant polysaccharides are often present in large amounts in fiber-rich by-products and wastes (e.g., cereal bran, fruit pomace, sugar-beet pulp, potato fibre and press cakes of oleaginous seeds or pulses). Specific glycanases could be used for production of novel oligosaccharides (Palframan, Gibson, & Rastall, 2003; Oosterveld, Beldman, & Voragen, 2002). Arabinogalactooligosaccharides can be made from soybeans by endogalactanases, arabinooligosaccharides can be made from sugar beet by endoarabinanases, arabinoxylooligosaccharides can be made from wheat by xylanases, and galacturonooligosaccharides can be made from polygalacturonic acid by endogalacturonases.

Bioactive Peptides

The research in the area of bioactive peptides is growing like never before, providing many new insights into their various physiological roles in the human body. Bioactive peptides are specific amino acid sequences which are inactive when present in their respective native protein but once released; they are capable of displaying a wide spectrum of biological activities. There is diverse range of plant and animal proteins which can serve as the sources of these peptides. These peptides can be released from their source proteins through enzymatic hydrolysis and fermentation. Various biological activities of the peptides on which research is focused include antihypertensive, opioid, antimicrobial, antioxidant, antithrombotic, hypocholesterolemic, immunomodulatory, and mineral binding activities. By products obtained from protein sources such as cheese whey, meat producing units, dairy industry, winery unit waste, olive mill industry, poultry industry, soybean, fish waste etc are subjected to controlled enzymatic hydrolysis to generate bioactive peptides (Toldra, Reig, Aristoy, & Mora, 2017). Bioactive peptides which are relevant nutritionally can be identified and quantified using a field of food peptidomics where Peptidomics has been used especially for milk peptides (milk peptidomics). However, nowadays Mass spectrometry is also being used with the technique for protein identification and quantification (Giacometti & Tomljanovic, 2017).

Enzymatic hydrolysis has been found to be most efficient and reliable method to produce a peptide with target functionalities such as anti-microbial, antioxidant, mineral binding, opioid activity etc. In some case, it has been observed that anti-oxidant activity depends on the source of proteins and protein substrate pre-treatment, type of proteases enzyme used and hydrolysis conditions applied.An enzyme in crude and pure form is being used for the production of anti-oxidant peptides, so for reducing the cost crude enzymes mixtures are used in case of pure enzymes. However, using microbial fermentation instead of crude enzymes are in trend in some countries. Natto, Douchii, and Tempeh are fermented soybean product which contains anti-oxidant peptide by the action of fungal proteases. Streptococcus Thermophilus 4F44 has been grown on bovine casein and proteases from lactic acid bacteria for milk protein are examples of bioactive peptide production. Similarly, a bioactive peptide with anti-oxidant and ACE inhibitory has been obtained from hydrolysis by two proteases on thornback ray skin gelatin. Gel-permeation chromatography and sub-fractionation of crude protein hydrolysates give various fractions of bioactive peptide and it has been noticed that smaller peptides show highest anti-oxidant activity as well as ACE inhibitory activity. Despite significant progress in the field of bioactive peptide production from natural sources, there are certain hurdles associated with it like their production at large scale without losing their biological activities. Challenging areas in case of bioactive peptides are identification, production, purification, chemical synthesis, identification of biological activities and their inculcation in various food items (Sanchez & Vazquez, 2017).

Phenolic Compounds/Flavonoids/Oil

Enzyme-based extraction approaches have been considered as a feasible alternative to chemical extraction for anti-oxidants/phenolic compounds from various agro-industrial wastes. Two approaches of enzyme-assisted extraction are: 1) Enzyme-assisted aqueous extraction (EAAE) and; 2) Enzyme-assisted cold pressing (EACP) (Azmir, Zaidul, Rahman, Sharif, Mohamed, Sahena, Jahurul, Ghafoor, Norulaini, & Omar, 2013). While the first method is used for extraction of oil from various seeds, the second method is used mainly for hydrolyzing cell wall. Various factors like enzyme concentration and composition, solid to water ratio, hydrolysis time, the moisture content of plant material and size/composition of plant material affects this entire process and are considered as key factors. EACP is considered suitable for extraction of bioactive compounds from oilseeds as it is non-toxic and non-inflammable in nature and extracted oil contains high amount of free fatty acids and phosphorus content as compared to traditional methods. EAAE is considered moreeco-friendly as it uses water in place of organic solvents.

Grapes are widely used for recovery of phenolic compounds like anthocyanin, flavonols etc. and it has been found that there is a positive correlation between the yield of total phenol and plant cell wall breakdown by enzymes.During wine making process, seeds and skin of red grapes are removed after maceration in contact with fermentation must while in white grapes these parts are removed just before fermentation process. Both these types of grapes are known to have a significant amount of flavonols, anthocyanins, flavonoids etc. Most of these useful compounds are present in inner layer of grape skin and are known to have number of useful biological properties like anti-fungal, anti-inflammatory, anti-oxidative, anti-microbial properties etc. their incorporation in food and medicine is desirable and are considered of great value. Organic solvent based extraction has a very negative environmental impact and is being replaced by supercritical fluid extraction,pressurised hot water extraction. However, food, pharmaceuticals and cosmetics industries are preferring green extraction method over these two procedures. One of these method is ''Enzyme assisted extraction'' procedure which involves disruption of cell wall by specific enzymes especially carbohydrases such as cellulases, hemicellulases and pectinases (Tomaz, Maslov, Stupic, Preiner, Asperger, & Kontic, 2016).

Various other studies have been reported as the improved release of phenolic compounds from *ribes nigrum* pomace using various enzymes. In another study release of total phenolic compounds from five citrus peel using EAAE method was compared and was found to be highest in case of celluzymeMX and in some studies phenolic anti-oxidants increased with enzyme concentration.A pectinolytic and cellulolytic enzyme in a ratio of 2:1 was used for extraction of bioactive compounds like anthocyanins, phenolic acids, non-anthocyanin flavonoids etc and yields were more as compared to non- enzymatic extraction. More phenolic anti-oxidants were recovered from raspberry wastes as compared to non -enzymatic extraction. In another study where various enzymes (Novoferm, pectinex, celluclast) were compared for extraction of phenolic compounds from grape wastes by EAAE method, Novoferm was found to be best for getting maximum yield.

Bioethanol

Although Bioethanol production from abundantly available lignocellulosic biomass (60 billion tons) from agrowaste is considered as the most viable alternative for fuel crisis it is difficult to achieve this due to several barriers associated with it. Genome engineering is found to best way to tackle these barriers as various proteins are expressed at the different level of bioethanol production from lignocellulosic biomass. Intolerance of host organisms to various temperature, PH, organic solvents stress can also be reduced by genome engineering (Ulaganathan, Goud, Reddy, & Kayalvili, 2017). Precision genome engineering facilitatesthe modification of gene directly inside the organism without isolating the gene, unlike traditional genetic engineering where the gene of interest is isolated first, manipulated *in vitro* and introduced back into the host. Precision genome engineering involves technique like: a) The bacterial immunity based CRISPR/Cas system; b) TALEN system (Xanthomonas transcription activator-like effector nuclease based); c) ZFN system which is based on zinc finger domain; d) nuclease-based meganuclease system which can recognise long regions; e)YOGE system which is oligonucleotide based. Change in one or more nucleotides can bring large changes for enhancing bioethanol production also enzyme specificities and host tolerance towards various stresses can also be altered by using precision genome engineering with traditional genetic engineering techniques.

1st and 2nd generation of biofuel has not been found promising for large-scale bioethanol production for various reasons associated with it (Sirajunnisa & Surendhiran, 2016). In first generation biofuel, bioethanol is produced from fermentation of starch from wheat, barley, corn, potato etc while in second generation lignocellulosic biomass from straw, wood, and grass is used for bioethanol production, These ways are not found to be encouraging in terms of economy and sustainability as they can push food prices up, can disturb food chain, occupies more cultivation land which can be used for food crops, depends on climate, low yield and high cost of transportation and hydrolysis for production of biofuel. Algae has been found to be a potential alternative for biofuel production overcoming all these disadvantages associated with 1st and 2nd generation biofuels. Algal cell walls are composed of cellulose, xylans, mannans and sulpfated glycans which can be hydrolyzed using chemicals and enzyme to produce sugar which is converted to ethanol. Yield is improved as compared to crops and requires less cultivation area than crops for ethanol production. Bioethanol production involves various steps like pre-treatment, hydrolysis, fermentation, and recovery of product, among these pre- treatment of biomass is most crucial and cost-intensive steps due to rigid algal cell wall which is hard to break. Physical, chemical and enzymatic methods are available for cells pre-treatment of which enzymatic method is considered to be environmentally friendly and higher glucose yield without production of any inhibitory product. Cellulase enzyme, pectinase and mixture of enzymes like β-glucosidase and cellulase has been found to be hydrolysing cell walls of algae for ethanol production (Sirajunnisa & Surendhiran, 2016). Commercially produced enzymes have also been tried for this purpose Lactozym, Spirizyme,Viscozyme and AMG for various red and green algae. In some studies, sequential treatment of enzyme and acid and vice-versa gave more yield of glucose for ethanol production. Bioethanol production from agricultural wastes has been summarised by Sarkar et al, (2012) and Jahnavi et al, (2017).

Industrial Enzymes Production

Various industrial enzymes which are used in food, chemicals, pharmaceuticals, textile and dye industries have been produced from microbial hydrolysis of agro-industrial wastes such as molasses, corncob, sugarcane bagasse, rice bran, wheat straw bran, rice bran etc (Table 3). Agricultural waste utilization, on one hand, reduces the cost of enzyme production as well as solve the massive problem of waste utilization and mobilizing biomass for a better purpose (Bharathiraja, Suriya, Krishanan, Manivasagan, & Kim, 2017).

BIOPROCESSING AND DOWNSTREAM PROCESSING

Various modes of bioprocesses are tried and tested for production of valuable compounds and undergo downstream processing for their removal and recovery after fermentation.Downstream processing processes are designed and optimized by keeping economic index (total annual cost), environmental function (eco-indicator 99) and control index (condition number) under consideration. Medium optimization by manual and using statistical software are some of the most common approaches which are used for this purpose. Various modes of fermentation like Batch, Fed-batch, continuous modes in both solid state and submerged fermentation mode etc are also tried for yield enhancement. On the other side, membrane filtration, column chromatography, precipitation etc are used for removal and recovery of valuable products like bioactive oligosaccharides and peptides, enzymes etc from fermented broth.

A hybrid separation based on liquid-liquid extraction (LLX) combined with dividing wall column (DWC) technology for purification of the ABE mixture. Multi-objective optimization (hybrid optimization algorithm) on the basis of an environmental factor, cost and controlling index was done and a configuration was proposed (Ramirez, Ramirez, Hernandez, Hernandez, & Kiss, 2017). Among the proposed four designs scheme 2 where only a reboiler is included showed best results in terms of cost, control-index, and environmental concern.

Bioactive peptides/carbohydrates and other compounds have been extracted traditionally using solvents and physical/chemical methods. However, these methods are time and energy intensive and also not fit for temperature sensitive molecules. Nowadays ultrasound, enzymes, microwaves, supercritical fluids and pressurized liquids are being researched for this purpose. Extraction of bioactive carbohydrates/peptides involve grinding, precipitation in acid/base medium and then it is followed by filtration and drying. New processing methods include novel cell disintegration methods with or without enzymes and extraction of bioactive compounds from complex matrix thereafter. In some cases, pre-treatment of samples also enhanced the desired products from fermentation broth (Hayes & Tiwari, 2015).

Most frequent method of COS fractionation in literature has been reported to be UFM reactor. Temperature, pH, the concentration of feed, type and pore size of membranes are some of the factors which affect the separation process (Sinha, Chand, & Tripathi, 2016). 1, 5 and 10 KDa UFM was connected to immobilized enzyme column reactor in which chitosanase enzyme from *Bacillus* sp was immobilized on chitin. 1% chitosan was passed through reactor at rate of 5 ml/min to produce PHC (partially hydrolysed chitosan) which is continuously added to UFM reactor system which is then converted to high molecular weight (12.6% yield), medium, medium molecular weight (9.4%) and low molecular weight COS(78%). Among these three fractions, COS of molecular weight 10 kDa and above were found to be having anti-microbial activity against pathogens (Jeon, Park, & Kim, 2001). UF has also been combined with nanofiltration and electrodialysis for better separation of COS. Column chromatography and solvent fractionation have also been reported for COS purification. Hurdles involved in ethanol/butanol produced as a result of ABE (acetone/butanol/ethanol) fermentation are expensive separation steps to purify these products. Continuous downstream processing for high-value biological products has been summarised recently (Zydney, 2016).

FUTURE RESEARCH DIRECTION

Future research will be directed towards followings points: 1) Finding novel biocatalysts through isolation and screening; 2) high throughput screening of biocatalysts; 3) bioreactor designing for large scale and continuous production of biomolecules from natural products having efficient enzymes; 4) enzyme engineering though genome engineering and over expression; 5) biocatalysts immobilization for improving the efficiency and stability; 6) downstream processing research for removal of products 7) fractionation, extraction/purification of desired products for improved yield; 8) exploration of novel bioactivity both *in vivo* and *in vitro* mode; 9) finding novel abundantly available biowaste as cheap substrates for biomolecule production.

CONCLUSION

Various trends and challenges related to the production of industrially valuable products from natural products have been summarised in Table 4. It is clear that future belongs to research emphasis on natural products which are abundantly available for industrial production of costly biomolecules.These biomoleculesare of great importance to various industrial sectors as mentioned in this chapter. Biocatalysis approaches for production has an edge over chemical route and biologically active molecules produced such are in demand due to their immensely useful bioactivity and eco-friendliness.However, more useful novel bioactivity will always be explored as it is the need of the hour. Need for finding novel biowaste for novel bioactive molecules production will drive the future of various industries. Biocatalysts need more attention about the isolation of robust enzyme as in case of thermostable and halotolerant GH5 cellulase from a culture isolated form Icelandic hot spring (Zarafeta, 2016). Gene overexpression using molecular biology techniques will be used for yield and activity improvement.Overexpressed Nitrile hydratase enzyme from *Rhodococcus rhodochrous*and *Pseudomonas putida* converted acrylonitrile to acrylamide upto 45% (W/W), showed improvement in thermal stability (3.5 fold) and catalytic activity (1.5 fold) which gave a maximum yield of conversion upto 93% (Choi, Han, & Kim, 2015). Some examples of immobilized biocatalysts use for industries are Aminoacylase, Nitrilase, Hydantoinase, Penicillin G acylase etc. However, the additional cost of enzyme immobilization should be justified over their advantages in the case of industrial use.

ACKNOWLEDGMENT

The corresponding author, Sujata Sinha is thankful to Department of Science and Technology (DST), Govt of India for providing grant (DST NO. SR/WOS-A/LS-129/2009)

REFERENCES

Ahmed, I. N., Chang, R., & Tsai, W. B. (2017). Poly (acrylic acid) nano gel as a substrate for cellulase immobilization for hydrolysis of cellulose. *Colloids and Surfaces. B, Biointerfaces*, *152*, 339–343. doi:10.1016/j.colsurfb.2017.01.040 PMID:28131958

Alander, M., Mättö, J., Kneifel, W., Johansson, M., Kögler, B., Crittenden, R., ... Saarela, M. (2001). Effect of galacto- oligosaccharide supplementation on human faecal microflora and on survival and persistence of *Bifidobacterium lactis* Bb-12 in the gastrointestinal tract. *International Dairy Journal*, *11*(10), 817–825. doi:10.1016/S0958-6946(01)00100-5

Alrumman, S. A. (2016). Enzymatic saccharification and fermentation of cellulosic date palm wastes to glucose and lactic acid. *Brazilian Journal of Microbiology*, *47*(1), 110–119. doi:10.1016/j.bjm.2015.11.015 PMID:26887233

Ambigaipalan, P., & Shahidi, F. (2017). Bioactive peptides form shrimp shell processing discard: Antioxidant and biological activities. *Journal of Functional Foods*, *34*, 7–17. doi:10.1016/j.jff.2017.04.013

Azmir, J., Zaidul, I. S. M., Rahman, M. M., Sharif, K. M., Mohamed, A., Sahena, F., ... Omar, A. K. M. (2013). Techniques for extraction of bioactive compounds from plant materials: A review. *Journal of Food Engineering*, *117*(4), 426–436. doi:10.1016/j.jfoodeng.2013.01.014

Bandhan, A. K., Chadha, B. S., Kaur, J., Saini, H. S., & Bhat, M. K. (2007). Production of xylanolytic and cellulolytic enzymes by thermophilic fungus *Myceliophthora* sp. IMI 387009. *Bioresource Technology*, *98*(3), 504–510. doi:10.1016/j.biortech.2006.02.009 PMID:16600593

Banerjee, J., Singh, R., Vijayraghvan, R., MacFarlane, D., Ptti, A. F., & Arora, A. (2017). Bioactives from fruit processing wastes: Green approaches to valuable chemicals. *Food Chemistry*, *225*, 10–22. doi:10.1016/j.foodchem.2016.12.093 PMID:28193402

Bankovic-Lilic, I. B., Stojkovic, I. J., Stamenkovic, O. S., & Veljkovic, V. B. (2014). Waste animal fats as feedstocks for biodiesel production. *Renewable & Sustainable Energy Reviews*, *32*, 238–254. doi:10.1016/j.rser.2014.01.038

Barros, F.F.C., Simiqueli, A.P.R., Andrade, C. J., & Pastore, G. M. (2013). Production of enzymes from agroindustrial wastes by biosurfactant-producing strains of *Bacillus subtilis. Biotechnology Research Journal*, 1-9.

Bharathiraja, S., Suriya, J., Krishanan, M., Manivasagan, P., & Kim, S. K. (2017). Chapter Six - Production of Enzymes From Agricultural Wastes and Their Potential Industrial Applications. *Advances in Food and Nutrition Research*, *80*, 125–148. doi:10.1016/bs.afnr.2016.11.003 PMID:28215322

Bilal, M., Asgher, M., Iqbal, H. M. N., Hu, H., & Zhang, X. (2017). Biotransformation of lignocellulosic materials into value-added prodcuts-A review. *International Journal of Biological Macromolecules*, *98*, 447–458. doi:10.1016/j.ijbiomac.2017.01.133 PMID:28163129

Chen, H. M., Zhen, L., Lin, W., & Yan, X. J. (2004). Product monitoring and oligosaccharide composition in agar hydrolysates by percolumn labeling by HPLC. *Talanta, 64*(3), 773–777. doi:10.1016/j.talanta.2004.04.002 PMID:18969671

Cherian, E., Dharmendirakumar, M., & Baskar, G. (2015). Immobilization of cellulase onto MnO_2n-anoparticles for bioethanol production by enhanced hydrolysis of agricultural waste. *Chinese Journal of Catalysis, 36*(8), 1223–1229. doi:10.1016/S1872-2067(15)60906-8

Choi, J. M., Han, S. S., & Kim, H. S. (2015). Industrial applications of enzyme biocatalysis: Current status and future aspects. *Biotechnology Advances, 33*(7), 1443–1454. doi:10.1016/j.biotechadv.2015.02.014 PMID:25747291

Courtois, J. (2009). Oligosaccharides from land plants and algae: Production and applications in therapeutics and biotechnology. *Current Opinion in Microbiology, 12*(3), 261–273. doi:10.1016/j.mib.2009.04.007 PMID:19467920

Crittenden, R. G., & Plyne, M. J. (1996). Production, properties and application of food grade oligosaccharides. *Trends in Food Science & Technology, 7*(11), 353–361. doi:10.1016/S0924-2244(96)10038-8

Crittenden, R. G., & Plyne, M. J. (2002). Production of food grade oligoscchrides using immobilized cells of *Zymomonas mobilis. Trends in Food Science & Technology, 58*, 297–302. PMID:11935179

Daniel, R. (2002). Construction of environmental libraries for functional screening of enzyme activity. In Directed molecular evolution of proteins. Wiley-VCH Verlag GmbH. doi:10.1002/3527600647.ch4

Del Valle, E. M. M. (2004). Cyclodextrins and their uses: A review. *Process Biochemistry, 39*, 1033–1046.

Delzenne, N. M. (2003). Oligosaccharide: State of Art. *The Proceedings of the Nutrition Society, 62*(01), 177–182. doi:10.1079/PNS2002225 PMID:12749343

Delzenne, N. M., & Roberfroid, M. R. (1994). Physiological effects of nondigestible oligosaccharides. *Lebensmittel-Wissenschaft + Technologie, 27*, 1–6.

DiCosimo, R., McAuliffe, J., Poulose, A. J., & Bohlmann, J. (2013). Industrial use of immobilized enzymes. *Chemical Society Reviews, 42*(15), 6437–6474. doi:10.1039/c3cs35506c PMID:23436023

Dorpe, V. S., Bronsealer, A., Nielandt, J., Stalmens, S., Wynendaele, E., Audenaert, K., ... Spiegeleer, B. D. (2012). Brainpeps: The blood-brain barrier peptide database. *Brain Structure & Function, 217*(3), 687–718. doi:10.100700429-011-0375-0 PMID:22205159

Enoki, T., Sagawa, H., Tominaga, T., Nishiyama, E., Koyama, N., Sakai, T., . . . Ikai, K. (2003). *US Patent, 0105029 A1*. US Patent Office.

Fadden, K., & Owen, R. W. (1992). Faecal steroids and colorectal cancer: The effect of lactulose on faecal bacterial metabolism in a continuous culture model of the large intestine. *European Journal of Cancer Prevention, 1*(2), 113–127. doi:10.1097/00008469-199202000-00004 PMID:1463973

Giacometti, J., & Tomljanovic, A. B. (2017). Peptidomics as a tool for characterizing bioactive milk peptides. *Food Chemistry, 23*, 91–98. doi:10.1016/j.foodchem.2017.03.016 PMID:28407976

Gibson, G. R., & Roberfroid, A. M. (1995). Dietary modulation of the human colonic microbiota: Introducing the concept of prebiotics. *The Journal of Nutrition*, *125*, 1401–1412. PMID:7782892

Gildberg, A. (2004). Enzymes and bioactive peptides from fish waste related to fish silage, fish feed and fish sauce production. *Journal of Aquatic Food Product Technology*, *13*(2), 3–11. doi:10.1300/J030v13n02_02

Hamed, I., Ozogul, F., & Regenstein, J. M. (2016). Industrial application of crustacean by-product: A review. *Trends in Food Science & Technology*, *48*, 40–50. doi:10.1016/j.tifs.2015.11.007

Harshvardhan, K., Suri, M., Goswami, A., & Goswami, T. (2017). Biological approach for the production of vanillin from lignocellulosic biomass (*Bambusa tulda*). *Journal of Cleaner Production*, *149*, 485–490. doi:10.1016/j.jclepro.2017.02.125

Hayes, M., & Tiwari, B. K. (2015). Bioactive carbohydrates and peptides in foods: An overview of source and downstream processing steps and associative bioactivities. *International Journal of Molecular Sciences*, *16*(9), 22485–22508. doi:10.3390/ijms160922485 PMID:26393573

Huey, W. C., Ichikawa, Y., & Shen, G. J. (1995). *Oligosaccharide enzyme substrates and inhibitors: methods and composition*. European Patent.

Ikeda, Y., Parashar, A., Chae, M., & Bressler, D. C. (2015). Reusability of immobilized cellulases with highly retained enzyme activity and their application for hydrolysis of model substrates and lignocellulosic biomass. *Journal of Thermodynamics & Catalysis.*, *6*(2), 149. doi:10.4172/2157-7544.1000149

Jahnavi, G., Prashanthi, G. S., Sarvanthi, K., & Rao, L. V. (2017). Status of availability of lignocellulosic feed stocks in India; Biotechnological strategies involved in the production of bioethanol. *Renewable & Sustainable Energy Reviews*, *73*, 798–820. doi:10.1016/j.rser.2017.02.018

Jeon, Y. J., Park, P. J., & Kim, S. K. (2001). Antimicrobial effect of chitooligosaccharides produced by bioreactor. *Carbohydrate Polymers*, *44*(1), 71–76. doi:10.1016/S0144-8617(00)00200-9

Kim, S. K., & Mendis, E. (2006). Bioactive compounds from marine processing byproducts – A review. *Food Research International*, *39*(4), 383–392. doi:10.1016/j.foodres.2005.10.010

Kim, S. K., & Rajapakse, N. (2005). Enzymatic production and biological activities of chitosan oligosaccharides (COS): A review. *Carbohydrate Polymers*, *62*(4), 357–368. doi:10.1016/j.carbpol.2005.08.012

Knietsch, A., Waschkowitz, T., Bowien, S., Henne, A., & Daniel, R. (2003). Construction and screening of metagenomic libraries derived from enrichment cultures: Generation of gene bank for genes conferring Alcohol Oxidoreductase activity of *Escherichia coli*. *Applied and Environmental Microbiology*, *69*(3), 408–1416. doi:10.1128/AEM.69.3.1408-1416.2003 PMID:12620823

Knob, A., Fortkamp, D., Prolo, T., Izidoro, C., & Almeida, J. M. (2014). Agroresidues as an alternative for xylanase production by filamentous fungi. *BioResource*, *9*(3), 5738–5773.

Kuroiwa, T., Ichikawa, S., Sato, S., Hiruta, O., Sato, S., & Mukataka, S. (2002). Factors affecting the composition of oligosaccharides produced in chitosan hydrolysis using immobilized chitosanases. *Biotechnology Progress*, *18*(5), 969–974. doi:10.1021/bp020011r PMID:12363347

Kuroiwa, T., Noguchi, Y., Nakajima, M., Sato, S., Sukekuni, M., & Ichikawa, S. (2008). Production of chitosan oligosaccharides using chitosanase immobilized on amylose-coated magnetic nanoparticles. *Process Biochemistry*, *43*(1), 62–69. doi:10.1016/j.procbio.2007.10.014

Li, B. B., Smith, B., & Hossain, M. M. (2006). Extraction of phenolics from citrus peels II. Enzyme assisted extraction method. *Separation and Purification Technology*, *48*(2), 189–196. doi:10.1016/j.seppur.2005.07.019

Li, J., Henriksson, G., & Gellerstedt, G. (2007). Lignin depolymerisation/repolymerisation and its critical role for delignification of aspen wood by steam explosion. *Bioresource Technology*, *98*(16), 3061–3068. doi:10.1016/j.biortech.2006.10.018 PMID:17141499

Li, P., Zeng, Y., Xie, Y., Li, X., Kang, Y., Wang, Y., ... Zhang, Y. (2017). Effect of pre-treatment on enzymatic hydrolysis of kitchen waste for xanthan production. *Bioresource Technology*, *223*, 84–90. doi:10.1016/j.biortech.2016.10.035 PMID:27788431

Liang, T. W., Hsieh, J. L., & Wang, S. L. (2012). Production and purification of a protease, a chitosanase and chitin oligosaccharides by Bacillus cereus TKU022 fermentation. *Carbohydrate Research*, *362*, 38–46. doi:10.1016/j.carres.2012.08.004 PMID:23079238

Lina, B. A. R., Jonker, D. K., & Kozianowski, G. (2002). Isomaltulose (Palatinose): A review of biological and toxicological studies. *Food and Chemical Toxicology*, *40*(10), 1375–1381. doi:10.1016/S0278-6915(02)00105-9 PMID:12387299

Maestri, E., Marmiroli, M., & Marmiroil, N. (2016). Bioactive peptides in plant-derived foodstuffs. *Journal of Proteomics*, *147*, 140–155. doi:10.1016/j.jprot.2016.03.048 PMID:27079980

Matsuo, K., & Miyazono, I. (1993). The influence of long-term administration of peptidoglycan on disease resistance and growth of juvenile rainbow trout. *Nippon Suisan Gakkaishi*, *59*(8), 1377–1379. doi:10.2331uisan.59.1377

McBain, A. J., & Macfarlane, G. T. (1997). Investigations of bifidobacterial ecology and oligosaccharide metabolism in a three-stage compound continuous culture system. *Scandinavian Journal of Gastroenterology*, *32*(sup222), 32–40. doi:10.1080/00365521.1997.11720715 PMID:9145444

Mizota, T. (1996). Functional and nutritional foods containing bifidogenic factors. *Bulletin of the International Dairy Foundation*, *313*, 31–35.

Moreno, A. D., Ibarra, D., Mialon, A., & Ballesteros, A. (2016). A bacterial laccase for enhancing saccharification and ethanol fermentation of steam pre treated biomass. *Fermentation*, *2*(2), 1–15. doi:10.3390/fermentation2020011

Moure, A., Gullon, P., Domı'nguez, H., & Parajo, J. C. (2006). Advances in the manufacture, purification and applications of xylo-oligosaccharides as food additives and nutraceuticals. *Process Biochemistry*, *41*(9), 1913–1923. doi:10.1016/j.procbio.2006.05.011

Mubarak, N. M., Wong, J. R., Tan, K. W., Sahu, J. N., Abdullah, E. C., Jayakumar, N. S., & Ganesan, P. (2014). Immobilization of cellulase enzyme on functionalized multiwall carbon nanotubes. *Journal of Molecular Catalysis. B, Enzymatic*, *107*, 124–131. doi:10.1016/j.molcatb.2014.06.002

Mussatto, S. I., & Mancilha, I. M. (2007). Non-digestible oligosaccharide: A Review. *Carbohydrate Polymers, 68*(3), 287–297. doi:10.1016/j.carbpol.2006.12.011

Nakakuki, T. (1993). Oligosaccharides: Production, Properties and Applications. Japanese Technology Reviews. *Gordon and Breach Switzearland, 3*(2), 45-52.

Nakakuki, T. (2002). Present status and future of functional oligosaccharide development in Japan. *Pure and Applied Chemistry, 74*(7), 1245–1251. doi:10.1351/pac200274071245

Oosterveld, A., Beldman, G., & Voragen, A. G. J. (2002). Enzymatic modification of pectic polysaccharides obtained from sugar beet pulp. *Carbohydrate Polymers, 48*(1), 73–81. doi:10.1016/S0144-8617(01)00216-8

Palframan, R., Gibson, G. R., & Rastall, R. A. (2003). Development of a quantitative tool for the comparison of the prebiotic effect of dietary oligosaccharides. *Letters in Applied Microbiology, 37*(4), 281–284. doi:10.1046/j.1472-765X.2003.01398.x PMID:12969489

Park, P., Je, J. Y., & Kim, S. K. (2003). Angiotensin I converting (ACE) inhibitory activity of hetero-chitooligosacchatides prepared from partially different deacetylated chitosans. *Journal of Agricultural and Food Chemistry, 51*(17), 4930–4934. doi:10.1021/jf0340557 PMID:12903948

Prado-Alves, F. H., Pavezzi, F. C., Leite, R. S. R., Oliveira, V. M., Sette, L. D., & Da Silva, R. (2016). Screening and production study of microbial xylanase producer from Brazilian cerado. *Applied Biochemistry and Biotechnology, 161*(1-8), 333–346. doi:10.100712010-009-8823-5

Ramirez, S. E., Ramirez, J. J. Q., Hernandez, S., Hernandez, J. G. S., & Kiss, A. A. (2017). Optimal hybrid separations for intensified downstream processing of biobutanol. *Separation and Purification Technology, 185*, 149–159. doi:10.1016/j.seppur.2017.05.011

Rao, M. S., Chander, R., & Sharma, A. (2008). Synergistic effect of chitooligosaccharides and lysozyme for meat preservation. *Lebensmittel-Wissenschaft + Technologie, 41*(10), 1995–2001. doi:10.1016/j.lwt.2008.01.013

Rastall, R. A., & Maitin, V. (2002). Prebiotics and synbiotics: Towards the next generation. *Current Opinion in Biotechnology, 13*(5), 490–496. doi:10.1016/S0958-1669(02)00365-8 PMID:12459343

Remaud-Simeon, M., Willemot, R., Sarçabal, P., Montalk, G. P., & Monsan, P. (2000). Glucansucrases: Molecular engineering and oligosaccharide synthesis. *Journal of Molecular Catalysis. B, Enzymatic, 10*(1-3), 117–128. doi:10.1016/S1381-1177(00)00119-3

Rivero-Urgell, M., & Santamaria-Orleans, A. (2001). Oligosaccharides: Application in infant food. *Early Human Development, 65*, S43–S52. doi:10.1016/S0378-3782(01)00202-X PMID:11755034

Roberfroid, M. J., & Slavin, M. (2000). Nondigestible oligosaccharides. *Critical Reviews in Food Science and Nutrition, 40*(6), 461–480. doi:10.1080/10408690091189239 PMID:11186236

Sako, T., Matsumoto, K., & Tanaka, R. (1999). Recent progress on research and applications of non-digestible galactooligosaccharides. *International Dairy Journal, 9*(1), 69–80. doi:10.1016/S0958-6946(99)00046-1

Salmen, S., Hoechstetter, J., Käsbauer, C., Paper, D. H., Bernhardt, G., & Buschauer, A. (2005). Sulphated oligosaccharides as inhibitors of hyaluronidases from bovine testis, bee venom and *Streptococcus agalactiae*. *Planta Medica, 71*(8), 727–732. doi:10.1055-2005-871255 PMID:16142636

Sanchez, A., & Vazquez, A. (2017), Bioactive Peptides: A review. *Food Quality and Safety, 1*(1), 29-46.

Santos, A. M .P. (2002). SÂ´ntese de oligossacarÂ´deos a partir da sacarosepor inulinase de kluyveromyces marxianus *var. bulgaricus*. Tese dedoutorado, FEA/UNICAMP, 2002.

Saratale, G. D., Kshirdsagar, S. D., Sampange, V. T., Saratale, R. G., Oh, S. E., Govindwar, S. P., & Oh, M. K. (2014). Cellulolytic enzyme production by utilizing agricultural waste under solid state fermentation and its application for Biohydrogen production. *Applied Biochemistry and Biotechnology, 174*(8), 2801–2817. doi:10.100712010-014-1227-1 PMID:25374139

Sarkar, N., Ghosh, S. K., Bannerjee, S., & Aikat, K. (2012). Bioethanol production from agricultural waste. *Renewable Energy, 37*(1), 19–27. doi:10.1016/j.renene.2011.06.045

Shahidi, F., Vidana, A. J. K., & Jeon, Y. J. (1999). Food applications of chitin and chitosans. *Trends in Food Science & Technology, 10*(2), 37–51. doi:10.1016/S0924-2244(99)00017-5

Sharma, K., Mahato, N., Cho, M. H., & Lee, Y. R. (2017). Converting citrus waste into value added products: Economic and environmentally friendly approaches. *Nutrition (Burbank, Los Angeles County, Calif.), 34*, 29–46. doi:10.1016/j.nut.2016.09.006 PMID:28063510

Singh, M., Sharma, R., & Banerjee, V. C. (2002). Biotechnological applications of cyclodextrins. *Biotechnology Advances, 20*(5-6), 341–359. doi:10.1016/S0734-9750(02)00020-4 PMID:14550020

Sinha, S., Chand, S., & Tripathi, P. (2016). Recent progress in chitosanase production of monomer-free chitooligosaccharides: Bioprocess strategies and future applications. *Applied Biochemistry and Biotechnology, 180*(5), 883–899. doi:10.100712010-016-2140-6 PMID:27206559

Sinha, S., Tripathi, P., & Chand, S. (2012). A New Bifunctional Chitosanase Enzyme from *Streptomyces* sp. and Its Application in Production of Antioxidant Chitooligosaccharides. *Applied Biochemistry and Biotechnology, 167*(5), 1029–1039. doi:10.100712010-012-9546-6 PMID:22322828

Sirajunnisa, A. R., & Surendhiran, D. (2016). Algae- A quintessential and positive resource of bioethanol production: A comprehensive review. *Renewable & Sustainable Energy Reviews, 66*, 248–267. doi:10.1016/j.rser.2016.07.024

Thadathil, N., & Velappan, S. P. (2014). Recent development in chitosanase research and its biotechnological applications: A review. *Food Chemistry, 150*, 392–399. doi:10.1016/j.foodchem.2013.10.083 PMID:24360467

Toldra, F., Reig, M., Aristoy, M. C., & Mora, L. (2017). Generation of Bioactive Peptides from food processing. *Food Chemistry*. doi:10.1016/j.foodchem.2017.06.119

Tomaz, I., Maslov, L., Stupic, D., Preiner, D., Asperger, D., & Kontic, J. K. (2016). Recovery of flavonoids from grape skins by enzyme-assisted extraction. *Separation Science and Technology, 51*(2), 255–268. doi:10.1080/01496395.2015.1085881

Uchiyama, T., Abe, T., Ikemura, T., & Watanabe, K. (2005). Substrate induced gene expression screening of environmental metagenome libraries for isolation of catabolic genes. *Nature Biotechnology*, *23*(1), 88–93. doi:10.1038/nbt1048 PMID:15608629

Ulaganathan, K., Goud, S., Reddy, M., & Kayalvili, U. (2017). Genome engineering for breaking barriers in lignocellulosic bioethanol production. *Renewable & Sustainable Energy Reviews*, *74*, 1080–1107. doi:10.1016/j.rser.2017.01.028

Uria, A.R., & Zilda, D.S. (2016). Metagenomics-guided mining of commercially useful biocatalysts from Marine micro-organisms. In *Advances in food and nutrition research.* Elsevier Publication.

Vergari, A., Tibuzzi, F., & Basile, G. (2010). An overview of the functional food market: from marketing issues and commercial players to future demand from life in space. *Biofarms for Nutraceuticals,* 308-321.

Villamiel, M., Corzo, N., Foda, M. I., Montes, F., & Olano, A. (2002). Lactulose formation catalysed by alkaline-substituted sepiolites in milk permeate. *Food Chemistry*, *76*(1), 7–11. doi:10.1016/S0308-8146(01)00239-4

Voragen, A. G. J. (1998). Technological aspects of functional food-related carbohydrates. *Trends in Food Science & Technology*, *9*(8-9), 328–335. doi:10.1016/S0924-2244(98)00059-4

Wu, S. C., Wen, T. N., & Pan, C. L. (2005). Agar oligosaccharides lysates prepared by two bacterial agarases stepwise hydrolyzed and their anti-oxidative properties. *Fisheries Science*, *71*(5), 1149–1159. doi:10.1111/j.1444-2906.2005.01075.x

Zarafeta, D., Kissas, D., Sayer, C., Gudbergsdottir, S. R., Ladoukakis, E., Isupov, M. N., ... Kolisis, F. N. (2016). Discovery and characterization of a thermostable and highly halotolerant GH5 Cellulase from an Icelandic hot spring isolate. *PLoS One*, *11*(1), e0146454. doi:10.1371/journal.pone.0146454 PMID:26741138

Zeng, J., & Zheng, L. Y. (2002). Studies on *Penicillium* sp. ZDZ1 chitosanase immobilized on chitin by cross-linking reaction. *Process Biochemistry*, *38*(4), 531–535. doi:10.1016/S0032-9592(02)00163-2

Zhang, Y., Zhou, X., Ji, L., & Che, F. (2017). Enzymatic single state preparation and anti-oxidant activity of hetero-chitooligosaccharides using non- pretreated housefly larvae powder. *Carbohydrate Polymers*, *172*, 113–119. doi:10.1016/j.carbpol.2017.05.037 PMID:28606517

Zydney, A. L. (2016). Continuous downstream processing for high value biological products: A review. *Biotechnology and Bioengineering*, *113*(3), 465–475. doi:10.1002/bit.25695 PMID:26153056

KEY TERMS AND DEFINITIONS

Bioactive: Any compound having effect on living organisms or biological organisms.

CLEA (Cross-Linked Enzyme Aggregates): Enzyme physically aggregated using various cross linker for enzyme immobilization purpose.

Downstream Processing: Purification and isolation of valuable products from fermentation broth after enzymatic production.

Enzyme Immobilization: Biocatalysts attached to any insoluble matrix or support by physical or chemical methods.

Industrially Valuable Products: Any compounds having useful activities applicable to industries like food, chemical, pharmaceuticals, and cosmetics, or any other industries.

Metagenomic: Isolation and extraction of DNA directly from ecological samples like soil, water, etc.

Oligosaccharides: Two or more sugar units joined together by glycosidic bonds to form polymers called oligosaccharides.

Peptides: Short chain of amino acids joined together by peptide bonds.

APPENDIX

Table 3. Production of various industrial enzymes from natural sources

Substrates	Enzyme Production	Remarks	References
Cassava Waste Water	Hydrolases (Amylases, proteases,lipases)	Cassava waste water was found to be better than synthetic medium used for induction of enzymes	Barros et al, (2013)
Wheat Bran, Corn Straw, Corn Cob, Cassava Bran, Sugarcane Bagasse	Xylanases	Both bacterial and fungal xylanase was obtained	Prado, 2016)
Pineapple Peel	Xylanase	Comparison of xylanase production was done with apple peel, rice peel,wheatbran,soybean peel etc but maximum production was obtained in pineapple peel.	Knob et al, (2014)
Agricultural Waste	Cellulolytic enzymes	Endoglucanase, exoglucanase, cellobiase, glucoamylase, xylanase, filter paperase	Saratale et al, (2014)
Rice Straw	Xylanase/ endoglucanase/ β-glucosidase	Multiple isoforms of these enzymes were produced under optimized culture conditions	Bandhan et al, (2007)

Table 4. Trends and challenges in various stages of enzymatic production of industrially valuable products from natural products

	Trends	Challenges
Natural Product	• Lignocellulosic waste/agricultural waste • Marine waste • Dairy waste • Domestic waste • Algal/fungal biomass	▪ Extraction ▪ Purification ▪ characterisation
Biocatalysis	• Isolation from ecological samples • Manual Screening • Metagenomic screening • Selection and statistical optimization • Immobilization/CLEA	▪ Novel biocatalysis ▪ Gene cloning & over expression at industrial scale ▪ Immobilization on nanomatrix ▪ Efficient biocatalysts ▪ Reusability ▪ Cost effectiveness
Biochemical Process	• Batch process • Hydrolysis by immobilized enzyme • Simultaneous production and purification	• Continuous process • Scale-up of process • Efficient Downstream processing
Industrial Product	• Bioactive molecules (oligosacchariddes/ peptides) • Industrial enzymes • Bioethanol • Phenolic compounds • Others	▪ Purification/characterisation of new bioactive molecules
Uses	• Bioactivity	▪ Finding more bioactivity to be explored in various areas ▪ *In vitro* and *In vivo* trial of bioactivity

Table 1. List of industrially valuable products from natural products using biocatalytic approach and their possible uses in various industry

Natural Product	Valuable Products	Biocatalysts	Bioactivities/ Possible Industrial Uses	References
Colloidal Housefly Larvae Chitin	Hetero-chito oligosaccharides (2-6)	*Chitinphilus*sp LZ32	Anti-oxidant activity/Hydroxyl scavenging ability	Zhang et al, (2017)
Ligno-Cellulosic Biomass	Biofuel, polymer	Laccases	Biofuel	Moreno et al, (2016)
Lignin	Vanillin	Natural bacterial consortium	Flavouring agent in food, beverages and pharmaceuticals	Harshvardhan et al, (2017)
Fish Wastes	Bioactive peptides, bioactive oligosaccharides	Pepsin	Stimulate non-specific host immune system/ nutraceuticals	Gildberg, 2004
Fat Wastes	Biodiesel	Lipases	Alternative for conventional fuel	Bankovick-llic et al, (2014)
Grape Wastes	Antioxidant phenolics	Carbohydrases (cellulase/ hemicellulase)	Anti-oxidants for foods/nutraceuticals	Tomaz et al, (2016)
Kitchen Waste	Xanthan gum	Amylase, α-Glucosidase	Oil recovery, food and pharmaceuticals	Li et al, (2017)
Agri Cellulosic Waste	Cellulase	Lactic acid, sugar	Food preservatives	Alrumman, 2016
Food Wastes	Bioactive peptides	Microbial peptidases	ACE Inhibitory, antioxidant, Antithrombotic, antimicrobial, opioid, immunomodulating	Toldra et al, (2017)
Raw Shrimp Shell Discard Protein	Bioactive peptides	Trypsin, chymotrypsin, pepsin, alcalase	Radical scavenging activity (ABTS, DPPH, Hydroxyl), reducing power, ferrous ion chelating activity	Ambigaipalan & Shahidi, 2017

Table 2. Various oligosaccharides from natural sources and their bioactivity with industrial uses

Natural Sources	Enzymes Used for Production	Oligosaccharides	Industry	Uses	References
Asparagus, Sugar Beet, Garlic, Chicory, Onion, Jerusalem Artichoke, Wheat, Honey, Banana, Barley, Tomato, Inulin and Rye	*Transfructosylase Aspergillus niger*	Fructooligosachharides	Food, pharma	Prebiotics	Delzenne, 2003
Honey, Sugarcane Juice or Food-Grade Molasses	Glycosyl transferase	Isomaltulose	Food	Artificial sweetener	Lina et al, (2002)
Bamboo Shoots, Fruits, Vegetables, Milk and Honey Cobs	Xylanase	Xylooligosaccharides	Nutraceuticals	Prebiotics	Crittenden and Plyne, 1996
Human Milk, Cow's Milk	β-galactosidase	Galactooligosaccharides	Food industry	Prebiotics	Alander et al, (2001)
Cell Wall of Marine Algae	Agarase	Agar oligosaccharides	Food, cosmetics, medical	Anti-oxdiants, hepatoprotective	Mussatto and Manchilla, 2007; Wu et al (2005)
Crab Shell, Shellfish Waste	chitosanase	Chitooligosaccharides	Food, pharma, cosmetics, waste water treatment, agriculture, environmental protection	Prebiotics Antioxidants Antidiabetic Antihypertensive Immunoprotective	Kim and Mendis, 2006
Mallow, Composite, Mustard, Soybean		Galactosylsucroses andstachyose			Sako et al, (1999)
Water Soluble Glucans	Cyclodextrin	Cyclodextrin	Food, Pharma, cosmetics, environment protection, packing and textile industry	Pharma: routes of drug administration. Solubilization of organic contaminants, removal of organic pollutants, heavy metals. In water treatment used for stabilization and adsorption of contaminants. In cosmetics to control release of fragrances	Courtois, 2009

Chapter 17
Industrial Enzyme Technology:
Potential Applications

Michael Bamitale Osho
McPherson University, Nigeria

ABSTRACT

Biotechnology, being the application of biological organisms and their components in pharmaceutical and other industrial processes, has emerged as the basic transformation tool for starch hydrolysis enzyme. Several advantages over chemical catalysts under mild environmental conditions with efficiency and high specificity have been accrued to this fact. Such include ingredient substitution through continuous fermentation, increased products yield and plant capacity, processing aid substitution, more efficient processing, less undesirable products with improved products. This chapter reports on the molecular properties of thermostable enzymes such as alpha-amylases, alpha-glucosidases, glucoamylases pullulanases as relates to pharmaceutical industries; highlights various technology development, continuous solid-state fermentation, metabolic engineering, sol-gel immobilized enzyme arrays often use in enzyme industries. The new modern biotechnology leads to improvement in the effects of various physiological conditions which may allow various industrial processes to carry out lower energy consumption, harmless to the environment, high efficiency, and the product's properties enhancement.

INTRODUCTION

History of enzyme technology dated back to 1874 began when Christian Hansen, a Danish Chemist extracted dried calves' stomachs with saline solution to produce rennet which was the first enzyme used for industrial purposes. French scientist, Louis Pasteur in 18th century discovered the fermentative activity of microorganisms. There were several experiments conducted by Eduard Buchner in 1897 at the University of Berlin, to study the ability of yeast extracts to ferment sugar and he found out that the sugar was fermented even when there were no living yeast cells in the mixture and the enzyme was named "zymase". Enzymes were used in 1930 in fruit juice clarification/manufacturing. Early 1960s starch industries witnessed a great advancement with the usage of alpha-amylases and glucoamylases which completely replaced traditional acid hydrolysis of starch.

DOI: 10.4018/978-1-5225-5237-6.ch017

Enzyme technology is a section of the biochemical science that is going through a stage of ontogenesis and significance recognitions in global industrialization. Biotechnology can be defined as the application of biological agents using scientific and engineering principles to the processing of materials to provide goods and services for human. It has a large impact and offers the potential for new industrial processes that are based on renewable raw materials and require less energy (Hamlyn, 1997). In industrial and analytical fields, biotechnology is currently considered as a useful alternative to process technology. Several advantages over chemical catalysts under mild environmental conditions with efficiency and high specificity have been accrued to this fact. Such include ingredient substitution through continuous fermentation, increased products yields and plant capacity, processing aid substitution, more efficient processing, less undesirable products with improved products (Souze, 1980). Biotechnology utilizes a wide range of enzymes produced on a commercial scale employing supposedly screened microorganisms. Such microorganisms have been characterized and optimized to synthesis a high-quality enzyme on large scales for industrial applications. Molecular biology techniques have allowed us to tailor a particular microorganism to produce enzyme with desired characteristics such as tolerance at high temperature and its stability in acidic or alkaline environment, thermostability, and high yields of an enzyme, but also retaining the enzyme activity under critical reaction conditions such as in presence of other metallic ions or compounds. However, the use of enzymes in industrial applications has been limited by several factors, mainly their instability, high cost of the enzymes, solubility in aqueous substrate and difficulty in recovery from bioreactor effluents. There have been several intense researches in the field of enzyme technology that has facilitated their practical applications (Mahmoud & Helmy, 2009). The main objective of this chapter is to extensively explore the great values of starch hydrolyzing enzymes as prospective tool for the many biotechnological opportunities they offer in pharmaceutical industry.

BACKGROUND

Biotechnology has influenced almost every sector of industrial activity- food, chemical feedstock, energy, feed, environment, and health care, these are directly driven by economic, environmental and social needs. It requires an understanding and application of a range of basic scientific and engineering disciplines, including microbiology, biochemistry, physics, chemistry, chemical and bioprocess engineering, besides molecular biology and genetics. Other features include their dependence on renewable feedstock, low energy consumption and environmentally favorable processing that can potentially lead to sustainable development. Although biological processes involving living cells and their constituents have been used by mankind, real break though has emerged with developments in both biological and engineering sciences, during the last fifty years. Table 1 revealed the impacts of emerging technologies on enzyme technology. All chemical reactions occurring in living cells (for breakdown of nutrients and synthesis of cellular constituents) are catalyzed by their group of molecules - enzymes. These are naturally evolved biocatalysts that are designed to perform their function in an efficient manner, and still providing a precise and suitable control mechanism to the cell for survival under range of environmental conditions with numerous applications in every sector. There exist more than 3,000 different known enzymes of which only 150 to 170 are used commercially. Currently only 5% of chemical products are produced using biotechnological methods. Enzymatic processes are fast becoming better financial and ecological alternatives to chemical-physical and mechanical processes and applications by virtue of being cost effective and more environmental friendly. Enzymatic processing of various application areas for industrial enzymes is classified as: i) Enzymes as final products; ii) Enzymes as processing aids; iii) Enzymes in food and beverage production; iv) Enzymes as industrial biocatalysts; and v) Enzyme in genetic engineering.

MAIN FOCUS OF THE CHAPTER

Fermentative Production

Traditionally, Solid State Fermentation (SSF) and submerged fermentation (SmF) have used for most enzymes production, although submerged cultures could be preferred SSF due to greater control of environmental factors such as pH and temperature and ease of handling. Low-cost medium has been recommended for the synthesis of amylases, as to meet the demand of the industries. SSF is defined as the growth of microorganisms on moist solid substrates with free water (Pandey, Soccol, Rodriguez-Leon, & Nigam, 2001). Synthetic media have been used for amylase production by various bacteria species through Submerged Fermentation (Haddaoui, Chambert, Petit-Glatron, Lindy, & Sarvas, 1999, McTigue, Kelly, Doyle, & Fogarty, 1995). Synthetic media such as soluble starch and nutrient broth are so expensive and could be replaced with cheap and readily agricultural by-products. Selection of suitable solid substrates for SSF has been dependent on agroindustrial residues due to their latent advantages for filamentous fungi, which are capable of penetrating into the hardest of these solid substrates and aided by the exertion of turgor pressure by the mycelium (Ramachandran, Patel, Nampoothiri, Francis, Nagy, Szakacs, & Pandey, 2004). SSF has been used to convert moist agricultural polymeric substrates such as maize bran, rice bran, wheat bran, rice husk, potato residue, groundnut oil cake, coconut oil cake, sugarcane bagasse, green gram bran, black gram bran etc. into fermented food products including industrial enzymes (Rahardjo, Sie, Weber, Tramper, & Rinzema, 2005). The solid substrates may provide both nutrition and support. Despite the advantages that SmF has over SSF, SSF technique is far preferred to SmF because of low capital investment, simple technique, low waste water output, better product recovery, high quality production and end-product inhibition (Lonsane, Ghildyal, Budiatman, & Ramakrishna, 1985). Appropriate selection of the solid substrate greatly increases the efficiency of solid state cultivation. Such substrate must meet the following requirements:

1. Should contain biodegradable polysaccharide,
2. Possess a porous solid matrix with a large surface area per unit volume,
3. Should retain added nitrogen sources,
4. Maintain gentle mixing and compression,
5. Should support microbial growth with relatively high water activity on the solid/gas interface (Orzua, Mussato, Contreras-Esquivel, Rodriguez, de la Garza, Teixeira, & Aguilar, 2009; Singhania, Patel, Soccol, & Pandey, 2009).

Components of Starch

Starch is a polymer of glucose linked to one another through the C1 oxygen by a glycosidic bond. This glycosidic bond hydrolyzes at low pH but stable at high pH. Latent aldehyde group known as the reducing end is present at the end of the polymeric chain.

Basically, there are two types of glucose polymers that are found in starch viz: amylose and amylopectin. Amylose is a linear polymer made up to 6000 D-glucose units with α, 1-4 glycosidic bonds (Figure 1) and composes around 20-30% of the total structure of starch depending on the source of starch (Marc, van der Maarel, Joost, Uitdehaag, Leemhuis, & Dijkhuizen, 2002). The number of glucose residues, also indicated with the term degree of polymerization (DP) varies with the source.

Amylopectin consists of short α, 1-4 glycosidic bond (linked linear chains) of 10 to 60 glucose units and α, 1-6 glycosidic bond (linked side chains) with 15 to 45 glucose units (Figure 2). The average number of branching points in amylopectin is 5% but varies with the botanical source. The complete amylopectin molecule contains about 2 000,000 glucose units, thereby being one of the largest molecules in nature (Marc, van der Maarel, Joost, Uitdehaag, Leemhuis, & Dijkhuizen, 2002). The most commonly accepted model of the structure of amylopectin is the cluster model, in which the side chains are ordered in clusters on the longer backbone.

Some of the profound differences between amylose and amylopectin are amylose is an unbranched structural component of starch while amylopectin is a branched component. Amylose is an insoluble component of starch while amylopectin is the soluble component.

Amylose is more used in cooking because of its easy separation from water while amylopectin tends to absorb water more. As amylopectin is soluble in water, amylose and the starch granule are insoluble in cold water. Amylose is a great storage system for energy while amylopectin only stores a small amount of energy.

Figure 1.

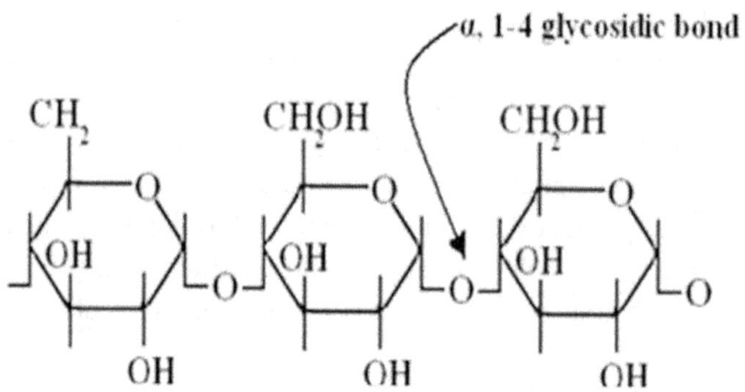

Figure 2.

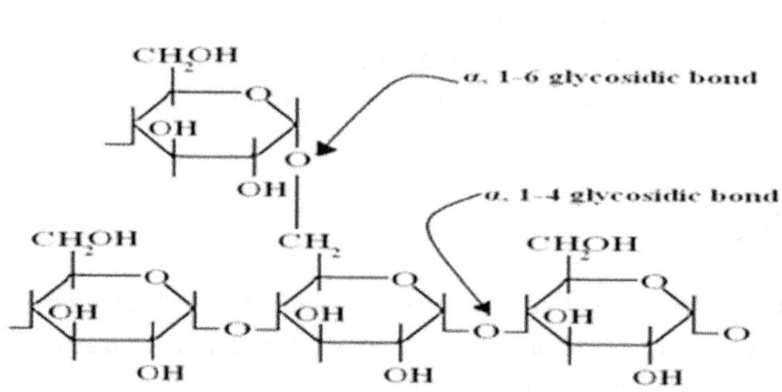

Starch Hydrolyzing Enzymes

The class of starch hydrolyzing enzyme is of great value for the many biotechnological opportunities they offer and their protein stability at extreme conditions. Potential noteworthy applications of amylase in many industries such as textile, pharmaceutical, food etc. has been established by α- Amylases (E.C.3.2.1.1) from Bacillus species which account for about 30% of the world's enzyme production (Gomes & Steiner, 2004). These significant enzymes are widely known for their specific industrial use in the starch conversion process (Mahmoud & Helmy, 2009). Amylolytic enzymes act on starch and related oligo- and polysaccharides (Pandey, Nigam, Soccol, Soccol, Singh, & Mohan, 2000a). The global research such as structural analysis catalytic mechanism and more importantly DNA sequence on starch hydrolyzing enzymes has led to the concept of one enzyme family—the alpha amylase. The three best known amylolytic enzymes viz: α-amylase, β-amylase and glucoamylase, form a large group of enzymes that have great biotechnological importance. Despite their quite closely related function of catalysis of hydrolysis of α-glucosidic bonds, they are quite different from structural and evolutionary points of view (Horvathova, Janeeek, & Sturdik, 2001). This chapter will report on the molecular properties of thermostable enzymes as relates to pharmaceutical industry.

1. Alpha-Amylase
2. Alpha-Glucosidase
3. Glucoamylase
4. Pullulanase

Alpha-Amylases: Molecular Properties and Applications

α-Amylase (then known as Taka diastase) from wheat bran koji culture of *Aspergillus* was the first enzymes to be commercially produced by Dr. Jhokichi Takamine *oryzae* in 1894 and was used as a digestive aid (Sivaramakrishnan, Gangadharan, Nampoothiri, Soccol, & Pandey, 2006). Late 1950s witnessed the industrial production of dextrose powder and crystals from starch using a-amylase and glucoamylase (Prasanna, 2005). Ever since, amylases have been used for various purposes most especially in the conversion of starch into sugar, syrups and dextrins. Starch hydrolysates are used as carbon sources in fermentation and as sources of sweetness in a range of manufactured food products and beverages. The world production of α -amylases from *Aspergillus* sp. and *B. licheniformis* was about 300 tonnes per year of pure enzyme protein (Lee, 1996) while the global market was about $2 billion in 2004 with an average annual growth rate of 3.3% and annual sale was estimated at $11 million (Kilara & Desai, 2002).

Amylases are among the most significant enzymes used for industrial purposes and applications. They are enzymes that hydrolyze glycosidic bonds in starch into glucose, maltose, maltriose and dextrin. Amylase occupies approximately 30% of the global enzyme production with great deal of variety of industries such as food, paper, textile; detergent etc. There are two categories types of amylase based on the degree of hydrolysis of the substrate (Fukumoto & Okada, 1963). These are saccharifying α-amylases which hydrolyze 50 to 60 percent and liquefying α-amylases cleave about 30 to 40 percent of the glycosidic linkages of starch. Enzymes belonging to amylases, endoamylases and exoamylases are able to hydrolyse starch. They are classified according to the manner in which the glycosidic bond is attacked. The starch degrading enzymes are found in the numerous glycoside hydrolase families (13, 14 and 15), but mainly family 13 (Henrissat & Bairoch, 1993). According to the manner of in which the glycosidic bond is attacked, amylases (EC 3.2.1.1) are classified as endoamylases having ability to hydrolyse starch α, 1-4 glycosidic bonds present in the inner part of the amylose or amylopectin chain. α- amylase is a well-known endoamylase. The end products of α-amylase action on linear oligosaccharides with α-configuration and α-limit dextrins, which constitute branched oligosaccharides. Exoamylases, either exclusively cleave α,1-4 glycosidic bonds such as β-amylase (EC 3.2.1.2) or cleave both α,1-4 and α,1-6 glycosidic bonds like glucoamylase (EC 3.2.1.3) or amyloglucosidase and α-glucosidase (Ikuta, Souza, Valencia, Castro, Schenberg, & Pizzirani-Kleiner, 1990).

In molecular biology, the presence of amylase can serve as an additional method to antibiotic resistance and selection of successful integration of a reporter construct. However, the insertion of foreign DNA into this gene result of amylolytic activity in the host cell that can be probed using a simple and inexpensive iodine staining technique. The enzyme, α-amylases would be prospectively useful in the pharmaceutical and fine chemicals industries for the treatment of digestive disorders. The study of gene regulatory elements and gene expression has been effective by Reporter gene assays procedures (Aubel, Morris, Lennon, Rimann, Kaufmann, & Folcher, 2001). Therefore, integration of amylase can serve as an additional method of selecting reporter gene assay as an important tool to study of gene regulatory elements and gene expression. An inexpensive and simple iodine staining procedure can be used to assay a loose of amylolytic activity of an inserted foreign DNA in the host cell (Ikuta, Souza, Valencia, Castro, Schenberg, & Pizzirani-Kleiner, 1990).

Enzyme stability is a crucial factor in the application of enzymes which can be achieved by screening stable enzymes intrinsically, adding stabilizing agents, chemical modification, immobilization, protein engineering, *etc.* Among the stabilizing fortifiers, calcium ions (Ca^{2+}) have been widely used to attain thermostability (Burhan, Nisa, Gokhan, Omer, Ashabil, & Osman, 2003; Mamo & Gessesse, 1999; Marchal, Jonkers, Franke, de Gooijer, & Tramper, 1999). It has also been reported that potassium ions (K^+), sodium (Na^+), ammonium (NH_4^+) and bovine serum albumin through their ionic interactions have protective effect on amylase and hence stabilized the enzyme against thermal denaturation (Janecek & Balaz, 1992; Brumm & Teague, 1989). *Geobacillus thermoleovorans* has been found to produce high maltose-forming, Ca^{2+} independent and hyperthermostable α- amylase (Malhotra, Noorvez, & Satyanarayana, 2000; Narang & Satyanarayana, 2001). Important characteristics of amylases such as hyperthermo- and pH-stability, high activity, high productivity, etc. have been improved by protein engineering, recombinant enzyme technology and enzyme immobilization.

Alpha-Glucosidase/Alpha-D-Glucoside Glucohydrolase

Alpha-glucosidase is a carbohydrate-hydrolase that releases terminal non-reducing (1→4)-linked alpha-glucose residues to alpha-glucose molecule. An example of alpha-glucosidase biocatalyzed reaction is conversion of maltotriose and water to alpha-glucose. Thermostable α-glucosidase has a number of potential applications due to their ability of catalyzing transglucosylation reactions and their wide range of substrate specificity (Cihan, Ozcan, Tekin, & Cokmus, 2010). *Geobacillus* sp. HTA-462 (Hung, Hatada, Goda, Lu, Hidaka, Li, Akita, Ohta. Watanabe, Matsui, Ito, & Horikoshi, 2005), *Bacillus thermoamyloliquefaciens* KP1071 (*Bacillus* sp. SAM1606 (Nakao, Nakayama, Harada, Kakudo, Ikemoto, Kobayashi, & Shibano, 1994), *G. stearotherothermophilus* ATCC 7953 (Albert, Davies, Woodson, & Soper, 1998), *G. thermodenitrificans* HR010 (Ezeji, Wolf, & Bahl, 2005), *Geobacillus* HTA-462 (Hung, Hatada, Goda, Lu, Hidaka, Li, Akita, Ohta. Watanabe, Matsui, Ito, & Horikoshi, 2005), and *Geobacillus* sp. A333 (Cihan, Cokmus, & Ozcan, 2009) and *Geobacillus* sp. A333 (Cihan, Ozcan, Tekin,

& Cokmus, 2010) were among the α-glucosidases of thermophilic and endospore-forming bacteria catalyzing transglycosylation reactions. Mohamed *et al.*, (2012) studied the effect of temperature and synthetic growth media on α-glucosidase production using *Geobacillus stearothermophilus* strain RM and its expression in *Escherichia coli*. The optimal temperatures for *G. stearothermophilus* strain RM was 55 °C and *Geobacillus* sp. was 60 °C, were due to their original temperatures where the samples was isolated. *Escherichia coli* system successfully increased the production of α-glucosidase as it was amplified from this identified *Geobacillus* sp. via degenerate primer and the complete gene was cloned and expressed into *E. coli* with 5-folds increase as compared to the wild type. Studies on *Geobacillus* sp. A333 (Cihan, Cokmus, & Ozcan, 2009) and *Geobacillus* HTA-462 (Hung, Hatada, Goda, Lu, Hidaka, Li, Akita, Ohta. Watanabe, Matsui, Ito, & Horikoshi, 2005) α-glucosidases revealed that these enzymes were active at pH values between 4.5-10.0 with optimum pH 6.8 and 5.5-9.5 with optimum pH 9.0) and at temperatures range of 40-70 °C and 40 - 80 °C with their optimum at 60 °C respectively. However, *G. stearothermophilus* ATCC 12016 α-glucosidase had a high optimal temperature of 70 °C and pH 6.4 among *Geobacillus* species (Takii, Takahashi, Yamamoto, & Suzuki, 1996). As a result ATCC 12016 α-glucosidase could only recover 7% of its activity at its optimal temperature within10 minutes. Cihan *et al.*, (2010) characterized a thermostable α-glucosidase from *Geobacillus thermodenitrificans* F84a. The F84a enzyme revealed activity and stability at a wide range of pH 4.5 to 11.0 with optimum 7.0 and attained 100% stability. The α-glucosidase showed maximum activity and stability at its optimal temperature 60 °C for growth. Therefore, it can be concluded that strain F84a α-glucosidase differed from the ATCC 12016 enzyme due to its high thermostability at its optimal temperature, from the HTA-462 enzyme due to its broad pH activity and stability values, and from the A333 enzyme due to its higher activity and stability values on these broad pH range.

Acarbose and luteolin have been found to be a strong inhibitor of α- glucosidase and are used as anti-diabetic drugs in combination with other anti-diabetic drugs and to stop developing diabetic symptoms respectively. Industrial production of glucoamylases from fungi source has certain limitations such as acidic pH requirement, non-extreme thermostability and protracted biocatalytic reaction that increase the process cost. Therefore, the quest for newer glucoamylases and bioengineering process to improve pH and temperature optimal that will alleviate the biocatalytic efficiency of the enzymes have been the main concern of today research.

Glucoamylase: Molecular Biology and Protein Engineering

Glucoamylase (E.C. 3.2.1.3) which secrete large quantities of the extracellular enzyme is one of the widely used biocatalysts involving in saccharification of partially starch/dextrin to glucose, which is an important substrate for fermentation products for range of food and beverage industries. That is glucoamylase hydrolyses single glucose units from the non-reducing ends of amylase and amyl pectin to produce D glucose as the sole end-product from starch and related polymers such as oligo- and polysaccharides in a successive manner (Nashiru, Koh, Lee, & Lee, 2001). Unlike α-amylase, glucoamylases are also able to hydrolyze the 1, 6- α -linkages (although at a lower rate than 1, 4-linkages) at the branching points of amyl pectin, (Fogarty, 1983; Parbat & Singhal, 2011). Glucoamylases are produced by bacteria, filamentous fungi and yeasts, but almost exclusively in fungi and then in bacteria and yeast (Parbat & Singhal, 2011). Some of the biotechnological significance of these industrially important hydrolytic enzymes is in the production of glucose syrup, high fructose corn syrup, and alcohol. These enzymes have found wide applications in processed food industry, fermentation industry, pharmaceutical industry and textile and paper industries (Vasudeo & Zambare, 2010; James & Lee, 1997),

Lam *et al.,* (2013) have highlighted the potential of using municipal pastry waste as feedstock for glucoamylase production. They reported that high glucoamylase yield of 253.7 U/g of the crude enzyme extract under the optimal process conditions of pH 5.5 and 55 °C was obtained without addition of nitrogen in comparison to other earlier reported waste substrates. Lakshmi and Jyothi (2014) carried out nutritional supplementation and laboratory scale optimization of some of fermentation conditions for glucoamylase production by *Aspergillus oryzae* NCIM 1212 in solid state fermentation. Incorporation of fructose as carbon source and peptone as organic nitrogen supplement showed an increase in glucoamylase activity of 42.32 U/ml and 45.21U/ml as compared to other carbon and nitrogen source respectively. The optimization parameters investigated have proved to be fruitful in enhancing scheme for enzymes of biotechnological importance.

The influence of host dependent post-translational modification on enzymatic and structural features, the recombinant protein produced in *P. pastoris* was described and compared to glucoamylase produced in the related hosts, *S. cerevisiae andA. niger*(Sauer, Sigurskjold, Christensen, Frandsen, Mirgorodskaya, Harrison, Roepstor¡, & Svensson, 2000). Recombinant glucoamylase produced from the three hosts (*Aspergillus awamori, A. niger* and *Saccharomyces cerevisiae*) showed essentially identical catalytic properties, but differs in thermostability (Fierobe, Mirgorodskaya, Frandsen, Roepstor, & Svensson, 1997).

Suzuki (1989) has developed an interesting approach to thermal stabilization of oligo-1,6-glucosidases (α-amylase family member). He proposed the so-called 'proline rule' that should be applicable to other proteins too, as a general strategy for their stabilization. The rule states that thermostability of a globular protein can be additively increased by increasing the frequency of the occurrence of proline residues at special positions on the surface of the protein and by clustering prolines around the flexible protein regions (Suzuki, 1999).

Based on a comparison of the amino acid sequences of three Bacillus α-amylases, differing in their thermostability, Janeèek (1993) calculated that one way leading to stabilization of a protein may be to decrease the hydrophobicity of the protein exterior and/or increase the hydrophobicity of the protein interior. Bacillus α-amylases were used also in studying the reasons responsible for their resistance to this process (Tomazic & Klibanov, 1988) and the mechanism of their irreversible thermal inactivation (Brosnan, Kelly, & Fogarty, 1992).

α-glucosidase from *Geobacillus* sp. A333 was recorded as an enzyme that have high transglycosylation activity on maltoligosaccharide substrates by using only some of its substrate in hydrolysis reaction. The substrates specificities in transglycosylation reactions differentiated the F84a enzyme from ATCC 12016 an HTA-462 α-glucosidases, but the F84a α-glucosidase seemed to be similar to the A333 enzyme in terms of transglycosylation catalysis (Cihan, A.C., Cokmus, C. & Ozcan, 2009). Table 1 summarized the engineering of elastomeric proteins with a new approach to improve the mechanical properties for the construction of biomaterials especially enzymes.

Pullulanase

Pullulanase (EC 3.2.1.41) is an extracellular carbohydrase first discovered from mesophilic *Klebsiella pneumoniae* in 1961 by Bender and Wallenfels. Pullulanase also known as de-branching enzyme hydrolyzes pullulan, polysaccharide and the extracellular yeast. Pullulanase attack on α-1, 4-glycosidic linkage and it also on α-1, 6-glycosidic linkage with other residues. As a result, pullulanase has been found to be a vital agent in structural studies of polysaccharides and oligosaccharide (Ling, Ling, Rosfarizan & Ariff, 2009; Saha, Mathupala, & Zeikus, 1988; Enlvoldsen, 1977). Pullulanase (EC. 3.2.1.41), isopullulanase (EC.3.2.1.57) and neopullulanase (EC.3.2.1.35) are the three kinds of bond that are cleaved by the enzyme. The pullulanase can be sub-divided into two types viz: Pullulanase Type-I and Pullulanase Type-II. The former hydrolyses only α-1, 6-glycocidic bond but the latter specifically attack on α-1,4-glycocidic linkage and α-1,6-glycocidic bond as obtained by Kriegshauser and Liebl, (2000).

Thermo stable pullulanase with unique specificities have been reported by many researchers. Extremophilic microorganisms that grow at 60-80 °C called thermophilic microorganisms were investigated by Rudiger *et al.,* (1995). Koch *et al.*, (1997) dicovered pullulanase of thermophilic microorganisms which attacks on both α-1, 4-glycosidic bond and α-1, 6-glycosidic bond in Amylopectin and malto-oligosaccharide respectively. Thermostable enzymes are essential for a quite number of industrial processes such as starch saccharification. An anaerobic thermophilic bacteria *Clostridium thermohydrosulfuricumis* strain E-39 ferments starch to ethanol and reported a glucoamylase and pullulanase activities from cell extract (Melashiemi, 1988). Pullulanase has improved the thermostability as compared with amylase used in the saccharification process (Madi, Antranikian, Ohmiya, & Gottschalk, 1987).

Some of the industrial applications of pullulanase enzyme can be found in starch processing industry. Pullulanase are used to complete the hydrolysis of starch initiated by α-amylases in production of sugar syrup (Malakar, Tiwari, & Malviya, 2010). Application of pullulanase with other amylolytic enzymes further increases the quality of sugar syrups. In baking industry, pullulanase was used as starch-modifying enzymes. The main challenge is the staling effect which seldom overcomes by increasing the crumb firmness and decreasing the moisture content of the crumb (Van Der Maarel, Van Der Veen, Uitdehaag, Leemhuis, & Dijkhuizen, 2002).

Pullulanase (EC 3.2.1.41), amylopectin 6-glucanohydrolase, debranching enzyme, alpha-dextrin endo-1,6-alpha-glucosidase, pullulan alpha-1,6-glucanohydrolase) is a specific type of amylolytic exoenzyme, that degrades pullulan. An extracellular enzyme is cell surface-anchored lipoprotein synthesized by the genus *Klebsiella*. It has been widely utilized to hydrolyze the α-1,6 glucosidic linkages in starch, amylopectin, pullulan, and related oligosaccharides, which enables an efficient and complete conversion of the branched polysaccharides into small sugars during saccharification process. Pullulanase is used as a processing aid in cereal processing biotechnology (production of ethanol and sweeteners). During saccharification process, pullulanase has been used to increase the final glucose concentration with reduced amount of glucoamylase. Therefore, preventing the reversion reaction that involves re-synthesis of saccharides from glucose molecules. The molecular cloning of genes and their expression in heterologous hosts is one of the possible approaches to overcome the problem of low activity of pullulanase enzymes by wild-type microorganisms. Large quantities of specific gene can be isolated in pure form by molecular cloning and the target DNA can be produced in large amounts under the control of the expression vector. Table 3 summarized the five groups of pullulan-degrading enzyme as it has been reported in the literature. These enzymes are classified based on the reaction specificities of substrate and end products.

Conclusively, industrial biocatalysts play significant transformation in enhancing novel applications of biotechnology at global commercial status. Variety of molecular strategies has been developed by extremophilic microbes in order to survive in harsh conditions. The use of natural polymeric substrates such as starch and other polysaccharides with number of these extremophiles of different taxonomic groups, produce starch hydrolyzing enzymes with protein stability that offers many biotechnological opportunities. Hence, thermostability should be considered as an important feature of starch hydrolyzing enzyme for industrial application. It is hoped that new applications will emerge in the biopharmaceutical sector.

REFERENCES

Ahmed, S., Riaz, S., & Jamil, A. (2009). Molecular cloning of fungal xylanases: An overview. *Applied Microbiology and Biotechnology, 84*(1), 19–35. doi:10.100700253-009-2079-4 PMID:19568746

Albert, H., Davies, D. J. G., Woodson, L. P., & Soper, C. J. (1998). Biological indicators for steam sterilization: Characterization of a rapid biological indicator utilizing *Bacillus stearothermophilus* spore-associated alpha-glucosidase enzyme. *Journal of Applied Microbiology, 85*(5), 865–874. doi:10.1046/j.1365-2672.1998.00607.x PMID:9830122

Anthonsen, H. W., Baptista, A., Drablos, F., Martel, P., & Petersen, S. B. (1994). The blind watch-maker and rational protein engineering. *Journal of Biotechnology, 36*(3), 185–220. doi:10.1016/0168-1656(94)90152-X PMID:7765263

Antikainen, N. M., & Martin, S. F. (2005). Altering protein specificity: Techniques and applications. *Bioorganic & Medicinal Chemistry, 13*(8), 2701–2716. doi:10.1016/j.bmc.2005.01.059 PMID:15781382

Ara, K., Saeki, K., Igarashi, K., Takaiwa, M., Uemura, T., Hagihara, H., ... Ito, S. (1995). Purification and characterization of an alkaline amylopullulanase with both α-1,4 and α-1,6 hydrolytic activity from alkalophilic *Bacillus sp.* KSM-1378. *Biochimica et Biophysica Acta, 1243*(3), 315–324. doi:10.1016/0304-4165(94)00148-Q PMID:7727505

Arnold, F. H. (1993). Engineering proteins for non-natural environments. *The FASEB Journal, 7*(9), 744–749. doi:10.1096/fasebj.7.9.8330682 PMID:8330682

Aubel, D., Morris, R., Lennon, B., Rimann, M., Kaufmann, H., Folcher, M., ... Fussenegger, M. (2001). Design of a novel mammalian screening system for the detection of bioavailable, non-cytotoxic strepto-gramin antibiotics. *The Journal of Antibiotics, 54*(1), 44–55. doi:10.7164/antibiotics.54.44 PMID:11269714

Bertoldo, C., & Antranikian, G. (2002). Starch-hydrolyzing enzymes from thermophilic archaea and bacteria. *Current Opinion in Chemical Biology, 6*(2), 151–160. doi:10.1016/S1367-5931(02)00311-3 PMID:12038998

Brosnan, M. P., Kelly, C. T., & Fogarty, W. M. (1992). Investigation of the mechanisms of irreversible thermoinactivation of *Bacillus stearothermophilus* α-amylase. *European Journal of Biochemistry, 203*(1-2), 225–231. doi:10.1111/j.1432-1033.1992.tb19850.x PMID:1730228

Brumm, P. J., & Teague, W. M. (1989). Effect of additives on the thermostability of *Bacillus stearothermophilus* α-amylase. *Biotechnology Letters, 11*(8), 541–544. doi:10.1007/BF01040032

Burhan, A., Nisa, U., Gokhan, C., Omer, C., Ashabil, A., & Osman, G. (2003). Enzymatic properties of a novel thermostable, thermophilic, alkaline and chelator resistant amylase from an alkaliphilic *Bacillus* sp. isolate ANT-6. *Process Biochemistry, 38*(10), 1397–1403. doi:10.1016/S0032-9592(03)00037-2

Carson, A. E., & Barker, T. H. (2009). Emerging concepts in engineering extracellular matrix variants for directing cell phenotype. *Regenerative Medicine, 4*(4), 593–600. doi:10.2217/rme.09.30 PMID:19580407

Chaput, J. C., Woodbury, N. W., Stearns, L. A., & Williams, B. A. R. (2008). Creating protein biocatalysts as tools for future industrial applications. *Expert Opinion on Biological Therapy*, *8*(8), 1087–1098. doi:10.1517/14712598.8.8.1087 PMID:18613761

Chirumamilla, R. R., Muralidhar, R., Marchant, R., & Nigam, P. (2001). Improving the quality of industrially important enzymes by directed evolution. *Molecular and Cellular Biochemistry*, *224*(1/2), 159–168. doi:10.1023/A:1011904405002 PMID:11693193

Chockalingam, K., Blenner, M., & Banta, S. (2007). Design and application of stimulusresponsive peptide systems. *Protein Engineering, Design & Selection*, *20*(4), 155–161. doi:10.1093/protein/gzm008 PMID:17376876

Chudasama, C. J., Jani, S. A., Jajda, H. M., & Pate, H. N. (2010). Optimization and production of alkaline protease from *Bacillus thuringiensis* CC7. *Journal of Cell Tissue Research*, *10*, 2257–2262.

Cihan, A. C., Cokmus, C., & Ozcan, B. (2009). Characterization of thermostable α-glucosidases from newly isolated *Geobacillus* sp. A333 and thermophilic bacterium A343. *World Journal of Microbiology & Biotechnology*, *25*(12), 2205–2217. doi:10.100711274-009-0127-y

Cihan, A. C., Ozcan, B., Tekin, N., & Cokmus, C. (2010). Characterization of a thermostable α-glucosidase from Geobacillus thermodenitrificans F84a. Current Research, Technology and Education Topics in Applied Microbiology and Microbial Technology, 945-955.

Collins, T., Gerday, C., & Feller, G. (2005). Xylanases, xylanase families and extremophilic xylanases. *FEMS Microbiology Reviews*, *29*(1), 3–23. doi:10.1016/j.femsre.2004.06.005 PMID:15652973

Duffner, F., Bertoldo, C., Andersen, J. T., Wagner, K., & Antranikian, G. (2000). A new thermoactive pullulanase from *Desulfurococcus mucosus*: Cloning, sequencing, purification, and characterization of the recombinant enzyme after expression in *Bacillus subtilis*. *Journal of Bacteriology*, *182*(22), 6331–6338. doi:10.1128/JB.182.22.6331-6338.2000 PMID:11053376

Duo, Y., Shi, P., Huang, H., Zhang, X., Luo, H., Wang, Y., & Yao, B. (2013). Characterization of three novel thermophilic xylanases from *Humicola insolens* Y1 with application potentials in the brewing industry. *Bioresource Technology*, *130*, 161–167. doi:10.1016/j.biortech.2012.12.067 PMID:23306124

El-Fallal, A., Abou Dobara, M., El-Sayed, A., & Omar, N. (2012). *Starch and Microbial α-Amylases: From Concepts to Biotechnological Applications*. Carbohydrates – Comprehensive Studies on Glycobiology and Glycotechnology. Intech. doi:10.5772/51571

Enevoldsen, B. S., Reimann, L., & Hansen, N. (1977). Biospecific Affinity Chromatography of Pullulanase. *FEBS Letters*, *79*(1), 121–124. doi:10.1016/0014-5793(77)80364-5 PMID:560992

Ezeji, T. C., Wolf, A., & Bahl, H. (2005). Isolation and identification of *Geobacillus thermodenitrificans* HRO10. An α-amylase and α- glucosidase producing thermophile. *Canadian Journal of Microbiology*, *51*, 685–693. doi:10.1139/w05-054 PMID:16234866

Fierobe, H. P., Mirgorodskaya, E., Frandsen, T. P., Roepstor, P., & Svensson, B. (1997). Overexpression and characterization of *Aspergillus awamori* wild-type and mutant glucoamylase secreted by the methylotrophic yeast *Pichia pastoris*: Comparison with wild-type recombinant glucoamylase produced using *Saccharomyces cerevisiae* and *Aspergillus niger* as hosts. *Protein Expression and Purification, 9*(2), 159–170. doi:10.1006/prep.1996.0689 PMID:9056481

Fogarty, W. M. (1983). Microbial amylases. In W. M. Fogarty (Ed.), *Microbial enzymes and biotechnology* (pp. 1–92). London: Applied Science Publications Ltd.

Fukumoto, J., & Okada, S. (1963). Studies on bacterial amylase, Amylase types of *Bacillus subtilis* species. *Journal of Fermentation Technology, 41*(1), 427–434.

Garg, A. P., Roberts, J. C., & McCarthy, A. (1998). Bleach boosting effect of cellulase free xylanase of *Streptomyces thermoviolaceus* and its comparison with two commercial enzyme preparations on birchwood Kraft pulp. *Enzyme Microbial Biotechnology, 22*(7), 594–598. doi:10.1016/S0141-0229(97)00250-0

Genckal, H., & Tari, C. (2006). Alkaline protease production from alkalophilic *Bacillus* sp. isolated from natural habitats. *Enzyme and Microbial Technology, 39*(4), 703–710. doi:10.1016/j.enzmictec.2005.12.004

Golynskiy, M. V., & Seelig, B. (2010). De novo enzymes: From computational design to mRNA display. *Trends in Biotechnology, 28*(7), 340–345. doi:10.1016/j.tibtech.2010.04.003 PMID:20483496

Gomes, J., & Steiner, W. (2004). The Biocatalytic potential of Extremophiles and Extremozymes. *Food Technology, 42*(4), 223–239.

Gupta, A., Joseph, B., Mani, A., & Thomas, G. (2008). Biosynthesis and properties of an extracellular thermostable serine alkaline protease from *Virgibacillus pantothenticus. World Journal of Microbiology & Biotechnology, 24*(2), 237–243. doi:10.100711274-007-9462-z

Gupta, R., Beg, Q. K., & Lorenz, P. (2002). Bacterial alkaline proteases: Molecular approaches and Industrial Applications. *Applied Microbiology and Biotechnology, 59*(1), 15–32. doi:10.100700253-002-0975-y PMID:12073127

Gushterova, A., Vasileva-Tonkova, E., Dimova, E., Nedkov, P., & Haertle, T. (2005). Keratinase production by newly isolated Antarctic actinomycete strains. *World Journal of Microbiology & Biotechnology, 21*(6-7), 831–834. doi:10.100711274-004-2241-1

Haddaoui, E., Chambert, R., Petit-Glatron, M. F., Lindy, O., & Sarvas, M. (1999). *Bacillus subtilis* k-amylase: The rate limiting step of secretion is growth phase-independent. *FEMS Microbiology Letters, 173*, 127–131. PMID:10400504

Hamlyn, P. F. (1997). Fungal Biotechnology. *NWFG Newsletter*.

Henrissat, B., & Bairoch, A. (1993). New families in the classification of glycosyl hydrolases based on amino acid sequence similarities. *The Biochemical Journal, 293*(3), 781–788. doi:10.1042/bj2930781 PMID:8352747

Horvathova, V., Janeeek, S., & Sturdik, E. (2001). Amylolytic Enzymes: Molecular Aspects of Their Properties. *General Physiology and Biophysics, 20*, 7–32. PMID:11508823

Hung, V. S., Hatada, Y., Goda, S., Lu, J., Hidaka, Y., Li, Z., ... Horikoshi, K. (2005). α- Glucosidase from a strain of deep-sea *Geobacillus*: A potential enzyme for the biosynthesis of complex carbohydrates. *Applied Microbial Biotechnology*, *68*(6), 757–765. doi:10.100700253-005-1977-3 PMID:15940457

Ikuta, N., Souza, M. B. N., Valencia, F. F., Castro, M. E. B., Schenberg, A. C. G., Pizzirani-Kleiner, A., & Astolfi-Filho, S. (1990). The α-amylase gene as a marker for gene cloning: Direct screening of recombinant clones. *Nature Biotechnology*, *8*(3), 241–242. doi:10.1038/nbt0390-241 PMID:1367421

Jackson, S. E. D., Craggs, T., & Huang, J. R. (2006). Understanding the folding of GFP using biophysical techniques. *Expert Review of Proteomics*, *3*(5), 545–559. doi:10.1586/14789450.3.5.545 PMID:17078767

James, J. A., & Lee, B. H. (1997). Glucoamylases: microbial sources, industrial applications and molecular biology: A review. *Journal of Food Biochemistry*, *21*(6), 1–52. doi:10.1111/j.1745-4514.1997.tb00223.x

Janecek, S., & Balaz, S. (1992). α-amylases and approaches leading to their enhanced stability. *FEBS Letters*, *304*(1), 1–3. doi:10.1016/0014-5793(92)80575-2 PMID:1618293

Janeèek, S. (1993). Does the increased hydrophobicity of the interior and hydrophilicity of the exterior of an enzyme structure reflect its increased thermostability? *International Journal of Biological Macromolecules*, *15*(5), 317–318. doi:10.1016/0141-8130(93)90033-I PMID:8251448

Johnvesly, B., & Naik, G. K. (2001). Studies on production of thermostable alkaline protease from thermophilic and alkaliphilic *Bacillus* sp. JB-99 in a chemical defined medium. *Process Biochemistry*, *37*(2), 139–144. doi:10.1016/S0032-9592(01)00191-1

Kilara, A., & Desai, M. (2002). Enzymes. In A. L. Branen, P. M. Davidson, S. Salminen, & J. H. Thorngate III, (Eds.), *Food Additives* (pp. 661–706). New York: Marcel Dekker Inc.

Kim, C. H., Nashiru, O., & Ko, J. H. (1996). Purification and biochemical characterization of pullulanase type I from *Thermus caldophilus* GK-24. *FEMS Microbiology Letters*, *138*(2-3), 147–152. doi:10.1111/j.1574-6968.1996.tb08148.x PMID:9026441

Koch, R., Canganella, F., Hippe, H., Jahnke, K. D., & Antranikian, G. (1997). Purification and Properties of a Thermostable Pullulanase from a Newly Isolated Thermophilic Anaerobic Bacterium *Fervidobacterium pennavorans* Ven5. *Applied and Environmental*, *63*, 1088–1094. PMID:16535541

Kriegshauser, G., & Liebl, W. (2000). Pullulanase from the Hyperthermophilic Bacterium Thermotoga maritime, Purification by β-Cyclodextrin Affinity Chromatography. *Journal of Chromatography. A*, *737*(1-2), 245–251. doi:10.1016/S0378-4347(99)00373-4 PMID:10681061

Kumar, C. G., & Takagi, H. (1999). Microbial alkaline proteases: From a bioindustrial viewpoint. *Biotechnology Advances*, *17*(7), 561–594. doi:10.1016/S0734-9750(99)00027-0 PMID:14538129

Kumar, J., Dahiya, J. S., Singh, D., & Nigam, P. (2000). Production of endo-1, 4- β-glucanase by a biocontrol fungus *Cladorrhinum foecundissimum*. *Bioresource Technology*, *75*(1), 95–97. doi:10.1016/S0960-8524(00)00037-7

Kuriki, T., Okada, S., & Imanaka, T. (1998). New type of pullulanase from *Bacillus stearothermophilus* and molecular cloning and expression of the gene in *Bacillus subtilis*. *Journal of Bacteriology*, *170*(4), 1554–1559. doi:10.1128/jb.170.4.1554-1559.1988 PMID:3127377

Lakshmi, M. V. V. C., & Jyothi, P. (2014). Production and optimization of Glucoamylase from *Aspergillus oryzae* NCIM1212 using wheat bran, varying chemical parameters under solid state fermentation. *International Journal of Current Microbiology and Applied Sciences*, *3*(5), 70–76.

Lam, W. C., Pleissner, D., & Lin, C. S. (2013). Production of Fungal Glucoamylase for Glucose Production from Food Waste. *Biomolecules*, *3*(4), 651–661. doi:10.3390/biom3030651 PMID:24970186

Lee, B. H. (1996). Other Microorganism Based Products. In *Fundamentals of Food Biotechnology* (pp. 291–352). New York: Wiley-VCH Inc.

Leveque, E., Janecek, S., Haye, B., & Belarbi, A. (2000). Thermophilic archaeal amylolytic enzymes. *Enzyme and Microbial Technology*, *26*(1), 3–14. doi:10.1016/S0141-0229(99)00142-8

Li, Y., Niu, D., Zhang, L., Wang, Z., & Shi, G. (2013). Purification, characterization and cloning of a thermotolerant isoamylase produced from Bacillus sp. CICIM 304. *Journal of Industrial Microbiology & Biotechnology*, *40*(5), 437–446. doi:10.100710295-013-1249-7 PMID:23494708

Ling, H. S., Ling, T. C., Mohamad, R., & Ariff, A. B. (2009). Characterization of Pullulanase type- I from *Bacillus cereus* H1.5. *American Journal of Biochemistry and Biotechnology*, *5*(4), 170–179. doi:10.3844/ajbbsp.2009.170.179

Lonsane, B. K., Ghildyal, N. P., Budiatman, S., & Ramakrishna, S. V. (1985). Engineering aspects of solid-state fermentation. *Enzyme and Microbial Technology*, *7*(6), 258–265. doi:10.1016/0141-0229(85)90083-3

Luo, H., Li, J., Yang, J., Wang, H., Yang, Y., Huang, H., ... Yao, B. A. (2009). Thermophilic and acid stable family-10 xylanase from the acidophilic fungus *Bispora* sp. MEY-1. *Extremophiles*, *13*(5), 849–857. doi:10.100700792-009-0272-0 PMID:19655217

Luo, H., Wang, K., Huang, H., Shi, P., Yang, P., & Yao, B. (2012). Gene cloning, expression and biochemical characterization of an alkali-tolerant b-mannanase from *Humicola insolens* Y1. *Journal of Industrial Microbiology & Biotechnology*, *39*(4), 547–555. doi:10.100710295-011-1067-8 PMID:22179540

Madi, E., Antranikian, G., Ohmiya, K., & Gottschalk, G. (1987). Thermostable Amylolytic Enzymes from a new Clostridium Isolate. *Applied and Environmental Microbiology*, *53*, 1661–1667. PMID:16347392

Mahmoud, D. A. R., & Helmy, W. A. (2009). Application of Immobilization Technology in Enzyme and Biomass Production (Review Article). *Journal of Applied Sciences Research*, *5*(12), 2466–2476.

Malakar, R., Tiwari, A., & Malviya, S. N. (2010). Pullulanase: A potential enzyme for industrial application. *International Journal of Biomedical Research*, *2*, 10–20.

Malhotra, R., Noorvez, S., & Satyanarayana, T. (2000). Production and partial characterization of thermostable and calcium-independent α-amylase of an extreme thermophile *Bacillus thermooleovorans* NP54. *Letters in Applied Microbiology*, *31*(5), 378–384. doi:10.1046/j.1472-765x.2000.00830.x PMID:11069641

Mamo, G., & Gessesse, A. (1999). Purification and characterization of two raw-starch-digesting thermostable a-amylases from a thermophilic *Bacillus*. *Enzyme and Microbial Technology*, *25*(3-5), 433–438. doi:10.1016/S0141-0229(99)00068-X

Mamo, G., Thunnissen, M., Hatti-Kaul, R., & Mattiasson, B. (2009). An alkaline active xylanase: Insights into mechanisms of high pH catalytic adaptation. *Biochimie*, *91*(9), 1187–1196. doi:10.1016/j.biochi.2009.06.017 PMID:19567261

Marc, J. E., van der Maarel, C., Joost, B. V., Uitdehaag, C. M., Leemhuis, H., & Dijkhuizen, L. (2002). Properties and applications of starch-converting enzymes of the α amylase family. *Journal of Biotechnology*, *94*(2), 137–155. doi:10.1016/S0168-1656(01)00407-2 PMID:11796168

Marchal, L. M., Jonkers, J., Franke, G. T., de Gooijer, C. D., & Tramper, J. (1999). The effect of process conditions on the a-amylolytic hydrolysis of amylopectin potato starch: An experimental design approach. *Biotechnology and Bioengineering*, *62*(3), 348–357. doi: PMID:10099546

McTigue, M. A., Kelly, C. T., Doyle, E. M., & Fogarty, W. M. (1995). The alkaline amylase of the alkalophilic *Bacillus* sp. IMD 370. *Enzyme and Microbial Technology*, *17*(6), 570–573. doi:10.1016/0141-0229(94)00098-C

Melasniemi, H. (1988). Purification and some Properties of the Extracellular α-amylase Pullulanase Produced by *Clostridium thermohydrosulfuricum*. *The Biochemical Journal*, *250*(3), 813–818. doi:10.1042/bj2500813 PMID:3260488

Mohamed, R. A., Salleh, A. B., Raja Abd Rahman, R. N. Z., Basri, M., & Leow, T. C. (2012). Isolation of the encoding gene for a thermostable α-glucosidase from *Geobacillus stearothermophilus* strain RM and its expression in *Escherichia coli*. *African Journal of Microbiological Research*, *6*(12), 2909–291.

Nakao, M., Nakayama, T., Harada, M., Kakudo, A., Ikemoto, H., Kobayashi, S., & Shibano, Y. (1994). Purification and characterization of thermostable alpha-glucosidase with transglucosylation activity. *Applied Microbiology and Biotechnology*, *41*, 337–343. doi:10.1007/BF00221229 PMID:7764968

Narang, S., & Satyanarayana, T. (2001). Thermostable a-amylase production by an extreme thermophile *Bacillus thermooleovorans*. *Letters in Applied Microbiology*, *32*(1), 31–35. doi:10.1046/j.1472-765x.2001.00849.x PMID:11169038

Nashiru, O., Koh, S., Lee, S., & Lee, D. (2001). Novel α-glucosidase from extreme thermophile *Thermus caldophilus* GK24. *Journal of Biochemistry and Molecular Biology*, *34*, 347–354.

Niehaus, F., Peters, A., Groudieva, T., & Antranikian, G. (2000). Cloning, expression and biochemical characterization of a unique thermostable pullulan-hydrolysing enzyme from the hyperthermophilic archaeon *Thermococcus aggregans*. *FEMS Microbiology Letters*, *190*(2), 223–229. doi:10.1111/j.1574-6968.2000.tb09290.x PMID:11034283

Nigam, P., & Pandey, A. (Eds.). (2009). Biotechnology for Agro-Industrial Residues Utilization. Springer Science Business Media B.V.

Nigam, P., Pandey, A., & Prabhu, K. A. (1987a). Ligninolytic activity of two Basidiomycetes moulds in the decomposition of bagasse. *Biological Wastes*, *21*, 1–10. doi:10.1016/0269-7483(87)90080-2

Nigam, P., Pandey, A., & Prabhu, K. A. (1987b). Cellulase and ligninase production by Basidiomycetes culture in solid-state fermentation. *Biological Wastes*, *20*(1), 1–9. doi:10.1016/0269-7483(87)90080-2

Orzua, M. C., Mussato, S. I., Contreras-Esquivel, J. C., Rodriguez, R., de la Garza, H., Teixeira, J. A., & Aguilar, C. N. (2009). Exploitation of agro industrial wastes as immobilization carrier for solid-state fermentation. *Industrial Crops and Products*, *30*(1), 24–27. doi:10.1016/j.indcrop.2009.02.001

Pandey, A., Nigam, P., Soccol, C. R., Soccol, V. T., Singh, D., & Mohan, R. (2000a). Advances in Microbial Amylases. *Biotechnology and Applied Biochemistry*, *31*(2), 135–152. doi:10.1042/BA19990073 PMID:10744959

Pandey, A., Soccol, C. R., Nigam, P., Soccol, V. T., Vandenberghe, L. P. S., & Mohan, R. (2000b). Biotechnological potential of agro-industrial residues, II-Cassava Bagasse. *Bioresource Technology*, *74*(1), 81–87. doi:10.1016/S0960-8524(99)00143-1

Pandey, A., Soccol, C. R., Rodriguez-Leon, J. A., & Nigam, P. (2001). *Solid- State Fermentation in Biotechnology: Fundamentals and Applications* (pp. 3–7). New Delhi, India: Asiatech Publishers Inc.

Parbat, R., & Singhal, B. (2011). Production of Glucoamylase by *Aspergillus oryzae* under solid state fermentation using agro industrial products. *International Journal of Microbiology Research*, *2*, 204–207.

Polizeli, M. L., Rizzatti, A. C., Monti, R., Terenzi, H. F., Jorge, J. A., & Amorim, D. S. (2005). Xylanases from fungi Properties and industrial applications. *Applied Microbiology and Biotechnology*, *67*(5), 577–591. doi:10.100700253-005-1904-7 PMID:15944805

Prasanna, V. A. (2005). Amylases and their applications. *African Journal of Biotechnology*, *4*(13), 1525–1529.

Rahardjo, Y. S. P., Sie, S., Weber, F. J., Tramper, J., & Rinzema, A. (2005). Effect of low oxygen concentrations on growth and a-amylase production of *Aspergillus oryzae* in model solid-state fermentation systems. *Biomolecular Engineering*, *21*(6), 163–172. doi:10.1016/j.bioeng.2005.01.001 PMID:15748690

Ramachandran, S., Patel, A. K., Nampoothiri, K. M., Francis, F., Nagy, V., Szakacs, G., & Pandey, A. (2004). Coconut oil cake – A potential raw material for the production of α-amylase. *Bioresource Technology*, *93*(2), 169–174. doi:10.1016/j.biortech.2003.10.021 PMID:15051078

Roy, A., Messaoud, E. B., & Bejar, S. (2003). Isolation and purification of an acidic pullulanase type II from newly isolated *Bacillus sp.* US149. *Enzyme and Microbial Technology*, *33*(5), 720–724. doi:10.1016/S0141-0229(03)00212-6

Rudiger, A., Jorgensen, P. L., & Antranikian, G. (1995). Isolation and Characterization of a Heat-stable Pullulanase from the Hyperthermophilic Archeon *Pyrococcus woesei* after cloning and expression of Its Gene in *Escherichia coli*. *Applied and Environmental Microbiology*, *61*, 567–575. PMID:7574598

Saha, B. C., Mathupala, S. P., & Zeikus, J. G. (1988). Purification and Characterization of a Highly Thermostable Novel Pullulanase from *Clostridium thermohydrosulfuricum*. *The Biochemical Journal*, *252*(2), 343–348. doi:10.1042/bj2520343 PMID:3415657

Shimizu, Y., Kuruma, Y., Ying, B. W., Umekage, S., & Ueda, T. (2006). Cell-free translation systems for protein engineering. *The FEBS Journal*, *273*(18), 4133–4140. doi:10.1111/j.1742-4658.2006.05431.x PMID:16930131

Sigurskjold, B.W., Christensen, U., Frandsen, T.P., Mirgorodskaya, E., Harrison, M., Roepstor, P., & Svensson, B. (2000). Glucoamylase: Structure/function relationships, and protein engineering. *Biochimica et Biophysica Acta*, 275–293.

Singh, D., Dahiya, J. S., & Nigam, P. (1995). Simultaneous raw starch hydrolysis and ethanol fermentation by glucoamylase from *Rhizoctonia solani* and *Saccharomyces cerevisiae*. *Journal of Basic Microbiology*, *35*(2), 117–121. doi:10.1002/jobm.3620350209 PMID:7783000

Singhania, R. R., Patel, A. K., Soccol, C. R., & Pandey, A. (2009). Recent advances in solid state fermentation. *Biochemical Engineering Journal*, *44*(1), 13–18. doi:10.1016/j.bej.2008.10.019

Sivaramakrishnan, S., Gangadharan, D., Nampoothiri, K. M., Soccol, C. R., & Pandey, A. (2006). α-Amylases from Microbial Sources – An Overview. *Recent Developments Food Technology and Biotechnology*, *44*(2), 173–184.

Souze. (1980). Immobilized enzyme bioprocess. *Biotechnology and Bioengineering*, *22*, 79-21.

Srinivasan, M. C., & Rele, M. V. (2005). Cellulase free xylanase from microorganisms and their applications to pulp and paper biotechnology: An overview. *Indian Journal of Microbiology*, *35*, 93–101.

Sunna, A., Moracci, M., Rossi, M., & Antranikian, G. (1997). Glycosyl hydrolases from hyperthermophiles. *Extremophiles*, *1*(1), 2–13. doi:10.1007007920050009 PMID:9680331

Suzuki Y. (1989). A general principle of increasing protein thermostability. *Proc. Jpn. Acad., Ser. B65*, 146-148.

Suzuki, Y. (1999). The proline rule - a strategy for protein thermal stabilization. Proc. Jpn. Acad. *Ser. B*, *75*, 133–137.

Takii, Y., Takahashi, K., Yamamoto, K., & Suzuki, Y. (1996). *Bacillus stearothermophilus* ATCC12016 α-glucosidase specific for α-1,4 bonds of maltosaccharides and α-glucans shows high amino acid sequence similarities to seven α-D-glucohydrolases with different substrate specificity. *Applied Microbiology and Biotechnology*, *44*(5), 629–634. doi:10.1007/BF00172496

Tomazic, S. J., & Klibanov, A. M. (1988). Why is one Bacillus α-amylase more resistant against irreversible thermoinactivation than another? *The Journal of Biological Chemistry*, *263*, 3092–3096. PMID:3125174

Van Der Maarel, M. J. E. C., Van Der Veen, B., Uitdehaag, J. C. M., Leemhuis, H., & Dijkhuizen, L. (2002). Properties and applications of starch-converting enzymes of the α-amylase family. *Journal of Biotechnology*, *94*(2), 137–155. doi:10.1016/S0168-1656(01)00407-2 PMID:11796168

Van der Sloot, A. M., Kiel, C., Serrano, L., & Stricher, F. (2009). Protein design in biological networks: From manipulating the input to modifying the output. *Protein Engineering, Design & Selection*, *22*(9), 537–542. doi:10.1093/protein/gzp032 PMID:19574296

Venkatesan, N., & Kim, B. H. (2002). Synthesis and enzyme inhibitory activities of novel peptide isosteres. *Current Medicinal Chemistry*, *9*(24), 2243–2270. doi:10.2174/0929867023368692 PMID:12470245

Wong, T. S., Zhurina, D., & Schwaneberg, U. (2006). The diversity challenge in directed protein evolution. *Combinatorial Chemistry & High Throughput Screening*, *9*(4), 271–288. doi:10.2174/138620706776843192 PMID:16724918

Yoshikuni, Y., & Keasling, J. D. (2007). Pathway engineering by designed divergent evolution. *Current Opinion in Chemical Biology*, *11*(2), 233–239. doi:10.1016/j.cbpa.2007.02.033 PMID:17353138

Zambare, V. (2010). Solid state fermentation of *Aspergillus oryzae* for glucoamylase production on agro residues. *International Journal of Life Sciences (Kathmandu)*, *4*(0), 16–25. doi:10.3126/ijls.v4i0.2892

APPENDIX

Table 1. Influence of Emerging Technologies on Enzyme Technology

New enzyme activities by chemical modification Singh *et al.*, (1995)
Global analysis of protein activities Mamo*et al.*, (2009)
Industrial biocatalysis Luo*et al.*, (2009)
Genetically modified organisms Hamlyn, 1997)
Enzyme immobilization Collins *et al.*, (2005)
Continuous solid-state fermentation Nigam and Pandey, 2009)
Sol-gel immobilized enzyme arrays Kumar *et al.*, (2000)
New developments in bioseparation Srinivasan and Rele, 2005)
Protein production and downstream processing Polizeli*et al.*, (2005)
Protein refolding for industrial processes Polizeli*et al.*, (2005)
Enzyme engineering Chudasama*et al.*, (2010)
Directed evolution to improve enantioselectivity Chudasama*et al.*, (2010)
Directed evolution and biocatalysis Johnvesly and Naik (2001)
Enzymic protection group techniques Nigam and Pandey (1987a)
Artificial enzymes Pandey*et al.*, (2000b)
High-throughput screening methods Gushterova*et al.*, (2005)
Pollution and waste reduction Kumar and Takagi, 1999)
Ahmed, *et al.*, (2009)
Bioelectro-catalysis Biosensors, bioreactors and biofuel cells Nigam and Pandey (1987b)
Enzymatic polymerization Duo *et al.*, (2013)
Metabolic engineering Li *et al.*, (2013)
Bioprocess engineering Chirumamilla*et al.*, (2001)
Dynamic kinetic resolution/ionic liquids Chirumamilla*et al.*, (2001)
Genomics for enzyme-based drug discovery Luo*et al.*, (2009)
Field-assisted protein purification Garg*et al.*, (1998)
Glycosylation of bioactive compounds Gupta*et al.*, (2002)
Combinatorial biocatalysts review Genckal and Tari, 2006)
Functional genomics Enzyme discovery Luo*et al.*, (2012)
Genome link with enzyme activities Pandey*et al.*, (2000a)
Regulatory aspects Single isomer pharmaceuticals Kumar and Takagi, 1999)
High-throughput methods Enzyme discovery and improvement Gupta *et al.*, (2008)

Table 2. Different Procedures Used in Protein Engineering

Techniques	Reference(s)
Engineering extracellular matrix variants	Carson and Barker, 2009
Peptidomimetics	Venkatesan and Kim, 2002
Cell-free translation systems	Shimizu *et al*., (2006)
De novo enzyme engineering	Golynskiy and Seelig, 2010)
Designed divergent evolution	Yoshikuni and Keasling, 2007
Site-directed mutagenesis	Arnold, 1993; Antikainen and Martin, 2005
Molecular dynamics	Anthonsen *et al*., (1994)
X-ray crystallography	Jackson *et al*., (2006)
Cell surface display technology	Antikainen and Martin, 2005; Chaput *et al*., (2008)
mRNA display	Golynskiy and Seelig, 2010
Computational methods (computational protein design)	Jackson *et al*., (2006); Van der Slootet *al*., (2009); Golynskiy and Seelig, 2010
Stimulus-responsive peptide systems	Chockalingam *et al*., (2007)
DNA shuffling	Antikainen and Martin, 2005; Jackson *et al*., (2006)
Random mutagenesis	Antikainen and Martin, 2005; Wong *et al*., (2006); Jackson *et al*., (2006)
Evolutionary methods/directed evolution	Arnold, 1993

Table 3. Groups of Pullulan-Hydrolyzing Enzyme and Their Action of Specificities

Enzymes Bonds Substrate Preffered End Products References
Pullulanase Type I α-(1,6) Oligosaccharides, Trimer (maltotriose) Bertoldo and
(3.2.1.41) Polysaccharides Antranikian, 2002;
Kim *et al.,* (1996)
Pullulanase Type II α-(1,6) Pullulan Trimer (maltotriose) Roy *et al.,* (2003);
(amylopullulanase) α-(1,4) Polysaccharides Mixture of glucose Duffner*et al.*,2000
3.2.1.41 Oligosaccharide maltose, Leveque *et al.*,2000
and maltotriose
Pullulanhydrolase α-(1,4) Pullulan Panose Ara*et al.,* (1995);
Type I Kuriki*et al.,* (1998)
(neopullulanase) Sunna*et al.,* (1997)
3.2.1.135
Pullulan hydrolase α-(1,4) Pullulan Isopanose Van Der Maarel
Type II *et al.,* (2002)
(isopullulanase)
3.2.1.57
Pullulanhydrolase α-(1,4) Pullulan Mixture of panose,
Type III α-(1,6) Starch, amylose, maltose and
3.2.1.— amylopectin maltotriose
Maltotriose and maltose Niehaus
et al., (2000)

Chapter 18
Role of Enzymes From Microbes in the Treatment of Recalcitrant From Industries

Veena Gayathri Krishnaswamy
Stella Maris College, India

ABSTRACT

The limited availability of fresh water is a global crisis. The growing consumption of fresh water due to anthropogenic activities has taken its toll on available water resources. Unfortunately, water bodies are still used as sinks for waste water from domestic and industrial sources. Azo dyes account for the majority of all dye stuffs, produced because they are extensively used in the textile, paper, food, leather, cosmetics, and pharmaceutical industries. Bacterial degradation of azo dyes under certain environmental conditions has gained momentum as a method of treatment, as these are inexpensive, eco-friendly, and can be applied to wide range of such complex dyes. The enzymatic approach has attracted much interest with regard to degradation of azo dyes from wastewater. The oxido-reductive enzymes are responsible for generating highly reactive free radicals that undergo complex series of spontaneous cleavage reactions, due to the susceptibility of enzymes to inactivation in the presence of the other chemicals. The oxidoreductive enzymes, such as lignin peroxidase, laccases, tyrosinase, azoreductase, riboflavin reductive, polyphenol oxidase, and aminopyrine n-demethylase, have been mainly utilized in the bacterial degradation of azo dye. Along with the reductive enzymes, some investigators have demonstrated the involvement in some other enzymes, such as Lignin peroxides and other enzymes. This chapter reviews the importance of enzymes in dye degradation.

INTRODUCTION

Industrial wastes are the wastes produced by industrial activities which includes any material that is rendered useless without a manufacturing process such as that of factories, industries, mills, and mining operations, which has existed on the start of the Industrial Revolution (Maczulak & Elizabeth, 2010). Examples of industrial wastesincludes chemical solvents, paints, sandpaper, paper products, industrial-byproducts, metals, and radioactive wastes, industrial wastes arenamed as toxic waste, chemicalwaste,

DOI: 10.4018/978-1-5225-5237-6.ch018

industrial solid waste andmunicipal solid waste. These industries use raw materials to producefinished goods duringthe manufacturing process, theseare materials which are rendered useless. They constitute the industrialwastes. Some examples of industrial wastes are metals, paints, sandpaper, slag, ash, radio-active wastes, etc.

Industrial wastes treatment deals with the mechanisms and processes used to treat wastewater that have been contaminated in some way by anthropogenic industrial or commercial activities prior to its release into the environment or its re-use. Most industries produce some wet waste although recent trends in the developed world have been to minimize such production or recycle such waste within the production process. However, many important industries remain dependent on processes that produce wastewater. Hence removal of waste has become the key issue from the industries.

Wastewater management is the key issue for many industries generating large volume of waste water in various processes. Effluent treatment plant is very important to manage wastewater treatment and achieve the norms given by government bodies. To avail the quality water, various advanced technology solutions are developed today. There is an array of technologies in industries wastewater treatment solutions to get economical yet effective results.

While devising the wastewater treatment plant for any industry, it is essential to assess the type of wastewater released from the industrial processes. This wastewater can be contaminated with heavy oil quantities, or hazardous chemicals, corrosive materials, metals or other harmful substances which make water unfit for its reuse. There are several factors to consider a right treatment processes, equipment or technologies.

The pollution of rivers and streams with chemical contaminants is one of the most crucial environmental problems. Waterborne chemical pollution entering rivers and streams causes tremendous amounts of destruction. Although some kinds of water pollution can occur through natural processes, it is mostly a result of human activities. We use water daily in our homes and industries. The water we use is taken from lakes and rivers and from underground (groundwater); and after we have used it and contaminated it most of it returns to these locations. This used water is called "wastewater". If it is not treated before being discharged into waterways, serious pollution is the result (Metcalf & Eddy, 2015).

Industrial wastewaters vary widely from composition, strength, flow and volume, depending on the specific industry or manufacturing establishment of the community. The specific composition and volume of the industrial waste will, of course, depend on the use to which the water has been put. Typical industries which produce significant volumes of wastewaters include paper and fiber plants, steel mills, refining and petrochemical operations, chemical and fertilizer plants, meat packers and poultry processors, vegetable and fruit packing operations and many more. Industrial discharges may consist of very strong organic wastewaters with a high oxygen demand or contain undesirable chemicals that can damage sewers and other structures. They may contain compounds, which resist biological degradation, or toxic components, which interfere with satisfactory operation of the wastewater treatment plant.

Conventional wastewater treatment consists of a combination of physical, chemical, and biological processes and operations to remove solids, organic matter and, sometimes, nutrients from wastewater. The conventional treatment processes have several shortcomings such as being unsuitable for use when the effluent contains high concentrations of the target pollutants, high running cost and low efficiency of removal (Stanisavljevic & Nedic, 2004).

According to Aitken (1993) enzymes were first proposed to the treatment of industrial waste in the 1930's but it wasn't until recent that enzyme technology received much attention (Whiteley & Lee, 2006) for the improvement in biological remediation for industrial effluents. The heterogeneous complexity of industrial waste has created gross uncertainty and deviation in predictions of suitable models for its bioremediation. Microorganisms can express specific xenobiotic metabolizing enzymes that would degrade even the most recalcitrant industrial waste but the limiting capacity for bioremediation emphasizes the fact that micro-organisms, by themselves, are insufficient as they lead to the generation of a considerable amount of biomass and have a very slow rate of degradation. Intrinsic bioremediation is the removal, transformation or detoxification of any contaminant to a less toxic form by any natural process and it is possible to enzymatically attack this complex industrial waste to recover valuable resources, remove toxic materials and recover the water.

Remarkable efforts have been made in the development and implementation of new efficient and eco-friendly treatment methods to reduce or eliminate toxic compounds in effluents prior to their disposal in to the environment. Enzymatic methods appear to be a most promising technology, with minor impact on ecosystems as compared to physicochemical methods. The enzyme technology has been explicitly demonstrated for a huge number of research work and patents. Number of redox enzymes, including peroxidases, tyrosinases and laccases have exhibited their potential for the remediation of a broad spectrum of recalcitrant organic compounds. Most important aspects about the current situation of the subject and future perspectives of the use of enzymes in industrial wastewater treatment are highlighted. The present chapter focuses on the application of enzymes in industrial wastewater treatment.

POTENTIAL USE OF ENZYMES WITH INDUSTRIAL EFFLUENTS

Industrial waste effluents contain a variable mixture of both organic and inorganic substances though the actually nature and concentrations will depend on the source. Much of the organic matter is contained within fatty acids, carbohydrates, proteins that can arise from the agricultural sector (crops, fruit vegetables, pesticides, herbicides, fertilizers, nitrates, phosphates); the fishing industry (oils, cleaning fluids); animal production processes (manure, nitrogen, phosphorous, microorganisms); paper and pulp industries (lignin, celluloses, colouring dyes, inks, sulphites, dioxins); abattoir effluents; wine distilling and brewing industries; meat canning and processing; dairy industries; petroleum industry (phenols, toluenes, benzenes, xylenes, sulphides) and in the tannery pond effluents. The inorganics contain predominantly the heavy metals, brine salts, sulphites and sulphides. The Different Role of enzymes in waste treatment is depicted in the Figure 1.

Industrial wastewaters from different industries express a range of heterogeneity, hence it has warranted extensive investigations and different strategies for their total bioremediation and water recovery. Though microorganisms can express enzymes that can degrade the most recalcitrant pollutants they are wrought with problems. Any approach to use microorganisms alone for the bioremediation and biodegradation of industrial wastewaters has limited potential to individual bacteria, capable of remediating a given pollutant, maybe inhibited by the presence of other pollutants. Moreover, bacteria exercise a very slow rate of degradation of polluted sites limiting their overall practical use for this purpose. Consequently, enzymology has been regarded as a key stone in the context of environmental biotechnology and biological remediation whether they are aerobic or anaerobic processes. The role of novel enzymes for use in bioremediation of industrial effluents remains a safe, economic and 'clean and green' alternative to redundant chemical processes. Another most important aspect about enzymes are they could be used in immobilized systems, hence even with the lowest quantity of the enzymes it could be reused. Table 1 shows the Industrial applications of microbial enzymes Chandrakant et al., (2011)

Figure 1. Role of enzymes in various field Waste treatment

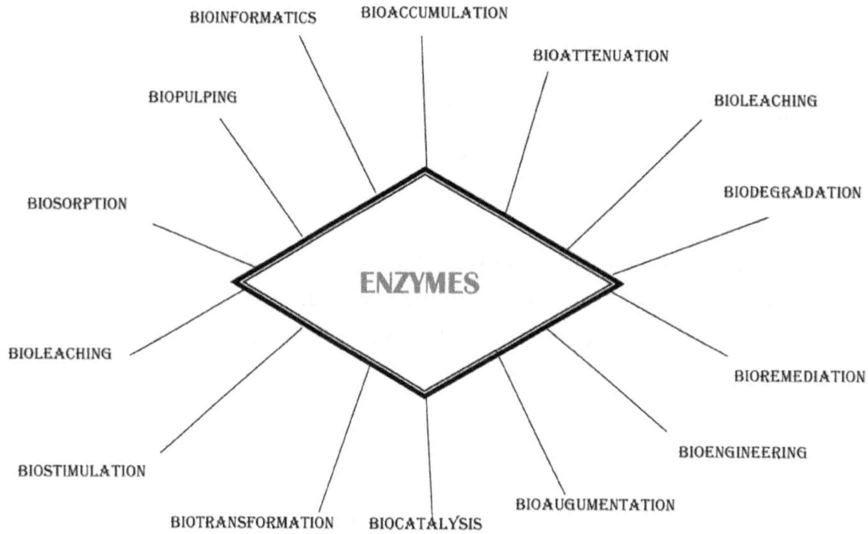

ENZYMES USED IN VARIOUS INDUSTRIES

Paper and Pulp

Enzymes can be used in the bio-deinking of newspaper before recycling which, in general, gives a better result as compared to any chemical method (Pelach, Pastor, Puig, Vilaseca, & Mutje, 2003). Wastepaper is a major constituent of solid waste and any enzymatic bioconversion of cellulose-rich waste into fermentable sugars (glucose) may limit environmental pollution and encourage reuse and sustainability (Van Wyk & Mohulatsi, 2003). Such effluents also contain significant amounts of coloured material, generated from quinone type products during the pulp bleaching process and a major challenge towards bioremediation is its ultimate removal. Bio-pulping is the enzymatic treatment of lignocellulosic material just before thermomechanical pulping to take place and is believed to increase the strength of the product while at the same time decrease environmental impact and chemical energy consumption. The enzymatic degradation of cellulose by cellulases has been the focus on several studies for their use for the bioconversions of agricultural wastes, in the improvement of the manufacture of recycled paper (Stork, Pereira, Wood, Dusterhoft, Toft, & Puls, 1995), in the production of food and fuel (Clarke, 1997), biopolishing of textiles (Cavaco-Paulo, 1998), additives in washing powder and animal feed, pulping, processing of fruit juices and beverages, baking and in bioethanol production (Tolan & Foody, 1999; Zaldivar, Nielsen, & Olsson, 2001). Endoglucanases cleaves the internal glycosidic bonds between cellulosic chains and act synergistically with exoglucanases and E-D-glucosidases during the solubilisation of crystalline cellulose. The current state of knowledge invites new and improved endoglucanases with varying pH and temperature optima, stability and substrate specificities for increased efficiency and economics of various biotechnological and bioremediation processes. Since endoglucanase and glucosidase enzymes have been found in the sulphidogenic bioreactor in our laboratories (Oyekola, Ngesi, & Whiteley, 2007) their role in the bioremediation of industrial waste was to be considered.

Olive Mill Industry

Olive oil production results from a large amount of wastewaters, which is dark liquid residues with high organic content composed mainly by sugars, tannins, polyphenols, polyalcohols, organic acids, proteins pectins and lipids. Cristiana Goncalves et al studied on the use of yeast and filamentous fungi such as *Yarrowia lipolytica, Candida rugose and Candida cylindracea* were used for the production of lipases in undiluted olive mill wastewater containing a COD of 30-261 g/L. In fed-batch cultures of yeasts it was possible to attain high values of lipase activity and simultaneously a higher level of organic matter degradation than the observed in batch operation. *Aspergillus ibericus* has shown to be a good lipase producer from olive mill wastewater of around 3000 U/L.

Mining Industry

Microorganisms offer a potentially large gene pool to choose from when searching for enzymes that may be potentially useful for the treatment, and recovery, of metals from contaminated wastewaters. In particular enzymes from SRB have received much attention due to their ability to enzymatically aid reductive precipitation of metal salts like mercury, palladium, chromium, technetium, arsenic (Lloyd Jr., Yong, & Macaskie, 1998; Lloyd Jr., Mabbet, Williams, & Macaskie, 2001).

Textile Industry

Synthetic dyes have a nearly exhaustible range of application for various types of industry including food, pharmaceutical, textile, printing, paper or chemical. It has been estimated that approximately 5-10% of dyes used in the industrial sector could remain persistent in wastewater (Dafale, Rao, Meshram, & Wate, 2008). Untreated dying effluents are a serious environmental problem with the twenty first century. The release of dyes into the environment is harmful due to toxicity, carcinogenic and/or mutagenic effects on living organisms. The presence of synthetic dyes in wastewater can cause an increase in BOD (Biochemical Oxygen Demand) and COD (Chemical Oxygen Demand) levels. Moreover, the chromophoric groups strongly absorb sunlight and therefore, photosynthetic activity of organisms is inhibited (da-Silva, de-Sa, Russo, Scio, & Ferreira-Leitao, 2010; Kagalkar, Jagtap, Jadhav, Govindwar, & Bapat, 2010). The negative impact on synthetic dyes has been observed in the oestrous cycle and reproductive system in rats (Nath, Sarkar, Mondal, & Paul, 2015), In biochemical markers of vital organs, such as the liver and kidney (Amin, Abdel, & Abd, 2010) and to fetal growth (Wan, Weng, Liang, Lu, & He, 2011 ; Gopinathan, Kanhere, & Banerjee, 2015) . Because of the aforementioned reasons, wastewater treatment containing synthetic dyes is a global problem requiring an immediate yet cost-effective solution. Degradation of synthetic dyes using biological methods is a promising, environmental friendly process and these methods are presented as a cheaper alternative to the expensive physicochemical methods producing large amount of sludge, which must be degraded by other processes. Using biological methods, the degradation of synthetic dyes can occur to a cost efficient, eco-friendly format with certain advantages. Not only are the products of enzyme-catalyzed reaction less toxic than synthetic dyes (Campos, Kandelbauer, Robra, Cavaco-Paulo, & Gubitz, 2001; Telke, Kagalkar, Jagtap, Desai, Bapat, & Govidwar, 2011; Adnan, Sathish Kumar, Yusoff, & Hadibarata, 2015), but the degradation products can be utilized by various natural organisms. These biological methods are founded on enzymes, which are produced by organisms. In the case of enzymatic degradation, azoreductases, peroxidases and phenol oxidases have a potential for dye biodegradation, but in certain aspects each of these enzyme falters in their biodegradation efficiency. Azoreductases (EC 1.7.1.6) requires additional co-factors such as NADH2, NADPH2 and FADH2 for the activation of the catalyst. Additionally, azoreductases can degrade only azo dyes via reductive cleavage of azo bonds (Pandey, Singh, & Iyengar, 2007). One of the main disadvantages of this enzyme-catalyzed degradation processes the formation of toxic products (Platzek, Lang, Grohmann, Gi, & Baltes, 1999). It is known that azoreductases are intracellular enzymes and their use of the pure form of production organisms is problematic due to issues of stability and necessary regeneration of co-factors. Furthermore, the direct application of microorganism produced azoreductases can result in additional problems of the inability of dye diffusion through cell membranes, due to the molecular weight of azo dyes (Robinson, Mcmullan, Marchant, & Nigam, 2001). Peroxidases (EC 1.11.1. X), such as horseradish peroxidase, chloroperoxidases, lignin peroxidases and manganese peroxidases, are other enzymes that could be useful for synthetic dye biodegradation but have a factor that limit their benefits in synthetic dye degradation. All of these enzymes belong to hemoproteins which catalyse the chemical reactions against the presence of hydrogen peroxides (Duran, Rosa, D'annibale, & Gianfreda, 2002). These reactions require specialized attention to because the presence of hydrogen peroxides at too high of a concentration can cause the inactivation of peroxidases (Aitken, Massey, Chen, & Heck, 1994). The last enzyme group described in the literature as useful for synthetic dye decolorization includes phenol oxidases. These enzymes belong to oxidases which catalyze the oxidation of phenolic compounds in the presence of oxygen without additional co-factors. One of these enzymes, laccase (EC 1.10.3.2), Is a member of the multi copper oxidase family (Telke et al, 2011). Laccases catalyses the removal of a hydrogen atom from the hydroxyl group of electron oxidation (Bollag, 1992), which generates non-toxic products of synthetic dye biodegradation (Campos, Kandelbauer, Robra, Cavaco-Paulo, & Gubitz, 2001; Telke, Kagalkar, Jagtap, Desai, Bapat, & Govidwar, 2011; Adnan, Sathish Kumar, Yusoff, & Hadibarata, 2015) . Although the biodegradation potential for laccases is mostly evaluated by the change of an absorption spectrum of dye (Suzuki, Timofei, Kurunczi, Dietze, & Schűűrmann, 2001; Bibi, Bhatti, & Asgher, 2011 ; Chmelova & Ondrejovic, 2015), the destruction of chromophores can lead to dye degradation (Chen & Ting, 2015).

USE OF ENZYMES IN THE DEGRADATION OF RECALCITRANT COMPOUNDS

Oxidoreductases

The removal of toxic organic compounds by both various species of bacteria and fungi and higher plants (Bollag, & Dec, 1998) through o*xidative coupling is mediated with oxidoreductases. Microbes extract energy via energy yielding biochemical reactions mediated by these enzymes to cleave chemical bonds and to assist the transfer of electrons from a reduced organic substrate (donor) to another chemical compound (acceptor). During such oxidation-reduction reactions, the contaminants are finally oxidized to harmless compounds (ITRC 2002). The oxidoreductases are main enzymes involved in the humification of various phenolic substances that are produced from the decomposition of lignin in a soil environment. In the same way, oxidoreductases can also detoxify most of the toxic xenobiotics, such as phenolic or anilinic compounds, through polymerization, copolymerization with other substrates, or binding to humic substances (Park 2006). Table 2 depicts the role of different enzymes in Wastewater.

Microbial enzymes have been exploited in the decolorization and degradation of azo dyes (Williams, 1977; Husain, 2006). Many bacteria reduce the radioactive metals from an oxidized soluble form of a reduced insoluble form. During the process of energy production, bacterium takes up electrons from organic compounds and use radioactive metal as the final electron acceptor. Some of these bacterial species reduce the radioactive metals indirectly with the help of an intermediate electron donor. Finally, precipitant can be seen as the result of redox reactions against the metal-reducing bacteria (Leung, 2004).

Chlorinated phenolic compounds are among the most abundant recalcitrant wastes found in the effluents generated by the paper and pulp industry. These compounds are produced upon the partial degradation of lignin during pulp bleaching process. Many fungal species are considered to be suitable for the removal of chlorinated phenolic compounds from the contaminated environments. The activity of fungi is mainly due to the action of extracellular oxidoreductase enzymes, like laccase, manganese peroxidase, and lignin peroxidase, which are released from fungal mycelium into their nearby environment. Being filamentous, fungi can reach the soil pollutants more effectively than bacteria Rubilar (2008).

Oxygenases

Oxygenases belongs to the oxidoreductase group of enzymes. They participate in oxidation of reduced substrates by transferring oxygen from molecular oxygen (O_2) utilizing FAD/NADH/NADPH as a co-substrate. Oxygenases has been grouped into two categories; the monooxygenases and dioxygenases on the basis of number of oxygen atoms used for oxygenation. They play a key role in the metabolism of organic compounds by increasing their reactivity or water solubility or bringing about cleavage of the aromatic ring. Oxygenases have a broad substrate range and are active against a wide range of compounds, including the chlorinated aliphatics. Generally, the introduction to O_2 atoms into the organic molecule by oxygenase results from the cleavage of the aromatic rings. Historically, the most studied enzymes in bioremediation are bacterial mono- or dioxygenases. There have been many reports on the role of oxygenases in biodegradation process (Arora & Lingens, 2009; Fetzner, 1998; Fetzner, 2003)

Halogenated organic compounds comprise the largest groups of environmental pollutants as a result of their widespread use as herbicides, insecticides, fungicides, hydraulic and heat transfer fluids, plasticizers, and intermediates for chemical synthesis. The degradation of these pollutants is achieved by specific oxygenases. Oxygenases mediates dehalogenation reactions of halogenated methanes, ethanes, and ethylenes in association with multifunctional enzymes (Fetzner & Lingens, 1994; Fetzner, 2001; Jones, 2001).

Monooxygenases

Monooxygenases incorporates one atom of the oxygen molecule into the substrate. Monooxygenases are classified into two subclasses based on the presence cofactor: flavin-dependent monooxygenases and P450 monooxygenases. Flavin-dependent monooxygenases. contain flavin as prosthetic group and require NADP or NADPH as coenzyme. P450 monooxygenases are heme containing oxygenases that exist on both eukaryotic and prokaryotic organisms. The monooxygenases comprise a versatile superfamily of enzymes that catalyzes oxidative reactions of substrates ranging from alkanes to complex endogenous molecules such as steroids and fatty acids. Monooxygenases acts as biocatalysts in bioremediation process and synthetic chemistry due to their highly region selectivity and stereo selectivity on wide range of substrates. Majority of mono-oxygenase studied previously are having cofactor, but there are certain monooxygenases which function independent as a cofactor. These enzymes require only molecular oxygen for their activities and utilize the substrate as reducing agent (Arora, 2009; Cirino & Arnold, 2002).

The desulfurization, dehalogenation, denitrification, ammonification, hydroxylation, biotransformation, and biodegradation of various aromatic and aliphatic compounds are catalyzed by monooxygenases. These properties have been explored for recent years for important application of biodegradation and biotransformation of aromatic compounds (Arora, 2010).

Under oxygen-rich conditions, mono-oxygenase catalyzes oxidative dehalogenation reactions, whereas under low oxygen levels, reductive dechlorination takes place. Oxidation of substrate can lead to de-halogenation as a result of the formation of labile products that undergo subsequent chemical decomposition Fetzner and Lingens (1994), Fetzner (2001), Jones (2001).

Dioxygenases

Dioxygenases are multi-component enzyme systems that introduce molecular oxygen into their substrate. Aromatic hydrocarbon dioxygenases, belong to a large family of Rieske non-heme iron oxygenases. These dioxygenases catalyze specifically the oxygenation of wide range of substrates. Dioxygenases primarily oxidized aromatic compounds and, therefore, have applications for environmental remediation. All members of this family have one or two electron transport proteins preceding their oxygenase components. The crystal structure of naphthalene dioxygenase has confirmed the presence of a Rieske (2Fe–2S) clustered and mononuclear iron in each alpha subunit (Dua, 2002).

The catechol dioxygenases serve as part of nature's strategy for degrading aromatic molecules in the Environment. They are found in the soil bacteria and involved in the transformation of aromatic The critical step in the metabolism of aromatic compounds is the destruction of the resonance structure by hydroxylation and fission of the benzoid ring which is achieved by dioxygenase-catalyzed reactions in the aerobic systems. Based on the substrate that is attacked by the ring cleaving enzyme dioxygenase, the aromatic metabolism can be grouped as catechol pathway, gentisate pathway, and proto catechaute pathway. In all these pathways, the ring activation by the introduction to hydroxyl groups is followed by the enzymatic ring cleavage. The ring fission products, then undergoes transformations leading to the general metabolic pathways of the organisms. Most of the aromatic catabolic pathways converge on catechol. Catechols are formed as intermediates in to a vast range of substituted and non-substituted mono and poly aromatic compounds (Gurujeyalakshmi & Oriel, 1989).

Phenol was degraded through ortho pathway and the crude extract showed the presence of ring cleaving enzyme catechol 1,2 dioxygenase (Nilotpala & Ingle, 2007). Catechols are cleaved either by ortho-fission (intradiol, that is, carbon bond between two hydroxyl groups or by a meta-fission (extra diol, that is, between one of the hydroxyl groups and a non-hydroxylated carbon) as given in Figure 2. Thus the ring is opened and the open ring is degraded (Cerniglia, 1984). As a general rule, most of the halo aromatics are degraded through the formation of the respective halocatechols, the ring fission of which takes place via ortho-mode. On the other hand, most of the non-halogenated aromatic compounds are degraded through meta pathway.

Streptomyces setonii (ATCC 39116) degraded aromatic compounds such as phenol or benzoate via an ortho cleavage pathway using catechol 1,2 dioxygenase (An, Park, & Kim, 2001). These dioxygenases are highly labile enzymes and there requires a detailed investigation into its structural properties. A bacterial strain, *Serratia plymuthica* was able to tolerate phenol up to a concentration of 1050 mg/L. Phenol was degraded through ortho pathway and the crude extract showed the presence of ring cleav-

Figure 2. The main pathways of phenol degradation under aerobic condition (ortho- and meta fission of the benzene ring)
(Adapted from Krug, Ziegler, & Straube, 1985) {Reaction: Phenol +O_2 + NADPH +H+ NADP+ + H2O + catechol (Mörtberg & Neujahr, 1987), catechol + O_2 ccMA + 2H+ (Ngai, Neidle, & Ornston, 1990).

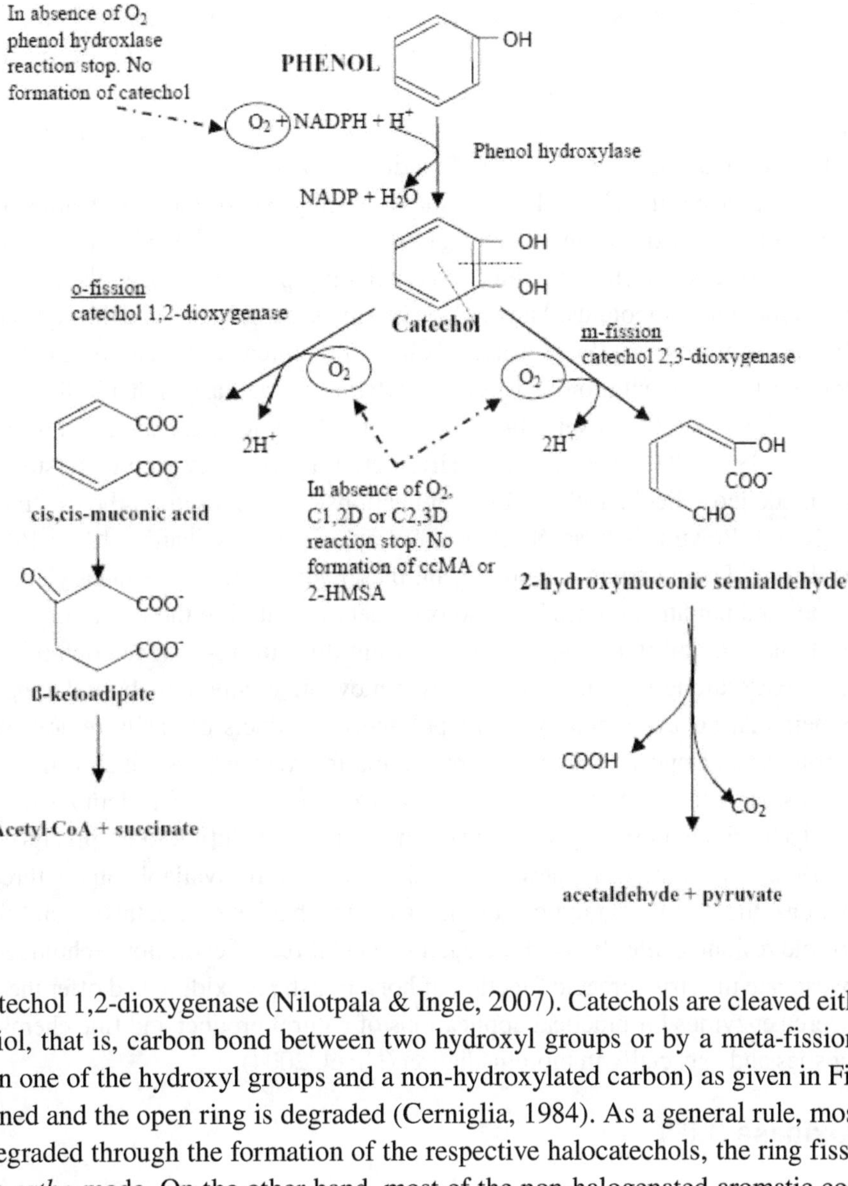

ing enzyme catechol 1,2-dioxygenase (Nilotpala & Ingle, 2007). Catechols are cleaved either by *ortho*-fission (intradiol, that is, carbon bond between two hydroxyl groups or by a meta-fission (extra diol, that is, between one of the hydroxyl groups and a non-hydroxylated carbon) as given in Figure 2. Thus the ring is opened and the open ring is degraded (Cerniglia, 1984). As a general rule, most of the halo aromatics is degraded through the formation of the respective halocatechols, the ring fission of which takes place via *ortho*-mode. On the other hand, most of the non-halogenated aromatic compounds are degraded through *meta* pathway.

Peroxidases

Peroxidases (donor: hydrogen peroxide oxidoreductases) is ubiquitous enzymes that catalyze the oxidation of lignin and other phenolic compounds. Peroxidases are oxidoreductases produced by a number of microorganisms and plants. They catalyze a variety of reactions but they all require the presence of peroxides such as hydrogen peroxide to activate them. (H_2O_2) Hydrogen peroxide first oxidizes the enzyme, which in turn oxidizes the substrate. Peroxidases that has been used for the laboratory-scale treatment of aqueous aromatic contaminants includes horseradish peroxidase, lignin peroxidase and a number of other peroxidases from different sources.

Halogenated organic compounds comprise the largest groups of environmental pollutants as a result of their widespread use as herbicides, insecticides, fungicides, hydraulic and heat transfer fluids, plasticizers, and intermediates for chemical synthesis. The degradation of these pollutants is achieved by specific oxygenases. Oxygenases also mediate dehalogenation reactions against halogenated methanes, ethanes, and ethylenes in association with multifunctional enzymes.

Horseradish peroxidase (HRP, EC 1.11.1.7) is undoubtedly one of the most studied enzymes in the relatively new area of enzymatic waste treatment. Once activated by HRP can catalyze the oxidation of H_2O_2, a wide variety of toxic aromatic compounds including phenols, biphenols, anilines, benzidines and related heteroaromatic compounds. The reaction products are polymerized through a non-enzymatic process which leads to the formation of water-insoluble precipitates that can be easily removed from water or wastewater by sedimentation or filtration. HRP is particularly suitable for wastewater treatment because it retains its activity over a broad range of pH and temperature (Nicell, Bewtra, Biswas, St. Pierre, & Taylor, 1993). The mechanism of HRP action is relatively well understood and has been mathematically modelled (Nicell, 1994). Most applications have focused on the treatment of phenolic contaminants. (Nicell, Bewtra, Biswas, St. Pierre, & Taylor, 1993; Klibanov, 1982; 1983). The use of HRP for the treatment of contaminants including anilines, hydroxyquinoline and arylamine carcinogens such as benzidines and naphthylamines has also been demonstrated in the laboratory (Kilanov, 1980). In addition, HRP has the ability to co-precipitate certain difficult-to-remove contaminants, including non-substrates of HRP, along with the more easily removable compounds by inducing the formation of mixed polymers that behave similarly to the polymeric products of easily removable compounds. This phenomenon has an important practical implication for wastewaters which usually contain many different pollutants. An extension of this principle of hazardous wastes was demonstrated when it was observed that polychlorinated biphenyls could be removed from solution to co-precipitation with phenols (Kilbanov 1982). In recent years' new information has become available on the three-dimensional structure of the enzyme and its catalytic intermediates, mechanisms of catalysis and the function of specific amino acid residues. Site-directed mutagenesis and directed evolution techniques are now used routinely to investigate the structure and function of horseradish peroxidase and offer the opportunity to develop engineered enzymes for practical applications of natural product and fine chemicals synthesis, medical diagnostics and especially in bioremediation (Nigel, 2004).

Lignin Peroxidase (LiP)

Lignin peroxidase (LiP; EC unknown), also known as ligninase or diarylopropane oxygenase was reported in 1983 (Aitken, 1993). It is part of the extracellular enzyme system of the white-rot fungus *Phanerochaete chrysosporium*. LiP was shown to mineralize a variety of recalcitrant aromatic compounds and to oxidize a number of polycyclic aromatic and phenolic compounds. LiPs role in lignin depolymerization has also been confirmed. (Aitken & Irvine, 1989; Venkatadri & Irvine, 1993). Its mechanism is very similar to that of HRP.19 LiPIs stability, which influences greatly the economic and technical feasibility of the enzymes application in waste treatment, was studied by Aitken and Irvine (1989). The authors reported that LiP was readily inactivated at low pH. The enzymes stability was improved by increasing the pH, increasing the enzyme concentration or incubating the enzyme in the presence of its substrate veratryl alcohol. It was also found that optimized conditions for phenolics removal included high enzyme concentration, a pH above 4.0 and controlled addition of Cornwell et al. (1990) reported an immobilization H_2O_2 (Aitken & Irvine, 1989) of LiP on porous ceramic supports did not adversely affect LiPs stability and showed a good potential for degradation of environmentally persistent aromatics (Cornwell, Tinland-Butez, Tardone, Cabasso, & Hammel, 1990). Venkatadri and Irvine (1993) developed reactor that could be used for hazardous waste treatment and LiP production.

Other Peroxidases

Chloroperoxidase (CPO; EC 1.11.1.10) from the fungus *Caldariomyces fumago* has been reported to oxidize several phenolic compounds. In addition, it has been shown to catalyze certain oxygen transfer reactions such as the oxidation of ethanol to acetaldehyde or the oxidation of chloride ions. This latter reaction might lead to the formation of a different range of products (which may be more toxic) when chloride ions are added to reaction mixtures containing chloroperoxidase (Aitken, Massey, Chen, & Heck, 1994). Manganese peroxidase (MnP; EC unknown), produced by *Phanerochaete chrysosporium* has also been observed to catalyze the oxidation of several mono-aromatic phenols, and aromatic dyes, but these reactions depend on the presence of both divalent manganese and certain types of buffers (Aitken & Irvine, 1995). In fact, MnP catalyzes the oxidation of Mn(II) to Mn(III) in the presence of Mn(III) stabilizing ligands. The resulting Mn(III) complexes can then carry out the oxidation of organic substrates (Aitken, Massey, Chen, & Heck, 1994). However, the enzymes requirement for high concentrations of Mn(III) makes its feasibility for wastewater treatment applications doubtful. The use of a microbial peroxidase from *Coprinus macrorhizus* as an alternative to HRP for the removal of aromatic compounds from wastewater was recently investigated. (Its performance was found to compare favorably to HRP in that it could catalyze the same reactions, although it was noticeably more easily inactivated. The use of hemoglobin (EC 1.14.99.3) as an HRP substitute was suggested by Chapsal et al. (1986) and was deemed feasible from a cost and efficiency point of view.

Polyphenol Oxidases

Polyphenol oxidases represent another family of oxidoreductases that have also been shown to catalyze oxidation reactions against phenolic compounds. They are subdivided into two subclasses: tyrosinases and laccases. Both enzyme groups require the presence of bimolecular oxygen for activity, but no coenzymes are needed (Bollag, 1998).

Tyrosinase

Tyrosinase (EC 1.14.18.1), also known as polyphenol oxidase, phenolase or catecholase, (Atlow, 1984) catalyzes two consecutive reactions: (1) the hydroxylation of monophenols with molecular oxygen to form o-diphenols; and (2) the dehydrogenation of o-diphenols with oxygen to form o-quinones. Quinones are mostly unstable and undergo non-enzymatic polymerization to yield water insoluble substances that can be easily removed by simple filtration (Atlow, Bonadonna-Aparo, & Klibanov, 1984; Wada, Ichikawa, & Tatsumi, 1993; Sun, 1992).

Laccase

Laccase (EC 1.10.3.2) is produced by several fungi and seems capable of decreasing the toxicity of phenolic compounds through a polymerization process (Bollag, 1988). Also, because of its relative non-specificity, it can induce the cross-coupling of pollutant phenols with naturally occurring phenols. In fact, laccase can oxidize phenolic compounds to their corresponding anionic free radicals which are highly reactive (Bollag, 1992).

In a study performed on laccase from the fungus *Rhizoctonia praticola*, Bollag et al. (1998) demonstrated the ability of the enzyme to detoxify some of the phenolic compounds tested. Detoxification of a particular phenol appeared to be dependent on the ability of the enzyme to transform the compound, as demonstrated by the disappearance of the parent phenol. However, the reaction products were not identified. Bollag et al. (1998) concluded that the ability of laccase to detoxify a solution containing phenols appe-ars to be function of the particular compound being treated, the source of the enzyme and other environmental factors.

Enzymes Used as Cell-Free Forms

Enzymes extracted from organisms producing them are also being employed in wastewater treatment. In many cases, such cell-free or isolated enzymes are preferred for use over the intact organism, especially when the effluent to be treated contains pollutants which cannot support growth. The isolated enzymes could be used in either the pure form or as a crude extract. Enzymes need not be acclimatized to the wastewater like microbial cultures and also do not require a supply of nutrients for their growth. When isolated enzymes are used, the growth rate of the source organism population does not affect the amount of enzyme available to treat the effluent. Besides, it is relatively easier to standardize optimum treatment conditions with isolated enzymes (Karam & Nicell, 1997). The use of isolated enzymes also has a definite advantage over microbial cultures in terms of ease of handling and storage. The delivery of cell-free crude enzyme extracted uses the least processed, yet functional form of the enzyme. The preparation of crude enzyme extracts typically includes simple processes such as grinding or homogenizing the source tissue in the presence of an appropriate buffer followed by filtration (Johnson & Pokora, 1994). Crude enzyme extracts are preferred for use over pure enzymes, primarily because they are relatively inexpensive. Considering the scale of operation, the use of pure enzymes in effluent treatment are not economically feasible. Crude enzyme extracts can also effectively remove pollutants from effluent. A crude extract from *Cladosporium cladosporioides* containing laccase was found to decolorize solutions of the azo dye acid blue 193 up to 47% in 8 hrs Vijayakumar et al (2006).

Enzyme Delivery of Immobilized Form

The functionality of enzymes depends largely on their conformation. Harsh reaction conditions like extreme temperature, very high or low pH, high ionic strength, high concentrations of reactants, and presence of inhibitors can alter the conformation of an enzyme (Karan & Nicell, 1997). Enzymes may not function optimally under such drastic conditions which are often encountered with effluent streams. Immobilization methods that increase the reusability of enzymes by preventing the loss of enzyme during the course of the reaction and minimizing the loss of activity of enzymes under harsh treatment conditions have been developed. An enzyme is said to be immobilized when it is physically confined to a certain region of space, retaining its catalytic activity and the capacity to be used repeatedly or continuously (Pescod, 1992). The use of immobilized enzymes in effluent treatment has many important advantages over the use of free enzymes including increased stability, localization, ease of handling, reusability and a consequent decrease in running cost (Karan & Nicell, 1997). The HRP enzyme has proved to be an adaptable molecule that can be used in the form of a cell-free crude extract (Kragl, 1996) or in an immobilized form entrapped in calcium alginate capsules at a laboratory scale (Maddhinni, Vurimindi, & Yerramilli, 2006; Costa, Azevedo, & Reis, 2005).

Fungal laccase immobilized using γ-aluminium oxide pellets, has been reported to decolorize solutions of azo dyes like Ponceau Red (65% decolorization), anthracinoid dyes like lanaset Blue 2R (100% decolorization) and a triphenyl methane dye like crystal violet (98% decolorization) after 24 hours (Omar, 2008).

However, not all enzymes are amenable to immobilization. Some of the methods of immobilization such as adsorption, covalent binding and chemical coupling can adversely affect the catalytic activity of certain enzymes. Adsorption is a widely used immobilization method which is preferred for its simplicity and ease of regeneration. But it has also been found that an immobilized enzyme that adsorbs too strongly to the supporting material may show a loss of functionality (Adlercreutz, 2006).

Enzymes immobilized by this method can operate in a relatively narrow range of pH, temperature and ionic strength. Drastic changes in reaction conditions cause desorption of the enzyme from the support material. Covalent binding of enzymes on to a support or matrix may modify the conformation of the enzyme. Considering that the functionality of enzymes depends largely on their conformation, such a change can result in the loss of enzymatic activity (Chou, 1998).

It has been shown that enzymes immobilized by covalent binding can retain their activities more effectively if they are immobilized in the presence of their substrate or a competitive inhibitor since the active site remains protected from conformational changes in the presence of a substrate or its structural analogue.

Similarly, the entrapment and encapsulation of enzymes has certain advantages such as large surface area for substrate-enzyme interaction in a relatively small volume. However, the drawbacks of this method are that a high concentration of enzyme is required and that there is occasional inactivation of the enzyme on entrapment. Additionally, the pore size of the (cross linked) polymer has to be very small to retain the enzyme within the capsule Wu et al (1998). HRP entrapped in calcium alginate gel showed lower decolorization (52%) of the azo dye direct yellow than the free enzyme (69%) Maddhinni et al (2006). Immobilization procedures need to be optimized to minimize the loss of enzyme activity and achieve maximum reusability. This method of enzyme delivery holds great potential for the continuous treatment of large volumes of effluent.

Enzymes Used with Different Nanoparticles Forms in Waste Treatment

Direct application of biologically active substances has several disadvantages, such as storage, operational stability and significant operational costs, because the active substance must be repeatedly added into cleaning process. These drawbacks can be reduced by immobilization of biologically active enzymes into suitable inert structure like nanofibers, thanks to their high specific surface. Immobilized enzymes are active, stabilized and thus advantageous to long-term operation in cleaning process.

Developments in nanoscale research have made it possible to invent economically feasible and environmentally stable treatment technologies for effectively treating water/wastewater meeting the ever increasing water quality standards. It is suggested that nanotechnology can adequately address many of the water quality issues by using different types of nanoparticles and/or nanofiber. Several commercials and noncommercial technological developments are employed on daily basis but nanotechnology has proved to be one of the advanced ways for water/wastewater treatment (Savage & Diallo, 2005).

The application of nanotechnology for the removal of toxic pollutants such as the pharmaceutical and personal care products, polycyclic aromatic hydrocarbons, polychlorinated biphenyls, phthalates, furans and dioxins, agrochemicals and pesticides, volatile organic compounds, viruses and bacteria, dyes, inorganic pollutants, etc. have been widely reported by several investigators in the field of nanotechnology (Olushola, 2014). Numerous scientists claim that nanotechnologies offer more affordable, effective, efficient and durable ways of achieving specific nanoparticles for water treatment will allow manufacturer to prepare less toxic particles using classical methods.

Nanotechnology has been cited, in different literatures, as one of the most advanced processes for wastewater treatment. It may be classified based on the nano-materials nature into three main categories: nano-adsorbents, nano-catalysts and nano-membranes. In nano-adsorption technology, numerous and effective works have been published recently with the aim to investigate the removal of pollutants from wastewater using nano-adsorbent materials (Zhang, Xu, Xu, Chen, He, Zhong, & Gu, 2014a; Zhang, Yan, Xu, Guo, Cui, Gao, Wei, & Du, 2014b; Tang, Zhang, Liu, Pan, Dong, Li, 2014; Shamsizadeh, Ghaedi, Ansari, Azizian, & Purkait, 2014; Kyzas & Matis, 2015). Nano-adsorbent can be produced using the atoms of those elements which are chemically active and have high adsorption capacity on the surface of the nano-material (Kyzas & Matis, 2015). The used materials for development of nano-adsorbents include activated carbon, silica, clay materials, metal oxides and modified compounds in the form of composites (El Saliby, Shon, Kandasamy, & Vigneswaran, 2008).

Nanomaterials are typically defined as materials smaller than 100 nms in at least one dimension. At this scale, materials often possess novel size-dependent properties different from their large counter parts which might already be explored for the water treatment purposes. These properties may relate to the high specific surface area, such as fast dissolution, high reactivity, and strong sorption, or to their discontinuous properties, such as super paramagnetism, localized surface plasmon resonance, and quantum confinement effect. Most applications are still in the stage of laboratory research.

Nanobiocatalyst advancements and its applications have been reported by Mission et al. (2015) has verified the significance and wide applications of nano-biocatalysts comprising of nanocarriers and enzymes (immobilized) and concluded that, it has been successfully employed in diverse bioprocess applications. Flow diagram demonstrating an integrated process of using advanced nano-biocatalyts for industrial bioprocessing applications (Mailin Misson, Hu Zhang, & Bo Jin, 2015).

Nanosorbents provided significant improvement over conventional adsorbents with their extremely high specific surface area and associated sorption sites, short intraparticle diffusion distance, and tunable pore size and surface chemistry.

The remediation of contaminated wastewater can be achieved by using a combination of enzyme technology and nanotechnology known as the SEN, i.e., Single Enzyme Nanoparticle (Watlington, 2005). An SEN may be described as an armored enzyme surrounded by a protective 'cage' which is a few nanometers thick. The enzyme chymotrypsin has been used to create an SEN. In this case, the enzyme molecule was 'caged' by a silicate shell which was linked with its surface. While the cage covered most of the enzyme, the active site was kept chemically accessible to maintain the functionality of the enzyme (Kim & Grate, 2003).

The technology used to create the chymotrypsin SEN can be applied to other enzymes as well. From the point of view of dye degradation in effluents, enzymes involved in redox reactions are of special interest for the synthesis of SENs. Enzymes involved in wastewater treatment that can be used for SEN synthesis include cell-free crude extracts or purified forms of enzymes like peroxidases, polyphenol oxidases (like laccases and tyrosinase), dehalogenases and organophosphorus hydrolases. These enzymes are capable of degrading a wide variety of recalcitrant organic contaminants such as phenols, polyaromatics, dyes, chlorinated compounds and pesticides (Kim & Grate, 2004).

SENs are able to withstand more drastic conditions of temperature, pH, contaminant concentration and salinity as compared to free enzymes. Another type of novel nanoparticle are nano-sponges. These are materials containing microscopic particles of nano-sized cavities. These particles can encapsulate or can be embedded with many types of substances and are capable of transporting them through an aqueous medium. Nano-sponges with complexes consisting of nano-polymers and enzymes have been synthesized recently. These nano-sponges are created by embedding the enzyme in a polymer matrix (Gitsov 2009, Regnnli & Bruns, 2010). Novel nanoparticles such as these could be synthesized using enzymes such as peroxidase and laccase. Such nano-sponges could find application in remediation of wash streams from dyeing and textile processing industries. At present, the utilization of nano-polymers in wastewater is primarily in the removal of heavy metals (Theron, Walker, & Cloete, 2008).

In recent times, research on carbon based nanotechnology such as the carbon nanotube is gaining momentum. The potential of these particles for use in remediation of soil, water and air is being evaluated. Carbon nanotubes carrying immobilized enzymes have been synthesized and incorporated into latex paints. The resulting materials can detect and eliminate hazardous chemical and biological agents.

Electrospun nanofiber membranes (ENMs) are the recently emerging membranes which give birth to a novel way to treat wastewater (Matsuura, Feng, Khulbe, Rana, Singh, Gopal, Kaur, Barhate, Ramakrishna, & Tabe, 2010; Botes & Eugene Cloete, 2010; Qu, Alvarez, & Li, 2013). The key features of this new emerging technique include less energy consumption, less expensive and lighter process as compared to the existing conventional techniques. Moreover, higher porosity and surface to volume ratio are the major advantages of this technique (Balamurugan, Sundarrajan, & Ramakrishna, 2011; Tabe, 2014). Electrospinning is advantageous over conventional nanofiber spinning techniques in being capable of producing fibers that are orders of magnitude thinner. The fiber diameter governs the surface area to volume ratio and affects membrane porosity. In electrospinning, fiber diameter can be adjusted by varying the process parameters such as solution concentration, applied voltage, surface tension, and spinning distance (Theron, Walker, & Cloete, 2004; Tabe, 2014).

Various types of natural and synthetic polymers have been electrospun into nanofibers. The reported number of these polymers has been more than 100. These include natural and synthetic polymers such as polystyrene (PS), poly (vinyl chloride) (PVC), polyvinylidene fluoride (PVDF), polybenzimidazole (PBI), poly (vinyl phenol) (PVP), Kevlar (poly (p-phenylene terephthalamide), or PPTA), polyurethanes (PUS), Nylon-6, poly (vinyl alcohol) (PVA), polycarbonates, poly(e-capro-lactone) (PCL), polysulfones, poly (ethylene terephthalate) (PET), and many others (Souhaimi & Matsuura, 2011; Feng, Khulbe, Matsuura, Tabe, & Ismail, 2013). The electrospun nanofiber integration with microbes can increase the potential of purification and filtration. Research has been conducted on the integration of microbe electrospun nanofibers containing algae or bacteria showed momentous impacts for environmental applications (San, Celebioglu, Tumtas, Uyar, & Tekinay, 2014).

A study the immobilization of ammonium oxidizing bacteria *Acinetobacter calcoaceticus* STB1 cells were performed on electrospun cellulose acetate nanofibrous webs (CA-NFW) for removal of ammonium from wastewater (Sarioglu., Yasa, Celebioglu, Uyar, & Tekinay, 2013). The study findings revealed that STB1/CA-NFW effectively removed the ammonium ions (98.5%) by converting them to nitrogen form that accumulated as bacterial biomass without loss of their reusability potential. Likewise, San et al. (2014) also carried out an experiment for decolorization of methylene blue (MB) dye by selecting electrospun CA-NFW for immobilization of dye decolorization bacteria species i.e. *Pseudomonas aeruginosa, Clavibacter michiganensis, Aeromonas eucrenophila* in wastewater. From the study, they noted that efficient decolorization of MB (95%) was attained in 24 hours. The fact behind decolorization of MB dye was the biodegradation carried out by bacteria. Moreover, the reusability of the nanofibrous biocomposite was also examined and it was found that bacteria immobilized NFW possess the capacity to decolorized dye at the end of 4th cycled. Consequently, due to its reusability, simple and porous characteristics this nanofibrous biocomposite can be utilized for treating industrial wastewater. As immobilized bacteria NFW required less space and volume for growth medium compared to free bacteria and hence it is more economical whereas the biofilm formed also possesses high resistance to metal toxicity, salinity and harsh environmental conditions.

CONCLUSION

A large number of enzymes from a variety of different microorganisms have been reported to play an important role in pollutant treatment. Enzymes have been employed in numerous fields primarily for their immense catalytic potential. In waste treatment, enzymes can be utilized to develop remediation processes that are environmentally less aggressive than conventional techniques. Their versatility and efficiency even in mild reaction conditions gives them an advantage over the conventional physico-chemical treatment methods. The biological origin of enzymes reduces the adverse impact on the environment there by making enzymatic wastewater treatment an ecologically sustainable technique. Enzymes can act on specific recalcitrant pollutants to remove them by precipitation or transformation to other innocuous products. They also can change the characteristics of a given waste to render it more amenable to treatment or aid in bio-converting waste material to value-added products. Enzymes seem to have a promising future. However, before the full potential for enzymes is realized, a number of issues still must be addressed including identification and characterization of reaction by-products, disposal of reaction residues and reduction of enzymatic treatment costs.

Actualization of enzyme applications in industrial processes requires high performance enzymes with specific characteristics, which will stimulate research to explore new avenues to overcome their weaknesses. Some of the strategies in the field are exploiting novel enzymes from nature, improving existing catalytic properties, broadening specialized enzymes to serve new functions, optimizing formulation of enzyme preparations, or de novo designing biocatalysts. These approaches have provided valuable candidates for the biocatalytic processes. Also, further research is needed to determine which enzyme is best suited in a particular situation and to optimize the enzymatic process as a whole.

REFERENCES

Adlercreutz, P. (2006). Immobilization of Enzymes for Use in Organic Media. In J. M. Guisan (Ed.), *Immobilization of Enzymes and Cells. Methods in Biotechnology™* (Vol. 22). Humana Press. doi:10.1007/978-1-59745-053-9_22

Adnan, L. A., Sathish Kumar, P., Yusoff, A. R. M., & Hadibarata, T. (2015). Metabolites characterization of laccase mediated Reactive Black 5 biodegradation by fast growing ascomycete fungus *Trichoderma atroviride* F03. *International Biodeterioration & Biodegradation*, *104*, 274–282. doi:10.1016/j.ibiod.2015.05.019

Aitken, M. D. (1993). Waste Treatment Applications of Enzymes: Opportunities and Obstacles. *Chemical Engineering Journal*, *52*(2), B49–B58. doi:10.1016/0300-9467(93)80057-U

Aitken, M. D., & Irvine, R. L. (1989). Stability testing of ligninase and Mn-peroxidase from *Phanerochaete chrysosporium*. *Biotechnology and Bioengineering*, *34*(10), 1251–1260. doi:10.1002/bit.260341003 PMID:18588065

Aitken, M. D., Massey, I. J., Chen, T., & Heck, P. E. (1994). Characterization of reaction products from the enzyme catalyzed oxidation of phenolic pollutants. *Water Research*, *28*(9), 1879–1889. doi:10.1016/0043-1354(94)90163-5

Al-Kassim, L., Taylor, K. E., Nicell, J. A., Bewtra, J. K., & Biswas, N. (1994). Enzymatic removal of selected aromatic contaminants from wastewater by a fungal peroxidase from *Coprinus macrorhizus* in batch reactors. *Journal of Chemical Technology and Biotechnology (Oxford, Oxfordshire)*, *61*(2), 179–182. doi:10.1002/jctb.280610214 PMID:7765416

Amin, K. A., Abdel, H. H., & Abd, E. A. H. (2010). Effect of food Azo dyes Tartrazine and Carmoisine On Biochemical Parameters Related to Renal, Hepatic Function and Oxidative Stress Biomarkers in Young Male Rats. *Food and Chemical Toxicology*, *48*(10), 2994–2999. doi:10.1016/j.fct.2010.07.039 PMID:20678534

An, H., Park, H., & Kim, E. (2001). Cloning and expression of thermophilic catechol 1,2 dioxygenase gene (cat A) from *Streptomyces setonii*. *FEMS Microbiology Letters*, *195*(1), 17–22. doi:10.1111/j.1574-6968.2001.tb10491.x PMID:11166989

Arora, P. K., Kumar, M., Chauhan, A., Raghava, G. P., & Jain, R. K. (2009). OxDBase: A database of oxygenases involved in biodegradation. *BMC Research Notes*, *2*(1), 67. doi:10.1186/1756-0500-2-67 PMID:19405962

Atlow, S.C., Bonadonna-Aparo, L., & Klibanov, A.M. (1984). Dephenolization of industrial wastewaters catalyzed by polyphenol oxidase. *Biotechnology and Bioengineering*, *26*, 599-603. DOI: 10.1002/bit.260260607

Balamurugan, R., Sundarrajan, S., & Ramakrishna, S. (2011). Recent trends in nanofibrous membranes and their suitability for air and water filtrations. *Membranes*, *1*(3), 232–248. doi:10.3390/membranes1030232 PMID:24957734

Bibi, I., Bhatti, H. N., & Asgher, M. (2011). Comparative study of natural and synthetic phenolic compounds as efficient laccase mediators for the transformation of cationic dye. *Biochemical Engineering Journal, 56*(3), 225–231. doi:10.1016/j.bej.2011.07.002

Bollag, J. M. (1992). Decontaminating soil with enzymes. *Environmental Science & Technology, 26*(10), 1876–1881. doi:10.1021/es00034a002

Bollag, J. M., & Dec, J. (1998). *Use of Plant material for the removal of pollutants by polymerization and binding to humic substances.* Tech. Rep. R-82092. Center for Bioremediation and Detoxification Environmental Resources Research Institute the Pennsylvania State University.

Bollag, J. M., Shuttleworth, K. L., & Anderson, D. H. (1988). Laccase-mediated detoxication of phenolic compounds. *Applied and Environmental Microbiology, 54*, 3086–3091. PMID:3223771

Botes, M., & Eugene Cloete, T. (2010). The potential of nanofibers and nanobiocides in water purification. *Critical Reviews in Microbiology, 36*(1), 68–81. doi:10.3109/10408410903397332 PMID:20088684

Campos, R., Kandelbauer, A., Robra, K. H., Cavaco-Paulo, A., & Gubitz, G. M. (2001). Indigo Degradation with Purified Laccases from Trametes Hirsuta and Sclerotium Rolfsii. *Journal of Biotechnology, 89*(2-3), 131–139. doi:10.1016/S0168-1656(01)00303-0 PMID:11500206

Cavaco-Paulo, A. (1998). Mechanism of Cellulose Action in Textile Processes. *Carbohydrate Polymers, 37*(3), 273–277. doi:10.1016/S0144-8617(98)00070-8

Cerniglia, C. E. (1984). Microbial transformations of aromatic hydrocarbons. In R. M. Atlas (Ed.), *Petroleum Microbiology* (pp. 99–128). New York: Macmillon.

Chapsal, J. M., Bourbigot, M. M., & Thomas, D. (1986). Oxidation of aromatic compounds by haemoglobin. *Water Research, 20*(6), 709–713. doi:10.1016/0043-1354(86)90094-1

Chen, S. H., & Ting, A. S. Y. (2015). Biodecolorization and biodegradation potential of recalcitrant triphenylmethane dyes by *Coriolopsis* sp. isolated from compost. *Journal of Environmental Management, 150*, 274–280. doi:10.1016/j.jenvman.2014.09.014 PMID:25527986

Chmelová, D., & Ondrejovič, M. (2015). Effect of metal ions on triphenylmethane dye decolorization by laccase from *Trametes versicolor. Nova Biotechnol. Chim., 14*, 191–200.

Chou, X., Wu, N., Lupher, D., & Davis, L. C. (1998). Benzotriazoles: Toxicity and Degradation. *Proceedings of the 1998 Conference on Hazardous Waste Research*, 374-382.

Cirino, P. C., & Arnold, F. H. (2002). Protein engineering of oxygenases for biocatalysis. *Current Opinion in Chemical Biology, 6*(2), 130–135. doi:10.1016/S1367-5931(02)00305-8 PMID:12038995

Clarke, Aj. (1997). *Biodegradation of Cellulose: Enzymology And Biotechnology*. Lancaster, PA: Technomic Publishing Co. Inc.

Colvin, V. L. (2003). The potential environmental impact of engineered nanomaterials. *Nature Biotechnology, 10*(10), 1166–1170. doi:10.1038/nbt875 PMID:14520401

Cornwell, R. L., Tinland-Butez, M.-F., Tardone, P. J., Cabasso, I., & Hammel, K. E. (1990). Lignin degradation and lignin peroxidase production in cultures of *Phanerochaete chrysosporium* immobilized on porous ceramic supports. *Enzyme and Microbial Technology*, *12*(12), 916–920. doi:10.1016/0141-0229(90)90109-4

Cornwell, R. L., Tinland-Butez, M.-F., Tardone, P. J., Cabasso, I., & Hammel, K. E. (1990). Lignin degradation and lignin peroxidase production in cultures of *Phanerochaete chrysosporium* immobilized on porous ceramic supports. *Enzyme and Microbial Technology*, *12*(12), 916–920. doi:10.1016/0141-0229(90)90109-4

Costa, S. A., Azevedo, H. S., & Reis, R. L. (2005). *Biodegradable Systems in Tissue Engineering and Regenerative Medicine*. Boca Raton, PA: C. R. C. Press.

da-Silva, M. R., de-Sa, L. R. V., Russo, C., Scio, E., & Ferreira-Leitao, V. S. (2010). The use of HRP in decolorization of reactive dyes and toxicological evaluation of their products. *Enzyme Research*, *2010*, 703824–703831. doi:10.4061/2010/703824 PMID:21318147

Dafale, N., Rao, N. N., Meshram, S. U., & Wate, S. R. (2008). Decolorization of Azo Dyes and Simulated Dye Bath Wastewater Using Acclimatized Microbial Consortium biostimulation and Halo Tolerance. *Bioresource Technology*, *99*(7), 2552–2558. doi:10.1016/j.biortech.2007.04.044 PMID:17566726

Dua, M., Singh, A., Sethunathan, N., & Johri, A. (2002). Biotechnology and bioremediation: Successes and limitations. *Applied Microbiology and Biotechnology*, *59*, 2–3, 143–152. PMID:12111139

Duran, N., Rosa, M. A., D'annibale, A., & Gianfreda, L. (2002). Applications of laccases and tyrosinases (phenol oxidases) immobilized on different supports: A review. *Enzyme and Microbial Technology*, *31*(7), 907–931. doi:10.1016/S0141-0229(02)00214-4

El Saliby, I. J., Shon, H., Kandasamy, J., & Vigneswaran, S. (2008). *Nanotechnology for wastewater treatment: in brief. Encyclopedia of Life Support Syst.* EOLSS.

Feng, C., Khulbe, K. C., Matsuura, T., Tabe, S., & Ismail, A. F. (2013). Preparation and characterization of electro-spun nanofiber membranes and their possible applications in water treatment. *Separation and Purification Technology*, *102*, 118–135. doi:10.1016/j.seppur.2012.09.037

Fetzner, S. (1998). Bacterial dehalogenation. *Applied Microbiology and Biotechnology*, *50*(6), 633–657. doi:10.1007002530051346 PMID:9891928

Fetzner, S. (2003). Oxygenases without requirement for cofactors or metal ions. *Applied Microbiology and Biotechnology*, *60*(3), 243–257. PMID:12436305

Fetzner, S., & Lingens, F. (1994). Bacterial dehalogenases: Biochemistry, genetics, and biotechnological applications. *Microbiological Reviews*, *58*(4), 641–685. PMID:7854251

Gopinathan, R., Kanhere, J., & Banerjee, J. (2015). Effect of Malachite Green Toxicity on Non Target Soil Organisms. *Chemosphere*, *120*, 637–644. doi:10.1016/j.chemosphere.2014.09.043 PMID:25462308

Gurujeyalakshmi, G., & Oriel, P. (1989). Isolation of phenol-degrading *Bacillus stearothermophilus* and partial characterization of the phenol hydroxylase. *Applied and Environmental Microbiology*, *55*(2), 500–502. PMID:2719481

Husain, Q. (2006). Potential applications of the oxidoreductive enzymes in the decolorization and detoxification of textile and other synthetic dyes from polluted water: A review. *Critical Reviews in Biotechnology*, *26*(4), 201–221. doi:10.1080/07388550600969936 PMID:17095432

Johnson, M. A., & Pokora, A. R. (1994). *Method for Deinking Printed Waste Using Soybean Peroxidase* (US Patent, 5370770). DOI: 10.1016/b978-1-4832-8381-4.50071-8

Jones, J. P., O'Hare, E. J., & Wong, L. L. (2001). Oxidation of polychlorinated benzenes by genetically engineered CYP101 (cytochrome P450cam). *European Journal of Biochemistry*, *268*(5), 1460–1467. doi:10.1046/j.1432-1327.2001.02018.x PMID:11231299

Kagalkar, A. N., Jagtap, U. B., Jadhav, J. P., Govindwar, S. P., & Bapat, S. A. (2010). Studies On Phytoremediation Potentiality Typhonium Flagelliforme for The Degradation of Brilliant Blue R. *Planta*, *232*(1), 271–285. doi:10.100700425-010-1157-2 PMID:20437182

Karam, J., & Nicell, J. A. (1997). Potential Applications of Enzymes in Waste Treatment. *Journal of Chemical Technology and Biotechnology (Oxford, Oxfordshire)*, *69*(2), 141–153. doi:

Karigar & Rao. (2011). Role of Microbial Enzymes in the Bioremediation of Pollutants: A Review. *Enzyme Research*. doi:10.4061/2011/805187

Kim, J., & Grate, J. (2003). Single Enzyme Nanoparticles Armored by a Nanometer-Scale Organic/Inorganic Network. *Nano Letters*, *3*(9), 1219–1222. doi:10.1021/nl034404b

Klibanov, A. M. (1982). Enzymatic removal of hazardous pollutants from industrial aqueous effluents. *Enzyme Engineering*, *6*, 319-23.

Klibanov, A. M. (1983). Peroxidase-catalyzed removal of phenols from coal-conversion waste waters. *Science*, *221*, 259-60.

Kragl, U. (1996). *Industrial Enzymology*. London: Macmillan Press.

Krug, M., Ziegler, H., & Straube, G. (1985). Degradation of phenolic compounds by the yeast *Candida tropicalis* HP-15. 1. mendocina NSYSU chloride release. *Journal of Hazardous Materials*, *124*, 68–73.

Kyzas, G. Z., & Matis, K. A. (2015). Nanoadsorbents for pollutants removal: A review. *Journal of Molecular Liquids*, *203*, 159–168. doi:10.1016/j.molliq.2015.01.004

Leung. (2004). Bioremediation: Techniques for cleaning up a mess. *Journal of Biotechnology*, *2*, 18–22.

Lloyd, Jr., Yong, & Macaskie. (1998). Enzymatic Recovery of Elemental Palladium by Using Sulphate Reducing Bacteria. *Applied and Environmental Microbiology*, *64*, 4607–4609. PMID:9797331

Lloyd, Jr., Mabbet, & Macaskie. (2001). Metal Reduction by Srb: Physiological Diversity and Metal Specificity. *Hydrometallurgy*, *59*, 327–337. doi:10.1016/S0304-386X(00)00175-4

Maczulak & Elizabeth. (2010). *Pollution: Treating Environmental Toxins*. New York: Infobase Publishing.

Maddhinni, V. L., Vurimindi, H. B., & Yerramilli, A. (2006). Degradation of Azo Dye with Horseradish Peroxidase (HRP). *J. Indian Inst. Sci.*, *86*, 507-514.

Matsuura, T., Feng, C., Khulbe, K. C., Rana, D., Singh, G., Gopal, R., ... Tabe, S. (2010). Development of novel membranes based on electrospun nanofibers and their application in liquid filtration, membrane distillation, and membrane adsorption. *Maku Japanese J. Membr.*, *35*, 119–127.

Misson, Zhang, & Jin. (2015). Nanobiocatalyst advancements and bioprocessing applications. *J R Soc Interface*, *12*(102). doi: 10.1098/rsif.2014.0891

Misson, M., Zhang, H., & Jin, B. (2015). Nanobiocatalyst advancements and bioprocessing applications. *Journal of the Royal Society, Interface*, *12*(102), 20140891. doi:10.1098/rsif.2014.0891 PMID:25392397

Mörtberg, M., & Neujahr, H. Y. (1987). In situ and in vitro kinetics of phenol hydroxylase. *Biochemical and Biophysical Research Communications*, *146*(1), 41–46. doi:10.1016/0006-291X(87)90687-5 PMID:3606624

Naghibi, F., Pourmorad, F., Honary, S., & Shamsi, M. (2003). Decontamination of Water Polluted with Phenol Using Raphanus Sativus Root. *Iranian Journal of Pharmaceutical Research*, *2*(1), 29–32.

Ayanda & Petrik. (2014). Nanotechnology: The Breakthrough in Water and Wastewater Treatment. *International Journal of Chemical Marine Environmental Research*, *1*(1), 1–2.

Nath, P. P., Sarkar, K., Mondal, M., & Paul, G. (2015). Metanil Yellow Impairs the Estrous Cycle Physiology and Ovarian Folliculogenesis in Female Rats. *Environmental Toxicology*. doi:10.1002/Tox.22205 PMID:26494366

Ngai, K. L., Neidle, E. L., & Ornston, L. N. (1990). Catechol and chlorocatechol 1,2-dioxygenase, Hydrocarbon and methylotrophy, Meth. Enzymol., 188, 122-126.

Nicell, J. A. (1994). Kinetics of horseradish peroxidase-catalysed polymerization and precipitation of aqueous 4- chlorophenol. *Journal of Chemical Technology and Biotechnology*, *60*, 203-15.

Nicell, J. A., Bewtra, J. K., Biswas, N., St. Pierre, C. & Taylor, K. E. (1993).Enzyme catalyzed polymerization and precipitation of aromatic compounds from aqueous solution. *Canadian Journal of Civil Engineering*, *20*, 725-35.

Nigel, C. (2004). Veitch Horseradish peroxidase: A modern view of a classic enzyme. *Phytochemistry*, *65*(3), 249–259. doi:10.1016/j.phytochem.2003.10.022 PMID:14751298

Nigel, C. (2004). Veitch Horseradish peroxidase: A modern view of a classic enzyme. *Phytochemistry*, *65*(3), 249–259. doi:10.1016/j.phytochem.2003.10.022 PMID:14751298

Nilotpala, P., & Ingle, A. O. (2007). Mineralization of phenol by a *Serratia plymuthica* strain GC isolated from sludge sample. *International Biodeterioration & Biodegradation*, *60*(2), 103–108. doi:10.1016/j.ibiod.2007.01.001

Omar, H. H. (2008). Algal Decolorization and Degradation of Monoazo and Diazo Dyes. *Pakistan Journal of Biological Sciences*, *11*(10), 1310–1316. doi:10.3923/pjbs.2008.1310.1316 PMID:18817261

Oyekola, O., Ngesi, N., & Whiteley, C. (2007). Isolation, Purification and Characterization of An Endoglucanase And B-Glucosidase From An Anaerobic Sulphidogenic Bioreactor. *Enz Microbiol Tech, 40*(8), 637–644. doi:10.1016/j.enzmictec.2006.05.020

Pandey, A., Singh, P., & Iyengar, L. (2007). Bacterial decolorization and degradation of azo dyes. *International Biodeterioration & Biodegradation, 59*(2), 73–84. doi:10.1016/j.ibiod.2006.08.006

Park, J. W., Park, B. K., & Kim, J. E. (2006). Remediation of soil contaminated with 2,4-dichlorophenol by treatment of minced shepherd's purse roots. *Archives of Environmental Contamination and Toxicology, 50*(2), 191–195. doi:10.100700244-004-0119-8 PMID:16392021

Pelach, M., Pastor, F., Puig, J., Vilaseca, F., & Mutje, P. (2003). Enzymatic Deinking of Old Newspapers With Cellulose. *Proc Biochem, 38*(7), 1063–1067. doi:10.1016/S0032-9592(02)00237-6

Pescod, M. B. (1992). *FAO Irrigation and Drainage, Food and Agriculture*. Rome, Italy: Organisation of United Nations.

Platzek, T., Lang, C., Grohmann, G., Gi, U. S., & Baltes, W. (1999). Formation of a carcinogenic aromatic amine from an azo dye by human skin bacteria in vitro. *Human and Experimental Toxicology, 18*(9), 552–559. doi:10.1191/096032799678845061 PMID:10523869

Qu, X., Alvarez, P. J., & Li, Q. (2013). Applications of nanotechnology in water and wastewater treatment. *Water Research, 47*(12), 3931–3946. doi:10.1016/j.watres.2012.09.058 PMID:23571110

Robinson, T., Mcmullan, G., Marchant, R., & Nigam, P. (2001). Remediation of dyes in textile effluent: A critical review on current treatment technologies with a proposed alternative. *Bioresource Technology, 77*(3), 247–255. doi:10.1016/S0960-8524(00)00080-8 PMID:11272011

Rubilar, O., Diez, M. C., & Gianfreda, L. (2008). Transformation of chlorinated phenolic compounds by white rot fungi. *Critical Reviews in Environmental Science and Technology, 38*(4), 227–268. doi:10.1080/10643380701413351

San, N., Celebioglu, O., Tumtas, A., Uyar, Y., & Tekinay, T. (2014). Reusable bacteria immobilized electrospun nano-fibrous webs for decolorization of methylene blue dye in wastewater treatment. *RSC Advances, 4*(61), 32249–32255. doi:10.1039/C4RA04250F

Sarioglu, O. F., Yasa, O., Celebioglu, A., Uyar, T., & Tekinay, T. (2013). Efficient ammonium removal from aquatic environments by Acinetobacter calcoaceticus STB1 immobilized on an electrospun cellulose acetate nanofibrous web. *Green Chemistry, 15*(9), 2566–2572. doi:10.1039/c3gc40885j

Savage, N., & Diallo, M. S. (2005). Nanomaterials and water purification: Opportunities and challenges. *Journal of Nanoparticle Research, 7*(4-5), 331–342. doi:10.100711051-005-7523-5

Shamsizadeh, A. A., Ghaedi, M., Ansari, A., Azizian, S., & Purkait, M. K. (2014). Tin oxide nanoparticle loaded on activated carbon as new adsorbent for efficient removal of malachite green-oxalate: Nonlinear kinetics and isotherm study. *Journal of Molecular Liquids, 195*, 212–218. doi:10.1016/j.molliq.2014.02.035

Souhaimi, M. K., & Matsuura, T. (2011). *Membrane distillation: principles and applications*. Elsevier.

Stanisavljević, M., & Nedić, L. (2004). Removal of Phenol from Industrial Wastewaters by Horseradish (*Cochlearia Armoracia* L.). *Working and Living Environmental Protection*, *2*(4), 345–349.

Stork, G., Pereira, H., Wood, Tm., Dusterhoft, Em., Toft, A., & Puls, J. (1995). Upgrading Recycled Pulps Using Enzymatic Treatment. *J Tech Assoc Pulp Pap Ind*, *78*, 79–188.

Sun, W. Q., Payne, G. F., Moas, M., Chu, J. H., & Wallace, K. K. (1992). Tyrosinase reaction/chitosan adsorption for removing phenols from wastewater. *Biotechnology Progress*, *8*(3), 179–186. doi:10.1021/bp00015a002

Suzuki, T., Timofei, S., Kurunczi, L., Dietze, U., & Schűűrmann, G. (2001). Correlation of aerobic biodegradability of sulfonated azo dyes with the chemical structure. *Chemosphere*, *45*(1), 1–19. doi:10.1016/S0045-6535(01)00074-1 PMID:11572582

Tabe, S. (2014). *Electrospun nanofiber membranes and their applications in water and wastewater treatment Nanotechnology for Water Treatment and Purification*. Springer International Publishing.

Tang, X., Zhang, Q., Liu, Z., Pan, K., Dong, Y., & Li, Y. (2014). Removal of Cu (II) by loofah fibers as a natural and low-cost adsorbent from aqueous solutions. *Journal of Molecular Liquids*, *199*, 401–407. doi:10.1016/j.molliq.2014.09.033

Telke, A. A., Kagalkar, A. N., Jagtap, U. B., Desai, N. S., Bapat, V. A., & Govidwar, S. P. (2011). Biochemical chracterization of laccase from hairy root culture of *Brassica juncea* L. and role of redox mediators to enhance its potential for the decolorization of textile dyes. *Planta*, *234*(6), 1137–1149. doi:10.100700425-011-1469-x PMID:21735196

Theron, J., Walker, J. A., & Cloete, E. (2008). Nanotechnology and Water Treatment, applications and Emerging Opportunities. *Critical Reviews in Microbiology*, *34*(1), 43–69. doi:10.1080/10408410701710442 PMID:18259980

Tolan, J. S., & Foody, B. (1999). Cellulase from Submerged Fermentation. *Advances in Biochemical Engineering/Biotechnology*, *65*, 41–67. doi:10.1007/3-540-49194-5_3

Van Wyk, J. (2003). Biodegradation of Wastepaper By Cellulose From *Trichoderma Viride*. *Bioresource Technology*, *86*(1), 21–23. doi:10.1016/S0960-8524(02)00130-X PMID:12421003

Venkatadri, R., & Irvine, R.L. (1993). Cultivation of *Phanerochaete chrysosporium* and production of lignin peroxidase in novel biofilm reactor systems: hollow fiber reactor and silicone membrane reactor. *Water Research*, *27*, 591-6. DOI: 10.1016/0043-1354(93)90168-H

Vijaykumar, M. H., Veeranagouda, Y., Neelkanteshwar, K., & Karegoudar, T. B. (2006). Degradation of 1:2 Metal Complex Dysse Acid Blue 193 by a Newly Isolated Fungus Cladosporium Cladosporioide. *World Journal of Microbiology & Biotechnology*, *22*(2), 157–162. doi:10.100711274-005-9013-4

Wada, S., Ichikawa, H., & Tatsumi, K. (1993). Removal of phenols from wastewater by soluble and immobilized tyrosinase. *Biotechnology and Bioengineering*, *42*, 854-8. DOI: 10.1002/bit.260420710

Wan, H., Weng, S., Liang, S., Lu, Q., & He, J. (2011). Evaluation of The Developmental Toxicity of Leucomalachite Green Administered Orally to Rats. *Food and Chemical Toxicology*, *49*(12), 3031–3037. doi:10.1016/j.fct.2011.10.003 PMID:22005260

Watlington, K. (2005). *Emerging Nanotechnologies for Site Remediation and Wastewater Treatment*. Washington, DC: U.S. Environmental Protection Agency.

Whiteley, C., & Lee, D. J. (2006). Enzyme Technology and Biological Remediation: A Review. *Enz. Microbial. Tech*, *38*(3-4), 291–316. doi:10.1016/j.enzmictec.2005.10.010

Williams, P. P. (1977). Metabolism of synthetic organic pesticides by anaerobic microorganisms. *Residue Reviews*, *66*, 63–135. doi:10.1007/978-1-4612-6352-4-3 PMID:335464

Wu, X., Chou, N., Lupher, D., & Davis, L. C. (1998). Benzotriazoles: Toxicity and Degradation. *Proceedings of the 1998 Conference on Hazardous Waste Research*, 374-382.

Zaldivar, J., Nielsen, J., & Olsson, L. (2001). Fuel Ethanol Production from Lignocellulose: A Challenge for Metabolic Engineering and Process Integration. *Applied Microbiology and Biotechnology*, *56*(1-2), 17–34. doi:10.1007002530100624 PMID:11499926

Zhang, Q., Xu, R., Xu, P., Chen, R., He, Q., Zhong, J., & Gu, X. (2014a). Performance study of ZrO2 ceramic micro-filtration membranes used in pretreatment of DMF wastewater. *Desalination*, *346*, 1–8. doi:10.1016/j.desal.2014.05.006

Zhang, Y., Yan, L., Xu, W., Guo, X., Cui, L., Gao, L., ... Du, B. (2014b). Adsorption of Pb (II) and Hg (II) from aqueous solution using magnetic CoFe2O4 reduced graphene oxide. *Journal of Molecular Liquids*, *191*, 177–182. doi:10.1016/j.molliq.2013.12.015

APPENDIX

Table 1. Industrial applications of microbial enzymes Chandrakant (2011)

S.No.	Enzyme	Substrate	Reaction	Applications
1	Oxidoreductase			
1.1	Oxygenase			
1.1.1	Monooxygenase	Alkane, steroids, fatty acid, and aromatic compounds	Incorporation of oxygen atom to substrate and utilize substrate as reducing agent. Desulfurization, dehalogenation, denitrification, ammonification, and hydroxylation of substrate	Protein engineering, bioremediation, synthetic chemistry, and so forth.
1.1.2	Dioxygenase	Aromatic compounds	Introduction of two oxygen atom to the substrate results in intradiol cleaving and extradiol cleaving with the formation of aliphatic product	Synthetic chemistry, pharmaceutical industry, bioremediation, and so forth.
1.2	Laccase	Ortho and paradiphenols, aminophenols, polyphenols, polyamines, lignins, and aryldiamines	Oxidation, decarboxylation and demethylation of substrate.	Food industry, paper and pulp industry, textile industry, nanotechnology, synthetic chemistry, bioremediation, cosmetics, and so forth.
1.3	Peroxidase			
1.3.1	Lignin peroxidase	Halogenated phenolic compounds, polycyclic aromatic compounds and other aromatic compounds	Oxidation of substrate in the presence of cosubstrate H_2O_2 and mediator like veratryl alcohol.	Food industry, paper and pulp industry, textile industry, pharmaceutical industry, bioremediation, and so forth.
1.3.2	Manganese peroxidase	Lignin and other phenolic compounds	In the presence of Mn^{2+} and H_2O_2 the co-substrate catalyses oxidation of Mn^{2+} to Mn^{3+} which results in an Mn^{3+} chelateoxalate, which in turn oxidizes the phenolic substrates.	Food industry, Paper and pulp industry, textile industry, pharmaceutical industry, bioremediation, and so forth.
1.3.3	Versatile peroxidase	Methoxybenzenes and phenolic aromatic	The enzyme catalyzes the electron transfer from an oxidizable substrate, with the formation and reduction of compound I and compound II intermediates.	Industrial biocatalyst, bioremediation, and so forth.
2	Hydrolase			
2.1	Lipase	Organic pollutants such as oil spill	The hydrolysis of triacylglycerols to glycerols and free-fatty acids	Control of oil spills, detergent production, baking industry, paper and pulp industry, personal care products, and so forth.
2.2	Cellulase	Cellulosic substance	Hydrolyses the substrate to simple carbohydrates.	Textile manufacturing. detergent production, paper and pulp industry, bioremediation, and so forth.
2.3	Protease	Proteins	Enzymes that hydrolyze peptide bonds in aqueous environment.	Leather, laundry, biocatalyst, bioremediation, and so forth

Table 2. Role of Different enzymes in Wastewater

ENZYMES	APPLICATIONS	REFERENCES
Horse Radish Peroxidase	Waste water treatment,Treatment of anilines, hydroxyquinoline and arylamine	Nicell et.al(1993),Kilbanoy et.al (1982,1983)
Lignin Peroxidase	Lignin depolymerisation,Catalyse aromatic compounds and polycyclic aromatic compounds	Atiken et.al(1989), Venkatadri et.al(1993), Conwell et al(1990)
Chloroperoxidase	Oxidizes several phenolic compounds	Atiken et al(1994)
Manganese Peroxidase	Catalyse the oxidation of mono aromatic phenols and aromatic dyes	Atiken et al (1994,1989), Chapsal et al (1986)
Tyrosinase	Removal of phenols	Atlow et al (1984), Wada et al (1993), Sun et al (1992)
Laccase	Decreasing the toxicity of phenolic compounds through polymerization, Bleaching plant effluent treatment,Degradation of dyes	Bollag et al (1992,1998), Pellimen et al (1988), Royer et al (1991), Lenkinen et al (1991)
Cellulolytic enzymes	Removal of sludges from pulp and drinking industries	Dut et al (1994,1995)
Phosphotriesterase	Pesticide detoxification	Munneck et al (1977), Caldwell et al (1991)
Cyanidase	Degrades cyanide to ammonia and formate	Basheer et al (1992,1993)
Cyanide hydratase	Hydrolyses cyanide to formamide	Basheer et al (1993), Nazly et al (1983)
Protease	Degrading insoluble proteins	Lagerkvist et al (1993)
Amylase	Treatment of starch containing food waste waters, Alcohol production from rice processing waste water, Enhances activated sludge waste water treatment	Blasheck et al (1992), Shoemaker et al (1986)
Azo reductase	Reduces azo dyes to aromatic amines	IM Banat et al (1996)
Dioxigenase	Oxidizes aromatic amines	Lawrence P.Wackett (2002)
Pectinestarase	Degrade pectin	Blasheck et al (1992)
Chitonase	Induce chitin degradation	Casio et al (1982)
Lipase	Reduction of total hydrocarbons in the soil	Thomas et al (1993)
Cellulase, bacterial lysozyme (muramidase)	Sludge dewatering	Utakulinen(1988)

Chapter 19
Moving into Nanotechnology Roles to Mimic and Boost Enzyme Activity

Maria Laura Soriano
University of Cordoba, Spain

ABSTRACT

A new tendency toward the design of artificial enzymes based on nanostructures (nanodots, nanofibers, mesoporous materials) has emerged. On one hand, nanotechnology bestows self-catalytic nanoparticles with a specific activity to achieve efficient reactions with low number of by-products. On other hand, the nanoparticles may behave as nanometric scaffolds for hosting enzymes, promoting their catalytic activity and stability. In this case, enzyme immobilization requires the preservation of the catalytic activity by preventing enzyme unfolding and avoiding its aggregation. These approaches render many other advantages like hosting/storing enzymes in nanotechnological solid, liquid, and gel-like media. This chapter focuses on the most up-to-date approaches to manipulate or mimic enzyme activity based on nanotechnology, and offers examples of their applications in the most promising fields. It also gives new insight into the creation of reusable nanotechnological tools for enzyme storage.

INTRODUCTION

Enzymes are biocatalysts of specific substrates with a strong potential in multiple reactions. They are known to have capability of modifying their functions upon environmental changes, which is the key of the adaptation of organisms, like bacteria resistance to pesticides or even to drugs. Although the concept of enzymes is known from the 19th century (Payen & Persoz, 1833), it was not until the next century when researchers perceived their great potential. There is a diversity in the type and functions of enzymes which also depend on the type organism hosting. Additionally, enzymes may have different domain composition, being of single domain or multidomain which evolve new functionalities. The discipline of enzymology is very wide but entails notably benefits not only for other fields of application, such as in technology and medicine, but also for the industrial sector. One of its main interests is to explore the structural changes affecting enzyme functions and to rationally control new functions by re-designing existing enzyme frameworks. Many are the advances in creating artificial active enzymes to mimic Nature;

DOI: 10.4018/978-1-5225-5237-6.ch019

thus, essential objectives are to stabilize enzymes and to control their catalytic performance in multistep processes of such complexity as those occurring in biological systems (e.g. metabolism, biosynthesis, signaling, channeling). However, the interest of the biotechnology is also to achieve a more efficient process control in large scale at low costs, very demanding in the industrial sector.

Some progress in the field has been made by supporting enzymes in different solid structures in order to maximize their catalytic activity, selectivity and recovery/reusability by incrementing their environmental tolerance. From the last two decades to now, the emerging advanced technology supported in nanoscale materials has attracted the attention of enzymologists. Nanotechnology offers a new perspective and opportunity not only to mimic but also to boost the enzymatic activity. Nanotechnology is considered as the manipulation of the nanomatter which brings unique physicochemical properties not observed for other materials, i.e., nanoparticles do not follow the fundamental rules known until now for micro and macroparticles; they can be explained thanks to the Nanoscience. Two main contributions of nanotechnology are highlighted. The first contribution is the use of those nanoparticles exhibiting catalytic activity by themselves; they are known as nanozymes. In the second approach, the nanostructures bring the possibility of hosting enzymes to improve both enzyme long-term stability against pH, heat and other external variables and reusability; it is possible thanks to their large surface area and unique physicochemical properties, which allow easy enzyme immobilization, in some cases with an improvement in catalytic activity, and in others, with high enzyme loading with minimum diffusion limitations, leading to great efficiencies in biocatalysts (Pundir, 2015).

It is important to note that enzyme immobilization requirements for achieving the best catalytic conditions are the preservation of the active conformational form of the enzyme and the prevention of any aggregation or unfolding. Thus, the main variables to consider are the type of enzyme, the nanoparticle surface, the mode of attachment and the matrix. With this aim in mind, many researchers have tried to understand the structural and conformational changes of enzymes in adequate micro- or nano- environments to create organized devices (Ariga, 2013). Thus, the convergence of nanotechnology and biotechnology lead to the nanobiotechnology, that has emerged for finding the foundations between the relationship of the enzyme structure and function in biological environments and evolution to a global industrial sector (Maine, 2014).

There exist a wide variety of analytical and biotechnological applications involving artificial enzymes coming from nanotechnology: nanozymes, enzyme-immobilized nanostructures and gels containing such enzyme-nanoparticles designated to the conversion of substrates to products, degradation of toxic substances, for sensing, delivering and/or imaging in biological systems, as well as creating more sophisticated self-assemblies.

This chapter focuses on the state-of-the-art of nanoenzymology by describing the recent scenarios of nanozyme and enzyme immobilization in nanoparticles and different gel matrices employed as well as the advantages presented for a variety of applications.

NANOPARTICLES WITH CATALYTIC ACTIVITY: NANOZYMES

Enzymes are of great interest in multitude of reactions; however, they have some drawbacks that limit their applications. These limitations are related to the difficulty on their isolation, high cost production, reusability and denaturation under certain conditions. Thus, researches have put much effort in the fabrication of highly stable artificial enzymes. With the discovery of nanotechnology, it was found that both nanoparticle and enzyme share several similar attributes as function of their surface charges and size dimensions. Nanozyme is the term coined by Manea et al. (2004) to refer to nanosize nanoparticles as powerful catalysts with enzymatic activity. Thus, nanoparticle attributes depend on their nature and structural characteristics of both inner and external cores, as occurered in enzymes. Thus, a wide variety of bare and coated inorganic and organic nanoparticles have been considered as artificial enzymes and its manipulation has enabled to mimic certain enzyme processes, as depicted in Figure 1. Recently, from 2007 in which intrinsic peroxidase-like activity was reported for ferromagnetic nanoparticles many are the authors who summarized the wide range of applications for these artificial enzymes, ranging from targeting and imaging to biosensing (Wang, 2016a-b; Golchin, 2017).

A variety of nanoparticles display catalase and peroxidase activities. These catalytic properties have been exploited in nanomedicine related to the diminution of free radicals, drug delivery and as biosensor. On the one hand, it is well known the catalytic activity of gold nanoparticles (AuNPs) that simulate peroxidase and glucose oxidase features. Cerium dioxide (CeO_2) nanoparticles also displayed catalase and the superoxide dismutase enzyme-like activities. On the other hand, platinum nanoparticles have catalysed reactions that occurred with superoxide dismutase. Besides, peroxidase activity of nanoenzyme derived from platinum are reported to be suitable as sensor for the detection of biothiols (Wang, 2016) as illustrated in Figure 2A.

Figure 1. Scheme of the nanozyme features, and the applications addressed for a variety of nanozyme with different catalytic activity as function of their core nature

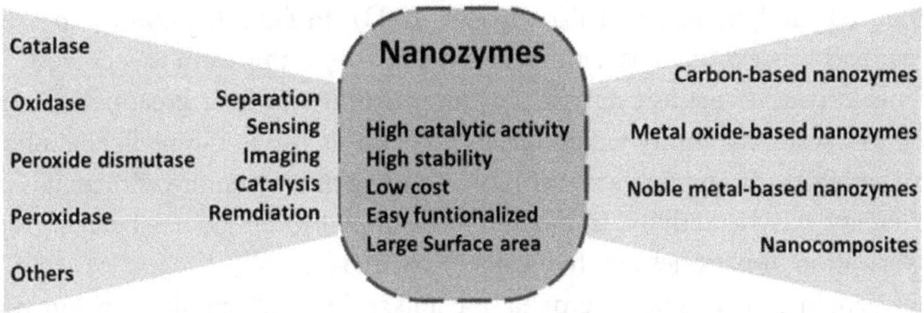

Figure 2. A) Schematic representation of a nanoparticle mimicking peroxidase activity for detecting target compounds. Used with permission from (Wang, X., Hua, Y., & Wei, H. (2016). Inorg. Chem. Front., 3, 41-60.) Copyright (2015) The Royal Society of Chemistry. B) Magnetic nanozyme coated with imprinting polymeric shell as recognition element for selectively sensing a specific compound, tetramethylbenzidine (TMB). Adapted from (Zhang, Z., Zhang, X., Liu, B., & Liu, J. (2017). J. Am. Chem. Soc. 139 (15), 5412-5419.). Copyright (2017) American Chemical Society.

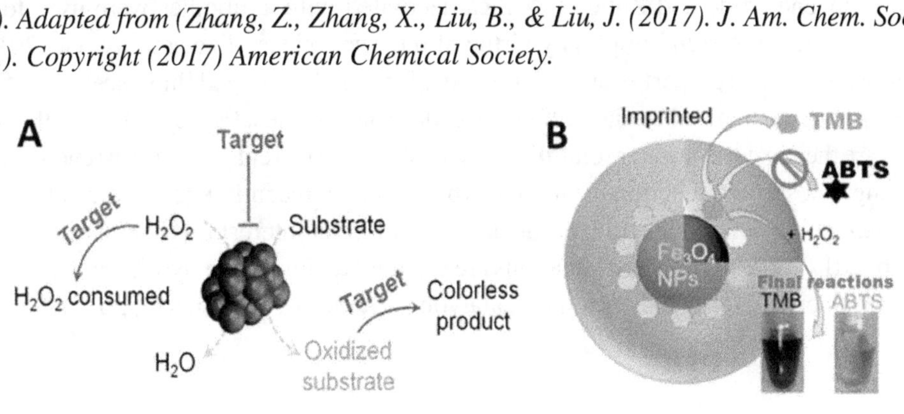

Additionally, magnetic nanoparticles (Huang, Hsiao, Chen, Chien, Yao, Chen, Ko, Hsu, Tai, Cheng, Wang, Yang, & Chen, 2009) like magnetite (Fe_3O_4) exhibited oxidase mimetic properties. Thus, this type of nanozyme is also of interest in other areas of applications related to environmental remediation and water treatments (Wu, 2014; Janoš, 2015).

Interestingly, recently Liu et al. (2017) demonstrated a hundred-fold enzymatic specificity of the peroxidase activity of nanozyme composed of Fe_3O_4 nanoparticles modified with molecularly imprinted polymers, as depicted in Figure 2B.

It was found that a number of molecules and ions may interfere in the catalytic activity of inorganic nanozymes by shifting the metal oxidation state. Thus, these molecules may act as activators or inhibitors when promoting or hindering their activity, respectively. This is the case of the catalase mimetic-activity of CeO_2 nnanoparticles which is altered by the presence of phosphate ions hindering the redox transformation of Ce(III)/Ce(IV) (Singh, 2016). The low peroxidase activity of AuNPs has been improved greatly by the formation of an alloy with mercury (Han, 2017). Doping of ferrite nanoparticles with other metal ions also affect the peroxidase activity, serving as sensing platform towards hydrogen peroxide (H_2O_2) and glucose (Shi, 2011). When adding a chitosan shell to those nanoparticles, the resulting nanozyme also oxidize luminol serving as chemiluminescent sensor (Fan, 2012). The peroxidase activity of AuNPs is also promoted by the presence of certain metal ions like mercury.

Carbon nanoparticles also display interesting enzyme mimetic activities, which depends in the carbon hybridisation and their morphology. In this family, fullerenes, carbon nanotubes and graphene derivatives are generally known by their catalytic activities. The first carbon nanozyme studied was fullerene for their light-induced catalytic behavior (Tokuyama, 1993). In fact, this hollow sphere resembling a nanoscopic soccer ball was explored as nanozyme even before of this term was coined.

Thus, fullerene derivatives act as catalytic scavengers, although their great potential as antioxidant is limited by heir low water-solubility which makes difficult its use in biomedical applications mainly. Apart from those variables, there exists a possibility of doping their structure with metals or heteroatoms to achieve a different catalytic activity. In fact, nitrogen doped fullerene has been considered of great interest in the fabrication of batteries and fuel cells (Noh, 2017).

Graphene has demonstrated to be a peroxidase catalyst, in which products resulted to display absorbance or emission bands. This makes useful its use as sensing platforms mainly for analytical and biomedical applications by virtue of their low toxicity. Graphene sensing property strongly depends on the groups located at the nanosheets edges and on doping. When decorating these nanosheets with AuNPs, the new hybrid nanostructure demonstrated the highest peroxidase catalytic effect if compared to both individual nanostructures by virtue of a synergistic effects. Immunosensors based on nanozyme composed of graphene oxide and $ZnFe_2O_4$@silica decorated with antibodies were used for sensing a specific cancer antigen in serum samples by virtue of their peroxidase-like activity (Ge, 2014).

With the amazing optical properties and low toxicity of the newly emerged fluorescent nanodots, carbon dots are starting to be recognized as alternative peroxidase nanozymes. Particularly valuable advantages of carbon dots are their excellent water solubility, low-cost effective preparation from renewable resources and easy doping and surface functionalization. Carbon dots are recently known to catalyse three types of substrates generating colored reacting products, being of great interest for their use as sensing probes (Garg & Bisth, 2017). The current investigations are directed to developed hybrid nanozymes of carbon dots with other nanoparticles to enhance their catalytic activities (Yousefinejad, 2017).

NANOPARTICLE AS THE IMMOBILIZATION MATRIX

From long time ago, it there is very challenging the fact of fabricating active artificial enzymes with highly catalytic activity in many processes. For that, nanotechnology emerged as a new type of nanoscale particles with fascinating optical, electrochemical and in some cases catalytic features capable of acting as enzymes.

From then, Nanotechnology has been used for enzyme immobilization which can be irreversible and reversible (Cipolatti, 2014). This involved the use of a variety of interactions ranging from reversible physical adsorption, affinity or ionic interactions and from irreversible covalent linkages, like amide, carbamate and ether bonds. There are several methods for enzyme immobilization (Johnson, 2011), and so many investigations demonstrated how size (Sanfins, 2014) and type of NP was affecting the catalytic activity of such enzymes in such conjugated systems (Ahmad & Sardar, 2015). However, only those reversible procedures lead to their recovery and reusability which is of great importance for a wide variety of applications. Ansari and Husain (2012) described their applicability in different fields ranging from biomedical, environmental analytics to synthetic purposes.

For the past two decades, many reports had reported the advantages of enzymes entrapped into nanomaterials. Regarding to nanomaterials with narrow size distribution, excellent biocompatibility and enzymatic activity was found for silica nanoparticles covalently modified with two enzymes, glutamate dehydrogenase and lactate dehydrogenase, and a staining reagent, as shown in Figure 3. This novel platform described by Qhobosheane et al. (2001) shows a great potential as novel biosensors and biomarkers.

After those studies, many reports demonstrated an increase of the catalytic activity of different enzymes associated to a wide variety of scaffolds of nano-size dimensions, such a variety of other metal oxide nanomaterials to noble metal nanoparticles, polymeric and carbon-based nanoparticles, as reported in many revision papers (Johnson, 2011; Cipolatti, 2014; Ahmad & Sardar, 2015). Noble metal nanoparticles resulted in good nanoplatforms for immobilize enzymes, as recently reported by Hondred et al. (2017).

Figure 3. Functionalization protocol of labelled-silica nanoparticles containing ruthenium bipyridyl complexes with enzymes via covalently linkage. Adapted from (Qhobosheane, M., Santra, S., Zhang, P., & Tan, W. (2001). Analyst,126,1274–1278.). Copyright (2001) Royal Society of Chemistry.

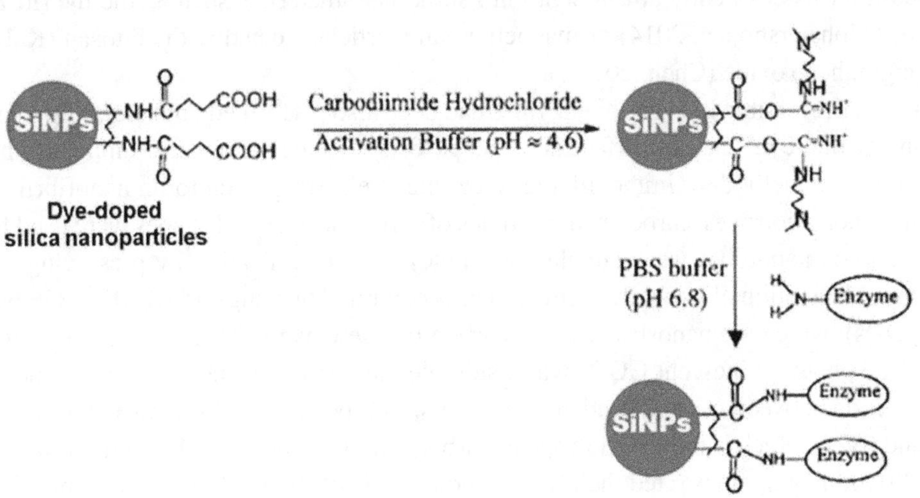

Phosphodiesterase trimer was immobilized in AuNPs of different nanosizes for the rapid detection of organophosphates via covalent attachment. Gómez-Anquela et al. (2015) reported electrodes based on AuNPs which were coordinated to the metal of the alcohol dehydrogenase; this electrode displayed a great potential for ethanol sensing with good performance and stable responses for a period of a month. Oxide nanoparticles were one of the most extensively used in enzyme nanoimmobilization (Ganesana, 2011; Ahmad, 2013). In fact, one of the most suitable matrix for enzyme immobilization is spherically silica-based nanomaterials; the large surface area and surface groups promotes high immobilization efficiencies. Exceptionally, silica nanotubes were also used to chemically bind enzymes and demonstrated they role as biocatalysts for the oxidation of glucose and also as analytical nanotool for the bioseparation (Mitchell, 2002).

Magnetic nanoparticles are of great interest for its superparamagnetism; thus, Jeong (2006) and Netto (2013) described different methodologies for coating those nanomagnets covered, starting with simple functionalization steps (covalent coupling reactions) and following with other more complex (involving polymeric or metallic coatings mainly with hydrophilic character, e.g. silica, chitosan or gold shells). Enzymes have an external hydrophilic character while its inner cavity used to be hydrophobic. Thus, it was found to be crucial that the selected nanoplatforms displayed hydrophilic shells to avoid enzyme unfolding of enzymes owing to their external hydrophilic character with inner hydrophobic cavities; so, a carefully selection of the nanoparticle surface is a prerequisite for enzyme immobilization without losing of enzymatic activity and selectivity. An analytical approach described by Kouassi group (2005) focused on cholesterol oxidase covalently bound to nanosized magnetite via carbodiimide activation for the analysis of the total cholesterol in serum samples.

Nickel oxide (Ni/NiO) nanoparticles were also explored to immobilize enzymes. Ganesana (2011) studied the acetylcholinesterase, and fabricated an electrochemical screen-printed biosensor suitable for detecting the organ phosphor pesticide paraoxon.

Those examples demonstrated the suitability of such enzymatic nanoplatform as biosensors, as a result of high sensitivity and fast responses with a low cost and enzyme losses.

Besides polymeric coatings, hybrid nanoparticles derived from metal oxide were also investigated. For instance, regarding to bioremediation purposes, laccase plays an important role in industrial applications for catalyzing the degradation of hazardous substances and dyes, so thus, its potential uses were demonstrated after its covalently attachment onto some nanomaterials such as the hybrid zinc ferrite nanoparticles (Mohajershojaei, 2014) or magnetite nanoparticles coated with chitosan (Kalkan, 2012) or even with graphene oxide (Chen, 2017).

Furthermore, irreversible linkages of enzymes had been also performed onto electrospun nanofibers, taking advantage of the increase of surface area and porosity of such nanometric supports. Thus, Park et al. (2013) reported a method for immobilizing lysozyme in electrospun chitosan nanofibrils.

Amongst carbon allotropes, carbon nanoparticles of different sizes and shapes were used for enzyme attachment. Carbon nanotubes were excellent examples of immobilization by preserving the enzyme conformation and retention of its catalytic functions, as compiled by Feng and Ji (2011). Graphene quantum dots (GQDs), which are nanosheets of sp^2 carbon of a few nanometers were used to immobilized peroxidase. The photoluminescent GQDs was a suitable platform to anchor horseradish peroxidase by amidation reaction between carboxyl and amine groups, as reported by Muthurasu and Ganesh (2014). Nanodiamond (ND), which is the sp^3 allotrope of carbon, has also been used for immobilized trypsin by Wei et al. (2010); they demonstrated the efficient proteolysis of myoglobin in just 5 min. In this case, the enzyme was covalently linked to the carboxyl groups of ND by amidation reaction.

Other elegant approach of enzyme immobilization onto nanoobjects is that reported by Ginet et al. (2011), which focuses on the enzymes anchored onto the bacterial magnetosome membranes, as illustrated in Figure 4. Those nano-sized magnetite particles are biomineralized by certain prokaryotes. The produced lipid-coated nanomagnet displayed excellent phosphohydrolase activity, and demonstrated a high stability over several cycles (Figure 4D), being very suitable for its applicability in the biodegradation of organophosphate pesticides.

Interestingly, it is known that enzymes are molecular machines that catalyse reactions that enable movement to accomplish tasks; thus, mimicking biological machinery by controlling micro and nano-engines is one of the hot topics nowadays (Ma, 2016). Sánchez group (Sánchez, 2010; Ma, 2015; 2017) developed different strategies to achieve the self-propulsion of nanoobjects. Because of the catalytic activity of platinum to trigger decomposition of hydrogen peroxide, motion of nanomotors of platinum metal was accomplished at only low H_2O_2 concentrations (Sánchez, 2010). Thus, promising bio-catalytic self-propulsion can be achieved with enzyme-powered Janus nanoparticles made of mesoporous silica; this is because it is well known that mesoporous silica materials with cylindrical pores can immobilized enzymes (Zhou, 2013; Masuda, 2014). In the design of the Janus nanomotors described by Ma and Sánchez (2017), a smooth silica layer is coated onto the nanoparticles, as indicated with red straight arrows (steps 2 and 4 of Figure 5). Thus, the left half containing carboxyl groups were the responsible of attaching the catalase enzyme via amidation reactions. In previous investigations, they reported the different self-propulsion motions of Janus nanoparticles when containing three different enzymes, catalase, glucose oxidase and urease; they have found that the highest speed was achieved by catalase (Ma, 2015)., making an important contribution for self-propelled nanomotors driven by non-toxic fuels.

However, enzyme-powered nanodevices are still upon investigation to better-understand the mechanisms of self-motion controlled by enzymatic reactions.

On the other hand, reversible immobilizations involve non-covalent attachments of the enzyme to the nanosystems. Although irreversible attachments prevent enzyme leaching, the reversible immobilization methods have the advantage of avoiding discarding the nanosystems containing the enzyme after inactivation. Many are the examples described from the last two decades, although new approaches are emerging in the recent years.

Figure 4. A-B) Micrographs of bacterial magnetosomes in Magnetospirillum magneticum AMB-1 cells. C) Illustration of the functionalized bacterial magnetosomes with an enzyme assisted by a membrane protein acting as anchor. D) Evaluation of batch-to-batch reproducibility and reusability of this novel catalyst (different batches) for bioremediation of ethylparaoxon at two different days (marked in red and blue). Adapted from (Ginet, N et al. (2011). PLoS ONE 6(6): e21442.). Copyright (2010) PLoS ONE.

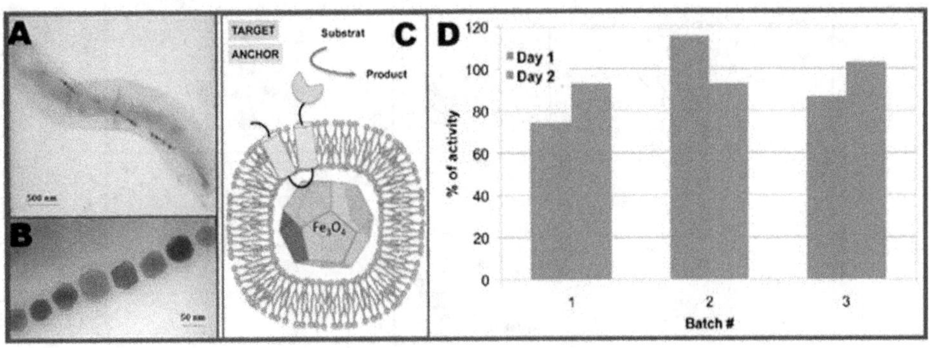

Figure 5. Design and TEM monograph of the biocatalytic Janus mesoporous silica nanoparticles acting as nanomotors. Used with permission from (Ma, X., & Sánchez, S. (2017). Tetrahedron 73, 4883-4886.) Copyright (2017) Elsevier Ltd.

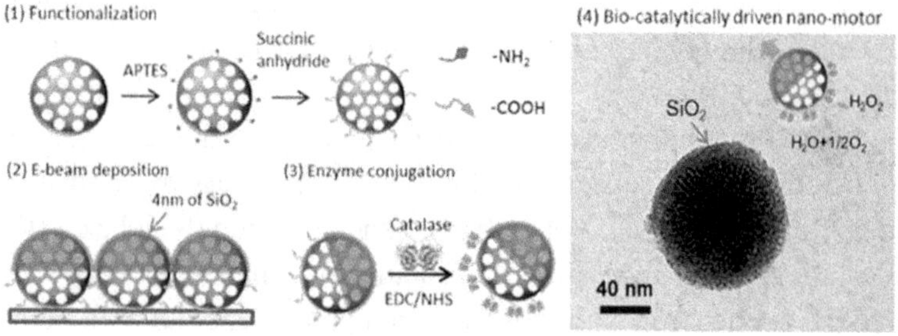

Rotello´s group (Hong, 2004a-b; You, 2006) studied the role of a digestive enzyme, chymotrypsin (ChT), in different nanoparticle types driven by electrostatic interactions. Firstly, Hong et al. (2004a) reported a conjugate formed by ChT electrostatically tagged to the surface of semiconductor quantum dots (SQDs) composed of cadmium selenide, resulting in the enzyme inhibition without denaturation. Secondly, they also described the enzymatic kinetics of the same enzyme attached to AuNPs; as reported by Hong et al., (2004 b) and You et al., (2006); an increase in both stability and selectivity of the same enzyme was achieved when attached to AuNPs containing superficial carboxylic and amine-acid groups (Figure 6). On the one hand, with ended carboxylic groups, the negatively charged conjugated ChT-AuNPs promotes the adsorption of only cationic oligopeptides as substrates having thus accessibility

Figure 6. A) Enzyme electrostatically linked onto AuNPs coated with carboxyl ended groups. Adapted with permission from (Hong, R., Emrick, T., & Rotello, V. M. (2004). J. Am. Chem. Soc., 126, 13572-13573. Copyright (2004) American Chemical Society. B) Influence of the diffusion and interaction of the substrate and the reaction product with the conjugated. Adapted from (You, C.-C., Agasti, S. S., De, M., Knapp, M. J., & Rotello, V. M. (2006). J. Am. Chem. Soc., 128, 14612-14618.). Copyright (2006) American Chemical Society. C) Mechanism of intracellular delivery of β-galactosidase (β -Gal) promoted by the peptide coated gold nanoparticles (NP_Pep), for its efficient enzymatic activity within the cell. D) Schematic illustration of the functionalized gold nanoparticles with a specific peptide and the enzyme. Adapted from (Ghosh, P. Yang, X. Arvizo, R. Zhu, Z.-J. Agasti, S. S. Mo, Z. & Rotello, V. M. (2010). J. Am. Chem. Soc., 132, 2642–2645.). Copyright (2010) American Chemical Society.

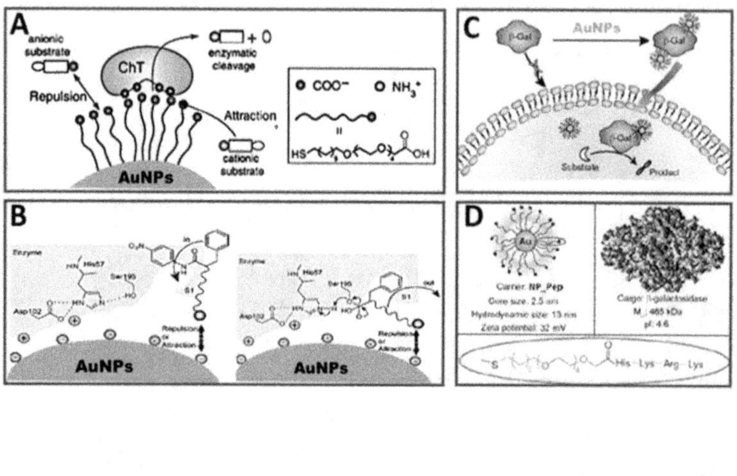

to the enzyme active sites, while negatively charged oligopeptides were hindered by electrostatic repulsion (Hong, 2004b). Afterwards, it was described the possibility of tuning the enzyme specificity using different amine-acid functionalized AuNPs associated to ChT. An enhancement of the catalytic activity when using AuNPs modified with anionic amino acids (e.g. glutamic and aspartic acids) was observed, owing to the electrostatic interactions which controlled the diffusion of the hydrolysed product, as illustrated in Figure 6B (You, 2006).

Effective delivery approaches were also described using metallic nanoparticles as enzyme/protein transporter. Rotello´s group (Ghosh, 2010) described the intracellular delivery of negatively charged proteins/enzymes by AuNPs previously conjugated with cell-penetrating peptides, as shown in Figure 6C. Membrane impermeable proteins such as β-galactosidase (Figure 6D) were used to demonstrate the potential of this investigation. In this way, the enzyme escapes from endosomes and maintains its activity within the cell. This is very promising for active-biomolecule delivery *in vivo* for therapeutic applications.

Polyion polymeric nanoparticles derived from carboxymethyl chitosan were also used for immobilizing enzymes for intracellular delivery of active molecules. Zhao et al. (2013) developed chitosan nanoparticles with a shell derived of polyamidoamine dendrimer (PAMAM) as novel nanocarrier of cationic lysozyme, enabling the control of the enzymatic activity by the influence of the pH as a sensitive off-on switch.

Recently, many researchers are trying to investigate the adsorption mechanisms of enzymes in the variety of nanomaterials. For instance, Adeogun et al. (2017) described an elegant work explaining the adsorption kinetics of amylase, lipase and protease onto zinc-ferrite ($ZnFe_2O_4$) nanoparticles, following a first- and a second-order kinetic models, respectively.

Adsorption of positively charged enzyme onto porous Co_3O_4 nanoparticles driven by electrostatic interaction seems to be ideal for preparing electrochemical biosensors. Thus, electrodes based on the horseradish peroxidase immobilized on the porous Co_3O_4 hexagonal nanosheets were reported by Liu et al. (2017); their attachment to the reduced graphene oxide sheets was suitable for sensing nitrite. The confine space for the substrate and enzyme promotes the probability of effective collisions, making possible to obtain really low limit of detection.

Interestingly, some adsorption methods led to some advantages for nanoparticles as promising renaturation tools. For instance, Vinogradov and Avnir (2015) reached a big challenge of renaturation of enzymes with alumina nanoparticles, which can be explained by the favorable conformation of the renature enzyme versus the denature conformations within the alumina nanocages. A recent approach of renaturation is demonstrated for only negatively charged denatured enzymes and the renaturation process is driven by electrostatic interactions, in which alumina nanoparticles acted as separation and refolding agents capable of release the renature enzyme (Volodina, 2017). This new promising method is fast, simple, efficient and recycling.

Moreover, of great importance is the mimicking of enzymatic cascade reactions as occurred in biological processes, and thus, Jia et al. (2013) claimed that nanomaterials provide an easy vehicle for immobilizing multi-enzymes. Recently the potential increment of the activities of multienzyme-nanodots supposes an important milestone in multi enzyme cascades and biological synthetic routes. Thus, Medintz´s group (Vranish, 2017) started with a preliminary study involving two enzymes, one bound to the surface of SQDs and the other enzyme free in the solution. It was demonstrated an increment in the catalytic activity of both enzymes possibly induced by SQDs. Some authors also performed multienzyme immobilization onto nanoparticles, as reported by the collaborators of Dong (2010), which synthesized in one-spot reaction a reusable low-cost magnetic nanoparticle functionalized with maltodextrin phosphorylase, glucose-1-phosphate thymidyl transferase and pyrophosphatase. Many other works had studied a variety of enzymes in different nanostructured systems with the intentions of exploiting their intrinsic electroactive properties, light harvesting or even recycling possibilities within the catalytic environments. This will open avenues to better understand the general phenomenon of multienzyme cascades based on this artificial nanoestructural scaffolds.

This section opens a new insight on future developments in this multidisciplinary field of nano-objects with enzymatic activity for a wider variety of applications.

NANOPARTICLES IN GELS AS THE IMMOBILIZATION MATRIX

It has been demonstrated the activation of enzymes inside the micro-compartments formed from certain gel matrices, which can be obtained by different methods. One of the most studied is the sol-gel encapsulation of enzymes or hydrogel formation driven by hydrogen bonds and electrostatic interactions.

Recently, new approaches for introducing the enzyme into nanoparticles have been developed.

Sol-gel processes has been involved to introduce nanoparticles into hydrogels by hydrolyzing certain precursors (e.g. glutaraldehyde, surfactants, ionic liquids, cyclodextrins, crown ethers) which give rise to a controlled polymerization; that is, cross-linking is an irreversible method that occur by formation of intermolecular bindings of the precursors. On the one hand, entrapment immobilization is one approach and involves occlusion of the enzyme within the polymeric matrix, fact that avoid denaturalization of the enzyme although it presents the disadvantage of diminishing of the mass transference. On the other hand, enzymes can be encapsulated in the hydrated pores of several sol-gel matrices; several investigations confirmed the increase of stabilities and catalytic activities.

A novel approach is the sol-gel encapsulation, which resulted to be a simple and workable method for the preparation of portable platforms containing active biomolecules (enzymes, proteins) for enhancing their stability over prolonged periods, which are assigned to perform synthetic and purification reactions and also as recycled analytical nanotools (solid-phase extraction, separation or sensing) in environmental and biomedical applications.

The sol-gel methods and host: guest chemistry are the responsible of the encapsulation of the lipase into calix[4]arenes coated Fe_3O_4 nanoparticles, as reported Yildiz et al. (2017). The advantages of this innovative nanoplatform, which was applied in two-phase solvent system for the hydrolysis reaction of racemic flurbiprofen methyl ester are the increment in the thermal and operational stabilities of the enzyme and also in their enzymatic activity and enantioselectivity if compared to free lipase.

Others investigated the immobilization a specific kind of glucanase, pullulanase, for degrading pullulan onto magnetic chitosan nanoparticles via sol-gel reaction. Firstly, Long et al. (2015) immobilized such enzyme onto a nanosized magnetite in a chitosan hydrogel by electrostatic adsorption and covalently bound. The investigations performed by Jiao and coworkers indicated that covalently linked the enzyme lead to a more stable system with higher enzymatic activity and reusability. Secondly, Long et al. (2017) compared to the immobilization of the same enzyme directly by simply sol-gel encapsulation and by encapsulating first the pullulanase onto the nanoparticle surface for further cross-linking, as depicted in Figure 7. Results showed an enhancement on the magnetic response of the sol-gel encapsulated enzyme into nanoparticles and even at higher temperatures (above 60 ºC) the enzyme activity resulted to be higher than those obtained for the free enzyme at the same conditions. However, better results were observed for the pullulanase immobilized firstly covalently to magnetite before the sol-gel process.

Figure 7. Schematic illustrations of the covalently binding of enzymes onto modified magnetite nanoparticles (A), and the sol-gel cross-linking processes for unbound and covalently bound enzyme and nanoparticles (B).

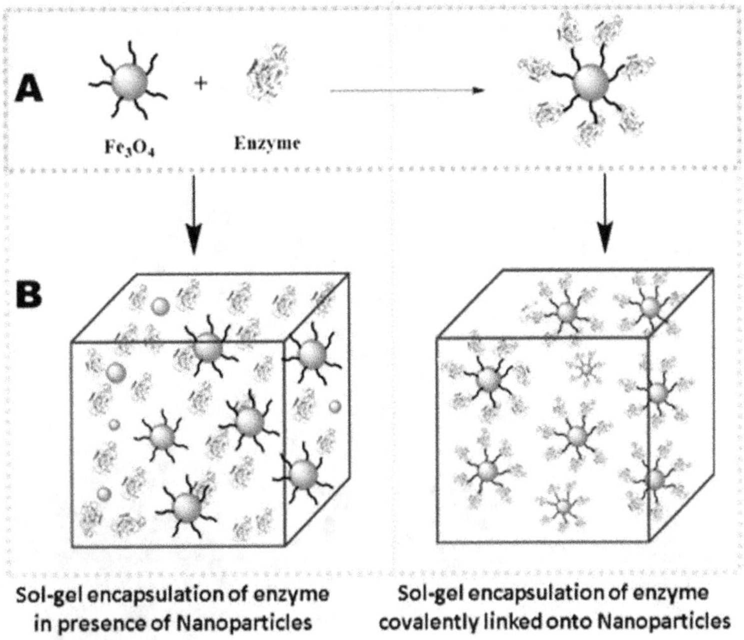

It is possible to combine an indicator of bioactivity (e.g. AuNPs) with an enzyme in a sol-gel process. This is the case described by Luckham and Brennan (2010) that managed to modify a paper substrate to promote the growth of AuNPs seeds, and thus, to be used as colorimetric biosensor of toxins. The group of Brennan reported a dipstick bioassay or test strips which used the bioactive paper containing acetylcholinesterase for the detection sub-lethal levels of paraoxon by visual observation of a color change, being able to register the concentration by images from a digital camera.

Interestingly, the use of silica matrices prepared by sol-gel processes introduces a new concept in chemical sensing by autonomously amplification of the signal. The foundations of this approach resulted in a self-acceleration of the catalytic process with an additional signal amplification, that is, it does not only prevent any interaction enzyme-substrate before the activation of the system but also activate in a cascade manner the whole process in the near nanogates. The material based on mesoporous silica nanoparticles displayed pores of suitable size to accommodate the certain enzymes, like porcine liver esterase (Xue & Zink, 2013). Thus, once the enzyme is accommodated in such pores, the functionalization of the silica surface with cyclodextrin cavities (via imine bonds) and its connection through an adamantane cluster with a star-like shape makes the gate of the novel catalytic system. Thus, the nanogate allow the separation of the substrates, represented as blue cubes, while the analyte target (blue ball) actuates as the nanogate for accessing to the enzyme and thus, generating the fluorescent triangular products.

Semi-wet supramolecular hydrogels bearing nanopores with ionic-exchange features are good candidates to encapsulate molecules for further analytical applications. It is expected a good mobility of the embedded molecules along the whole gel network. A promising photoluminescent sensor reported by Wada et al. (2010) entails the use of enzyme, mesoporous silica particles and hybrid supramolecular

Figure 8. A) Assemblage of the enzymatic chemical amplifier, consisting in the first star-like cluster formation with ended adamantane moieties, which selectively linked to the β-cyclodextrin (β-CD) functionalized silica surface via host:guest chemistry. The enzyme, represented as a green knot is trapped into the pore. B) Mechanism proposed of the sensing process, in which a specific target analyte acting as nanogate (ball) allows the access of the substrates (cubes) to the enzyme (knot) for the catalytic generation of fluorescent products (triangles). Adapted from (Xue, M., & Zink, J. I. (2013). J. Am. Chem. Soc. 135, 17659–17662.). Copyright (2013) American Chemical Society.

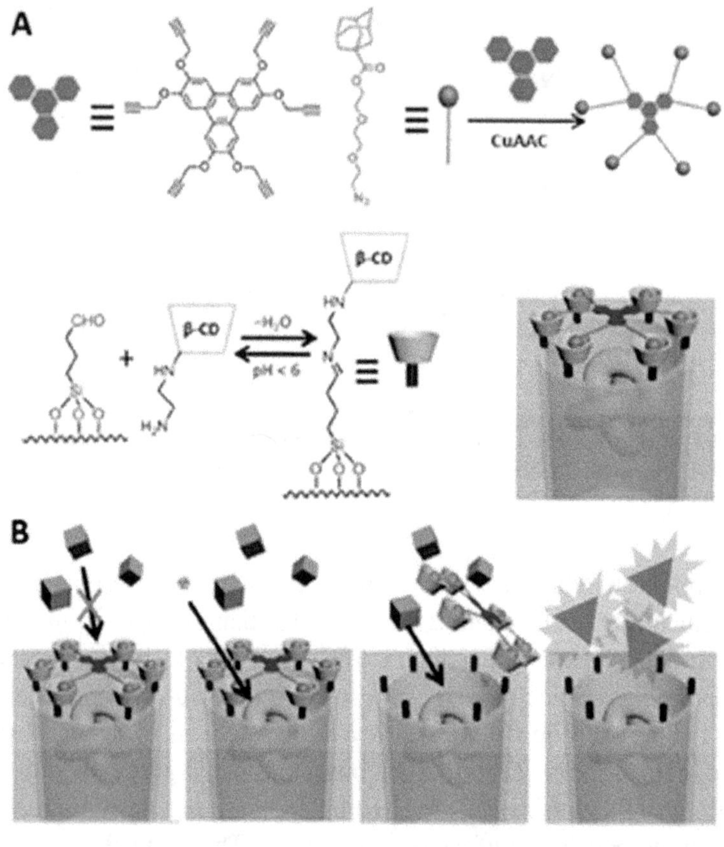

hydrogels consisting of cationic nanopores and hydrophobic nanofibers in an aqueous gel bulk phase. In this example, there exists a cooperative sensing mechanism capable of distinguished the polyanions as result of a better discrimination of the polysulfates versus polyphosphates. Nanocellulose (NC) is a newly emerged low-cost nanoparticle, ecofriendly and renewable, displaying fascinating properties which expected to revolutionize the future nanotechnology. Their fascinating behaviour in aqueous media provided liquid-to-gel transition and chiral inversions (Dufresne, 2013); thus, over different superficial charges and addition or suitable amines, those nanofibers can form reversible hydrogels and organogels (Ruiz-Palomero, 2017a; 2016, respectively). Analytical applications involved the use of a novel hydrogel media composed of cellulose nanofibrils and GQDs for the determination of laccase. Soriano´s group (Ruiz-Palomero, Benítez-Martínez, Soriano, & Valcárcel, 2017b) described a new sensing media for the determination of enzyme laccase and also proposed the hydrogel as a suitable environmental media for the stabilization and storage of laccase, previously extracted from shampoos, as depicted in Figure 9A.

Enzymes in Nanocompartments

As observed in Nature, enzymes are localized in nanoscale compartments, like inside cells. This restriction of the enzymes compartmentalized in nanocages guarantees the enzyme stabilization and also the well-controlled reactions with the optimize stoichiometry (Zhao, 2016).

Recently, nanogels are on the top of the top of enzyme immobilization by wrapping them on synthetic polymers to form in this case nanoscopic gels, as illustrated in Figure 9B. In this research line, the methods for complexation entails covalent and non-covalent interactions. These are of great importance in therapeutic formulations. The encapsulation with amphiphilic polymers with self-cross linkages was suitable for wrapping capsase-3 enzyme in nanogels whilst α-glucosidase enzyme is immersed in nanogel by inverse inversion emulsion polymerization (Raghupathi & Thayumanavan, 2017). It was demonstrated that for electrostatic immobilization of enzymes, their release from the nanogel lead that almost the entire catalytic activities prevails. Others designed a manner to protect the enzyme lignin peroxidase by microemulsion polymerization to form a 90-nm size nanoparticle displaying high catalytic activity towards tetramethylbenzidine and high reusability (Tay, 2016).

From investigations similar to those mentioned above, there is a new avenue in the functionalization of such gel fibers to create more sophisticated sensors with intelligent characteristics applicable in other complex fields like drug delivery or even in self-healing and prosthesis.

Figure 9. A) Schematic illustration of the optical properties of two macroscopic hydrogels composed of nanocellulose (NC) and graphene quantum dots (GQDs) in presence (A1) and absence (A2) of laccase. B) Nanogels containing polymer-wrapped enzymes.

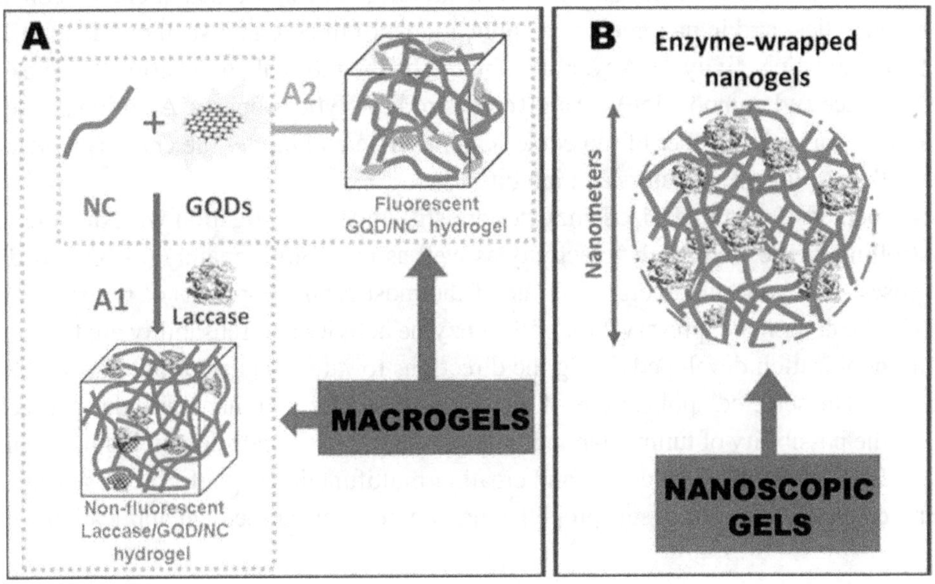

FUTURE TRENDS

Nanozymes is a subject widely explored; there is a recently interest in studying the biocompatibility of inorganic and organic nanoenzymes and in the exploitation of these multienzyme-like properties in biological and environmental systems. These investigations provided reliable information for their potential advances in nanomedicine and energy conversion technologies. However, the current trend is directed to the green chemistry, using green nanofabrication techniques of stable inorganic nanozymes or synthesizing biocompatible low-toxic nanozymes like graphene and carbon dots. More investigation is needed to evaluate synergistic nanozymetic activity of hybrids and nanocomposites composed of two or more nanoparticles in order to enable its integration in portable low-cost sensing platforms (Cheng, 2017).

On the other hand, enzyme immobilization onto multifunctional nanoparticles or nanostructured materials offer many advantages in environmental and biomedical fields. Many are the nanoparticles explore in that subject, although the trend moves towards the exploitation of enzyme immobilization in renewable nanoparticles or tridimensional nanostructures (nanogels, nanoemulsions) which are obtained by low -cost and green methods, like carbon dots, graphene quantum dots and nanocellulose. These innovative materials has huge potential in nanomedicine as reporting agents and tissue repairing tools

CONCLUSION

The easy-to-understand aspects of the recent advances in the catalytic activities of nanoobjects are described in this chapter. The most exciting topics are nanozyme, enzymes immobilized into nanoparticles and within nanostructured gels. Nanozymes still surprise scientists for their fascinating catalytic properties that mimic bioactive molecules and nowadays are of great importance in energy conversion and storage, in bioanalytical and in nanomedicine applications. Both core and shell of nanoparticles play a key role in their catalytic activity. However, it is limited by the rational engineering of nanozyme (size, morphology, surface and composition) for a given desired catalytic property. Additionally, as occurred in natural enzymes, a wide number of molecules can promote and hinder the enzymatic activity of the nanoparticles, thus acting as activators and inhibitors.

Innovative methods for immobilized enzymes are still under investigation with the intention of rationally controlling the desired catalytic activity as well as reusability, mainly for bioremediation and sensing purposes. However, nanomotors are one of the most promising areas of research for this type of systems. On the other hand, preservation of the enzyme activity and reusability are the main aspects of enzyme immobilization developed, being the directions towards the use of polymers, gels and even within nanogels. The selected applications of enzymes immobilized in nanoparticles and gels opens a new insight in the possibility of tuning nanoparticles as central nanoscaffolds for helping to understand their great potential in catalytic processes and creating multifunctional platforms in sensing, imaging, drug delivery, catalysis and even in self-propelled nanomotors for biomedical applications.

ACKNOWLEDGMENT

It is acknowledged the support from the European Commission within the Seventh Research Framework Programme EU-FP7 for the funding project (FP7-NMP-2007-2013-SME5-280550).

REFERENCES

Adeogun, A. I., Kareeem, S. O., Adebayo, O. S., & Balogun, S. A. (2017). Comparative adsorption of amylase, protease and lipase on $ZnFe_2O_4$: Kinetics, isothermal and thermodynamics studies. *Three Biotech*, *7*(3), 198. doi:10.100713205-017-0859-6 PMID:28667640

Ahmad, R., & Sardar, M. (2015). Enzyme immobilization: An overview on nanoparticles as immobilization matrix. *Biochemistry and Analytical Biochemistry*, *4*, 178. doi:10.4172/2161-1009.1000178

Ahmad, R., Khatoon, N., & Sardar, M. (2013). Biosynthesis, Characterization and Application of TiO_2 nanoparticles in biocatalysis and protein folding. *Journal of Proteins and Proteomics*, *4*, 115–121.

Ansari, S. A., & Husain, Q. (2012). Potential applications of enzymes immobilized on/in nano materials: A review. *Biotechnology Advances*, *30*(3), 512–523. doi:10.1016/j.biotechadv.2011.09.005 PMID:21963605

Ariga, K., Ji, Q., Mori, T., Naito, M., Yamauchi, Y., Abe, H., & Hill, J. P. (2013). Enzyme nanoarchitectonics: Organization and device application. *Chemical Society Reviews*, *42*(15), 6322–6345. doi:10.1039/c2cs35475f PMID:23348617

Chen, J., Leng, J., Yang, X., Liao, L., Liu, L., & Xiao, A. (2017). Enhanced performance of magnetic graphene oxide-immobilized laccase and its application for the decolorization of dyes. *Molecules, 22*, E221. doi: 10.3390/molecules22020221

Cheng, N., & Song, Y. (2017). *Nanozyme mediated dual-immunoassay integrated with smartphone for use in simultaneous detection of pathogens*. ACS Applied Matererials & Interfaces; doi:10.1021/acsami.7b12734

Cipolatti, E. P., Silva, M. J. A., Klein, M., Feddern, V., Feltes, M. M. C., Oliveira, J. V., ... Oliveira, D. (2014). Current status and trends in enzymatic nanoimmobilization. *Journal of Molecular Catalysis. B, Enzymatic*, *99*, 56–67. doi:10.1016/j.molcatb.2013.10.019

Dong, Q., Ouyang, L.-M., Yu, H.-L., & Xu, J.-H. (2010). Efficient biosynthesis of uridine diphosphate glucose from maltodextrin by multiple enzymes immobilized on magnetic nanoparticles. *Carbohydrate Research*, *345*(11), 1622–1626. doi:10.1016/j.carres.2010.04.025 PMID:20627237

Dufresne, A. (2013). Nanocellulose: A new ageless bionanomaterial. *Materials Today*, *16*(6), 220–227. doi:10.1016/j.mattod.2013.06.004

Fan, Y., & Huang, Y. (2012). The effective peroxidase-like activity of chitosan-functionalized CoFe2O4 nanoparticles for chemiluminescence sensing of hydrogen peroxide and glucose. *Analyst (London)*, *137*(5), 1225–1231. doi:10.1039/c2an16105b PMID:22258160

Feng, W., & Ji, P. (2011). Enzymes immobilized on carbon nanotubes. *Biotechnology Advances*, *29*(6), 889–895. doi:10.1016/j.biotechadv.2011.07.007 PMID:21820044

Ganesana, M., Istarnboulie, G., Marty, J.-L., Noguer, T., & Andreescu, S. (2011). Site-specific immobilization of a (His) 6-tagged acetylcholinesterase on nickel nanoparticles for highly sensitive toxicity biosensors. *Biosensors & Bioelectronics*, *30*(1), 43–48. doi:10.1016/j.bios.2011.08.024 PMID:21937214

Ganesana, M., Istarnboulie, G., Marty, J. L., Noguer, T., & Andreescu, S. (2011). Site-specific immobilization of a (His)6-tagged acetylcholinesterase on nickel nanoparticles for highly sensitive toxicity biosensors. *Biosensors & Bioelectronics*, *30*(1), 43–48. doi:10.1016/j.bios.2011.08.024 PMID:21937214

Garg, B., & Bisht, T. (2016). Carbon Nanodots as Peroxidase Nanozymes for Biosensing. *Molecules (Basel, Switzerland)*, *21*, 1653.

Ge, S. (2014). Disposable electrochemical immunosensor based on peroxidase-like magnetic silica-graphene oxide composites for detection of cancer antigen 153. *Sensors and Actuators. B, Chemical*, *192*, 317–326.

Ghosh, P., Yang, X., Arvizo, R., Zhu, Z. J., Agasti, S. S., Mo, Z., & Rotello, V. M. (2010). Intracellular delivery of a membrane-impermeable enzyme in active form using functionalized gold nanoparticles. *Journal of the American Chemical Society*, *132*(8), 2642–2645. doi:10.1021/ja907887z PMID:20131834

Ginet, N., Pardoux, R., Adryanczyk, G., Garcia, D., Brutesco, C., & Pignol, D. (2011). Single-step production of a recyclable nanobiocatalyst for organophosphate pesticides biodegradation using functionalized bacterial magnetosomes. *PLoS One*, *6*(6), e21442. doi:10.1371/journal.pone.0021442 PMID:21738665

Golchin, J., Golchin, K., Alidadian, N., Ghaderi, S., Eslamkhah, S., Eslamkhah, M., & Akbarzadeh, A. (2017). Nanozyme applications in biology and medicine: an overview. *Artificial Cells, Nanomedicine, and Biotechnology, an International Journal, 45*(6), 1069-1076.

Gómez-Anquela, C., García-Mendiola, T., Abad, J. M., Pita, M., Pariente, F., & Lorenzo, E. (2015). Scaffold electrodes based on thioctic acid-capped gold nanoparticles coordinated Alcohol Dehydrogenase and Azure A films for high performance biosensor. *Bioelectrochemistry (Amsterdam, Netherlands)*, *106*, 335–342. doi:10.1016/j.bioelechem.2015.06.009 PMID:26152878

Han, K. N., Choi, J.-S., & Kwon, J. (2017). Gold nanozyme-based paper chip for colorimetric detection of mercury ions. *Scientific Reports*, *7*(1), 2806. doi:10.103841598-017-02948-x PMID:28584238

Hondred, J. A., Breger, J. C., Garland, N. T., Oh, E., Susumu, K., Walper, S. A., ... Claussen, J. C. (2017). Enhanced enzymatic activity from phosphotriesterase trimer gold nanoparticle bioconjugates for pesticide detection. *Analyst (London)*, *142*(17), 3261–3271. doi:10.1039/C6AN02575G PMID:28765846

Hong, R., Fischer, N. O., Verma, A., Goodman, C. M., Emrick, T., & Rotello, V. M. (2014a). Control of protein structure and function through surface recognition by tailored nanoparticle scaffolds. *Journal of the American Chemical Society*, *126*(3), 739–743. doi:10.1021/ja037470o PMID:14733547

Hong, R., Emrick, T., & Rotello, V. M. (2004b). Monolayer-controlled substrate selectivity using noncovalent enzyme-nanoparticle conjugates. *Journal of the American Chemical Society*, *126*(42), 13572–13573. doi:10.1021/ja0461163 PMID:15493887

Huang, D.-M., Hsiao, J.-K., Chen, Y.-C., Chien, L. Y., Yao, M., Chen, Y. K., ... Chen, Y. C. (2009). The promotion of human mesenchymal stem cell proliferation by superparamagnetic iron oxide nanoparticles. *Biomaterials*, *30*(22), 3645–3651. doi:10.1016/j.biomaterials.2009.03.032 PMID:19359036

Janoš, P., Kuráň, P., Pilařová, V., Trögl, J., Šťastný, M., Pelant, O., ... Skoumal, M. (2015). Magnetically separable reactive sorbent based on the CeO_2/γ-Fe_2O_3 composite and its utilization for rapid degradation of the organophosphate pesticide parathion methyl and certain nerve agents. *Chemical Engineering Journal*, *262*, 747–755. doi:10.1016/j.cej.2014.10.016

Jeong, J., Ha, T. H., & Chung, B. H. (2006). Enhanced reusability of hexa-arginine-tagged esterase immobilized on gold-coated magnetic nanoparticles. *Analytica Chimica Acta*, *569*(1-2), 203–209. doi:10.1016/j.aca.2006.03.102

Jia, F., Narasimhan, B., & Mallapragada, S. K. (2013). Biomimetic multienzyme complexes based on nanoscale platforms. *AIChE Journal. American Institute of Chemical Engineers*, *59*(2), 355–360. doi:10.1002/aic.13992

Johnson, P. A., Park, H. J., & Driscoll, A. J. (2011). Enzyme nanoparticle fabrication: Magnetic nanoparticle synthesis and enzyme immobilization. *Methods in Molecular Biology (Clifton, N.J.)*, *679*, 183–191. doi:10.1007/978-1-60761-895-9_15 PMID:20865397

Kalkan, N. A., Aksoy, S., Aksoy, E. A., & Hasirci, N. (2012). Preparation of chitosan-coated magnetite nanoparticles and application for immobilization of laccase. *Journal of Applied Polymer Science*, *123*(2), 707–716. doi:10.1002/app.34504

Kouassi, G. K., Irudayaraj, J., & McCarty, G. (2005). Examination of cholesterol oxidase attachment to magnetic nanoparticles. *Journal of Nanobiotechnology*, *3*(1), 1–9. doi:10.1186/1477-3155-3-1 PMID:15661076

Liu, H., Guo, K., Lv, J., Gao, Y., Duan, C., Deng, L., & Zhu, Z. (2017). A novel nitrite biosensor based on the direct electrochemistry of horseradish peroxidase immobilized on porous Co_3O_4 nanosheets and reduced graphene oxide composite modified electrode. *Sensors and Actuators. B, Chemical*, *238*, 249–256. doi:10.1016/j.snb.2016.07.073

Long, J., Li, X., Wu, Z., Xu, E., Xu, X., Jin, Z., & Jiao, A. (2015). Immobilization of pullulanase onto activated magnetic chitosan/Fe_3O_4 nanoparticles prepared by in situ mineralization and effect of surface functional groups on the stability. *Colloids & Surfaces A. Physicochemical & Engineering*, *472*, 69–77. doi:10.1016/j.colsurfa.2015.02.038

Long, J., Li, X., Zhan, X., Xu, X., Tian, Y., Xie, Z., & Jin, Z. (2017). Sol–gel encapsulation of pullulanase in the presence of hybrid magnetic (Fe3O4–chitosan) nanoparticles improves thermal and operational stability. *Bioprocess and Biosystems Engineering*, *40*(6), 821–831. doi:10.100700449-017-1747-5 PMID:28243738

Luckham, R. E., & Brennan, J. D. (2010). Bioactive paper dipstick sensors for acetylcholinesterase inhibitors based on sol–gel/enzyme/gold nanoparticle composites. *Analyst (London)*, *135*(8), 2028–2035. doi:10.1039/c0an00283f PMID:20593080

Ma, X., & Sánchez, S. (2017). Bio-catalytic mesoporous Janus nano-motors powered by catalase enzyme. *Tetrahedron*, *73*(33), 4883–4886. doi:10.1016/j.tet.2017.06.048

Ma, X., Hortelão, A. C., Patiño, T., & Sánchez, S. (2016). Enzyme catalysis to power micro/nanomachines. *ACS Nano*, *10*(10), 9111–9122. doi:10.1021/acsnano.6b04108 PMID:27666121

Ma, X., Jannasch, A., Albrecht, U. R., Hahn, K., Miguel-López, A., Schäffer, E., & Sánchez, S. (2015). Enzyme-powered hollow mesoporous Janus nanomotors. *Nano Letters*, *15*(10), 7043–7050. doi:10.1021/acs.nanolett.5b03100 PMID:26437378

Maine, E., Thomas, V. J., Bliemel, M., Murira, A., & James Utterback, J. (2014). The emergence of the nanobiotechnology industry. *Nature Nanotechnology*, *9*(1), 2–5. doi:10.1038/nnano.2013.288 PMID:24390555

Manea, F., Houillon, F. B., Pasquato, L., & Scrimin, P. (2004). Nanozymes: Gold-Nanoparticle-Based Transphosphorylation Catalysts. *Angewandte Chemie International Edition*, *43*(45), 6165–6169. doi:10.1002/anie.200460649 PMID:15549744

Masuda, Y., Jugimiya, S.-I., Kawachi, Y., & Kato, K. (2014). Interparticle mesoporous silica as an effective support for enzyme immobilisation. *RSC Advances*, *4*(7), 3573–3580. doi:10.1039/C3RA46122J

Mitchell, D. T., Lee, S. B., Trofin, L., Li, N., Nevanen, T. K., Soderlund, H., & Martin, C. R. (2002). Smart nanotubes for bioseparations and biocatalysis. *Journal of the American Chemical Society*, *124*(40), 11864–11865. doi:10.1021/ja027247b PMID:12358526

Mohajershojaei, K., Khosravi, A., & Mahmoodi, N. M. (2014). Decolorization of Dyes using immobilized laccase enzyme on zinc ferrite nanoparticle from single and binary systems. *Fibers and Polymers*, *15*(10), 2139–2145. doi:10.100712221-014-2139-y

Muthurasu, A., & Ganesh, V. (2014). Horseradish peroxidase enzyme immobilized graphene quantum dots as electrochemical biosensors. *Applied Biochemistry and Biotechnology*, *174*(3), 945–959. doi:10.100712010-014-1019-7 PMID:24984603

Netto, C. G. C. M., Toma, H. E., & Andrade, L. H. (2013). Superparamagnetic nanoparticles as versatile carriers and supporting materials for enzymes. *Journal of Molecular Catalysis B: Enzymatic, 85– 86*, 71– 92.

Noh, S. H., Kwon, C., Hwang, J., Ohsaka, T., Kim, B.-J., Kim, T.-Y., ... Han, B. (2017). Self-assembled nitrogen-doped fullerenes and their catalysis for fuel cell and rechargeable metal–air battery applications. *Nanoscale*, *9*(22), 7373–7379. doi:10.1039/C7NR00930E PMID:28405666

Park, J. M., Kim, M., Park, H. S., Jang, A., Min, J., & Kim, Y. H. (2013). Immobilization of lysozyme-CLEA onto eletrospun chitosan nanofiber for effective antimicrobial applications. *International Journal of Biological Macromolecules*, *54*, 37–43. doi:10.1016/j.ijbiomac.2012.11.025 PMID:23201775

Payen, A., & Persoz, J. (1833). Mémoire sur la Diastase, les principaux produits de ses réactions, et leurs applications aux arts industriels. *Annales de Chimie et de Physique*, *53*, 73–92.

Pundir, C. S. (2015). Enzyme Nanoparticles: Preparation, Characterisation, Properties and Applications. Elsevier.

Qhobosheane, M., Santra, S., Zhang, P., & Tan, W. (2001). Biochemically functionalized silica nanoparticles. *Analyst (London)*, *126*(8), 1274–1278. doi:10.1039/b101489g PMID:11534592

Raghupathi, K., & Thayumanavan, S. (2017). NanoArmoring of enzymes: rational design of polymer-wrapped enzymes. In C. V. Kumar (Ed.), *Methods in Enzymology* (Vol. 590, pp. 381–411). Elsevier.

Ruiz-Palomero, C., Kennedy, S. R., Soriano, M. L., Jones, C. D., Valcárcel, M., & Steed, J. W. (2016). Pharmaceutical Crystallization with Nanocellulose Organogels. *Chemical Communications*, *52*(50), 7782–7785. doi:10.1039/C6CC03088B PMID:27168091

Ruiz-Palomero, C., Soriano, M. L., & Valcárcel, M. (2017a). Nanocellulose as analyte and analytical tool: Opportunities and challenges. *TrAC Trends in Analytical Chemistry*, *87*, 1–18. doi:10.1016/j.trac.2016.11.007

Ruiz-Palomero, C., Benítez-Martínez, S., Soriano, M. L., & Valcárcel, M. (2017b). Fluorescent nanocellulosic hydrogels based on graphene quantum dots for sensing laccase. *Analytica Chimica Acta*, *974*, 93–99. doi:10.1016/j.aca.2017.04.018 PMID:28535886

Sanchez, S., Solovev, A. A., Mei, Y., & Schmidt, O. G. (2010). Dynamics of Biocatalytic Microengines Mediated by Variable Friction Control. *Journal of the American Chemical Society*, *132*(38), 13144–13145. doi:10.1021/ja104362r PMID:20860367

Sanfins, E., Augustsson, C., Dahlbäck, B., Linse, S., & Cedervall, T. (2014). Size-dependent effects of nanoparticles on enzymes in the blood coagulation cascade. *Nano Letters*, *14*(8), 4736–4744. doi:10.1021/nl501863u PMID:25025946

Singh, S., (2016). Cerium oxide based nanozymes: Redox phenomenon at biointerfaces. *Biointerphases,11*(40), 04B202.

Shi, W., Zhang, X., He, S., & Huang, Y. (2011). CoFe2O4 magnetic nanoparticles as a peroxidase mimic mediated chemiluminescence for hydrogen peroxide and glucose. *Chemical Communications*, *47*(38), 10785–10787. doi:10.1039/c1cc14300j PMID:21892508

Tay, T., Köse, E., Keçili, R., & Say, R. (2016). Design and preparation of nano-lignin peroxidase (NanoLiP) by protein block copolymerization approach. *Polymers*, *8*(6), 223. doi:10.3390/polym8060223

Tokuyama, H., Yamago, S., Nakamura, E., Shiraki, T., & Sugiura, Y. (1993). Photoinduced biochemical-activity of fullerene carboxylic-acid. *Journal of the American Chemical Society*, *115*(17), 7918–7919. doi:10.1021/ja00070a064

Vinogradov, V. V., & Avnir, D. (2015). Enzyme renaturation to higher activity driven by the sol-gel transition: Carbonic anhydrase. *Scientific Reports*, *5*(1), 14411. doi:10.1038rep14411 PMID:26394694

Volodina, K. V., Avnir, D., & Vinogradov, V. V. (2017). Alumina nanoparticle-assisted enzyme refolding: A versatile methodology for proteins renaturation. *Scientific Reports*, *7*(1), 1458. doi:10.103841598-017-01436-6 PMID:28469182

Vranish, J. N., Ancona, M. G., Oh, E., Susumu, K., & Medintz, I. L. (2017). Enhancing coupled enzymatic activity by conjugating one enzyme to a nanoparticle. *Nanoscale*, *9*(16), 5172–5187. doi:10.1039/C7NR00200A PMID:28393943

Wada, A., Tamaru, S., Ikeda, M., & Hamachi, I. (2009). MCM-enzyme-supramolecular hydrogel hybrid as a fluorescence sensing material for polyanions of biological significance. *Journal of the American Chemical Society, 131*(14), 5321–5330. doi:10.1021/ja900500j PMID:19351208

Wang, X., Guo, W., Hu, Y., Wu, J., & Wei, H. (2016a). Introduction to Nanozymes. In *Nanozymes: Next Wave of Artificial Enzymes. Springer Briefs in Molecular Science*. Berlin: Springer. doi:10.1007/978-3-662-53068-9_1

Wang, X., Hua, Y., & Wei, H. (2016b). Nanozymes in bionanotechnology: From sensing to therapeutics and beyond. *Inorganic Chemistry Frontiers, 3*(1), 41–60. doi:10.1039/C5QI00240K

Wei, L., Zhang, W., Lu, H., & Yang, P. (2010). Immobilization of enzyme on detonation nanodiamond for highly efficient proteolysis. *Talanta, 80*(3), 1298–1304. doi:10.1016/j.talanta.2009.09.029 PMID:20006091

Wu, X. C., Zhang, Y., Han, T., Wu, H. X., Guo, S. W., & Zhang, J. Y. (2014). Composite of graphene quantum dots and Fe3O4 nanoparticles: Peroxidase activity and application in phenolic compound removal. *RSC Advances, 4*(7), 3299–3305. doi:10.1039/C3RA44709J

Xue, M., & Zink, J. I. (2013). Enzymatic Chemical Amplifier Based on Mechanized Nanoparticles. *Journal of the American Chemical Society, 135*(47), 17659–17662. doi:10.1021/ja4066317 PMID:24224766

Yildiz, H., Ozyilmaz, E., Bhatti, A. A., & Yilmaz, M. (2017). Enantioselective resolution of racemic flurbiprofen methyl ester by lipase encapsulated mercapto calix[4]arenes capped Fe_3O_4 nanoparticles. *Bioprocess and Biosystems Engineering, 40*(8), 1189–1196. doi:10.100700449-017-1779-x PMID:28488138

You, C.-C., Agasti, S. S., De, M., Knapp, M. J., & Rotello, V. M. (2006). Modulation of the catalytic behavior of r-chymotrypsin at monolayer-protected nanoparticle surfaces. *Journal of the American Chemical Society, 128*(45), 14612–14618. doi:10.1021/ja064433z PMID:17090046

Yousefinejad, S., Rasti, H., Hajebi, M., Kowsari, M., Sadravi, S., & Honarasa, F. (2017). Design of C-dots/ Fe_3O_4 magnetic nanocomposite as an efficient new nanozyme and its application for determination of H_2O_2 in nanomolar level. *Sensors and Actuators. B, Chemical, 247*, 691–696. doi:10.1016/j.snb.2017.02.145

Zhang, X., Zhao, J., Wen, Y., Zhu, C., Yang, J., & Yao, F. (2013). Carboxymethyl chitosan-poly(amidoamine) dendrimer core–shell nanoparticles for intracellular lysozyme delivery. *Carbohydrate Polymers, 98*(2), 1326–1334. doi:10.1016/j.carbpol.2013.08.005 PMID:24053810

Zhang, Z., Zhang, X., Liu, B., & Liu, J. (2017). Molecular imprinting on inorganic nanozymes for hundred-fold enzyme specificity. *Journal of the American Chemical Society, 139*(15), 5412–5419. doi:10.1021/jacs.7b00601 PMID:28345903

Zhao, Z., Fu, J., Dhakal, S., Johnson-Buck, A., Liu, M., Zhang, T., ... Yan, H. (2016). Nanocaged enzymes with enhanced catalytic activity and increased stability against protease digestion. *Nature Communications, 7*, 10619. doi:10.1038/ncomms10619 PMID:26861509

Zhou, Z., & Hartmann, M. (2013). Progress in enzyme immobilization in ordered mesoporous materials and related applications. *Chemical Society Reviews, 42*(9), 3894–3912. doi:10.1039/c3cs60059a PMID:23570038

Compilation of References

Abdel-Mohsen, H. T., Conrad, J., & Beifuss, U. (2014). Laccase-catalyzed synthesis of catechol thioethers by reaction of catechols with thiols using air as an oxidant. *Green Chemistry*, *16*(1), 90–95. doi:10.1039/C3GC41968A

Abdel-Mohsen, H. T., Conrad, J., Harms, K., Nohr, D., & Beifuss, U. (2017). Laccase-catalyzed green synthesis and cytotoxic activity of novel pyrimidobenzothiazoles and catechol thioethers. *RSC Advances*, *7*(28), 17427–17441. doi:10.1039/C6RA28102H

Abdel-Naby, M. A., & Kwon, D.Y. (1992). Production of xylanase and xylosidase by *Aspergillus niger* NRC 107. *Korean Journal of Applied Microbial Biotechnology*, *20*(5), 543–550.

Abdeshahian, P., Samat, N., & Wan, Y. W. M. (2009). Xylanase production by *Aspergillus niger* FTCC 5003 using palm kernel cake in fermentative bioprocess. *Pakistan Journal of Biological Sciences*, *12*(15), 1049–1055. doi:10.3923/pjbs.2009.1049.1055 PMID:19943460

Abou-Okeil, A., El-Shafie, A., & El Zawahry, M. M. (2010). Ecofriendly laccase–hydrogen peroxide/ultrasound-assisted bleaching of linen fabrics and its influence on dyeing efficiency. *Ultrasonics Sonochemistry*, *17*(2), 383–390. doi:10.1016/j.ultsonch.2009.08.007 PMID:19748814

Accinelli, C., Saccà, M. L., Batisson, I., Fick, J., Mencarelli, M., & Grabic, R. (2010). Removal of oseltamivir (Tamiflu) and other selected pharmaceuticals from wastewater using a granular bioplastic formulation entrapping propagules of *Phanerochaete chrysosporium*. *Chemosphere*, *81*(3), 436–443. doi:10.1016/j.chemosphere.2010.06.074 PMID:20673959

Acuña, C. P. S., González, G. H. J., & Aristizábal-Torres, I. D. (2012). Physicochemical characteristics and functional properties of vitabosa (*Mucuna deeringiana*) and soybean (*Glycine max*). *Food Science and Technology (Campinas)*, *32*(1), 98–105. doi:10.1590/S0101-20612012005000007

Adelakun, O. E., Kudanga, T., Green, I. R., le Roes-Hill, M., & Burton, S. G. (2012b). Enzymatic modification of 2, 6-dimethoxyphenol for the synthesis of dimers with high antioxidant capacity. *Process Biochemistry*, *47*(12), 1926–1932. doi:10.1016/j.procbio.2012.06.027

Adelakun, O. E., Kudanga, T., Parker, A., Green, I. R., le Roes-Hill, M., & Burton, S. G. (2012a). Laccase-catalyzed dimerization of ferulic acid amplifies antioxidant activity. *Journal of Molecular Catalysis. B, Enzymatic*, *74*(1-2), 29–35. doi:10.1016/j.molcatb.2011.08.010

Adeogun, A. I., Kareeem, S. O., Adebayo, O. S., & Balogun, S. A. (2017). Comparative adsorption of amylase, protease and lipase on $ZnFe_2O_4$: Kinetics, isothermal and thermodynamics studies. *Three Biotech*, *7*(3), 198. doi:10.100713205-017-0859-6 PMID:28667640

Adibi, H., Rashidi, A., Khodaei, M. M., Alizadeh, A., Majnooni, M. B., Pakravan, N., ... Nematollahi, D. (2011). Catecholthioether derivatives: Preliminary study of in-vitro antimicrobial and antioxidant activities. *Chemical & Pharmaceutical Bulletin*, *59*(9), 1149–1152. doi:10.1248/cpb.59.1149 PMID:21881259

Adjonu, R., Doran, G., Torley, P., & Agboola, S. (2013). Screening of whey protein isolate hydrolysates for their dual functionality: Influence of heat pre-treatment and enzyme specificity. *Food Chemistry*, *136*(3-4), 1435–1443. doi:10.1016/j.foodchem.2012.09.053 PMID:23194546

Adlercreutz, P. (2006). Immobilization of Enzymes for Use in Organic Media. In J. M. Guisan (Ed.), *Immobilization of Enzymes and Cells. Methods in Biotechnology*™ (Vol. 22). Humana Press. doi:10.1007/978-1-59745-053-9_22

Adnan, L. A., Sathish Kumar, P., Yusoff, A. R. M., & Hadibarata, T. (2015). Metabolites characterization of laccase mediated Reactive Black 5 biodegradation by fast growing ascomycete fungus *Trichoderma atroviride* F03. *International Biodeterioration & Biodegradation*, *104*, 274–282. doi:10.1016/j.ibiod.2015.05.019

Agarwal, S., Kushwaha, A., Verma, V., & Singh, M. P. (2017). Nutritional attributes of *Pleurotus* mushroom. In V. V. Singh & A. K. Singh (Eds.), *Incredible World of Biotechnology, M.P* (pp. 13–24). New York, USA: Nova Science Publishers.

Agarwal, S., & Singh, M. P. (2015). Oyster Mushroom: An Effective Tool of Bioremediation and Sustainable Development. In Insight. *International Journal of Sciences*, *2*, 14–25.

Agematu, H., Kominato, K., Shibamoto, N., Yoshioka, T., Nishida, H., Okamoto, R., ... Murao, S. (1993a). Transformation of 7-(4-hydroxyphenylacetamido) cephalosporanic acid into a new cephalosporin antibiotic, 7-[1-oxaspiro (2.5) octa-6-oxo-4, 7-diene-2-carboxamido] cephalosporanic acid, by laccase. *Bioscience, Biotechnology, and Biochemistry*, *57*(8), 1387–1388. doi:10.1271/bbb.57.1387

Agematu, H., Tsuchida, T., Kominato, K., Shibamoto, N., Yoshioka, T., Nishida, H., ... Murao, S. (1993b). Enzymatic dimerization of penicillin X. *The Journal of Antibiotics*, *46*(1), 141–148. doi:10.7164/antibiotics.46.141 PMID:8436547

Aghaie-Khouzani, M., Forootanfar, H., Moshfegh, M., Khoshayand, M. R., & Faramarzi, M. A. (2012). Decolorization of some synthetic dyes using optimized culture broth of laccase producing ascomycete *Paraconiothyrium variabile. Biochemical Engineering Journal*, *60*, 9–15. doi:10.1016/j.bej.2011.09.002

Ahel, M., Giger, W., & Koch, M. (1994). Behaviour of alkylphenolpolyethoxylate surfactants in the aquatic environment I Occurrence and transformation in sewage treatment. *Water Research*, *28*(5), 1131–1142. doi:10.1016/0043-1354(94)90200-3

Ahmad, R., Khatoon, N., & Sardar, M. (2013). Biosynthesis, Characterization and Application of TiO_2 nanoparticles in biocatalysis and protein folding. *Journal of Proteins and Proteomics*, *4*, 115–121.

Ahmad, R., & Sardar, M. (2015). Enzyme immobilization: An overview on nanoparticles as immobilization matrix. *Biochemistry and Analytical Biochemistry*, *4*, 178. doi:10.4172/2161-1009.1000178

Ahmed, I. N., Chang, R., & Tsai, W. B. (2017). Poly (acrylic acid) nano gel as a substrate for cellulase immobilization for hydrolysis of cellulose. *Colloids and Surfaces. B, Biointerfaces*, *152*, 339–343. doi:10.1016/j.colsurfb.2017.01.040 PMID:28131958

Ahmed, S., Riaz, S., & Jamil, A. (2009). Molecular cloning of fungal xylanases: An overview. *Applied Microbiology and Biotechnology*, *84*(1), 19–35. doi:10.100700253-009-2079-4 PMID:19568746

Aitken, M. D. (1993). Waste Treatment Applications of Enzymes: Opportunities and Obstacles. *Chemical Engineering Journal*, *52*(2), B49–B58. doi:10.1016/0300-9467(93)80057-U

Aitken, M. D., & Irvine, R. L. (1989). Stability testing of ligninase and Mn-peroxidase from *Phanerochaete chrysosporium. Biotechnology and Bioengineering*, *34*(10), 1251–1260. doi:10.1002/bit.260341003 PMID:18588065

Aitken, M. D., Massey, I. J., Chen, T., & Heck, P. E. (1994). Characterization of reaction products from the enzyme catalyzed oxidation of phenolic pollutants. *Water Research*, *28*(9), 1879–1889. doi:10.1016/0043-1354(94)90163-5

Akacha, N. B., & Gargouri, M. (2015). *Food and Bioproducts Processing*. Academic Press. doi:10.1016/j.fbp.2014.09.011

Akiba, S., Kimura, Y., Yamamoto, K., & Kumagai, H. (1995). Purification and characterization of a protease-resistant cellulase from Aspergillus niger. *Journal of Fermentation and Bioengineering*, *79*(2), 125–130. doi:10.1016/0922-338X(95)94078-6

Alama, M. Z., Mansora, M. F., & Jalal, K. C. A. (2009). Optimization of decolorization of methylene blue by lignin peroxidase enzyme produced from sewage sludge with *Phanerocheate chrysosporium*. *Journal of Hazardous Materials*, *162*(2-3), 708–715. doi:10.1016/j.jhazmat.2008.05.085 PMID:18599210

Alander, M., Mättö, J., Kneifel, W., Johansson, M., Kögler, B., Crittenden, R., ... Saarela, M. (2001). Effect of galacto-oligosaccharide supplementation on human faecal microflora and on survival and persistence of *Bifidobacterium lactis* Bb-12 in the gastrointestinal tract. *International Dairy Journal*, *11*(10), 817–825. doi:10.1016/S0958-6946(01)00100-5

Albersheim, P., Neukom, H., & Deuel, H. (1960). Splitting of pectin chain molecules in neutral solutions. *Archives of Biochemistry and Biophysics*, *90*(1), 46–51. doi:10.1016/0003-9861(60)90609-3 PMID:13682085

Albert, H., Davies, D. J. G., Woodson, L. P., & Soper, C. J. (1998). Biological indicators for steam sterilization: Characterization of a rapid biological indicator utilizing *Bacillus stearothermophilus* spore-associated alpha-glucosidase enzyme. *Journal of Applied Microbiology*, *85*(5), 865–874. doi:10.1046/j.1365-2672.1998.00607.x PMID:9830122

Alcalde, M., Bulter, T., Zumárraga, M., García-Arellano, H., Mencía, M., Plou, F. J., & Ballesteros, A. (2005). Screening mutant libraries of fungal laccases in the presence of organic solvents. *Journal of Biomolecular Screening*, *10*(6), 624–631. doi:10.1177/1087057105277058 PMID:16103414

Alder-Nissen, J. (1986). *Enzymatic hydrolysis of food proteins*. London: Elsevier Applied Science.

Ali, M. R., Salim Hossain, M., & Islam, M. A. (2014). *Aspect of thrombolytic therapy: a review*. Academic Press.

Ali, S., Haq, I., Qadeer, M. A., & Iqbal, J. (2002). Production of citric acid by *Aspergillus niger* using cane molasses in a stirred fermentor. *Electronic Journal of Biotechnology*, *5*(3), 258–271. doi:10.2225/vol5-issue3-fulltext-3

Aljawish, A., Chevalot, I., Jasniewski, J., Paris, C., Scher, J., & Muniglia, L. (2014b). Laccase-catalysed oxidation of ferulic acid and ethyl ferulate in aqueous medium: A green procedure for the synthesis of new compounds. *Food Chemistry*, *145*, 1046–1054. doi:10.1016/j.foodchem.2013.07.119 PMID:24128582

Aljawish, A., Chevalot, I., Jasniewski, J., Revol-Junelles, A. M., Scher, J., & Muniglia, L. (2014a). Laccase-catalysed functionalisation of chitosan by ferulic acid and ethyl ferulate: Evaluation of physicochemical and biofunctional properties. *Food Chemistry*, *161*, 279–287. doi:10.1016/j.foodchem.2014.03.076 PMID:24837951

Al-Kassim, L., Taylor, K. E., Nicell, J. A., Bewtra, J. K., & Biswas, N. (1994). Enzymatic removal of selected aromatic contaminants from wastewater by a fungal peroxidase from *Coprinus macrorhizus* in batch reactors. *Journal of Chemical Technology and Biotechnology (Oxford, Oxfordshire)*, *61*(2), 179–182. doi:10.1002/jctb.280610214 PMID:7765416

Alkorta, I., Garbisu, C., Llama, M. J., & Serra, J. L. (1998). Industrial applications of pectic enzymes: A review. *Process Biochemistry*, *33*(1), 21–28. doi:10.1016/S0032-9592(97)00046-0

Allan, R., & Roxon, J. (1974). Metabolism by intestinal bacteria: The effect of bile salts on tartrazine azo reduction. *Xenobiotica*, *4*(10), 637–643. doi:10.1080/00498257409169766 PMID:4611051

Allen, T. M. (1998). Liposomal drug formulations. Rationale for development and what we can expect for the future. *Drugs*, *56*(5), 747–756. doi:10.2165/00003495-199856050-00001 PMID:9829150

Allen, T. M., & Cullis, P. R. (2013). Liposomal drug delivery systems: From concept to clinical applications. *Advanced Drug Delivery Reviews*, *65*(1), 36–48. doi:10.1016/j.addr.2012.09.037 PMID:23036225

Alrumman, S. A. (2016). Enzymatic saccharification and fermentation of cellulosic date palm wastes to glucose and lactic acid. *Brazilian Journal of Microbiology*, *47*(1), 110–119. doi:10.1016/j.bjm.2015.11.015 PMID:26887233

Amare, G., & Gashaw, M. (1999). High-level xylanase production by an alkaliphilic *Bacillus* sp. by using solid-state fermentation. *Enzyme and Microbial Technology*, *25*(1-2), 68–72. doi:10.1016/S0141-0229(99)00006-X

Ambigaipalan, P., & Shahidi, F. (2017). Bioactive peptides form shrimp shell processing discard: Antioxidant and biological activities. *Journal of Functional Foods*, *34*, 7–17. doi:10.1016/j.jff.2017.04.013

Amin, K. A., Abdel, H. H., & Abd, E. A. H. (2010). Effect of food Azo dyes Tartrazine and Carmoisine On Biochemical Parameters Related to Renal, Hepatic Function and Oxidative Stress Biomarkers in Young Male Rats. *Food and Chemical Toxicology*, *48*(10), 2994–2999. doi:10.1016/j.fct.2010.07.039 PMID:20678534

Aminov, R. I. (2010). A brief history of the antibiotic era: Lessons learned and challenges for the future. *Frontiers in Microbiology*, *1*, 134. doi:10.3389/fmicb.2010.00134 PMID:21687759

Anastas, P. T., Kirchhoff, M. M., & Williamson, T. C. (2001). Catalysis as a foundational pillar of green chemistry. *Applied Catalysis A, General*, *221*(1-2), 3–13. doi:10.1016/S0926-860X(01)00793-1

Anastas, P. T., & Warner, J. C. (1998). Principles of green chemistry. *Green chemistry. Theory into Practice*, 29–56.

Andersen, L. D., & Dalbøge, H. D. (1999). Future Trend of Feed Enzymes. *Proceedings of the Roche-American Soybean Association Joint Seminar.*

An, H., Park, H., & Kim, E. (2001). Cloning and expression of thermophilic catechol 1,2 dioxygenase gene (cat A) from *Streptomyces setonii. FEMS Microbiology Letters*, *195*(1), 17–22. doi:10.1111/j.1574-6968.2001.tb10491.x PMID:11166989

Annabi, N., Mithieux, S. M., Camci-Unal, G., Dokmeci, M. R., Weiss, A. S., & Khademhosseini, A. (2013). Elastomeric recombinant protein-based biomaterials. *Biochemical Engineering Journal*, *77*, 110–118. doi:10.1016/j.bej.2013.05.006 PMID:23935392

Annamalai, N., Thavasi, R., Jayalakshmi, S., & Balasubramanian, T. (2009). Production and optimization of cellulase-free, alkali-stable xylanase by *Bacillus pumilus* SV-85S in submerged fermentation. *Indian Journal of Biotechnology*, *8*, 291–297.

Ansari, S. A., & Husain, Q. (2012). Potential applications of enzymes immobilized on/in nano materials: A review. *Biotechnology Advances*, *30*(3), 512–523. doi:10.1016/j.biotechadv.2011.09.005 PMID:21963605

Anthonsen, H. W., Baptista, A., Drablos, F., Martel, P., & Petersen, S. B. (1994). The blind watchmaker and rational protein engineering. *Journal of Biotechnology*, *36*(3), 185–220. doi:10.1016/0168-1656(94)90152-X PMID:7765263

Anthony, T. K., Chandra, R., Rajendran, A., & Gunasekaran, P. (2003). High molecular weight cellulase-free xylanase from alkali-tolerant *Aspergillus fumigatus* AR1. *Enzyme and Microbial Technology*, *32*(6), 647–654. doi:10.1016/S0141-0229(03)00050-4

Antikainen, N. M., & Martin, S. F. (2005). Altering protein specificity: Techniques and applications. *Bioorganic & Medicinal Chemistry*, *13*(8), 2701–2716. doi:10.1016/j.bmc.2005.01.059 PMID:15781382

Anuradha, P., Vijayalakshmi, K. N. D. P., & Sridevi, K. (2007). Production and properties of alkaline xylanase from *Bacillus* sp. isolated from sugarcane fields. *Current Science*, *92*, 1283–1286.

Anyanwutaku, I. O., Petroski, R. J., & Rosazza, J. P. (1994). Oxidative coupling of mithramycin and hydroquinone catalyzed by copper oxidases and benzoquinone.Implications for the mechanism of action of aureolic acid antibiotics. *Bioorganic & Medicinal Chemistry*, *2*(6), 543–551. doi:10.1016/0968-0896(94)80025-1 PMID:8000877

AOAC. (1990). *Association of the Official Analytical Chemists* (K. Elrich, Ed.). Washington, DC: AOAC.

Ara, K., Saeki, K., Igarashi, K., Takaiwa, M., Uemura, T., Hagihara, H., ... Ito, S. (1995). Purification and characterization of an alkaline amylopullulanase with both α-1,4 and α-1,6 hydrolytic activity from alkalophilic *Bacillus sp.* KSM- 1378. *Biochimica et Biophysica Acta*, *1243*(3), 315–324. doi:10.1016/0304-4165(94)00148-Q PMID:7727505

Araujo, R., Fernandes, M., Cavaco-Paulo, A., & Gomes, A. (2011). Biology of human hair: Know your hair to control it. *Advances in Biochemical Engineering/Biotechnology*, *13*, 121–143. PMID:21072698

Archana, A., & Satyanarayan, T. (1997). Xylanase production by thermophilic *Bacillus licheniformis* A99 in solid state fermentation. *Enzyme and Microbial Technology*, *21*(1), 12–17. doi:10.1016/S0141-0229(96)00207-4

Ariga, K., Ji, Q., Mori, T., Naito, M., Yamauchi, Y., Abe, H., & Hill, J. P. (2013). Enzyme nanoarchitectonics: Organization and device application. *Chemical Society Reviews*, *42*(15), 6322–6345. doi:10.1039/c2cs35475f PMID:23348617

Arnold, F. H. (1993). Engineering proteins for non-natural environments. *The FASEB Journal*, *7*(9), 744–749. doi:10.1096/fasebj.7.9.8330682 PMID:8330682

Aroclor. (1956). Advertisement by Monsanto Chemical Company. *Chem. Eng.*, 386.

Arora, D. S., & Sharma, R. K. (2010). Ligninolytic fungal laccases and their biotechnological applications. *Applied Biochemistry and Biotechnology*, *160*(6), 1760–1788. doi:10.100712010-009-8676-y PMID:19513857

Arora, P. K., Kumar, M., Chauhan, A., Raghava, G. P., & Jain, R. K. (2009). OxDBase: A database of oxygenases involved in biodegradation. *BMC Research Notes*, *2*(1), 67. doi:10.1186/1756-0500-2-67 PMID:19405962

Arthur, W. H., & Elmer, L. (1950). Oxygen transfer in submerged fermentation. *Industrial & Engineering Chemistry Research*, *42*(9), 1792–1801. doi:10.1021/ie50489a031

Artik, N., Karhan, M., & Aydar, G. (2004). Effects of polyphenoloxidase (LACCASE) application on clarity stability of sour cherry juice. *Journal of Food Technology*, *2*(4), 237–243.

Asadgol, Z., Forootanfar, H., Rezaei, S., Mahvi, A. H., & Faramarzi, M. A. (2014). Removal of phenol and bisphenol-A catalyzed by laccase in aqueous solution. *Journal of Environmental Health Science & Engineering*, *12*(1), 93. doi:10.1186/2052-336X-12-93 PMID:25031840

Asano, K., Sriprang, R., Gobsuk, J., Eurwilaichitr, L., Tanapongpipat, S., & Kirtikara, K. (2005). Endo-1,4-β-xylanase B from *Aspergillusniger* BCC14405 isolated in Thailand: Purification, characterization and gene isolation. *Journal of Biochemistry and Molecular Biology*, *38*(1), 17–23. PMID:15715941

Ashford, M., Fell, J., Attwood, D., Sharma, H., & Woodhead, P. (1994). Studies on pectin formulations for colonic drug delivery. *Journal of Controlled Release*, *30*(3), 225–232. doi:10.1016/0168-3659(94)90028-0

Ashrafi, S. D., Nasseri, S., Alimohammadi, M., Mahvi, A. H., & Faramarzi, M. A. (2015). Optimization of the enzymatic elimination of flumequine by laccase-mediated system using response surface methodology. *Desalination and Water Treatment*, *57*(31), 14478–14487. doi:10.1080/19443994.2015.1063462

Astolfi, P., Brandi, P., Galli, C., Gentili, P., Gerini, M. F., Greci, L., & Lanzalunga, O. (2005). New mediators for the enzyme laccase: Mechanistic features and selectivity in the oxidation of non-phenolic substrates. *New Journal of Chemistry, 29*(10), 1308–1317. doi:10.1039/b507657a

Atlow, S.C., Bonadonna-Aparo, L., & Klibanov, A.M. (1984). Dephenolization of industrial wastewaters catalyzed by polyphenol oxidase. *Biotechnology and Bioengineering, 26*, 599-603. DOI: 10.1002/bit.260260607

Aubel, D., Morris, R., Lennon, B., Rimann, M., Kaufmann, H., Folcher, M., ... Fussenegger, M. (2001). Design of a novel mammalian screening system for the detection of bioavailable, non-cytotoxic streptogramin antibiotics. *The Journal of Antibiotics, 54*(1), 44–55. doi:10.7164/antibiotics.54.44 PMID:11269714

Auriol, M., Filali-Meknassi, Y., Adams, C. D., & Tyagi, R. D. (2006). Natural and synthetic hormone removal using the horseradish peroxidase enzyme: Temperature and pH effects. *Water Research, 40*(15), 2847–2856. doi:10.1016/j.watres.2006.05.032 PMID:16849026

Auriol, M., Filali-Meknassi, Y., Adams, C. D., Tyagi, R. D., Noguerol, T.-N., & Piña, B. (2008). Removal of estrogenic activity of natural and synthetic hormones from a municipal wastewater: Efficiency of horseradish peroxidase and laccase from *Trametes versicolor. Chemosphere, 70*(3), 445–452. doi:10.1016/j.chemosphere.2007.06.064 PMID:17897698

Auriol, M., Filali-Meknassi, Y., Tyagi, R. D., & Adams, C. D. (2007). Laccase-catalyzed conversion of natural and synthetic hormones from a municipal wastewater. *Water Research, 41*(15), 3281–3288. doi:10.1016/j.watres.2007.05.008 PMID:17585984

Auriol, M., Filali-Meknassi, Y., Tyagi, R. D., Adams, C. D., & Surampalli, R. Y. (2006). Endocrine disrupting compounds removal from wastewater, a new challenge. *Process Biochemistry, 41*(3), 525–539. doi:10.1016/j.procbio.2005.09.017

Aust, S. D. (1990). Degradation of environmental pollutants by *Phanerochaete chrysosporium. Microbial Ecology, 20*(1), 197–209. doi:10.1007/BF02543877 PMID:24193974

Aviezer, D., Brill-Almon, E., Shaaltiel, Y., Hashmueli, S., Bartfeld, D., Mizrachi, S., & Galun, E. (2009). A plant-derived recombinant human glucocerebrosidase enzyme--a preclinical and phase I investigation. *PLoS One, 4*(3), e4792. doi:10.1371/journal.pone.0004792 PMID:19277123

Awad, D., Bartok, M., Mostaghimi, F., Schrader, I., Sudumbrekar, N., Schaffran, T., ... Gabel, D. (2015). Halogenated dodecaborate clusters as agents to trigger release of liposomal contents. *ChemPlusChem, 80*(4), 656–664. doi:10.1002/cplu.201402286

Ayala, M., Roman, R., & Vazquez-Duhalt, R. (2007). A catalytic approach to estimate the redox potential of heme-peroxidases. *Biochemical and Biophysical Research Communications, 357*(3), 804–808. doi:10.1016/j.bbrc.2007.04.020 PMID:17442271

Ayanda & Petrik. (2014). Nanotechnology: The Breakthrough in Water and Wastewater Treatment. *International Journal of Chemical Marine Environmental Research, 1*(1), 1–2.

Ayyub, O. B., & Kofinas, P. (2015). Enzyme induced stiffening of nanoparticle–hydrogel composites with structural color. *ACS Nano, 9*(8), 8004–8011. doi:10.1021/acsnano.5b01514 PMID:26196060

Azarikia, F., Wu, B. C., Abbasi, S., & McClements, D. J. (2015). Stabilization of biopolymer microgels formed by electrostatic complexation: Influence of enzyme (laccase) cross-linking on pH, thermal, and mechanical stability. *Food Research International, 78*, 18–26. doi:10.1016/j.foodres.2015.11.013 PMID:28433280

Azimi, M., Nafissi-Varcheh, N., Mogharabi, M., Faramarzi, M. A., & Aboofazeli, R. (2016). Study of laccase activity and stability in the presence of ionic and non-ionic surfactants and the bioconversion of indole in laccase-TX-100 system. *Journal of Molecular Catalysis. B, Enzymatic*, *126*, 69–75. doi:10.1016/j.molcatb.2016.02.001

Azin, M., Moravej, R., & Zareh, D. (2007). Production of xylanase by *Trichoderma longibrachiatum* on a mixture of wheat bran and wheat straw: Optimization of culture condition by Taguchi method. *Enzyme and Microbial Technology*, *40*(4), 801–805. doi:10.1016/j.enzmictec.2006.06.013

Azmir, J., Zaidul, I. S. M., Rahman, M. M., Sharif, K. M., Mohamed, A., Sahena, F., ... Omar, A. K. M. (2013). Techniques for extraction of bioactive compounds from plant materials: A review. *Journal of Food Engineering*, *117*(4), 426–436. doi:10.1016/j.jfoodeng.2013.01.014

Baba, M., Shigeta, S., Yuasa, S., Takashima, H., Sekiya, K., Ubasawa, M., ... De Clercq, E. (1994). Preclinical evaluation of MKC-442, a highly potent and specific inhibitor of human immunodeficiency virus type 1 in vitro. *Antimicrobial Agents and Chemotherapy*, *38*(4), 688–692. doi:10.1128/AAC.38.4.688 PMID:7518216

Baba, M., Tanaka, H., De Clercq, E., Pauwels, R., Balzarini, J., Schols, D., ... Miyasaka, T. (1989). Highly specific inhibition of human immunodeficiency virus type 1 by a novel 6-substituted acyclouridine derivative. *Biochemical and Biophysical Research Communications*, *165*(3), 1375–1381. doi:10.1016/0006-291X(89)92756-3 PMID:2575380

Babior, B. M. (1999). NADPH oxidase: An update. *Blood*, *93*(5), 1464–1476. PMID:10029572

Bagshawe, K. D. (1987). Antibody directed enzymes revive anti-cancer prodrugs concept. *British Journal of Cancer*, *56*(5), 531–532. doi:10.1038/bjc.1987.237 PMID:3426915

Bailey, M. J., Biely, P., & Poutanen, K. (1992). Interlaboratory testing of methods for assay of xylanase activity. *Journal of Biotechnology*, *23*(3), 257–270. doi:10.1016/0168-1656(92)90074-J

Baiocco, P., Barreca, A. M., Fabbrini, M., Galli, C., & Gentili, P. (2003). Promoting laccase activity towards non-phenolic substrates: A mechanistic investigation with some laccase–mediator systems. *Organic & Biomolecular Chemistry*, *1*(1), 191–197. doi:10.1039/B208951C PMID:12929410

Bajaj, P. (2002). Finishing of textile materials. *Journal of Applied Polymer Science*, *83*(3), 631–659. doi:10.1002/app.2262

Bakir, U., Yavascaoglu, S., Guvenc, F., & Ersayin, A. (2001). An endo-β-1,4-xylanase from *Rhizopus oryzae*: Production, partial purification and biochemical characterisation. *Enzyme and Microbial Technology*, *29*(6-7), 328–334. doi:10.1016/S0141-0229(01)00379-9

Bakker, B. M., Michels, P. A. M., Opperdoes, F. R., & Westerhoff, H. V. (1999). What controls glycolysis in bloodstream form Trypanosoma brucei? *The Journal of Biological Chemistry*, *274*(21), 14551–14559. doi:10.1074/jbc.274.21.14551 PMID:10329645

Bakker, B. M., Walsh, M. C., Ter Kuile, B. H., Mensonides, F. I. C., Michels, P. A. M., Opperdoes, F. R., & Westerhoff, H. V. (1999). Contribution of glucose transport to the control of the glycolytic flux in Trypanosoma brucei. *Proceedings of the National Academy of Sciences of the United States of America*, *96*(18), 10098–10103. doi:10.1073/pnas.96.18.10098 PMID:10468568

Bakri, Al-Jazairi, M., & Al-Kayat, G. (2008). Xylanase production by a newly isolated *Aspergillus niger* SS7 in submerged culture. *Polish Journal of Microbiology*, *57*(3), 249–251. PMID:19004246

Balamurugan, R., Sundarrajan, S., & Ramakrishna, S. (2011). Recent trends in nanofibrous membranes and their suitability for air and water filtrations. *Membranes*, *1*(3), 232–248. doi:10.3390/membranes1030232 PMID:24957734

Baldo, B. A. (2015). Enzymes approved for human therapy: Indications, mechanisms and adverse effects. *BioDrugs*, *29*(1), 31–55. doi:10.100740259-015-0116-7 PMID:25648140

Baldo, B. A. (2016). Enzymes Approved for Therapy. In B. A. Baldo (Ed.), *Safety of Biologics Therapy: Monoclonal Antibodies, Cytokines, Fusion Proteins, Hormones, Enzymes, Coagulation Proteins, Vaccines, Botulinum Toxins* (pp. 419–477). Cham: Springer International Publishing. doi:10.1007/978-3-319-30472-4_9

Baldrian, P. (2006). Fungal laccases-occurrence and properties. *FEMS Microbiology Reviews*, *30*(2), 215–242. doi:10.1111/j.1574-4976.2005.00010.x PMID:16472305

Baldrian, P., Valášková, V., Merhautová, V., & Gabriel, J. (2005). Degradation of lignocellulose by *Pleurotus ostreatus* in the presence of copper, manganese, lead and zinc. *Research in Microbiology*, *156*(5-6), 670–676. doi:10.1016/j.resmic.2005.03.007 PMID:15921894

Baldwin, E. A., Hagenmaier, R., & Bai, J. (Eds.). (2011). *Edible coatings and films to improve food quality*. CRC Press.

Baltimore, D. (1970). Rna-dependent dna polymerase¡ n virions of rous sarcoma virus. *Nature*, *226*(121), 1. PMID:16057031

Balti, R., Bougatef, A., El-Hadj, A. N., Zekri, D., Barkia, A., & Nasri, M. (2010). Influence of degree of hydrolysis on functional properties and angiotensin I-converting enzyme inhibitory activity of protein hydrolysates from cuttlefish (*Sepia officinalis*) by-products. *Journal of the Science of Food and Agriculture*, *90*, 2006–2014. PMID:20583200

Bandhan, A. K., Chadha, B. S., Kaur, J., Saini, H. S., & Bhat, M. K. (2007). Production of xylanolytic and cellulolytic enzymes by thermophilic fungus *Myceliophthora* sp. IMI 387009. *Bioresource Technology*, *98*(3), 504–510. doi:10.1016/j.biortech.2006.02.009 PMID:16600593

Banerjee, J., Singh, R., Vijayraghvan, R., MacFarlane, D., Ptti, A. F., & Arora, A. (2017). Bioactives from fruit processing wastes: Green approaches to valuable chemicals. *Food Chemistry*, *225*, 10–22. doi:10.1016/j.foodchem.2016.12.093 PMID:28193402

Bangham, A. D., Standish, M. M., & Watkins, J. C. (1965). Diffusion of univalent ions across the lamellae of swollen phospholipids. *Journal of Molecular Biology*, *13*(1), 238–252. doi:10.1016/S0022-2836(65)80093-6 PMID:5859039

Bankovic-Lilic, I. B., Stojkovic, I. J., Stamenkovic, O. S., & Veljkovic, V. B. (2014). Waste animal fats as feedstocks for biodiesel production. *Renewable & Sustainable Energy Reviews*, *32*, 238–254. doi:10.1016/j.rser.2014.01.038

Bansal, N., & Kanwar, S. S. (2013). Peroxidase(s) in Environment Protection. *The Scientific World Journal*, *2013*, 1–9. doi:10.1155/2013/714639 PMID:24453894

Barilli, A., Belinghieri, F., Passarella, D., Lesma, G., Riva, S., Silvani, A., & Danieli, B. (2004). Enzyme assisted enantioselective synthesis of the alkaloid (+)-aloperine. *Tetrahedron, Asymmetry*, *15*(18), 2921–2925. doi:10.1016/j.tetasy.2004.06.056

Barnard, M. L., Baker, R. R., & Matalon, S. (1993). Mitigation of oxidant injury to lung microvasculature by intratracheal instillation of antioxidant enzymes. *The American Journal of Physiology*, *265*(4 Pt 1), L340–L345. PMID:8238368

Barneto, A. G., Aracri, E., Andreu, G., & Vidal, T. (2012). Investigating the structure–effect relationships of various natural phenols used as laccase mediators in the biobleaching of kenaf and sisal pulps. *Bioresource Technology*, *112*, 327–335. doi:10.1016/j.biortech.2012.02.136 PMID:22437048

Barreca, A. M., Fabbrini, M., Galli, C., Gentili, P., & Ljunggren, S. (2003). Laccase/mediated oxidation of a lignin model for improved delignification procedures. *Journal of Molecular Catalysis. B, Enzymatic*, *26*(1-2), 105–110. doi:10.1016/j.molcatb.2003.08.001

Barros, F.F.C., Simiqueli, A.P.R., Andrade, C. J., & Pastore, G. M. (2013). Production of enzymes from agroindustrial wastes by biosurfactant-producing strains of *Bacillus subtilis*. *Biotechnology Research Journal*, 1-9.

Barr-Sinoussi, F., Chermann, J. C., Rey, F., Nugeyre, M. T., & Chamaret, S. (1983). Isolation of a T-lymphotropic retrovirus from a patient at risk of acquired immune deficiency syndrome (AIDS). *Science*, *220*, 868. doi:10.1126cience.6189183 PMID:6189183

Barth, R. F., Soloway, A. H., Fairchild, R. G., & Brugger, R. M. (1992). Boron neutron capture therapy for cancer. Realities and prospects. *Cancer*, *70*(12), 2995–3007. doi: PMID:1451084

Barton, N. W., Furbish, F. S., Murray, G. J., Garfield, M., & Brady, R. O. (1990). Therapeutic response to intravenous infusions of glucocerebrosidase in a patient with Gaucher disease. *Proceedings of the National Academy of Sciences of the United States of America*, *87*(5), 1913–1916. doi:10.1073/pnas.87.5.1913 PMID:2308952

Ba, S., Haroune, L., Cruz-Morató, C., Jacquet, C., Touahar, I. E., Bellenger, J.-P., & Cabana, H. (2014). Synthesis and characterization of combined cross-linked laccase and tyrosinase aggregates transforming acetaminophen as a model phenolic compound in wastewaters. *The Science of the Total Environment*, *487*, 748–755. doi:10.1016/j.scitotenv.2013.10.004 PMID:24867811

Ba, S., Jones, J. P., & Cabana, H. (2014). Hybrid bioreactor (HBR) of hollow fiber microfilter membrane and cross-linked laccase aggregates eliminate aromatic pharmaceuticals in wastewaters. *Journal of Hazardous Materials*, *280*, 662–670. doi:10.1016/j.jhazmat.2014.08.062 PMID:25218263

Bassetto, M., Massarotti, A., Coluccia, A., & Brancale, A. (2016). Structural biology in antiviral drug discovery. *Current Opinion in Pharmacology*, *30*, 116–130. doi:10.1016/j.coph.2016.08.014 PMID:27611878

Bauer, C. G., Kuhn, A., Gajovic, N., Skorobogatko, O., Holt, P. J., Bruce, N. C., ... Scheller, F. W. (1999). New enzyme sensors for morphine and codeine based on morphine dehydrogenase and laccase. *Fresenius' Journal of Analytical Chemistry*, *364*(1-2), 179–183. doi:10.1007002160051320

Bax, B. E., Bain, M. D., Fairbanks, L. D., Webster, A. D., & Chalmers, R. A. (2000). In vitro and in vivo studies with human carrier erythrocytes loaded with polyethylene glycol-conjugated and native adenosine deaminase. *British Journal of Haematology*, *109*(3), 549–554. doi:10.1046/j.1365-2141.2000.02059.x PMID:10886203

Becker, D., Rodriguez-Mozaz, S., Insa, S., Schoevaart, R., Barceló, D., de Cazes, M., ... Wagner, M. (2017). Removal of endocrine disrupting chemicals in wastewater by enzymatic treatment with fungal laccases. *Organic Process Research & Development*, *21*(4), 480–491. doi:10.1021/acs.oprd.6b00361

Bedard, K., & Krause, K.-H. (2007). The NOX family of ROS-generating NADPH oxidases: Physiology and pathophysiology. *Physiological Reviews*, *87*(1), 245–313. doi:10.1152/physrev.00044.2005 PMID:17237347

Beede, D. N., & Bloom, D. E. (1995). The Economics of Municipal Solid Waste. *The World Bank Research Observer*, *10*(2), 113–150. doi:10.1093/wbro/10.2.113

Begum, M. F., & Alimon, A. R. (2011). Bioconversion and saccharification of some lignocellulosic wastes by *Aspergillus oryzae* ITCC-4857.01 for fermentable sugar production. *Electronic Journal of Biotechnology*, *14*(5), 1–9. doi:10.2225/vol14-issue5-fulltext-3

Belancic, A., Scarpa, J., Peirano, A., Diaz, R., Steiner, J., & Eyzayuirre, J. (1995). *Penicillium purpurogenum* produces several xylanases: Purification and properties of two of the enzymes. *Journal of Biotechnology*, *41*(1), 71–79. doi:10.1016/0168-1656(95)00057-W PMID:7640003

Bendary, E., Francis, R. R., Ali, H. M. G., Sarwat, M. I., & El Hady, S. (2013). Antioxidant and structure–activity relationships (SARs) of some phenolic and anilines compounds. *Annals of Agricultural Science*, *58*(2), 173–181. doi:10.1016/j.aoas.2013.07.002

Bento, I., Silva, C. S., Chen, Z., Martins, L. O., Lindley, P. F., & Soares, C. M. (2010). Mechanisms underlying dioxygen reduction in laccases. Structural and modelling studies focusing on proton transfer. *BMC Structural Biology*, *10*(1), 28. doi:10.1186/1472-6807-10-28 PMID:20822511

Bergmans, M. E. F., Beldman, G., Gruppen, H., & Voragen, A. G. J. (1996). Optimisation of the selective extraction of (glucurono) arabinoxylans from wheat bran: Use of barium and calcium hydroxide solution at elevated temperatures. *Journal of Cereal Science*, *23*(3), 235–245. doi:10.1006/jcrs.1996.0024

Berlanga-Reyes, C. M., Carvajal-Millán, E., Juvera, G. C., Rascón-Chu, A., Marquez-Escalante, J. A., & Martínez-López, A. L. (2009). Laccase induced maize bran arabinoxylan gels. *Food Science and Biotechnology*, *18*(4), 1027–1029.

Berthelot, A. (1911). Recherches sur la flore intestinale. Isolement des microbes qui attaquent specialement les produits ultimes de la digestion proteiques. *Compt. rend. Acad. d. sc, 153*, 306.

Bertoldo, C., & Antranikian, G. (2002). Starch-hydrolyzing enzymes from thermophilic archaea and bacteria. *Current Opinion in Chemical Biology*, *6*(2), 151–160. doi:10.1016/S1367-5931(02)00311-3 PMID:12038998

Bertrand, G. (1894). Sur le latex de l'arbre à laque. *Comptes Rendus Hebdomadaires des Séances de l'Académie des Sciences*, *118*, 1215–1218.

Bertrand, G. (1985). Sur la laccase et sur le pouvoir oxydant de cette diastase. *Comp Rendus L'Acad Sci.*, *120*, 266–269.

Bertrand, T., Jolivalt, C., Briozzo, P., Caminade, E., Joly, N., Madzak, C., & Mougin, C. (2002). Crystal structure of a four-copper laccase complexed with an arylamine: Insights into substrate recognition and correlation with kinetics. *Biochemistry*, *41*(23), 7325–7333. doi:10.1021/bi0201318 PMID:12044164

Betancor, L., Johnson, G. R., & Luckarift, H. R. (2013). Stabilized laccases as heterogeneous bioelectrocatalysts. *ChemCatChem*, *5*(1), 46–60. doi:10.1002/cctc.201200611

Bevilaqua, J. V., Cammarota, M. C., Freire, D. M. G., & Sant'Anna, G. L. Jr. (2002). Phenol removal through combined biological and enzymatic treatments. *Brazilian Journal of Chemical Engineering*, *19*(2), 151–158. doi:10.1590/S0104-66322002000200010

Bharathiraja, S., Suriya, J., Krishanan, M., Manivasagan, P., & Kim, S. K. (2017). Chapter Six - Production of Enzymes From Agricultural Wastes and Their Potential Industrial Applications. *Advances in Food and Nutrition Research*, *80*, 125–148. doi:10.1016/bs.afnr.2016.11.003 PMID:28215322

Bhat, M. K., & Hazlewood, G. P. (2001). Enzymology and other characteristics of cellulases and xylanases. In M. R. Bedford & G. G. Partridge (Eds.), Enzymes in Farm Animal Nutrition. Academic Press.

Bhatnagar, D., & Imelda-Joseph, P. R. R. (2010). Amylase and acid protease production by solid state fermentation using *Aspergillus niger* from mangrove swamp. *Indian Journal of Fisheries*, *57*(1), 45–51.

Bhattacharya, A., Sood, P., & Citovsky, V. (2010). The roles of plant phenolics in defence and communication during *Agrobacterium* and *Rhizobium* infection. *Molecular Plant Pathology*, *11*, 705–719. PMID:20696007

Bibi, I., Bhatti, H. N., & Asgher, M. (2011). Comparative study of natural and synthetic phenolic compounds as efficient laccase mediators for the transformation of cationic dye. *Biochemical Engineering Journal*, *56*(3), 225–231. doi:10.1016/j.bej.2011.07.002

Bilal, M., Asgher, M., Iqbal, H. M. N., Hu, H., & Zhang, X. (2017). Biotransformation of lignocellulosic materials into value-added prodcuts-A review. *International Journal of Biological Macromolecules*, *98*, 447–458. doi:10.1016/j.ijbiomac.2017.01.133 PMID:28163129

Blanco, D. C., Gonza'lez, D. M., Monmany, D. M., & Tzanov, T. (2009). Dyeing properties, synthesis, isolation and characterization of an in situ generated phenolic pigment, covalently bound to cotton. *Enzyme and Microbial Technology*, *44*(6-7), 380–385. doi:10.1016/j.enzmictec.2009.02.003

Blánquez, P., & Guieysse, B. (2008). Continuous biodegradation of 17β-estradiol and 17α-ethynylestradiol by *Trametes versicolor. Journal of Hazardous Materials*, *150*(2), 459–462. doi:10.1016/j.jhazmat.2007.09.085 PMID:17996366

Blume, G., & Cevc, G. (1990). Liposomes for the sustained drug release in vivo. *Biochimica et Biophysica Acta*, *1029*(1), 91–97. doi:10.1016/0005-2736(90)90440-Y PMID:2223816

Blume-Jensen, P., & Hunter, T. (2001). Oncogenic kinase signalling. *Nature*, *411*(, 6835), 355–365. doi:10.1038/35077225 PMID:11357143

Bocchini, D. A., Oliveira, O. M. M. F., Gomes, E., & Silva, R. (2005). Use of sugarcane bagasse and grass hydrolysates as carbon sources for xylanase production by *Bacillus circulans* D1 in submerged fermentation. *Process Biochemistry*, *40*(12), 3653–3659. doi:10.1016/j.procbio.2005.02.021

Bocharova, V., Sharp, D., Jones, A., Cheng, S., Griffin, P. J., Agapov, A. L., ... Sokolov, A. P. (2015). Enzyme induced formation of monodisperse hydrogel nanoparticles tunable in size. *Chemistry of Materials*, *27*(7), 2557–2565. doi:10.1021/acs.chemmater.5b00187

Boesch, F. H., Madinda, C., Robbins, N. F., Gray, M. M., Cournil, M., Ooms, A., ... Peeters, M. (2015). Origin of the HIV-1 group O epidemic in western lowland gorillas. *Proceedings of the National Academy of Sciences of the United States of America*, *112*(11), E1343–E1352. doi:10.1073/pnas.1502022112 PMID:25733890

Bollag, J. M., & Dec, J. (1998). *Use of Plant material for the removal of pollutants by polymerization and binding to humic substances*. Tech. Rep. R-82092. Center for Bioremediation and Detoxification Environmental Resources Research Institute the Pennsylvania State University.

Bollag, J. M. (1992). Decontaminating soil with enzymes. *Environmental Science & Technology*, *26*(10), 1876–1881. doi:10.1021/es00034a002

Bollag, J. M. (1992). Enzymes catalyzing oxidative coupling reactions of pollutants. *Metal Ions in Biological Systems*, *28*, 205–217.

Bollag, J. M., Shuttleworth, K. L., & Anderson, D. H. (1988). Laccase-mediated detoxication of phenolic compounds. *Applied and Environmental Microbiology*, *54*, 3086–3091. PMID:3223771

Botes, M., & Eugene Cloete, T. (2010). The potential of nanofibers and nanobiocides in water purification. *Critical Reviews in Microbiology*, *36*(1), 68–81. doi:10.3109/10408410903397332 PMID:20088684

Bourbonnais, R., & Paice, M. G. (1990). Oxidation of non-phenolic substrates: An expanded role for laccase in lignin biodegradation. *FEBS Letters*, *267*(1), 99–102. doi:10.1016/0014-5793(90)80298-W PMID:2365094

Bourne, H. R., Sanders, D. A., & McCormick, F. (1990). The GTPase superfamily: A conserved switch for diverse cell functions. *Nature*, *348*(6297), 125–132. doi:10.1038/348125a0 PMID:2122258

Bourne, H. R., Sanders, D. A., & McCormick, F. (1991). The GTPase superfamily: Conserved structure and molecular mechanism. *Nature*, *349*(6305), 117–127. doi:10.1038/349117a0 PMID:1898771

Bouwens, E. M., Trivedi, K., Vliet, C. V., & Winkel, C. (1997). *Method of enhancing color in a tea based foodstuff European patent application.* EP 760213 A1.

Bouwens, E.M., Trivedi, K., Vliet, C.V., & Winkel, C. (1999). *Method of enhancing color in a tea based foodstuff.* US5879730 A.

Božič, M., Gorgieva, S., & Kokol, V. (2012a). Homogeneous and heterogeneous methods for laccase-mediated functionalization of chitosan by tannic acid and quercetin. *Carbohydrate Polymers, 89*(3), 854–864. doi:10.1016/j.carbpol.2012.04.021 PMID:24750872

Božič, M., Gorgieva, S., & Kokol, V. (2012b). Laccase-mediated functionalization of chitosan by caffeic and gallic acids for modulating antioxidant and antimicrobial properties. *Carbohydrate Polymers, 87*(4), 2388–2398. doi:10.1016/j.carbpol.2011.11.006

Božič, M., Štrancar, J., & Kokol, V. (2013). Laccase-initiated reaction between phenolic acids and chitosan. *Reactive & Functional Polymers, 73*(10), 1377–1383. doi:10.1016/j.reactfunctpolym.2013.01.005

Božić, T., Novaković, I., Gašić, M. J., Juranić, Z., Stanojković, T., Tufegdžić, S., ... Sladić, D. (2010). Synthesis and biological activity of derivatives of the marine quinone avarone. *European Journal of Medicinal Chemistry, 45*(3), 923–929. doi:10.1016/j.ejmech.2009.11.033 PMID:19995673

Brausch, J. M., Connors, K. A., Brooks, B. W., & Rand, G. M. (2012). Human pharmaceuticals in the aquatic environment: A review of recent toxicological studies and considerations for toxicity testing. *Reviews of Environmental Contamination and Toxicology, 218*, 1–99. doi:10.1007/978-1-4614-3137-4_1 PMID:22488604

Brayfield, A. (Ed.). (2014). *"Indometacin". Martindale: The Complete Drug Reference.* London, UK: Pharmaceutical Press.

Brenna, O., & Bianchi, E. (1994). Immobilized laccase for phenolic removal in must and wine. *Biotechnology Letters, 16*(1), 35–40. doi:10.1007/BF01022620

Bresnick, A. R. (2013). Chemical agents for the prevention of inhibition or tumor metastasis *U.S. Patent US20130090355A1.*

Breuil, C., & Saddler, J. N. (1985). Comparison of the 3,5-dinitrosalicylic acid and Nelson-Somogyi methods of assaying for reducing sugars and determining cellulase activity. *Enzyme and Microbial Technology, 7*(7), 327–332. doi:10.1016/0141-0229(85)90111-5

Brienzo, M., Siqueira, A. F., & Milagres, A. M. F. (2009). Search for optimum conditions of sugarcane bagasse hemicelluloses extraction. *Biochemical Engineering Journal, 46*(2), 199–204. doi:10.1016/j.bej.2009.05.012

Brijwani, K., Oberoi, H. S., & Vadlani, P. V. (2010). Production of a cellulolytic enzyme system in mixed-culture solid-state fermentation of soybean hulls supplemented with wheat bran. *Process Biochemistry, 45*(1), 120–128. doi:10.1016/j.procbio.2009.08.015

Brijwani, K., & Vadlani, P. V. (2010). *Solid state fermentation of soybean hulls for cellulolytic enzymes production. In Soybean: Application and Technology* (pp. 305–322). InTech.

Brinton, C., Wichmann, H., Willaman, J., Wilson, C., & Dore, W. (1927). Definitions written by the Committee on Nomenclature of Pectin of the Agriculture-Food Division. *Journal of the American Chemical Society, 49*, 38–40.

Brittes, J., Lúcio, M., Nunes, C., Lima, J. L. F. C., & Reis, S. (2010). Effects of resveratrol on membrane biophysical properties: Relevance for its pharmacological effects. *Chemistry and Physics of Lipids, 163*(8), 747–754. doi:10.1016/j.chemphyslip.2010.07.004 PMID:20691168

Brosnan, M. P., Kelly, C. T., & Fogarty, W. M. (1992). Investigation of the mechanisms of irreversible thermoinactivation of *Bacillus stearothermophilus* α-amylase. *European Journal of Biochemistry*, *203*(1-2), 225–231. doi:10.1111/j.1432-1033.1992.tb19850.x PMID:1730228

Brown, J.E., Partlow, B.P., Berman, A.M., House, M.D., & Kaplan, D.L. (2016). Injectable silk-based biomaterials for cervical tissue augmentation: An in vitro study. *American Journal of Obstetrics and Gynecology, 214*(1), 118. e1- e9.

Brumm, P. J., & Teague, W. M. (1989). Effect of additives on the thermostability of *Bacillus stearothermophilus* α-amylase. *Biotechnology Letters*, *11*(8), 541–544. doi:10.1007/BF01040032

Buchdunger, E., Matter, A., & Druker, B. J. (2001). Bcr-Abl inhibition as a modality of CML therapeutics. *Biochimica et Biophysica Acta (BBA)-. Revue Canadienne*, *1551*(1), M11–M18.

Bugg, T. D. H., Ahmad, M., Hardiman, E. M., & Rahmanpour, R. (2011). Pathways for degradation of lignin in bacteria and fungi. *Natural Product Reports*, *28*(12), 1883–1896. doi:10.1039/c1np00042j PMID:21918777

Burhan, A., Nisa, U., Gokhan, C., Omer, C., Ashabil, A., & Osman, G. (2003). Enzymatic properties of a novel thermostable, thermophilic, alkaline and chelator resistant amylase from an alkaliphilic *Bacillus* sp. isolate ANT-6. *Process Biochemistry*, *38*(10), 1397–1403. doi:10.1016/S0032-9592(03)00037-2

Burton, S. G., & Davids, L. M. (2014). *U.S. Patent No. 8,663,608.* Washington, DC: U.S. Patent and Trademark Office.

Burton, S. (2003). Laccases and phenol oxidases in organic synthesis. *Current Organic Chemistry*, *7*(13), 1317–1331. doi:10.2174/1385272033486477

Cabana, H., Jones, J. P., & Agathos, S. N. (2007). Elimination of endocrine disrupting chemicals using white rot fungi and their lignin modifying enzymes: A review. *Engineering in Life Sciences*, *7*(5), 429–456. doi:10.1002/elsc.200700017

Cabrita, J. F., Abrantes, L. M., & Viana, A. S. (2005). N-Hydroxysuccinimide-terminated self-assembled monolayers on gold for biomolecules immobilisation. *Electrochimica Acta*, *50*(10), 2117–2124. doi:10.1016/j.electacta.2004.09.019

Calabrese, B. S., Pickard, M., Vazquez-Duhalt, R., & Heller, A. (2002). Electroreduction of O_2 to water at 0.6 V (NHE) at pH 7 on the 'wired' *Pleurotus ostreatus* Laccase Cathode. *Biosensors & Bioelectronics*, *17*(11-12), 1071–1074. doi:10.1016/S0956-5663(02)00100-8 PMID:12392957

Calcaterra, N. E., & Barrow, J. C. (2014). Classics in chemical neuroscience: Diazepam (valium). *ACS Chemical Neuroscience*, *5*(4), 253–260. doi:10.1021/cn5000056 PMID:24552479

Call, H. P., & Mücke, I. (1997). History, overview and applications of mediated lignolytic systems, especially laccase-mediator systems (LignozymR process). *Journal of Biotechnology*, *53*(2-3), 163–202. doi:10.1016/S0168-1656(97)01683-0

Camarero, S., Garcia, O., Vidal, T., Colom, J., del Río, J. C., Gutiérrez, A., ... Martínez, Á. T. (2004). Efficient bleaching of non-wood high-quality paper pulp using laccase-mediator system. *Enzyme and Microbial Technology*, *35*(2-3), 113–120. doi:10.1016/j.enzmictec.2003.10.019

Camarero, S., Ibarra, D., Martínez, M. J., & Martínez, Á. T. (2005). Lignin-derived compounds as efficient laccase mediators for decolorization of different types of recalcitrant dyes. *Applied and Environmental Microbiology*, *71*(4), 1775–1784. doi:10.1128/AEM.71.4.1775-1784.2005 PMID:15812000

Campos, R., Kandelbauer, A., Robra, K. H., Cavaco-Paulo, A., & Gubitz, G. M. (2001). Indigo Degradation with Purified Laccases from Trametes Hirsuta and Sclerotium Rolfsii. *Journal of Biotechnology*, *89*(2-3), 131–139. doi:10.1016/S0168-1656(01)00303-0 PMID:11500206

Cañas, A. I., & Camarero, S. (2010). Laccases and their natural mediators: Biotechnological tools for sustainable eco-friendly processes. *Biotechnology Advances*, *28*(6), 694–705. doi:10.1016/j.biotechadv.2010.05.002 PMID:20471466

Cannatelli, M. D., & Ragauskas, A. J. (2015a). Laccase-catalyzed synthesis of 2, 3-ethylenedithio-1, 4-quinones. *Journal of Molecular Catalysis. B, Enzymatic*, *119*, 85–89. doi:10.1016/j.molcatb.2015.05.016

Cannatelli, M. D., & Ragauskas, A. J. (2015b). Laccase-catalyzed α-arylation of benzoylacetonitrile with substituted hydroquinones. *Chemical Engineering Research & Design*, *97*, 128–134. doi:10.1016/j.cherd.2014.08.021

Cannatelli, M. D., & Ragauskas, A. J. (2017). Two decades of laccases: Advancing sustainability in the chemical industry. *Chemical Record (New York, N.Y.)*, *17*(1), 122–140. doi:10.1002/tcr.201600033 PMID:27492131

Cantarella, G., Galli, C., & Gentili, P. (2003). Free radical versus electron-transfer routes of oxidation of hydrocarbons by laccase/mediator systems: Catalytic or stoichiometric procedures. *Journal of Molecular Catalysis. B, Enzymatic*, *22*(3-4), 135–144. doi:10.1016/S1381-1177(03)00014-6

Cantarelli, C. (1986). Trattamenti enzimatici sui costituenti fenolici dei mosti come prevenzione della maderizzazione. *Vini d'Italia.*, *3*, 87–98.

Cantarelli, C., & Giovanelli, G. (1990). Stabilization of pome and grape juice against phenolic deterioration by enzymic treatments. *Inst Fruchtsaft Union Wiss Tech Komm.*, *21*, 35–57.

Cantarelli, C., & Giovanelli, G. (1991). La stabilizzazione dei vini bianchi mediante ossidazione enzimatica. *Vini d'Italia.*, *33*, 21–32.

Carballa, M., Omil, F., Lema, J. M., Llompart, M., García-Jares, C., Rodríguez, I., & Ternes, T. (2004). Behavior of pharmaceuticals, cosmetics and hormones in a sewage treatment plant. *Water Research*, *38*(12), 2918–2926. doi:10.1016/j.watres.2004.03.029 PMID:15223286

Carson, A. E., & Barker, T. H. (2009). Emerging concepts in engineering extracellular matrix variants for directing cell phenotype. *Regenerative Medicine*, *4*(4), 593–600. doi:10.2217/rme.09.30 PMID:19580407

Carvalho, J., Gonçalves, C., Gil, A. M., & Gama, F. M. (2007). Production and characterization of a new dextrin based hydrogel. *European Polymer Journal*, *43*(7), 3050–3059. doi:10.1016/j.eurpolymj.2007.02.046

Cass, A. E. G., Davis, G., Francis, G. D., Hill, H. A. O., Aston, W. J., Higgins, I. J., ... Turner, A. P. F. (1984). Ferrocene-mediated enzyme electrode for amperometric determination of glucose. *Analytical Chemistry*, *56*(4), 667–671. doi:10.1021/ac00268a018 PMID:6721151

Castagna, M., Takai, Y., Kaibuchi, K., Sano, K., Kikkawa, U., & Nishizuka, Y. (1982). Direct activation of calcium-activated, phospholipid-dependent protein kinase by tumor-promoting phorbol esters. *The Journal of Biological Chemistry*, *257*(13), 7847–7851. PMID:7085651

Castiglioni, S., Thomas, K. V., Kasprzyk-Hordern, B., Vandam, L., & Griffiths, P. (2014). Testing wastewater to detect illicit drugs: State of the art, potential and research needs. *The Science of the Total Environment*, *487*, 613–620. doi:10.1016/j.scitotenv.2013.10.034 PMID:24963530

Cavaco-Paulo, A. (1998). Mechanism of Cellulose Action in Textile Processes. *Carbohydrate Polymers*, *37*(3), 273–277. doi:10.1016/S0144-8617(98)00070-8

Cerniglia, C. E. (1984). Microbial transformations of aromatic hydrocarbons. In R. M. Atlas (Ed.), *Petroleum Microbiology* (pp. 99–128). New York: Macmillon.

Ceron-Litvoc, D., Soares, B. G., Geddes, J., Litvoc, J., & De Lima, M. S. (2009). Comparison of carbamazepine and lithium in treatment of bipolar disorder, a systematic review of randomized controlled trials. *Human Psychopharmacology, 24*(1), 19–28. doi:10.1002/hup.990 PMID:19053079

Cesar, T., & Mrsa, V. (1996). Purification and properties of xylanase produced by *Thermomyces lanuginosus. Enzyme and Microbial Technology, 19*(4), 289–296. doi:10.1016/0141-0229(95)00248-0

Chaidarun, S. S., Eggo, M. C., Sheppard, M. C., & Stewart, P. M. (1994). Expression of epidermal growth factor (EGF), its receptor, and related oncoprotein (erbB-2) in human pituitary tumors and response to EGF in vitro. *Endocrinology, 135*(5), 2012–2021. doi:10.1210/endo.135.5.7956924 PMID:7956924

Chakroun, H., Bouaziz, M., Yangui, T., Blibech, I., Dhouib, A., & Sayadi, S. (2013). Enzymatic transformation of tyrosol by *Trametes trogii* laccases: Identification of the product and study of its biological activities. *Journal of Molecular Catalysis. B, Enzymatic, 87*, 11–17. doi:10.1016/j.molcatb.2012.10.006

Chandra, R. P., & Ragauskas, A. J. (2002). Evaluating laccase-facilitated coupling of phenolic acids to high-yield kraft pulps. *Enzyme and Microbial Technology, 30*(7), 855–861. doi:10.1016/S0141-0229(02)00020-0

Chang, H., Hu, J., Wang, L., & Shao, B. (2008). Occurrence of sulfonamide antibiotics in sewage treatment plants. *Chinese Science Bulletin, 53*(4), 514–520. doi:10.100711434-008-0123-x

Chan, R. P., Pope, D. J., Gilbert, A. P., Sacra, P. J., Baron, J., & Lennard-Jones, J. E. (1983). Studies of two novel sulfasalazine analogs, ipsalazide and balsalazide. *Digestive Diseases and Sciences, 28*(7), 609–615. doi:10.1007/BF01299921 PMID:6345112

Chapla, D., Divecha, J., Madamwar, D., & Shah, A. (2010). Utilisation of agro-industrial waste for xylanase production by*Aspergillus foetidus*MTCC 4898 under solid state fermentation and its application in saccharification. *Biochemical Engineering Journal, 49*(3), 361–369. doi:10.1016/j.bej.2010.01.012

Chapsal, J. M., Bourbigot, M. M., & Thomas, D. (1986). Oxidation of aromatic compounds by haemoglobin. *Water Research, 20*(6), 709–713. doi:10.1016/0043-1354(86)90094-1

Chaput, J. C., Woodbury, N. W., Stearns, L. A., & Williams, B. A. R. (2008). Creating protein biocatalysts as tools for future industrial applications. *Expert Opinion on Biological Therapy, 8*(8), 1087–1098. doi:10.1517/14712598.8.8.1087 PMID:18613761

Charrois, G. J., & Allen, T. M. (2004). Drug release rate influences the pharmacokinetics, biodistribution, therapeutic activity, and toxicity of pegylated liposomal doxorubicin formulations in murine breast cancer. *Biochimica et Biophysica Acta, 1663*(1-2), 167–177. doi:10.1016/j.bbamem.2004.03.006 PMID:15157619

Chaurasia, P. K., Bharati, S. L., & Sarma, C. (2016). *Laccases in Pharmaceutical Chemistry: A Comprehensive Appraisal*. Academic Press.

Chaurasia, , P. K., Bharati, S. L., & Singh, S. K. (2013). Comparative Studies on the Blue and Yellow Laccases. *Research in Plant Sciences., 1*(2), 32–37.

Chaurasia, P. K., Bharati, S. L., Sharma, M., Singh, S. K., Yadav, S. S., & Yadava, S. (2015). Fungal Laccases and their Biotechnological Significances in the Current Perspective. *RE:view, 19*(19), 1916–1934.

Chedeville, O., Debacq, M., Almanza, M. F., & Porte, C. (2007). Use of an ejector for phenol containing water treatment by ozonation. *Separation and Purification Technology, 57*(2), 201–208. doi:10.1016/j.seppur.2007.04.004

Chedeville, O., Debacq, M., & Porte, C. (2009). Removal of phenolic compounds present in olive mill wastewaters by ozonation. *Desalination*, *249*(2), 865–869. doi:10.1016/j.desal.2009.04.014

Cheftel, J. C., Cuq, J. L., & Lorient, D. (1985). *Protéines alimentaires: Biochimie - propriétés fonctionnelles, valeur nutritionnelle - Modifications chimiques*. Technique et Documentation, Lavoisier-Paris-France.

Chen, J., Leng, J., Yang, X., Liao, L., Liu, L., & Xiao, A. (2017). Enhanced performance of magnetic graphene oxide-immobilized laccase and its application for the decolorization of dyes. *Molecules*, *22*, E221. doi: 10.3390/molecules22020221

Chen, B. (2016). HIV capsid assembly, mechanism, and structure. *Biochemistry*, *55*(18), 2539–2552. doi:10.1021/acs.biochem.6b00159 PMID:27074418

Chen, C. S., Mrkisch, M., Huang, S., Whistesides, G. M., & Ingber, D. E. (1998). Micropatterned surfaces for control of cell shape, position, and function. *Biotechnology Progress*, *14*(3), 356–363. doi:10.1021/bp980031m PMID:9622515

Cheng, N., & Song, Y. (2017). *Nanozyme mediated dual-immunoassay integrated with smartphone for use in simultaneous detection of pathogens*. ACS Applied Matererials & Interfaces; doi:10.1021/acsami.7b12734

Chen, H. M., Zhen, L., Lin, W., & Yan, X. J. (2004). Product monitoring and oligosaccharide composition in agar hydrolysates by percolumn labeling by HPLC. *Talanta*, *64*(3), 773–777. doi:10.1016/j.talanta.2004.04.002 PMID:18969671

Chenite, A., Gori, S., Shive, M., Desrosiers, E., & Buschmann, M. (2006). Monolithic gelation of chitosan solutions via enzymatic hydrolysis of urea. *Carbohydrate Polymers*, *64*(3), 419–424. doi:10.1016/j.carbpol.2005.12.010

Chen, J., Blevins, W. E., Park, H., & Park, K. (2000). Gastric retention properties of super porous hydrogel composites. *Journal of Controlled Release*, *64*(1-3), 39–51. doi:10.1016/S0168-3659(99)00139-X PMID:10640644

Chen, S. C., Chen, C. H., Chern, C. L., Hsu, L. S., Huang, Y. C., Chung, K. T., & Chye, S. M. (2006). p-Phenylenediamine induces p53-mediated apoptosis in Mardin-Darby canine kidney cells. *Toxicology In Vitro*, *20*(6), 801–807. doi:10.1016/j.tiv.2005.12.003 PMID:16469477

Chen, S. H., & Ting, A. S. Y. (2015). Biodecolorization and biodegradation potential of recalcitrant triphenylmethane dyes by *Coriolopsis* sp. isolated from compost. *Journal of Environmental Management*, *150*, 274–280. doi:10.1016/j.jenvman.2014.09.014 PMID:25527986

Chen, T., Barton, S. C., Binyamin, G., Gao, Z., Zhang, Y., Kim, H. H., & Heller, A. (2001). A miniature biofuel cell. *Journal of the American Chemical Society*, *123*(35), 8630–8631. doi:10.1021/ja0163164 PMID:11525685

Chen, T., Embree, H. D., Wu, L. Q., & Payne, G. F. (2002). In vitro protein–polysaccharide conjugation: Tyrosinase-ecatalyzed conjugation of gelatin and chitosan. *Biopolymers*, *64*(6), 292–302. doi:10.1002/bip.10196 PMID:12124847

Cherian, E., Dharmendirakumar, M., & Baskar, G. (2015). Immobilization of cellulase onto MnO_2nanoparticles for bioethanol production by enhanced hydrolysis of agricultural waste. *Chinese Journal of Catalysis*, *36*(8), 1223–1229. doi:10.1016/S1872-2067(15)60906-8

Cheynier, V., Arellano, H., Souquet, J. M., & Moutounet, M. (1997). Estimation of the oxidative changes in phenolic compounds of Carignane during winemaking. *American Journal of Enology and Viticulture*, *48*, 225–228.

Chiacchierini, E., Restuccia, D., & Vinci, G. (2004). Bioremediation of food industry effluents: Recent applications of free and immobilised polyphenoloxidases. *Food Science & Technology International*, *10*(6), 373–382. doi:10.1177/1082013204049388

Chi, C. F., Wang, B., Wang, Y. M., Zhang, B., & Deng, S. G. (2015). Isolation and characterization of three antioxidant peptides from protein hydrolysate of bluefin leatherjacket (*Navodon septentrionalis*) heads. *Journal of Functional Foods*, *12*, 1–10. doi:10.1016/j.jff.2014.10.027

Chirivì, C., Fontana, G., Monti, D., Ottolina, G., Riva, S., & Danieli, B. (2012). The quest for new mild and selective modifications of natural structures: Laccase-catalysed oxidation of ergot alkaloids leads to unexpected stereoselective C-4 hydroxylation. *Chemistry (Weinheim an der Bergstrasse, Germany)*, *18*(33), 10355–10361. doi:10.1002/chem.201201076 PMID:22777708

Chirumamilla, R. R., Muralidhar, R., Marchant, R., & Nigam, P. (2001). Improving the quality of industrially important enzymes by directed evolution. *Molecular and Cellular Biochemistry*, *224*(1/2), 159–168. doi:10.1023/A:1011904405002 PMID:11693193

Chmelová, D., & Ondrejovič, M. (2015). Effect of metal ions on triphenylmethane dye decolorization by laccase from *Trametes versicolor. Nova Biotechnol. Chim.*, *14*, 191–200.

Chockalingam, K., Blenner, M., & Banta, S. (2007). Design and application of stimulusresponsive peptide systems. *Protein Engineering, Design & Selection*, *20*(4), 155–161. doi:10.1093/protein/gzm008 PMID:17376876

Choi, J. M., Han, S. S., & Kim, H. S. (2015). Industrial applications of enzyme biocatalysis: Current status and future aspects. *Biotechnology Advances*, *33*(7), 1443–1454. doi:10.1016/j.biotechadv.2015.02.014 PMID:25747291

Choi, S. M., Yoo, S. D., & Lee, B. M. (2004). Toxicological characteristics of endocrine-disrupting chemicals: Developmental toxicity, carcinogenicity, and mutagenicity. *Journal of Toxicology and Environmental Health, Part B. Critical Review*, *7*, 1–24. PMID:14681080

Chou, X., Wu, N., Lupher, D., & Davis, L. C. (1998). Benzotriazoles: Toxicity and Degradation. *Proceedings of the 1998 Conference on Hazardous Waste Research*, 374-382.

Christakopoulos, P., Nerinckx, W., Kekos, D., Marcis, B., & Claeyssens, M. (1996). Purification and characterization of two low molecular mass alkaline xylanases from *Fusarium oxysporum* F3. *Journal of Biotechnology*, *51*(2), 181–189. doi:10.1016/0168-1656(96)01619-7 PMID:8987884

Chronopoulou, L., Lorenzoni, S., Masci, G., Dentini, M., Togna, A. R., Togna, G., ... Palocci, C. (2010). Lipase-supported synthesis of peptidic hydrogels. *Soft Matter*, *6*(11), 2525–2532. doi:10.1039/c001658f

Chudasama, C. J., Jani, S. A., Jajda, H. M., & Pate, H. N. (2010). Optimization and production of alkaline protease from *Bacillus thuringiensis* CC7. *Journal of Cell Tissue Research*, *10*, 2257–2262.

Chung, C., Degner, B., Decker, E. A., & McClements, D. J. (2013). Oil –filled hydrogel particles for reduced for reduced-fat food applications: Fabrication, characterization, and properties. *Innovative Food Science & Emerging Technologies*, *20*, 324–334. doi:10.1016/j.ifset.2013.08.006

Chung, J. E., Kurisawa, M., Uyama, H., & Kobayashi, S. (2003). Enzymatic synthesis and antioxidant property of gelatin-catechin conjugates. *Biotechnology Letters*, *25*(23), 1993–1997. doi:10.1023/B:BILE.0000004391.27564.8e PMID:14719812

Cihan, A. C., Ozcan, B., Tekin, N., & Cokmus, C. (2010). Characterization of a thermostable α-glucosidase from Geobacillus thermodenitrificans F84a. Current Research, Technology and Education Topics in Applied Microbiology and Microbial Technology, 945-955.

Cihan, A. C., Cokmus, C., & Ozcan, B. (2009). Characterization of thermostable α-glucosidases from newly isolated *Geobacillus* sp. A333 and thermophilic bacterium A343. *World Journal of Microbiology & Biotechnology*, *25*(12), 2205–2217. doi:10.100711274-009-0127-y

Cihlar, T., & Ray, A. S. (2010). Nucleoside and nucleotide HIV reverse transcriptase inhibitors: 25 years after zidovudine. *Antiviral Research*, *85*(1), 39–58. doi:10.1016/j.antiviral.2009.09.014 PMID:19887088

Cipolatti, E. P., Silva, M. J. A., Klein, M., Feddern, V., Feltes, M. M. C., Oliveira, J. V., ... Oliveira, D. (2014). Current status and trends in enzymatic nanoimmobilization. *Journal of Molecular Catalysis. B, Enzymatic*, *99*, 56–67. doi:10.1016/j.molcatb.2013.10.019

Cirino, P. C., & Arnold, F. H. (2002). Protein engineering of oxygenases for biocatalysis. *Current Opinion in Chemical Biology*, *6*(2), 130–135. doi:10.1016/S1367-5931(02)00305-8 PMID:12038995

Civas, A., Eberhard, R., Le Dizet, P., & Petek, F. (1984). Glycosidases induced in Aspergillus tamarii. Secreted α-d-galactosidase and β-d-mannanase. *The Biochemical Journal*, *219*(3), 857–863. doi:10.1042/bj2190857 PMID:6331399

Claesson-Welsh, L. (2015). Vascular permeability-the essentials. *Upsala Journal of Medical Sciences*, *120*(3), 135–143. doi:10.3109/03009734.2015.1064501 PMID:26220421

Clara, M., Strenn, B., & Kreuzinger, N. (2004). Carbamazepine as a possible anthropogenic marker in the aquatic environment: Investigations on the behaviour of carbamazepine in wastewater treatment and during groundwater infiltration. *Water Research*, *38*(4), 947–954. doi:10.1016/j.watres.2003.10.058 PMID:14769414

Clarke, Aj. (1997). *Biodegradation of Cellulose: Enzymology And Biotechnology*. Lancaster, PA: Technomic Publishing Co. Inc.

Claus, H. (2003). Laccases and their occurrence in prokaryotes. *Archives of Microbiology*, *179*(3), 145–150. doi:10.100700203-002-0510-7 PMID:12610719

Claus, H. (2004). Laccases: Structure, reactions, distribution. *Micron (Oxford, England)*, *35*(1-2), 93–96. doi:10.1016/j.micron.2003.10.029 PMID:15036303

Codner, R. (1971). Pectinolytic and cellulolytic enzymes in the microbial modification of plant tissues. *Journal of Applied Microbiology*, *34*(1), 147–160. PMID:5564376

Cohen, P., & Frame, S. (2001). The renaissance of GSK3. *Nature Reviews. Molecular Cell Biology*, *2*(10), 769–776. doi:10.1038/35096075 PMID:11584304

Coiffier, B., Mounier, N., Bologna, S., Fermé, C., Tilly, H., Sonet, A., & Herbrecht, R. (2003). Efficacy and Safety of Rasburicase (recombinant urate oxidase) for the Prevention and Treatment of Hyperuricemia During Induction Chemotherapy of Aggressive Non-Hodgkin's Lymphoma: Results of the GRAAL1 (Groupe d'Etude des Lymphomes de l'Adulte Trial on Rasburicase Activity in Adult Lymphoma) Study. *Journal of Clinical Oncology*, *21*(23), 4402–4406. doi:10.1200/JCO.2003.04.115 PMID:14581437

Colborn, T., vom Saal, F. S., & Soto, A. M. (1993). Developmental effects of endocrine-disrupting chemicals in wildlife and humans. *Environmental Health Perspectives*, *101*(5), 378–384. doi:10.1289/ehp.93101378 PMID:8080506

Coleman, T., & Brines, M. (2004). Science review: recombinant human erythropoietin in critical illness: a role beyond anemia? *Critical Care (London, England)*, *8*(5), 337–341. doi:10.1186/cc2897 PMID:15469595

Collett, M. S., & Erikson, R. L. (1978). Protein kinase activity associated with the avian sarcoma virus src gene product. *Proceedings of the National Academy of Sciences of the United States of America*, *75*(4), 2021–2024. doi:10.1073/pnas.75.4.2021 PMID:205879

Collins, T., Gerday, C., & Feller, G. (2005). Xylanases, xylanase families and extremophilic xylanases. *FEMS Microbiology Reviews*, *29*(1), 3–23. doi:10.1016/j.femsre.2004.06.005 PMID:15652973

Colvin, V. L. (2003). The potential environmental impact of engineered nanomaterials. *Nature Biotechnology*, *10*(10), 1166–1170. doi:10.1038/nbt875 PMID:14520401

Conchie, J., & Macdonald, D. (1959). Glycosidases in the mammalian alimentary tract. *Nature*, *184*(4694), 1233–1233. doi:10.1038/1841233a0 PMID:13811463

Congreve, M., & Andrews, S. (2017). Perspective in Medicinal Chemistry: Structure-Based Drug Design. *Current Topics in Medicinal Chemistry*, *17*(2), 93–94. doi:10.2174/156802661702161122204251 PMID:28093978

Conrad, L. S., Sponholz, W. R., & Berker, O. (2000). *Treatment of cork with a phenol oxidizing enzyme*. US Patent 6152966.

Constable, D. J., Dunn, P. J., Hayler, J. D., Humphrey, G. R., Leazer, J. L. Jr, Linderman, R. J., ... Zaks, A. (2007). Key green chemistry research areas—a perspective from pharmaceutical manufacturers. *Green Chemistry*, *9*(5), 411–420. doi:10.1039/B703488C

Constantin, M. A., Conrad, J., & Beifuss, U. (2012). Laccase-catalyzed oxidative phenolic coupling of vanillidene derivatives. *Green Chemistry*, *14*(9), 2375–2379. doi:10.1039/c2gc35848d

Contado, C. (2015). Nanomaterials in consumer products: A challenging analytical problem. *Frontiers in Chemistry*, *3*, 48. doi:10.3389/fchem.2015.00048 PMID:26301216

Cooper, E. R., Siewicki, T. C., & Phillips, K. (2008). Preliminary risk assessment database and risk ranking of pharmaceuticals in the environment. *The Science of the Total Environment*, *398*(1-3), 26–33. doi:10.1016/j.scitotenv.2008.02.061 PMID:18440598

Copeland, R. A. (2013). *Evaluation of enzyme inhibitors in drug discovery: a guide for medicinal chemists and pharmacologists*. John Wiley & Sons.

Copeland, R. A. (2000). *Enzymes : a practical introduction to structure, mechanism, and data analysis*. Wiley. doi:10.1002/0471220639

Coral, G., Arikan, B., Unaldi, M. N., & Guvenmez, H. K. (2002). Some properties of thermostable xylanase from an *Aspergillus niger* strain. *Annals of Microbiology*, *52*, 299–306.

Cornwell, R. L., Tinland-Butez, M.-F., Tardone, P. J., Cabasso, I., & Hammel, K. E. (1990). Lignin degradation and lignin peroxidase production in cultures of *Phanerochaete chrysosporium* immobilized on porous ceramic supports. *Enzyme and Microbial Technology*, *12*(12), 916–920. doi:10.1016/0141-0229(90)90109-4

Corvo, M. L., Boerman, O. C., Oyen, W. J. G., Bloois, L. V., Cruz, M. E. M., Crommelin, D. J. A., & Storm, G. (1999). Intravenous administration of superoxide dismutase entrapped in long circulating liposomes. *In vivo* fate in a rat model of adjuvant arthritis. *Biochimica et Biophysica Acta*, *1419*(2), 325–334. doi:10.1016/S0005-2736(99)00081-4 PMID:10407083

Cosgrove, D. J. (1997). Assembly and enlargement of the primary cell wall in plants. *Annual Review of Cell and Developmental Biology*, *13*(1), 171–201. doi:10.1146/annurev.cellbio.13.1.171 PMID:9442872

Costa, S. A., Azevedo, H. S., & Reis, R. L. (2005). *Biodegradable Systems in Tissue Engineering and Regenerative Medicine*. Boca Raton, PA: C. R. C. Press.

Costerousse, O., Allegrini, J., Huang, H., Bounhik, J., & Alhenc-Gelas, F. (1994). Regulation of ACE gene expression and plasma levels during rat postnatal development. *American Journal of Physiology. Endocrinology and Metabolism*, *267*(5), E745–E753. doi:10.1152/ajpendo.1994.267.5.E745 PMID:7977726

Coughlan, M. (1991). Towards an understanding of the mechanism of action of main chain-hydrolyzing xylanases. *Progress in Biotechnology*.

Coughlan, M. P., Tuohy Filho, M., Exf, P. J., Claeyssens, M., Vrsanska, M., & Hughes, M. (1993). Enzymological aspects of microbial hemicellulases with emphasis on fungal systems. Hemicellulose and Hemicellulases, 53-84.

Courtin, C. M., & Delcour, J. A. (2001). Relative activity of endoxylanases towards water-extractable and water-unextractable arabinoxylan. *Journal of Cereal Science*, *33*(3), 301–312. doi:10.1006/jcrs.2000.0354

Courtois, J. (2009). Oligosaccharides from land plants and algae: Production and applications in therapeutics and biotechnology. *Current Opinion in Microbiology*, *12*(3), 261–273. doi:10.1016/j.mib.2009.04.007 PMID:19467920

Couto, S. G., Toca, J. L., & Herrera, J. L. (2006). Industrial and biotechnological applications of laccases: A review. *Biotechnology Advances*, *24*(5), 500–513. doi:10.1016/j.biotechadv.2006.04.003 PMID:16716556

Cowan, W. D. (2000). Review and Preview of The Development of Geographical Feed Enzyme Markets. *Proceedings of the 3rd European Symposium on Feed Enzymes*.

Crittenden, R. G., & Plyne, M. J. (1996). Production, properties and application of food grade oligosaccharides. *Trends in Food Science & Technology*, *7*(11), 353–361. doi:10.1016/S0924-2244(96)10038-8

Crittenden, R. G., & Plyne, M. J. (2002). Production of food grade oligoscchrides using immobilized cells of *Zymomonas mobilis*. *Trends in Food Science & Technology*, *58*, 297–302. PMID:11935179

Crueger, W., & Crueger, A. (Eds.). (1989). Industrial Microbiology. Sinauer Association.

Cruz-Morató, C., Jelić, A., Perez, S., Petrović, M., Barceló, D., Marco-Urrea, E., & Vicent, T. (2013). Continuous treatment of clofibric acid by *Trametes versicolor* in a fluidized bed bioreactor: Identification of transformation products and toxicity assessment. *Biochemical Engineering Journal*, *75*, 79–85. doi:10.1016/j.bej.2013.03.020

D'Abrosca, B., Fiorentino, A., Izzo, A., Cefarelli, G., Pascarella, M. T., Uzzo, P., & Monaco, P. (2008). Phytotoxicity evaluation of five pharmaceutical pollutants detected in surface water on germination and growth of cultivated and spontaneous plants. *Journal of Environmental Science and Health. Part A, Toxic/Hazardous Substances & Environmental Engineering*, *43*(3), 285–294. doi:10.1080/10934520701792803 PMID:18205060

d'Acunzo, F., Baiocco, P., Fabbrini, M., Galli, C., & Gentili, P. (2002). A mechanistic survey of the oxidation of alcohols and ethers with the enzyme laccase and its mediation by TEMPO. *European Journal of Organic Chemistry*, *2002*(24), 4195–4201. doi:

Dabek, M., McCrae, S. I., Stevens, V. J., Duncan, S. H., & Louis, P. (2008). Distribution of β-glucosidase and β-glucuronidase activity and of β-glucuronidase gene gus in human colonic bacteria. *FEMS Microbiology Ecology*, *66*(3), 487–495. doi:10.1111/j.1574-6941.2008.00520.x PMID:18537837

Dafale, N., Rao, N. N., Meshram, S. U., & Wate, S. R. (2008). Decolorization of Azo Dyes and Simulated Dye Bath Wastewater Using Acclimatized Microbial Consortium biostimulation and Halo Tolerance. *Bioresource Technology*, *99*(7), 2552–2558. doi:10.1016/j.biortech.2007.04.044 PMID:17566726

Dai, X. J., Liu, M. Q., Jin, H. X., & Jing, M. Y. (2011). Optimisation of solid-state fermentation of *Aspergillus niger* JL-15 for xylanase production and xylo-oligosaccharides preparation. *Czech Journal of Food Sciences*, *29*(5), 557–567.

Daley, G. Q., Van Etten, R. A., & Baltimore, D. (1990). Induction of Chronic Myelogenous Leukemia in Mice by the P210 ^bcr/abl Gene of the Philadelphia Chromosome. *Science*, *247*(4944), 824–830. doi:10.1126cience.2406902 PMID:2406902

Dan, A., Yang, Y., Dai, Y. N., Chen, C. X., Wang, S. Y., & Tao, R. (2013). Removal and factors influencing removal of sulfonamides and trimethoprim from domestic sewage in constructed wetlands. *Bioresource Technology*, *146*, 363–370. doi:10.1016/j.biortech.2013.07.050 PMID:23954243

Daniel, R. (2002). Construction of environmental libraries for functional screening of enzyme activity. In Directed molecular evolution of proteins. Wiley-VCH Verlag GmbH. doi:10.1002/3527600647.ch4

Darah, I., Sumathi, G., Jain, K., & Lim, S. H. (2011). Influence of agitation speed on tannase production and morphology of *Aspergillus niger* FETL FT3 in submerged fermentation. *Applied Biochemistry and Biotechnology*, *165*(7-8), 7–8. doi:10.100712010-011-9387-8 PMID:21947762

Das, A. K., Maity, I., Parmar, H. S., McDonald, T. O., & Konda, M. (2015). Lipase-catalyzed dissipative self-assembly of a thixotropic peptide bolaamphiphile hydrogel for human umbilical cord stem-cell proliferation. *Biomacromolecules*, *16*(4), 1157–1168. doi:10.1021/bm501835v PMID:25714334

Das, D., & Goyal, A. (2014). Pharmaceutical Enzymes. In S. K. Brar, G. S. Dhillon, & C. R. Soccol (Eds.), *Biotransformation of Waste Biomass into High Value Biochemicals* (pp. 367–387). New York, NY: Springer New York. doi:10.1007/978-1-4614-8005-1_15

da-Silva, M. R., de-Sa, L. R. V., Russo, C., Scio, E., & Ferreira-Leitao, V. S. (2010). The use of HRP in decolorization of reactive dyes and toxicological evaluation of their products. *Enzyme Research*, *2010*, 703824–703831. doi:10.4061/2010/703824 PMID:21318147

Davaran, S., Hanaee, J., & Khosravi, A. (1999). Release of 5-amino salicylic acid from acrylic type polymeric prodrugs designed for colon-specific drug delivery. *Journal of Controlled Release*, *58*(3), 279–287. doi:10.1016/S0168-3659(98)00167-9 PMID:10099153

Davidsen, J., Jorgensen, K., Andresen, T. L., & Mouritsen, O. G. (2003). Secreted phospholipase A_2 as a new enzymatic trigger mechanism for localised liposomal drug release and absorption in diseased tissue. *Biochimica et Biophysica Acta*, *1609*(1), 95–101. doi:10.1016/S0005-2736(02)00659-4 PMID:12507763

Davis, J. L., Paris, H. L., Beals, J. W., Binns, S. E., Giordano, G. R., Scalzo, R. L., ... Bell, C. (2016). Liposomal-encapsulated Ascorbic Acid: Influence on Vitamin C Bioavailability and Capacity to Protect Against Ischemia–Reperfusion Injury. *Nutrition and Metabolic Insights*, *9*, 25–30. doi:10.4137/NMI.S39764 PMID:27375360

De Béthune, M. P. (2010). Non-nucleoside reverse transcriptase inhibitors (NNRTIs), their discovery, development, and use in the treatment of HIV-1 infection: A review of the last 20 years (1989–2009). *Antiviral Research*, *85*(1), 75–90. doi:10.1016/j.antiviral.2009.09.008 PMID:19781578

De Clercq, E., & Li, G. (2016). Approved antiviral drugs over the past 50 years. *Clinical Microbiology Reviews*, *29*(3), 695–747. doi:10.1128/CMR.00102-15 PMID:27281742

de Souza, C. G. M., & Peralta, R. M. (2003). Purification and characterization of the main laccase produced by the white-rot fungus *Pleurotus pulmonarius* on wheat bran solid state medium. *Journal of Basic Microbiology*, *43*(4), 278–286. doi:10.1002/jobm.200390031 PMID:12872309

De Vries, J., Rombouts, F., Voragen, A., & Pilnik, W. (1982). Enzymic degradation of apple pectins. *Carbohydrate Polymers*, *2*(1), 25–33. doi:10.1016/0144-8617(82)90043-1

Debbab, A., Aly, A. H., Lin, W. H., & Proksch, P. (2010). Bioactive compounds from marine bacteria and fungi. *Microbial Biotechnology*, *3*(5), 544–563. doi:10.1111/j.1751-7915.2010.00179.x PMID:21255352

Debyser, W., Derdelinckx, G., & Delcour, J. A. (1997a). Arabinoxylans and arabinoxylan hydrolysing activities in barley malts and worts derived from them. *Journal of Cereal Science*, *26*(1), 67–74. doi:10.1006/jcrs.1996.0107

Debyser, W., Derdelinckx, G., & Delcour, J. A. (1997b). Arabinoxylan solubilization and inhibition of the barley malt xylanolytic system by wheat during brewing with wheat whole meal adjunct: Evidence for a new class of enzyme inhibitors. *Journal of the American Society of Brewing Chemists*, *55*, 153–156.

Dehghanifard, E., Jonidi Jafari, A., Rezaei Kalantary, R., Mahvi, A. H., Faramarzi, M. A., & Esrafili, A. (2013). Biodegradation of 2,4-dinitrophenol with laccase immobilized on nano-porous silica beads. *Iranian Journal of Environmental Health Sciences & Engineering*, *10*(1), 25. doi:10.1186/1735-2746-10-25 PMID:23547870

Del Valle, E. M. M. (2004). Cyclodextrins and their uses: A review. *Process Biochemistry*, *39*, 1033–1046.

Del Zoppo, G. J., Saver, J. L., Jauch, E. C., & Adams, H. P. Jr. (2009). Expansion of the time window for treatment of acute ischemic stroke with intravenous tissue plasminogen activator: A science advisory from the American Heart Association/American Stroke Association. *Stroke*, *40*(8), 2945–2948. doi:10.1161/STROKEAHA.109.192535 PMID:19478221

Delzenne, N. M. (2003). Oligosaccharide: State of Art. *The Proceedings of the Nutrition Society*, *62*(01), 177–182. doi:10.1079/PNS2002225 PMID:12749343

Delzenne, N. M., & Roberfroid, M. R. (1994). Physiological effects of nondigestible oligosaccharides. *Lebensmittel-Wissenschaft + Technologie*, *27*, 1–6.

Desai, S. S., & Nityanand, C. (2011). Microbial laccases and their applications:a review. *Asian J Biotechnol*, *3*(2), 98–124. doi:10.3923/ajbkr.2011.98.124

Deshpande, P. P., Biswas, S., & Torchilin, V. P. (2013). Current trends in the use of liposomes for tumor targeting. *Nanomedicine (London)*, *8*(9), 1509–1528. doi:10.2217/nnm.13.118 PMID:23914966

Devi, N. A., & Rao, A. A. (1996). Fractionation, purification, and preliminary characterization of polygalacturonases produced by Aspergillus carbonarius. *Enzyme and Microbial Technology*, *18*(1), 59–65. doi:10.1016/0141-0229(96)00055-5

Dhiman, S. S., Garg, G., Sharma, J., & Mahajan, R. (2011). Characterisation of statistically produced xylanase for enrichment of fruit juice clarification process. *New Biotechnology*, *28*(6), 746–755. doi:10.1016/j.nbt.2010.11.004 PMID:21093618

Diaconu, M., Litescu, S. C., & Radu, G. L. (2010). Laccase–MWCNT–chitosan biosensor—A new tool for total polyphenolic content evaluation from in vitro cultivated plants. *Sensors and Actuators. B, Chemical*, *145*(2), 800–806. doi:10.1016/j.snb.2010.01.064

Díaz-Rodríguez, A., Lavandera, I., Kanbak-Aksu, S., Sheldon, R. A., Gotor, V., & Gotor-Fernández, V. (2012). From diols to lactones under aerobic conditions using a laccase/TEMPO catalytic system in aqueous medium. *Advanced Synthesis & Catalysis*, *354*(18), 3405–3408. doi:10.1002/adsc.201200892

DiCosimo, R., McAuliffe, J., Poulose, A. J., & Bohlmann, J. (2013). Industrial use of immobilized enzymes. *Chemical Society Reviews*, *42*(15), 6437–6474. doi:10.1039/c3cs35506c PMID:23436023

Dietmar, H., Bernd, N., Kulbe, K. D., Walter, S., & Silvia, Z. (1996). Production of fungal xylanases. *Journal of Bioresource Technology*, *58*(2), 137–161. doi:10.1016/S0960-8524(96)00094-6

Disease, A. W. G. o. M. o. P., Kishnani, P. S., Steiner, R. D., Bali, D., Berger, K., Byrne, B. J., & Watson, M. S. (2006). Pompe disease diagnosis and management guideline. *Genetics in Medicine*, *8*(5), 267-288.

Dittmer, N. T., Suderman, R. J., Jiang, H., Zhu, Y.-C., Gorman, M. J., Kramer, K. J., & Kanost, M. R. (2004). Characterization of cDNAs encoding putative laccase-like multicopper oxidases and developmental expression in the tobacco hornworm, Manduca sexta, and the malaria mosquito, Anopheles gambiae. *Insect Biochemistry and Molecular Biology*, *34*(1), 29–41. doi:10.1016/j.ibmb.2003.08.003 PMID:14723895

Djekrif-Dakhmouche, S., Gheribi-Aoulmi, S., Meraihi, Z., & Bennamoun, L. (2006). Application of a statistical design to the optimisation of culture medium for α-amylase production by *Aspergillus niger* ATCC 16404 grown on orange waste powder. *Journal of Food Engineering*, *73*(2), 190–197. doi:10.1016/j.jfoodeng.2005.01.021

Dobrev, G. T., Pishtiyski, I. G., Stanchev, V. S., & Mircheva, R. (2007). Optimisation of nutrient medium containing agricultural wastes for xylanase production by *Aspergillus niger* B03 using optimal composite experimental design. *Bioresource Technology*, *28*(14), 2671–2678. doi:10.1016/j.biortech.2006.09.022 PMID:17092711

Domaradzka, D., Guzik, U., & Wojcieszyńska, D. (2015). Biodegradation and biotransformation of polycyclic non-steroidal anti-inflammatory drugs. *Reviews in Environmental Science and Biotechnology*, *14*(2), 229–239. doi:10.100711157-015-9364-8

Domeradzka, N. E., Werten, M. W., de Wolf, F. A., & de Vries, R. (2016). Protein cross-linking tools for the construction of nanomaterials. *Current Opinion in Biotechnology*, *39*, 61–67. doi:10.1016/j.copbio.2016.01.003 PMID:26871735

Doner, L. W., & Hicks, K. B. (1997). Isolation of hemicelluloses from corn fiber by alkaline hydrogen peroxide extraction. *Cereal Chemistry*, *74*(2), 176–181. doi:10.1094/CCHEM.1997.74.2.176

Dong, Q., Ouyang, L.-M., Yu, H.-L., & Xu, J.-H. (2010). Efficient biosynthesis of uridine diphosphate glucose from maltodextrin by multiple enzymes immobilized on magnetic nanoparticles. *Carbohydrate Research*, *345*(11), 1622–1626. doi:10.1016/j.carres.2010.04.025 PMID:20627237

Donkor, O. N., Henriksson, A., Vasiljevic, T., & Shah, N. (2007). α-Galactosidase and proteolytic activities of selected probiotic and dairy cultures in fermented soymilk. *Food Chemistry*, *104*(1), 10–20. doi:10.1016/j.foodchem.2006.10.065

Doolittle, D. J., Sherrill, J. M., & Butterworth, B. E. (1983). Influence of intestinal bacteria, sex of the animal, and position of the nitro group on the hepatic genotoxicity of nitrotoluene isomers in vivo. *Cancer Research*, *43*(6), 2836–2842. PMID:6850595

Doralice, S. L. B., & Regina, T. R. M. (2001). Decolorization of textile indigo dye by ligninolytic fungi. *Journal of Biotechnology*, *89*(2-3), 141–145. doi:10.1016/S0168-1656(01)00304-2 PMID:11500207

Dorpe, V. S., Bronsealer, A., Nielandt, J., Stalmens, S., Wynendaele, E., Audenaert, K., ... Spiegeleer, B. D. (2012). Brainpeps: The blood-brain barrier peptide database. *Brain Structure & Function*, *217*(3), 687–718. doi:10.100700429-011-0375-0 PMID:22205159

Drasar, B. S., & Hill, M. J. (1974). *Human intestinal flora: Academic Press (London) Ltd., 24/28 Oval Road*. London: NWI.

Driss, D., Brhiri, F., Elleuch, L., Bouly, N., Stals, I., Miled, N., ... Chaabouni, S. E. (2011). Purification and properties of an extracellular acidophilic endo-1,4-β-xylanase, naturally deleted in the "thumb", from *Penicillium occitanis* Pol6. *Process Biochemistry*, *46*(6), 1299–1306. doi:10.1016/j.procbio.2011.02.022

Druker, B. J. (2002). STI571 (Gleevec™) as a paradigm for cancer therapy. *Trends in Molecular Medicine*, *8*(4), S14–S18. doi:10.1016/S1471-4914(02)02305-5 PMID:11927282

Drury, J. L., & Mooney, D. J. (2003). Hydrogels for tissue engineering: Scaffold design variables and applications. *Biomaterials*, *24*(24), 4337–4351. doi:10.1016/S0142-9612(03)00340-5 PMID:12922147

Dua, M., Singh, A., Sethunathan, N., & Johri, A. (2002). Biotechnology and bioremediation: Successes and limitations. *Applied Microbiology and Biotechnology*, *59*, 2–3, 143–152. PMID:12111139

Ducros, V., Brzozowski, A. M., Wilson, K. S., Brown, S. H., Stergard, P., Schneider, P., ... Davies, G. J. (1998). Crystal structure of the type-2 Cu depleted laccase from *Coprinus cinereus* at 2.2A° resolution. *Nature Structural Biology*, *5*(4), 310–316. doi:10.1038/nsb0498-310 PMID:9546223

Duffner, F., Bertoldo, C., Andersen, J. T., Wagner, K., & Antranikian, G. (2000). A new thermoactive pullulanase from *Desulfurococcus mucosus*: Cloning, sequencing, purification, and characterization of the recombinant enzyme after expression in *Bacillus subtilis*. *Journal of Bacteriology*, *182*(22), 6331–6338. doi:10.1128/JB.182.22.6331-6338.2000 PMID:11053376

Dufresne, A. (2013). Nanocellulose: A new ageless bionanomaterial. *Materials Today*, *16*(6), 220–227. doi:10.1016/j.mattod.2013.06.004

Duo, Y., Shi, P., Huang, H., Zhang, X., Luo, H., Wang, Y., & Yao, B. (2013). Characterization of three novel thermophilic xylanases from *Humicola insolens* Y1 with application potentials in the brewing industry. *Bioresource Technology*, *130*, 161–167. doi:10.1016/j.biortech.2012.12.067 PMID:23306124

Dur´an, Rosa, D'Annibale, & Gianfreda. (2002). Applications of laccases and tyrosinases (phenoloxidases) immobilized on different supports: a review. *Enzyme and Microbial Technology, 31*(7), 907–931.

Duran, N., Rosa, M. A., D'annibale, A., & Gianfreda, L. (2002). Applications of laccases and tyrosinases (phenol oxidases) immobilized on different supports: A review. *Enzyme and Microbial Technology*, *31*(7), 907–931. doi:10.1016/S0141-0229(02)00214-4

During, A., Debouche, C., Raas, T., & Larondelle, Y. (2012). Among plant lignans, pinoresinol has the strongest anti-inflammatory properties in human intestinal Caco-2 cells. *The Journal of Nutrition*, *142*(10), 1798–1805. doi:10.3945/jn.112.162453 PMID:22955517

Dusterhoft, E. M., Posthumus, M. A., & Voragen, A. G. J. (1992). Non-starch polysaccharides from sunflower (*Helianthus annuus*) meal and palm kernel (*Elaeis guineensis*) meal preparation of cell wall material and extraction of polysaccharide fractions. *Journal of the Science of Food and Agriculture*, *59*, 151–160. doi:10.1002/jsfa.2740590204

Dutta, T., Sengupta, R., Sahoo, R., Ray, S. S., Bhattacharjee, A., & Ghosh, S. (2007). A novel cellulase free alkaliphilic xylanase from alkali tolerant *Penicillium citrinum*: Production, purification and characterization. *Letters in Applied Microbiology*, *44*(2), 206–211. doi:10.1111/j.1472-765X.2006.02042.x PMID:17257262

Dwivedi, U. N., Singh, P., Pandey, V. P., & Kumar, A. (2011). Structure–function relationship among bacterial, fungal and plant laccases. *Journal of Molecular Catalysis. B, Enzymatic*, *68*(2), 117–128. doi:10.1016/j.molcatb.2010.11.002

Dzidic, S., & Bedeković, V. (2003). Horizontal gene transfer-emerging multidrug resistance in hospital bacteria. *Acta Pharmacologica Sinica*, *24*, 519–526. PMID:12791177

Edagwa, B., McMillan, J., Sillman, B., & Gendelman, H. E. (2017). Long-acting slow effective release antiretroviral therapy. *Expert Opinion on Drug Delivery*, 1–11. doi:10.1080/17425247.2017.1288212 PMID:28128004

Eggert, C. (1997). Laccase-catalyzed formation of cinnabarinic acid is responsible for antibacterial activity of *Pycnoporus cinnabarinus*. *Microbiological Research*, *152*(3), 315–318. doi:10.1016/S0944-5013(97)80046-8 PMID:9352667

Eggert, C., Temp, U., Dean, J. F., & Eriksson, K. E. L. (1996). A fungal metabolite mediates degradation of non-phenolic lignin structures and synthetic lignin by laccase. *FEBS Letters*, *391*(1-2), 144–148. doi:10.1016/0014-5793(96)00719-3 PMID:8706903

Eibes, G., Debernardi, G., Feijoo, G., Moreira, M. T., & Lema, J. M. (2011). Oxidation of pharmaceutically active compounds by a ligninolytic fungal peroxidase. *Biodegradation, 22*(3), 539–550. doi:10.100710532-010-9426-0 PMID:20972884

Eisenman, H. C., Mues, M., Weber, S. E., Frases, S., Chaskes, S., Gerfen, G., & Casadevall, A. (2007). *Cryptococcus neoformans* laccase catalyses melanin synthesis from both D- and L-DOPA. *Microbiology, 153*(12), 3954–3962. doi:10.1099/mic.0.2007/011049-0 PMID:18048910

El Saliby, I. J., Shon, H., Kandasamy, J., & Vigneswaran, S. (2008). *Nanotechnology for wastewater treatment: in brief. Encyclopedia of Life Support Syst.* EOLSS.

Elavarasan, K., & Shamasundar, B. A. (2015). Effect of oven drying and freeze drying on the antioxidant and functional properties of protein hydrolysates derived from freshwater fish (*Cirrhinus mrigala*) using papain enzyme. *Journal of Food Science and Technology*. doi:10.100713197-015-2084-9 PMID:27162411

El-Fakharany, E. M., Haroun, B. M., Ng, T. B., & Redwan, E. R. (2010). Oyster mushroom laccase inhibits hepatitis C virus entry into Peripheral blood cells and hepatoma cells. *Protein and Peptide Letters, 17*, 1031–1039. doi:10.2174/092986610791498948 PMID:20156183

El-Fallal, A., Abou Dobara, M., El-Sayed, A., & Omar, N. (2012). *Starch and Microbial α-Amylases: From Concepts to Biotechnological Applications*. Carbohydrates – Comprehensive Studies on Glycobiology and Glycotechnology. Intech. doi:10.5772/51571

Eliaz, R. E., & Szoka, F. C. Jr. (2001). Liposome-encapsulated doxorubicin targeted to CD44: A strategy to kill CD44-overexpressing tumor cells. *Cancer Research, 61*(6), 2592–2601. PMID:11289136

Ellouze, O., Fattouch, S., Mestiri, F., Mohammed, R. A., & Mohammed, N. M. (2008). Optimisation of extracellular xylanase production by *Sclerotinia sclerotiorum* S2 using factorial design. *Indian Journal of Biochemistry & Biophysics, 45*, 404–409. PMID:19239127

Elmacı, S. B., Gülgör, G., Tokatlı, M., Erten, H., İşci, A., & Özçelik, F. (2015). Effectiveness of chitosan against wine-related microorganisms. *Antonie van Leeuwenhoek, 107*(3), 675–686. doi:10.100710482-014-0362-6 PMID:25528342

Emirdağ-Öztürk, S., Hajdok, S., Conrad, J., & Beifuss, U. (2013). Laccase-catalyzed reaction of 3-tert-butyl-1H-pyrazol-5 (4H)-one with substituted catechols using air as an oxidant. *Tetrahedron, 69*(18), 3664–3668. doi:10.1016/j.tet.2013.03.023

Enevoldsen, B. S., Reimann, L., & Hansen, N. (1977). Biospecific Affinity Chromatography of Pullulanase. *FEBS Letters, 79*(1), 121–124. doi:10.1016/0014-5793(77)80364-5 PMID:560992

Eng, C. M., Guffon, N., Wilcox, W. R., Germain, D. P., Lee, P., Waldek, S., ... Desnick, R. J. (2001). Safety and efficacy of recombinant human α-galactosidase A replacement therapy in Fabry's disease. *The New England Journal of Medicine, 345*(1), 9–16. doi:10.1056/NEJM200107053450102 PMID:11439963

Engelmann, C., Illner, S., & Kragl, U. (2015). Laccase initiated C-C couplings: Various techniques for reaction monitoring. *Process Biochemistry, 50*(10), 1591–1599. doi:10.1016/j.procbio.2015.05.011

Enguita, F. J., Marçal, D., Martins, L. O., Grenha, R., Henriques, A. O., Lindley, P. F., & Carrondo, M. A. (2004). Substrate and dioxygen binding to the endospore coat laccase from *Bacillus subtilis*. *The Journal of Biological Chemistry, 279*(22), 23472–23476. doi:10.1074/jbc.M314000200 PMID:14764581

Enoki, T., Sagawa, H., Tominaga, T., Nishiyama, E., Koyama, N., Sakai, T., . . . Ikai, K. (2003). *US Patent, 0105029 A1*. US Patent Office.

Eremia, S. A., Vasilescu, I., Radoi, A., Litescu, S. C., & Radu, G. L. (2013). Disposable biosensor based on platinum nanoparticles-reduced graphene oxide-laccase biocomposite for the determination of total polyphenolic content. *Talanta, 110*, 164–170. doi:10.1016/j.talanta.2013.02.029 PMID:23618190

Espinoza, F. M. C., Morel, M. H., & Rouau, X. (1998). Effect of lysine, tyrosine, cysteine and glutathione on the oxidative crosslinking of feruloylated arabinoxylans by a fungal laccase. *Journal of Agricultural and Food Chemistry, 46*, 2589.

Espinoza, F. M. C., Morel, M. H., Surget, A., Asther, M., Moukha, S., Sigoillot, J. C., & Rouau, X. (1999). Attempt to crosslinkferuloylated arabinoxylans and proteins with a fungal laccase. *Food Hydrocolloids, 13*(1), 65–71. doi:10.1016/S0268-005X(98)00072-1

Espinoza, F. M. C., & Rouau, X. (1999). Effect of cysteinyl caffeic acid, caffeic acid, and l-dopa on the oxidative cross-linking of feruloylated arabinoxylans by a fungal laccase. *Journal of Agricultural and Food Chemistry, 47*(2), 497–503. doi:10.1021/jf980723q PMID:10563923

Ezeji, T. C., Wolf, A., & Bahl, H. (2005). Isolation and identification of *Geobacillus thermodenitrificans* HRO10. An α-amylase and α- glucosidase producing thermophile. *Canadian Journal of Microbiology, 51*, 685–693. doi:10.1139/w05-054 PMID:16234866

Fabbrini, M., Galli, C., & Gentili, P. (2002). Comparing the catalytic efficiency of some mediators of laccase. *Journal of Molecular Catalysis. B, Enzymatic, 16*(5-6), 231–240. doi:10.1016/S1381-1177(01)00067-4

Fabbrini, M., Galli, C., Gentili, P., & Macchitella, D. (2001). An oxidation of alcohols by oxygen with the enzyme laccase and mediation by TEMPO. *Tetrahedron Letters, 42*(43), 7551–7553. doi:10.1016/S0040-4039(01)01463-0

Fabbro, D., Parkinson, D., & Matter, A. (2002). Protein tyrosine kinase inhibitors: New treatment modalities? *Current Opinion in Pharmacology, 2*(4), 374–381. doi:10.1016/S1471-4892(02)00179-0 PMID:12127869

Faccin, G. L., Viera, L. D. N., Miotto, L. A., Barreto, P. L. M., & Amante, E. R. (2009). Chemical, sensorial and rheological properties of a new organic rice bran beverage. *Rice Science, 16*(3), 226–234. doi:10.1016/S1672-6308(08)60083-9

Fadden, K., & Owen, R. W. (1992). Faecal steroids and colorectal cancer: The effect of lactulose on faecal bacterial metabolism in a continuous culture model of the large intestine. *European Journal of Cancer Prevention, 1*(2), 113–127. doi:10.1097/00008469-199202000-00004 PMID:1463973

Fadel, M., & Fouda, M. S. (1993). Physiological studies on xylanase production by *Penicillium fumiculosum* on some agricultural wastes. *Microbiology, 148*, 304–312. PMID:8368028

Fang, H. Y., Changa, S. M., Hsieh, M. C., & Fang, T. J. (2007). Production, optimization growth conditions and properties of the xylanase from *Aspergillus carneus* M34. *Journal of Molecular Catalysis. B, Enzymatic, 49*(1-4), 36–42. doi:10.1016/j.molcatb.2007.08.002

Fan, Y., & Huang, Y. (2012). The effective peroxidase-like activity of chitosan-functionalized CoFe2O4 nanoparticles for chemiluminescence sensing of hydrogen peroxide and glucose. *Analyst (London), 137*(5), 1225–1231. doi:10.1039/c2an16105b PMID:22258160

FAO. (2004). *Laccase from* Myceliophthora thermophila *expressed in* Aspergillus oryzae. Chemical and Technical Assessment (CTA). FAO.

Faramarzi, M. A., & Forootanfar, H. (2011). Biosynthesis and characterization of gold nanoparticles produced by laccase from *Paraconiothyrium variabile. Colloids and Surfaces. B, Biointerfaces, 87*(1), 23–27. doi:10.1016/j.colsurfb.2011.04.022 PMID:21616647

Faria, N. R., Rambaut, A., Suchard, M. A., Baele, G., Bedford, T., Ward, M. J., ... Posada, D. (2014). The early spread and epidemic ignition of HIV-1 in human populations. *Science*, *346*(6205), 56–61. doi:10.1126cience.1256739 PMID:25278604

Feijoo, G., Dosoretz, C., & Lema, J. M. (1995). Production of lignin-peroxidase by *Phanerochaete chrysosporium* in a packed bed bioreactor operated in semicontinuous mode. *Journal of Biotechnology*, *42*(3), 247–253. doi:10.1016/0168-1656(95)00085-5

Feng, C., Khulbe, K. C., Matsuura, T., Tabe, S., & Ismail, A. F. (2013). Preparation and characterization of electro-spun nanofiber membranes and their possible applications in water treatment. *Separation and Purification Technology*, *102*, 118–135. doi:10.1016/j.seppur.2012.09.037

Feng, W., & Ji, P. (2011). Enzymes immobilized on carbon nanotubes. *Biotechnology Advances*, *29*(6), 889–895. doi:10.1016/j.biotechadv.2011.07.007 PMID:21820044

Fenice, M., Sermanni, G. G., Federici, F., & D'Annibale, A. (2003). Submerged and solid-state production of laccase and Mn-peroxidase by *Panustigrinus* on olive mill wastewater-based media. *Journal of Biotechnology*, *100*(1), 77–85. doi:10.1016/S0168-1656(02)00241-9 PMID:12413788

Fernandez-Espinar, M. T., Ramon, D., Pinaga, F., & Valles, S. (1992). Xylanase production by *Aspergillus nidulans*. *FEMS Microbiology*, *91*, 91–96. doi:10.1111/j.1574-6968.1992.tb05190.x

Ferraroni, M., Myasoedova, N. M., Schmatchenko, V., Leontievsky, A. A., Golovleva, L. A., Scozzafava, A., & Briganti, F. (2007). Crystal structure of a blue laccase from Lentinus tigrinus: Evidences for intermediates in the molecular oxygen reductive splitting by multicopper oxidases. *BMC Structural Biology*, *7*(1), 60. doi:10.1186/1472-6807-7-60 PMID:17897461

Ferreira, L., Carvalho, R., Gil, M. H., & Dordick, J. S. (2002). Enzymatic synthesis of inulin-containing hydrogels. *Biomacromolecules*, *3*(2), 333–341. doi:10.1021/bm010150h PMID:11888320

Ferreira, L., Gil, M. H., Cabrita, A. M., & Dordick, J. S. (2005). Biocatalytic synthesis of highly ordered degradable dextran-based hydrogels. *Biomaterials*, *26*(23), 4707–4716. doi:10.1016/j.biomaterials.2004.11.051 PMID:15763250

Ferreira, L., Gil, M. H., & Dordick, J. S. (2002). Enzymatic synthesis of dextran-containing hydrogels. *Biomaterials*, *23*(19), 3957–3967. doi:10.1016/S0142-9612(02)00132-1 PMID:12162328

Ferry, Y., & Leech, D. (2005). Amperometric detection of catecholamine neurotransmitters using electrocatalytic substrate recycling at a laccase electrode. *Electroanalysis*, *17*(2), 2113–2119. doi:10.1002/elan.200403069

Fetzner, S. (1998). Bacterial dehalogenation. *Applied Microbiology and Biotechnology*, *50*(6), 633–657. doi:10.1007002530051346 PMID:9891928

Fetzner, S. (2003). Oxygenases without requirement for cofactors or metal ions. *Applied Microbiology and Biotechnology*, *60*(3), 243–257. PMID:12436305

Fetzner, S., & Lingens, F. (1994). Bacterial dehalogenases: Biochemistry, genetics, and biotechnological applications. *Microbiological Reviews*, *58*(4), 641–685. PMID:7854251

Fierobe, H. P., Mirgorodskaya, E., Frandsen, T. P., Roepstor, P., & Svensson, B. (1997). Overexpression and characterization of *Aspergillus awamori* wild-type and mutant glucoamylase secreted by the methylotrophic yeast *Pichia pastoris*: Comparison with wild-type recombinant glucoamylase produced using *Saccharomyces cerevisiae* and *Aspergillus niger* as hosts. *Protein Expression and Purification*, *9*(2), 159–170. doi:10.1006/prep.1996.0689 PMID:9056481

Filby, A. L., Neuparth, T., Thorpe, K. L., Owen, R., Galloway, T. S., & Tyler, C. R. (2007). Health impacts of estrogens in the environment, considering complex mixture effects. *Environmental Health Perspectives*, *115*(12), 1704–1710. doi:10.1289/ehp.10443 PMID:18087587

Fini, L., Hotchkiss, E., Fogliano, V., Graziani, G., Romano, M., De Vol, E. B., ... Ricciardiello, L. (2007). Chemopreventive properties of pinoresinol-rich olive oil involve a selective activation of the ATM–p53 cascade in colon cancer cell lines. *Carcinogenesis*, *29*(1), 139–146. doi:10.1093/carcin/bgm255 PMID:17999988

Firozjaee, T. T., Najafpour, G. D., Asgari, A., & Khavarpour, M. (2013). Biological treatment of phenolicwastewater in an anaerobiccontinuous stirred tank reactor. *Chemical Industry & Chemical Engineering Quarterly*, *19*(2), 173–179. doi:10.2298/CICEQ120216052F

Fleming, F. F., Yao, L., Ravikumar, P. C., Funk, L., & Shook, B. C. (2010). Nitrile-containing pharmaceuticals: Efficacious roles of the nitrile pharmacophore. *Journal of Medicinal Chemistry*, *53*(22), 7902–7917. doi:10.1021/jm100762r PMID:20804202

Fogarty, W. M. (1983). Microbial amylases. In W. M. Fogarty (Ed.), *Microbial enzymes and biotechnology* (pp. 1–92). London: Applied Science Publications Ltd.

Fogarty, W. M., & Kelly, C. T. (1990). In W. M. Forgarty & C. T. Kelly (Eds.), *Microbial Enzymes and Biotechnology* (2nd ed.; pp. 71–132). London: Elsevier Science Publishers. doi:10.1007/978-94-009-0765-2_3

Foh, M. B. K., Amadou, I., Foh, B. M., Kamara, M. K., & Xia, W. (2010). Functionality and Antioxidant Properties of Tilapia (*Oreochromis niloticus*) as Influenced by the degree of hydrolysis. *International Journal of Molecular Sciences*, *11*(12), 1851–1869. doi:10.3390/ijms11041851 PMID:20480046

Forbath, T. (1957). Flexible processing keys enzymes' future. *Chemical Engineering (Albany, N.Y.)*, *64*, 226–229.

Fore, H., Walker, R., & Golberg, L. (1967). Studies on Brown FK. II. Degradative changes undergone in vitro and in vivo. *Food and Cosmetics Toxicology*, *5*, 459–473. doi:10.1016/S0015-6264(67)83147-X PMID:4295788

Forootanfar, H., & Faramarzi, M. A. (2015). Insights into laccase producing organisms, fermentation states, purification strategies, and biotechnological applications. *Biotechnology Progress*, *31*(6), 1443–1463. doi:10.1002/btpr.2173 PMID:26399693

Forootanfar, H., Faramarzi, M. A., Shahverdi, A. R., & Yazdi, M. T. (2011). Purification and biochemical characterization of extracellular laccase from the ascomycete *Paraconiothyrium variabile*. *Bioresource Technology*, *102*(2), 1808–1814. doi:10.1016/j.biortech.2010.09.043 PMID:20933400

Forootanfar, H., Moezzi, A., Aghaie-Khozani, M., Mahmoudjanlou, Y., Ameri, A., Niknejad, F., & Faramarzi, M. A. (2012). Synthetic dye decolorization by three sources of fungal laccase. *Iranian Journal of Environmental Health Sciences & Engineering*, *9*(1), 27. doi:10.1186/1735-2746-9-27 PMID:23369690

Forootanfar, H., Movahednia, M. M., Yaghmaei, S., Tabatabaei-Sameni, M., Rastegar, H., Sadighi, A., & Faramarzi, M. A. (2012). Removal of chlorophenolic derivatives by soil isolated ascomycete of *Paraconiothyrium variabile* and studying the role of its extracellular laccase. *Journal of Hazardous Materials*, *209–210*, 199–203. doi:10.1016/j.jhazmat.2012.01.012 PMID:22277342

Forootanfar, H., Rezaei, S., Zeinvand-Lorestani, H., Tahmasbi, H., Mogharabi, M., Ameri, A., & Faramarzi, M. A. (2016). Studies on the laccase-mediated decolorization, kinetic, and microtoxicity of some synthetic azo dyes. *Journal of Environmental Health Science & Engineering*, *14*(1), 7. doi:10.118640201-016-0248-9 PMID:27182441

Frederick, M. M., Kiang, C., Frederick, J. R., & Reilly, P. J. (1985). Purification and characterization of endo-xylanases from *Aspergillus niger*. I. Two isozymes active on xylan backbones near branch points. *Biotechnology and Bioengineering, 27*(4), 525–532. doi:10.1002/bit.260270420 PMID:18553703

Freire, R. S., Durán, N., & Kubota, L. T. (2001). Effects of fungal laccase immobilization procedures for the development of a biosensor for phenol compounds. *Talanta, 54*(4), 681–686. doi:10.1016/S0039-9140(01)00318-6 PMID:18968290

Friend, D. R., & Chang, G. W. (1984). A colon-specific drug-delivery system based on drug glycosides and the glycosidases of colonic bacteria. *Journal of Medicinal Chemistry, 27*(3), 261–266. doi:10.1021/jm00369a005 PMID:6699871

Friend, D. R., & Tozer, T. N. (1992). Drug glycosides in oral colon-specific drug delivery. *Journal of Controlled Release, 19*(1-3), 109–120. doi:10.1016/0168-3659(92)90069-4

Fu, J., Nyanhongo, G. S., Gubitz, G. M., Cavaco-Paulo, A., & Kim, S. (2012). Enzymatic colouration with laccase and peroxidases: Recent progress. *Biocatalysis and Biotransformation, 30*(1), 125–140. doi:10.3109/10242422.2012.649563

Fukumoto, J., & Okada, S. (1963). Studies on bacterial amylase, Amylase types of *Bacillus subtilis* species. *Journal of Fermentation Technology, 41*(1), 427–434.

Fukuoka, M., Yano, S., Giaccone, G., Tamura, T., Nakagawa, K., Douillard, J.-Y., ... Baselga, J. (2003). Multi-institutional randomized phase II trial of gefitinib for previously treated patients with advanced non--small-cell lung cancer. *Journal of Clinical Oncology, 21*(12), 2237–2246. doi:10.1200/JCO.2003.10.038 PMID:12748244

Fusco, D. (2010). Laccase based biosensor for the determination of polyphenol index in wine. *Talanta, 81*(1-2), 235–240. doi:10.1016/j.talanta.2009.11.063 PMID:20188914

Gaanappriya, M., Guhankumar, P., & Balakrishnan, V. (2011). Isolation of xylan degrading enzyme from *Trichoderma* spp. *Research in Plant Biology, 1*(3), 15–20.

Gabel, D., Awad, D., Schaffran, T., Radovan, D., Daraban, D., Damian, L., ... Edwards, K. (2007). The Anionic Boron Cluster (B(12)H(11)SH)(2-) as a means to trigger release of liposome contents. *ChemMedChem, 2*(1), 51–53. doi:10.1002/cmdc.200600227 PMID:17154431

Gabel, D., Foster, S., & Fairchild, R. G. (1987). The Monte Carlo simulation of the biological effect of the 10B(n, alpha)7Li reaction in cells and tissue and its implication for boron neutron capture therapy. *Radiation Research, 111*(1), 14–25. doi:10.2307/3577018 PMID:3602351

Gabizon, A., Horowitz, A. T., Goren, D., Tzemach, D., Shmeeda, H., & Zalipsky, S. (2003). In vivo fate of folate-targeted polyethylene-glycol liposomes in tumor-bearing mice. *Clinical Cancer Research, 9*(17), 6551–6559. PMID:14695160

Gabizon, A., & Papahadjopoulos, D. (1988). Liposome formulations with prolonged circulation time in blood and enhanced uptake by tumors. *Proceedings of the National Academy of Sciences of the United States of America, 85*(18), 6949–6953. doi:10.1073/pnas.85.18.6949 PMID:3413128

Gadipelly, C., Pérez-González, A., Yadav, G. D., Ortiz, I., Ibáñez, R., Rathod, V. K., & Marathe, K. V. (2014). Pharmaceutical industry waste water: Review of the technologies for water treatment and reuse. *Industrial & Engineering Chemistry Research, 53*(29), 11571–11592. doi:10.1021/ie501210j

Gagné, F., Blaise, C., & André, C. (2006). Occurrence of pharmaceutical products in a municipal effluent and toxicity to rainbow trout (Oncorhynchus mykiss) hepatocytes. *Ecotoxicology and Environmental Safety, 64*(3), 329–336. doi:10.1016/j.ecoenv.2005.04.004 PMID:15923035

Gale, E. F. (2006). The bacterial amino acid decarboxylases. *Advances in Enzymology and Related Areas of Molecular Biology*, *6*, 1–32.

Gallo, R. C., & Montagnier, L. (2003). The discovery of HIV as the cause of AIDS. *The New England Journal of Medicine*, *349*(24), 2283–2285. doi:10.1056/NEJMp038194 PMID:14668451

Gane, P. J., & Dean, P. M. (2000). Recent advances in structure-based rational drug design. *Current Opinion in Structural Biology*, *10*(4), 401–404. doi:10.1016/S0959-440X(00)00105-6 PMID:10981625

Ganesana, M., Istarnboulie, G., Marty, J.-L., Noguer, T., & Andreescu, S. (2011). Site-specific immobilization of a (His) 6-tagged acetylcholinesterase on nickel nanoparticles for highly sensitive toxicity biosensors. *Biosensors & Bioelectronics*, *30*(1), 43–48. doi:10.1016/j.bios.2011.08.024 PMID:21937214

Garbelli, A., Riva, V., Crespan, E., & Maga, G. (2017). How to win the HIV-1 drug resistance hurdle race: Running faster or jumping higher? *The Biochemical Journal*, *474*(10), 1559–1577. doi:10.1042/BCJ20160772 PMID:28446620

García-Galán, M. J., Rodríguez-Rodríguez, C. E., Vicent, T., Caminal, G., Díaz-Cruz, M. S., & Barceló, D. (2011). Biodegradation of sulfamethazine by *Trametes versicolor*: Removal from sewage sludge and identification of intermediate products by UPLC-QqTOF-MS. *The Science of the Total Environment*, *409*(24), 5505–5512. doi:10.1016/j.scitotenv.2011.08.022 PMID:21944200

Garg, A. P., Roberts, J. C., & McCarthy, A. (1998). Bleach boosting effect of cellulase free xylanase of *Streptomyces thermoviolaceus* and its comparison with two commercial enzyme preparations on birchwood Kraft pulp. *Enzyme Microbial Biotechnology*, *22*(7), 594–598. doi:10.1016/S0141-0229(97)00250-0

Garg, B., & Bisht, T. (2016). Carbon Nanodots as Peroxidase Nanozymes for Biosensing. *Molecules (Basel, Switzerland)*, *21*, 1653.

Gashaw, I., Ellinghaus, P., Sommer, A., & Asadullah, K. (2012). What makes a good drug target? *Drug Discovery Today*, *17*, S24–S30. doi:10.1016/j.drudis.2011.12.008 PMID:22155646

Gassara-Chatti, F., Brar, S. K., Ajila, C. M., Verma, M., Tyagi, R. D., & Valero, J. R. (2013). Encapsulation of ligninolytic enzymes and its application in clarification of juice. *Food Chemistry*, *137*(1-4), 18–24. doi:10.1016/j.foodchem.2012.09.083 PMID:23199985

Gasser, C. A., Ammann, E. M., Shahgaldian, P., & Corvini, P. F.-X. (2014). Laccases to take on the challenge of emerging organic contaminants in wastewater. *Applied Microbiology and Biotechnology*, *98*(24), 9931–9952. doi:10.100700253-014-6177-6 PMID:25359481

Gavezzotti, P., Vavříková, E., Valentová, K., Fronza, G., Kudanga, T., Kuzma, M., ... Křen, V. (2014). Enzymatic oxidative dimerization of silymarin flavonolignans. *Journal of Molecular Catalysis. B, Enzymatic*, *109*, 24–30. doi:10.1016/j.molcatb.2014.07.012

Gawande, P. V., & Kamat, M. Y. (1999). Production of *Aspergillus* xylanase by lignocellulosic waste fermentation and its application. *Journal of Applied Microbiology*, *87*(4), 511–519. doi:10.1046/j.1365-2672.1999.00843.x PMID:10583678

Gbogouri, G. A., Linder, M., Fanni, J., & Parmentier, M. (2004). Influence of hydrolysis degree on the functional properties of salmon byproducts hydrolysates. *Journal of Food Science*, *69*(8), 615–622. doi:10.1111/j.1365-2621.2004.tb09909.x

Genckal, H., & Tari, C. (2006). Alkaline protease production from alkalophilic *Bacillus* sp. isolated from natural habitats. *Enzyme and Microbial Technology*, *39*(4), 703–710. doi:10.1016/j.enzmictec.2005.12.004

Gernjak, W., Krutzler, T., Glaser, A., Malato, S., Caceres, T., Bauer, R., & Fernandez-Alba, A. R. (2003). Photo-Fenton treatment of water containing natural phenolic pollutants. *Chemosphere*, *50*(1), 71–78. doi:10.1016/S0045-6535(02)00403-4 PMID:12656231

Geronikaki, A., Eleftheriou, P., & Poroikov, V. (2016). *Anti-HIV Agents: Current Status and Recent Trends*. Academic Press.

Ge, S. (2014). Disposable electrochemical immunosensor based on peroxidase-like magnetic silica-graphene oxide composites for detection of cancer antigen 153. *Sensors and Actuators. B, Chemical*, *192*, 317–326.

Gessesse, A., & Gashe, B. A. (1997). Production of alkaline xylanase by an alkaliphilic *Bacillus spp*. isolated from an alkaline soda lake. *Journal of Applied Microbiology*, *83*(4), 402–406. doi:10.1046/j.1365-2672.1997.00242.x

Geyer, H., Jacobi, I., Linder, D., Stirm, S., Bialojan, S., Strube, K. H., & Geyer, R. (1996). Glycosylation of recombinant ancrod from Agkistrodon rhodostoma after expression in mouse epithelial cells. *European Journal of Biochemistry*, *237*(1), 113–127. doi:10.1111/j.1432-1033.1996.0113n.x PMID:8620863

Ghasemi, F., Tabandeh, F., Bambai, B., & Rao, K. R. S. (2010). Decolorization of different azo dyes by *Phanerochaete chrysosporium* RP78 under optimal condition. *International Journal of Environmental Science and Technology*, *7*(3), 457–464. doi:10.1007/BF03326155

Ghindilis, A. L., Gavrilova, V. P., & Yaropolov, A. I. (1992). Laccase based biosensor for determination of polyphenols: Determination of catechols in tea. *Biosensors & Bioelectronics*, *7*(2), 127–131. doi:10.1016/0956-5663(92)90017-H PMID:1637524

Ghosh, D., Saha, M., Sana, B., & Mukherjee, J. (2005). Article. *J. Marine Enzymes. Adv. Biochem. Eng. Biotechnol.*, *96*, 189–218.

Ghosh, M., & Nanda, G. (1994). Purification and some properties of xylanase from *Aspergillus sydowii* MG 49. *Applied and Environmental Microbiology*, *60*, 4620–4623. PMID:7811101

Ghosh, P., Yang, X., Arvizo, R., Zhu, Z. J., Agasti, S. S., Mo, Z., & Rotello, V. M. (2010). Intracellular delivery of a membrane-impermeable enzyme in active form using functionalized gold nanoparticles. *Journal of the American Chemical Society*, *132*(8), 2642–2645. doi:10.1021/ja907887z PMID:20131834

Giacometti, J., & Tomljanovic, A. B. (2017). Peptidomics as a tool for characterizing bioactive milk peptides. *Food Chemistry*, *23*, 91–98. doi:10.1016/j.foodchem.2017.03.016 PMID:28407976

Gianfreda, L., Xu, F., & Bollag, J. M. (1999). Laccases: A useful group of oxidoreductive enzymes. *Bioremediation Journal*, *3*(1), 125. doi:10.1080/10889869991219163

Gibbs, J. B. (1991). Ras C-terminal processing enzymes-New drug targets? *Cell*, *65*(1), 1–4. doi:10.1016/0092-8674(91)90352-Y PMID:2013090

Gibson, J. F. (1997). Chemistry in the vineyards. *Education in Chemistry*, *34*, 123–126.

Gibson, G. R., & Roberfroid, A. M. (1995). Dietary modulation of the human colonic microbiota: Introducing the concept of prebiotics. *The Journal of Nutrition*, *125*, 1401–1412. PMID:7782892

Gifford, A. H., Lahey, T., & Reyn, C. F. V. (2006). Fatal hemoptysis from invasive *Aspergillus niger* in a patient with cavitary lung disease and *Mycobacterium avium complex* infection. *Medical Mycology*, *44*(6), 557–560. doi:10.1080/13693780600575635 PMID:16966174

Gildberg, A. (2004). Enzymes and bioactive peptides from fish waste related to fish silage, fish feed and fish sauce production. *Journal of Aquatic Food Product Technology, 13*(2), 3–11. doi:10.1300/J030v13n02_02

Ginet, N., Pardoux, R., Adryanczyk, G., Garcia, D., Brutesco, C., & Pignol, D. (2011). Single-step production of a recyclable nanobiocatalyst for organophosphate pesticides biodegradation using functionalized bacterial magnetosomes. *PLoS One, 6*(6), e21442. doi:10.1371/journal.pone.0021442 PMID:21738665

Gingell, R., & Walker, R. (1971). Mechanisms of azo reduction by Streptococcus faecalis II. The role of soluble flavins. *Xenobiotica, 1*(3), 231–239. doi:10.3109/00498257109033172 PMID:4341449

Giovanelli, G. (1989). Enzymic treatment of malt polyphenols for beer stabilization. *Industrie delle Bevande, 18*, 497–502.

Giovanelli, G., & Ravasini, G. (1993). Apple juice stabilization by combined enzyme- membrane filtration process. *Lebensmittel-Wissenschaft + Technologie, 26*(1), 1–7. doi:10.1006/fstl.1993.1001

Girio, F. M., Fonseca, C., Carvalheiro, F., Duarte, L. C., Marques, S., & Bogel-Lukasik, R. (2010). Hemicelluloses for fuel ethanol: A review. *Bioresource Technology, 101*(13), 4775–4800. doi:10.1016/j.biortech.2010.01.088 PMID:20171088

Giugliani, R., Lampe, C., Guffon, N., Ketteridge, D., Leao-Teles, E., Wraith, J. E., & Harmatz, P. (2014). Natural history and galsulfase treatment in mucopolysaccharidosis VI (MPS VI, Maroteaux-Lamy syndrome)-10-year follow-up of patients who previously participated in an MPS VI Survey Study. *American Journal of Medical Genetics. Part A, 164a*(8), 1953–1964. doi:10.1002/ajmg.a.36584 PMID:24764221

Gkioni, K., Leeuwenburgh, S. C., Douglas, T. E., Mikos, A. G., & Jansen, J. A. (2010). Mineralization of hydrogels for bone regeneration. *Tissue Engineering. Part B, Reviews, 16*(6), 577–585. doi:10.1089/ten.teb.2010.0462 PMID:20735319

Glenn, J. K., Lakshmi, A., & Gold, M. H. (1986). Mn(II) oxidation is the principal function of the extracellular Mn-Peroxidase from *Phanerochaete chrysosporium. Archives of Biochemistry and Biophysics, 251*(2), 688–696. doi:10.1016/0003-9861(86)90378-4 PMID:3800395

Gliko-Kabir, I., Yagen, B., Baluom, M., & Rubinstein, A. (2000). Phosphated crosslinked guar for colon-specific drug delivery: II. In vitro and in vivo evaluation in the rat. *Journal of Controlled Release, 63*(1), 129–134. doi:10.1016/S0168-3659(99)00180-7 PMID:10640586

Gliko-Kabir, I., Yagen, B., Penhasi, A., & Rubinstein, A. (1998). Low swelling, crosslinked guar and its potential use as colon-specific drug carrier. *Pharmaceutical Research, 15*(7), 1019–1025. doi:10.1023/A:1011921925745 PMID:9688054

Gnanasambandam, R., Mathias, M., & Proctor, A. (1998). Structure and performance of soy hull carbon adsorbents as affected by pyrolysis temperature. *Journal ofthe American Oil Chemists'. Society, 75*(5), 615–621.

Goga, A., McLaughlin, J., Afar, D. E. H., Saffran, D. C., & Witte, O. N. (1995). Alternative signals to RAS for hematopoietic transformation by the BCR-ABL oncogene. *Cell, 82*(6), 981–988. doi:10.1016/0092-8674(95)90277-5 PMID:7553858

Gokmen, V., Borneman, Z., & Nijhuis, H. H. (1998). Improved ultrafiltration for color reduction and stabilization of apple juice. *Journal of Food Science, 63*(3), 504–507. doi:10.1111/j.1365-2621.1998.tb15773.x

Golan-Rozen, N., Chefetz, B., Ben-Ari, J., Geva, J., & Hadar, Y. (2011). Transformation of the recalcitrant pharmaceutical compound carbamazepine by *Pleurotus ostreatus*: Role of cytochrome P450 monooxygenase and manganese peroxidase. *Environmental Science & Technology, 45*(16), 6800–6805. doi:10.1021/es200298t PMID:21744850

Golchin, J., Golchin, K., Alidadian, N., Ghaderi, S., Eslamkhah, S., Eslamkhah, M., & Akbarzadeh, A. (2017). Nanozyme applications in biology and medicine: an overview. *Artificial Cells, Nanomedicine, and Biotechnology, an International Journal, 45*(6), 1069-1076.

Gold, M. H., & Alic, M. (1993). Molecular biology of the lignin-degrading Basidiomycete *Phanerochaete chrysosporium*. *Microbiological Reviews*, *57*, 605–622. PMID:8246842

Golynskiy, M. V., & Seelig, B. (2010). De novo enzymes: From computational design to mRNA display. *Trends in Biotechnology*, *28*(7), 340–345. doi:10.1016/j.tibtech.2010.04.003 PMID:20483496

Golz-Berner, K., Walze, B., Zastrow, L., & Doucet, O. (2004). *Cosmetic or dermatological preparation with skin-lightening proteins*. WO2004017931.

Gomes, J., & Steiner, W. (2004). The Biocatalytic potential of Extremophiles and Extremozymes. *Food Technology*, *42*(4), 223–239.

Gómez-Anquela, C., García-Mendiola, T., Abad, J. M., Pita, M., Pariente, F., & Lorenzo, E. (2015). Scaffold electrodes based on thioctic acid-capped gold nanoparticles coordinated Alcohol Dehydrogenase and Azure A films for high performance biosensor. *Bioelectrochemistry (Amsterdam, Netherlands)*, *106*, 335–342. doi:10.1016/j.bioelechem.2015.06.009 PMID:26152878

Gómez-Estaca, J., de Lacey, A. L., López-Caballero, M., Gómez-Guillén, M., & Montero, P. (2010). Biodegradable gelatin–chitosan films incorporated with essential oils as antimicrobial agents for fish preservation. *Food Microbiology*, *27*(7), 889–896. doi:10.1016/j.fm.2010.05.012 PMID:20688230

Gómez-Guillén, M., Giménez, B., López-Caballero, M., & Montero, M. (2011). Functional and bioactive properties of collagen and gelatin from alternative sources: A review. *Food Hydrocolloids*, *25*(8), 1813–1827. doi:10.1016/j.foodhyd.2011.02.007

Gomez-Toribio, V., Garcıa-Martın, A. B., Martınez, M. J., Martınez, A. T., & Guillen, F. (2009). Enhancing the production of hydroxyl radicals by *Pleurotus eryngii* via quinone redox cycling for pollutant removal. *Applied and Environmental Microbiology*, *75*(12), 3954–3962. doi:10.1128/AEM.02138-08 PMID:19376890

Goodman, L. E., Judd, S. R., Farnsworth, C. C., Powers, S., Gelb, M. H., Glomset, J. A., & Tamanoi, F. (1990). Mutants of Saccharomyces cerevisiae defective in the farnesylation of Ras proteins. *Proceedings of the National Academy of Sciences of the United States of America*, *87*(24), 9665–9669. doi:10.1073/pnas.87.24.9665 PMID:2124698

Goodman, L. S. (1996). *Goodman and Gilman's the pharmacological basis of therapeutics* (Vol. 1549). New York: McGraw-Hill.

Gopinathan, R., Kanhere, J., & Banerjee, J. (2015). Effect of Malachite Green Toxicity on Non Target Soil Organisms. *Chemosphere*, *120*, 637–644. doi:10.1016/j.chemosphere.2014.09.043 PMID:25462308

Gorrasi, G., Bugatti, V., & Vittoria, V. (2012). Pectins filled with LDH-antimicrobial molecules: Preparation, characterization and physical properties. *Carbohydrate Polymers*, *89*(1), 132–137. doi:10.1016/j.carbpol.2012.02.061 PMID:24750614

Gottlieb, M. S., Schroff, R., Schanker, H. M., Weisman, J. D., Fan, P. T., Wolf, R. A., & Saxon, A. (1981). Pneumocystis carinii pneumonia and mucosal candidiasis in previously healthy homosexual men: Evidence of a new acquired cellular immunodeficiency. *The New England Journal of Medicine*, *305*(24), 1425–1431. doi:10.1056/NEJM198112103052401 PMID:6272109

Goyal, M., Kalra, K. L., Sareen, V. K., & Soni, G. (2008). Xylanase production with xylan rich lignocellulosic wastes by a local soil isolate of *Trichoderma viride*. *Brazilian Journal of Microbiology*, *39*(3), 535–541. doi:10.1590/S1517-83822008000300025 PMID:24031262

Grabowski, G. A. (2005). Recent clinical progress in Gaucher disease. *Current Opinion in Pediatrics*, *17*(4), 519–524. doi:10.1097/01.mop.0000172702.33128.19 PMID:16012266

Grabowski, G. A., Barton, N. W., Pastores, G., Dambrosia, J. M., Banerjee, T. K., McKee, M. A., & Brady, R. O. (1995). Enzyme therapy in type 1 Gaucher disease: Comparative efficacy of mannose-terminated glucocerebrosidase from natural and recombinant sources. *Annals of Internal Medicine, 122*(1), 33–39. doi:10.7326/0003-4819-122-1-199501010-00005 PMID:7985893

Greenway, F., Liu, Z., Yu, Y., & Gupta, A. (2011). A clinical trial testing the safety and efficacy of a standardized *Eucommia ulmoides*oliver bark extract to treat hypertension. *Alternative Medicine Review, 16*, 338–347. PMID:22214253

Grim, Y., & Kopecek, J. (1991). Bioadhesive water-soluble polymeric drug carriers for site-specific oral drug delivery. Synthesis, characterization, and release of 5-aminosalicylic acid by Streptococcus faecium in vitro. *New Polymeric Materials, 3*(1), 49-59.

Gros, M., Petrović, M., Ginebreda, A., & Barceló, D. (2010). Removal of pharmaceuticals during wastewater treatment and environmental risk assessment using hazard indexes. *Environment International, 36*(1), 15–26. doi:10.1016/j.envint.2009.09.002 PMID:19819553

Gruber, A. J., Hudson, J. I., & Pope, H. G. Jr. (1996). The management of treatment-resistant depression in disorders on the interface of psychiatry and medicine. *The Psychiatric Clinics of North America, 19*(2), 351–369. doi:10.1016/S0193-953X(05)70292-6 PMID:8827194

Guaadaoui, A., Benaicha, S., Elmajdoub, N., Bellaoui, M., & Hamal, A. (2014). What is a bioactive compound? A combined definition for a preliminary consensus. *International Journal of Nutrition and Food Sciences, 3*, 174–179.

Gummadi, S. N., & Panda, T. (2003). Purification and biochemical properties of microbial pectinases—a review. *Process Biochemistry, 38*(7), 987–996. doi:10.1016/S0032-9592(02)00203-0

Guo, X. L., Zhu, Z. W., & Li, H. L. (2014). Biodegradation of sulfamethoxazole by *Phanerochaete chrysosporium. Journal of Molecular Liquids, 198*, 169–172. doi:10.1016/j.molliq.2014.06.017

Guo, X., MacKay, J. A., & Szoka, F. C. Jr. (2003). Mechanism of pH-triggered collapse of phosphatidylethanolamine liposomes stabilized by an ortho ester polyethyleneglycol lipid. *Biophysical Journal, 84*(3), 1784–1795. doi:10.1016/S0006-3495(03)74986-8 PMID:12609880

Gupta, A., Joseph, B., Mani, A., & Thomas, G. (2008). Biosynthesis and properties of an extracellular thermostable serine alkaline protease from *Virgibacillus pantothenticus. World Journal of Microbiology & Biotechnology, 24*(2), 237–243. doi:10.100711274-007-9462-z

Gupta, R., Beg, Q. K., & Lorenz, P. (2002). Bacterial alkaline proteases: Molecular approaches and Industrial Applications. *Applied Microbiology and Biotechnology, 59*(1), 15–32. doi:10.100700253-002-0975-y PMID:12073127

Gupta, U., & Kar, R. (2009). Xylanase production by a thermo-tolerant *Bacillus* species under solid-state and submerged fermentation. *Brazilian Archives of Biology and Technology, 52*(6), 1363–1371. doi:10.1590/S1516-89132009000600007

Gurujeyalakshmi, G., & Oriel, P. (1989). Isolation of phenol-degrading *Bacillus stearothermophilus* and partial characterization of the phenol hydroxylase. *Applied and Environmental Microbiology, 55*(2), 500–502. PMID:2719481

Gurung, N., Ray, S., Bose, S., & Rai, V. (2013). A broader view: Microbial enzymes and their relevance in industries, medicine, and beyond. *BioMed Research International*. PMID:24106701

Gushterova, A., Vasileva-Tonkova, E., Dimova, E., Nedkov, P., & Haertle, T. (2005). Keratinase production by newly isolated Antarctic actinomycete strains. *World Journal of Microbiology & Biotechnology, 21*(6-7), 831–834. doi:10.100711274-004-2241-1

Gu, Z., & Tang, Y. (2010). Enzyme-assisted photolithography for spatial functionalization of hydrogels. *Lab on a Chip*, *10*(15), 1946–1951. doi:10.1039/c001335h PMID:20436969

Haddaoui, E., Chambert, R., Petit-Glatron, M. F., Lindy, O., & Sarvas, M. (1999). *Bacillus subtilis* k-amylase: The rate limiting step of secretion is growth phase-independent. *FEMS Microbiology Letters*, *173*, 127–131. PMID:10400504

Hadzhiyska, H., Calafell, M., Gibert, J. M., Daga', J. M., & Tzanov, T. (2006). Laccase-assisted dyeing of cotton. *Biotechnology Letters*, *28*(10), 755–759. doi:10.100710529-006-9043-5 PMID:16791731

Haeberlin, B., Rubas, W., Nolen, H. W. III, & Friend, D. R. (1993). In vitro evaluation of dexamethasone-β-d-glucuronide for colon-specific drug delivery. *Pharmaceutical Research*, *10*(11), 1553–1562. doi:10.1023/A:1018956232628 PMID:8290467

Haghighi, B., Gorton, L., Ruzgas, T., & Jönsson, L. J. (2003). Characterization of graphite electrodes modified with laccase from Trametes versicolor and their use for bioelectrochemical monitoring of phenolic compounds in flow injection analysis. *Analytica Chimica Acta*, *487*(1), 314. doi:10.1016/S0003-2670(03)00077-1

Hahn, V., Mikolasch, A., Manda, K., Gördes, D., Thurow, K., & Schauer, F. (2009b). Laccase-catalyzed carbon–nitrogen bond formation: Coupling and derivatization of unprotected L-phenylalanine with different para-hydroquinones. *Amino Acids*, *37*(2), 315–321. doi:10.100700726-008-0154-2 PMID:18695937

Hahn, V., Mikolasch, A., Wende, K., Bartrow, H., Lindequist, U., & Schauer, F. (2009a). Synthesis of model morpholine derivatives with biological activities by laccase-catalysed reactions. *Biotechnology and Applied Biochemistry*, *54*(4), 187–195. doi:10.1042/BA20090219 PMID:19788410

Hain, Q., & Qayyum, S. (2013). Biological and enzymatic treatment of bisphenolA and other endocrine disrupting compounds: A review. *Critical Reviews in Biotechnology*, *33*(3), 260–292. doi:10.3109/07388551.2012.694409 PMID:22712546

Hakim, A. B. R. (2006). *Study on the effect of substrates concentration in the production of xylanase from Aspergillus niger using palm kernel cake.* University College of Engineering and Technology.

Hakulinen, N., Kiiskinen, L.-L., Kruus, K., Saloheimo, M., Paananen, A., Koivula, A., & Rouvinen, J. (2002). Crystal structure of a laccase from *Melanocarpus albomyces* with an intact trinuclear copper site. *Nature Structural Biology*, *9*, 601–605. PMID:12118243

Hakulinen, N., & Rouvinen, J. (2015). Three-dimensional structures of laccases. *Cellular and Molecular Life Sciences*, *72*(5), 857–868. doi:10.100700018-014-1827-5 PMID:25586561

Hall, A. (1990). The cellular functions of small GTP-binding proteins. *Science*, *249*(4969), 635–640. doi:10.1126cience.2116664 PMID:2116664

Hamed, I., Ozogul, F., & Regenstein, J. M. (2016). Industrial application of crustacean by-product: A review. *Trends in Food Science & Technology*, *48*, 40–50. doi:10.1016/j.tifs.2015.11.007

Hamlyn, P. F. (1997). Fungal Biotechnology. *NWFG Newsletter*.

Hammel, K. E., & Cullen, D. (2008). Role of fungal peroxidases in biological ligninolysis. *Current Opinion in Plant Biology*, *11*(3), 349–355. doi:10.1016/j.pbi.2008.02.003 PMID:18359268

Hammel, K. E., Jensen, K. A. Jr, Mozuch, M. D., Landucci, L. L., Tien, M., & Pease, E. A. (1993). Ligninolysis by a purified lignin peroxidase. *The Journal of Biological Chemistry*, *268*, 12274–12281. PMID:8509364

Hammel, K. E., & Moen, M. A. (1991). Depolymerization of a synthetic lignin in vitro by lignin peroxidise. *Enzyme and Microbial Technology*, 13.

Hammond, P. T., & Whitesides, G. M. (1995). Formation of polymer microstructures by selective deposition of poly ion multilayers using patterned self assembled self assembled monolayers as a template. *Macromolecules, 28*(22), 7569–7571. doi:10.1021/ma00126a040

Ham, S. S., Kim, S. H., Moon, S. Y., Chung, M. J., Cui, C. B., Han, E. K., ... Choe, M. (2009). Antimutagenic effects of subfractions of Chaga mushroom (*Inonotus obliquus*) extract. *Mutation Research/Genetic Toxicology and Environmental Mutagenesis*, *672*(1), 55–59. doi:10.1016/j.mrgentox.2008.10.002 PMID:18992843

Hamzeh, A., Benjakul, S., & Senphan, T. (2016). Comparative study on antioxidant activity of hydrolysates from splendid squid *(Loligo formosana)* gelatin and protein isolate prepared using protease from hepatopancreas of Pacific white shrimp *(Litopenaeus vannamei)*. *Journal of Food Science and Technology*, *53*(6), 3615–3623. doi:10.100713197-016-2348-z PMID:27777469

Han, K. N., Choi, J.-S., & Kwon, J. (2017). Gold nanozyme-based paper chip for colorimetric detection of mercury ions. *Scientific Reports*, *7*(1), 2806. doi:10.103841598-017-02948-x PMID:28584238

Hanke, M. T., & Koessler, K. K. (1924). Studies on proteinogenous amines XVII. On the faculty of normal intestinal bacteria to form toxic amines. *The Journal of Biological Chemistry*, *59*(3), 835–853.

Hansch, C., Sammes, P. G., & Taylor, J. B. (1989). *Comprehensive medicinal chemistry: the rational design, mechanistic study & therapeutic applications of chemical compounds* (Vol. 5). Pergamon Pr.

Haq, I., Tasneem, M., Raana, K., Khan, A., Mukhtar, H., & Javed, M. (2004). Optimization of cultural conditions for the production of xylanase by chemically mutated strain of *A. niger* GCBCX-20. *International Journal of Agriculture and Biotechnology, 6,* 1115-1118.

Harada, K. M., Tanaka, K., Fukuda, Y., Hashimoto, W., & Murata, K. (2005). Degradation of rice bran hemicellulose by *Paenibacillus* sp. strain HC1: Gene cloning, characterization and function of β-D-glucosidase as an enzyme involved in degradation. *Archives of Microbiology*, *184*(4), 215–224. doi:10.100700203-005-0038-8 PMID:16205911

Harayama, S. (1997). Polycyclic aromatic hydrocarbon bioremediation design. *Current Opinion in Biotechnology*, *8*(3), 268–273. doi:10.1016/S0958-1669(97)80002-X PMID:9206005

Harford-Cross, C. F., Carmichael, A. B., Allan, F. K., England, P. A., Rouch, D. A., & Wong, L. (2000). Protein engineering of cytochrome P458 (cam) (CYP101) for the oxidation of polycyclic aromatic hydrocarbons. *Protein Engineering*, *13*(2), 121–128. doi:10.1093/protein/13.2.121 PMID:10708651

Harmatz, P., Garcia, P., Guffon, N., Randolph, L. M., Shediac, R., Braunlin, E., & Decker, C. (2014). Galsulfase (Naglazyme(R)) therapy in infants with mucopolysaccharidosis VI. *Journal of Inherited Metabolic Disease*, *37*(2), 277–287. doi:10.100710545-013-9654-7 PMID:24108527

Harmatz, P., Giugliani, R., Schwartz, I., Guffon, N., Teles, E. L., Miranda, M. C., & Swiedler, S. J. (2006). Enzyme replacement therapy for mucopolysaccharidosis VI: A phase 3, randomized, double-blind, placebo-controlled, multinational study of recombinant human N-acetylgalactosamine 4-sulfatase (recombinant human arylsulfatase B or rhASB) and follow-on, open-label extension study. *The Journal of Pediatrics*, *148*(4), 533–539. doi:10.1016/j.jpeds.2005.12.014 PMID:16647419

Harris, R. C., McKanna, J. A., Akai, Y., Jacobson, H. R., Dubois, R. N., & Breyer, M. D. (1994). Cyclooxygenase-2 is associated with the macula densa of rat kidney and increases with salt restriction. *The Journal of Clinical Investigation*, *94*(6), 2504–2510. doi:10.1172/JCI117620 PMID:7989609

Harris, Z. L., Davis-Kaplan, S. R., Gitlin, J. D., & Kaplan, J. (2004). A fungal multicopper oxidase restores iron homeostasis in aceruloplasminemia. *Blood*, *103*(12), 4672–4673. doi:10.1182/blood-2003-11-4060 PMID:14739215

Harshvardhan, K., Suri, M., Goswami, A., & Goswami, T. (2017). Biological approach for the production of vanillin from lignocellulosic biomass (*Bambusa tulda*). *Journal of Cleaner Production*, *149*, 485–490. doi:10.1016/j.jclepro.2017.02.125

Hata, T., Kawai, S., Okamura, H., & Nishida, T. (2010). Removal of diclofenac and mefenamic acid by the white rot fungus *Phanerochaete sordida* YK-624 and identification of their metabolites after fungal transformation. *Biodegradation*, *21*(5), 681–689. doi:10.100710532-010-9334-3 PMID:20127144

Hata, T., Shintate, H., Kawai, S., Okamura, H., & Nishida, T. (2010). Elimination of carbamazepine by repeated treatment with laccase in the presence of 1-hydroxybenzotriazole. *Journal of Hazardous Materials*, *181*(1-3), 1175–1178. doi:10.1016/j.jhazmat.2010.05.103 PMID:20619797

Havlir, D. V., Tierney, C., Friedland, G. H., Pollard, R. B., Smeaton, L., Sommadossi, J. P., ... Richman, D. D. (2000). In vivo antagonism with zidovudine plus stavudine combination therapy. *The Journal of Infectious Diseases*, *182*(1), 321–325. doi:10.1086/315683 PMID:10882616

Hawksworth, G., Drasar, B., & Hili, M. (1971). Intestinal bacteria and the hydrolysis of glycosidic bonds. *Journal of Medical Microbiology*, *4*(4), 451–459. doi:10.1099/00222615-4-4-451 PMID:5002686

Hayes, M., & Hyatt, A. (1974). The decarboxylation of amino acids by bacteria derived from human dental plaque. *Archives of Oral Biology*, *19*(5), 361–369. doi:10.1016/0003-9969(74)90177-0 PMID:4528022

Hayes, M., & Tiwari, B. K. (2015). Bioactive carbohydrates and peptides in foods: An overview of source and downstream processing steps and associative bioactivities. *International Journal of Molecular Sciences*, *16*(9), 22485–22508. doi:10.3390/ijms160922485 PMID:26393573

Heberer, T., Reddersen, K., & Mechlinski, A. (2002). From municipal sewage to drinking water: fate and removal of pharmaceutical residues in the aquatic environment in urban areas. *Water Science and Technology: A Journal of the International Association on Water Pollution Research*, *46*, 81–88.

Heberer, T. (2002). Occurrence, fate and removal of pharmaceutical residues in the aquatic environment: A review of recent research data. *Toxicology Letters*, *131*(1-2), 5–17. doi:10.1016/S0378-4274(02)00041-3 PMID:11988354

Heberer, T. (2002). Tracking persistent pharmaceutical residues from municipal sewage to drinking water. *Journal of Hydrology (Amsterdam)*, *266*(3-4), 175–189. doi:10.1016/S0022-1694(02)00165-8

Heck, J. X., Hertz, P. F., & Ayub, M. A. Z. (2002). Cellulase and xylanase production by isolated amazon *Bacillus* strains using soybean industrial residue based solid-state cultivation. *Brazilian Journal of Microbiology*, *33*(3), 213–218. doi:10.1590/S1517-83822002000300005

Hehre, E. J., & Sery, T. W. (1952). Dextran-splitting anaerobic bacteria from the human intestine. *Journal of Bacteriology*, *63*(3), 424. PMID:14927577

Heidary, M., Khoobi, M., Ghasemi, S., Habibi, Z., & Faramarzi, M. A. (2014). Synthesis of quinazolinones from alcohols *via* laccase-mediated tandem oxidation. *Advanced Synthesis & Catalysis*, *356*(8), 1789–1794. doi:10.1002/adsc.201400103

Hela, W., Brandtner, M., Widek, R., & Schuh, R. (2003). Determination of sulfonamides in animal tissues using cation exchange reversed phase sorbent for sample cleanup and HPLC-DAD for detection. *Food Chemistry*, *83*(4), 601–608. doi:10.1016/S0308-8146(03)00254-1

Heller, M. C., & Keoleian, G. A. (2003). Assessing the sustainability of the US food system: A life cycle perspective. *Agricultural Systems*, *76*(3), 1007–1041. doi:10.1016/S0308-521X(02)00027-6

Henrissat, B., & Bairoch, A. (1993). New families in the classification of glycosyl hydrolases based on amino acid sequence similarities. *The Biochemical Journal*, *293*(3), 781–788. doi:10.1042/bj2930781 PMID:8352747

Hernandez, C. P., Morrow, K., Lopez-Barcons, L. A., Zabaleta, J., Sierra, R., Velasco, C., & Rodriguez, P. C. (2010). Pegylated arginase I: A potential therapeutic approach in T-ALL. *Blood, 115*(25), 5214–5221. doi:10.1182/blood-2009-12-258822 PMID:20407034

Hernández-Luna, C. E., Gutiérrez-Soto, G., & Salcedo-Martínez, S. M. (2008). Screening for decolorizing basidiomycetes in Mexico. *World Journal of Microbiology & Biotechnology, 24*(4), 465–473. doi:10.100711274-007-9495-3

Hershfield, M. S., Buckley, R. H., Greenberg, M. L., Melton, A. L., Schiff, R., Hatem, C., ... Abuchowski, A. (1987). Treatment of adenosine deaminase deficiency with polyethylene glycol-modified adenosine deaminase. *The New England Journal of Medicine, 316*(10), 589–596. doi:10.1056/NEJM198703053161005 PMID:3807953

Herter, S., Michalik, D., Mikolasch, A., Schmidt, M., Wohlgemuth, R., Bornscheuer, U., & Schauer, F. (2013). Laccase-mediated synthesis of 2-methoxy-3-methyl-5-(alkylamino)- and 3-methyl-2,5- is(alkylamino)-[1,4]-benzoquinones. *Journal of Molecular Catalysis. B, Enzymatic, 90*, 91–97. doi:10.1016/j.molcatb.2013.01.020

Hidaka, H., Inagaki, M., Kawamoto, S., & Sasaki, Y. (1984). Isoquinolinesulfonamides, novel and potent inhibitors of cyclic nucleotide-dependent protein kinase and protein kinase C. *Biochemistry, 23*(21), 5036–5041. doi:10.1021/bi00316a032 PMID:6238627

Hillman, B. (2004). *Role of gut bacteria in human toxicology and pharmacology*. CRC Press.

Ho, H. T., & Hitchcock, M. J. (1989). Cellular pharmacology of 2',3'-dideoxy-2',3'-didehydrothymidine, a nucleoside analog active against human immunodeficiency virus. *Antimicrobial Agents and Chemotherapy, 33*(6), 844–849. doi:10.1128/AAC.33.6.844 PMID:2764535

Hölker, U., Höfer, M., & Lenz, J. (2004). Biotechnological advantages of laboratory-scale solid-state fermentation with fungi. *Applied Microbiology and Biotechnology, 64*, 175–186. doi:10.100700253-003-1504-3 PMID:14963614

Holker, U., & Lenz, J. (2005). Solid-state fermentation are there any biotechnological advantages? *Current Opinion in Microbiology, 8*(3), 301–306. doi:10.1016/j.mib.2005.04.006 PMID:15939353

Hondred, J. A., Breger, J. C., Garland, N. T., Oh, E., Susumu, K., Walper, S. A., ... Claussen, J. C. (2017). Enhanced enzymatic activity from phosphotriesterase trimer gold nanoparticle bioconjugates for pesticide detection. *Analyst (London), 142*(17), 3261–3271. doi:10.1039/C6AN02575G PMID:28765846

Hong, L. S., Ibrahim, D., & Omar, I. C. (2011). Lignocellulolytic materials-as a raw material for the production of fermentable sugars via solid state fermentation. *Asian Journal of Scientific Research, 4*(1), 53–61. doi:10.3923/ajsr.2011.53.61

Hong, R., Emrick, T., & Rotello, V. M. (2004b). Monolayer-controlled substrate selectivity using noncovalent enzyme-nanoparticle conjugates. *Journal of the American Chemical Society, 126*(42), 13572–13573. doi:10.1021/ja0461163 PMID:15493887

Hong, R., Fischer, N. O., Verma, A., Goodman, C. M., Emrick, T., & Rotello, V. M. (2014a). Control of protein structure and function through surface recognition by tailored nanoparticle scaffolds. *Journal of the American Chemical Society, 126*(3), 739–743. doi:10.1021/ja037470o PMID:14733547

Hope, M. J., Bally, M. B., Webb, G., & Cullis, P. R. (1985). Production of large unimellar vesicles by a rapid extrusion procedure. Characterization of size distribution, trapped volume and ability to maintain a membrane potential. *Biochimica et Biophysica Acta, 812*(1), 55–65. doi:10.1016/0005-2736(85)90521-8 PMID:23008845

Hopkins, A. L., & Groom, C. R. (2002). Opinion: The druggable genome. *Nature Reviews. Drug Discovery, 1*(9), 727–730. doi:10.1038/nrd892 PMID:12209152

Horovitz, D. D., Magalhaes, T. S., Acosta, A., Ribeiro, E. M., Giuliani, L. R., Palhares, D. B., & Llerena, J. C. Jr. (2013). Enzyme replacement therapy with galsulfase in 34 children younger than five years of age with MPS VI. *Molecular Genetics and Metabolism, 109*(1), 62–69. doi:10.1016/j.ymgme.2013.02.014 PMID:23535281

Horvathova, V., Janeeek, S., & Sturdik, E. (2001). Amylolytic Enzymes: Molecular Aspects of Their Properties. *General Physiology and Biophysics, 20*, 7–32. PMID:11508823

Hossain, Kh. M. G., Gonza'lez, M. D., Juan, A. R., & Tzanov, T. (2010). Enzyme-mediated coupling of a bi-functional phenolic compound onto wool to enhance its physical, mechanical and functional properties. *Enzyme and Microbial Technology, 46*(3-4), 326–330. doi:10.1016/j.enzmictec.2009.12.008

Hossain, Kh. M. G., Gonza'lez, M. D., Lozano, G. R., & Tzanov, T. (2009). Multifunctional modification of wool using an enzymatic process in aqueous-organic media. *Journal of Biotechnology, 141*(1-2), 58–63. doi:10.1016/j.jbiotec.2009.02.011 PMID:19428731

Howard, P. C., Beland, F. A., & Cerniglia, C. E. (1983a). Reduction of the carcinogen 1-nitropyrene to 1-aminopyrene by rat intestinal bacteria. *Carcinogenesis, 4*(8), 985–990. doi:10.1093/carcin/4.8.985 PMID:6347427

Howard, P. C., Heflich, R. H., Evans, F. E., & Beland, F. A. (1983b). Formation of DNA adducts in vitro and in Salmonella typhimurium upon metabolic reduction of the environmental mutagen 1-nitropyrene. *Cancer Research, 43*(5), 2052–2058. PMID:6339047

Huang, D.-M., Hsiao, J.-K., Chen, Y.-C., Chien, L. Y., Yao, M., Chen, Y. K., ... Chen, Y. C. (2009). The promotion of human mesenchymal stem cell proliferation by superparamagnetic iron oxide nanoparticles. *Biomaterials, 30*(22), 3645–3651. doi:10.1016/j.biomaterials.2009.03.032 PMID:19359036

Huang, Q., & Weber, W. J. (2005). Transformation and removal of bisphenol A from aqueous phase via peroxidase-mediated oxidative coupling reactions: Efficacy, products, and pathways. *Environmental Science & Technology, 39*(16), 6029–6036. doi:10.1021/es050036x PMID:16173560

Huang, S. L., & MacDonald, R. C. (2004). Acoustically active liposomes for drug encapsulation and ultrasound-triggered release. *Biochimica et Biophysica Acta, 1665*(1-2), 134–141. doi:10.1016/j.bbamem.2004.07.003 PMID:15471579

Huang, Y. C., Hung, W. C., Kang, W. Y., Chen, W. T., & Chai, C. Y. (2007). p-Phenylenediamine induced DNA damage in SV-40 immortalized human uroepithelial cells and expression of mutant p53 and COX-2 proteins. *Toxicology Letters, 170*(2), 116–123. doi:10.1016/j.toxlet.2007.02.011 PMID:17403587

Hu, B. H., & Messersmith, P. B. (2003). Rational design of transglutaminase substrate peptides for rapid enzymatic formation of hydrogels. *Journal of the American Chemical Society, 125*(47), 14298–14299. doi:10.1021/ja038593b PMID:14624577

Hu, D. D., Zhang, R. Y., Zhang, G. Q., Wang, H. X., & Ng, T. B. (2011). A laccase with antiproliferative activity against tumor cells from an edible mushroom, white common Agrocybe cylindracea. *Phytomedicine, 18*(5), 374–379. doi:10.1016/j.phymed.2010.07.004 PMID:20739163

Huey, W. C., Ichikawa, Y., & Shen, G. J. (1995). *Oligosaccharide enzyme substrates and inhibitors: methods and composition.* European Patent.

Hung, V. S., Hatada, Y., Goda, S., Lu, J., Hidaka, Y., Li, Z., ... Horikoshi, K. (2005). α- Glucosidase from a strain of deep-sea *Geobacillus*: A potential enzyme for the biosynthesis of complex carbohydrates. *Applied Microbial Biotechnology, 68*(6), 757–765. doi:10.100700253-005-1977-3 PMID:15940457

Hunter, T. (2000). Signaling—2000 and beyond. *Cell*, *100*(1), 113–127. doi:10.1016/S0092-8674(00)81688-8 PMID:10647936

Hunt, J. A., Chen, R., van Veen, T., & Bryan, N. (2014). Hydrogels for tissue engineering and regenerative medicine. *Journal of Materials Chemistry. B, Materials for Biology and Medicine*, *2*(33), 5319–5338. doi:10.1039/C4TB00775A

Hu, P., & Tirelli, N. (2012). Scavenging ROS: Superoxide dismutase/catalase mimetics by the use of an oxidation-sensitive nanocarrier/enzyme conjugate. *Bioconjugate Chemistry*, *23*(3), 438–449. doi:10.1021/bc200449k PMID:22292618

Husain, Q. (2006). Potential applications of the oxidoreductive enzymes in the decolorization and detoxification of textile and other synthetic dyes from polluted water: A review. *Critical Reviews in Biotechnology*, *26*(4), 201–221. doi:10.1080/07388550600969936 PMID:17095432

Hu, W. S., & Hughes, S. H. (2012). HIV-1 reverse transcription. *Cold Spring Harbor Perspectives in Medicine*, *2*(10), a006882. doi:10.1101/cshperspect.a006882 PMID:23028129

Hu, X., Lingyun, H., Wang, H., Yang, X., Zhang, G., Wang, G., & … . (2011). Hydrogel Contact Lens for Extended Delivery of Ophthalmic Drugs. *International Journal of Polymer Science*, *2011*, 1–9. doi:10.1155/2011/814163

Ibarra-Escutia, P., Gomez, J. J., Calas-Blanchard, C., Marty, J. L., & Ramirez-Silva, M. T. (2010). Amperometric biosensor based on a high resolution photopolymer deposited onto a screen-printed electrodefor phenolic compounds monitoring in tea infusions. *Talanta*, *81*(4-5), 1636–1642. doi:10.1016/j.talanta.2010.03.017 PMID:20441951

Ihler, G., Glew, R. H., & Schnure, F. W. (1973). Enzyme Loading of Erythrocytes. *Proceedings of the National Academy of Sciences of the United States of America*, *70*(9), 2663–2666. doi:10.1073/pnas.70.9.2663 PMID:4354859

Ihler, G., Lantzy, A., Purpura, J., & Glew, R. H. (1975). Enzymatic degradation of uric acid by uricase-loaded human erythrocytes. *The Journal of Clinical Investigation*, *56*(3), 595–602. doi:10.1172/JCI108129 PMID:1159075

Ihssen, J., Schubert, M., Meyer, L. T., & Richter, M. (2014). Laccase Catalyzed Synthesis of Iodinated Phenolic Compounds with Antifungal Activity. *PLoS One*, *9*(3), 89924. doi:10.1371/journal.pone.0089924 PMID:24594755

Ikeda, Y., Parashar, A., Chae, M., & Bressler, D. C. (2015). Reusability of immobilized cellulases with highly retained enzyme activity and their application for hydrolysis of model substrates and lignocellulosic biomass. *Journal of Thermodynamics & Catalysis.*, *6*(2), 149. doi:10.4172/2157-7544.1000149

Ikehata, K., Jodeiri Naghashkar, N., & Gamal El-Din, M. (2006). Degradation of aqueous pharmaceuticals by ozonation and advanced oxidation processes: A review. *Ozone Science and Engineering*, *28*(6), 353–414. doi:10.1080/01919510600985937

Ikuta, N., Souza, M. B. N., Valencia, F. F., Castro, M. E. B., Schenberg, A. C. G., Pizzirani-Kleiner, A., & Astolfi-Filho, S. (1990). The α-amylase gene as a marker for gene cloning: Direct screening of recombinant clones. *Nature Biotechnology*, *8*(3), 241–242. doi:10.1038/nbt0390-241 PMID:1367421

Iluyemi, F. B., Hanafi, M. M., Radziah, O., & Kamarudin, M. S. (2006). Fungal solid state culture of palm kernel cake. *Bioresource Technology*, *97*(3), 477–482. doi:10.1016/j.biortech.2005.03.005 PMID:16216731

Imaizumi, S., Woolworth, V., Kinouchi, H., Chen, S. F., Fishman, R. A., & Chan, P. H. (1990). Liposome-entrapped superoxide dismutase ameliorates infarct volume in focal cerebral ischaemia. *Acta Neurochirurgica*, *51*, 236–238. PMID:2089905

Ingale, S., Joshi, S. J., & Gupta, A. (2014). Production of Bioethanol using agricultural waste: Banana pseudo stem. *Brazilian Journal of Microbiology*, *45*(3), 885–892. doi:10.1590/S1517-83822014000300018 PMID:25477922

Inoue, N., Kawashima, S., Kanazawa, K., Yamada, S., Akita, H., & Yokoyama, M. (1998). Polymorphism of the NADH/NADPH oxidase p22 phox gene in patients with coronary artery disease. *Circulation*, *97*(2), 135–137. doi:10.1161/01.CIR.97.2.135 PMID:9445163

Intarasirisawat, R., Benjakul, S., Visessanguan, W., & Wu, J. (2012). Antioxidative and functional properties of protein hydrolysate from defatted skipjack (*Katsuwonous pelamis*) roe. *Food Chemistry*, *135*(4), 3039–3048. doi:10.1016/j.foodchem.2012.06.076 PMID:22980906

Irfan, M., Nadeem, M., Syed, Q., & Baig, S. (2012). Effect of medium composition on xylanase production by *Bacillus subtilis* using various agricultural wastes. *American-Eurasian Journal of Agricultural and Environmental Sciences*, *12*(5), 561–565.

Ito, K., Ogasawara, H., Sugimoto, T., & Ishikawa, T. (1992). Purification and properties of acid stable xylanases from *Aspergillus kawachii*. *Bioscience, Biotechnology, and Biochemistry*, *56*(4), 547–550. doi:10.1271/bbb.56.547 PMID:27280644

Ivanova, V., Yankov, D., Kabaivanova, L., & Pashkkoulov, D. (2001). Simultaneous biosynthesis and purification of two extra cellular *Bacillus hydrolases* in aqueous two alpha amylase. *Journal of Biochemical Engineering*, *8*, 61–81.

Iwata, H., Hata, Y., Matsuda, T., & Ikada, Y. (1991). Initiation of radical polymerization by glucose oxidase utilizing dissolved oxygen. *Journal of Polymer Science. Part A, Polymer Chemistry*, *29*(8), 1217–1218. doi:10.1002/pola.1991.080290818

Izawa, H., Nawaji, M., Kaneko, Y., & Kadokawa, J. (2009). Preparation of glycogenbased polysaccharide materials by phosphorylase catalyzed chain elongation of glycogen. *Macromolecular Bioscience*, *9*(11), 1098–1104. doi:10.1002/mabi.200900106 PMID:19637153

Izydorczyk, M., Biliaderis, C., & Bushuk, W. (1990). Oxidative gelation studies of water-soluble pentosans from wheat. *Journal of Cereal Science*, *11*(2), 153–169. doi:10.1016/S0733-5210(09)80117-3

Jabbari, E., Yaszemski, M. J., & Currier, B. L. (2006). *European Patent: 1 664 168 B1*. European Patent Office.

Jackson, S. E. D., Craggs, T., & Huang, J. R. (2006). Understanding the folding of GFP using biophysical techniques. *Expert Review of Proteomics*, *3*(5), 545–559. doi:10.1586/14789450.3.5.545 PMID:17078767

Jadhav, S. B., & Singhal, R. S. (2014). Laccase–gum Arabic conjugate for preparation of water-soluble oligomer of catechin with enhanced antioxidant activity. *Food Chemistry*, *150*, 9–16. doi:10.1016/j.foodchem.2013.10.127 PMID:24360412

Jafari, N., Rezaei, S., Rezaie, R., Dilmaghani, H., Khoshayand, M. R., & Faramarzi, M. A. (2017). Improved production and characterization of a highly stable laccase from the halophilic bacterium *Chromohalobacter salexigens* for the efficient delignification of almond shell bio-waste. *International Journal of Biological Macromolecules*, *105*, 489–498. doi:10.1016/j.ijbiomac.2017.07.055 PMID:28709895

Jahnavi, G., Prashanthi, G. S., Sarvanthi, K., & Rao, L. V. (2017). Status of availability of lignocellulosic feed stocks in India; Biotechnological strategies involved in the production of bioethanol. *Renewable & Sustainable Energy Reviews*, *73*, 798–820. doi:10.1016/j.rser.2017.02.018

Jain, S. K., & Jain, A. (2008). Target-specific drug release to the colon. *Expert Opinion on Drug Delivery*, *5*(5), 483–498. doi:10.1517/17425247.5.5.483 PMID:18491977

Jamdar, S. N., Rajalakshmi, V., Pednekar, M. D., Juan, F., Yardi, V., & Sharma, A. (2010). Influence of degree of hydrolysis on functional properties, antioxidant activity and ACE inhibitory activity of peanut protein hydrolysate. *Food Chemistry*, *121*(1), 178–184. doi:10.1016/j.foodchem.2009.12.027

James, J. A., & Lee, B. H. (1997). Glucoamylases: microbial sources, industrial applications and molecular biology: A review. *Journal of Food Biochemistry*, *21*(6), 1–52. doi:10.1111/j.1745-4514.1997.tb00223.x

Janecek, S., & Balaz, S. (1992). α-amylases and approaches leading to their enhanced stability. *FEBS Letters*, *304*(1), 1–3. doi:10.1016/0014-5793(92)80575-2 PMID:1618293

Janeček, S. (1993). Does the increased hydrophobicity of the interior and hydrophilicity of the exterior of an enzyme structure reflect its increased thermostability? *International Journal of Biological Macromolecules*, *15*(5), 317–318. doi:10.1016/0141-8130(93)90033-I PMID:8251448

Janoš, P., Kuráň, P., Pilařová, V., Trögl, J., Šťastný, M., Pelant, O., ... Skoumal, M. (2015). Magnetically separable reactive sorbent based on the CeO_2/γ-Fe_2O_3 composite and its utilization for rapid degradation of the organophosphate pesticide parathion methyl and certain nerve agents. *Chemical Engineering Journal*, *262*, 747–755. doi:10.1016/j.cej.2014.10.016

Jarosz-Wilkołazka, A., Ruzgas, T., & Gorton, L. (2004). Use of laccase-modified electrode for amperometric detection of plant flavonoids. *Enzyme and Microbial Technology*, *35*(2-3), 238–241. doi:10.1016/j.enzmictec.2004.04.016

Jasuja, R., Passam, F. H., Kennedy, D. R., Kim, S. H., van Hessem, L., Lin, L., ... Furie, B. C. (2012). Protein disulfide isomerase inhibitors constitute a new class of antithrombotic agents. *The Journal of Clinical Investigation*, *122*(6), 2104–2113. doi:10.1172/JCI61228 PMID:22565308

Jayani, R. S., Saxena, S., & Gupta, R. (2005). Microbial pectinolytic enzymes: A review. *Process Biochemistry*, *40*(9), 2931–2944. doi:10.1016/j.procbio.2005.03.026

Jayant, M., Rashmi, J., Shailendra, M., & Deepesh, Y. (2011). Production of cellulase by different co-culture of *Aspergillus niger* and *Penicillium chrysogenum* from waste paper, cotton waste and baggase. *Journal of Yeast and Fungal Research*, *2*(2), 24–27.

Jayaram, H. N., Ahluwalia, G. S., & Cooney, D. A. (2000). *Enzyme Applications, Therapeutic Kirk-Othmer Encyclopedia of Chemical Technology*. John Wiley & Sons, Inc.

Jee, E., Bánsági, T. Jr, Taylor, A. F., & Pojman, J. A. (2016). Temporal control of gelation and polymerization fronts driven by an autocatalytic enzyme reaction. *Angewandte Chemie International Edition*, *55*(6), 2127–2131. doi:10.1002/anie.201510604 PMID:26732469

Jelic, A., Cruz-Morató, C., Marco-Urrea, E., Sarrà, M., Perez, S., Vicent, T., & Barcelo, D. (2012). Degradation of carbamazepine by *Trametes versicolor* in an air pulsed fluidized bed bioreactor and identification of intermediates. *Water Research*, *46*(4), 955–964. doi:10.1016/j.watres.2011.11.063 PMID:22178304

Jeong, J., Ha, T. H., & Chung, B. H. (2006). Enhanced reusability of hexa-arginine-tagged esterase immobilized on gold-coated magnetic nanoparticles. *Analytica Chimica Acta*, *569*(1-2), 203–209. doi:10.1016/j.aca.2006.03.102

Jeon, Y. J., Park, P. J., & Kim, S. K. (2001). Antimicrobial effect of chitooligosaccharides produced by bioreactor. *Carbohydrate Polymers*, *44*(1), 71–76. doi:10.1016/S0144-8617(00)00200-9

Ji, Z. P., & Su, Y. Q. (2008). Study on Antimicrobial Activities of Extracts from *Eucommia ulmoides* Oliv. Leaves. *Chemistry and Industry of Forest Products, 2*, 16.

Jia, F., Narasimhan, B., & Mallapragada, S. K. (2013). Biomimetic multienzyme complexes based on nanoscale platforms. *AIChE Journal. American Institute of Chemical Engineers*, *59*(2), 355–360. doi:10.1002/aic.13992

Jiang, Z., Li, X., Yang, S., Li, L., & Tan, S. (2005). Improvement of bread making quality of wheat flour by the hyperthermophilic xylanase B from *Thermotoga maritime. Food Research International*, *38*(1), 37–43. doi:10.1016/j.foodres.2004.07.007

Jiménez, J. P., Serrano, J., Tabernero, M., Arranz, S., Díaz-Rubio, M. E., García-Diz, L., ... Saura-Calixto, F. (2008). Effects of grape antioxidant dietary fiber in cardiovascular disease risk factors. *Nutrition (Burbank, Los Angeles County, Calif.)*, *24*(7-8), 646–653. doi:10.1016/j.nut.2008.03.012 PMID:18485668

Joana Gil-Chávez, G., Villa, J. A., Fernando Ayala-Zavala, J., Basilio Heredia, J., Sepulveda, D., Yahia, E. M., & González-Aguilar, G. A. (2013). Technologies for extraction and production of bioactive compounds to be used as nutraceuticals and food ingredients: An overview. *Comprehensive Reviews in Food Science and Food Safety*, *12*(1), 5–23. doi:10.1111/1541-4337.12005

Jobling, S., Beresford, N., Nolan, M., Rodgers-Gray, T., Brighty, G. C., Sumpter, J. P., & Tyler, C. R. (2002). Altered sexual maturation and gamete production in wild roach (*Rutilus rutilus*) living in rivers that receive treated sewage effluents. *Biology of Reproduction*, *66*(2), 272–281. doi:10.1095/biolreprod66.2.272 PMID:11804939

Johannes, C., & Majcherczyk, A. (2000). Laccase activity tests and laccase inhibitors. *Journal of Biotechnology*, *78*(2), 193–199. doi:10.1016/S0168-1656(00)00208-X PMID:10725542

Johannes, C., & Majcherczyk, A. (2000). Natural mediators in the oxidation of polycyclic aromatic hydrocarbons by laccase mediator systems. *Applied and Environmental Microbiology*, *66*(2), 524–528. doi:10.1128/AEM.66.2.524-528.2000 PMID:10653713

John, A. M., Thomas, N. S. B., Mufti, G. J., & Padua, R. A. (2004). Targeted therapies in myeloid leukemia. *Seminars in Cancer Biology*, *14*(1), 41–62. doi:10.1016/j.semcancer.2003.11.006 PMID:14757535

Johnson, M. A., & Pokora, A. R. (1994). *Method for Deinking Printed Waste Using Soybean Peroxidase* (US Patent, 5370770). DOI: 10.1016/b978-1-4832-8381-4.50071-8

Johnson, L. M., DeForest, C. A., Pendurti, A., Anseth, K. S., & Bowman, C. N. (2010). Formation of three-dimensional hydrogel multilayers using enzyme-mediated redox chain initiation. *ACS Applied Materials & Interfaces*, *2*(7), 1963–1972. doi:10.1021/am100275n PMID:20586415

Johnson, L. M., Fairbanks, B. D., Anseth, K. S., & Bowman, C. N. (2009). Enzyme-mediated redox initiation for hydrogel generation and cellular encapsulation. *Biomacromolecules*, *10*(11), 3114–3121. doi:10.1021/bm900846m PMID:19821604

Johnson, P. A., Park, H. J., & Driscoll, A. J. (2011). Enzyme nanoparticle fabrication: Magnetic nanoparticle synthesis and enzyme immobilization. *Methods in Molecular Biology (Clifton, N.J.)*, *679*, 183–191. doi:10.1007/978-1-60761-895-9_15 PMID:20865397

Johnvesly, B., & Naik, G. K. (2001). Studies on production of thermostable alkaline protease from thermophilic and alkaliphilic *Bacillus* sp. JB-99 in a chemical defined medium. *Process Biochemistry*, *37*(2), 139–144. doi:10.1016/S0032-9592(01)00191-1

Jonckheere, H., Anné, J., & De Clercq, E. (2000). The HIV-1 Reverse Transcription (RT) Process as Target for RT Inhibitors. *Medicinal Research Reviews*, *20*(2), 129–154. doi: PMID:10723025

Jones, J. P., O'Hare, E. J., & Wong, L. L. (2001). Oxidation of polychlorinated benzenes by genetically engineered CYP101 (cytochrome P450cam). *European Journal of Biochemistry*, *268*(5), 1460–1467. doi:10.1046/j.1432-1327.2001.02018.x PMID:11231299

Jones, S.M., van Dyk, J. S., & Pletschke, B.I. (2012). *Bacillus subtilis* SJ01 produces hemicellulose degrading multi-enzyme complexes. *BioResources*, *7*(1), 1294–1309.

Jonker, A. M., Borrmann, A., van Eck, E. R., van Delft, F. L., Löwik, D. W., & van Hest, J. (2015). A fast and activatable crosslinking strategy for hydrogel formation. *Advanced Materials*, *27*(7), 1235–1240. doi:10.1002/adma.201404448 PMID:25535032

Joss, A., Keller, E., Alder, A. C., Göbel, A., McArdell, C. S., Ternes, T., & Siegrist, H. (2005). Removal of pharmaceuticals and fragrances in biological wastewater treatment. *Water Research*, *39*(14), 3139–3152. doi:10.1016/j.watres.2005.05.031 PMID:16043210

Jubeh, T. T., Antler, S., Haupt, S., Barenholz, Y., & Rubinstein, A. (2005). Local prevention of oxidative stress in the intestinal epithelium of the rat by adhesive liposomes of superoxide dismutase and tempamine. *Molecular Pharmaceutics*, *2*(1), 2–11. doi:10.1021/mp0499095 PMID:15804172

Jung, H., Xu, F., & Li, K. (2002). Purification and characterization of laccase from wood-degrading fungus *Trichophyton rubrum* LKY-7. *Enzyme and Microbial Technology*, *30*(2), 161–168. doi:10.1016/S0141-0229(01)00485-9

Jung, J., & Wicker, L. (2012). Laccase mediated conjugation of heat treated beta-lactoglobulin and sugar beet pectin. *Carbohydrate Polymers*, *89*(4), 1244–1249. doi:10.1016/j.carbpol.2012.04.028 PMID:24750938

Junker, B., Moore, J., Sturr, M., & McLoughlin, K. (2001). Article. *Bioprocess and Biosystems Engineering*, *24*, 39–49. doi:10.1007004490100230

Jun, S. Y., Park, P. J., Jung, W. K., & Kim, S. K. (2004). Purification and characterization of an antioxidative peptide from enzymatic hydrolysate of yellowfin sole (*Limanda aspera*) frame protein. *European Food Research and Technology*, *219*(1), 20–26. doi:10.100700217-004-0882-9

Justus, E., Awad, D., Hohnholt, M., Schaffran, T., Edwards, K., Karlsson, G., ... Gabel, D. (2007). Synthesis, liposomal preparation, and in vitro toxicity of two novel dodecaborate cluster lipids for boron neutron capture therapy. *Bioconjugate Chemistry*, *18*(4), 1287–1293. doi:10.1021/bc070040t PMID:17569498

Juwaied, A. A., Al-amiery, A. A. H., Abdumuniem, Z., & Anaam, U. (2011). Optimization of cellulase production by *Aspergillus niger* and *Tricoderma viride* using sugar cane waste. *Journal of Yeast and Fungal Research*, *2*(2), 19–23.

Kadokawa, J. I., & Kaneko, Y. (2013). Phosphorylase-catalyzed enzymatic polymerization. In *Advances in the Engineering of Polysaccharide Materials: by Phosphorylase-Catalyzed Enzymatic Chain-Elongation* (p. 43). CRC Press.

Kagalkar, A. N., Jagtap, U. B., Jadhav, J. P., Govindwar, S. P., & Bapat, S. A. (2010). Studies On Phytoremediation Potentiality Typhonium Flagelliforme for The Degradation of Brilliant Blue R. *Planta*, *232*(1), 271–285. doi:10.100700425-010-1157-2 PMID:20437182

Kalkan, N. A., Aksoy, S., Aksoy, E. A., & Hasirci, N. (2012). Preparation of chitosan-coated magnetite nanoparticles and application for immobilization of laccase. *Journal of Applied Polymer Science*, *123*(2), 707–716. doi:10.1002/app.34504

Kamble, R. D., & Jadhav, A. R. (2012). Isolation, purification and characterization of xylanase produced by a new species of *Bacillus* in solid state fermentation. *International Journal of Microbiology*. PMID:22315613

Kanabar, D., Randhawa, M., & Clayton, P. (2001). Improvement of symptoms in infant colic following reduction of lactose load with lactase. *Journal of Human Nutrition and Dietetics*, *14*(5), 359–363. doi:10.1046/j.1365-277X.2001.00304.x PMID:11906576

Kanbi, L. D., Antonyuk, S., Hough, M. A., Hall, J. F., Dodd, F. E., & Hasnain, S. S. (2002). Crystal structures of the Met148Leu and Ser86Asp mutants of Rusticyanin from *Thiobacillus ferrooxidans*: Insights into the structural relationship with the cupredoxins and the multi copper proteins. *Journal of Molecular Biology*, *320*(2), 263–275. doi:10.1016/S0022-2836(02)00443-6 PMID:12079384

Kang, T. S., & Stevens, R. C. (2009). Structural aspects of therapeutic enzymes to treat metabolic disorders. *Human Mutation*, *30*(12), 1591–1610. doi:10.1002/humu.21111 PMID:19790257

Kansoh, A. L., & Gammel, A. (2001). Xylanolytic activities of *Streptomyces sp.1*, taxonomy production, partial purification and utilization of agricultural wastes. *Acta Microbiologica et Immunologica Hungarica*, *48*(1), 39–52. doi:10.1556/AMicr.48.2001.1.5 PMID:11233699

Kantelinen, A., Hortling, B., Sundquist, J., Linko, M., & Viikari, L. (1993). Proposed mechanism of the enzymatic bleaching of kraft pulp with xylanases. *Holzforschung*, *47*(4), 318–324. doi:10.1515/hfsg.1993.47.4.318

Kantor, L.S., Lipton, K., Manchester, A. and Oliveira V. (1997). Estimating and Addressing America's Food Losses. *Food Review*, 1-12.

Kapilan, R., & Arasaratnam, V. (2011). Paddy husk as support for solid state fermentation to produce xylanase from *Bacillus pumilus*. *Rice Science*, *18*(1), 36–45. doi:10.1016/S1672-6308(11)60006-1

Karam, J., & Nicell, J. A. (1997). Potential Applications of Enzymes in Waste Treatment. *Journal of Chemical Technology and Biotechnology (Oxford, Oxfordshire)*, *69*(2), 141–153. doi:

Kargman, S., Charleson, S., Cartwright, M., Frank, J., Riendeau, D., Mancini, J., ... O'Neill, G. (1996). Characterization of prostaglandin G/H synthase 1 and 2 in rat, dog, monkey, and human gastrointestinal tracts. *Gastroenterology*, *111*(2), 445–454. doi:10.1053/gast.1996.v111.pm8690211 PMID:8690211

Karigar & Rao. (2011). Role of Microbial Enzymes in the Bioremediation of Pollutants: A Review. *Enzyme Research*. doi:10.4061/2011/805187

Karmen, A., Wróblewski, F., & LaDue, J. S. (1955). Transaminase activity in human blood. *The Journal of Clinical Investigation*, *34*(1), 126–133. doi:10.1172/JCI103055 PMID:13221663

Kar, P. K., Tandon, V., & Saha, N. (2002). Anthelmintic efficacy of *Flemingia vestita*: Genistein-induced effect on the activity of nitric oxide synthase and nitric oxide in the trematode parasite, Fasciolopsis buski. *Parasitology International*, *51*(3), 249–257. doi:10.1016/S1383-5769(02)00032-6 PMID:12243779

Kasinathan, N., Volety, S. M., & Josyula, V. R. (2016). Chondroitinase: A promising therapeutic enzyme. *Critical Reviews in Microbiology*, *42*(3), 474–484. PMID:25319196

Kasprzyk-Hordern, B., Dinsdale, R. M., & Guwy, A. J. (2009). The removal of pharmaceuticals, personal care products, endocrine disruptors and illicit drugs during wastewater treatment and its impact on the quality of receiving waters. *Water Research*, *43*(2), 363–380. doi:10.1016/j.watres.2008.10.047 PMID:19022470

Kaur, B., Kumar, B., Garg, N., & Kaur, N. (2015). Statistical optimization of conditions for decolorization of synthetic dyes by *Cordyceps militaris* MTCC 3936 Using RSM. *BioMed Research International*, *2015*, 1–17. doi:10.1155/2015/536745 PMID:25722980

Kaur, H., Dutt, D., & Tyagi, C. H. (2011). Production of novel alkali-thermo-tolerant cellulose-poor xylanases from *Coprinopsis cinerea* HK-1 NFCCI-2032. *BioResources*, *6*(2), 1376–1391.

Kavya, V., & Padmavathi, T. (2009). Optimization of growth conditions for xylanase production from *Aspergillus niger* in solid state fermentation. *Polish Journal of Microbiology*, *58*(2), 125–130. PMID:19824396

Kawai, S., Umezawa, T., Shimada, M., Higuchi, T., Koide, K., Nishida, T., ... Haraguchi, T. (1987). C-alpha-C-beta cleavage of phenolic beta-1 lignin substructure model compound by laccase of *Coriolus versicolor*. *Journal of the Japan Wood Research Society (Japan)*, *33*, 792–797.

Kempf, D. J., Norbeck, D. W., Sham, H. L., & Zhao, C. (1996). *Retroviral protease inhibiting compounds*. Google Patents.

Kermani, M., Pourmoghaddas, H., Bina, B., & Khazaei, Z. (2006). Removal of phenol from aqueous solutions by rice husk ash and activated carbon. *Pakistan Journal of Biological Sciences*, *9*(10), 1905–1910. doi:10.3923/pjbs.2006.1905.1910

Kermanshai, R., McCarry, B. E., Rosenfeld, J., Summers, P. S., Weretilnyk, E. A., & Sorger, G. J. (2001). Benzyl isothiocyanate is the chief or sole anthelmintic in papaya seed extracts. *Phytochemistry*, *57*(3), 427–435. doi:10.1016/S0031-9422(01)00077-2 PMID:11393524

Kersten, P. J., Kalyanaraman, B., Hammel, K. E., Reinhammar, B., & Kirk, T. K. (1990). Comparison of lignin peroxidase, horseradish peroxidase and laccase in the oxidation of methoxybenzenes. *The Biochemical Journal*, *268*(2), 475–480. doi:10.1042/bj2680475 PMID:2163614

Khan, A. A., Piris, J., & Truelove, S. (1977). An experiment to determine the active therapeutic moiety of sulphasalazine. *Lancet*, *310*(8044), 892–895. doi:10.1016/S0140-6736(77)90831-5 PMID:72239

Khandeparkar, R., & Bhosle, N. B. (2007). Application of thermoalkalophilic xylanase from *Arthrobacter* sp. MTCC 5214 in biobleaching of kraft pulp. *Bioresource Technology*, *98*(4), 897–903. doi:10.1016/j.biortech.2006.02.037 PMID:16617015

Khanna, S., & Gauri, S. (1993). Regulation, purification, and properties of xylanase from *Cellulomonas fimi*. *Enzyme and Microbial Technology*, *15*(11), 990–995. doi:10.1016/0141-0229(93)90177-4

Kheadr, E. E., Vuillemard, J. C., & El-Deeb, S. A. (2003). Impact of liposome-encapsulated enzyme cocktails on Cheddar cheese ripening. *Food Research International*, *36*(3), 241–252. doi:10.1016/S0963-9969(02)00166-7

Khetan, S. K., & Collins, T. J. (2007). Human pharmaceuticals in the aquatic environment: A challenge to Green Chemistry. *Chemical Reviews*, *107*(6), 2319–2364. doi:10.1021/cr020441w PMID:17530905

Kidwai, M., Jain, A., Sharma, A., & Kuhad, R. C. (2013a). Laccase—a natural source for the synthesis of benzofuro [2, 3-c] pyrazolin-5-ones. *Catalysis Science & Technology*, *3*(1), 230–234. doi:10.1039/C2CY20452E

Kidwai, M., Jain, A., Sharma, A., & Kuhad, R. C. (2013b). Laccase-catalysed reaction between Meldrum's acid and catechols/hydroquinones–An investigation. *Comptes Rendus. Chimie*, *16*(8), 728–735. doi:10.1016/j.crci.2013.04.006

Kiese, M. (1974). *Methemoglobinemia: a Comprehensive Treatise: Causes, Consequences, and Correction of Increased Contents of Ferrihemoglobin in Blood*. CRC Press.

Kilara, A., & Desai, M. (2002). Enzymes. In A. L. Branen, P. M. Davidson, S. Salminen, & J. H. Thorngate III, (Eds.), *Food Additives* (pp. 661–706). New York: Marcel Dekker Inc.

Kim, C. H., Nashiru, O., & Ko, J. H. (1996). Purification and biochemical characterization of pullulanase type I from *Thermus caldophilus* GK-24. *FEMS Microbiology Letters*, *138*(2-3), 147–152. doi:10.1111/j.1574-6968.1996.tb08148.x PMID:9026441

Kim, C. J., & Lee, P. I. (1992). Composite poly (vinyl alcohol) beads for controlled drug delivery. *Pharmaceutical Research*, *9*(1), 10–16. doi:10.1023/A:1018963223484 PMID:1589392

Kim, G. N., Shin, J. G., & Jang, H. D. (2009). Antioxidant and antidiabetic activity of Dangyuja (*Citrus grandis Osbeck*) extract treated with *Aspergillus saitoi*. *Food Chemistry*, *117*(1), 35–41. doi:10.1016/j.foodchem.2009.03.072

Kim, J., & Grate, J. (2003). Single Enzyme Nanoparticles Armored by a Nanometer-Scale Organic/Inorganic Network. *Nano Letters*, *3*(9), 1219–1222. doi:10.1021/nl034404b

Kim, J., Park, M. R., Kim, D. S., Lee, J. O., Maeng, S. H., Cho, S. Y., & Jin, D. K. (2013). IgE-mediated anaphylaxis and allergic reactions to idursulfase in patients with Hunter syndrome. *Allergy*, *68*(6), 796–802. doi:10.1111/all.12155 PMID:23621439

Kim, S. K., & Mendis, E. (2006). Bioactive compounds from marine processing byproducts – A review. *Food Research International*, *39*(4), 383–392. doi:10.1016/j.foodres.2005.10.010

Kim, S. K., & Rajapakse, N. (2005). Enzymatic production and biological activities of chitosan oligosaccharides (COS): A review. *Carbohydrate Polymers*, *62*(4), 357–368. doi:10.1016/j.carbpol.2005.08.012

Kim, S. W., Kang, S. W., & Lee, J. S. (1997). Cellulase and xylanase production by *Aspergillus niger* KKS in various bioreactors. *Bioresource Technology*, *59*(1), 63–67. doi:10.1016/S0960-8524(96)00127-7

Kim, S. Y., Lopez, C., Guebitz, G. M., & Cavaco-Paulo, A. (2008). Biological coloration of flax fabrics with flavonoids using laccase from *Trametes hirsuta*. *Engineering in Life Sciences*, *8*(3), 324–330. doi:10.1002/elsc.200700061

Kim, S. Y., Moldes, D., & Cavaco-Paulo, A. (2007). Laccase for enzymatic coloration of unbleached cotton. *Enzyme and Microbial Technology*, *40*(7), 1788–1793. doi:10.1016/j.enzmictec.2007.01.002

Kim, S., & Aga, D. S. (2007). Potential ecological and human health impacts of antibiotics and antibiotic-resistant bacteria from wastewater treatment plants. *Journal of Toxicology and Environmental Health. Part B, Critical Reviews*, *10*(8), 559–573. doi:10.1080/15287390600975137 PMID:18049923

Kimura, I., Sasahar, H., & Tajima, S. (1995). Purification and characterization of two xylanases and an arabinofuranosidase from *Aspergillus sojae*. *Journal of Fermentation and Bioengineering*, *80*(4), 334–339. doi:10.1016/0922-338X(95)94200-B

Kim, Y. K., Lee, S. C., Cho, Y. Y., Oh, H. J., & Ko, Y. H. (2012). *Isolation of cellulolytic Bacillus subtilis strains from agricultural environments*. Microbiology.

Kirk, O., Borchert, T. V., & Fuglsang, C. C. (2002). Industrial enzyme applications. *Current Opinion in Biotechnology*, *13*(4), 345–351. doi:10.1016/S0958-1669(02)00328-2 PMID:12323357

Klibanov, A. M. (1982). Enzymatic removal of hazardous pollutants from industrial aqueous effluents. *Enzyme Engineering*, *6*, 319-23.

Klibanov, A. M. (1983). Peroxidase-catalyzed removal of phenols from coal-conversion waste waters. *Science*, *221*, 259-60.

Klompong, V., Benjakul, S., Kantachote, D., & Shahidi, F. (2007). Antioxidative activity and functional properties of protein hydrolysate of yellow stripe trevally (*Selaroides leptolepis*) as influenced by the degree of hydrolysis and enzyme type. *Food Chemistry*, *102*(4), 1317–1327. doi:10.1016/j.foodchem.2006.07.016

Knietsch, A., Waschkowitz, T., Bowien, S., Henne, A., & Daniel, R. (2003). Construction and screening of metagenomic libraries derived from enrichment cultures: Generation of gene bank for genes conferring Alcohol Oxidoreductase activity of *Escherichia coli*. *Applied and Environmental Microbiology*, *69*(3), 408–1416. doi:10.1128/AEM.69.3.1408-1416.2003 PMID:12620823

Knob, A., Fortkamp, D., Prolo, T., Izidoro, C., & Almeida, J. M. (2014). Agroresidues as an alternative for xylanase production by filamentous fungi. *BioResource*, *9*(3), 5738–5773.

Koch, R., Canganella, F., Hippe, H., Jahnke, K. D., & Antranikian, G. (1997). Purification and Properties of a Thermostable Pullulanase from a Newly Isolated Thermophilic Anaerobic Bacterium *Fervidobacterium pennavorans* Ven5. *Applied and Environmental*, *63*, 1088–1094. PMID:16535541

Ko, E. M., Leem, Y. E., & Choi, H. (2001). Purification and characterization of laccase isozymes from the white-rot basidiomycete *Ganoderma lucidum*. *Applied Microbiology and Biotechnology*, *57*, 98–102. doi:10.1007002530100727 PMID:11693941

Kohli, V., Nigam, P., Singh, D., & Chaudhary, K. (2001). Thermostable, alkalophilic and cellulose free xylanase production by *Thermoactinoyces thalophillus Subgroup C. Enzyme and Microbial Technology*, *28*(7-8), 606–610. doi:10.1016/S0141-0229(01)00320-9 PMID:11339942

Kola, I., & Landis, J. (2004). Can the pharmaceutical industry reduce attrition rates? *Nature Reviews. Drug Discovery*, *3*(August), 1–5. doi:10.1038/nrd1470 PMID:15286737

Kolibaba, K. S., & Druker, B. J. (1997). Protein tyrosine kinases and cancer. *Biochimica et Biophysica Acta (BBA)-. Revue Canadienne*, *1333*(3), F217–F248.

Kono, H., Rusyn, I., Yin, M., Gäbele, E., Yamashina, S., Dikalova, A., ... Thurman, R. G. (2000). NADPH oxidase--derived free radicals are key oxidants in alcohol-induced liver disease. *The Journal of Clinical Investigation*, *106*(7), 867–872. doi:10.1172/JCI9020 PMID:11018074

Kontopidis, G., Holt, C., & Sawyer, L. (2004). Beta-lactoglobulin:binding properties, structure, and function. *Journal of Dairy Science*, *87*(4), 785–796. doi:10.3168/jds.S0022-0302(04)73222-1 PMID:15259212

Kouassi, G. K., Irudayaraj, J., & McCarty, G. (2005). Examination of cholesterol oxidase attachment to magnetic nanoparticles. *Journal of Nanobiotechnology*, *3*(1), 1–9. doi:10.1186/1477-3155-3-1 PMID:15661076

Kragl, U. (1996). *Industrial Enzymology*. London: Macmillan Press.

Kramer, K. J., Kanost, M. R., Hopkins, T. L., Jiang, H., Zhu, Y. C., Xu, R., ... Turecek, F. (2001). Oxidative conjugation of catechols with proteins in insect skeletal systems. *Tetrahedron*, *57*(2), 385–392. doi:10.1016/S0040-4020(00)00949-2

Krause, K. H., Lambeth, D., & Krönke, M. (2012). NOX enzymes as drug targets. *Cellular and Molecular Life Sciences*, *69*(14), 2279–2282. doi:10.100700018-012-1006-5 PMID:22585058

Krawczyk, H., Persson, T., Andersson, A., & Jonsson, A. S. (2007). Isolation of hemicelluloses from barley husks. *Food and Bioproducts Processing*, *86*(1), 31–36. doi:10.1016/j.fbp.2007.10.018

Kresse, G. B. (2004). Nonindustrial Enzyme Usage: Therapeutic Enzymes. In Enzymes in Industry (pp. 295-317). Wiley-VCH Verlag GmbH & Co. KGaA.

Kresse, G.-B. (2000). Enzymes, 6. Enzymes in Therapy. In Ullmann's Encyclopedia of Industrial Chemistry. Wiley-VCH Verlag GmbH & Co. KGaA.

Kriegshauser, G., & Liebl, W. (2000). Pullulanase from the Hyperthermophilic Bacterium Thermotoga maritime, Purification by β-Cyclodextrin Affinity Chromatography. *Journal of Chromatography. A*, *737*(1-2), 245–251. doi:10.1016/S0378-4347(99)00373-4 PMID:10681061

Krisana, A., Rutchadaporn, S., Jarupan, G., Lily, E., Sutipa, T., & Kanyawim, K. (2005). Endo-1,4-xylanase B from *Aspergillus* cf. *niger* BCC14405 isolated in Thailand: Purification, characterisation and gene isolation. *Journal of Biochemistry and Molecular Biology*, *38*(1), 17–23. PMID:15715941

Kris-Etherton, P. M., Hecker, K. D., Bonanome, A., Coval, S. M., Binkoski, A. E., Hilpert, K. F., ... Etherton, T. D. (2002). Bioactive compounds in foods: Their role in the prevention of cardiovascular disease and cancer. *The American Journal of Medicine*, *113*(9), 71–88. doi:10.1016/S0002-9343(01)00995-0 PMID:12566142

Kristl, J., Teskač, K., Caddeo, C., Abramović, Z., & Šentjurc, M. (2009). Improvements of cellular stress response on resveratrol in liposomes. *European Journal of Pharmaceutics and Biopharmaceutics*, *73*(2), 253–259. doi:10.1016/j.ejpb.2009.06.006 PMID:19527785

Krug, M., Ziegler, H., & Straube, G. (1985). Degradation of phenolic compounds by the yeast *Candida tropicalis* HP-15. 1. mendocina NSYSU chloride release. *Journal of Hazardous Materials*, *124*, 68–73.

Kudanga, T., Burton, S., Nyanhongo, G. S., & Guebitz, G. M. (2012). Versatility of oxidoreductases in the remediation of environmental pollutants. *Frontiers in Bioscience (Elite Edition)*, *4*(3), 1127–1149. doi:10.2741/e446 PMID:22201941

Kudanga, T., Nemadziva, B., & Le Roes-Hill, M. (2017). Laccase catalysis for the synthesis of bioactive compounds. *Applied Microbiology and Biotechnology*, *101*(1), 13–33. doi:10.100700253-016-7987-5 PMID:27872999

Kudanga, T., Nyanhongo, G. S., Guebitz, G. M., & Burton, S. (2011). Potential applications of laccase-mediated coupling and grafting reactions: A review. *Enzyme and Microbial Technology*, *48*(3), 195–208. doi:10.1016/j.enzmictec.2010.11.007 PMID:22112901

Kuhad, R. C., Kuhad, S., Kapoor, M., Sharma, K. K., & Singh, A. (2007). Lignocellulolytic microorganisms, their enzymes and possible biotechnologies based on lignocellulolytic microorganisms and their enzymes. In R. C. Kuhad & A. Singh (Eds.), Lignocellulose Biotechnology: Future Prospects. Academic Press.

Kuhad, R. C., Singh, A., & Eriksson, K. E. L. (1997). Microorganisms and enzymes involved in the degradation of plant fiber cell wall. In K. E. L. Eriksson (Ed.), *Biotechnology in the Pulp and paper industry.Advances in biochemical engineering biotechnology*. Berlin: Springer Verlag. doi:10.1007/BFb0102072

Kulkarni, N., Shendye, A., & Rao, M. (1999). Molecular and biotechnological aspects of xylanases. *FEMS Microbiology Reviews*, *23*(4), 411–456. doi:10.1111/j.1574-6976.1999.tb00407.x PMID:10422261

Kumar, A., Rishi, G., Bhuvnesh, S., Yogender, P. K., & Ramesh, C. K. (2012). Xylanase production from an alkalophilic actinomycete isolate *Streptomyces* sp. RCK-2010, its characterization and application in saccharification of second generation biomass. *Journal of Molecular Catalysis. B, Enzymatic*, *74*(3-4), 170–177. doi:10.1016/j.molcatb.2011.10.001

Kumar, C. G., & Takagi, H. (1999). Microbial alkaline proteases: From a bioindustrial viewpoint. *Biotechnology Advances*, *17*(7), 561–594. doi:10.1016/S0734-9750(99)00027-0 PMID:14538129

Kumari, G., Modi, M., Gupta, S. K., & Singh, R. K. (2011). Rhodium (II) acetate-catalyzed stereoselective synthesis, SAR and anti-HIV activity of novel oxindoles bearing cyclopropane ring. *European Journal of Medicinal Chemistry*, *46*(4), 1181–1188. doi:10.1016/j.ejmech.2011.01.037 PMID:21339030

Kumari, G., & Singh, K, R. (. (2013). Anti-HIV drug development: Structural features and limitations of present day drugs and future challenges in the successful HIV/AIDS treatment. *Current Pharmaceutical Design*, *19*(10), 1767–1783. doi:10.2174/13816128113199990295 PMID:23092282

Kumari, G., & Singh, R. K. (2012). Highly active antiretroviral therapy for treatment of HIV/AIDS patients: Current status and future prospects and the Indian scenario. *HIV & AIDS Review*, *11*(1), 5–14. doi:10.1016/j.hivar.2012.02.003

Kumari, G., & Singh, R. K. (2016). Molecular Modeling, Synthesis, and Anti-HIV Activity of Novel Isoindolinedione Analogues as Potent Non-nucleoside Reverse Transcriptase Inhibitors. *Chemical Biology & Drug Design*, *87*(2), 200–212. doi:10.1111/cbdd.12620 PMID:26212217

Kumar, J., Dahiya, J. S., Singh, D., & Nigam, P. (2000). Production of endo-1, 4- β-glucanase by a biocontrol fungus *Cladorrhinum foecundissimum*. *Bioresource Technology*, *75*(1), 95–97. doi:10.1016/S0960-8524(00)00037-7

Kunamneni, A., Camarero, S., García-Burgos, C., Plou, F. J., Ballesteros, A., & Alcalde, M. (2008). Engineering and applications of fungal laccases for organic synthesis. *Microbial Cell Factories*, *7*(1), 32. doi:10.1186/1475-2859-7-32 PMID:19019256

Kunamneni, A., Plou, F. J., Ballesteros, A., & Alcalde, M. (2008). Laccases and their applications: A patent review. *Recent Patents on Biotechnology*, *2*(1), 10–24. doi:10.2174/187220808783330965 PMID:19075849

Kuriki, T., Okada, S., & Imanaka, T. (1998). New type of pullulanase from *Bacillus stearothermophilus* and molecular cloning and expression of the gene in *Bacillus subtilis*. *Journal of Bacteriology*, *170*(4), 1554–1559. doi:10.1128/jb.170.4.1554-1559.1988 PMID:3127377

Kurisawa, M., Chung, J. E., Uyama, H., & Kobayashi, S. (2003a). Laccase-catalyzed synthesis and antioxidant property of poly (catechin). *Macromolecular Bioscience*, *3*(12), 758–764. doi:10.1002/mabi.200300038

Kurisawa, M., Chung, J. E., Uyama, H., & Kobayashi, S. (2003b). Enzymatic synthesis and antioxidant properties of poly (rutin). *Biomacromolecules*, *4*(5), 1394–1399. doi:10.1021/bm034136b PMID:12959611

Kurniawati, S., & Nicell, J. A. (2007). Efficacy of mediators for enhancing the laccase-catalyzed oxidation of aqueous phenol. *Enzyme and Microbial Technology*, *41*(3), 353–361. doi:10.1016/j.enzmictec.2007.03.003

Kuroiwa, T., Ichikawa, S., Sato, S., Hiruta, O., Sato, S., & Mukataka, S. (2002). Factors affecting the composition of oligosaccharides produced in chitosan hydrolysis using immobilized chitosanases. *Biotechnology Progress*, *18*(5), 969–974. doi:10.1021/bp020011r PMID:12363347

Kuroiwa, T., Noguchi, Y., Nakajima, M., Sato, S., Sukekuni, M., & Ichikawa, S. (2008). Production of chitosan oligosaccharides using chitosanase immobilized on amylose-coated magnetic nanoparticles. *Process Biochemistry*, *43*(1), 62–69. doi:10.1016/j.procbio.2007.10.014

Kurosawa, Y., Nirengi, S., Homma, T., Esaki, K., Ohta, M., Clark, J. F., & Hamaoka, T. (2015). A single-dose of oral nattokinase potentiates thrombolysis and anti-coagulation profiles. *Scientific Reports*, *5*(1), 11601. doi:10.1038rep11601 PMID:26109079

Kushwaha, A., Agarwal, S., Gupta, K. K., Maurya, S., Chaurasia, P. K., Singh, A. K., & Singh, M. P. (2017). Laccase Enzyme from white rot fungi: An overview and its Application. In Incredible world of Biotechnology. Nova Science Publishers Inc.

Kuuva, T., Lantto, R., Reinikainen, T., Buchert, J., & Autio, K. (2003). Rheological properties of laccase-induced sugar beet pectin gels. *Food Hydrocolloids*, *17*(5), 679–684. doi:10.1016/S0268-005X(03)00034-1

Kuznetsov, B. A., Shumakovich, G. P., Koroleva, O. V., & Yaropolov, A. I. (2001). On applicability of laccase as label in the mediated and mediatorless electro immunoassay: Effect of distance on the direct electron transfer between laccase and electrode. *Biosensors & Bioelectronics*, *16*(1-2), 73–84. doi:10.1016/S0956-5663(00)00135-4 PMID:11261856

Kyzas, G. Z., & Matis, K. A. (2015). Nanoadsorbents for pollutants removal: A review. *Journal of Molecular Liquids*, *203*, 159–168. doi:10.1016/j.molliq.2015.01.004

Labat, E., Morel, M. H., & Rouau, X. (2000). Effects of laccase and ferulic acid on wheat flour doughs. *Cereal Chemistry*, *77*(6), 823–828. doi:10.1094/CCHEM.2000.77.6.823

Labat, E., Morel, M. H., & Rouau, X. (2000). Wheat gluten phenolic acids: Occurrence and fate upon mixing. *Journal of Agricultural and Food Chemistry*, *48*(12), 6280–6283. doi:10.1021/jf0006271 PMID:11312800

LaBella, F., Waykole, P., & Queen, G. (1968). Formation of insoluble gels and dityrosine by the action of peroxidase on soluble collagens. *Biochemical and Biophysical Research Communications*, *30*(4), 333–338. doi:10.1016/0006-291X(68)90746-8 PMID:5637039

Ladisch, M., & Kohlmann, K. (1992). Recombinant human insulin. *Biotechnology Progress*, *8*(6), 469–478. doi:10.1021/bp00018a001 PMID:1369033

Lakshmi, M. V. V. C., & Jyothi, P. (2014). Production and optimization of Glucoamylase from *Aspergillus oryzae* NCIM1212 using wheat bran, varying chemical parameters under solid state fermentation. *International Journal of Current Microbiology and Applied Sciences*, *3*(5), 70–76.

Lakshmi, P., & Pradeepthi, B. (2014). Formulation and in-vitro evaluation of colon targeted aceclofenac-guargum microspheres by ionic gelation process. *Pharmanest*, *5*(3), 2163–2168.

Lambeth, J. D. (2004). NOX enzymes and the biology of reactive oxygen. *Nature Reviews. Immunology*, *4*(3), 181–189. doi:10.1038/nri1312 PMID:15039755

Lam, W. C., Pleissner, D., & Lin, C. S. (2013). Production of Fungal Glucoamylase for Glucose Production from Food Waste. *Biomolecules*, *3*(4), 651–661. doi:10.3390/biom3030651 PMID:24970186

Lantto, R., Scha¨nberg, C., & Buchert, J. (2004). Effects of laccase mediator combinations on wool. *Textile Research Journal*, *74*(8), 713–717. doi:10.1177/004051750407400809

Lavanya, C., Dhankar, R., Chhikara, S., & Sheoran, S. (2014). Degradation of toxic dyes: A review. *International Journal of Current Microbiology and Applied Sciences*, *3*(6), 189–199.

Lawton, T. J., Sayavedra-Soto, L. A., Arp, D. J., & Rosenzweig, A. C. (2009). Crystal structure of a two-domain multicopper oxidase: Implications for the evolution of multicopper blue proteins. *The Journal of Biological Chemistry*, *284*(15), 10174–10180. doi:10.1074/jbc.M900179200 PMID:19224923

Lee, B. H. (1996). Other Microorganism Based Products. In *Fundamentals of Food Biotechnology* (pp. 291–352). New York: Wiley-VCH Inc.

Lee, W. C., & Tsai, T. H. (2010). Preparation and characterization of liposomal coenzyme Q10 for *in vivo* topical application. *International Journal of Pharmaceutics*, *395*(1-2), 78–83. doi:10.1016/j.ijpharm.2010.05.006 PMID:20635514

Lee, Y.-Y., Kim, T. G., & Cho, K.-S. (2015). Isolation and characterization of a novel electricity-producing yeast, *Candida* sp. IR11. *Bioresource Technology*, *192*, 556–563. doi:10.1016/j.biortech.2015.06.038 PMID:26092068

Leite, O. D., Lupetti, K. O., Fatibello-Filho, O., Vieira, I. C., & de Barbosa, A. M. (2003). Synergic effect studies of the bi-enzymatic system laccaseperoxidase in a voltammetric biosensor for catecholamines. *Talanta*, *59*(5), 889–896. doi:10.1016/S0039-9140(02)00650-1 PMID:18968977

Lemons, M. L., Howland, D. R., & Anderson, D. K. (1999). Chondroitin sulfate proteoglycan immunoreactivity increases following spinal cord injury and transplantation. *Experimental Neurology*, *160*(1), 51–65. doi:10.1006/exnr.1999.7184 PMID:10630190

Leonetti, C., Scarsella, M., Semple, S. C., Molinari, A., D'Angelo, C., Stoppacciaro, A., ... Zupi, G. (2004). *In vivo* administration of liposomal vincristine sensitizes drug-resistant human solid tumors. *International Journal of Cancer*, *110*(5), 767–774. doi:10.1002/ijc.20174 PMID:15146568

Leroy, E., Jacquet, P., Coativy, G., Reguerre, A., & Lourdin, D. (2012). Compatibilization of starch–zein melt processed blends by an ionic liquid used as plasticizer. *Carbohydrate Polymers*, *89*(3), 955–963. doi:10.1016/j.carbpol.2012.04.044 PMID:24750886

Leung. (2004). Bioremediation: Techniques for cleaning up a mess. *Journal of Biotechnology*, *2*, 18–22.

Leveque, E., Janecek, S., Haye, B., & Belarbi, A. (2000). Thermophilic archaeal amylolytic enzymes. *Enzyme and Microbial Technology*, *26*(1), 3–14. doi:10.1016/S0141-0229(99)00142-8

Lewis, K. (2013). Platforms for antibiotic discovery. *Nature Reviews. Drug Discovery*, *12*(5), 371–387. doi:10.1038/nrd3975 PMID:23629505

Liang, T. W., Hsieh, J. L., & Wang, S. L. (2012). Production and purification of a protease, a chitosanase and chitin oligosaccharides by Bacillus cereus TKU022 fermentation. *Carbohydrate Research*, *362*, 38–46. doi:10.1016/j.carres.2012.08.004 PMID:23079238

Li, B. B., Smith, B., & Hossain, M. M. (2006). Extraction of phenolics from citrus peels II. Enzyme assisted extraction method. *Separation and Purification Technology*, *48*(2), 189–196. doi:10.1016/j.seppur.2005.07.019

Li, C. J., & Trost, B. M. (2008). Green chemistry for chemical synthesis. *Proceedings of the National Academy of Sciences of the United States of America*, *105*(36), 13197–13202. doi:10.1073/pnas.0804348105 PMID:18768813

Lienert, J., Güdel, K., & Escher, B. I. (2007). Screening method for ecotoxicological hazard assessment of 42 pharmaceuticals considering human metabolism and excretory routes. *Environmental Science & Technology*, *41*(12), 4471–4478. doi:10.1021/es0627693 PMID:17626454

Li, F., & Yang, X. W. (2012). Analysis of anti-inflammatory dehydrodiisoeugenol and metabolites excreted in rat feces and urine using HPLC-UV. *Biomedical Chromatography*, *26*(6), 703–707. doi:10.1002/bmc.1717 PMID:21932389

Li, J., Henriksson, G., & Gellerstedt, G. (2007). Lignin depolymerisation/repolymerisation and its critical role for delignification of aspen wood by steam explosion. *Bioresource Technology*, *98*(16), 3061–3068. doi:10.1016/j.biortech.2006.10.018 PMID:17141499

Li, J., Ye, Q., & Gan, J. (2014). Degradation and transformation products of acetaminophen in soil. *Water Research*, *49*, 44–52. doi:10.1016/j.watres.2013.11.008 PMID:24316789

Li, K., Horanyi, P. S., Collins, R., Phillips, R. S., & Eriksson, K. E. L. (2001). Investigation of the role of 3-hydroxyanthranilic acid in the degradation of lignin by white-rot fungus *Pycnoporus cinnabarinus*. *Enzyme and Microbial Technology*, *28*(4-5), 301–307. doi:10.1016/S0141-0229(00)00332-X PMID:11240183

Lina, B. A. R., Jonker, D. K., & Kozianowski, G. (2002). Isomaltulose (Palatinose): A review of biological and toxicological studies. *Food and Chemical Toxicology*, *40*(10), 1375–1381. doi:10.1016/S0278-6915(02)00105-9 PMID:12387299

Liney, K. E., Hagger, J. A., Tyler, C. R., Depledge, M. H., Galloway, T. S., & Jobling, S. (2005). Health effects in fish of long-term exposure to effluents from wastewater treatment works. *Environmental Health Perspectives*, *114*(S-1), 81–89. doi:10.1289/ehp.8058 PMID:16818251

Ling, H. S., Ling, T. C., Mohamad, R., & Ariff, A. B. (2009). Characterization of Pullulanase type- I from *Bacillus cereus* H1.5. *American Journal of Biochemistry and Biotechnology*, *5*(4), 170–179. doi:10.3844/ajbbsp.2009.170.179

Lin, J., Ndlovu, L. M., Singh, S., & Pillay, B. (1999). Purification and biochemical characteristics of β-D-xylanase from a thermophilic fungus, *Thermomyces lanuginosus*-SSBP. *Biotechnology and Applied Biochemistry*, *30*, 73–79. PMID:10467122

Li, P., Zeng, Y., Xie, Y., Li, X., Kang, Y., Wang, Y., ... Zhang, Y. (2017). Effect of pre-treatment on enzymatic hydrolysis of kitchen waste for xanthan production. *Bioresource Technology*, *223*, 84–90. doi:10.1016/j.biortech.2016.10.035 PMID:27788431

Li, S., Yang, X., Yang, S., Zhu, M., & Wang, X. (2012). Technology prospecting on Enzymes: Application, Marketing and Engineering. *Computational and Structural Biotechnology Journal*, *2*(3), e201209017. doi:10.5936/csbj.201209017 PMID:24688658

Lisdat, F., Wollenberger, U., Makower, A., Hortnagl, H., Pfeiffer, D., & Scheller, F. W. (1997). Catecholamine detection using enzymatic amplification. *Biosensors & Bioelectronics*, *12*(12), 1199–1211. doi:10.1016/S0956-5663(97)00098-5 PMID:9474769

Littoz, F., & McClements, D. J. (2008). Bio-mimetic approach to improving emulsion stability: Cross-linking adsorbed beet pectin layers using laccase. *Food Hydrocolloids*, *22*(7), 1203–1211. doi:10.1016/j.foodhyd.2007.06.009

Liu, H., Guo, K., Lv, J., Gao, Y., Duan, C., Deng, L., & Zhu, Z. (2017). A novel nitrite biosensor based on the direct electrochemistry of horseradish peroxidase immobilized on porous Co_3O_4 nanosheets and reduced graphene oxide composite modified electrode. *Sensors and Actuators. B, Chemical*, *238*, 249–256. doi:10.1016/j.snb.2016.07.073

Liu, W., Ye, A., Liu, W., Liu, C., & Singh, H. (2013). Liposomes as food ingredients and nutraceutical delivery systems. *Agro Food Industry Hi-Tech*, *24*(2), 68–71.

Liu, Y., Javvaji, V., Raghavan, S. R., Bentley, W. E., & Payne, G. F. (2012). Glucose oxidase-mediated gelation: A simple test to detect glucose in food products. *Journal of Agricultural and Food Chemistry*, *60*(36), 8963–8967. doi:10.1021/jf301376b PMID:22906038

Li, X. J., & Zhang, H. Y. (2008). Western-medicine-validated anti-tumor agents and traditional Chinese medicine. *Trends in Molecular Medicine*, *14*(1), 1–2. doi:10.1016/j.molmed.2007.11.002 PMID:18054521

Li, X., Gao, Y., Kuang, Y., & Xu, B. (2010). Enzymatic formation of a photo responsive supramolecular hydrogel. *Chemical Communications*, *46*(29), 5364–5366. doi:10.1039/c0cc00163e PMID:20582376

Li, X., Sun, B. Z., Lv, J., Li, Y. H. S. E., & Zhu, Y. (2011). Production and improved bleaching abilities of a thermostable xylanase from a newly isolated *Streptomyces chartreusis* strain. *African Journal of Biotechnology*, *10*(64), 14132–14142. doi:10.5897/AJB10.2360

Li, X., Xu, Q.-M., Cheng, J.-S., & Yuan, Y.-J. (2016). Improving the bioremoval of sulfamethoxazole and alleviating cytotoxicity of its biotransformation by laccase producing system under coculture of *Pycnoporus sanguineus* and *Alcaligenes faecalis*. *Bioresource Technology*, *220*, 333–340. doi:10.1016/j.biortech.2016.08.088 PMID:27591519

Li, Y., Niu, D., Zhang, L., Wang, Z., & Shi, G. (2013). Purification, characterization and cloning of a thermotolerant isoamylase produced from Bacillus sp. CICIM 304. *Journal of Industrial Microbiology & Biotechnology*, *40*(5), 437–446. doi:10.100710295-013-1249-7 PMID:23494708

Llorca, M., Rodríguez-Mozaz, S., Couillerot, O., Panigoni, K., de Gunzburg, J., Bayer, S., ... Barceló, D. (2014). Identification of new transformation products during enzymatic treatment of tetracycline and erythromycin antibiotics at laboratory scale by an on-line turbulent flow liquid-chromatography coupled to a high resolution mass spectrometer LTQ-Orbitrap. *Chemosphere*, *119C*, 90–98. doi:10.1016/j.chemosphere.2014.05.072 PMID:24972175

Lloret, L., Eibes, G., Feijoo, G., Moreira, M. T., & Lema, J. M. (2012a). Continuous operation of a fluidized bed reactor for the removal of estrogens by immobilized laccase on Eupergit supports. *Journal of Biotechnology*, *162*(4), 404–406. doi:10.1016/j.jbiotec.2012.04.007 PMID:22546624

Lloret, L., Eibes, G., Feijoo, G., Moreira, M. T., & Lema, J. M. (2012b). Degradation of estrogens by laccase from *Myceliophthora thermophila* in fed-batch and enzymatic membrane reactors. *Journal of Hazardous Materials, 213–214*, 175–183. doi:10.1016/j.jhazmat.2012.01.082 PMID:22342900

Lloret, L., Eibes, G., Lú-Chau, T., Moreira, M. T., Feijoo, G., & Lema, J. M. (2010). Laccase-catalyzed degradation of anti-inflammatories and estrogens. *Biochemical Engineering Journal, 51*(3), 124–131. doi:10.1016/j.bej.2010.06.005

Lloret, L., Eibes, G., Moreira, M. T. T., Feijoo, G., & Lema, J. M. M. (2013). On the use of a high-redox potential laccase as an alternative for the transformation of non-steroidal anti-inflammatory drugs (NSAIDs). *Journal of Molecular Catalysis. B, Enzymatic, 97*, 233–242. doi:10.1016/j.molcatb.2013.08.021

Lloyd, A. W., Martin, G. P., & Soozandehfar, S. H. (1994). Azopolymers: A means of colon specific drug delivery? *International Journal of Pharmaceutics, 106*(3), 255–260. doi:10.1016/0378-5173(94)90011-6

Lloyd, Jr., Mabbet, & Macaskie. (2001). Metal Reduction by Srb: Physiological Diversity and Metal Specificity. *Hydrometallurgy, 59*, 327–337. doi:10.1016/S0304-386X(00)00175-4

Lloyd, Jr., Yong, & Macaskie. (1998). Enzymatic Recovery of Elemental Palladium by Using Sulphate Reducing Bacteria. *Applied and Environmental Microbiology, 64*, 4607–4609. PMID:9797331

Loera, O., & Cordova, J. (2003). Improvement of xylanase production by a parasexual cross between *Aspergillus niger* strains. *Brazilian Archives of Biology and Technology, 46*(2), 177–181. doi:10.1590/S1516-89132003000200006

Long, J., Li, X., Wu, Z., Xu, E., Xu, X., Jin, Z., & Jiao, A. (2015). Immobilization of pullulanase onto activated magnetic chitosan/Fe_3O_4 nanoparticles prepared by in situ mineralization and effect of surface functional groups on the stability. *Colloids & Surfaces A. Physicochemical & Engineering, 472*, 69–77. doi:10.1016/j.colsurfa.2015.02.038

Long, J., Li, X., Zhan, X., Xu, X., Tian, Y., Xie, Z., & Jin, Z. (2017). Sol–gel encapsulation of pullulanase in the presence of hybrid magnetic (Fe3O4–chitosan) nanoparticles improves thermal and operational stability. *Bioprocess and Biosystems Engineering, 40*(6), 821–831. doi:10.100700449-017-1747-5 PMID:28243738

Longo, N., Harding, C. O., Burton, B. K., Grange, D. K., Vockley, J., Wasserstein, M., & Sile, S. (2014). Single-dose, subcutaneous recombinant phenylalanine ammonia lyase conjugated with polyethylene glycol in adult patients with phenylketonuria: An open-label, multicentre, phase 1 dose-escalation trial. *Lancet, 384*(9937), 37–44. doi:10.1016/S0140-6736(13)61841-3 PMID:24743000

Lonsane, B. K., Ghildyal, N. P., Budiatman, S., & Ramakrishna, S. V. (1985). Engineering aspects of solid state fermentation. *Enzyme and Microbial Technology, 7*(6), 256–265. doi:10.1016/0141-0229(85)90083-3

Lonsane, B. K., & Ramesh, M. V. (1990). Production of bacterial thermostable α-amylase by solid state fermentation: A potential tool for achieving economy in enzyme production and starch hydrolysis. *Advances in Applied Microbiology, 35*, 1–56. doi:10.1016/S0065-2164(08)70242-9 PMID:2205080

Luckham, R. E., & Brennan, J. D. (2010). Bioactive paper dipstick sensors for acetylcholinesterase inhibitors based on sol–gel/enzyme/gold nanoparticle composites. *Analyst (London), 135*(8), 2028–2035. doi:10.1039/c0an00283f PMID:20593080

Lu, F. X., Lu, M., Lu, Z., Bie, X. M., Zhao, H. Z., & Wang, Y. (2008). Purification and characterization of xylanase from *Aspergillus ficuum* AF-98. *Bioresource Technology, 99*(13), 5938–5941. doi:10.1016/j.biortech.2007.10.051 PMID:18068974

Lund, M., & Ragauskas, A. J. (2001). Enzymatic modification of kraft lignin through oxidative coupling with water-soluble phenols. *Applied Microbiology and Biotechnology, 55*(6), 699–703. doi:10.1007002530000561 PMID:11525617

Luo, H., Li, J., Yang, J., Wang, H., Yang, Y., Huang, H., ... Yao, B. A. (2009). Thermophilic and acid stable family-10 xylanase from the acidophilic fungus *Bispora* sp. MEY-1. *Extremophiles*, *13*(5), 849–857. doi:10.100700792-009-0272-0 PMID:19655217

Luo, H., Wang, K., Huang, H., Shi, P., Yang, P., & Yao, B. (2012). Gene cloning, expression and biochemical characterization of an alkali-tolerant b-mannanase from *Humicola insolens* Y1. *Journal of Industrial Microbiology & Biotechnology*, *39*(4), 547–555. doi:10.100710295-011-1067-8 PMID:22179540

Luo, X., Ma, M., Chen, B., Yao, S., Wan, Z., Yang, D., & Hang, H. (2004). Analysis of Nine Bioactive Compounds in *Eucommia ulmoides* Oliv.and Their Preparation by HPLC-Photodiode Array Detection and Mass Spectrometry. *Journal of Liquid Chromatography & Related Technologies*, *27*(1), 63–81. doi:10.1081/JLC-120027086

Lu, W., Li, D., & Wu, Y. (2003). Influence of water activity and temperature on xylanase biosynthesis in pilot-scale solid-state fermentation by *Aspergillus sulphureus*. *Enzyme and Microbial Technology*, *32*(2), 305–311. doi:10.1016/S0141-0229(02)00292-2

Macellaro, G., Pezzella, C., Cicatiello, P., Sannia, G., & Piscitelli, A. (2014). Fungal laccases degradation of endocrine disrupting compounds. *BioMed Research International*, *2014*, 614038. doi:10.1155/2014/614038 PMID:24829908

Mach, C. M., Mathew, L., Mosley, S. A., Kurzrock, R., & Smith, J. A. (2009). Determination of minimum effective dose and optimal dosing schedule for liposomal curcumin in a xenograft human pancreatic cancer model. *Anticancer Research*, *29*(6), 1895–1899. PMID:19528445

Macheix, J. J., Sapis, J. C., Fleuriet, A., & Lee, C. Y. (1991). Phenolic compounds and polyphenoloxidases in relation to browning in grapes and wines. *Critical Reviews in Food Science and Nutrition*, *30*(4), 441–486. doi:10.1080/10408399109527552 PMID:1910524

Maciel, G.M., de Souza, Vandenberghe, L. P., Windson, C., Haminiuk, I., Fendrich, R.C., ... Soccol, C.R. (2008). Xylanase production by *Aspergillus niger* LPB 326 in solid state fermentation using statistical experimental design. *Food Technology and Biotechnology*, *46*(2), 183–189.

MacMurrough, I., Madigan, D., Kelly, R., & O'Rourke, T. (1999). Haze formation. Shelf-life prediction for lager beer. *Food Technology*, *53*, 58–62.

Maczulak & Elizabeth. (2010). *Pollution: Treating Environmental Toxins*. New York: Infobase Publishing.

Maddhinni, V. L., Vurimindi, H. B., & Yerramilli, A. (2006). Degradation of Azo Dye with Horseradish Peroxidase (HRP). *J. Indian Inst. Sci.*, *86*, 507-514.

Madhavi, V., & Lele, S. S. (2009). Laccase: Properties and applications. *Bioresource Technology*, *4*(4), 1694–1717.

Madi, E., Antranikian, G., Ohmiya, K., & Gottschalk, G. (1987). Thermostable Amylolytic Enzymes from a new Clostridium Isolate. *Applied and Environmental Microbiology*, *53*, 1661–1667. PMID:16347392

Maestri, E., Marmiroli, M., & Marmiroil, N. (2016). Bioactive peptides in plant-derived foodstuffs. *Journal of Proteomics*, *147*, 140–155. doi:10.1016/j.jprot.2016.03.048 PMID:27079980

Magnani, M., Mancini, U., Bianchi, M., & Fazi, A. (1992). Comparison of Uricase-Bound and Uricase-Loaded Erythrocytes as Bioreactors for Uric Acid Degradation. In M. Magnani & J. R. DeLoach (Eds.), *The Use of Resealed Erythrocytes as Carriers and Bioreactors* (pp. 189–194). Boston, MA: Springer US. doi:10.1007/978-1-4615-3030-5_23

Mahmoud, D. A. R., & Helmy, W. A. (2009). Application of Immobilization Technology in Enzyme and Biomass Production (Review Article). *Journal of Applied Sciences Research*, *5*(12), 2466–2476.

Maier, G., Mayer, P., & Dietrich, H. (1990). Application of a polyphenoloxidase to stabilization of apple juices. *Deutsche Lebensmittel-Rundschau*, *86*, 137–142.

Maine, E., Thomas, V. J., Bliemel, M., Murira, A., & James Utterback, J. (2014). The emergence of the nanobiotechnology industry. *Nature Nanotechnology*, *9*(1), 2–5. doi:10.1038/nnano.2013.288 PMID:24390555

Majeau, J. A., Brar, S. K., & Tyagi, R. D. (2010). Laccases for removal of recalcitrant and emerging pollutants. *Bioresource Technology*, *101*(7), 2331–2350. doi:10.1016/j.biortech.2009.10.087 PMID:19948398

Malakar, R., Tiwari, A., & Malviya, S. N. (2010). Pullulanase: A potential enzyme for industrial application. *International Journal of Biomedical Research*, *2*, 10–20.

Maldonado, M., & De Saad, A. S. (1998). Production of pectinesterase and polygalacturonase by Aspergillus niger in submerged and solid state systems. *Journal of Industrial Microbiology & Biotechnology*, *20*(1), 34–38. doi:10.1038j.jim.2900470 PMID:9523455

Malhotra, R., Noorvez, S., & Satyanarayana, T. (2000). Production and partial characterization of thermostable and calcium-independent α-amylase of an extreme thermophile *Bacillus thermooleovorans* NP54. *Letters in Applied Microbiology*, *31*(5), 378–384. doi:10.1046/j.1472-765x.2000.00830.x PMID:11069641

Mallett, A. K., Wise, A., & Rowland, I. R. (1983). Effect of dietary cellulose on the metabolic activity of the rat caecal microflora. *Archives of Toxicology*, *52*(4), 311–317. doi:10.1007/BF00316499 PMID:6307219

Mamo, G., & Gessesse, A. (1997). Article. *Biotechnology Techniques*, *11*(6), 447–450. doi:10.1023/A:1018437310837

Mamo, G., & Gessesse, A. (1999). Purification and characterization of two raw-starch-digesting thermostable a-amylases from a thermophilic *Bacillus*. *Enzyme and Microbial Technology*, *25*(3-5), 433–438. doi:10.1016/S0141-0229(99)00068-X

Mamo, G., Thunnissen, M., Hatti-Kaul, R., & Mattiasson, B. (2009). An alkaline active xylanase: Insights into mechanisms of high pH catalytic adaptation. *Biochimie*, *91*(9), 1187–1196. doi:10.1016/j.biochi.2009.06.017 PMID:19567261

Manda, K. (2006). *Biotransformation von Aminosäuren, Peptiden und dihydroxylierten Aromaten durch Laccasen aus Pilzen zur Herstellung von Hybridmolekülen* (Unpublished doctoral dissertation). University of Greifswald, Germany.

Manda, K., Hammer, E., Mikolasch, A., Gördes, D., Thurow, K., & Schauer, F. (2006). Laccase-induced derivatization of unprotected amino acid L-tryptophan by coupling with p-hydroquinone 2, 5-dihydroxy-N-(2-hydroxyethyl)-benzamide. *Amino Acids*, *31*(4), 409–419. doi:10.100700726-005-0276-8 PMID:16583315

Manea, F., Houillon, F. B., Pasquato, L., & Scrimin, P. (2004). Nanozymes: Gold-Nanoparticle-Based Transphosphorylation Catalysts. *Angewandte Chemie International Edition*, *43*(45), 6165–6169. doi:10.1002/anie.200460649 PMID:15549744

Manley, P. W., Cowan-Jacob, S. W., Buchdunger, E., Fabbro, D., Fendrich, G., Furet, P., ... Zimmermann, J. (2002). Imatinib: A selective tyrosine kinase inhibitor. *European Journal of Cancer*, *38*, S19–S27. doi:10.1016/S0959-8049(02)80599-8 PMID:12528769

Marchal, L. M., Jonkers, J., Franke, G. T., de Gooijer, C. D., & Tramper, J. (1999). The effect of process conditions on the a-amylolytic hydrolysis of amylopectin potato starch: An experimental design approach. *Biotechnology and Bioengineering*, *62*(3), 348–357. doi: PMID:10099546

Marc, J. E., van der Maarel, C., Joost, B. V., Uitdehaag, C. M., Leemhuis, H., & Dijkhuizen, L. (2002). Properties and applications of starch-converting enzymes of the α amylase family. *Journal of Biotechnology*, *94*(2), 137–155. doi:10.1016/S0168-1656(01)00407-2 PMID:11796168

Marco-urrea, E., Pérez-trujillo, M., Blánquez, P., Vicent, T., & Caminal, G. (2010). Biodegradation of the analgesic naproxen by *Trametes versicolor* and identification of intermediates using HPLC-DAD-MS and NMR. *Bioresource Technology*, *101*(7), 2159–2166. doi:10.1016/j.biortech.2009.11.019 PMID:20004093

Marco-Urrea, E., Pérez-Trujillo, M., Cruz-Morató, C., Caminal, G., & Vicent, T. (2010a). Degradation of the drug sodium diclofenac by *Trametes versicolor* pellets and identification of some intermediates by NMR. *Journal of Hazardous Materials*, *176*(1-3), 836–842. doi:10.1016/j.jhazmat.2009.11.112 PMID:20031320

Marco-Urrea, E., Pérez-Trujillo, M., Cruz-Morató, C., Caminal, G., & Vicent, T. (2010b). White-rot fungus-mediated degradation of the analgesic ketoprofen and identification of intermediates by HPLC–DAD–MS and NMR. *Chemosphere*, *78*(4), 474–481. doi:10.1016/j.chemosphere.2009.10.009 PMID:19913277

Marco-Urrea, E., Pérez-Trujillo, M., Vicent, T., & Caminal, G. (2009). Ability of white-rot fungi to remove selected pharmaceuticals and identification of degradation products of ibuprofen by *Trametes versicolor. Chemosphere*, *74*(6), 765–772. doi:10.1016/j.chemosphere.2008.10.040 PMID:19062071

Mares, D. J., & Stone, B. A. (1973). Studies on wheat endosperm-II. Properties of wall components and studies on organisation in the wall. *Australian Journal of Biological Sciences*, *26*(4), 813–830. doi:10.1071/BI9730813

Margot, J., Bennati-Granier, C., Maillard, J., Blánquez, P., Barry, D., & Holliger, C. (2013). Bacterial versus fungal laccase: Potential for micropollutant degradation. *AMB Express*, *3*(1), 63. doi:10.1186/2191-0855-3-63 PMID:24152339

Margot, J., Maillard, J., Rossi, L., Barry, D. A., & Holliger, C. (2013). Influence of treatment conditions on the oxidation of micropollutants by *Trametes versicolor* laccase. *New Biotechnology*, *30*(6), 803–813. doi:10.1016/j.nbt.2013.06.004 PMID:23831273

Markom, M., Hasan, M., Daud, W. R. W., Singh, H., & Jahim, J. M. (2007). Extraction of hydrolysable tannins from *Phyllanthus niruri*Linn.: Effects of solvents and extraction methods. *Separation and Purification Technology*, *52*(3), 487–496. doi:10.1016/j.seppur.2006.06.003

Marrot, B., Barrios-Martinez, A., Moulin, P., & Roche, N. (2006). Biodegradation of high phenol concentration by activated sludge in an immersed membrane bioreactor. *Biochemical Engineering Journal*, *30*(2), 174–183. doi:10.1016/j.bej.2006.03.006

Martin, B. D., Ampofo, S. A., Linhardt, R. J., & Dordick, J. S. (1992). Biocatalytic synthesis of sugar-containing polyacrylate-based hydrogels. *Macromolecules*, *25*(26), 7081–7085. doi:10.1021/ma00052a001

Martin, S. C., Forsberg, P. O., & Eriksson, S. D. (1991). The effects of in vitro phosphorylation and dephosphorylation on the thrombin-induced gelation and plasmin degradation of fibrinogen. *Thrombosis Research*, *61*(3), 243–252. doi:10.1016/0049-3848(91)90100-B PMID:1827546

Martins, S., Aguilar, C. N., Garza-Rodriguez, I. D. L., Mussatto, S. I., & Teixeira, J. A. (2010). Kinetic study of nordihydroguaiaretic acid recovery from *Larrea tridentata* by microwave-assisted extraction. *Journal of Chemical Technology and Biotechnology (Oxford, Oxfordshire)*, *85*(8), 1142–1147. doi:10.1002/jctb.2412

Martins, S., Mussatto, S. I., Martínez-Avila, G., Montañez-Saenz, J., Aguilar, C. N., & Teixeira, J. A. (2011). Bioactive phenolic compounds: Production and extraction by solid-state fermentation. A review. *Biotechnology Advances*, *29*(3), 365–373. doi:10.1016/j.biotechadv.2011.01.008 PMID:21291993

Martins, V. G., Costa, J. A. V., & Prentice-Hernández, C. (2009). Hidrolisado protéico de pescado obtido por vias química enzimática a partir de corvina (*Micropogonias furnieri*). *Quimica Nova*, *32*(1), 61–66. doi:10.1590/S0100-40422009000100012

Masuda, Y., Jugimiya, S.-I., Kawachi, Y., & Kato, K. (2014). Interparticle mesoporous silica as an effective support for enzyme immobilisation. *RSC Advances*, *4*(7), 3573–3580. doi:10.1039/C3RA46122J

Masuyer, G., Schwager, S. L. U., Sturrock, E. D., Isaac, R. E., & Acharya, K. R. (2012). Molecular recognition and regulation of human angiotensin-I converting enzyme (ACE) activity by natural inhibitory peptides. *Scientific Reports*, *2*(1), 717. doi:10.1038rep00717 PMID:23056909

Matern, U., & Kneusel, R. E. (1988). Phenolic compounds in plant disease resistance. *Phytoparasitica*, *16*(2), 153–170. doi:10.1007/BF02980469

Mathiasen, T.E. (1995). *Laccase and beer storage. PCT international application.* WO 9521240 A2.

Mathioni, S. M., Patel, N., Riddick, B., Sweigard, J. A., Czymmek, K. J., Caplan, J. L., & Donofrio, N. M. (2013). Transcriptomics of the rice blast fungus *Magnaporthe oryzae* in response to the bacterial antagonist Lysobacter enzymogenes reveals candidate fungal defense response genes. *PLoS One*, *8*(10), e76487. doi:10.1371/journal.pone.0076487 PMID:24098512

Matsuo, K., & Miyazono, I. (1993). The influence of long-term administration of peptidoglycan on disease resistance and growth of juvenile rainbow trout. *Nippon Suisan Gakkaishi*, *59*(8), 1377–1379. doi:10.2331uisan.59.1377

Matsuura, T., Feng, C., Khulbe, K. C., Rana, D., Singh, G., Gopal, R., ... Tabe, S. (2010). Development of novel membranes based on electrospun nanofibers and their application in liquid filtration, membrane distillation, and membrane adsorption. *Maku Japanese J. Membr.*, *35*, 119–127.

Matsuura, T., & Ohkatsu, Y. (2000). Phenolic antioxidants: Effect of o-benzyl substituents. *Polymer Degradation & Stability*, *70*(1), 59–63. doi:10.1016/S0141-3910(00)00088-4

Mattinen, M. L., Kruus, K., Buchert, J., Nielsen, J. H., Andersen, H. J., & Steffensen, C. L. (2005). Laccase-catalysed polymerization of tyrosine-containing peptides. *The FEBS Journal*, *272*(14), 3640–3650. doi:10.1111/j.1742-4658.2005.04786.x PMID:16008563

Maugh, T. H. (1984). Semisynthetic enzymes are new catalysts; enzymes may be the ultimate catalysts, especially if their specificities can be changed by chemical modification. *Science*, *223*(4632), 154–157. doi:10.1126cience.223.4632.154 PMID:17733792

Ma, X., Hortelão, A. C., Patiño, T., & Sánchez, S. (2016). Enzyme catalysis to power micro/nanomachines. *ACS Nano*, *10*(10), 9111–9122. doi:10.1021/acsnano.6b04108 PMID:27666121

Ma, X., Jannasch, A., Albrecht, U. R., Hahn, K., Miguel-López, A., Schäffer, E., & Sánchez, S. (2015). Enzyme-powered hollow mesoporous Janus nanomotors. *Nano Letters*, *15*(10), 7043–7050. doi:10.1021/acs.nanolett.5b03100 PMID:26437378

Ma, X., & Sánchez, S. (2017). Bio-catalytic mesoporous Janus nano-motors powered by catalase enzyme. *Tetrahedron*, *73*(33), 4883–4886. doi:10.1016/j.tet.2017.06.048

Mayer, A. M., & Staples, R. C. (2002). Laccase: New functions for an old enzyme. *Phytochemistry*, *60*(6), 551–565. doi:10.1016/S0031-9422(02)00171-1 PMID:12126701

McBain, A. J., & Macfarlane, G. T. (1997). Investigations of bifidobacterial ecology and oligosaccharide metabolism in a three-stage compound continuous culture system. *Scandinavian Journal of Gastroenterology*, *32*(sup222), 32–40. doi:10.1080/00365521.1997.11720715 PMID:9145444

McCleary, B. V. (1991). *Comparison of endolytic hydrolases that depolymerize 1, 4-β-d-mannan, 1, 5-α-l-arabinan, and 1, 4-β-d-galactan.* ACS Publications. doi:10.1021/bk-1991-0460.ch034

McCleary, B., & Matheson, N. (1986). Polysaccharides having a β-D-xylan backbone. *Advances in Carbohydrate Chemistry and Biochemistry*, *44*, 158–164.

McClements, D. J. (2017). Recent progress in hydrogel delivery systems for improving nutraceuticalbioavilability. *Food Hydrocolloids*, *68*, 238–245. doi:10.1016/j.foodhyd.2016.05.037

McClintock, S. D., Hoesel, L. M., Das, S. K., Till, G. O., Neff, T., Kunkel, R. G., ... Ward, P. A. (2006). Attenuation of half sulfur mustard gas-induced acute lung injury in rats. *Journal of Applied Toxicology*, *26*(2), 126–131. doi:10.1002/jat.1115 PMID:16252256

McLeod, A. D., Friend, D. R., & Tozer, T. N. (1994). Glucocorticoid–dextran conjugates as potential prodrugs for colon-specific delivery: Hydrolysis in rat gastrointestinal tract contents. *Journal of Pharmaceutical Sciences*, *83*(9), 1284–1288. doi:10.1002/jps.2600830919 PMID:7530303

McTigue, M. A., Kelly, C. T., Doyle, E. M., & Fogarty, W. M. (1995). The alkaline amylase of the alkalophilic *Bacillus* sp. IMD 370. *Enzyme and Microbial Technology*, *17*(6), 570–573. doi:10.1016/0141-0229(94)00098-C

Meerum Terwogt, J. M., Groenewegen, G., Pluim, D., Maliepaard, M., Tibben, M. M., Huisman, A., ... Schellens, J. M. (2002). Phase I and pharmacokinetic study of SPI-77, a liposomal encapsulated dosage form of cisplatin. *Cancer Chemotherapy and Pharmacology*, *49*(3), 201–210. doi:10.1007002800100371 PMID:11935212

Meinlschmidt, P., Sussmann, D., Schweiggert-Weisz, U., & Eisner, P. (2016). Enzymatic treatment of soy protein isolates: Effects on the potential allergenicity, technofunctionality, and sensory properties. *Food Science & Nutrition*, *4*(1), 11–23. doi:10.1002/fsn3.253 PMID:26788306

Melasniemi, H. (1988). Purification and some Properties of the Extracellular α-amylase Pullulanase Produced by *Clostridium thermohydrosulfuricum*. *The Biochemical Journal*, *250*(3), 813–818. doi:10.1042/bj2500813 PMID:3260488

Melikoglu, M., Lin, C. S. K., & Webb, C. (2013). Analysing global food waste problem: Pinpointing the facts and estimating the energy content. *Cent. Eur. J. Eng.*, *3*(2), 157–164.

Melo, J. S., Kholi, S., Patwardhan, A. W., & D'Souza, S. F. (2005). Effect of oxygen transfer limitations in phenol biodegradation. *Process Biochemistry*, *40*(2), 625–628. doi:10.1016/j.procbio.2004.01.049

Mensah, C. A., Adamafio, N. A., Kwarteng, A. K., & Rodrigues, F. K. (2012). Reduced tannin content of laccase-treated cocoa (Theobroma cacao) pod husk. *Int J Biol Chem.*, *6*(1), 31–36. doi:10.3923/ijbc.2012.31.36

Messerschmidt, A. (Ed.). (1997). *Multi-copper oxidases*. Singapore: World Scientific. doi:10.1142/3146

Mezajoug-Kenfack, L. B., Arab-Tehrany, E., Tchiégang, C., & Linder, M. (2011). Compositional and nutritional studies of two defatted flours obtained from *Ricinodendron heudelotii* (Bail.) and *Tetracarpidium conophorum* (Müll. Arg.). *Food Global Science Books*, *5*(1), 52–57.

Mezajoug-Kenfack, L. B., Ngangoum, E. S., Tchiégang, C., & Linder, M. (2014). Proteolytic enzymes from *Abrus precatorius* Linn. *Burnatia enneandra* Micheli. and *Ziziphus mauritiana* Lam: Optimization of the extraction process by Response Surface Methodology. *The Journal of Bioprocess Technology*, *100*, 453–461.

Mezajoug-Kenfack, L. B., & Tchiégang, C. (2016). Physico chemical properties of defatted flours from *Ricinodendron heudelotii* (Bail.) and *Tetracarpidium conophorum* (Müll. Arg.) with respect to particle sizes. *Global Advanced Research Journal of Agricultural Science*, *5*(2), 94–102.

Miao, X.-S., & Metcalfe, C. D. (2003). Determination of carbamazepine and its metabolites in aqueous samples using liquid chromatography-electrospray tandem mass spectrometry. *Analytical Chemistry*, *75*(15), 3731–3738. doi:10.1021/ac030082k PMID:14572037

Micheli, F. (2001). Pectin methylesterases: Cell wall enzymes with important roles in plant physiology. *Trends in Plant Science*, *6*(9), 414–419. doi:10.1016/S1360-1385(01)02045-3 PMID:11544130

Mielgo, I. C., Lopez, C., Moreira, M. T., Feijoo, G., & Lema, J. M. (2003). Oxidative Degradation of Azo Dyes by Manganese Peroxidase under Optimized Condition. *Biotechnology Progress*, *19*(2), 325–331. doi:10.1021/bp020136w PMID:12675567

Migliore, L., Fiori, M., Spadoni, A., & Galli, E. (2012). Biodegradation of oxytetracycline by *Pleurotus ostreatus* mycelium: A mycoremediation technique. *Journal of Hazardous Materials*, *215–216*, 227–232. doi:10.1016/j.jhazmat.2012.02.056 PMID:22436341

Mihara, H., Sumi, H., Yoneta, T., Mizumoto, H., Ikeda, R., Seiki, M., & Maruyama, M. (1991). A novel fibrinolytic enzyme extracted from the earthworm, Lumbricus rubellus. *Japanese Journal of Physiology*, *41*(3), 461–472. doi:10.2170/jjphysiol.41.461 PMID:1960890

Mikolasch, A., Hammer, E., Jonas, U., Popowski, K., Stielow, A., & Schauer, F. (2002). Synthesis of 3-(3, 4-dihydroxyphenyl)-propionic acid derivatives by N-coupling of amines using laccase. *Tetrahedron*, *58*(38), 7589–7593. doi:10.1016/S0040-4020(02)00872-4

Mikolasch, A., Hessel, S., Salazar, M. G., Neumann, H., Manda, K., Gōrdes, D., ... Beller, M. (2008b). Synthesis of new N-analogous corollosporine derivatives with antibacterial activity by laccase-catalyzed amination. *Chemical & Pharmaceutical Bulletin*, *56*(6), 781–786. doi:10.1248/cpb.56.781 PMID:18520080

Mikolasch, A., Manda, K., Schlüter, R., Lalk, M., Witt, S., Seefeldt, S., ... Lindequist, U. (2012). Comparative analyses of laccase-catalyzed amination reactions for production of novel β-lactam antibiotics. *Biotechnology and Applied Biochemistry*, *59*(4), 295–306. doi:10.1002/bab.1026 PMID:23586863

Mikolasch, A., Niedermeyer, T. H. J., Lalk, M., Witt, S., Seefeldt, S., Hammer, E., ... Lindequist, U. (2006). Novel penicillins synthesized by biotransformation using laccase from *Trametes* species. *Chemical & Pharmaceutical Bulletin*, *54*(5), 632–638. doi:10.1248/cpb.54.632 PMID:16651757

Mikolasch, A., Niedermeyer, T. H. J., Lalk, M., Witt, S., Seefeldt, S., Hammer, E., ... Lindequist, U. (2007). Novel cephalosporins synthesized by amination of 2, 5-dihydroxybenzoic acid derivatives using fungal laccases II. *Chemical & Pharmaceutical Bulletin*, *55*(3), 412–416. doi:10.1248/cpb.55.412 PMID:17329882

Mikolasch, A., & Schauer, F. (2009). Fungal laccases as tools for the synthesis of new hybrid molecules and biomaterials. *Applied Microbiology and Biotechnology*, *82*(4), 605–624. doi:10.100700253-009-1869-z PMID:19183983

Mikolasch, A., Wurster, M., Lalk, M., Witt, S., Seefeldt, S., Hammer, E., ... Lindequist, U. (2008a). Novel β-lactam antibiotics synthesized by amination of catechols using fungal laccase. *Chemical & Pharmaceutical Bulletin*, *56*(7), 902–907. doi:10.1248/cpb.56.902 PMID:18591799

Milagres, A. M. F., Laci, L. S., & Prade, R. A. (1993). Characterisation of xylanase production by a local isolate of *Penicillium janthinellum. Enzyme and Microbial Technology*, *15*(3), 248–253. doi:10.1016/0141-0229(93)90145-R

Milagres, A. M. F., Santos, E., Piovan, T., & Roberto, I. C. (2004). Production of xylanase by *Thermoascus aurantiacus* from sugar cane bagasse in an aerated growth fermentor. *Process Biochemistry*, *39*(11), 1387–1391. doi:10.1016/S0032-9592(03)00272-3

Miller, G. L. (1959). Use of Dinitrosalicyclic acid reagent for determination of reducing sugar. *Analytical Chemistry*, *31*(3), 426–428. doi:10.1021/ac60147a030

Mimura, H., & Shinichi, N. (1999). Physiological characteristics of *Bacillus* sp. *Biocontrol Science and Technology*, *4*(2), 105–108. doi:10.4265/bio.4.105

Min, K.-L., Kim, Y.-H., Kim, Y. W., Jung, H. S., & Hah, Y. C. (2001). Characterization of a novel laccase produced by the wood-rotting fungus *Phellinus ribis*. *Archives of Biochemistry and Biophysics*, *392*(2), 279–286. doi:10.1006/abbi.2001.2459 PMID:11488603

Minussi, R. C., Pereira, D. G., Pastore, G. M., & Duran, N. (1998).Enzymatic oxidation of ferulic acid by laccase. *IV National Congress of Biotechnology, Talca-Chile, Resume'n.*

Minussi, R.C., Pastore, G.M., & Duran, N. (2002). Potential applications of laccase in the food industry. *Trends in Food Science and Technology, 13*(6-7), 205–216.

Minussi, R. C., Pastore, G. M., & Duran, N. (1998). Phenol removal and stabilization of red wine in the presence of laccase from *Trametes versicolor*. In *Brazilian Meeting on the Chemistry of Food and Beverages*, Sa˜o Carlos, Brazil.

Mirsalis, J. C., Hamm, T. E., Sherrill, J. M., & Butterworth, B. E. (1982). Role of gut flora in the genotoxicity of dinitrotoluene. *Nature*, *295*(5847), 322–323. doi:10.1038/295322a0 PMID:7057895

Mirzadeh, S.-S., Khezri, S.-M., Rezaei, S., Forootanfar, H., Mahvi, A. H., & Faramarzi, M. A. (2014). Decolorization of two synthetic dyes using the purified laccase of *Paraconiothyrium variabile* immobilized on porous silica beads. *Journal of Environmental Health Science & Engineering*, *12*(1), 6. doi:10.1186/2052-336X-12-6 PMID:24393474

Misson, Zhang, & Jin. (2015). Nanobiocatalyst advancements and bioprocessing applications. *J R Soc Interface*, *12*(102). doi: 10.1098/rsif.2014.0891

Mitchell, D. T., Lee, S. B., Trofin, L., Li, N., Nevanen, T. K., Soderlund, H., & Martin, C. R. (2002). Smart nanotubes for bioseparations and biocatalysis. *Journal of the American Chemical Society*, *124*(40), 11864–11865. doi:10.1021/ja027247b PMID:12358526

Mitchell, M. D., & Miller, D. M. (2012). Looking at ocriplasmin as a new option in eye disease. *Drugs of Today (Barcelona, Spain)*, *48*(8), 519–524. PMID:22916339

Mizota, T. (1996). Functional and nutritional foods containing bifidogenic factors. *Bulletin of the International Dairy Foundation*, *313*, 31–35.

Mogharabi, M., & Faramarzi, M. A. (2014). Laccase and laccase-mediated systems in the synthesis of organic compounds. *Advanced Synthesis & Catalysis*, *356*(5), 897–927. doi:10.1002/adsc.201300960

Mogharabi, M., Nassiri-Koopaei, N., Bozorgi-Koushalshahi, M., Nafissi-Varcheh, N., Bagherzadeh, G., & Faramarzi, M. A. (2012). Immobilization of laccase in alginate-gelatin mixed gel and decolorization of synthetic dyes. *Bioinorganic Chemistry and Applications*, *2012*, 1–6. doi:10.1155/2012/823830 PMID:22899898

Mohajershojaei, K., Khosravi, A., & Mahmoodi, N. M. (2014). Decolorization of Dyes using immobilized laccase enzyme on zinc ferrite nanoparticle from single and binary systems. *Fibers and Polymers*, *15*(10), 2139–2145. doi:10.100712221-014-2139-y

Mohamed, R. A., Salleh, A. B., Raja Abd Rahman, R. N. Z., Basri, M., & Leow, T. C. (2012). Isolation of the encoding gene for a thermostable α-glucosidase from *Geobacillus stearothermophilus* strain RM and its expression in *Escherichia coli*. *African Journal of Microbiological Research*, *6*(12), 2909–291.

Mohammadi, M., Ghanbarzadeh, B., & Hamishehkar, H. (2014). Formulation of nanoliposomal vitamin D3 for potential application in beverage fortification. *Advanced Pharmaceutical Bulletin*, *4*(2), 569–575. PMID:25671191

Mohammadi, R., Mahmoudzade, M., Atefi, M., Khosravi-Darani, K., & Mozafari, M. R. (2015). Applications of nanoliposomes in cheese technology. *International Journal of Dairy Technology*, *68*(1), 11–23. doi:10.1111/1471-0307.12174

Mohan, R. (2000). *Advances in microbial amylases*. Biotechnology and Applied.

Montazer, M., & Maryan, S. A. (2010). Influences of different enzymatic treatment on denim garment. *Applied Biochemistry and Biotechnology*, *160*(7), 2114–2128. doi:10.100712010-009-8727-4 PMID:19728171

Montereali, M. R., Seta, L. D., Vastarella, W., & Pilloton, R. (2010). A disposable Laccase-Tyrosinase based biosensor for amperometric detection of phenolic compounds in must andWine. *Journal of Molecular Catalysis. B, Enzymatic*, *64*(3-4), 189–194. doi:10.1016/j.molcatb.2009.07.014

Moore, A., Martinez-Munoz, I., & Hoseney, R. (1990). Factors affecting the oxidative gelation of wheat water-solubles. *Cereal Chemistry*, *67*(1), 81–84.

Moore, W., & Holdeman, L. (1975). Discussion of current bacteriological investigations of the relationships between instestinal flora, diet, and colon cancer. *Cancer Research*, *35*(11 Part 2), 3418–3420.

Moreira, L., & Filho, E. X. F. (2008). An overview of mannan structure and mannan-degrading enzyme systems. *Applied Microbiology and Biotechnology*, *79*(2), 165–178. doi:10.100700253-008-1423-4 PMID:18385995

Moreno, A. D., Ibarra, D., Mialon, A., & Ballesteros, A. (2016). A bacterial laccase for enhancing saccharification and ethanol fermentation of steam pre treated biomass. *Fermentation*, *2*(2), 1–15. doi:10.3390/fermentation2020011

Moriyama, K., Minamihata, K., Wakabayashi, R., Goto, M., & Kamiya, N. (2014). Enzymatic preparation of a redox-responsive hydrogel for encapsulating and releasing living cells. *Chemical Communications*, *50*(44), 5895–5898. doi:10.1039/C3CC49766F PMID:24761434

Morozova, O. V., Shumakovich, G. P., Gorbacheva, M. A., Shleev, S. V., & Yaropolov, A. I. (2007a). "Blue" laccases. *Biochemistry (Moscow)*, *72*(10), 1136–1150. doi:10.1134/S0006297907100112 PMID:18021071

Morozova, O. V., Shumakovich, G. P., Shleev, S. V., & Yaropolov, Y. I. (2007b). Laccase-mediator systems and their applications: A review. *Applied Biochemistry and Microbiology*, *43*(5), 523–535. doi:10.1134/S0003683807050055 PMID:18038679

Morrow, K., Hernandez, C. P., Raber, P., Valle, L. D., Wilk, A. M., Majumdar, S., & Rodriguez, P. C. (2013). Anti-leukemic mechanisms of pegylated arginase I in acute lymphoblastic T-cell leukemia. *Leukemia*, *27*(3), 569–577. doi:10.1038/leu.2012.247 PMID:22926702

Mörtberg, M., & Neujahr, H. Y. (1987). In situ and in vitro kinetics of phenol hydroxylase. *Biochemical and Biophysical Research Communications*, *146*(1), 41–46. doi:10.1016/0006-291X(87)90687-5 PMID:3606624

Moure, A., Gullon, P., Domı'nguez, H., & Parajo, J. C. (2006). Advances in the manufacture, purification and applications of xylo-oligosaccharides as food additives and nutraceuticals. *Process Biochemistry*, *41*(9), 1913–1923. doi:10.1016/j.procbio.2006.05.011

Mourez, T., Simon, F., & Plantier, J.-C. (2013, July 01). Non-M Variants of Human Immunodeficiency Virus Type 1. *Clinical Microbiology Reviews*, *26*(3), 448–461. doi:10.1128/CMR.00012-13

Mubarak, N. M., Wong, J. R., Tan, K. W., Sahu, J. N., Abdullah, E. C., Jayakumar, N. S., & Ganesan, P. (2014). Immobilization of cellulase enzyme on functionalized multiwall carbon nanotubes. *Journal of Molecular Catalysis. B, Enzymatic*, *107*, 124–131. doi:10.1016/j.molcatb.2014.06.002

Mukherjee, S., Sunil Kumar, S., Misra, A. K., & Fan, M. (2007). Removal of phenols from water environment by activated carbon,bagasse ash and wood charcoal. *Chemical Engineering Journal*, *129*(1-3), 133–142. doi:10.1016/j.cej.2006.10.030

Mullai, P., & Rene, E. R. (2008). A simple multi layered neural network model for predicting bacterial xylanase production by *Bacillus* sp. using *Avena sativa* as substrate. *Biotechnology (Faisalabad)*, *7*(3), 499–504. doi:10.3923/biotech.2008.499.504

Müller, C. M., Laurindo, J. B., & Yamashita, F. (2009). Effect of cellulose fibers addition on the mechanical properties and water vapor barrier of starch-based films. *Food Hydrocolloids*, *23*(5), 1328–1333. doi:10.1016/j.foodhyd.2008.09.002

Müller, W. E. G., Grebenjuk, V. A., Le Pennec, G., Schröder, H. C., Brümmer, F., Hentschel, U., ... Breter, H. J. (2004). Sustainable production of bioactive compounds by sponges—cell culture and gene cluster approach: A review. *Marine Biotechnology (New York, N.Y.)*, *6*, 105–117. doi:10.100710126-002-0098-6 PMID:15085406

Munro, N. H., Green, D. W., Dangerfield, A., & McGrath, K. M. (2011). Biomimetic mineralisation of polymeric scaffolds using a combined soaking and Kitano approach. *Dalton Transactions (Cambridge, England)*, *40*(36), 9259–9268. doi:10.1039/c1dt11056j PMID:21829841

Munzig, E., Eckert, K., Harrach, T., Graf, H., & Maurer, H. R. (1994). Bromelain protease F9 reduces the CD44 mediated adhesion of human peripheral blood lymphocytes to human umbilical vein endothelial cells. *FEBS Letters*, *351*(2), 215–218. doi:10.1016/0014-5793(94)00860-4 PMID:7521849

Murakami, Y., Shoji, M., Hirata, A., Tanaka, S., Yokoe, I., & Fujisawa, S. (2005). Dehydrodiisoeugenol, an isoeugenol dimer, inhibits lipopolysaccharide-stimulated nuclear factor kappa B activation and cyclooxygenase-2 expression in macrophages. *Archives of Biochemistry and Biophysics*, *434*(2), 326–332. doi:10.1016/j.abb.2004.11.013 PMID:15639233

Murphy, P. S., & Evans, G. R. D. (2012). Advances in Wound Healing: A Review of Current Wound Healing Products. *Plastic Surgery International*, *2*, 1–8. doi:10.1155/2012/190436 PMID:22567251

Murugan, S., Arnold, D., Pongiya, U. D., & Narayanan, P. M. (2011). *Production of xylanase from Arthrobacter sp. MTCC 6915 using saw dust as substrate under solid state fermentation*. Enzyme Research.

Mussatto, S. I., & Mancilha, I. M. (2007). Non-digestible oligosaccharide: A Review. *Carbohydrate Polymers*, *68*(3), 287–297. doi:10.1016/j.carbpol.2006.12.011

Mussatto, S. I., & Teixeira, J. A. (2010). Lignocellulose as raw material in fermentation processes. In A. Méndez-Vilas (Ed.), *Current research, technology and education topics in applied microbiology and microbial biotechnology* (Vol. 2). Badajoz, Spain: Formatex Research Center.

Muthurasu, A., & Ganesh, V. (2014). Horseradish peroxidase enzyme immobilized graphene quantum dots as electrochemical biosensors. *Applied Biochemistry and Biotechnology*, *174*(3), 945–959. doi:10.100712010-014-1019-7 PMID:24984603

NACO-Annual Report 2015-2016. (n.d.). Retrieved from http://naco.gov.in/ 202015-16_NACO.pdf

Nagar, S., Mittal, A., & Gupta, V. K. (2012). A cost effective method for screening and isolation of xylan degrading bacteria using agro waste material. *Asian Journal of Biological Sciences*, *5*(8), 384–394. doi:10.3923/ajbs.2012.384.394

Naghibi, F., Pourmorad, F., Honary, S., & Shamsi, M. (2003). Decontamination of Water Polluted with Phenol Using Raphanus Sativus Root. *Iranian Journal of Pharmaceutical Research, 2*(1), 29–32.

Nakakuki, T. (1993). Oligosaccharides: Production, Properties and Applications. Japanese Technology Reviews. *Gordon and Breach Switzearland, 3*(2), 45-52.

Nakakuki, T. (2002). Present status and future of functional oligosaccharide development in Japan. *Pure and Applied Chemistry, 74*(7), 1245–1251. doi:10.1351/pac200274071245

Nakao, M., Nakayama, T., Harada, M., Kakudo, A., Ikemoto, H., Kobayashi, S., & Shibano, Y. (1994). Purification and characterization of thermostable alpha-glucosidase with transglucosylation activity. *Applied Microbiology and Biotechnology, 41*, 337–343. doi:10.1007/BF00221229 PMID:7764968

Nandy, S. K. (2016). Bioprocess Technology Governs Enzyme use an production in Industrial Biotechnology: An Overview. *Enzyme Engg., 5*(01), 1. doi:10.4172/2329-6674.1000144

Naqash, S. Y., & Nazeer, R. A. (2013). Antioxidant and functional properties of protein hydrolysates from pink perch (*Nemipterus japonicus*) muscle. *Journal of Food Science and Technology, 50*(5), 972–978. doi:10.100713197-011-0416-y PMID:24426005

Narang, S., & Satyanarayana, T. (2001). Thermostable a-amylase production by an extreme thermophile *Bacillus thermooleovorans. Letters in Applied Microbiology, 32*(1), 31–35. doi:10.1046/j.1472-765x.2001.00849.x PMID:11169038

Nascimento, R. P., Coelho, R. R. R., Marques, S., Alves, L., Girio, F. M., Bon, E. P. S., & Amaral-Collaco, M. T. (2002). Production and partial characterization of xylanase from *Streptomyces* sp. strain AMT-3 isolated from Brazilian cerrado soil. *Enzyme and Microbial Technology, 31*(4), 549–555. doi:10.1016/S0141-0229(02)00150-3

Nashiru, O., Koh, S., Lee, S., & Lee, D. (2001). Novel α-glucosidase from extreme thermophile *Thermus caldophilus* GK24. *Journal of Biochemistry and Molecular Biology, 34*, 347–354.

Nath, P. P., Sarkar, K., Mondal, M., & Paul, G. (2015). Metanil Yellow Impairs the Estrous Cycle Physiology and Ovarian Folliculogenesis in Female Rats. *Environmental Toxicology*. doi:10.1002/Tox.22205 PMID:26494366

Ncanana, S., Baratto, L., Roncaglia, L., Riva, S., & Burton, S. G. (2007). Laccase mediated oxidation of totarol. *Advanced Synthesis & Catalysis, 349*(8-9), 1507–1513. doi:10.1002/adsc.200700005

Nchez, A. S., Ysunza, F., Garciäa, M. J. B., & Esqueda, M. (2002). Biodegradation of Viticulture Wastes by *Pleurotus*: A Source of Microbial and Human Food and Its Potential Use in Animal Feeding. *Journal of Agricultural and Food Chemistry*, 50. PMID:11958618

Ncube, T., Howard, R. L., Abotsi, E. K., van Rensburg, E. L. J., & Ncube, I. (2012). *Jatropha curcas* seed cake as substrate for production of xylanase and cellulase by *Aspergillus niger* FGSA733 in solid-state fermentation. *Industrial Crops and Products, 37*(1), 118–123. doi:10.1016/j.indcrop.2011.11.024

Needham, D., & Dewhirst, M. W. (2001). The development and testing of a new temperature-sensitive drug delivery system for the treatment of solid tumors. *Advanced Drug Delivery Reviews, 53*(3), 285–305. doi:10.1016/S0169-409X(01)00233-2 PMID:11744173

Negrón-Encarnación, I., & Arce, R. (2007). Light-induced transformations of aza-aromatic pollutants adsorbed on models of atmospheric particulate matter: Acridine and 9(10-H) acridone. *Atmospheric Environment, 41*(32), 6771–6783. doi:10.1016/j.atmosenv.2007.04.062 PMID:18836521

Neifar, M., Ghorbel, E. R., Kamoun, A., Baklouti, S., Mokni, A., Jaouani, A., & Chaabouni, E. S. (2011). Effective clarification of pomegranate juice using laccase treatment optimized by response surface methodology followed by ultrafiltration. *Journal of Food Process Engineering*, *34*(4), 1199–1219. doi:10.1111/j.1745-4530.2009.00523.x

Nematollahi, D., & Forooghi, Z. (2002). Electrochemical oxidation of catechols in the presence of 4-hydroxy-6-methyl-2-pyrone. *Tetrahedron*, *58*(24), 4949–4953. doi:10.1016/S0040-4020(02)00422-2

Netto, C. G. C. M., Toma, H. E., & Andrade, L. H. (2013). Superparamagnetic nanoparticles as versatile carriers and supporting materials for enzymes. *Journal of Molecular Catalysis B: Enzymatic, 85– 86*, 71– 92.

Ngai, K. L., Neidle, E. L., & Ornston, L. N. (1990). Catechol and chlorocatechol 1,2-dioxygenase, Hydrocarbon and methylotrophy, Meth. Enzymol., 188, 122-126.

Ngangoum, E. S., Mezajoug Kenfack, L. B., Sanjit, K., Tchiégang, C., & Mookambeswaran, V. M. (2017). Extraction, purification, and biochemical characterization of serine protease from leaves of Abrus precatorius. *Preparative Biochemistry & Biotechnology*, 1–9. doi:10.1080/2017/1373289 PMID:28857663

Nguyen, L. N., Hai, F. I., Price, W. E., Leusch, F. D. L., Roddick, F., Mcadam, E. J., & Nghiem, L. D. (2014). Continuous biotransformation of bisphenol A and diclofenac by laccase in an enzymatic membrane reactor. *International Biodeterioration & Biodegradation*, *95*, 25–32. doi:10.1016/j.ibiod.2014.05.017

Nicell, J. A. (1994). Kinetics of horseradish peroxidase-catalysed polymerization and precipitation of aqueous 4- chlorophenol. *Journal of Chemical Technology and Biotechnology*, *60*, 203-15.

Nicell, J. A., Bewtra, J. K., Biswas, N., St. Pierre, C. & Taylor, K. E. (1993).Enzyme catalyzed polymerization and precipitation of aromatic compounds from aqueous solution. *Canadian Journal of Civil Engineering*, *20*, 725-35.

Nicell, J. A., Wagner, M., & Sakurai, A. (2003). Peroxidase-catalyzed oxidation of phenols in industrial wastewaters., 93–124.

Nicotra, S., Cramarossa, M. R., Mucci, A., Pagnoni, U. M., Riva, S., & Forti, L. (2004a). Biotransformation of resveratrol: Synthesis of trans-dehydrodimers catalyzed by laccases from *Myceliophtora thermophila* and from *Trametes pubescens. Tetrahedron*, *60*(3), 595–600. doi:10.1016/j.tet.2003.10.117

Nicotra, S., Intra, A., Ottolina, G., Riva, S., & Danieli, B. (2004b). Laccase-mediated oxidation of the steroid hormone 17β-estradiol in organic solvents. *Tetrahedron, Asymmetry*, *15*(18), 2927–2931. doi:10.1016/j.tetasy.2004.06.034

Niehaus, F., Peters, A., Groudieva, T., & Antranikian, G. (2000). Cloning, expression and biochemical characterization of a unique thermostable pullulan-hydrolysing enzyme from the hyperthermophilic archaeon *Thermococcus aggregans. FEMS Microbiology Letters*, *190*(2), 223–229. doi:10.1111/j.1574-6968.2000.tb09290.x PMID:11034283

Niek, N., Stefaan, C., & Demeester, J. (2006). Deoxyribonuclease I. In B. M & G. W (Eds.), Directory of thepraeutic enzymes. Taylor & Francis Group.

Nielsen, J. E., & Borchert, T. V. (2000). Protein engineering of bacterial $α$-amylases. *Biochimica et Biophysica Acta (BBA)-. Protein Structure and Molecular Enzymology*, *1543*(2), 253–274. doi:10.1016/S0167-4838(00)00240-5

Nigam, P., & Pandey, A. (Eds.). (2009). Biotechnology for Agro-Industrial Residues Utilization. Springer Science Business Media B.V.

Nigam, P., Pandey, A., & Prabhu, K. A. (1987a). Ligninolytic activity of two Basidiomycetes moulds in the decomposition of bagasse. *Biological Wastes*, *21*, 1–10. doi:10.1016/0269-7483(87)90080-2

Nigam, P., & Singh, D. (1995a). Enzyme and microbial systems involved in starch processing. *Enzyme and Microbial Technology*, *17*(9), 770–778. doi:10.1016/0141-0229(94)00003-A

Nigam, P., & Singh, D. (1995b). Processes for fermentative production of xylitol-a sugar substitute. *Process Biochemistry*, *30*, 117–124.

Nigel, C. (2004). Veitch Horseradish peroxidase: A modern view of a classic enzyme. *Phytochemistry*, *65*(3), 249–259. doi:10.1016/j.phytochem.2003.10.022 PMID:14751298

Nilotpala, P., & Ingle, A. O. (2007). Mineralization of phenol by a *Serratia plymuthica* strain GC isolated from sludge sample. *International Biodeterioration & Biodegradation*, *60*(2), 103–108. doi:10.1016/j.ibiod.2007.01.001

Ninawe, S., & Kuhad, R. C. (2005). Use of xylan-rich cost effective agro-residues in the production of xylanase by *Streptomyces cyaneus* SN32. *Journal of Applied Microbiology*, *99*(5), 1141–1148. doi:10.1111/j.1365-2672.2005.02711.x PMID:16238744

Nishikawa, R., Ji, X.-D., Harmon, R. C., Lazar, C. S., Gill, G. N., Cavenee, W. K., & Huang, H. J. (1994). A mutant epidermal growth factor receptor common in human glioma confers enhanced tumorigenicity. *Proceedings of the National Academy of Sciences of the United States of America*, *91*(16), 7727–7731. doi:10.1073/pnas.91.16.7727 PMID:8052651

No, H. K., Meyers, S. P., Prinyawiwatkul, W., & Xu, Z. (2007). Applications of chitosan for improvement of quality and shelf life of foods: A review. *Journal of Food Science*, *72*(5), 87–100. doi:10.1111/j.1750-3841.2007.00383.x PMID:17995743

Noh, S. H., Kwon, C., Hwang, J., Ohsaka, T., Kim, B.-J., Kim, T.-Y., ... Han, B. (2017). Self-assembled nitrogen-doped fullerenes and their catalysis for fuel cell and rechargeable metal–air battery applications. *Nanoscale*, *9*(22), 7373–7379. doi:10.1039/C7NR00930E PMID:28405666

Norsker, M., Jensen, M., & Adler-Nissen, J. (2000). Enzymatic gelation of sugar beet pectin in food products. *Food Hydrocolloids*, *14*(3), 237–243. doi:10.1016/S0268-005X(00)00016-3

Nugroho Prasetyo, E., Kudanga, T., Østergaard, L., Rencoret, J., Gutiérrez, A., del Río, J. C., & Guebitz, G. M. (2010). Polymerization of lignosulfonates by the laccase-HBT (1-hydroxybenzotriazole) system improves dispersibility. *Bioresource Technology*, *101*(14), 5054–5062. doi:10.1016/j.biortech.2010.01.048 PMID:20176477

Nutracea. (2012). *Rice bran facts*. Retrieved from http://www.nutracea.com/RiceBranFacts

Oelofse, S. H. H., & Nahman, A. (2013). Estimating the magnitude of food waste generated in South Africa. *Waste Management & Research*, *31*(1), 80–86. doi:10.1177/0734242X12457117 PMID:22878934

Ogata, M., Oka, K., Seki, M., Hoshi, M., Takatsu, H., Mashino, T., ... Endo, T. (2007). Solubilization of dipropofol, an antibacterial agent, using saccharide and ascorbic acid. *Biological & Pharmaceutical Bulletin*, *30*(8), 1565–1568. doi:10.1248/bpb.30.1565 PMID:17666822

Ogata, M., Sato, K. T., Kunikane, T., Oka, K., Seki, M., Urano, S., ... Endo, T. (2005). Antibacterial activity of dipropofol and related compounds. *Biological & Pharmaceutical Bulletin*, *28*(6), 1120–1122. doi:10.1248/bpb.28.1120 PMID:15930760

Ohta, K., Moriyama, S., Tanaka, H., Shige, T., & Akimoto, H. (2001). Purification and characterisation of an acidophilic xylanase from *Aureobasidium pullulans* var. *melanigenum* and sequence analysis of the encoding gene. *Journal of Bioscience and Bioengineering*, *92*(3), 262–270. doi:10.1016/S1389-1723(01)80260-7 PMID:16233094

Okafor, U. A., Emezue, T. N., Okochi, V. I., Onyegeme-okerenta, B. M., & Nwodo-Chinedu, S. (2007a). Xylanase production by *Penicillium chrysogenum* PCL501 fermented on cellulosic wastes. *African Journal of Biochemistry Research, 1*(4), 48–53.

Okafor, U. A., Okochi, V. I., Onyegeme-okerenta, B. M., & Nwodo-Chinedu, S. (2007b). Xylanase production by *Aspergillus niger* ANL 301 using agro-wastes. *African Journal of Biotechnology, 6*(14), 1710–1714.

Okezie, U., Akanbi, C. T., Otunola, E. T., & Adeyemi, I. A. (2003). Effect of addition of ripe bananas on some physico-chemical properties of maize extract. *International Journal of Food Sciences and Nutrition, 54*(6), 437–445. doi:10.1080/09637480220138106 PMID:14522689

Omar, H. H. (2008). Algal Decolorization and Degradation of Monoazo and Diazo Dyes. *Pakistan Journal of Biological Sciences, 11*(10), 1310–1316. doi:10.3923/pjbs.2008.1310.1316 PMID:18817261

Onaca, O., Hughes, D. W., Balasubramanian, V., Grzelakowski, M., Meier, W., & Palivan, C. G. (2010). SOD antioxidant nanoreactors: Influence of block copolymer composition on the nanoreactor efficiency. *Macromolecular Bioscience, 10*(5), 531–538. PMID:20112239

Ong, S. T., Keng, P. S., Lee, W. N., Ha, S. T., & Hung, W. T. (2011). Dye waste treatment. *Water (Basel), 3*(1), 157–176. doi:10.3390/w3010157

Oosterveld, A., Beldman, G., & Voragen, A. G. J. (2002). Enzymatic modification of pectic polysaccharides obtained from sugar beet pulp. *Carbohydrate Polymers, 48*(1), 73–81. doi:10.1016/S0144-8617(01)00216-8

Opperdoes, F. R., & Michels, P. A. (2001). Enzymes of carbohydrate metabolism as potential drug targets. *International Journal for Parasitology, 31*(5–6), 482–490. doi:10.1016/S0020-7519(01)00155-2 PMID:11334933

Orzua, M. C., Mussato, S. I., Contreras-Esquivel, J. C., Rodriguez, R., de la Garza, H., Teixeira, J. A., & Aguilar, C. N. (2009). Exploitation of agro industrial wastes as immobilization carrier for solid-state fermentation. *Industrial Crops and Products, 30*(1), 24–27. doi:10.1016/j.indcrop.2009.02.001

Osiadacz, J., Al-Adhami, A. J., Bajraszewska, D., Fischer, P., & Peczyñska-Czoch, W. (1999). On the use of *Trametes versicolor* laccase for the conversion of 4-methyl-3-hydroxyanthranilic acid to actinocin chromophore. *Journal of Biotechnology, 72*(1-2), 141–149. doi:10.1016/S0168-1656(99)00100-5

Osma, J. F., Herrera, J. L. T., & Couto, S. R. (2010). Uses of laccases in the food industry. *Enzyme Research, 2010*, 918761. doi:10.4061/2010/918761 PMID:21048873

Ostadhadi-Dehkordi, S., Tabatabaei-Sameni, M., Forootanfar, H., Kolahdouz, S., Ghazi-Khansari, M., & Faramarzi, M. A. (2012). Degradation of some benzodiazepines by a laccase-mediated system in aqueous solution. *Bioresource Technology, 125*, 344–347. doi:10.1016/j.biortech.2012.09.039 PMID:23069616

Osti, E., & Osti, F. (2004). Treatment of cutaneous burns with burnishield (Hydrogel) and a semi-permeable adhesive film. *Annals of Burns and Fire Disasters, 3*, 137–141.

Othman, A. M., Elshafei, A., Hassan, M., Haroun, B., Elsayed, M., & Farrag, A. (2014). Purification, Biochemical Characterization and Applications of *Pleurotus ostreatus* ARC280. Laccase. *British Microbiology Research Journal, 4*(12), 1418–1439. doi:10.9734/BMRJ/2014/11218

Otto, B., Schlosser, D., & Reisser, W. (2010). First description of a laccase-like enzyme in soil algae. *Archives of Microbiology, 192*(9), 759–768. doi:10.100700203-010-0603-7 PMID:20623267

Oyekola, O., Ngesi, N., & Whiteley, C. (2007). Isolation, Purification and Characterization of An Endoglucanase And B-Glucosidase From An Anaerobic Sulphidogenic Bioreactor. *Enz Microbiol Tech*, *40*(8), 637–644. doi:10.1016/j.enzmictec.2006.05.020

Ozgun, D., Basak, S., & Cinar, O. (2014). The inhibition effect of three different pharmaceuticals in domestic wastewater. *World Journal of Environmental Research*, *4*, 1–6.

Padmanabhan, R. V., Gudapaty, R., Liener, I. E., Schwartz, B. A., & Hoidal, J. R. (1985). Protection against pulmonary oxygen toxicity in rats by the intratracheal administration of liposome-encapsulated superoxide dismutase or catalase. *The American Review of Respiratory Disease*, *132*(1), 164–167. PMID:4014861

Palframan, R., Gibson, G. R., & Rastall, R. A. (2003). Development of a quantitative tool for the comparison of the prebiotic effect of dietary oligosaccharides. *Letters in Applied Microbiology*, *37*(4), 281–284. doi:10.1046/j.1472-765X.2003.01398.x PMID:12969489

Palma, M. B., Milagres, A. M. F., Prata, A. M. R., & Manicilha, D. I. M. (1996). Influence of aeration and agitation on xylanase production. *Brazilian Journal of Biochemistry*, *31*(2), 141–145.

Palmer, A. M., & Alavijeh, M. S. (2012). Translational CNS medicines research. *Drug Discovery Today*, *17*(19-20), 1068–1078. doi:10.1016/j.drudis.2012.05.001 PMID:22580061

Palmieri, G., Giardina, P., Bianco, C., Scaloni, A., Capassoi, A., & Sannia, G. (1997). A novel white laccase from *Pleurotus ostreatus*. *The Journal of Biological Chemistry*, *272*(50), 31301–31307. doi:10.1074/jbc.272.50.31301 PMID:9395457

Pandey, A., Nigam, P., Soccol, C. R., Soccol, V. T., Singh, D., & Mohan, R. (2000). Advances in microbial amylases. *Biotechnology and Applied Biochemistry*, *31*(2), 135–152. doi:10.1042/BA19990073 PMID:10744959

Pandey, A., Selvakumar, P., Soccol, C. R., & Nigam, P. (1999). Solid state fermentation for the production of industrial enzymes. *Current Science*, *77*, 149–162.

Pandey, A., Singh, P., & Iyengar, L. (2007). Bacterial decolorization and degradation of azo dyes. *International Biodeterioration & Biodegradation*, *59*(2), 73–84. doi:10.1016/j.ibiod.2006.08.006

Pandey, A., Soccol, C. R., Nigam, P., Soccol, V. T., Vandenberghe, L. P. S., & Mohan, R. (2000b). Biotechnological potential of agro-industrial residues, II-Cassava Bagasse. *Bioresource Technology*, *74*(1), 81–87. doi:10.1016/S0960-8524(99)00143-1

Pandey, A., Soccol, C. R., Rodriguez-Leon, J. A., & Nigam, P. (2001). *Solid- State Fermentation in Biotechnology: Fundamentals and Applications* (pp. 3–7). New Delhi, India: Asiatech Publishers Inc.

Pang, P. K., & Ibrahim, C. O. (2005). Xylanase production by a local fungal isolate, *Aspergillus niger* USM AI 1 via solid state fermentation using palm kernel cake (PKC) as substrate. *Songklanakarin Journal of Science and Technology*, *27*(2), 325–336.

Parbat, R., & Singhal, B. (2011). Production of Glucoamylase by *Aspergillus oryzae* under solid state fermentation using agro industrial products. *International Journal of Microbiology Research*, *2*, 204–207.

Parkes, M., Myant, C., Dini, D., & Cann, P. (2015). Tribology-optimised silk protein hydrogels for articular cartilage repair. *Tribology International*, *89*, 9–18. doi:10.1016/j.triboint.2014.11.024

Park, J. M., Kim, M., Park, H. S., Jang, A., Min, J., & Kim, Y. H. (2013). Immobilization of lysozyme-CLEA onto eletrospun chitosan nanofiber for effective antimicrobial applications. *International Journal of Biological Macromolecules*, *54*, 37–43. doi:10.1016/j.ijbiomac.2012.11.025 PMID:23201775

Park, J. W., Hong, K., Carter, P., Asgari, H., Guo, L. Y., Keller, G. A., ... Benz, C. C. (1995). Development of anti-p185HER2 immunoliposomes for cancer therapy. *Proceedings of the National Academy of Sciences of the United States of America*, *92*(5), 1327–1331. doi:10.1073/pnas.92.5.1327 PMID:7877976

Park, J. W., Park, B. K., & Kim, J. E. (2006). Remediation of soil contaminated with 2,4-dichlorophenol by treatment of minced shepherd's purse roots. *Archives of Environmental Contamination and Toxicology*, *50*(2), 191–195. doi:10.100700244-004-0119-8 PMID:16392021

Park, P., Je, J. Y., & Kim, S. K. (2003). Angiotensin I converting (ACE) inhibitory activity of hetero-chitooligosacchatides prepared from partially different deacetylated chitosans. *Journal of Agricultural and Food Chemistry*, *51*(17), 4930–4934. doi:10.1021/jf0340557 PMID:12903948

Parmar, H. S., Bhinchar, M. K., Bhatia, M., Chordia, N., Raval, I., Chauhan, D. S., … Kumar, A. (2014). Study on gluco-regulatory potential of glimepiride sulfonamide using in silico, in vitro and in vivo approaches. *Current Pharmaceutical Design*, *20*(32), 5212–7. Retrieved from http://www.ncbi.nlm.nih.gov/pubmed/24641189

Parvathy, K. S., Negi, P. S., & Srinivas, P. (2009). Antioxidant, antimutagenic and antibacterial activities of curcumin-β-diglucoside. *Food Chemistry*, *115*(1), 265–271. doi:10.1016/j.foodchem.2008.12.036

Pasha, I., Saeed, F., Waqas, K., Anjum, F. M., & Arshad, M. U. (2013). Nutraceutical and functional scenario of wheat straw. *Critical Reviews in Food Science and Nutrition*, *53*(3), 287–295. doi:10.1080/10408398.2010.528080 PMID:23216000

Pastores, G. M., Weinreb, N. J., Aerts, H., Andria, G., Cox, T. M., Giralt, M., & Tylki-Szymanska, A. (2004). Therapeutic goals in the treatment of Gaucher disease. *Seminars in Hematology*, *41*(4Suppl 5), 4–14. doi:10.1053/j.seminhematol.2004.07.009 PMID:15468045

Patani, G. A., & LaVoie, E. J. (1996, January). Bioisosterism: A Rational Approach in Drug Design. *Chemical Reviews*, *96*(8), 3147–3176. doi:10.1021/cr950066q

Patel, H., & Gupte, A. (2016). Optimization of different culture conditions for enhanced laccase production and its purification from *Tricholoma giganteum* AGHP. *Bioresources and Bioprocessing*, *3*(1), 11. doi:10.118640643-016-0088-6

Paul, M. K., & Mukhopadhyay, A. K. (2004). Tyrosine kinase – Role and significance in Cancer. *International Journal of Medical Sciences*, *1*(283), 101–115. doi:10.7150/ijms.1.101 PMID:15912202

Pauwels, R., Andries, K., Debyser, Z., Kukla, M. J., Schols, D., Breslin, H. J., ... De Clercq, E. (1994, December 01). New tetrahydroimidazo [4, 5, 1-jk][1, 4]-benzodiazepin-2 (1H)-one and-thione derivatives are potent inhibitors of human immunodeficiency virus type 1 replication and are synergistic with 2′, 3′-dideoxynucleoside analogs. *Antimicrobial Agents and Chemotherapy*, *38*(12), 2863–2870. doi:10.1128/AAC.38.12.2863

Payen, A., & Persoz, J. (1833). Mémoire sur la Diastase, les principaux produits de ses réactions, et leurs applications aux arts industriels. *Annales de Chimie et de Physique*, *53*, 73–92.

Pazarlıog˘lu, N. K., Sariişik, M., & Telefoncu, A. (2005). Laccase: Production by *Trametes versicolor* and application to denim washing. *Process Biochemistry (Barking, London, England)*, *40*(5), 1673–1678. doi:10.1016/j.procbio.2004.06.052

Pearce, K. N., & Kinsella, J. E. (1978). Emulsifying properties of proteins: Evaluation of turbidimetric technique. *Journal of Agricultural and Food Chemistry*, *26*(3), 716–723. doi:10.1021/jf60217a041

Pei, L. X., Li, Y. M., Bu, X. Z., Gu, L. Q., & Chan, A. S. (2006). One-pot synthesis of 5, 6-dihydroxylated benzo [b] furan derivatives. *Tetrahedron Letters*, *47*(15), 2615–2618. doi:10.1016/j.tetlet.2006.02.055

Pelach, M., Pastor, F., Puig, J., Vilaseca, F., & Mutje, P. (2003). Enzymatic Deinking of Old Newspapers With Cellulose. *Proc Biochem*, *38*(7), 1063–1067. doi:10.1016/S0032-9592(02)00237-6

Pel, H. J., de Winde, J. H., Archer, D. B., Dyer, P. S., Hofmann, G., Schaap, P. J., ... Stam, H. (2007). Genome sequencing and analysis of the versatile cell factory Aspergillus niger CBS 513.88. *Nature Biotechnology*, *25*(2), 221–231. doi:10.1038/nbt1282 PMID:17259976

Peng, C. H., Chang, C. H., Peng, R. Y., & Chyau, C. C. (2010). Improved membrane transport of astaxanthine by liposomal encapsulation. *European Journal of Pharmaceutics and Biopharmaceutics*, *75*(2), 154–161. doi:10.1016/j.ejpb.2010.03.004 PMID:20226858

Peng, F., Bian, J., Peng, P., Guan, Y., Xu, F., & Sun, R. C. (2012). Fractional separation and structural features of hemicelluloses from sweet sorghum leaves. *BioResources*, *7*(4), 4744–4759. doi:10.15376/biores.7.4.4744-4759

Peppas, N. A. (1997). Hydrogels and drug delivery. *Current Opinion in Colloid & Interface Science*, *2*(5), 531–537. doi:10.1016/S1359-0294(97)80103-3

Pereira, L., Bastos, C., Tzanov, T., Cavaco-Paulo, A., & Guebitz, G. M. (2005). Environmentally friendly bleaching of cotton using laccases. *Environmental Chemistry Letters*, *3*(2), 66–69. doi:10.100710311-005-0004-3

Perez, J., Munoz-Dorado, J., De la Rubia, T., & Martinez, J. (2002). Biodegradation and biological treatments of cellulose, hemicellulose and lignin: An overview. *International Microbiology*, *5*(2), 53–63. doi:10.100710123-002-0062-3 PMID:12180781

Pescod, M. B. (1992). *FAO Irrigation and Drainage, Food and Agriculture*. Rome, Italy: Organisation of United Nations.

Petersen, B. R., Mathiasen, T. E., Peelen, B., & Andersen, H. (1996). Use of laccase for deoxygenation of oil-containing product such as salad dressing. *PCT Int. Appl. WO*, *9635768*, A1.

Peters, R. T., Toby, G., Lu, Q., Liu, T., Kulman, J. D., Low, S. C., & Pierce, G. F. (2013). Biochemical and functional characterization of a recombinant monomeric factor VIII-Fc fusion protein. *Journal of Thrombosis and Haemostasis*, *11*(1), 132–141. doi:10.1111/jth.12076 PMID:23205847

Petrikovics, I., Budai, M., Baskin, S. I., Rockwood, G. A., Childress, J., Budai, L., & Szilasi, M. (2009). Characterization of liposomal vesicles encapsulating rhodanese for cyanide antagonism. *Drug Delivery*, *16*(6), 312–319. doi:10.1080/10717540903003711 PMID:19606945

Petrikovics, I., Jayanna, P., Childress, J., Budai, M., Martin, S., Kuzmitcheva, G., & Rockwood, G. (2011). Optimization of Liposomal Lipid Composition for a New, Reactive Sulfur Donor, and In Vivo Efficacy Studies on Mice to Antagonize Cyanide Intoxication. *Journal of Drug Delivery*, *2011*, 7. doi:10.1155/2011/928626 PMID:22187653

Pezzella, C., Guarino, L., & Piscitelli, A.A. (2015). How to enjoy laccases Cell. *Mol. Life Sci.*

Pezzella, C., Guarino, L., & Piscitelli, A. (2015). How to enjoy laccases. *Cellular and Molecular Life Sciences*, *72*(5), 923–940. doi:10.100700018-014-1823-9 PMID:25577278

Pham, N. H., Hollmann, F., Kracher, D., Preims, M., Haltrich, D., & Ludwig, R. (2015). Engineering an enzymatic regeneration system for NAD(P)H oxidation. *Journal of Molecular Catalysis. B, Enzymatic*, *120*, 38–46. doi:10.1016/j.molcatb.2015.06.011

Philpott, G. W., Bower, R. J., Parker, K. L., Shearer, W. T., & Parker, C. W. (1974). Affinity Cytotoxicity of Tumor Cells with Antibody-Glucose Oxidase Conjugates, Peroxidase, and Arsphenamine. *Cancer Research*, *34*(9), 2159–2164. PMID:4858425

Phuong, N. T., Ho, V. A., Nguyen, D. H., Khoa, N. C., Quyen, T. N., & Lee, Y. (2015). Enzyme-mediated fabrication of an oxidized chitosan hydrogel as a tissue sealant. *Journal of Bioactive and Compatible Polymers*, *30*(4), 1–12. doi:10.1177/0883911515578760

Phupong, V., Shuangshoti, S., Sutthiruangwong, P., Maneesri, S., Nuayboonma, P., & Shotelersuk, V. (2005). Prenatal diagnosis of Pompe disease by electron microscopy. *Archives of Gynecology and Obstetrics*, *271*(3), 260–262. doi:10.100700404-004-0620-3 PMID:15146345

Pietruszka, J., & Wang, C. (2012). Laccase-catalysed α-arylation of cyclic β-dicarbonyl compounds. *Green Chemistry*, *14*(9), 2402–2409. doi:10.1039/c2gc35662g

Pita, M., Gutierrez-Sanchez, C., Olea, D., Velez, M., Garcia-Diego, C., Shleev, S., & De Lacey, A. L. (2011). High Redox Potential Cathode Based on Laccase Covalently Attached to Gold Electrode. *The Journal of Physical Chemistry C*, *115*(27), 13420–13428. doi:10.1021/jp203643h

Pitt, J. I., & Hocking, A. D. (2009). *Fungi and food spoilage*. Springer, US. doi:10.1007/978-0-387-92207-2

Platzek, T., Lang, C., Grohmann, G., Gi, U. S., & Baltes, W. (1999). Formation of a carcinogenic aromatic amine from an azo dye by human skin bacteria in vitro. *Human and Experimental Toxicology*, *18*(9), 552–559. doi:10.1191/096032799678845061 PMID:10523869

Poeckel, D., Niedermeyer, T. H., Pham, H. T., Mikolasch, A., Mundt, S., Lindequist, U., ... Werz, O. (2006). Inhibition of human 5-lipoxygenase and anti-neoplastic effects by 2-amino-1, 4-benzoquinones. *Medicinal Chemistry (Shariqah, United Arab Emirates)*, *2*(6), 591–595. doi:10.2174/157340641060206059l PMID:17105440

Pointing, S. B. (2001). Feasibility of bioremediation by white-rot fungi. *Applied Microbiology and Biotechnology*, *57*(1-2), 20–33. doi:10.1007002530100745 PMID:11693920

Pokusaeva, K., Fitzgerald, G. F., & van Sinderen, D. (2011). Carbohydrate metabolism in Bifidobacteria. *Genes & Nutrition*, *6*(3), 285–306. doi:10.100712263-010-0206-6 PMID:21484167

Polak, J., & Jarosz-Wilkolazka, A. (2012). Fungal laccases as green catalysts for dye synthesis. *Process Biochemistry*, *47*(9), 1295–1307. doi:10.1016/j.procbio.2012.05.006

Polizeli, M. L., Rizzatti, A. C., Monti, R., Terenzi, H. F., Jorge, J. A., & Amorim, D. S. (2005). Xylanases from fungi Properties and industrial applications. *Applied Microbiology and Biotechnology*, *67*(5), 577–591. doi:10.100700253-005-1904-7 PMID:15944805

Pomati, F., Orlandi, C., Clerici, M., Luciani, F., & Zuccato, E. (2008). Effects and interactions in an environmentally relevant mixture of pharmaceuticals. *Toxicological Sciences : An Official Journal of the Society of Toxicology*, *102*(1), 129–137. doi:10.1093/toxsci/kfm291 PMID:18048368

Potthast, A., Rosenau, T., Chen, C. L., & Gratzl, J. S. (1995). Selective enzymic oxidation of aromatic methyl groups to aldehydes. *The Journal of Organic Chemistry*, *60*(14), 4320–4321. doi:10.1021/jo00119a006

Potthast, A., Rosenau, T., Chen, C. L., & Gratzl, J. S. (1996). A novel method for conversion of benzyl alcohols to benzaldehydes by laccase-catalysed oxidation. *Journal of Molecular Catalysis A Chemical*, *108*(1), 5–9. doi:10.1016/1381-1169(95)00251-0

Pownall, H. J., Hickson, D. L., & Smith, L. C. (1983). Transport of biological lipophiles: Effect of lipophile structure. *Journal of the American Chemical Society*, *105*(8), 2440–2445. doi:10.1021/ja00346a055

Pozdnyakova, N., Rodakiewicz-Nowak, J., & Turkovskaya, O. (2004). Catalytic properties of yellow laccase from *Pleurotus ostreatus* D1. *Journal of Molecular Catalysis. B, Enzymatic*, *30*(1), 19–24. doi:10.1016/j.molcatb.2004.03.005

Prade, R. A., Zhan, D., Ayoubi, P., & Mort, A. J. (1999). Pectins, pectinases and plant-microbe interactions. *Biotechnology & Genetic Engineering Reviews*, *16*(1), 361–392. doi:10.1080/02648725.1999.10647984 PMID:10819085

Prado-Alves, F. H., Pavezzi, F. C., Leite, R. S. R., Oliveira, V. M., Sette, L. D., & Da Silva, R. (2016). Screening and production study of microbial xylanase producer from Brazilian cerado. *Applied Biochemistry and Biotechnology, 161*(1-8), 333–346. doi:10.100712010-009-8823-5

Prasad, N. K., Vindal, V., Narayana, S. L., Ramakrishna, V., Kunal, S. P., & Srinivas, M. (2012). In silico analysis of *Pycnoporus cinnabarinus* laccase active site with toxic industrial dyes. *Journal of Molecular Modeling, 18*(5), 2013–2019. doi:10.100700894-011-1215-0 PMID:21877154

Prasanna, V. A. (2005). Amylases and their applications. *African Journal of Biotechnology, 4*(13), 1525–1529.

Prasanth, V., Jayaprakash, R., & Mathew, S. T. (2012). *Colon Specific Drug Delivery Systems: A Review on Various Pharmaceutical Approaches*. Academic Press.

Prasetyo, E. N., Kudanga, T. W., Steiner, W., Murkovic, M., Nyanhongo, G. S., & Guebitz, G. M. (2010). Laccase-generated tetramethoxyazobismethylene quinone (TMAMQ) as a tool for antioxidant activity measurement. *Food Chemistry, 118*(2), 437–444. doi:10.1016/j.foodchem.2009.04.102

Pringsheim, T., Davenport, W. J., & Dodick, D. (2008). Acute Treatment and Prevention of Menstrually Related Migraine Headache: Evidence-Based. *Revista de Neurología, 70*(17), 1555–1563. doi:10.1212/01.wnl.0000310638.54698.36 PMID:18427072

Prudhviraj, G., Vaidya, Y., Singh, S. K., Yadav, A. K., Kaur, P., Gulati, M., &Gowthamarajan, K. (2015). Effect of co-administration of probiotics with polysaccharide based colon targeted delivery systems to optimize site specific drug release. *European Journal of Pharmaceutics and Biopharmaceutics, 97*, 164-172.

Puls, J. (1993). Chemistry of hemicelluloses: relationship between hemicellulose structure and enzymes required for hydrolysis. *Hemicellulose and Hemicellulases.*

Pundir, C. S. (2015). Enzyme Nanoparticles: Preparation, Characterisation, Properties and Applications. Elsevier.

Purich, D. L., & Allison, R. D. (2002). *The enzyme reference : a comprehensive guidebook to enzyme nomenclature, reactions, and methods*. Academic Press. doi:10.1016/B978-012568041-7/50002-2

Purkarthofer, H., Lischnig, T., & Steiner, W. (1993). Thermostability of endo-β-xylanase from the thermophilic fungus *Thermomyces lanuginosus. Biotechnology Letters, 15*(4), 411–414. doi:10.1007/BF00128286

Purnomo, A. S., Mori, T., Kamei, I., Nishii, T., & Kondo, R. (2010). Application of mushroom waste medium from *Pleurotus ostreatus* for bioremediation of DDT-contaminated soil. *International Biodeterioration & Biodegradation, 64*(5), 397–402. doi:10.1016/j.ibiod.2010.04.007

Purwanto, L. A., Ibrahim, D., & Sudrajat, H. (2009). Effect of agitation speed on morphological changes in *Aspergillus niger* hyphae during production of tannase. *World Journal of Chemistry, 4*(1), 34–38.

Qhobosheane, M., Santra, S., Zhang, P., & Tan, W. (2001). Biochemically functionalized silica nanoparticles. *Analyst (London), 126*(8), 1274–1278. doi:10.1039/b101489g PMID:11534592

Quaratino, D., Federici, F., Petruccioli, M., Fenice, M., & D'Annibale, A. (2007). Production, purification and partial characterisation of a novel laccase from the white-rot fungus *Panus tigrinus* CBS 577.79. *Antonie van Leeuwenhoek, 91*(1), 57–69. doi:10.100710482-006-9096-4 PMID:17013550

Quinn, D. M. (1987). Acetylcholinesterase: Enzyme structure, reaction dynamics, and virtual transition states. *Chemical Reviews, 87*(5), 955–979. doi:10.1021/cr00081a005

Qu, X., Alvarez, P. J., & Li, Q. (2013). Applications of nanotechnology in water and wastewater treatment. *Water Research, 47*(12), 3931–3946. doi:10.1016/j.watres.2012.09.058 PMID:23571110

Raafat, D., Von Bargen, K., Haas, A., & Sahl, H. G. (2008). Insights into the mode of action of chitosan as an antibacterial compound. *Applied and Environmental Microbiology, 74*(12), 3764–3773. doi:10.1128/AEM.00453-08 PMID:18456858

Radjenovic, J., Petrovic, M., & Barceló, D. (2007). Analysis of pharmaceuticals in wastewater and removal using a membrane bioreactor. *Analytical and Bioanalytical Chemistry, 387*(4), 1365–1377. doi:10.100700216-006-0883-6 PMID:17115140

Radomski, J., & Mellinger, T. (1962). The absorption, fate and excretion in rats of the water-soluble azo dyes, FD&C Red No. 2, FD&C Red No. 4, and FD&C Yellow No. 6. *The Journal of Pharmacology and Experimental Therapeutics, 136*(2), 259–266. PMID:14489780

Rafil, F., Franklin, W., Heflich, R., & Cerniglia, C. (1991). Reduction of nitroaromatic compounds by anaerobic bacteria isolated from the human gastrointestinal tract. *Applied and Environmental Microbiology, 57*(4), 962–968. PMID:2059053

Raghupathi, K., & Thayumanavan, S. (2017). NanoArmoring of enzymes: rational design of polymer-wrapped enzymes. In C. V. Kumar (Ed.), *Methods in Enzymology* (Vol. 590, pp. 381–411). Elsevier.

Rahardjo, Y. S. P., Sie, S., Weber, F. J., Tramper, J., & Rinzema, A. (2005). Effect of low oxygen concentrations on growth and a-amylase production of *Aspergillus oryzae* in model solid-state fermentation systems. *Biomolecular Engineering, 21*(6), 163–172. doi:10.1016/j.bioeng.2005.01.001 PMID:15748690

Rahmani, K., Faramarzi, M. A., Mahvi, A. H., Gholami, M., Esrafili, A., Forootanfar, H., & Farzadkia, M. (2015). Elimination and detoxification of sulfathiazole and sulfamethoxazole assisted by laccase immobilized on porous silica beads. *International Biodeterioration & Biodegradation, 97*, 107–114. doi:10.1016/j.ibiod.2014.10.018

Raj, K. C., & Chandra, T. S. (1996). Purification and characterization of xylanase from alkalitolerant *Aspergillus fischeri* Fxn 1. *FEMS Microbiology Letters, 145*(3), 457–461. doi:10.1111/j.1574-6968.1996.tb08616.x PMID:8978101

Ramachandran, S., Patel, A. K., Nampoothiri, K. M., Francis, F., Nagy, V., Szakacs, G., & Pandey, A. (2004). Coconut oil cake – A potential raw material for the production of α-amylase. *Bioresource Technology, 93*(2), 169–174. doi:10.1016/j.biortech.2003.10.021 PMID:15051078

Ramanan, R. M. K., Chellamuthu, P., Tang, L., & Nguyen, K. T. (2006). Development of a temperature-sensitive composite hydrogel for drug delivery applications. *Biotechnology Progress, 22*(1), 118–125. doi:10.1021/bp0501367 PMID:16454501

Ramírez, P., Mano, N., Andreu, R., Ruzgas, T., Heller, A., Gorton, L., & Shleev, S. (2008). Direct electron transfer from graphite and functionalized gold electrodes to T1 and T2/T3 copper centers of bilirubin oxidase. *Biochimica et Biophysica Acta, 1777*(10), 1364–1369. doi:10.1016/j.bbabio.2008.06.010 PMID:18639515

Ramirez, S. E., Ramirez, J. J. Q., Hernandez, S., Hernandez, J. G. S., & Kiss, A. A. (2017). Optimal hybrid separations for intensified downstream processing of biobutanol. *Separation and Purification Technology, 185*, 149–159. doi:10.1016/j.seppur.2017.05.011

Ramos-Cormenzana, A., Monteoliva-Sanchez, M. and Lopez, M.J. (1995). Bioremediation of Alpechin. *Intonational Biodeterioration & Biodegradation*, 249-268.

Ramos, R., Carvalho, V., & Gama, M. (2006). Novel hydrogel obtained by chitosan and dextrin-VA co-polymerization. *Biotechnology Letters, 28*(16), 127912–127984. doi:10.100710529-006-9087-6 PMID:16802098

Rao, M. S., Chander, R., & Sharma, A. (2008). Synergistic effect of chitooligosaccharides and lysozyme for meat preservation. *Lebensmittel-Wissenschaft + Technologie*, *41*(10), 1995–2001. doi:10.1016/j.lwt.2008.01.013

Rashid, A. N. (1999). Characterization of xylanase produced by *Bacillus pumilus* strain PJ19. *Journal of Microbiology and Biotechnology*, *9*(2), 157–162.

Rastall, R. A., & Maitin, V. (2002). Prebiotics and synbiotics: Towards the next generation. *Current Opinion in Biotechnology*, *13*(5), 490–496. doi:10.1016/S0958-1669(02)00365-8 PMID:12459343

Rauner, N., Buenger, L., Schuller, S., & Tiller, J. C. (2015). Post polymerization of urease induced calcified, polymer hydrogels. *Macromolecular Rapid Communications*, *36*(2), 224–230. doi:10.1002/marc.201400426 PMID:25284027

Ray, A. S. (2005). Intracellular interactions between nucleos(t)ide inhibitors of HIV reverse transcriptase. *AIDS Reviews*, *7*, 113–125. PubMed

Reddy, C. A. (1995). The potential for white-rot fungi in the treatment of pollutants. *Current Opinion in Biotechnology*, *6*(3), 320–328. doi:10.1016/0958-1669(95)80054-9

Reeve, J. G., Morgan, J., Schwander, J., & Bleehen, N. M. (1993). Role for membrane and secreted insulin-like growth factor-binding protein-2 in the regulation of insulin-like growth factor action in lung tumors. *Cancer Research*, *53*(19), 4680–4685. PMID:7691401

Regina, G. L., Coluccia, A., & Silvestri, R. (2010). Looking for an active conformation of the future HIV type-1 non-nucleoside reverse transcriptase inhibitors. *Antiviral Chemistry & Chemotherapy*, *20*(6), 213–237. PubMed

Remaud-Simeon, M., Willemot, R., Sarçabal, P., Montalk, G. P., & Monsan, P. (2000). Glucansucrases: Molecular engineering and oligosaccharide synthesis. *Journal of Molecular Catalysis. B, Enzymatic*, *10*(1-3), 117–128. doi:10.1016/S1381-1177(00)00119-3

Rengaraj, S., Moon, S. H., Sivabalan, R., Arabindoo, B., & Murugesan, V. (2002). Agricultural solid waste for the removal of organics: Adsorption of phenol from water and wastewater by palm seed coat activated carbon. *Waste Management (New York, N.Y.)*, *22*(5), 543–548. doi:10.1016/S0956-053X(01)00016-2 PMID:12092764

Renner, G. (1980). Metabolic studies on pentachloronitrobenzene (PCNB) in rats. *Xenobiotica*, *7*(8), 537–550. doi:10.3109/00498258009033788 PMID:7445521

Ren, X., & Diller, B. G. (2007). Oxidoreductases for modification of linen fibers. *Colloid Surface A*, *299*(1-3), 15–21. doi:10.1016/j.colsurfa.2006.11.011

Renzetti, S., Courtin, C. M., Delcour, J. A., & Arendt, E. K. (2010). Oxidative and proteolytic enzyme preparations as promising improvers for oat bread formulations: Rheological, biochemical and microstructural background. *Food Chemistry*, *119*(4), 1465–1473. doi:10.1016/j.foodchem.2009.09.028

Resnik, J. L., Reichart, D. B., Huey, K., Webster, N. J. G., & Seely, B. L. (1998). Elevated insulin-like growth factor I receptor autophosphorylation and kinase activity in human breast cancer. *Cancer Research*, *58*(6), 1159–1164. PMID:9515800

Reust, C. E. (2011). Common adverse effects of antiretroviral therapy for HIV disease. *American Family Physician*, *83*(12).

Rezaei, S., Shahverdi, A. R., & Faramarzi, M. A. (2014). An extremely halophilic laccase from the saline water isolate *Chromohalobacter* sp. *Molecular Biology Research Communications*, *3*, 260.

Rezaei, S., Shahverdi, A. R., & Faramarzi, M. A. (2017). Isolation, one-step affinity purification, and characterization of a polyextremotolerant laccase from the halophilic bacterium *Aquisalibacillus elongatus* and its application in the delignification of sugar beet pulp. *Bioresource Technology*, *230*, 67–75. doi:10.1016/j.biortech.2017.01.036 PMID:28161622

Rezaei, S., Tahmasbi, H., Mogharabi, M., Ameri, A., Forootanfar, H., Khoshayand, M. R., & Faramarzi, M. A. (2015). Laccase-catalyzed decolorization and detoxification of Acid Blue 92: Statistical optimization, microtoxicity, kinetics, and energetics. *Journal of Environmental Health Science & Engineering*, *13*(1), 31. doi:10.118640201-015-0183-1 PMID:25908991

Rezaei, S., Tahmasbi, H., Mogharabi, M., Firuzyar, S., Ameri, A., Khoshayand, M. R., & Faramarzi, M. A. (2015). Efficient decolorization and detoxification of reactive orange 7 using laccase isolated from *Paraconiothyrium variabile*, kinetics and energetics. *Journal of the Taiwan Institute of Chemical Engineers*, *56*, 113–121. doi:10.1016/j.jtice.2015.04.008

Rezaie, R., Rezaei, S., Jafari, N., Forootanfar, H., Khoshayand, M. R., & Faramarzi, M. A. (2017). Delignification and detoxification of peanut shell bio-waste using an extremely halophilic laccase from an Aquisalibacillus elongatus isolate. *Extremophiles*, *21*(6), 993–1004. doi:10.100700792-017-0958-7 PMID:28871494

Rezende, M. I., Barbosa, A. M., Vasconcelos, A. F. D., & Endo, A. S. (2002). Xylanase production by *Trichoderma harzianum rifai* by solid state fermentation on sugarcane bagasse. *Brazilian Journal of Microbiology*, *33*(1), 67–72. doi:10.1590/S1517-83822002000100014

Rida, S. M., El-Hawash, S. A., Fahmy, H. T., Hazza, A. A., & El-Meligy, M. M. (2006). Synthesis and in vitro evaluation of some novel benzofuran derivatives as potential anti-HIV-1, anticancer, and antimicrobial agents. *Archives of Pharmacal Research*, *29*(1), 16–25. doi:10.1007/BF02977463 PMID:16491838

Rifaat, H. M., Nagieb, Z. A., & Ahmedv, Y. M. (2005). Production of xylanase by *Streptomyces* species and the bleaching effect on rice straw pulp. *Applied Ecology and Environmental Research*, *4*(1), 151–160. doi:10.15666/aeer/0401_151160

Ritter, G., Maier, G., Schoepplein, E., & Dietrich, H. (1992). The application of polyphenoloxidase in the processing of apple juice. *Bulletin de Liaison-Groupe Polyphenols.*, *16*, 209–212.

Ritter, L., Solomon, K., Sibley, P., Hall, K., Keen, P., Mattu, G., & Linton, B. (2002). Sources, pathways, and relative risks of contaminants in surface water and groundwater: A perspective prepared for the Walkerton inquiry. *Journal of Toxicology and Environmental Health. Part A.*, *65*, 1–142. doi:10.1080/152873902753338572 PMID:11809004

Riva, S. (2006). Laccases: Blue enzymes for green chemistry. *Trends in Biotechnology*, *24*(5), 219–226. doi:10.1016/j.tibtech.2006.03.006 PMID:16574262

Rivero-Urgell, M., & Santamaria-Orleans, A. (2001). Oligosaccharides: Application in infant food. *Early Human Development*, *65*, S43–S52. doi:10.1016/S0378-3782(01)00202-X PMID:11755034

Rob, A., Ball, A. S., Tuncer, M., & Wilson, M. T. (1997). Novel Applications of Peroxidase. *Journal of Chemical Education*, *74*(2), 212. doi:10.1021/ed074p212

Robbins, R. J., Larson, S. M., Sinha, N., Shaha, A., Divgi, C., Pentlow, K. S., ... Tuttle, R. M. (2002). A retrospective review of the effectiveness of recombinant human TSH as a preparation for radioiodine thyroid remnant ablation. *Journal of Nuclear Medicine : Official Publication, Society of Nuclear Medicine*, *43*(11), 1482–1488. PMID:12411552

Roberfroid, M. J., & Slavin, M. (2000). Nondigestible oligosaccharides. *Critical Reviews in Food Science and Nutrition*, *40*(6), 461–480. doi:10.1080/10408690091189239 PMID:11186236

Robinson, T., Mcmullan, G., Marchant, R., & Nigam, P. (2001). Remediation of dyes in textile effluent: A critical review on current treatment technologies with a proposed alternative. *Bioresource Technology*, *77*(3), 247–255. doi:10.1016/S0960-8524(00)00080-8 PMID:11272011

Rocasalbas, G., Francesko, A., Touriño, S., Fernández-Francos, X., Guebitz, G. M., & Tzanov, T. (2013). Laccase-assisted formation of bioactive chitosan/gelatin hydrogel stabilized with plant polyphenols. *Carbohydrate Polymers*, *92*(2), 989–996. doi:10.1016/j.carbpol.2012.10.045 PMID:23399119

Rodríguez Couto, S., Sanromán, M., & Gübitz, G. M. (2005). Influence of redox mediators and metal ions on synthetic acid dye decolourization by crude laccase from *Trametes hirsuta*. *Chemosphere*, *58*(4), 417–422. doi:10.1016/j.chemosphere.2004.09.033 PMID:15620733

Rodríguez-Rodríguez, C. E., Marco-Urrea, E., & Caminal, G. (2010b). Naproxen degradation test to monitor *Trametes versicolor* activity in solid-state bioremediation processes, *179*, 1152–1155.

Rodríguez-Rodríguez, C. E., Marco-Urrea, E., & Caminal, G. (2010a). Degradation of naproxen and carbamazepine in spiked sludge by slurry and solid-phase *Trametes versicolor* systems. *Bioresource Technology*, *101*(7), 2259–2266. doi:10.1016/j.biortech.2009.11.089 PMID:20031398

Romanowska, I., Polak, J., Janowska, K., & Bieleckim, S. (2003). The application of fungal endoxylanase in bread making. *Communications in Agricultural and Applied Biological Sciences*, *68*, 317–320. PMID:15296186

Rooklidge, S. J. (2004). Environmental antimicrobial contamination from terraccumulation and diffuse pollution pathways. *The Science of the Total Environment*, *325*(1-3), 1–13. doi:10.1016/j.scitotenv.2003.11.007 PMID:15144773

Rosenblat, M., Volkova, N., Coleman, R., & Aviram, M. (2007). Anti-oxidant and anti-atherogenic properties of liposomal glutathione: Studies *in vitro*, and in the atherosclerotic apolipoprotein E-deficient mice. *Atherosclerosis*, *195*(2), e61–e68. doi:10.1016/j.atherosclerosis.2007.05.012 PMID:17588583

Rossi, L., Bianchi, M., & Magnani, M. (1992). Increased glucose metabolism by enzyme-loaded erythrocytes in vitro and in vivo normalization of hyperglycemia in diabetic mice. *Biotechnology and Applied Biochemistry*, *15*(2), 207–216. PMID:1586460

Rossi, M., Giovanelli, G., Cantarelli, C., & Brenna, O. (1988). Effects of laccase and other enzymes on barley wort phenolics as a possible treatment to prevent haze in beer. *Bulletin de Liaison-Groupe Polyphenols.*, *14*, 85–88.

Rossi, S. (Ed.). (2013). *Australian Medicines Handbook*. Adelaide: The Australian Medicines Handbook Unit Trust.

Rowland, I. (1988). Factors affecting metabolic activity of the intestinal microflora. *Drug Metabolism Reviews*, *19*(3-4), 243–261. doi:10.3109/03602538808994135 PMID:3068030

Roy, A., Messaoud, E. B., & Bejar, S. (2003). Isolation and purification of an acidic pullulanase type II from newly isolated *Bacillus sp.* US149. *Enzyme and Microbial Technology*, *33*(5), 720–724. doi:10.1016/S0141-0229(03)00212-6

Roy, J. J., Abraham, T. E., Abhijith, K. S., Kumar, P. V. S., & Thakur, M. S. (2005). Biosensor for the determination of phenols based on Cross-Linked Enzyme Crystals (CLEC) of laccase. *Biosensors & Bioelectronics*, *21*(1), 206–211. doi:10.1016/j.bios.2004.08.024 PMID:15967371

Rubilar, O., Diez, M. C., & Gianfreda, L. (2008). Transformation of chlorinated phenolic compounds by white rot fungi. *Critical Reviews in Environmental Science and Technology*, *38*(4), 227–268. doi:10.1080/10643380701413351

Rubinstein, A. (1995). Approaches and opportunities in colon-specific drug delivery. *Critical Reviews™ in Therapeutic Drug Carrier Systems, 12*(2-3).

Rudiger, A., Jorgensen, P. L., & Antranikian, G. (1995). Isolation and Characterization of a Heat-stable Pullulanase from the Hyperthermophilic Archeon *Pyrococcus woesei* after cloning and expression of Its Gene in *Escherichia coli*. *Applied and Environmental Microbiology*, *61*, 567–575. PMID:7574598

Ruiz-Palomero, C., Benítez-Martínez, S., Soriano, M. L., & Valcárcel, M. (2017b). Fluorescent nanocellulosic hydrogels based on graphene quantum dots for sensing laccase. *Analytica Chimica Acta*, *974*, 93–99. doi:10.1016/j.aca.2017.04.018 PMID:28535886

Ruiz-Palomero, C., Kennedy, S. R., Soriano, M. L., Jones, C. D., Valcárcel, M., & Steed, J. W. (2016). Pharmaceutical Crystallization with Nanocellulose Organogels. *Chemical Communications*, *52*(50), 7782–7785. doi:10.1039/C6CC03088B PMID:27168091

Ruiz-Palomero, C., Soriano, M. L., & Valcárcel, M. (2017a). Nanocellulose as analyte and analytical tool: Opportunities and challenges. *TrAC Trends in Analytical Chemistry*, *87*, 1–18. doi:10.1016/j.trac.2016.11.007

Rutschmann, J., Stadler, P. A., Berde, B., & Schild, H. O. (1978). Ergot alkaloids and related compounds. *Handbook of Experimental Pharmacology*, *49*, 61–62.

Rüttimann-Johnson, C. A. R. M. E. N., & Lamar, R. T. (1996). Polymerization of pentachlorophenol and ferulic acid by fungal extracellular lignin-degrading enzymes. *Applied and Environmental Microbiology*, *62*, 3890–3893. PMID:8967777

Ryan, D., Leukes, W., & Burton, S. (2007). Improving the bioremediation of phenolic wastewaters by *Trametes versicolor*. *Bioresource Technology*, *98*(3), 579–587. doi:10.1016/j.biortech.2006.02.001 PMID:16545562

S, O. C. (2006). Additional Therapeutic Enzymes. In B. M & G. W (Eds.), *Directory of therapeutic enzymes*. Taylor and Francis.

Sadighi, A., & Faramarzi, M. A. (2013). Congo red decolorization by immobilized laccase through chitosan nanoparticles on the glass beads. *Journal of the Taiwan Institute of Chemical Engineers*, *44*(2), 156–162. doi:10.1016/j.jtice.2012.09.012

Saffran, M., Bedra, C., Kumar, G. S., & Neckers, D. C. (1988). Vasopressin: A model for the study of effects of additives on the oral and rectal administration of peptide drugs. *Journal of Pharmaceutical Sciences*, *77*(1), 33–38. doi:10.1002/jps.2600770107 PMID:3346821

Saffran, M., Field, J., Pena, J., Jones, R., & Okuda, Y. (1991). Oral insulin in diabetic dogs. *The Journal of Endocrinology*, *131*(2), 267–278. doi:10.1677/joe.0.1310267 PMID:1744572

Saffran, M., Kumar, S., Savariar, C., Burnham, J. C., Williams, F., & Neckers, D. C. (1986). A new approach to the oral administration of insulin and other peptide drugs. *Science*, *233*(4768), 1081–1085. doi:10.1126cience.3526553 PMID:3526553

Sagui, F., Chirivì, C., Fontana, G., Nicotra, S., Passarella, D., Riva, S., & Danieli, B. (2009). Laccase-catalyzed coupling of catharanthine and vindoline: An efficient approach to the bisindole alkaloid anhydrovinblastine. *Tetrahedron*, *65*(1), 312–317. doi:10.1016/j.tet.2008.10.064

Saha, B. C., Mathupala, S. P., & Zeikus, J. G. (1988). Purification and Characterization of a Highly Thermostable Novel Pullulanase from *Clostridium thermohydrosulfuricum*. *The Biochemical Journal*, *252*(2), 343–348. doi:10.1042/bj2520343 PMID:3415657

Sakai, T., Sakamoto, T., Hallaert, J., & Vandamme, E. J. (1993). ⌈ Pectin, Pectinase, and Protopectinase: Production,⌈ Properties, and Applications. *Advances in Applied Microbiology*, *39*, 213–294. doi:10.1016/S0065-2164(08)70597-5 PMID:8213306

Sakamoto, T., Hours, R. A., & Sakai, T. (1994). Purification, characterization, and production of two pectic transeliminases with protopectinase activity from Bacillus subtilis. *Bioscience, Biotechnology, and Biochemistry*, *58*(2), 353–358. doi:10.1271/bbb.58.353 PMID:7764545

Sako, T., Matsumoto, K., & Tanaka, R. (1999). Recent progress on research and applications of non-digestible galactooligosaccharides. *International Dairy Journal*, *9*(1), 69–80. doi:10.1016/S0958-6946(99)00046-1

Sakurai, T., & Kataoka, K. (2007). Structure and function of type I copper in multicopper oxidases. *Cellular and Molecular Life Sciences*, *64*(19-20), 2642–2656. doi:10.100700018-007-7183-y PMID:17639274

Salmen, S., Hoechstetter, J., Käsbauer, C., Paper, D. H., Bernhardt, G., & Buschauer, A. (2005). Sulphated oligosaccharides as inhibitors of hyaluronidases from bovine testis, bee venom and *Streptococcus agalactiae. Planta Medica*, *71*(8), 727–732. doi:10.1055-2005-871255 PMID:16142636

Sammartino, M., Piacquadio, P., De Stefano, G., & Sciancalepore, V. (1998). Apple juice stabilization by conventional and innovative methods. *Industrie delle Bevande*, *27*, 367–369.

Sanborn, T. J., Messersmith, P. B., & Barron, A. E. (2002). In situ crosslinking of a biomimetic peptide-PEG hydrogel via thermally triggered activation of factor XIII. *Biomaterials*, *23*(13), 2703–2710. doi:10.1016/S0142-9612(02)00002-9 PMID:12059019

Sanchez, A., & Vazquez, A. (2017), Bioactive Peptides: A review. *Food Quality and Safety, 1*(1), 29-46.

Sánchez-Moreno, C., Larrauri, J. A., & Saura-Calixto, F. (1998). A procedure to measure the antiradical efficiency of polyphenols. *Journal of the Science of Food and Agriculture*, *76*(2), 270–276. doi:

Sanchez, S., Solovev, A. A., Mei, Y., & Schmidt, O. G. (2010). Dynamics of Biocatalytic Microengines Mediated by Variable Friction Control. *Journal of the American Chemical Society*, *132*(38), 13144–13145. doi:10.1021/ja104362r PMID:20860367

Sandra, A., & Pagano, R. E. (1979). Liposome-cell interactions. Studies of lipid transfer using isotopically asymmetric vesicles. *The Journal of Biological Chemistry*, *254*(7), 2244–2249. PMID:429283

Sandrim, V. C., Rizzatti, A. C. S., Terenzi, H. F., Jorge, J. A., Milagres, A. M. F., & Polizeli, M. L. T. M. (2005). Purification and biochemical characterization of two xylanases produced by *Aspergillus caespitosus* and their potential for kraft pulp bleaching. *Process Biochemistry*, *40*(5), 1823–1828. doi:10.1016/j.procbio.2004.06.061

Sanfins, E., Augustsson, C., Dahlbäck, B., Linse, S., & Cedervall, T. (2014). Size-dependent effects of nanoparticles on enzymes in the blood coagulation cascade. *Nano Letters*, *14*(8), 4736–4744. doi:10.1021/nl501863u PMID:25025946

Sanghi, A., Garg, N., Kuhar, K., Kuhad, R. C., & Gupta, V. K. (2009). Enhanced production of cellulase-free xylanase by alkalophilic *Bacillus subtilis* ash and its application in biobleaching of kraft pulp. *BioResources*, *4*(3), 1109–1129.

San, N., Celebioglu, O., Tumtas, A., Uyar, Y., & Tekinay, T. (2014). Reusable bacteria immobilized electrospun nanofibrous webs for decolorization of methylene blue dye in wastewater treatment. *RSC Advances*, *4*(61), 32249–32255. doi:10.1039/C4RA04250F

Santhanam, N., Vivanco, J. M., Decker, S. R., & Reardon, K. F. (2011). Expression of industrially relevant laccases: Prokaryotic style. *Trends in Biotechnology*, *29*(10), 480–489. doi:10.1016/j.tibtech.2011.04.005 PMID:21640417

Santiago, L. G., Maldonado-Valderrama, J., Martin-Molina, A., Haro-Pérez, C., Maldonado-Valderrama, J., Martin-Molina, A., & Gàlvez-Ruiz, M. J. (2008). Adsorption of soy protein isolate at air–water and oil–water interfaces. *Colloids and Surfaces. A, Physicochemical and Engineering Aspects*, *323*(1-3), 155–162. doi:10.1016/j.colsurfa.2007.11.001

Santos, A. M .P. (2002). SÂ´ntese de oligossacarÂ´deos a partir da sacarosepor inulinase de kluyveromyces marxianus *var. bulgaricus*. Tese dedoutorado, FEA/UNICAMP, 2002.

Santosa, I. J. S., Grossmana, M. J., Sartorattob, A., Ponezib, A. N., & Durrant, L. R. (2012). Degradation of the Recalcitrant Pharmaceuticals Carbamazepine and 17α-Ethinylestradiol by Ligninolytic Fungi. *Chemical Engineering Transactions*, *27*, 169–174.

Saratale, G. D., Kshirdsagar, S. D., Sampange, V. T., Saratale, R. G., Oh, S. E., Govindwar, S. P., & Oh, M. K. (2014). Cellulolytic enzyme production by utilizing agricultural waste under solid state fermentation and its application for Biohydrogen production. *Applied Biochemistry and Biotechnology*, *174*(8), 2801–2817. doi:10.100712010-014-1227-1 PMID:25374139

Sarioglu, O. F., Yasa, O., Celebioglu, A., Uyar, T., & Tekinay, T. (2013). Efficient ammonium removal from aquatic environments by Acinetobacter calcoaceticus STB1 immobilized on an electrospun cellulose acetate nanofibrous web. *Green Chemistry*, *15*(9), 2566–2572. doi:10.1039/c3gc40885j

Sarkar, N., Ghosh, S. K., Bannerjee, S., & Aikat, K. (2012). Bioethanol production from agricultural waste. *Renewable Energy*, *37*(1), 19–27. doi:10.1016/j.renene.2011.06.045

Sarmah, A. K., Meyer, M. T., & Boxall, A. B. A. (2006). A global perspective on the use, sales, exposure pathways, occurrence, fate and effects of veterinary antibiotics (VAs) in the environment. *Chemosphere*, *65*(5), 725–759. doi:10.1016/j.chemosphere.2006.03.026 PMID:16677683

Sarnoski-Brocavich, S., & Hilas, O. (2013). Canagliflozin (invokana), a novel oral agent for type-2 diabetes. *P&T*, *38*(11), 656. PMID:24391386

Sathe, S. K., & Salunkhe, D. K. (1981). Functional properties of the Great Northern bean (*Phaseolus vulgaris* L.) protei:s: emulsion, foaming, viscosity and gelation properties. *Journal of Food Science*, *46*(1), 71–74. doi:10.1111/j.1365-2621.1981.tb14533.x

Sathishkumar, P., Chae, J.-C. C., Unnithan, A. R., Palvannan, T., Kim, H. Y., Lee, K.-J. J., & Oh, B.-T. T. (2012). Laccase-poly(lactic-co-glycolic acid) (PLGA) nanofiber: Highly stable, reusable, and efficacious for the transformation of diclofenac. *Enzyme and Microbial Technology*, *51*(2), 113–118. doi:10.1016/j.enzmictec.2012.05.001 PMID:22664196

Saurav, K., & Kannabiran, K. (2010). Diversity and optimization of process parameters for the growth of *Streptomyces* VITSVK9 spp. isolated from Bay of Begal, India. *Journal of Natural and Environmental Sciences*, *1*(2), 56–65.

Savage, N., & Diallo, M. S. (2005). Nanomaterials and water purification: Opportunities and challenges. *Journal of Nanoparticle Research*, *7*(4-5), 331–342. doi:10.100711051-005-7523-5

Saw, H. Y., Janaun, J., & Subbarao, D. (2008). Hydration properties of palm kernel cake. *Journal of Food Engineering*, *89*(2), 227–231. doi:10.1016/j.jfoodeng.2008.04.027

Schaber, M. D., O'hara, M. B., Garsky, V. M., Mosser, S. C., Bergstrom, J. D., Moores, S. L., ... Gibbs, J. B. (1990). Polyisoprenylation of Ras in vitro by a farnesyl-protein transferase. *The Journal of Biological Chemistry*, *265*(25), 14701–14704. PMID:2203759

Schaetzle, O., Barrière, F., & Baronian, K. (2008). Bacteria and yeasts as catalysts in microbial fuel cells: Electron transfer from micro-organisms to electrodes for green electricity. *Energy & Environmental Science*, *1*(6), 607–620. doi:10.1039/b810642h

Schaffran, T., Jiang, N., Bergmann, M., Küstermann, E., Süss, R., Schubert, R., ... Gabel, D. (2014). Hemorrhage in mouse tumors induced by dodecaborate cluster lipids intended for boron neutron capture therapy. *International Journal of Nanomedicine*, *9*, 3583–3590. PMID:25114527

Scheline, R. R. (1973). Metabolism of foreign compounds by gastrointestinal microorganisms. *Pharmacological Reviews*, *25*(4), 451–523. PMID:4587548

Scheline, R., Nygaard, R., & Longberg, B. (1970). Enzymatic reduction of the azo dye, acid yellow, by extracts of Streptococcus faecalis isolated from rat intestine. *Food and Cosmetics Toxicology*, *8*(1), 55–58. doi:10.1016/S0015-6264(70)80223-1 PMID:4392274

Schlessinger, J. (2000). Cell signaling by receptor tyrosine kinases. *Cell*, *103*(2), 211–225. doi:10.1016/S0092-8674(00)00114-8 PMID:11057895

Schmitt, C., & Turgeon, S. L. (2011). Protein/polysaccharide complexes and coacervates in food systems. *Advances in Colloid and Interface Science*, *167*(1-2), 63–70. doi:10.1016/j.cis.2010.10.001 PMID:21056401

Schroder, M., Aichernig, N., Gu¨bitz, G. M., & Kokol, V. (2007). Enzymatic coating of lignocellulosic surfaces with polyphenols. *Biotechnology Journal*, *2*(3), 334–341. doi:10.1002/biot.200600209 PMID:17260331

Schultz, A., Jonas, U., Hammer, E., & Schauer, F. (2001). Dehalogenation of chlorinated hydroxybiphenyls by fungal laccase. *Applied and Environmental Microbiology*, *67*(9), 4377–4381. doi:10.1128/AEM.67.9.4377-4381.2001 PMID:11526052

Schunkert, H., Ingelfinger, J. R., Hirsch, A. T., Pinto, Y., Remme, W. J., Jacob, H., & Dzau, V. J. (1993). Feedback regulation of angiotensin converting enzyme activity and mRNA levels by angiotensin II. *Circulation Research*, *72*(2), 312–318. doi:10.1161/01.RES.72.2.312 PMID:8380358

Scrinis, G., & Lyons, K. (2010). *Nanotechnology and the techno-corporate agri-food paradigm, Food security, nutrition and sustainability*. Earthscan Publisher.

Selinheimo, E., Kruus, K., Buchert, J., Hopia, A., & Autio, K. (2006). Effects of laccase, xylanase and their combination on the rheological properties of wheat doughs. *Journal of Cereal Science*, *43*(2), 152–159. doi:10.1016/j.jcs.2005.08.007

Semdé, R., Amighi, K., Devleeschouwer, M. J., & Moës, A. J. (2000). Effect of pectinolytic enzymes on the theophylline release from pellets coated with water insoluble polymers containing pectin HM or calcium pectinate. *International Journal of Pharmaceutics*, *197*(1), 169–179. doi:10.1016/S0378-5173(99)00465-2 PMID:10704804

Senthivelan, T., Kanagaraj, J., & Panda, R. (2016). Recent trends in fungal laccase for various industrial applications: An eco-friendly approach-A review. *Biotechnology and Bioprocess Engineering; BBE*, *21*(1), 19–38. doi:10.100712257-015-0278-7

Seo, H.-W., Seo, J. K., & Yang, H. S. (2016). Supplementation of Pork Patties with Bovine Plasma Protein Hydrolysates Augments Antioxidant Properties and Improves Quality. *Korean J. Food Sci. An.*, *36*(2), 198–205. doi:10.5851/kosfa.2016.36.2.198 PMID:27194928

Seyis, I., & Aksoz, N. (2005). Effect of carbon and nitrogen sources on xylanase production by *Trichoderma harzianum* 1073 D3. *International Biodeterioration & Biodegradation*, *55*(2), 115–119. doi:10.1016/j.ibiod.2004.09.001

Shaaltiel, Y., Bartfeld, D., Hashmueli, S., Baum, G., Brill-Almon, E., Galili, G., & Aviezer, D. (2007). Production of glucocerebrosidase with terminal mannose glycans for enzyme replacement therapy of Gaucher's disease using a plant cell system. *Plant Biotechnology Journal*, *5*(5), 579–590. doi:10.1111/j.1467-7652.2007.00263.x PMID:17524049

Shah, K. R., Kydonieus, A., Jamshidi, K., Decker, S. C., & Chang, T. (1996). *US Patent 5,527,271*. US Patent Office.

Shah, A. K., Sidid, S. S., Ahmad, A., & Rele, M. V. (1999). Treatment of bagasse pulp with cellulase-free xylanases from an alkalophilic *Bacillus spp*. Sam-3. *Bioresource Technology*, *68*(2), 133–140. doi:10.1016/S0960-8524(98)00133-3

Shah, A. R., & Madamwar, D. (2005a). Xylanase production by a newly isolated *Aspergillus foetidus* strain and its characterisation. *Process Biochemistry*, *40*(5), 1763–1771. doi:10.1016/j.procbio.2004.06.041

Shah, A. R., & Madamwar, D. (2005b). Xylanase production under solid-state fermentation and its characterization by an isolated strain of *Aspergillus foetidus* in India. *World Journal of Microbiology & Biotechnology*, *21*(3), 233–243. doi:10.100711274-004-3622-1

Shahidi, F., Vidana, A. J. K., & Jeon, Y. J. (1999). Food applications of chitin and chitosans. *Trends in Food Science & Technology*, *10*(2), 37–51. doi:10.1016/S0924-2244(99)00017-5

Shak, S., Capon, D. J., Hellmiss, R., Marsters, S. A., & Baker, C. L. (1990). Recombinant human DNase I reduces the viscosity of cystic fibrosis sputum. *Proceedings of the National Academy of Sciences of the United States of America*, *87*(23), 9188–9192. doi:10.1073/pnas.87.23.9188 PMID:2251263

Shamsizadeh, A. A., Ghaedi, M., Ansari, A., Azizian, S., & Purkait, M. K. (2014). Tin oxide nanoparticle loaded on activated carbon as new adsorbent for efficient removal of malachite green-oxalate: Nonlinear kinetics and isotherm study. *Journal of Molecular Liquids*, *195*, 212–218. doi:10.1016/j.molliq.2014.02.035

Sharma, D., & Satyanarayana, T. (2006). A marked enhancement in the production of a highly alkaline and thermostable pectinase by Bacillus pumilus dcsr1 in submerged fermentation by using statistical methods. *Bioresource Technology*, *97*(5), 727–733. doi:10.1016/j.biortech.2005.04.012 PMID:15936940

Sharma, K., Mahato, N., Cho, M. H., & Lee, Y. R. (2017). Converting citrus waste into value added products: Economic and environmentally friendly approaches. *Nutrition (Burbank, Los Angeles County, Calif.)*, *34*, 29–46. doi:10.1016/j.nut.2016.09.006 PMID:28063510

Sharma, P., Goel, R., & Capalash, N. (2007). Bacterial laccases. *World Journal of Microbiology & Biotechnology*, *23*(6), 823–832. doi:10.100711274-006-9305-3

Sharp, P. M., & Hahn, B. H. (2011). Origins of HIV and the AIDS pandemic. *Cold Spring Harbor Perspectives in Medicine*, *1*(1), a006841.

Sheldon, R. A. (1992). Organic synthesis-past, present and future. *Chemistry & Industry*, *23*, 903–906.

Shenoy, R., & Bowman, C. N. (2012). Kinetics of interfacial radical polymerization initiated by a glucose-oxidase mediated redox system. *Biomaterials*, *33*(29), 6909–6914. doi:10.1016/j.biomaterials.2012.06.014 PMID:22795542

Shih, M. C., Yang, K. T., & Kuo, S. J. (2002). Quality and antioxidative activity of black soybean tofu as affected by Bean Cultivar. *Journal of Food Science*, *67*(2), 480–484. doi:10.1111/j.1365-2621.2002.tb10623.x

Shimizu, Y., Kuruma, Y., Ying, B. W., Umekage, S., & Ueda, T. (2006). Cell-free translation systems for protein engineering. *The FEBS Journal*, *273*(18), 4133–4140. doi:10.1111/j.1742-4658.2006.05431.x PMID:16930131

Shin, J. S., & Kim, B. G. (1997). Kinetic resolution of α-methylbenzylamine with ω-transaminase screened fromsoil microorganisms: Application of a biphasic system to overcome product inhibition. B*iotechnology and Bioengineering*, 55*(2)*, 348–335. doi: PMID:18636493

Shi, W., Zhang, X., He, S., & Huang, Y. (2011). CoFe2O4 magnetic nanoparticles as a peroxidase mimic mediated chemiluminescence for hydrogen peroxide and glucose. C*hemical Communications*, 47*(*38), 10785–10787. doi:10.1039/c1cc14300j PMID:21892508

Shleev, S. V., Morozova, O.V., & Nikitina O.V. (2004). Comparison of physico-chemical characteristics of four laccases from different basidiomycetes. *Biochimie*, *86*(9-10), 693–703.

Shleev, S. V., Morozova, O. V., Nikitina, O. V., Gorshina, E. S., Rusinova, T. V., Serezhenkov, V. A., & Yaropolov, A. I. (2004). Comparison of physico-chemical characteristics of four laccases from different basidiomycetes. *Biochimie*, *86*(9-10), 693–703. doi:10.1016/j.biochi.2004.08.005 PMID:15556280

Shleev, S., Christenson, A., Serezhenkov, V., Burbaev, D., Yaropolov, A., Gorton, L., & Ruzgas, T. (2005). Electrochemical redox transformations of T1 and T2 copper sites in native Trametes hirsuta laccase at gold electrode. *The Biochemical Journal*, *385*(3), 745–754. doi:10.1042/BJ20041015 PMID:15453829

Shleev, S., Tkac, J., Christenson, A., Ruzgas, T., Yaropolov, A. I., Whittaker, J. W., & Gorton, L. (2005). Direct electron transfer between copper-containing proteins and electrodes. *Biosensors & Bioelectronics*, *20*(12), 2517–2554. doi:10.1016/j.bios.2004.10.003 PMID:15854824

Shraddha, S. (2011). R., Sehgal, S., Kamthania, M., & Kumar, A. (2011). Laccase: Microbial sources, production, purification, and potential biotechnological applications. *Enzyme Research*, ▪▪▪, 217861. PMID:21755038

Shrivastava, R., Christian, V., & Vyas, B. R. M. (2005). Enzymatic decolorization of sulfonphthalein dyes. *Enzyme and Microbial Technology*, *36*(2-3), 333–337. doi:10.1016/j.enzmictec.2004.09.004

Si, J. Q. (1994). *Use of laccase in baking industry*. International patent application PCT/DK94/00232.

Si, J.Q. (1994). *Use of laccase in baking*. PCT Int. Appl. WO 9428728A1.

Sigal, G. B., Mrksich, M., & Whitesides, G. M. (1998). Effect of surface wett ability on the adsorption of proteins of proteins and detergents. *Journal of the American Chemical Society*, *120*(14), 3464–3473. doi:10.1021/ja970819l

Sigurskjold, B.W., Christensen, U., Frandsen, T.P., Mirgorodskaya, E., Harrison, M., Roepstor, P., & Svensson, B. (2000). Glucoamylase: Structure/function relationships, and protein engineering. *Biochimica et Biophysica Acta*, 275–293.

Si, H., Liu, D., Si, H., & Liu, D. (2007). Phytochemical genistein in the regulation of vascular function: New insights. *Current Medicinal Chemistry*, *14*(24), 2581–2589. doi:10.2174/092986707782023325 PMID:17979711

Sikri, N., & Bardia, A. (2007). A History of Streptokinase Use in Acute Myocardial Infarction. *Texas Heart Institute Journal*, *34*(3), 318–327. PMID:17948083

Silva, R., Fabry, B., & Boccaccini, A. R. (2014). Fibrous protein-based hydrogels for cell encapsulation. *Biomaterials*, *35*(25), 6727–6738. doi:10.1016/j.biomaterials.2014.04.078 PMID:24836951

Silverman, R. B., & Holladay, M. W. (2014). *The organic chemistry of drug design and drug action*. Academic Press.

Simoes, M. L. G., Tauk-Tornisielo, S. M., & Tapia, D. M. P. (2009). Screening of culture condition for xylanase production by filamentous fungi. *African Journal of Biotechnology*, *8*(22), 6317–6326. doi:10.5897/AJB2009.000-9479

Simone, E. A., Dziubla, T. D., Arguiri, E., Vardon, V., Shuvaev, V. V., Christofidou-Solomidou, M., & Muzykantov, V. R. (2009). Loading PEG-Catalase into Filamentous and Spherical Polymer Nanocarriers. *Pharmaceutical Research*, *26*(1), 250–260. doi:10.100711095-008-9744-7 PMID:18956141

Simon, G. L., & Gorbach, S. L. (1982). Intestinal microflora. *The Medical Clinics of North America*, *66*(3), 557–574. doi:10.1016/S0025-7125(16)31407-9 PMID:7043123

Simon, G. L., Gorbach, S. L., & Bustos-Fernandez, L. (1983). *Bacteriology of the colon. In Colon Structure and Function* (p. 103). Plenum Medical New York.

Sim, W.-J., Lee, J.-W., & Oh, J.-E. (2010). Occurrence and fate of pharmaceuticals in wastewater treatment plants and rivers in Korea. *Environmental Pollution*, *158*(5), 1938–1947. doi:10.1016/j.envpol.2009.10.036 PMID:19913963

Singh, S., (2016). Cerium oxide based nanozymes: Redox phenomenon at biointerfaces. *Biointerphases,11*(40), 04B202.

Singh, A., Yadav, D., Yadav, M., Dhamanage, A., Kulkarni, S., & Singh, R. K. (2015, March). Modeling, Synthesis and Biological Evaluation of N-Heteroaryl Compounds as Reverse Transcriptase Inhibitors Against HIV-1. *Chemical Biology & Drug Design*, *85*(3), 336–347. doi:10.1111/cbdd.12397 PMID:25055732

Singh, A., Yadav, M., Srivastava, R., Singh, N., Kaur, R., Gupta, S. K., & Singh, R. K. (2016, December). Design and anti-HIV activity of arylsulphonamides as non-nucleoside reverse transcriptase inhibitors. *Medicinal Chemistry Research*, *25*(12), 2842–2859. doi:10.100700044-016-1707-7

Singhania, R. R., Patel, A. K., Soccol, C. R., & Pandey, A. (2009). Recent advances in solid state fermentation. *Biochemical Engineering Journal*, *44*(1), 13–18. doi:10.1016/j.bej.2008.10.019

Singhania, R. R., Sukumaran, R. K., Patel, A. K., Larroche, C., & Pandey, A. (2010). Advancement and comparative profiles in the production technologies using solid-state and submerged fermentation for microbial cellulases. *Enzyme and Microbial Technology*, *46*(7), 541–549. doi:10.1016/j.enzmictec.2010.03.010

Singh, D., Dahiya, J. S., & Nigam, P. (1995). Simultaneous raw starch hydrolysis and ethanol fermentation by glucoamylase from *Rhizoctonia solani* and *Saccharomyces cerevisiae*. *Journal of Basic Microbiology*, *35*(2), 117–121. doi:10.1002/jobm.3620350209 PMID:7783000

Singh, M. P., & Srivastava, A. K. (2016). Decolorization of Synthetic Textile Dye and Enzymes Production by Improved Strains of *Pleurotus* Species. *Cellular and Molecular Biology*, *62*, 145. doi:10.4172/1165-158X.1000128

Singh, M., Sharma, R., & Banerjee, V. C. (2002). Biotechnological applications of cyclodextrins. *Biotechnology Advances*, *20*(5-6), 341–359. doi:10.1016/S0734-9750(02)00020-4 PMID:14550020

Singh, R. K., Miazga, A., Dąbrowska, A., Lipniacki, A., Piasek, A., Kulikowski, T., & Shugar, D. (2014). Myristoylated Derivatives of 2, 3-didehydro-2, 3-dideoxythymidine (stavudine) bi-Functional Prodrugs with Potent Anti-HIV-1 Activity and Low Cytotoxicity. *Antiviral Chemistry & Chemotherapy*, *23*(6), 231–235.

Singh, R. K., Yadav, D., Rai, D., Kumari, G., Pannecouque, C., & De Clercq, E. (2010, September). Synthesis, structure–activity relationship and antiviral activity of 3′-N, N-dimethylamino-2′, 3′-dideoxythymidine and its prodrugs. *European Journal of Medicinal Chemistry*, *45*(9), 3787–3793. doi:10.1016/j.ejmech.2010.05.028 PMID:20538384

Singh, S. A., & Appu Rao, A. (2002). A simple fractionation protocol for, and a comprehensive study of the molecular properties of, two major endopolygalacturonases from Aspergillus niger. *Biotechnology and Applied Biochemistry*, *35*(2), 115–123. doi:10.1042/BA20010077 PMID:11916453

Singh, S. K., Sczakas, G., Soccol, C. R., & Pandey, A. (2008). Production of enzymes by solid-state fermentation. In A. Pandey, C. R. Soccol, & C. Larroche (Eds.), *Current Development in Solid-state Fermentation* (pp. 183–204). New York, NY: Springer. doi:10.1007/978-0-387-75213-6_9

Singh, S. K., Yadav, A. K., Prudhviraj, G., Gulati, M., Kaur, P., & Vaidya, Y. (2015). A novel dissolution method for evaluation of polysaccharide based colon specific delivery systems: A suitable alternative to animal sacrifice. *European Journal of Pharmaceutical Sciences*, *73*, 72–80. doi:10.1016/j.ejps.2015.03.012 PMID:25829049

Singh, S., Topuz, F., Hahn, K., Albrecht, K., & Groll, J. (2013). Embedding of active proteins and living cells in redox-sensitive hydrogels and nanogels through enzymatic crosslinking. *Angewandte Chemie International Edition*, *52*(10), 3000–3003. doi:10.1002/anie.201206266 PMID:23386357

Singh, S., Tyagi, C. H., Datt, D., & Upadhyaya, J. S. (2009). Production of high level of cellulose-poor xylanases by wild strains of white-rot fungus *Coprinellus disseminatus* in solid-state fermentation. *New Biotechnology, 26*(3/4). PMID:19761879

Singh, U. P., Bhat, H. R., Verma, A., Kumawat, M. K., Kaur, R., Gupta, S. K., & Singh, R. K. (2013). discovery of a novel class of non-nucleoside reverse transcriptase inhibitors (NNRTIs) against HIV-1 and their antibacterial properties. *RSC Advances, 3*(38), 17335–17348. doi:10.1039/c3ra41604f

Singh, U., & Singh, R. K. (2011). Implication for rational drug design. *Retrovirology, 8*(S2), 82. doi:10.1186/1742-4690-8-S2-P82

Sinha, S., Chand, S., & Tripathi, P. (2016). Recent progress in chitosanase production of monomer-free chitooligosaccharides: Bioprocess strategies and future applications. *Applied Biochemistry and Biotechnology, 180*(5), 883–899. doi:10.100712010-016-2140-6 PMID:27206559

Sinha, S., Tripathi, P., & Chand, S. (2012). A New Bifunctional Chitosanase Enzyme from *Streptomyces* sp. and Its Application in Production of Antioxidant Chitooligosaccharides. *Applied Biochemistry and Biotechnology, 167*(5), 1029–1039. doi:10.100712010-012-9546-6 PMID:22322828

Sinha, V., & Kumria, R. (2001). Colonic drug delivery: Prodrug approach. *Pharmaceutical Research, 18*(5), 557–564. doi:10.1023/A:1011033121528 PMID:11465408

Sirajunnisa, A. R., & Surendhiran, D. (2016). Algae- A quintessential and positive resource of bioethanol production: A comprehensive review. *Renewable & Sustainable Energy Reviews, 66*, 248–267. doi:10.1016/j.rser.2016.07.024

Sivaramakrishnan, S., Gangadharan, D., Nampoothiri, K. M., Soccol, C. R., & Pandey, A. (2006). α-Amylases from Microbial Sources – An Overview. *Recent Developments Food Technology and Biotechnology, 44*(2), 173–184.

Slomczynski, D., Nakas, J., & Tanenbaum, S. W. (1995). Production and Characterization of Laccase from *Botrytis cinerea* 61-34. *Applied and Environmental Microbiology, 61*, 907–912. PMID:16534974

Sluis-Cremer, N., & Tachedjian, G. (2008). Mechanisms of inhibition of HIV replication by non-nucleoside reverse transcriptase inhibitors. *Virus Research, 134*(1), 147–156. PubMed

Soliman, H. M., Abdel-Dayem, A. S., & El-Tanash, A. B. (2012). Production of xylanase by *Aspergillus niger* and *Trichoderma viride* using some agriculture residues. *International Journal of Agricultural Research, 7*, 46–57. doi:10.3923/ijar.2012.46.57

Solís-Oba, M., Ugalde-Saldívar, V. M., González, I., & Viniegra-González, G. (2005). An electrochemical–spectrophotometrical study of the oxidized forms of the mediator 2, 2′-azino-bis-(3-ethylbenzothiazoline-6-sulfonic acid) produced by immobilized laccase. *Journal of Electroanalytical Chemistry, 579*(1), 59–66. doi:10.1016/j.jelechem.2005.01.025

Solomon, E. I., Augustine, A. J., & Yoon, J. (2008). O_2 Reduction to H_2O by the multicopper oxidases. *Dalton Transactions (Cambridge, England), 30*(30), 3921–3932. doi:10.1039/b800799c PMID:18648693

Solomon, E. I., Sundaram, U. M., & Machonkin, T. E. (1996). Multicopper oxidases and oxygenases. *Chemical Reviews, 96*(7), 2563–2606. doi:10.1021/cr950046o PMID:11848837

Songulashvili, G., Elisashvili, V., Wasser, S., Nevo, E., & Hadar, Y. (2006). Laccase and manganese peroxidase activities of *Phellinus robustus* and *Ganoderma adspersum* grown on food industry wastes in submerged fermentation. *Biotechnology Letters, 28*(18), 1425–1429. doi:10.100710529-006-9109-4 PMID:16823599

Sonia, K. G., Chadha, B. S., & Saini, H. S. (2005). Sorghum straw for xylanase hyper-production by *Thermomyces lanuginosus* (D_2W_3) under solid-state fermentation. *Bioresource Technology*, *96*(14), 1561–1569. doi:10.1016/j.biortech.2004.12.037 PMID:15978988

Souhaimi, M. K., & Matsuura, T. (2011). *Membrane distillation: principles and applications*. Elsevier.

Souze. (1980). Immobilized enzyme bioprocess. *Biotechnology and Bioengineering, 22*, 79-21.

Spada, M., Pagliardini, S., Yasuda, M., Tukel, T., Thiagarajan, G., Sakuraba, H., & Desnick, R. J. (2006). High Incidence of Later-Onset Fabry Disease Revealed by Newborn Screening. *American Journal of Human Genetics*, *79*(1), 31–40. doi:10.1086/504601 PMID:16773563

Sperinde, J. J., & Griffith, L. G. (2000). Control and prediction of gelation kinetics in enzymatically cross-linked poly (ethylene glycol) hydrogels. *Macromolecules*, *33*(15), 5476–5480. doi:10.1021/ma000459d

Spoerke, E. D., Anthony, S. G., & Stupp, S. I. (2009). Enzyme directed templating of artificial bone mineral. *Advanced Materials*, *21*(4), 425–430. doi:10.1002/adma.200802242 PMID:22068437

Srinivasan, M. C., & Rele, M. V. (2005). Cellulase free xylanase from microorganisms and their applications to pulp and paper biotechnology: An overview. *Indian Journal of Microbiology*, *35*, 93–101.

Srinivasan, M., Sudheer, A. R., & Menon, V. P. (2007). Ferulic acid: Therapeutic potential through its antioxidant property. *Journal of Clinical Biochemistry and Nutrition*, *40*(2), 92–100. doi:10.3164/jcbn.40.92 PMID:18188410

Stalmans, P., Benz, M. S., Gandorfer, A., Kampik, A., Girach, A., Pakola, S., & Haller, J. A. (2012). Enzymatic vitreolysis with ocriplasmin for vitreomacular traction and macular holes. *The New England Journal of Medicine*, *367*(7), 606–615. doi:10.1056/NEJMoa1110823 PMID:22894573

Stamatopoulos, K., Batchelor, H. K., & Simmons, M. J. (2016). Dissolution profile of theophylline modified release tablets, using a biorelevant Dynamic Colon Model (DCM). *European Journal of Pharmaceutics and Biopharmaceutics*, *108*, 9–17. doi:10.1016/j.ejpb.2016.08.004 PMID:27531624

Stanic, D., Monogioudi, E., Dilek, E., Radosavljevic, J., Markovic, A. M., Vuckovic, O., ... Cirkovic, T. (2010). Digestibility and allergenicity assessment of enzymatically crosslinked beta-casein. *Molecular Nutrition & Food Research*, *54*(9), 1273–1284. doi:10.1002/mnfr.200900184 PMID:20205191

Stanisavljević, M., & Nedić, L. (2004). Removal of Phenol from Industrial Wastewaters by Horseradish (*Cochlearia Armoracia* L.). *Working and Living Environmental Protection*, *2*(4), 345–349.

Staples, C. A., Dorn, P. B., Klecka, G. M., O'Block, S. T., & Harris, L. R. (1998). A review of the environmental fate, effects, and exposures ofbisphenol A. *Chemosphere*, *36*(10), 2149–2173. doi:10.1016/S0045-6535(97)10133-3 PMID:9566294

Stashak, T. S., Farstvedt, E., & Othic, A. (2004). Update on wound dressings: Indications and best use. *Clinical Techniques in Equine Practice*, *3*(2), 148–163. doi:10.1053/j.ctep.2004.08.006

Steiert, J. G., & Crawford, R. L. (1985). Microbial degradation of chlorinated phenols. *Trends in Biotechnology*, *3*(12), 300–305. doi:10.1016/0167-7799(85)90032-0

Stichnothe, H., & Schuchardt, F. (2011). Life cycle assessment of two palm oil production systems. *Biomass and Bioenergy*, *35*(9), 3976–3984. doi:10.1016/j.biombioe.2011.06.001

Stork, G., Pereira, H., Wood, Tm., Dusterhoft, Em., Toft, A., & Puls, J. (1995). Upgrading Recycled Pulps Using Enzymatic Treatment. *J Tech Assoc Pulp Pap Ind*, *78*, 79–188.

Strong, P. J., & Claus, H. (2011). Laccase: A Review of Its Past and Its Future in Bioremediation. *Critical Reviews in Environmental Science and Technology*, *41*(4), 373–434. doi:10.1080/10643380902945706

Suárez, S., Carballa, M., Omil, F., & Lema, J. M. (2008). How are pharmaceutical and personal care products (PPCPs) removed from urban wastewaters? *Reviews in Environmental Science and Biotechnology*, *7*(2), 125–138. doi:10.100711157-008-9130-2

Subramaniyan, S., & Prema, P. (2002). Biotechnology of microbial xylanases: Enzymology, molecular biology, and application. *Critical Reviews in Biotechnology*, *22*(1), 33–64. doi:10.1080/07388550290789450 PMID:11958335

Suda, T., Hata, T., Kawai, S., Okamura, H., & Nishida, T. (2012). Treatment of tetracycline antibiotics by laccase in the presence of 1-hydroxybenzotriazole. *Bioresource Technology*, *103*(1), 498–501. doi:10.1016/j.biortech.2011.10.041 PMID:22071243

Sugarman, B. J., Aggarwal, B. B., Hass, P. E., Figari, I. S., Palladino, M. A., & Shepard, H. M. (1985). Recombinant human tumor necrosis factor-alpha: Effects on proliferation of normal and transformed cells in vitro. *Science*, *230*(4728), 943–945. doi:10.1126cience.3933111 PMID:3933111

Suk, K., Lee, J. Y., & Kim, S. H. (1997). Synergistic stimulation of fibrinogen gelation by casein kinase II and polycationic compounds. *IUBMB Life*, *42*(3), 487–495. doi:10.1080/15216549700202891 PMID:9247706

Sumathi, S., Chai, S. P., & Mohamed, A. R. (2008). Oil palm as a source of renewable energy in Malaysia. *Renewable & Sustainable Energy Reviews*, *12*(9), 2404–2421. doi:10.1016/j.rser.2007.06.006

Sumpter, J. P. (1998). Xenoendorine disrupters--environmental impacts. *Toxicology Letters*, *102–103*, 337–342. doi:10.1016/S0378-4274(98)00328-2 PMID:10022275

Sunna, A., Moracci, M., Rossi, M., & Antranikian, G. (1997). Glycosyl hydrolases from hyperthermophiles. *Extremophiles*, *1*(1), 2–13. doi:10.1007007920050009 PMID:9680331

Sun, R. C., Tomkinson, J., Wang, Y. X., & Xiao, B. (2000). Physico-chemical and structural characterisation of hemicelluloses from wheat straw by an alkaline peroxide extraction. *Polymer*, *41*(7), 2647–2656. doi:10.1016/S0032-3861(99)00436-X

Suntres, Z. E. (2011). Liposomal Antioxidants for Protection against Oxidant-Induced Damage. *Journal of Toxicology*, *2011*, 1–16. doi:10.1155/2011/152474 PMID:21876690

Suntres, Z. E., & Shek, P. N. (1994). Incorporation of α-tocopherol in liposomes promotes the retention of liposome-encapsulated glutathione in the rat lung. *The Journal of Pharmacy and Pharmacology*, *46*(1), 23–28. doi:10.1111/j.2042-7158.1994.tb03714.x PMID:8201524

Sun, W. Q., Payne, G. F., Moas, M., Chu, J. H., & Wallace, K. K. (1992). Tyrosinase reaction/chitosan adsorption for removing phenols from wastewater. *Biotechnology Progress*, *8*(3), 179–186. doi:10.1021/bp00015a002

Suparno, O., Covington, A. D., & Evans, C. S. (2007). Application of diphenols for dyeing. *J Soc Leath Tech Ch.*, *91*, 139–141.

Suri, R., Metcalfe, C., Lees, B., Grieve, R., Flather, M., Normand, C., & Wallis, C. (2001). Comparison of hypertonic saline and alternate-day or daily recombinant human deoxyribonuclease in children with cystic fibrosis: A randomised trial. *Lancet*, *358*(9290), 1316–1321. doi:10.1016/S0140-6736(01)06412-1 PMID:11684212

Su, T., Tang, Z., He, H., Li, W., Wang, X., Liao, C., ... Wang, Q. (2014). Glucose oxidase triggers gelation of N-hydroxyimide–heparin conjugates to form enzyme-responsive hydrogels for cell-specific drug delivery. *Chemical Science (Cambridge)*, *5*(11), 4204–4209. doi:10.1039/C4SC01603C

Sutar, R. S., & Rathod, V. K. (2015). Ultrasonics sonochemistry ultrasound assisted enzyme catalyzed degradation of Cetirizine dihydrochloride. *Ultrasonics Sonochemistry, 24*, 80–86. doi:10.1016/j.ultsonch.2014.10.016 PMID:25515235

Suzuki Y. (1989). A general principle of increasing protein thermostability. *Proc. Jpn. Acad., Ser. B65*, 146-148.

Suzuki, T., Timofei, S., Kurunczi, L., Dietze, U., & Schüürmann, G. (2001). Correlation of aerobic biodegradability of sulfonated azo dyes with the chemical structure. *Chemosphere, 45*(1), 1–19. doi:10.1016/S0045-6535(01)00074-1 PMID:11572582

Suzuki, Y. (1999). The proline rule - a strategy for protein thermal stabilization. Proc. Jpn. Acad. *Ser. B, 75*, 133–137.

Svarachorn, A. (1999). Production of fungal-xylanase using agricultural waste by solid state fermentation. *The Journal of Scientific Research Chulalongkorn University, 24*(1).

Syed, Y. Y., & Dhillon, S. (2013). Ocriplasmin: A review of its use in patients with symptomatic vitreomacular adhesion. *Drugs, 73*(14), 1617–1625. doi:10.100740265-013-0124-1 PMID:24062204

Szendefy, J., Szakacs, G., & Christopher, L. (2006). Potential of solid-state fermentation enzymes of *Aspergillus oryzae* in biobleaching of paper pulp. *Enzyme and Microbial Technology, 39*(6), 1354–1360. doi:10.1016/j.enzmictec.2006.06.016

Tabe, S. (2014). *Electrospun nanofiber membranes and their applications in water and wastewater treatment Nanotechnology for Water Treatment and Purification.* Springer International Publishing.

Taha, F. S., Yamamah, G. A., Mohamed, G. F., Mohamed, S. S., & Wagdy, S. M. (2014). Biological Activities of Sunflower Protein Hydrolysate. *World Applied Sciences Journal, 30*(11), 462–1470.

Tahmasbi, H., Khoshayand, M. R., Bozorgi-Koushalshahi, M., Heidary, M., Ghazi-Khansari, M., & Faramarzi, M. A. (2016). Biocatalytic conversion and detoxification of imipramine by the laccase-mediated system. *International Biodeterioration & Biodegradation, 108*, 1–8. doi:10.1016/j.ibiod.2015.11.029

Takada, K., Nakamura, A., Matsuo, N., Inoue, A., Someya, K., & And Shimogaki, H. (2003). Influence of oxidative and/or reductive treatment on human hair (I): Analysis of hair-damage after oxidative and/or reductive treatment. *Journal of Oleo Science, 52*(10), 541–548. doi:10.5650/jos.52.541

Takahashi, M., Uechi, S., Takara, K., Asikin, Y., & Wada, K. (2009). Evaluation of an oral carrier system in rats: Bioavailability and antioxidant properties of liposome-encapsulated curcumin. *Journal of Agricultural and Food Chemistry, 57*(19), 9141–9146. doi:10.1021/jf9013923 PMID:19757811

Takata, Y., Yamamoto, K., & Kadokawa, J. (2015). Preparation of pHresponsive amphoteric glycogen hydrogels by αglucanphosphorylasecatalyzed successive enzymatic reactions. *Macromolecular Chemistry and Physics, 216*(13), 1415–1420. doi:10.1002/macp.201500151

Takei, T., Sugihara, K., Ijima, H., & Kawakami, K. (2011). In situ gellable sugar beet pectin via enzyme-catalyzed coupling reaction of feruloyl groups for biomedical applications. *Journal of Bioscience and Bioengineering, 112*(5), 491–494. doi:10.1016/j.jbiosc.2011.07.006 PMID:21807557

Takei, T., Sugihara, K., Yoshida, M., & Kawakami, K. (2013). Injectable and biodegradable sugar beet pectin/gelatin hydrogels for biomedical applications. *Journal of Biomaterials Science. Polymer Edition, 24*(11), 1333–1342. doi:10.1080/09205063.2012.757727 PMID:23796034

Takemori, T., Ito, Y., Ito, M., & Yoshama, M. (1992). Flavor and taste improvement of cacao nib by enzymatic treatment. *Japan Kokai Tokkyo Koho JP, 04126037*, A2.

Takii, Y., Takahashi, K., Yamamoto, K., & Suzuki, Y. (1996). *Bacillus stearothermophilus* ATCC12016 α-glucosidase specific for α-1,4 bonds of maltosaccharides and α-glucans shows high amino acid sequence similarities to seven α-D-glucohydrolases with different substrate specificity. *Applied Microbiology and Biotechnology*, *44*(5), 629–634. doi:10.1007/BF00172496

Tamagawa, Y., Hirai, H., Kawai, S., & Nishida, T. (2005). Removal of estrogenic activity of endocrine-disrupting genistein by ligninolytic enzymes from white rot fungi. *FEMS Microbiology Letters*, *244*(1), 93–98. doi:10.1016/j.femsle.2005.01.023 PMID:15727826

Tandon, V. K., Yadav, D. B., Singh, R. V., Vaish, M., Chaturvedi, A. K., & Shukla, P. K. (2005). Synthesis and biological evaluation of novel 1, 4-naphthoquinone derivatives as antibacterial and antiviral agents. *Bioorganic & Medicinal Chemistry Letters*, *15*(14), 3463–3466. doi:10.1016/j.bmcl.2005.04.075 PMID:15950468

Tang, X., Zhang, Q., Liu, Z., Pan, K., Dong, Y., & Li, Y. (2014). Removal of Cu (II) by loofah fibers as a natural and low-cost adsorbent from aqueous solutions. *Journal of Molecular Liquids*, *199*, 401–407. doi:10.1016/j.molliq.2014.09.033

Tan, I. A. W., Ahmad, A. L., & Hameed, B. H. (2009). Fixed-bed adsorption performance of oil palm shell-based activated carbon for removal of 2,4,6-trichlorophenol. *Bioresource Technology*, *100*(3), 1494–1496. doi:10.1016/j.biortech.2008.08.017 PMID:18809316

Tan, L. U. L., Wong, K. K. Y., Yu, E. K. C., & Saddler, J. N. (1985). Purification and characterization of two D-xylanases from *Trichodermaharzianum*. *Enzyme and Microbial Technology*, *7*(7), 425–430. doi:10.1016/0141-0229(85)90041-9

Tan, Q. Y., Wang, N., Yang, H., Chen, L., Xiong, H. R., Zhang, L. K., & Zhang, J. Q. (2010). Preparation and characterization of lipid vesicles containing uricase. *Drug Delivery*, *17*(1), 28–37. doi:10.3109/10717540903508953 PMID:22747073

Tan, Q. Y., Wang, N., Yang, H., Zhang, L. K., Liu, S., Chen, L., & Zhang, J. Q. (2010). Characterization, stabilization and activity of uricase loaded in lipid vesicles. *International Journal of Pharmaceutics*, *384*(1), 165–172. doi:10.1016/j.ijpharm.2009.09.036 PMID:19782743

Tantoush, Z., Stanic, D., Stojadinovic, M., Ognjenovic, J., Mihajlovic, L., Markovic, A. M., & Cirkovic, T. (2011). Digestibility and allergenicity of b-lactoglobulin following laccase-mediated cross-linking in the presence of sour cherry phenolics. *Food Chemistry*, *125*(1), 84–91. doi:10.1016/j.foodchem.2010.08.040

Taspinar, A., & Kolankaya, N. (1998). Optimization of enzymatic chlorine removal from kraft pulp. *Bulletin of Environmental Contamination and Toxicology*, *61*, 15–21. doi:10.1007001289900723 PMID:9657825

Tateishi, M., Ishida, T., Mitsudomi, T., Kaneko, S., & Sugimachi, K. (1990). Immunohistochemical evidence of autocrine growth factors in adenocarcinoma of the human lung. *Cancer Research*, *50*(21), 7077–7080. PMID:2208175

Tavares, A. P. M., Cristóvão, R. O., Gamelas, J. A. F., Loureiro, J. M., Boaventura, R. A. R., & Macedo, E. A. (2009). Sequential decolourization of reactive textile dyes by laccase mediator system. *Journal of Chemical Technology and Biotechnology (Oxford, Oxfordshire)*, *84*(3), 442–446. doi:10.1002/jctb.2060

Taverna, R. D., & Langdon, R. G. (1973). Glucose transport in white erythrocyte ghosts and membrane-derived vesicles. *Biochimica et Biophysica Acta (BBA) -. Biomembranes*, *298*(2), 422–428. doi:10.1016/0005-2736(73)90369-6 PMID:4719139

Tay, T., Köse, E., Keçili, R., & Say, R. (2016). Design and preparation of nano-lignin peroxidase (NanoLiP) by protein block copolymerization approach. *Polymers*, *8*(6), 223. doi:10.3390/polym8060223

Tchiégang, C., Kapseu, C., Ndjouenkeu, R., & Ngassoum, M. B. (1997). Amandes de *Ricinodendron heudelotii* (Bail): Matière première potentielle pour les industries agro-alimentaires tropicales. *Journal of Food Engineering*, *32*(1), 1–10. doi:10.1016/S0260-8774(97)00094-0

Tebit, D.M., & Arts, E.J. (2011). Tracking a century of global expansion and evolution of HIV to drive understanding and to combat disease. *Lancet Infect Dis, 11*, 45–56. (10)70186-910.1016/S1473-3099

Teixeira, L. S. M., Feijen, J., van Blitterswijk, C. A., Dijkstra, P. J., & Karperien, M. (2012). Enzyme-catalyzed crosslinkable hydrogels: Emerging strategies for tissue engineering. *Biomaterials*, *33*(5), 1281–1290. doi:10.1016/j.biomaterials.2011.10.067 PMID:22118821

Telke, A. A., Kagalkar, A. N., Jagtap, U. B., Desai, N. S., Bapat, V. A., & Govidwar, S. P. (2011). Biochemical chracterization of laccase from hairy root culture of *Brassica juncea* L. and role of redox mediators to enhance its potential for the decolorization of textile dyes. *Planta*, *234*(6), 1137–1149. doi:10.100700425-011-1469-x PMID:21735196

Tenkanen, H., Puls, J., & Poutanen, K. (1992). Two major xylanases of *Trichoderma reesei. Enzyme and Microbial Technology*, *14*(7), 566–574. doi:10.1016/0141-0229(92)90128-B

Ternes, T. A. (1998). Occurrence of drugs in German sewage treatment plants and rivers1Dedicated to Professor Dr. Klaus Haberer on the occasion of his 70th birthday. *Water Research*, *32*(11), 3245–3260. doi:10.1016/S0043-1354(98)00099-2

Tetsch, L., Bend, J., & Hölker, U. (2006). Molecular and enzymatic characterisation of extra-and intracellular laccases from the acidophilic ascomycete *Hortaea acidophila. Antonie van Leeuwenhoek*, *90*(2), 183–194. doi:10.100710482-006-9064-z PMID:16871425

Thadathil, N., & Velappan, S. P. (2014). Recent development in chitosanase research and its biotechnological applications: A review. *Food Chemistry*, *150*, 392–399. doi:10.1016/j.foodchem.2013.10.083 PMID:24360467

Thanaporn, L., Parichart, K., Nakul, R., Phungjai, T., Saowanee, A., Paemanee, A., ... Eurwiilaichitr, L. (2011). Production of multi-fiber modifying enzyme from *Mamillisphaeria* sp. for refining of recycled paper pulp. *Bioscience, Biotechnology, and Biochemistry*, *75*(12), 2297–2303. doi:10.1271/bbb.110431 PMID:22146707

Theron, J., Walker, J. A., & Cloete, E. (2008). Nanotechnology and Water Treatment, applications and Emerging Opportunities. *Critical Reviews in Microbiology*, *34*(1), 43–69. doi:10.1080/10408410701710442 PMID:18259980

Thomas, D. A., Myers, M. A., Wichert, B., Schreier, H., & Gonzalez-Rothi, R. J. (1991). Acute effects of liposome aerosol inhalation on pulmonary function in healthy human volunteers. *Chest*, *99*(5), 1268–1270. doi:10.1378/chest.99.5.1268 PMID:2019191

Thurston, C. F. (1994). The structure and function of fungal laccases. *Microbiology*, *140*(1), 19–26. doi:10.1099/13500872-140-1-19

Tian, L., Branford-White, C., Wang, W., Nie, H., & Zhu, L. (2012). Laccase-mediated system pretreatment to enhance the effect of hydrogen peroxide bleaching of cotton fabric. *International Journal of Biological Macromolecules*, *50*(3), 782–787. doi:10.1016/j.ijbiomac.2011.11.025 PMID:22155402

Tien, M., Kirk, T. K., Morgan, M. A., Maytield, M. B., Kuwahara, M., & Gold, M. H. (1988). Lignin Peroxidase of *Phanerochaete chrysosporium. Methods in Enzymology*, *161*, 238–249. doi:10.1016/0076-6879(88)61025-1

Tokuyama, H., Yamago, S., Nakamura, E., Shiraki, T., & Sugiura, Y. (1993). Photoinduced biochemical-activity of fullerene carboxylic-acid. *Journal of the American Chemical Society*, *115*(17), 7918–7919. doi:10.1021/ja00070a064

Tolan, J. S., & Foody, B. (1999). Cellulase from Submerged Fermentation. *Advances in Biochemical Engineering/Biotechnology, 65*, 41–67. doi:10.1007/3-540-49194-5_3

Toldra, F., Reig, M., Aristoy, M. C., & Mora, L. (2017). Generation of Bioactive Peptides from food processing. *Food Chemistry*. doi:10.1016/j.foodchem.2017.06.119

Tomaz, I., Maslov, L., Stupic, D., Preiner, D., Asperger, D., & Kontic, J. K. (2016). Recovery of flavonoids from grape skins by enzyme-assisted extraction. *Separation Science and Technology, 51*(2), 255–268. doi:10.1080/01496395.2015.1085881

Tomazic, S. J., & Klibanov, A. M. (1988). Why is one Bacillus α-amylase more resistant against irreversible thermoinactivation than another? *The Journal of Biological Chemistry, 263*, 3092–3096. PMID:3125174

Tomlin, J., & Read, N. (1988). The relation between bacterial degradation of viscous polysaccharides and stool output in human beings. *British Journal of Nutrition, 60*(3), 467–475. doi:10.1079/BJN19880119 PMID:3219318

Torchilin, V., & Weissig, V. (2003). *Liposomes: a practical approach*. Oxford, UK: Oxford University Press.

Torres, E., Bustos-Jaimes, I., & Le Borgne, S. (2003). Potential use of oxidative enzymes for the detoxification of organic pollutants. *Applied Catalysis B: Environmental, 46*(1), 1–15. doi:10.1016/S0926-3373(03)00228-5

Torronen, A., Harkki, A., & Rouvinen, J. (1994). Three dimensional structure of endo-1,4-β-xylanase II from *Trichoderma reesei*: Two conformational states in the active site. *The EMBO Journal, 13*(11), 2493–2501. PMID:8013449

Touahar, I. E., Haroune, L., Ba, S., Bellenger, J.-P. P., & Cabana, H. (2014). Characterization of combined cross-linked enzyme aggregates from laccase, versatile peroxidase and glucose oxidase, and their utilization for the elimination of pharmaceuticals. *The Science of the Total Environment, 481*, 90–99. doi:10.1016/j.scitotenv.2014.01.132 PMID:24589758

Trivić, S., Leskovac, V., Zeremski, J., Vrvić, M., & Winston, G. W. (2002). Bioorganic mechanisms of the formation of free radicals catalyzed by glucose oxidase. *Bioorganic Chemistry, 30*(2), 95–106. doi:10.1006/bioo.2001.1229 PMID:12020134

Turkoglu, M., & Ugurlu, T. (2002). In vitro evaluation of pectin–HPMC compression coated 5-aminosalicylic acid tablets for colonic delivery. *European Journal of Pharmaceutics and Biopharmaceutics, 53*(1), 65–73. doi:10.1016/S0939-6411(01)00225-9 PMID:11777754

Tychanowicz, G. K., Zilly, A., de Souza, C. G. M., & Peralta, R. M. (2001). Decolourisation of industrial dyes by solid-state cultures of *Pleurotus pulmonarius. Process Biochemistry, 39*(7), 855–859. doi:10.1016/S0032-9592(03)00194-8

Tyler, C. R., Jobling, S., & Sumpter, J. P. (1998). Endocrine disruption in wildlife: A critical review of the evidence. *Critical Reviews in Toxicology, 28*(4), 319–361. doi:10.1080/10408449891344236 PMID:9711432

Tzanov, T., Basto, C., Gu¨bitz, G. M., & Cavaco-Paulo, A. (2003 a). Laccases to improve the whiteness in a conventional bleaching of Cotton. *Macromolecular Materials and Engineering, 288*(10), 807–810. doi:10.1002/mame.200300100

Tzanov, T., Silva, C., Zille, A., Oliveira, J., & Cavaco-Paulo, A. (2003 b). Effect of some process parameters in enzymatic dyeing of wool. *Applied Biochemistry and Biotechnology, 111*(1), 1–14. doi:10.1385/ABAB:111:1:1 PMID:14566065

Uchida, H., Fukuda, T., Miyamoto, H., Kawabata, T., Suzuki, M., & Uwajima, T. (2001). Polymerization of Bisphenol A by purified laccase from *Trametes villosa. Biochemical and Biophysical Research Communications, 287*(2), 355–358. doi:10.1006/bbrc.2001.5593 PMID:11554734

Uchiyama, T., Abe, T., Ikemura, T., & Watanabe, K. (2005). Substrate induced gene expression screening of environmental metagenome libraries for isolation of catabolic genes. *Nature Biotechnology, 23*(1), 88–93. doi:10.1038/nbt1048 PMID:15608629

Ulaganathan, K., Goud, S., Reddy, M., & Kayalvili, U. (2017). Genome engineering for breaking barriers in lignocellulosic bioethanol production. *Renewable & Sustainable Energy Reviews*, *74*, 1080–1107. doi:10.1016/j.rser.2017.01.028

Ullah, F., Othman, M. B. H., Javed, F., Ahmad, Z., & Akil, H. M. (2015). Classification, processing and applications of hydrogels: A review. *Materials Science and Engineering C*, *57*(1), 414–433. doi:10.1016/j.msec.2015.07.053 PMID:26354282

Umesh, U. J., Vishal, V. D., Amar, A. T., & Sanjay, P. G. (2008). Decolorization of Direct Blue GLL with enhanced lignin peroxidase enzyme production in *Comamonas* sp UVS. *Journal of Chemical Technology and Biotechnology (Oxford, Oxfordshire)*, *84*, 126–132.

UNAIDS Fact Sheet. (2016). *Global statistics-2015*. Retrieved from http://www.unaids.org/sites/ default/

Underkofler, L., Barton, R., & Rennert, S. (1958). Production of microbial enzymes and their applications. *Applied Microbiology*, *6*(3), 212. PMID:13534309

Underkofler, L., Severson, G., Goering, K., & Christensen, L. (1947). Commercial production and use of mold bran. *Cereal Chemistry*, *24*(1), 1–22.

Upadhayay, P., Shrivastava, R., & Agrawal, P.K. (2016).Bioprospecting and Biotechnological applications of fungal laccase. *Biotech.*

Uria, A.R., & Zilda, D.S. (2016). Metagenomics-guided mining of commercially useful biocatalysts from Marine microorganisms. In *Advances in food and nutrition research.* Elsevier Publication.

Uxa, R., Baczyk, D., Kingdom, J. C. P., Viero, S., Casper, R., & Keating, S. (2010). Genetic Polymorphisms in the Fibrinolytic System of Placentas with Massive Perivillous Fibrin Deposition. *Placenta*, *31*(6), 499–505. doi:10.1016/j.placenta.2010.03.013 PMID:20447686

Vacca, J. P., Dorsey, B. D., Guare, J. P., Holloway, M. K., Hungate, R. W., & Levin, R. B. (1995). *HIV protease inhibitors useful for the treatment of AIDS*. Google Patents.

Vahabzadeh, F., Mehranian, M., & Saatari, A. R. (2004). Color removal ability of *Phanerochaete chrysosporium* in relation to lignin peroxidase and manganese peroxidase produced in molasses wastewater. *World Journal of Microbiology & Biotechnology*, *20*(8), 859–864. doi:10.100711274-004-9005-9

Van den Mooter, G., & Kinget, R. (1995a). Oral colon-specific drug delivery: A review. *Drug Delivery*, *2*(2), 81–93. doi:10.3109/10717549509031355

Van den Mooter, G., Samyn, C., & Kinget, R. (1995b). In vivo evaluation of a colon-specific drug delivery system: An absorption study of theophylline from capsules coated with azo polymers in rats. *Pharmaceutical Research*, *12*(2), 244–247. doi:10.1023/A:1016283027139 PMID:7784340

van der Heijden, R., Jacobs, D. I., Snoeijer, W., Hallard, D., & Verpoorte, R. (2004). The Catharanthus alkaloids: Pharmacognosy and biotechnology. *Current Medicinal Chemistry*, *11*(5), 607–628. doi:10.2174/0929867043455846 PMID:15032608

Van der Sloot, A. M., Kiel, C., Serrano, L., & Stricher, F. (2009). Protein design in biological networks: From manipulating the input to modifying the output. *Protein Engineering, Design & Selection*, *22*(9), 537–542. doi:10.1093/protein/gzp032 PMID:19574296

Van Wyk, J. (2003). Biodegradation of Wastepaper By Cellulose From *Trichoderma Viride. Bioresource Technology*, *86*(1), 21–23. doi:10.1016/S0960-8524(02)00130-X PMID:12421003

Vane, J. R., Bakhle, Y. S., & Botting, R. M. (1998). Cyclooxygenases 1 and 2. *Annual Review of Pharmacology and Toxicology*, *38*(1), 97–120. doi:10.1146/annurev.pharmtox.38.1.97 PMID:9597150

Vasquez, M. I., Lambrianides, A., Schneider, M., Kümmerer, K., & Fatta-Kassinos, D. (2014). Environmental side effects of pharmaceutical cocktails: What we know and what we should know. *Journal of Hazardous Materials*, *279*, 169–189. doi:10.1016/j.jhazmat.2014.06.069 PMID:25061892

Vellard, M. (2003). The enzyme as drug: Application of enzymes as pharmaceuticals. *Current Opinion in Biotechnology*, *14*(4), 444–450. doi:10.1016/S0958-1669(03)00092-2 PMID:12943856

Venkatadri, R., & Irvine, R.L. (1993). Cultivation of *Phanerochaete chrysosporium* and production of lignin peroxidase in novel biofilm reactor systems: hollow fiber reactor and silicone membrane reactor. *Water Research*, *27*, 591-6. DOI: 10.1016/0043-1354(93)90168-H

Venkatesan, N., & Kim, B. H. (2002). Synthesis and enzyme inhibitory activities of novel peptide isosteres. *Current Medicinal Chemistry*, *9*(24), 2243–2270. doi:10.2174/0929867023368692 PMID:12470245

Venkatesh, S., Sizemore, S. P., & Byrne, M. E. (2007). Biomimetic hydrogels for enhanced loading and extended release of ocular therapeutics. *Biomaterials*, *28*(4), 717–724. doi:10.1016/j.biomaterials.2006.09.007 PMID:17007923

Vergari, A., Tibuzzi, F., & Basile, G. (2010). An overview of the functional food market: from marketing issues and commercial players to future demand from life in space. *Biofarms for Nutraceuticals,* 308-321.

Veugelers, P. J., & Schechter, M. T. (1997). The incubation period of HIV infection. *Current Opinion in Infectious Diseases*, *10*(1), 7–11. doi:10.1097/00001432-199702000-00003

Vijaykumar, M. H., Veeranagouda, Y., Neelkanteshwar, K., & Karegoudar, T. B. (2006). Degradation of 1:2 Metal Complex Dysse Acid Blue 193 by a Newly Isolated Fungus Cladosporium Cladosporioide. *World Journal of Microbiology & Biotechnology*, *22*(2), 157–162. doi:10.100711274-005-9013-4

Villadsen, J., Nielsen, J., & Liden, G. (2011). *Bioreaction engineering principles* (3rd ed.). Springer publishers. doi:10.1007/978-1-4419-9688-6

Villamiel, M., Corzo, N., Foda, M. I., Montes, F., & Olano, A. (2002). Lactulose formation catalysed by alkaline-substituted sepiolites in milk permeate. *Food Chemistry*, *76*(1), 7–11. doi:10.1016/S0308-8146(01)00239-4

Viniegra-Gonzalez, G., Favela-Torres, E., Aguilar, C. N., Romero-Gomez, S., Diaz-Godinez, G., & Augur, C. (2003). Advantages of fungal enzyme production in solid state over liquid fermentation system. *Biochemical Engineering Journal*, *13*(2-3), 157–167. doi:10.1016/S1369-703X(02)00128-6

Vinogradov, V. V., & Avnir, D. (2015). Enzyme renaturation to higher activity driven by the sol-gel transition: Carbonic anhydrase. *Scientific Reports*, *5*(1), 14411. doi:10.1038rep14411 PMID:26394694

Vollman, J., Fritz, N. C., Wagentristi, H., & Ruckenbauer, P. (2000). Environmental and genetic variation of soybean seed protein content under Central European growing conditions. *Journal of the Science of Food and Agriculture*, *80*(9), 1300–1306. doi:

Volodina, K. V., Avnir, D., & Vinogradov, V. V. (2017). Alumina nanoparticle-assisted enzyme refolding: A versatile methodology for proteins renaturation. *Scientific Reports*, *7*(1), 1458. doi:10.103841598-017-01436-6 PMID:28469182

Voogt, D., & Brinkman, P. (1989). U. A. T. Production, properties and usage of polychlorinated biphenyls in Halogenated Biphenyls, Terphenyls, Naphthalenes, Dibenzodioxins, and Related Products. *Science*, 3–45.

Voragen, A. G. J. (1998). Technological aspects of functional food-related carbohydrates. *Trends in Food Science & Technology*, *9*(8-9), 328–335. doi:10.1016/S0924-2244(98)00059-4

Vorauer-Uhl, K., Fürnschlief, E., Wagner, A., Ferko, B., & Katinger, H. (2001). Topically applied liposome encapsulated superoxide dismutase reduces postburn wound size and edema formation. *European Journal of Pharmaceutical Sciences*, *14*(1), 63–67. doi:10.1016/S0928-0987(01)00149-X PMID:11457651

Vranish, J. N., Ancona, M. G., Oh, E., Susumu, K., & Medintz, I. L. (2017). Enhancing coupled enzymatic activity by conjugating one enzyme to a nanoparticle. *Nanoscale*, *9*(16), 5172–5187. doi:10.1039/C7NR00200A PMID:28393943

Wada, S., Ichikawa, H., & Tatsumi, K. (1993). Removal of phenols from wastewater by soluble and immobilized tyrosinase. *Biotechnology and Bioengineering*, *42*, 854-8. DOI: 10.1002/bit.260420710

Wada, A., Tamaru, S., Ikeda, M., & Hamachi, I. (2009). MCM-enzyme-supramolecular hydrogel hybrid as a fluorescence sensing material for polyanions of biological significance. *Journal of the American Chemical Society*, *131*(14), 5321–5330. doi:10.1021/ja900500j PMID:19351208

Wagner, J. R., Lee, C. T., Durrant, J. D., Malmstrom, R. D., Feher, V. A., & Amaro, R. E. (2016, June 8). Emerging computational methods for the rational discovery of allosteric drugs. *Chemical Reviews*, *116*(11), 6370–6390. doi:10.1021/acs.chemrev.5b00631 PMID:27074285

Wakerly, Z., Fell, J., Attwood, D., & Parkins, D. (1996). Pectin/ethylcellulose film coating formulations for colonic drug delivery. *Pharmaceutical Research*, *13*(8), 1210–1212. doi:10.1023/A:1016016404404 PMID:8865314

Wallerstein, L. (1939). Enzyme Preparations from Microorganisms COMMERCIAL PRODUCTION AND INUSTRIAL APPLICATION. *Industrial & Engineering Chemistry*, *31*(10), 1218–1224. doi:10.1021/ie50358a012

Walsh, G., & Walsh, G. (2015). Therapeutic Proteins: Blood Products, Vaccines and Enzymes. *Proteins*, 177–203.

Walther, F. J., David-Cu, R., & Lopez, S. L. (1995). Antioxidant-surfactant liposomes mitigate hyperoxic lung injury in premature rabbits. *The American Journal of Physiology*, *269*(5 Pt 1), L613–L617. PMID:7491979

Wanga, H. X., & Ng, T. B. (2004). Purification of a novel low-molecular-mass laccase with HIV-1 reverse transcriptase inhibitory activity from the mushroom *Tricholoma giganteum. Biochemical and Biophysical Research Communications*, *315*(2), 450–454. doi:10.1016/j.bbrc.2004.01.064 PMID:14766229

Wang, C. Y., Lee, M.-S., King, C. M., & Warner, P. O. (1980). Evidence for nitroaromatics as direct-acting mutagens of airborne particulates. *Chemosphere*, *9*(2), 83–87. doi:10.1016/0045-6535(80)90093-4

Wang, H., Yang, C., Wang, L., Kong, D., Zhang, Y., & Yang, Z. (2011). Self-assembled nanospheres as a novel delivery system for taxol: A molecular hydrogel with nanosphere morphology. *Chemical Communications*, *47*(15), 4439–4441. doi:10.1039/c1cc10506j PMID:21387062

Wang, J., Majima, N., Hirai, H., & Kawagishi, H. (2012). Effective Removal of Endocrine-Disrupting Compounds by Lignin Peroxidase from the White-Rot Fungus *Phanerochaete sordida* YK-624. *Current Microbiology*, *64*(3), 300–303. doi:10.100700284-011-0067-2 PMID:22203416

Wang, L., Waltenberger, B., Pferschy-Wenzig, E.-M., Blunder, M., Liu, X., Malainer, C., & Atanasov, A. G. (2014). Natural product agonists of peroxisome proliferator-activated receptor gamma (PPARγ): A review. *Biochemical Pharmacology*, *92*(1), 73–89. doi:10.1016/j.bcp.2014.07.018 PMID:25083916

Wang, L., Ying, G. G., Chen, F., Zhang, L. J., Zhao, J. L., Lai, H. J., ... Tao, R. (2012). Monitoring of selected estrogenic compounds and estrogenic activity in surface water and sediment of the Yellow River in China using combined chemical and biological tools. *Environmental Pollution*, *165*, 241–249. doi:10.1016/j.envpol.2011.10.005 PMID:22015335

Wang, L., Ying, G. G., Zhao, J. L., Liu, S., Yang, B., Zhou, L. J., ... Su, H. C. (2011). Assessing estrogenic activity in surface water and sediment of the Liao Riversystem in northeast China using combined chemical and biological tools. *Environmental Pollution*, *159*(1), 148–156. doi:10.1016/j.envpol.2010.09.017 PMID:20951481

Wang, R., Xu, D. I., Liang, L., Xu, T. T., Liu, W., Ouyang, P. K., ... Xu, H. (2016). Enzymatically crosslinked epsilon-poly-L-lysine hydrogels with inherent antibacterial properties for wound infection prevention. *RSC Advances*, *6*(11), 8620–8627. doi:10.1039/C5RA15616E

Wang, S. L., Yen, Y. H., Shih, I. L., Chang, A. C., Chang, W. T., Wu, W. C., & Chai, Y. D. (2003). Production of xylanases from rice bran by *Streptomyces actuosus* A-151. *Enzyme and Microbial Technology*, *33*(7), 917–925. doi:10.1016/S0141-0229(03)00246-1

Wang, X., Guo, W., Hu, Y., Wu, J., & Wei, H. (2016a). Introduction to Nanozymes. In *Nanozymes: Next Wave of Artificial Enzymes. Springer Briefs in Molecular Science*. Berlin: Springer. doi:10.1007/978-3-662-53068-9_1

Wang, X., Hua, Y., & Wei, H. (2016b). Nanozymes in bionanotechnology: From sensing to therapeutics and beyond. *Inorganic Chemistry Frontiers*, *3*(1), 41–60. doi:10.1039/C5QI00240K

Wang, Z., Chen, W., Cui, Z., He, K., & Yu, W. (2015). Studies on photoyellowing of silk fibroin and alteration of its tyrosine content. *Journal of the Textile Institute*, *2015*, 1–7.

Wan, H., Weng, S., Liang, S., Lu, Q., & He, J. (2011). Evaluation of The Developmental Toxicity of Leucomalachite Green Administered Orally to Rats. *Food and Chemical Toxicology*, *49*(12), 3031–3037. doi:10.1016/j.fct.2011.10.003 PMID:22005260

Wan, Y., Lu, R., Akiyama, K., Miyakoshi, T., & Du, Y. (2007). Enzymatic synthesis of bioactive compounds by *Rhus* laccase from Chinese *Rhus vernicifera*. *Science in China. Series B, Chemistry*, *50*(2), 179–182. doi:10.100711426-007-0022-z

Ward, O. P. (2012). Production of recombinant proteins by filamentous fungi. *Biotechnology Advances*, *30*(5), 1119–1139. doi:10.1016/j.biotechadv.2011.09.012 PMID:21968147

Watlington, K. (2005). *Emerging Nanotechnologies for Site Remediation and Wastewater Treatment*. Washington, DC: U.S. Environmental Protection Agency.

Watts, G. (2008). How clean is your water? *BMJ (Clinical Research Ed.)*, *337*(7661), a237. doi:10.1136/bmj.a237 PMID:18595920

Weber, R., Gaus, C., Tysklind, M., Johnston, P., Forter, M., Hollert, H., ... Zennegg, M. (2008). Dioxin- and POP contaminated sites-contemporary and future relevance and challenges: Overview on background aims and scope of the series. *Environmental Science and Pollution Research International*, *15*(5), 363–393. doi:10.100711356-008-0024-1 PMID:18597132

Wei, L., Zhang, W., Lu, H., & Yang, P. (2010). Immobilization of enzyme on detonation nanodiamond for highly efficient proteolysis. *Talanta*, *80*(3), 1298–1304. doi:10.1016/j.talanta.2009.09.029 PMID:20006091

Weinreb, N. J., Barranger, J. A., Charrow, J., Grabowski, G. A., Mankin, H. J., & Mistry, P. (2005). Guidance on the use of miglustat for treating patients with type 1 Gaucher disease. *American Journal of Hematology*, *80*(3), 223–229. doi:10.1002/ajh.20504 PMID:16247743

Wellington, K. W., & Kolesnikova, N. I. (2012). A laccase-catalysed one-pot synthesis of aminonaphthoquinones and their anticancer activity. *Bioorganic & Medicinal Chemistry*, *20*(14), 4472–4481. doi:10.1016/j.bmc.2012.05.028 PMID:22682920

Wellington, K. W., Qwebani-Ogunleye, T., Kolesnikova, N. I., Brady, D., & de Koning, C. B. (2013). One-pot laccase-catalysed synthesis of 5, 6-dihydroxylated benzo [b] furans and catechol derivatives, and their anticancer activity. *Archiv der Pharmazie*, *346*(4), 266–277. doi:10.1002/ardp.201200413 PMID:23447437

Wells, A., Teria, M., & Eve, T. (2006). Green oxidations with laccase–mediator systems. *Biochemical Society Transactions*, *34*(2), 304–308. doi:10.1042/BST0340304 PMID:16545100

Weng, S.-S., Liu, S.-M., & Lai, H.-T. (2013). Application parameters of laccase-mediator systems for treatment of sulfonamide antibiotics. *Bioresource Technology*, *141*, 152–159. doi:10.1016/j.biortech.2013.02.093 PMID:23561949

Wen, X., Jia, Y., & Li, J. (2009). Degradation of tetracycline and oxytetracycline by crude lignin peroxidase prepared from *Phanerochaete chrysosporium* – A white rot fungus. *Chemosphere*, *75*(8), 1003–1007. doi:10.1016/j.chemosphere.2009.01.052 PMID:19232429

Wesenberg, D., Kyriakides, I., & Agathos, S. N. (2003). White-rot fungi and their enzymes for the treatment of industrial dye effluents. *Biotechnology Advances*, *22*(2), 161–187. doi:10.1016/j.biotechadv.2003.08.011 PMID:14623049

Whitaker, J. R. (1984). Pectic substances, pectic enzymes and haze formation in fruit juices. *Enzyme and Microbial Technology*, *6*(8), 341–349. doi:10.1016/0141-0229(84)90046-2

Whiteley, C., & Lee, D. J. (2006). Enzyme Technology and Biological Remediation: A Review. *Enz.Microbial. Tech*, *38*(3-4), 291–316. doi:10.1016/j.enzmictec.2005.10.010

Williams, P. P. (1977). Metabolism of synthetic organic pesticides by anaerobic microorganisms. *Residue Reviews*, *66*, 63–135. doi:10.1007/978-1-4612-6352-4-3 PMID:335464

Wilson, N., & Shah, N. P. (2007). Microencapsulation of Vitamins. *ASEAN Food Journal*, *14*(1), 1–14.

Wilson, R., & Turner, A. (1992). Glucose oxidase: An ideal enzyme. *Biosensors & Bioelectronics*, *7*(3), 165–185. doi:10.1016/0956-5663(92)87013-F

Wiriyaphan, C., Chitsomboon, B., & Yongsawadigul, J. (2012). Antioxidant activity of protein hydrolysates derived from threadfin bream surimi byproducts. *Food Chemistry*, *132*(1), 104–111. doi:10.1016/j.foodchem.2011.10.040 PMID:26434269

Wise, A., Mallett, A. K., & Rowland, I. R. (1982). *Dietary protein and cecal microbial metabolism in the rat*. Academic Press.

Witayakran, S. (2008). *Laccase in organic synthesis and its applications* (Doctoral dissertation). Georgia Institute of Technology.

Witayakran, S., & Ragauskas, A. J. (2007). One-pot synthesis of 1, 4-naphthoquinones and related structures with laccase. *Green Chemistry*, *9*(5), 475–480. doi:10.1039/b606686k

Witschi, A., Reddy, S., Stofer, B., & Lauterburg, B. H. (1992). The systemic availability of oral glutathione. *European Journal of Clinical Pharmacology*, *43*(6), 667–669. doi:10.1007/BF02284971 PMID:1362956

Wong, K. K., & Saddler, J. N. (1993). Applications of hemicellulases in the food, feed, and pulp and paper industries. Hemicellulose and Hemicellulases, 127-143.

Wong, D. W. S. (2009). Structure and action mechanism of ligninolytic enzymes. *Applied Biochemistry and Biotechnology*, *157*(2), 174–209. doi:10.100712010-008-8279-z PMID:18581264

Wong, J. H., Ng, T. B., Jiang, Y., Liu, F., Sze, S. C., & Zhang, K. Y. (2010). Purification and characterization of a laccase with inhibitory activity toward HIV-1 reverse transcriptase and tumor cells from an edible mushroom (*Pleurotus cornucopiae*). *Protein and Peptide Letters*, *17*(8), 1040–1047. doi:10.2174/092986610791498966 PMID:19807674

Wong, K. K. Y., Tan, L. U. L., & Saddler, J. N. (1988). Multiplicity of α-1,4 xylanase in microorganisms functions and applications. *Microbiological Reviews*, *52*, 305–317. PMID:3141761

Wong, T. S., Zhurina, D., & Schwaneberg, U. (2006). The diversity challenge in directed protein evolution. *Combinatorial Chemistry & High Throughput Screening*, *9*(4), 271–288. doi:10.2174/138620706776843192 PMID:16724918

World Farmers Organisation. (2014). Retrieved from http://www.wfo-oma.com/news/global-production-of-fruits-and-vegetables-increased-by-9-4.html

Wu, X., Chou, N., Lupher, D., & Davis, L. C. (1998). Benzotriazoles: Toxicity and Degradation. *Proceedings of the 1998 Conference on Hazardous Waste Research*, 374-382.

Wu, D.-C., Teismann, P., Tieu, K., Vila, M., Jackson-Lewis, V., Ischiropoulos, H., & Przedborski, S. (2003). NADPH oxidase mediates oxidative stress in the 1-methyl-4-phenyl-1, 2, 3, 6-tetrahydropyridine model of Parkinson's disease. *Proceedings of the National Academy of Sciences of the United States of America*, *100*(10), 6145–6150. doi:10.1073/pnas.0937239100 PMID:12721370

Wu, H. C., Chen, H. M., & Shiau, C. Y. (2003). Free amino acids and peptides as related to antioxidant properties in protein hydrolysates of mackerel (*Scomber austriasicus*). *Food Research International*, *36*(9-10), 949–957. doi:10.1016/S0963-9969(03)00104-2

Wu, S. C., Wen, T. N., & Pan, C. L. (2005). Agar oligosaccharides lysates prepared by two bacterial agarases stepwise hydrolyzed and their anti-oxidative properties. *Fisheries Science*, *71*(5), 1149–1159. doi:10.1111/j.1444-2906.2005.01075.x

Wu, X. C., Zhang, Y., Han, T., Wu, H. X., Guo, S. W., & Zhang, J. Y. (2014). Composite of graphene quantum dots and Fe3O4 nanoparticles: Peroxidase activity and application in phenolic compound removal. *RSC Advances*, *4*(7), 3299–3305. doi:10.1039/C3RA44709J

Wyss, M., Brugger, R., Kronenberger, A., Remy, R., Fimbel, R., Oesterhelt, G., & … van Loon, A.P.G.M. (1999). Biochemical Characterization of Fungal Phytases (myo-Inositol Hexakisphosphate Phosphohydrolases): Catalytic properties. *Applied and Environmental Microbiology*, *65*(2), 367–373. PMID:9925555

Xue, M., & Zink, J. I. (2013). Enzymatic Chemical Amplifier Based on Mechanized Nanoparticles. *Journal of the American Chemical Society*, *135*(47), 17659–17662. doi:10.1021/ja4066317 PMID:24224766

Xu, F. (1996). Oxidation of phenols, anilines, and benzenethiols by fungal laccases: Correlation between activity and redox potentials as well as halide inhibition. *Biochemistry*, *35*(23), 7608–7614. doi:10.1021/bi952971a PMID:8652543

Xu, F. (1997). Effects of redox potential and hydroxide inhibition on the pH activity profile of fungal laccases. *The Journal of Biological Chemistry*, *272*(2), 924–928. doi:10.1074/jbc.272.2.924 PMID:8995383

Xu, F., Berka, R. M., Wahleithner, J. A., Nelson, B. A., Shuster, J. R., Brown, S. H., & Solomon, E. I. (1998). Site-directed mutations in fungal laccase: Effect on redox potential, activity and pH profile. *The Biochemical Journal*, *334*(1), 63–70. doi:10.1042/bj3340063 PMID:9693103

Xu, F., Kulys, J. J., Duke, K., Li, K., Krikstopaitis, K., Deussen, H. J. W., ... Schneider, P. (2000). Redox chemistry in laccase-catalyzed oxidation of N-hydroxy compounds. *Applied and Environmental Microbiology*, *66*(5), 2052–2056. doi:10.1128/AEM.66.5.2052-2056.2000 PMID:10788380

Xu, J., Li, X., & Sun, F. (2010). Cyclodextrin-containing hydrogels for contact lenses as a platform for drug incorporation and release. *Acta Biomaterialia*, *6*(2), 486–493. doi:10.1016/j.actbio.2009.07.021 PMID:19619677

Xu, R., Tang, R., Zhou, Q., Li, F., & Zhang, B. (2015). Enhancement of catalytic activity of immobilized laccase for diclofenac biodegradation by carbon nanotubes. *Chemical Engineering Journal*, *262*, 88–95. doi:10.1016/j.cej.2014.09.072

Xu, Y. M., Li, Y. L., Xu, S. C., Liu, Y., Wang, X., & Tang, J. W. (2008). Improvement of xylanase production by *Aspergillus niger* XY-1 using respond surface methodology for optimizing the medium composition. *Journal of Zhejiang University. Science*, *9*(7), 558–566. doi:10.1631/jzus.B0820038 PMID:18600786

Yadav, M., Singh, A., & Singh, R. K. (2017). Docking studies on novel bisphenylbenzimidazoles (BPBIs) as non-nucleoside inhibitors of HIV-1 reverse transcriptase. *Indian Journal of Chemistry B*, *56*(7), 714-723.

Yadav, A., Amit, A., Chaudhary, R., Chandel, A. S., Mahantesh, V., Suman, S. S., & Pandey, K. (2015). Leishmania donovani: Impairment of the cellular immune response against recombinant ornithine decarboxylase protein as a possible evasion strategy of Leishmania in visceral leishmaniasis. *International Journal for Parasitology*, *45*(1), 33–42. doi:10.1016/j.ijpara.2014.08.013 PMID:25449949

Yagi, K. (2006). *Mechanism of Enzyme Action-an Approach Through the Study of Slow Reactions* (pp. 1–36). John Wiley & Sons, Inc.

Yang, J., Li, W., Ng, T. B., Deng, X., Lin, J., & Ye, X. (2017). Laccases: Production, expression regulation, and applications in pharmaceutical biodegradation. *Frontiers in Microbiology*, *8*, 832. doi:10.3389/fmicb.2017.00832 PMID:28559880

Yang, L. P., & Keating, G. M. (2009). Fenofibric acid: In combination therapy in the treatment of mixed dyslipidemia. *American Journal of Cardiovascular Drugs*, *9*(6), 401–409. doi:10.2165/11203920-000000000-00000 PMID:19929038

Yang, L., Yu, L. E., & Ray, M. B. (2008). Degradation of paracetamol in aqueous solutions by TiO_2 photocatalysis. *Water Research*, *42*(13), 3480–3488. doi:10.1016/j.watres.2008.04.023 PMID:18519147

Yang, R., Xu, S., Wang, Z., & Yang, W. (2004). Aqueous extraction of corncob xylan and production of xylooligosacchrids. *Lebensmittel-Wissenschaft + Technologie*, *38*(6), 677–682. doi:10.1016/j.lwt.2004.07.023

Yang, S. Q., Yan, Q. J., Jiang, Z. Q., Li, L. T., Tian, H. M., & Wang, Y. Z. (2006). High-level of xylanase production by the thermophilic *Paecilomyces thermophila* J18 on wheat straw in solid-state fermentation. *Bioresource Technology*, *97*(15), 1794–1800. doi:10.1016/j.biortech.2005.09.007 PMID:16230011

Yang, S., Hai, F. I., Nghiem, L. D., Nguyen, L. N., Roddick, F., & Price, W. E. (2013). Removal of bisphenol A and diclofenac by a novel fungal membrane bioreactor operated under non-sterile conditions. *International Biodeterioration & Biodegradation*, *85*, 483–490. doi:10.1016/j.ibiod.2013.03.012

Yang, S.-S., & Wang, J.-Y. (1999). Protease and amylase production of Streptomyces rimosus in submerged and solid state cultivations. *Botanical Bulletin of Academia Sinica*, 40.

Yang, Z., Kevin, P., Molino, B. F., Haydar, S. N., Yeyu, C., Bois, F., ... Michels, P. C. (2009). Novel Oxidation of Cyclosporin A: Preparation of Cyclosporin Methyl Vinyl Ketone (Cs-MVK). *Synlett*, *18*, 2935–2938.

Yang, Z., Liang, G., Wang, L., & Xu, B. (2006). Using a kinase/phosphatase switch to regulate a supramolecular hydrogel and forming the supramolecular hydrogel in vivo. *Journal of the American Chemical Society*, *128*(9), 3038–3043. doi:10.1021/ja057412y PMID:16506785

Yaropolov, A. I., Skorobogatko, O. V., Vartanov, S. S., & Varfolomeyev, S. D. (1994). Laccase: Properties, catalytic mechanism and applicability. *Applied Biochemistry and Biotechnology*, *49*(3), 257–280. doi:10.1007/BF02783061

Yaver, D. S., Xu, F., Golightly, E. J., Brown, K. M., Brown, S. H., Rey, M. W., ... Dalboge, H. (1996). Purification, characterization, molecular cloning, and expression of two laccase genes from the white rot basidiomycete *Trametes villosa. Applied and Environmental Microbiology*, *62*, 834–841. PMID:8975613

Yildiz, H., Ozyilmaz, E., Bhatti, A. A., & Yilmaz, M. (2017). Enantioselective resolution of racemic flurbiprofen methyl ester by lipase encapsulated mercapto calix[4]arenes capped Fe_3O_4 nanoparticles. *Bioprocess and Biosystems Engineering*, *40*(8), 1189–1196. doi:10.100700449-017-1779-x PMID:28488138

Yin, W., Su, R., Qi, W., & He, Z. (2012). A casein-polysaccharide hybrid hydrogel cross-linked by transglutaminase for drug delivery. *Journal of Materials Science*, *47*(4), 2045–2055. doi:10.100710853-011-6005-7

Ylikahri, R. (1979). Metabolic and nutritional aspects of xylitol. *Advances in Food Research*, *25*, 159–180. doi:10.1016/S0065-2628(08)60237-2 PMID:391001

Yoon, M. Y. (1998). *Process for improved shrink resistance in wool.* WO Patent 98/27264.

Yoshida, H. (1883). LXIII.—chemistry of lacquer (Urushi). Part I. communication from the chemical society of Tokyo. *Journal of the Chemical Society Transactions*, *43*(0), 472–486. doi:10.1039/CT8834300472

Yoshikuni, Y., & Keasling, J. D. (2007). Pathway engineering by designed divergent evolution. *Current Opinion in Chemical Biology*, *11*(2), 233–239. doi:10.1016/j.cbpa.2007.02.033 PMID:17353138

Yoshitake, A., Katayama, Y., Nakamura, M., Iimura, Y., Kawai, S., & Morohoshi, N. (1993). N-linked carbohydrate chains protect laccase III from proteolysis in *Coriolus versicolor. Microbiology*, *139*, 179–185.

You, C.-C., Agasti, S. S., De, M., Knapp, M. J., & Rotello, V. M. (2006). Modulation of the catalytic behavior of r-chymotrypsin at monolayer-protected nanoparticle surfaces. *Journal of the American Chemical Society*, *128*(45), 14612–14618. doi:10.1021/ja064433z PMID:17090046

Yousefi-Ahmadipour, A., Bozorgi-Koshalshahi, M., Mogharabi, M., Amini, M., Ghazi-Khansari, M., & Faramarzi, M. A. (2016). Laccase-catalyzed treatment of ketoconazole, identification of biotransformed metabolites, determination of kinetic parameters, and evaluation of micro-toxicity. *Journal of Molecular Catalysis. B, Enzymatic*, *133*, 77–84. doi:10.1016/j.molcatb.2016.07.015

Yousefinejad, S., Rasti, H., Hajebi, M., Kowsari, M., Sadravi, S., & Honarasa, F. (2017). Design of C-dots/Fe_3O_4 magnetic nanocomposite as an efficient new nanozyme and its application for determination of H_2O_2 in nanomolar level. *Sensors and Actuators. B, Chemical*, *247*, 691–696. doi:10.1016/j.snb.2017.02.145

Yuan, Q. P., Wang, J. D., Huai, Z., & Qian, Z. M. (2005). Effect of temperature shift on production of xylanase by *Aspergillus niger. Process Biochemistry*, *40*(10), 3255–3257. doi:10.1016/j.procbio.2005.03.020

Yung, C., Wu, L., Tullman, J., Payne, G., Bentley, W., & Barbari, T. (2007). Transglutaminasecrosslinked gelatin as a tissue engineering scaffold. *Journal of Biomedical Materials Research. Part A*, *83*(4), 1039–1046. doi:10.1002/jbm.a.31431 PMID:17584898

Zaldivar, J., Nielsen, J., & Olsson, L. (2001). Fuel Ethanol Production from Lignocellulose: A Challenge for Metabolic Engineering and Process Integration. *Applied Microbiology and Biotechnology*, *56*(1-2), 17–34. doi:10.1007002530100624 PMID:11499926

Zambare, V. (2010). Solid state fermentation of *Aspergillus oryzae* for glucoamylase production on agro residues. *International Journal of Life Sciences (Kathmandu)*, *4*(0), 16–25. doi:10.3126/ijls.v4i0.2892

Zamorani, A. (1989). Enzymatic processing of musts and wines. In Biotechnology applications in beverage production. Elsevier Applied Science. doi:10.1007/978-94-009-1113-0_14

Zamorani, A., Spettoli, P., Lante, A., Crapisi, A., & Pasini, G. (1993). Immobilized laccase and tyrosinase: An approach for wine stabilization. *Italian Journal of Food Science*, *4*, 409–414.

Zarafeta, D., Kissas, D., Sayer, C., Gudbergsdottir, S. R., Ladoukakis, E., Isupov, M. N., ... Kolisis, F. N. (2016). Discovery and characterization of a thermostable and highly halotolerant GH5 Cellulase from an Icelandic hot spring isolate. *PLoS One*, *11*(1), e0146454. doi:10.1371/journal.pone.0146454 PMID:26741138

Zavada, S., McHardy, N., & Scott, T. (2014). Oxygen-mediated enzymatic polymerization of thiol–ene hydrogels. *Journal of Materials Chemistry. B, Materials for Biology and Medicine*, *2*(17), 2598–2605. doi:10.1039/C3TB21794A PMID:24995128

Zeinvand-Lorestani, H., Sabzevari, O., Setayesh, N., Amini, M., Nili-Ahmadabadi, A., & Faramarzi, M. A. (2015). Comparative study of in vitro prooxidative properties and genotoxicity induced by aflatoxin B1 and its laccase-mediated detoxification products. *Chemosphere*, *135*, 1–6. doi:10.1016/j.chemosphere.2015.03.036 PMID:25876029

Zeng, J., & Zheng, L. Y. (2002). Studies on *Penicillium* sp. ZDZ1 chitosanase immobilized on chitin by cross-linking reaction. *Process Biochemistry*, *38*(4), 531–535. doi:10.1016/S0032-9592(02)00163-2

Zeng, Z., She, Y., Peng, Z., Wei, J., & He, X. (2016). Enzyme-mediated in situ formation of pH-sensitive nanogels for proteins delivery. *RSC Advances*, *6*(10), 8032–8042. doi:10.1039/C5RA25133H

Zhang, Q., Su, Y., & Zhang, J. (2013). Seasonal difference in antioxidant capacity and active compounds contents of *Eucommia ulmoides* Oliver leaf. *Molecules (Basel, Switzerland)*, *18*(12), 1857–1868. doi:10.3390/molecules18021857 PMID:23377129

Zhang, Q., Xu, R., Xu, P., Chen, R., He, Q., Zhong, J., & Gu, X. (2014a). Performance study of ZrO2 ceramic microfiltration membranes used in pretreatment of DMF wastewater. *Desalination*, *346*, 1–8. doi:10.1016/j.desal.2014.05.006

Zhang, X., Liu, M., Li, Y., Dong, Y., Pingguan-Murphy, B., Lu, T. J., & Xu, F. (2015). Engineering cell microenvironment using novel functional hydrogels. *European Polymer Journal*, *72*, 590–601. doi:10.1016/j.eurpolymj.2015.03.019

Zhang, X., Zhao, J., Wen, Y., Zhu, C., Yang, J., & Yao, F. (2013). Carboxymethyl chitosan-poly(amidoamine) dendrimer core–shell nanoparticles for intracellular lysozyme delivery. *Carbohydrate Polymers*, *98*(2), 1326–1334. doi:10.1016/j.carbpol.2013.08.005 PMID:24053810

Zhang, Y., & Geißen, S. U. (2012). Elimination of carbamazepine in a non-sterile fungal bioreactor. *Bioresource Technology*, *112*, 221–227. doi:10.1016/j.biortech.2012.02.073 PMID:22409982

Zhang, Y., & Geissen, S.-U. (2010). In vitro degradation of carbamazepine and diclofenac by crude lignin peroxidase. *Journal of Hazardous Materials*, *176*(1-3), 1089–1092. doi:10.1016/j.jhazmat.2009.10.133 PMID:19945218

Zhang, Y., Geissen, S.-U., & Gal, C. (2008). Carbamazepine and diclofenac: Removal in wastewater treatment plants and occurrence in water bodies. *Chemosphere*, *73*(8), 1151–1161. doi:10.1016/j.chemosphere.2008.07.086 PMID:18793791

Zhang, Y., Yan, L., Xu, W., Guo, X., Cui, L., Gao, L., ... Du, B. (2014b). Adsorption of Pb (II) and Hg (II) from aqueous solution using magnetic CoFe2O4 reduced graphene oxide. *Journal of Molecular Liquids*, *191*, 177–182. doi:10.1016/j.molliq.2013.12.015

Zhang, Y., Zhou, X., Ji, L., & Che, F. (2017). Enzymatic single state preparation and anti-oxidant activity of hetero-chitooligosaccharides using non- pretreated housefly larvae powder. *Carbohydrate Polymers*, *172*, 113–119. doi:10.1016/j.carbpol.2017.05.037 PMID:28606517

Zhang, Z., Zhang, X., Liu, B., & Liu, J. (2017). Molecular imprinting on inorganic nanozymes for hundred-fold enzyme specificity. *Journal of the American Chemical Society*, *139*(15), 5412–5419. doi:10.1021/jacs.7b00601 PMID:28345903

Zhao, J., & Kwan, H. S. (1999). Characterization, molecular cloning, and differential expression analysis of laccase genes from the edible mushroom *Lentinula edodes*. *Applied and Environmental Microbiology*, *65*, 4908–4913. PMID:10543802

Zhao, Z., Fu, J., Dhakal, S., Johnson-Buck, A., Liu, M., Zhang, T., ... Yan, H. (2016). Nanocaged enzymes with enhanced catalytic activity and increased stability against protease digestion. *Nature Communications*, *7*, 10619. doi:10.1038/ncomms10619 PMID:26861509

Zhong, M., Sun, J., Wei, D., Zhu, Y., Guo, L., Wei, Q., ... Zhang, X. (2014). Establishing a cell-affinitive interface and spreading space in a 3D hydrogel by introduction of microcarriers and an enzyme. *Journal of Materials Chemistry. B, Materials for Biology and Medicine*, *2*(38), 6601–6610. doi:10.1039/C4TB00887A

Zhou, Z., Lin, X., & Madura, J. D. (2006). HIV-1 RT nonnucleoside inhibitors and their interaction with RT for antiviral drug development. *Infectious Disorders-Drug Targets, 6*(4), 391-413.

Zhou, Z., & Hartmann, M. (2013). Progress in enzyme immobilization in ordered mesoporous materials and related applications. *Chemical Society Reviews*, *42*(9), 3894–3912. doi:10.1039/c3cs60059a PMID:23570038

Zimran, A., Brill-Almon, E., Chertkoff, R., Petakov, M., Blanco-Favela, F., Munoz, E. T., & Aviezer, D. (2011). Pivotal trial with plant cell-expressed recombinant glucocerebrosidase, taliglucerase alfa, a novel enzyme replacement therapy for Gaucher disease. *Blood*, *118*(22), 5767–5773. doi:10.1182/blood-2011-07-366955 PMID:21900191

Zulfiqar, A., Masood, S. B., Anjum, F. M., & Asgher, M. (2009). Effect of wheat bran concentration on xylanase biosynthesis by *Aspergillus niger. International Journal of Agriculture and Biology*, *11*, 571–576.

Zwick, E., Bange, J., & Ullrich, A. (2002). Receptor tyrosine kinases as targets for anticancer drugs. *Trends in Molecular Medicine*, *8*(1), 17–23. doi:10.1016/S1471-4914(01)02217-1 PMID:11796262

Zydney, A. L. (2016). Continuous downstream processing for high value biological products: A review. *Biotechnology and Bioengineering*, *113*(3), 465–475. doi:10.1002/bit.25695 PMID:26153056

About the Contributors

Shashi Lata Bharati is presently working as Assistant Professor in the Chemistry Department of "North Eastern Regional Institute of Science and Technology, Arunachal Pradesh, India". She has PhD degree in the subject chemistry. She has been awarded UGC-DSA (New Delhi) fellowship for meritorious student during her research period. She was further awarded by UGC as Post Doctoral Fellow for women. She has published more than 15 research papers and one book in journal/publishers of repute. She has also published one chapter. She has also worked as reviewer. She has expertise in the field of inorganic complex synthesis, organo-metallic chemistry, bio-inorganic chemistry and enzymology.

Pankaj Kumar Chaurasia is presently working as "Assistant Professor" in PG Department of Chemistry, LS College Muzaffarpur, India (Under BRA Bihar University, Bihar). He has also taught in "Central University of Allahabad" as "Guest Faculty" during 2016-2017. He has been awarded by different research fellowship like CSIR-JRF (NET), CSIR-SRF (NET) and CSIR-RA. He has published 22 research papers in journals of repute and 03 book chapter in books of international publishers. He has also worked as reviewer and authored an international book. His research areas of expertise are enzymology, bioorganic chemistry, enzymatic organic synthesis, organo-metallic chemistry, biochemistry and biotechnology.

* * *

Doaa Awad is a Ph.D. Student at University of Bremen, Bremen, Germany, 2003 - 2006. Postdoctoral Fellow at University of Bremen, Bremen, Germany 2009 - 2011. Postdoctoral Fellow at Jacobs University, Bremen, Germany 2015 - 2016. Assistant Professor at Alexandria University, Faculty of Science, Biochemistry Department, Alexandria, Egypt, 2014-now.

Mahmoud Balbaa has received his PhD in Hokkaido University, Japan during the period of 1984-1988. Currently, he is working as professor of biochemistry in Alexandria University, Egypt. He was appointed as head of Biochemistry Department, Alexandria University, Egypt from 2007 to 2009. His research has included the study of enzyme characterization, enzyme inhibition, cell signaling and the biochemical parameters in diseases. Based on this research and fellowship training, he has received several awards and honors, such as post-doctoral fellowship from Medical Research Council, Canada, and post-doctoral fellowship from AIEJ, Japan, the prize of Scientific Encouragement of Alexandria University and the prize of Graduate Studies of Hokkaido University School of medicine. He is serving as an editorial member of several reputed journals and expert Reviewer for many journals. He have authorized more than 50 peer-reviewed research articles. He is a member of Egyptian Biochemical Society and Egyptian Society of Clinical Chemistry.

Mezajoug Kenfack Laurette Blandine is an academic researcher in food unit departement of University Institute of Technology. She works on protein extraction and concentration, nutritional evaluation of some local wild crops.

Arvind Kumar Singh Chandel has done his graduation (B. Pharm.) from the department of Pharmacy Barkatullah University Bhopal and post graduation (M.S. Pharm.) from National Institute of Pharmaceutical Education and Research (NIPER) Hajipur, He has awarded Four times GPAT (AICTE New Delhi) and CSIR-JRF-GATE and CSIR-SRF-GATE from the CSIR-HRDG, Currently pursuing his PhD from CSIR CSMCRI.He has published 9 research papers and currently he is working on Design and Synthesis of amphiphilic gels for biomedical applications.

Mohammad Ali Faramarzi got his Pharm D (1995) and PhD (2002) from Tehran University of Medical Sciences, Tehran, Iran. This was followed by postdoctoral studies in Environmental Biotechnology at University of Zürich, Zürich, Switzerland. Currently, he is the Professor of Pharmaceutical Biotechnology at Tehran University of Medical Sciences. His major research interest is discovery, characterization, and applications of biocatalysts that have potential uses for biotechnological purposes using techniques such as organic synthesis, biochemistry, microbiology, and industrial enzymology. His specific interests include enzymatic reactions in the synthesis of pharmacological active compounds using laccase, lipase, chitinase, peroxidase, etc.

Vijay Kumar Garlapati is working as an Assistant Professor in Biotechnology and Bioinformatics dept. of the Jaypee University of Information Technology, HP, India. Dr.Garlapati holds a Ph.D. from IIT Kharagpur, India and M.Tech (Biotechnology) from Andhra University, India. He has done his Post Doctoral study at Universitat Autonoma de Barcelona, Spain. His research interests include Bioprocess engineering / Industrial Biotechnology, Fermentation technology, Biocatalysis and Microbial Biotechnology. He has been credited with Filing of one International & three National patents, publishing of 33 research articles in peer-reviewed International Journals, 10 Book chapters, guided one Ph.D. student and has communicated/presented at several conferences either as oral presentations or posters. He is a recipient of A4U Postdoctoral Fellowship, Spain, 2011-2013; Institute Research Fellowship, IIT Kharagpur, 2006-2010; GATE - MHRD Fellowship, 2002-2003.

Akshaya Gupte attained his M.Sc. and Ph. D degrees from Sardar Patel University. At present working as the Head of the Dept. of Microbiology since 2003 at Natubhai V Patel College of Pure and Applied Sciences. He is a recipient of the Globol Impact of Applied Microbiologist Fellowship for three years and also a DAAD Fellowship. He visited University of Gottingen, Germany and University of Lund, Sweden during the year 1995-1996. He has published more than 45 research papers in reputed journals. Eight students have obtained Doctoral degree under his supervision. He has to his credit seven projects completed from various Govt. of India funding agencies such as DBT and UGC.

Hooi Ling Ho was graduated with B.Sc. (Hons) Microbial Technology & Bioprocessing and M.Sc. Fermentation Technology & Bioprocess Engineering from University Putra Malaysia. In 2011, she obtained PhD from Monash University (Australia) in Biochemistry & Molecular Cell Biology. She has been involved in teaching and research of xylanase production and molecular studies of protein kinases. She is also an academic writer, reviewer and proofreader for national and international peer review journals.

K. Veena Gayathri, assistant professor, is from Department of Biotechnology at Stella Maris College, Chennai(India). Her main area of research is"Biodegradation of Recalcitrant Compounds using Extremophilic Organisms She completed her PhD from Centre for Environmental Studies, Anna University, Chennai, during her research period, she was awarded the "UGC- meritorious student in science fellowship". She had established a method for degradation of chlorophenols by a halotolerant bacterial strain, for which she was granted a patent at the national level by Intellectual Property Rights, India on 27th April 2017. She had received UGC-minor research project on October 2016. She has presented her research findings in 27 international peer reviewed journals and in various national and international conferences. She has received 7 best paper awards and has guided around 30 M.sc graduate projects in different aspects of environmental and food biotechnology.

Nafisa Lakhdawala has studied B.Tech in Food from Techno India, Salt Lake, Kolkata. She loves writing and has an eye for detail. Later, she underwent Lead Auditor Training in SGS. She was also a part of the editing team for her departmental magazine in college. She loves physics & chemistry and enjoys reading. She is extremely passionate about teaching and has been teaching chemistry part-time at a tutorial for three years now.

Subir Kumar Nandy is a Chemical Engineer (Ph.D. from IIT Bombay) having professional experiences in the biotechnology field. Subir has gained a deep interest in Microbiology, Synthetic Biology, Metabolic Engineering, Fermentation optimization through more than ten years of experiences together in academia and industry. Subir holds a broad range of research competencies and the ability to combine an innovative mindset with a systematic and result driven approach.

Michael Bamitale Osho bagged his Ph.D. (April 2013) in Food & Industrial Microbiology from the Federal University of Agriculture, Abeokuta, Nigeria. A seasoned Lecturer in the Department of Biological Sciences, McPherson University, Seriki Sotayo, Nigeria. Also, an associate Lecturer in the Department of Microbiology, Olabisi Onabanjo University, Ago-Iwoye, Nigeria (2015 – Till date). He has supervised many undergraduate studentsand three M.Sc. students. He received a special grant from Federal Ministry of Education, Abuja under STEP-B project in 2012 during his doctoral programme He was Acting. Head of Department, Biological Sciences (March, 2014- Sept. 2016); Senate Member, McPherson University (March, 2014- Till date). He is a member of many academic societies. Dr. Osho has over fifteen scholarly articles published to his credit in national and international journals of high repute. His areas of specialization as an Industrial Microbiology and Biotechnology include immobilization of cells on novel matrix from organic materials for enzyme production, production of second generation biofuels, biosynthesis of nanoparticles from microbes and their applications.

Helina Patel received her B.Sc. degree from Veer Narmad South Gujarat University, Surat, Gujarat, in 2011 and M.Sc degree from Sardar Patel University, Vallabh Vidyanagar, Gujarat, in 2013. She is currently a Ph.D. scholar in the Department of Microbiology at Natubhai V. Patel College of Pure and Applied Sciences, Vallabh Vidyanagar. Her research work focuses on production and characterization of laccase enzyme for its application in organic synthesis, conducted under the supervision of Dr. Akshaya Gupte.

Ishan H. Raval has done MSc Biotechnology from DAVV Indore got (Gold Medal) and DST INSPIRE Fellowship for PhD. He is doing his PhD from CSMCRI Bhavnagar on Microbiology and Molecular Biology field. He has published number of International Paper in peer reviewed journals.

Anuradha Singh, a PhD graduate from the University of Allahabad, is working as Principal Investigator in a Department of Science and Technology (DST) funded research project at the University of Allahabad. She is well versed in synthetic chemistry, computer-aided drug design and other techniques used in the field of medicinal chemistry. Her research findings have been published in journals of international repute. She has been awarded with gold medal during her post graduate studies and Swarnajayanti Fellowships by National Academy of Sciences, India.

Mohan Prasad Singh is presently working as Professor of Biotechnology at University of Allahabad (India). He is also heading Centre of Bioinformatics as Coordinator at University of Allahabad (India). He did M.Sc., M. Phil. and Ph.D. from Panjab University, Chandigarh, India. He also qualified prestigious NET and JRF (Junior Research Fellow) examination. He joined as Lecturer (Assistant Professor) in HNB Garhwal University in Botany in 1992. In 2000 he moved to the Department of Biotechnology at VBS Purvanchal University, Jaunpur where he became Reader (Associate Professor) in 2000 and Professor in 2008. Prof. Singh has been carrying out his research on various aspects of edible oyster mushroom over last two decades. He has improved production and nutritional content of mushroom and suggested the effective way to get rid of environmental problems through mushroom. He has published 47 research papers, edited four books and has been serving as an editorial board member of repute. His research area is Mushroom nutraceuticals, biodegradation & bioremediation, molecular biology.

Ramendra K. Singh, Professor of Chemistry, is Director of Bioorganic Research Laboratory in the University of Allahabad, India. He has about 30 years of experience in the field of nucleic acids research and more than a decade in the field of antiretroviral research with wide range of training in various national and international institutes. His research interests also include developing natural compound curcumin as anti-cancer agent and fluorescently labeled nucleosides/ oligonucleotides as markers in drug delivery. He is member of several national bodies, editor to different international journals and referee to more than a dozen national/ international journals. He has been recipient of various awards/fellowships, like, Young Scientists' Award, India; XVI IUBMB Fellowship, India; Jawaharlal Nehru Visiting Fellowship, India; UNESCO Fellowship, Japan; Post-Doc Fellowship, Japan; INSA International Exchange Fellowship, Poland and Fulbright Fellowship, USA. Dr. Singh has published more than 70 research papers/ book chapters, in addition to several articles, in journals/books of national and international repute.

Sujata Sinha is working as Scientist at IIT Delhi. She has done M Tech (Gold Medalist) in Biochemical Engg from IIT BHU and PhD in Life Sciences. She has varied work experience in the field of nanotech, biocatalysis and Molecular Biology and also in grant management. She has got number of manuscript published out of these research experience in international peer reviewed journals of repute and has been cited by various authors working in the same field. She has got various fellowships and travel grant from DST.

M. Laura Soriano received her Ph.D. degree from the Universidad de Castilla La-Mancha (UCLM, Spain) in 2007. Then, she worked in the chemistry department at Durham University (England 2007-2009) and in the pharmaceutical department at the University of Trieste (Italy, 2009-2011) as Postdoctoral Fellows. In 2012, she joined the analytical department at Córdoba University. Dr. Soriano has published an ebook, diverse book chapter, 30 articles and two front covers in high impact factor international peer-reviewed journals in diverse categories of chemistry: her h-index is 13 and the total citations are over 500. Her actual research interests lie on fabricating nanostructures and gel systems for analytical and pharmaceutical purposes.

Index

A

Abrus precatorius 326-327
Adverse 1-2, 4-5, 7-12, 14, 17, 24, 26, 30-32, 34-41, 47, 49-50, 52-53, 56, 58-60, 131, 136, 179, 352, 410
agricultural extracts 294-296, 304-307, 309, 312
AIDS 14, 16, 54-55, 100-102, 123, 151, 153, 299-300, 377
alpha-glucosidases 375
Antibiotics 55, 91-92, 179-180, 184, 187-188, 215, 219, 221-222, 267, 279-280
anticancerous 253
antigenicity 1-2, 29, 46-47, 49
Anti-Inflammatory 46-47, 55-56, 142, 149, 179, 184-185, 226, 229, 250, 278, 282, 353, 358, 361
antineoplasic 46
Antioxidants 128, 134, 204, 253, 266, 352, 359
Aspergillus spp 294-295, 301, 306, 310-312

B

bacterial enzymes 71-73, 77, 89-91, 93
Bioactive 178-179, 184-185, 187-189, 204, 304, 352-354, 357-358, 360-361, 363-364, 371, 431, 434
Bioactive Molecules 187, 204, 352, 357, 364, 434
Biocatalyst 178-180, 204, 222, 232, 255, 265, 341
Biochemical 62, 119, 123, 141, 178, 204, 264, 341, 343, 352, 356-357, 376, 399-400
Biodegradation 183, 214, 228, 232, 250, 279, 282-283, 285-286, 397, 400-402, 410, 427
Bioengineering 46-47, 351, 354, 381
Bioprocess 279, 341-342, 346, 348, 354, 376, 408
Bioreactor 224-226, 228, 232-233, 308, 346, 351, 364, 376, 398
Bioremediation 180, 182, 214, 216, 219, 221, 224, 229-230, 234, 250, 265, 279, 283-284, 300, 397-398, 401, 404, 426-427, 434
Biotechnology 1-2, 17, 23, 190, 294, 298, 305, 337, 341-342, 375-376, 383, 397, 422
Biotransformation 13, 60, 183, 204, 214, 231, 250, 279, 402
Blood 2, 5, 10-11, 14, 23-27, 29-31, 33, 37, 40, 49, 54, 56-57, 59-61, 105, 128-129, 134, 144, 147-149, 151-153, 169, 171, 185, 354, 357-358
blue oxidase 253
boron 131-133, 136
Burnatia enneandra 327

C

Cancer 24, 47-51, 58-61, 72, 91, 109, 129-130, 142, 145-147, 151, 153, 179, 185-186, 266, 424
capture therapy 131, 136
Catalyst 165, 170, 184, 204, 254, 263, 283, 341, 343, 355, 400, 424, 427
Catalytic Activity 26, 28-29, 62, 182, 364, 407, 421-425, 427, 429, 433-434
Characterization 29, 264, 296, 327, 346, 351, 410
CLEA (Cross-Linked Enzyme Aggregates) 355, 371
Cloning 1-3, 7, 25, 152, 346, 351, 355, 383
colonic microflora 71-72, 92
Conventional 39, 59, 76, 87-88, 151, 161, 168, 224, 227, 254, 257, 261-263, 279, 284, 396, 408-410
Coupling 162, 178, 183-184, 186-188, 210, 217, 262-264, 267, 400, 407, 426
Cross-Linked Enzyme Aggregate 250, 355

D

Deficiency 1-3, 6-10, 12-13, 38-41, 50, 77, 100, 144, 151
Delivery 17, 46-47, 51, 54, 57, 61, 71-73, 77-78, 82, 88, 92-93, 113, 128-129, 133-136, 166, 168-171, 406-407, 423, 428-429, 433-434
Delivery Systems 71-73, 77, 82, 88, 93, 129, 168-170
Detoxification 47, 61-62, 189, 213, 221, 223, 227, 229, 231-233, 250, 397, 406
Diseases 1-3, 17, 23-25, 28, 33, 35, 41, 46-47, 51-52, 56, 58, 61, 109, 141-142, 144, 148, 150, 179,

184, 219, 266-267
downstream processing 304, 351-352, 356, 363-364, 371
Drugs 14, 17, 24-25, 41, 46-47, 51, 57-59, 61, 72, 75, 77, 92, 100, 102, 107, 109, 112, 120-123, 128-130, 136, 142, 146, 149, 151, 169, 180, 184, 187, 215, 224, 226, 229, 233-234, 250, 253, 264, 266, 278, 280-281, 381, 421

E

endocrine disrupting chemicals 214-215, 250, 278-280, 282
enzymatic removal 221
enzyme immobilization 355-356, 364, 371-372, 380, 421-422, 425-427, 433-434
Events 1-2, 4-5, 7-8, 12, 24, 34-36, 38-39, 47, 49-50, 52, 58-60, 111, 216-217

F

Fabrication 159, 162, 164, 422, 424
fibrinolytic 23-25, 27, 31-33
Food 3, 16, 24, 32-33, 74, 128, 133, 151, 159, 162, 167, 169, 171, 178-180, 189, 214, 253, 255, 258-260, 267, 278-279, 283-284, 294, 296-297, 299-301, 305, 308, 327, 337, 341-344, 351-354, 357-362, 372, 376-377, 379, 381-382, 395, 398-399
Fractionation 23-24, 363-364
fruit juice industry 253
Functional Properties 25, 37, 129, 259, 261, 326-327, 329, 333, 336-337
Fungi 13, 16, 48, 52, 151, 178, 180, 183, 217, 224, 250, 253-254, 264, 266-267, 279, 284-286, 294-295, 297-298, 300-301, 304, 307-308, 310-312, 341, 346-347, 377, 381, 399-401, 406

G

glucoamylases 375, 381
Green Chemistry 178-179, 204, 434
growth conditions 295-296, 303, 309, 312

H

HAART 112, 119, 123
hereditary 1-3, 15, 23-24
HIV 100-102, 104-109, 112, 114-115, 123, 151, 267
Hydrogel 159-162, 164-167, 169-170, 257, 430, 432

I

Immobilization 166, 221, 231, 351, 353, 355-356, 364, 371-372, 380, 405-407, 409-410, 421-422, 425-427, 429-430, 433-434
immunogenicity 5, 10, 13, 26, 33-34, 40-41, 46-48, 51, 56, 58-59, 62, 135
in vitro drug release 71, 87, 92-93
Industrial Applications 180, 255, 294-296, 299, 308, 312, 358, 376, 383, 397, 426
industrially valuable products 351-352, 364, 372
Industry 23, 41, 133, 141, 151, 169, 171, 179, 184, 189, 253, 255, 257, 259-260, 263, 265, 267, 278-280, 283-284, 295-296, 299-301, 306-308, 342-344, 348, 352-354, 357-360, 376, 379, 382-383, 396-397, 399, 401
industry wastewater 280
Inhibitors 28, 30, 35-36, 38-39, 100, 107, 112-113, 116-117, 120, 122-123, 142, 145-146, 148, 150-151, 358, 406, 424, 434

K

Kinetics 166, 169, 181, 215, 227-228, 343, 428-429
kLa 343

L

Laccase 166, 178-190, 205, 207-214, 216-222, 224, 226-229, 231-234, 253-267, 282-283, 352, 400-401, 406-407, 409, 426, 432-433
laccase enzyme 253, 255, 258-261, 264
laccase mediator 220
Lignin Peroxidase 226, 281, 284-286, 395, 401, 404, 433
Lysosomal 1-3, 5-7, 9-10, 77, 130, 144

M

Manganese Peroxidase 222, 283-286, 401, 405
Mechanism 23-24, 32, 79-80, 100, 112, 114, 118, 146, 151, 171, 180-183, 186, 206-207, 211, 213, 217, 224, 255, 286, 376, 379, 382, 404, 428, 432
Metagenomic 351, 355, 372
microbially triggered 71-73, 88, 93
Microencapsulation 46-47
Microorganisms 72, 179, 188, 214, 234, 250, 284, 295, 300-301, 303-307, 309, 311-312, 341, 344, 347, 375-377, 383, 397, 399, 403, 410

N

N glycosylated 253
nanoenzyme 423
nanogel 433
nanomotors 427-428, 434
Nanoparticles 166, 170, 260, 356, 407-409, 421-431, 434
NNRTIs 112, 116-121, 123
non-deficiency 46-47
Nonsteroidal Anti-Inflammatory Drugs (NSAIDs) 250
NRTIs 112-116, 123
Nutrients 15, 17, 47, 72, 128, 133, 283, 295, 299, 303-304, 312, 376, 396, 406

O

Oligosaccharides 15, 57, 77, 79, 352, 354, 357-360, 363, 372, 380, 383
Oxidation 59, 61, 116, 134-135, 149-150, 162, 169, 181-188, 205-206, 212, 215, 217-218, 220, 222, 229, 253-260, 262-263, 266-267, 279-280, 282-283, 285, 307, 400-405, 424, 426
Oxidoreductases 214, 263, 279, 286, 400, 403, 405

P

Peptides 35, 40, 166, 260, 327, 332-333, 335, 352, 354, 357, 360, 363, 372, 429
Peroxidases 182, 216, 261, 279-280, 283-286, 397, 399-400, 403-405, 409
Pharmaceutical 17, 23, 41, 71, 128, 141, 144, 151, 159, 167-169, 178-180, 184-185, 189, 213-216, 219, 221, 223, 231, 233-234, 253, 255, 265-267, 278, 280-281, 299, 342, 346, 358, 375-376, 379-380, 382, 395, 399, 408
Pharmaceutical Industry 41, 141, 151, 169, 184, 265, 267, 280, 342, 358, 376, 379, 382
Phenolic Compounds 179, 184-186, 189, 214, 255-257, 260, 262, 264, 280, 284, 297, 354, 361, 400-401, 403-406
Polymeric 72-73, 159-160, 182, 184, 189, 262, 264, 284-285, 377, 383, 404, 423, 425-426, 429-430
Polymerization 110, 115, 118, 162, 164-166, 170, 178, 184, 189, 262, 279, 297, 356-357, 378, 400, 405-406, 430, 433
Polysaccharides 14-15, 71-72, 77-78, 164, 169, 297, 300, 306-308, 312, 353, 359, 379, 381, 383
probiotics 16, 71, 88, 90-93
Processes 1-2, 59, 61-62, 73, 109, 114, 141-142, 144, 146, 162, 166, 178-179, 190, 214, 221, 224, 258, 260-262, 283-284, 294, 300, 303, 307, 342, 347-348, 352, 355, 363, 375-376, 383, 396-400, 406, 408, 410, 422-423, 425, 429-431, 434
prodrugs 51, 77, 83
protein stability 379, 383
pullulanases 375
Purified 2-3, 9-10, 23-24, 27, 29-30, 35, 48, 58, 180-181, 185, 221, 226, 228-229, 232, 265, 296, 309, 327-328, 331, 333, 336, 346, 409

R

rational drug design 100, 120-121
Recalcitrant Compounds 182, 222, 400
Recombinant 1-3, 5-11, 13-14, 23-26, 28-41, 47, 50, 56, 58-60, 104, 141, 144, 152-153, 267, 380, 382
Recombinant DNA Technology 5, 9, 24, 29-30, 34, 141, 144, 152, 267
Remediation 234, 280, 286, 396-397, 402, 408-410, 424
Replacement 1-2, 16, 23-24, 31, 113, 144, 263, 295

S

Sources 1-2, 13, 15-16, 23-25, 48, 90, 141-142, 165, 231, 295, 304-306, 309, 312, 327, 341, 345-347, 354, 357-360, 379, 395, 404
Steric Stabilization 129
Storage 1-3, 6, 9, 133, 144, 167, 256-257, 297, 300, 307, 378, 406-407, 421, 432, 434
Sustainability 204, 342, 362, 398

T

textile dye bleaching 253
Therapeutics 2, 23-24, 33, 41, 46, 51, 145
Therapy 1, 4, 7, 11-12, 17, 23-26, 28, 31-34, 38-39, 46-52, 56-57, 59-61, 70, 100, 102, 112, 114-115, 129-131, 134, 136, 144, 153, 219
thermophiles 300-301
thrombolytic 23-31, 33
Trypsin 10, 16, 47, 53, 55, 326-328, 331-333, 335-336, 426

U

uricase 59, 135

V

versatile peroxidase 282

W

waste treatment 397-398, 404-405, 407, 410
Wastewater 183, 214-215, 231-234, 257, 280, 283, 285, 304, 395-397, 399-400, 404-410
white rot fungi 253-254, 264, 266-267, 279, 284, 286
Wound 46, 52-54, 57, 168, 189

X

Xenobiotic 182, 250, 284, 397
xylan 294-301, 305-309, 311-312, 359
xylanase 294-301, 303-312, 355
xylanase activity 296, 298-301, 303, 305-311

www.ingramcontent.com/pod-product-compliance
Lightning Source LLC
LaVergne TN
LVHW080619300926
849162LV00005B/1212